Electronic Communications Systems

A Complete Course

WILLIAM SCHWEBER
Analog Devices, Inc.

Prentice-Hall, Englewood Cliffs, N.J. 07632

Library of Congress Cataloging-in-Publication Data

Schweber, William L.
    Electronic communications systems : a complete course / William L. Schweber.
      p.   cm.
    Includes index.
    ISBN 0-13-590092-1
    1. Telecommunication.   I. Title.
TK5101.S33   1991
621.382—dc20
                                                                                                                         90-7408
                                                                                                                           CIP

Acquisitions Editor: **Sharon Jacobus**
Editorial/production supervision: **Lillian Glennon**
Interior design: **Nancy Field**
Cover design: **Bruce Kenselaar**
Manufacturing buyer: **Mary McCartney/Ed O'Dougherty**
Technical checker: **Paul Perletti**
Photo credit: **Chuck O'Rear/West Light**

© 1991 by Prentice-Hall, Inc.
A Division on Simon & Schuster
Englewood Cliffs, New Jersey 07632

All rights reserved. No part of this book may be reproduced, in any form or by any means, without permission in writing from the publisher.

Printed in the United States of America

10 9 8 7 6 5 4 3 2 1

ISBN 0-13-590092-1

Prentice-Hall International (UK) Limited, *London*
Prentice-Hall of Australia Pty. Limited, *Sydney*
Prentice-Hall Canada Inc., *Toronto*
Prentice-Hall Hispanoamericana, S.A., *Mexico*
Prentice-Hall of India Private Limited, *New Delhi*
Prentice-Hall of Japan, Inc., *Tokyo*
Simon & Schuster Asia Pte. Ltd., *Singapore*
Editora Prentice-Hall do Brasil, Ltds., *Rio de Janeiro*

To my wife,

who is always there

Contents

Preface xi

PART A: SPECTRUM, NOISE, AND MODULATION

1 The Electromagnetic Spectrum 1

    1.1 Introduction to Modern Communications Systems **2**
    1.2 Electromagnetic Waves and Energy **7**
    1.3 The Electromagnetic Spectrum and Allocations **11**
    1.4 Bandwidth and Information Capacity **14**
    1.5 Simplex, Duplex, and Half Duplex Systems **17**

2 Fourier and Spectrum Analysis 22

    2.1 Time and Frequency Domains **23**
    2.2 The Spectrum Analyzer **25**
    2.3 Fourier Analysis Examples **27**
    2.4 Modulation and the Frequency Spectrum **32**
    2.5 Spectra of Digital Signals **34**
    2.6 Superposition **38**
    2.7 Power and Energy Spectra **42**

3 Decibels and Noise — 46

- 3.1 Signal Magnitudes and Ranges 46
- 3.2 dB Calculation Examples 48
- 3.3 dB Reference Values 50
- 3.4 System Measurements with dB 53
- 3.5 Charting dB 58
- 3.6 dB and Bandwidth 61
- 3.7 Noise and its Effects 63
- 3.8 Sources and Types of Noise 66
- 3.9 Noise Measurements 70

4 Amplitude Modulation — 78

- 4.1 Need for Modulation 79
- 4.2 Basics of AM 80
- 4.3 Modulation Index and Signal Power 84
- 4.4 AM Circuits 91
- 4.5 Suppressed Carriers and Single Sideband 95
- 4.6 SSB Transmitter Circuits 102 *(omit)*
- 4.7 Baseband AM 107
- 4.8 Continuous Wave AM 108
- 4.9 Transmitter Functions and Testing 109

5 Receivers for AM — 119

- 5.1 Role of the Receiver 119
- 5.2 Receiver Techniques and Stages 121
- 5.3 RF Stage 125
- 5.4 Mixer and Local Oscillator 129
- 5.5 IF Stage 134
- 5.6 AM Demodulation and Audio Stages 140
- 5.7 SSB and CW Demodulation 143
- 5.8 Complete Receivers 145
- 5.9 Amplitude Modulation Features and Drawbacks 153
- 5.10 AM Receiver Testing 154

6 Frequency and Phase Modulation — 159

- 6.1 The Concept of Frequency Modulation 160
- 6.2 FM Spectrum and Bandwidth 161
- 6.3 Transmitters 167
- 6.4 Receiver Functions 175
- 6.5 FM Demodulators 178
- 6.6 The Phase-Locked Loop and Stereo Demodulation 182
- 6.7 Phase Modulation 190
- 6.8 Comparison of AM, FM, and PM 192

    6.9 FM Receiver Systems **195**
    6.10 FM Testing and Equipment **198**

PART B: MEDIA, TRANSMISSION LINES, AND WAVE PROPAGATION

7 Wire and Cable Media   203

    7.1 Wire and Cable Parameters **204**
    7.2 Balanced and Unbalanced Lines **205**
    7.3 Line Drivers and Receivers **209**
    7.4 Twisted Pair and Coaxial Cable **214**
    7.5 Time Domain Reflectometry **216**

8 Transmission Lines   223

    8.1 Impedance and Line Fundamentals **224**
    8.2 Microstrip Lines and Striplines **228**
    8.3 Waveguides **231**
    8.4 Line and Load Matching **237**
    8.5 S Parameters; the Smith Chart **247**
    8.6 Test Equipment **254**

9 Propagation and Antennas   263

    9.1 Propagation and the Function of Antennas **264**
    9.2 Propagation Modes **265**
    9.3 Antenna Characterization **273**
    9.4 Antenna Fundamentals **280**
    9.5 Elementary Antennas **284**
    9.6 Advanced Multiple Element Antennas **290**
    9.7 Advanced Single Element Antennas **298**

PART C: DIGITAL SYSTEMS

10 Digital Information   308

    10.1 Digital Information in Communications **309**
    10.2 Digital Specifications **315**
    10.3 Sampling, Bandwidth, and Bit Rates **323**
    10.4 Digital Testing **325**

11 Digital Communication Fundamentals — 332

- 11.1 Analog to Digital/Digital to Analog Converters **333**
- 11.2 Pulse Code Modulation **335**
- 11.3 Synchronization **344**
- 11.4 Delta Modulation **348**
- 11.5 Troubleshooting **354**

12 Digital Communication Systems — 358

- 12.1 The Complexity of Digital Communications **359**
- 12.2 Coding **362**
- 12.3 Format **366**
- 12.4 The Physical Interface and Throughput **369**
- 12.5 Protocol and State Diagrams **376**
- 12.6 Asynchronous and Synchronous Systems; Effective Throughput **380**
- 12.7 Error Detection and Correction **387**

13 Digital Modulation and Testing — 402

- 13.1 Basic Modulation and Demodulation **403**
- 13.2 Quadrature Amplitude Modulation **410**
- 13.3 Loopbacks, Error Rates, and Eye Patterns **414**
- 13.4 Random Bit Generation and Data Encryption **421**

PART D: COMMUNICATIONS SYSTEMS AND APPLICATIONS

14 TV/Video and Facsimile — 430

- 14.1 Imaging Basics **431**
- 14.2 The TV Signal **434**
- 14.3 Color TV **439**
- 14.4 TV Receivers **442**
- 14.5 Facsimile **448**

15 Frequency Synthesizers — 456

- 15.1 Direct and Indirect Synthesis **457**
- 15.2 Basic Indirect Synthesis **460**
- 15.3 Extending Synthesizers **466**
- 15.4 Synthesizers and Microprocessor Systems **471**

16 The Telephone System — 479

16.1 An Overview of the System 480
16.2 The Telephone Instrument and the Local Loop 483
16.3 The Central Office and Loop Supervision 492
16.4 The Central Office and Switching 496
16.5 Electronic Switching Systems 502
16.6 Echoes and Echo Cancellation 508
16.7 Digital Signals and Switching 512

17 The RS-232 Interface Standard and Modems — 518

17.1 The Role of the Interface Standard 519
17.2 RS-232 Operation 521
17.3 RS-232 ICs 531
17.4 RS-232 Examples and Troubleshooting 538
17.5 Modem Functions 544
17.6 Standard Modems 552
17.7 Other Communications Standards 554

18 Networks — 560

18.1 Network Applications 561
18.2 Topologies 564
18.3 Protocols and Access 568
18.4 Network Examples 573
18.5 Wide-Area Networks and Packet Switching 582
18.6 Integrated Services Digital Networks 588

19 Satellite Communication and Navigation — 593

19.1 Communications and Orbits 594
19.2 Satellite Design 599
19.3 Ground Stations 606
19.4 LORAN Navigation 610
19.5 Satellite Navigation 615

20 Cellular Telephone Systems — 622

20.1 The Cellular Concept 623
20.2 Cellular System Implementation 628
20.3 Cellular System Protocol and Testing 636

21 Radar Systems — 642

21.1 Radar Concepts and Display 643
21.2 Pulse Shapes 648
21.3 Radar System Circuitry and Components 657
21.4 Advanced Radar Systems 661

PART E: BROADBAND SYSTEMS

22 Multiplexing — 669

22.1 An Introduction to Multiplexing 670
22.2 Space Division Multiplexing 672
22.3 Frequency Division Multiplexing 674
22.4 Time Division Multiplexing 679
22.5 Multiple Stage Multiplexing 688

23 Microwave Equipment and Devices — 693

23.1 Test Instruments and Methods 694
23.2 Vacuum Tube Devices 704
23.3 Semiconductor Devices 715
23.4 Surface Acoustic Waves 720

24 Fiber Optics — 726

24.1 Fiber Optic System Characteristics 727
24.2 The Optical Fiber 729
24.3 Sources and Detectors 736
24.4 Complete Systems 743
24.5 Fiber Optic Testing 753

Answers to Selected Problems — 765

Appendices — 785

A Electromagnetic Spectrum 785
B dB Chart 786
C ASCII Code Chart 787

Index — 791

Preface

Electronic Communication Systems: A Complete Course is the first book to cover the traditional aspects of communications, yet recognize and explore the three developments which have radically changed communication systems. These developments—which have changed the way that technicians and engineers must deal with these systems—are:

- the widespread use of *integrated circuits* (*ICs*) to provide system functions in a single compact, high-performance device, thus replacing circuitry which previously required many discrete components.

- the use of *microprocessors and software* to manage and improve the operation of traditional analog communication systems.

- the use of *digital techniques and signals* in the communication system itself, to supplement or virtually replace analog techniques.

These three factors have shaped major changes in the way that communication systems are designed, implemented, and maintained. *Electronic Communication Systems: A Complete Course* explores communication systems with this perspective. Due to the impact of these three factors, students must learn to approach and comprehend systems in a new manner, rather than just learn more details of the same basic circuitry. In addition, this book provides a presentation both of the way that systems are commonly implemented, along with a discussion of the tradeoffs that exist in any system design: speed, power, performance, errors, complexity. The book takes a circuitry plus systems viewpoint, discussing circuits and their resultant systems (or equipment) with equal emphasis.

Over half the book is devoted to digital communications, actual communication systems (video, facsimile, telephone, modems, RS-232, cellular phones, computer networks, satellites, radar, fiber optics) following the necessary basic topics such as bandwidth, AM, FM, antennas, transmitters, receivers, and microwaves. Whenever possible, there are sections which specifically discuss troubleshooting goals, techniques, and instrumentation.

In-Text Learning Aids

Electronic Communication Systems: A Complete Course was designed with the beginning student in mind. Each chapter opens with Chapter Objectives contained in a blue box and an Introduction. Review Questions, also shaded in blue, follow each chapter section and serve to reinforce what the student has just learned. Color is also used to highlight important elements in the illustrations. Within each chapter, clearly marked, worked-out solutions are provided for numerical examples. Every chapter concludes with a Chapter Summary (corresponding to the Chapter Objectives), Summary Questions, and numerical Practice Problems (broken out by chapter section).

To provide historical perspective, explore a subject in more depth, or go slightly off the main-line path of a chapter, the book uses sidebars to present additional material. Sidebars provide the student with interesting additional information that is not essential to the basic flow of the chapter, but which add new facets to the student's understanding of the topic.

The Supplement Package

Careful thought was put into developing a comprehensive and useful set of supplementary teaching aids for instructors and students. A brief description of each item in this package follows.

- A LAB MANUAL by Ralph Folger, et al. (Hudson Valley Community College, Troy, NY) contains 24 lab experiments keyed to the text. Its features include lists of parts readily available from any electronics store. To purchase, contact your local bookstore. ISBN: 0-13-590373-4

- The INSTRUCTOR'S RESOURCE MANUAL provides teachers with additional numerical problems and applications, worked-out solutions to these additional problems as well as to those found in the text, chapter outlines for each chapter, and black-line Transparency Masters of over 75 key illustrations from the text. Available to instructors only. ISBN 0-13-590340-8

- A TEST ITEM FILE, compiled by Paul Perletti (Mt. Hood Community College, Portland, OR), offers approximately 100 problems for each chapter to be used in preparing exams. Available to instructors only. ISBN: 0-13-590357-2

A DEDICATION

In communications, one person stands out through his far-reaching, long-lasting contributions. Major Edward H. Armstrong (1890–1954) conceived, analyzed, produced, and perfected several major innovations. Two of these—the superhetrodyne receiver and FM transmission—are still vital to communications systems in this advanced age of digital electronics, microprocessors, and ICs; his first invention—the regenerative amplifier—allowed vacuum tubes in amplifiers to have much greater effective gain for weak signals (although subsequent developments have made it obsolete).

This book encompasses the world of digitally-driven, microprocessor-based analog and/or digital communication while recognizing the efforts and legacy of persons like Major Armstrong.

ACKNOWLEDGEMENTS

A book is created by the author but with the guidance of others. The reviews and comments of the following individuals were essential in making sure that this book's topics and focus, as well as depth, stayed balanced on the dual path defined by traditional circuits and systems combined with the radical changes effected by digital signals, microprocessors, and ICs, for these are changes which now permeate both new and older systems.

George Borchers—ITT Technical Institute, Salt Lake City, UT
Paul Cary—Lincoln Technical Institute, Allentown, PA
Robert E. Greenwood—Ryerson Polytechnic Institute, Toronto, Ontario
Dwight Holtman—ITT Technical Institute, St. Louis, MO
David Johnson—Pennsylvania College of Technology, Williamsport, PA
George Johnson—ITT Technical Institute, St. Louis, MO
Charles M. Killman—DeVry Institute of Technology, Kansas City, MO
Eugene L. Larchar—SUNY College of Agriculture & Technology at Morrisville, NY
Russell Puckett—Texas A & M University, College Station, TX
Arlyn I. Smith—Alfred State College, Wellsville, NY
James Stewart—DeVry Institute of Technology, Woodbridge, NJ
David Tancig—Parkland College, Champaign, IL

Paul Perletti—Mt. Hood Community College, Portland, OR deserves special recognition for his meticulous attention to the technical accuracy of the text.

Bill Schweber
Sharon, Massachusetts 1990

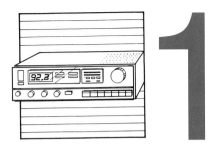

The Electromagnetic Spectrum

CHAPTER OBJECTIVES

When you have completed this chapter, you will understand:

- The broad issues in modern communications systems
- The relationship among frequency, wavelength, and propagation velocity
- The wide span and appearance of the overall electromagnetic spectrum
- The meaning of bandwidth and how it affects system information capacity
- The basic differences between simplex, half-duplex, and full-duplex communication systems

INTRODUCTION

The goal of a communications system is to transfer information from one place to another. This is done by sending the information as electromagnetic energy through vacuum, air, wire, or strands of glass and plastic fiber. The extremely wide range of energy frequencies and wavelengths that are used for this take up a large part of the electromagnetic spectrum. In this chapter we discuss some of the basic physical principles that define communications systems: electromagnetic frequencies, wavelengths, and velocities; energy; energy bandwidth and information that the energy can carry, and communications in one and two directions.

The electromagnetic spectrum is divided into subsections, or bands, each with some technical characteristics and peculiarities unique to that band. To avoid interference, users are assigned specific frequencies within the bands. Sending information requires a span of frequencies, and to send information at a higher rate, a greater span is needed than for a lower rate of information transfer. Communications systems are also divided into groups that allow information transfer in one direction only, in both directions simultaneously, and in both directions but only one direction at a time.

1.1 INTRODUCTION TO MODERN COMMUNICATIONS SYSTEMS

Communications systems and modern electronics have made it possible to send messages over great distances easily and reliably. They have also made it possible to send large amounts of data quickly from one point to another. Communications capability is now so common and available that it is hard to imagine the difficulty and uncertainty that people used to accept as the best that could be achieved with the technology they had available.

Modern communications systems use a wide range and variety of electronic equipment to meet the needs of users. Hand-held radios (Figure 1.1) allow direct contact with nearby base stations. Small satellite dish antennas can communicate with orbiting satellites, while larger dishes allow contact with space vehicles and space satellites millions of miles away (Figure 1.2). Commercial radio and television stations use powerful transmitters and large antennas to reach an audience within hundreds of miles. Radar can locate ships and planes regardless of weather or darkness (Figure 1.3). Increasingly, communications systems are being used to transmit digital messages, either directly from a computer or from some other source whose signal has been converted to computer-compatible digital format. The communications distance in a system can be to another planet, around the world, or as short as the distance from one part of an electronic chassis to another.

All communications systems have at least two endpoints. At one end, there is a source of signals (a voice, someone typing at a keyboard, or computer data, for example), circuitry for converting the signal source into a signal that is compatible with the rest of the system, and a transmitter which puts the converted signal onto the communications pathway or link (wire, air, or light-carrying fiber). After the transmitted signal passes through this link, it is received by the far end, converted in signal format as needed, and finally passed to the user (Figure 1.4).

In any communications system, both the user at the sending end and the user at the receiving send must agree on many factors for successful and meaningful communications to occur. These factors include what signals are used, the way the message is coded (such as what alphabet), the meaning of symbols within the message (the language: English, French, Spanish, etc.), and the type of modulation used. Unless this is done, the message may be received but meaningless, like getting a perfect copy of message in a language you do not understand, with strange symbols. You would not even know if the specks on the paper were mere dirt spots or part of the message. In communications systems, both parties must know the rules of the conversation or the overall communications effort will be wasted. Each of the key factors is discussed in a subsequent chapter.

Electronic communications began with *copper wire* as the only type of link between the sender and the receiver. Later, *broadcasting* was developed, which allowed signals to be transmitted through the air, without any wires at all. In recent years, *fiber optics* has become a common link between users. The principle

1.1 Introduction to Modern Communications Systems

Figure 1.1 Hand-held radio provides direct two-way communication to a nearby station (courtesy of Tandy Corp.).

behind this type of link is to use electromagnetic energy in the form of light to convey the information. Regular electrical signals in the system are converted to lightwaves, send through hair-thin optical fibers made of glass or plastic, and then received as light. The receiver reconverts the light into conventional electrical signals for further use.

What This Book Covers

We look at the concepts, implementations, and applications of communications systems in the five parts of this book. In Chapter 1 we discuss some of the basic factors that define the potential of a communications system, including how electromagnetic energy in the form of radio transmissions, signals in wire, or light are

Figure 1.2 Dish antennas are needed for communications at larger distances [courtesy of Electronic Space Systems Corp. (ESSCO)].

actually used to convey the message; the wide range of frequencies that make up the electromagnetic spectrum; the need for a span of frequencies, called *bandwidth*, to pass the desired amount of information in the time available; and the basic modes of single-direction and two-direction communications.

The remainder of Part A deals with two ways of looking at and measuring signals. There is the traditional *signal versus time* and the equally important and valid *signal versus frequency components* of spectrum analysis, along with signal and noise magnitudes, and powers and decibel units of measurements. In the second half of this section we examine how an *information* signal with the desired message is used to affect another signal, called a *carrier,* which has more power or a more desired frequency. This *modulation* is a key part of a complete communications system. Modulation can change the amplitude, frequency, or even the phase of the carrier signal, and each type of modulation has advantages in performance, simplicity, and use of the available frequencies. Of course, modulation at the transmitter requires corresponding demodulation at the receiver, and all aspects are studied.

The active components of the transmitter and receiver—the tubes, transistors, and integrated circuits—play a significant role in the overall system performance. But the passive components described in Part B, such as wire and cable,

1.1 Introduction to Modern Communications Systems

Figure 1.3 Radar screen, showing use of communications system for conveying location (courtesy of Lockheed Sanders, Inc.).

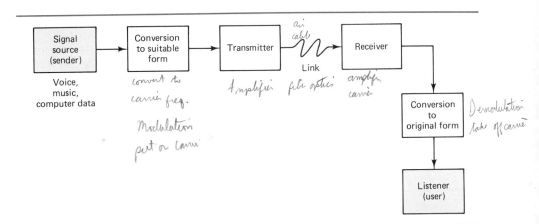

Figure 1.4 Basic block diagram of the functions of communications system, showing information signal source, conversion, transmitter, link, receiver, conversion, and listener.

the transmission line that carries the signal to or from the antenna, and the antenna itself, are vital elements of the system. Each has its own characteristics with respect to electronic communications. For antennas, the way the signal travels through air and space at various frequencies and the many antenna designs used determine some of the primary capabilities of the communications system.

In Part C we look at *digital* systems, a special but very important type of communication. Digital signals can represent *analog* information (where the signal can have any value in an overall range) such as voice or video, as well as conveying information that is inherently digital, such as numerical information. This part looks at how digital signals represent analog information, how analog signals are *sampled* to capture their essential information, and the similarities between digital signals and the even simpler case of binary (two-value) digital signals. For digital systems, the transmitter and receiver must be synchronized so that the receiver knows which digital signals represent which parts of the original signal.

Once a system uses digital rather than analog signals to represent the information of the message, the entire nature of the communications system changes. The information to be sent must be encoded with a standard scheme to represent letters and numbers, the message can be preceded by information about the message (intended address, message length), and the receiver can actually decide if the message was received properly or with an error. *Protocols* designed into the system define all the rules of the conversation between users. Among other functions, when an error is detected, these protocols allow the receiver to request the message again, and the transmitter to send it—all without getting the overall sequence of messages confused. The many aspects of digital signals and communications are studied in the second half of Part C.

In Part D we look at specific applications and implementations of communications systems. These include TV (video) and facsimile (fax), which transmit images; frequency synthesizers, which easily provide precise tuning of signals; the telephone system, which is the most common communications system and used for everything from voice to computer data and images; the basic and popular RS-232 interface standard, which allows many computer and digital systems to be connected to each other and phone systems; and networks, which are complex and interconnected systems with many users who must send information to other users in the network.

Part D concludes with a study of the use of satellites for communications, time measurement, and high-precision navigation anywhere on the globe; radiotelephone and cellular phone systems, which provide many channels for users of mobile phones; and radar, which uses the reflections of electronic signals to measure distances to other objects, as well as their movement.

In Part E we look at some communications topics that provide the highest levels of performance. *Multiplexing* allows any users to share the same communications channel, while *microwaves* (extremely high frequencies) provide the wide range of frequencies needed to carry large amounts of information through air or vacuum. Finally, in many applications, copper wire and cables are being replaced by fiber optics. The hair-thin glass or plastic fibers carry tremendous amounts of information on a beam of electronically controlled light, with the light guided by the fiber acting as a *light pipe*.

The five major parts of this book divide the overall sphere of communications into the principles, practices, and applications that drive the way systems are designed, built, and used. Wherever practical, we also show the kinds of test equipment that communication requires. Most of this equipment is very different from the traditional and common voltmeter and oscilloscope of electronics, and is designed to do certain specific tasks. We also explain why a certain technical approach is used compared to another approach in the design or application. In electronic communications there is no single ''right'' answer to producing a system that meets the needs of the users. Instead, there are many choices, and the most correct choice is a balance of technology, price, reliability, and convenience. The final decision on which approach to use is a trade-off among the many factors involved. As technology changes or the needs change, the second-best choice may become a first choice in the next generation of systems and equipment.

Review Questions

1. What is the range of distances over which communications systems are used?
2. Give three examples of communications system equipment.
3. What are the basic block functions of any communications system?
4. Explain why the users at either end of the communications system must agree on various technical factors. What are some of these factors?
5. What are three physical paths for the communications signal?

1.2 ELECTROMAGNETIC WAVES AND ENERGY

All electronic communications systems send information from one point to another by transmitting *electromagnetic* energy from the sender to the intended receiver. This electromagnetic energy can travel in various modes: as a voltage or current through wires, as radio emissions through air or the vacuum of space, or as light. For all modes, the same basic laws of physics apply. The simplest and most important of these laws relates the *wavelength*, the distance between peaks of the oscillations of the energy wave or, equivalently, the distance traveled in the time to complete one cycle); the *velocity*, the speed at which the energy travels through the wire, air, vacuum, or optical fiber; and the *frequency*, the number of oscillations or cycles per second (hertz, abbreviated Hz units 1/3) of the waveform:

$$\text{wavelength} = \frac{\text{velocity}}{\text{frequency}}$$

Wavelength can also be measured as the distance between the negative peaks, zero points, or any point on the waveform, as long as the same point is used on each adjacent cycle for measurement.

Propagation Velocity

The speed at which a wave travels depends on the substance, or *medium*, through which it is traveling and is called the *propagation velocity*. The fastest propagation velocity possible occurs in vacuum, and is symbolized by the letter c (often referred to as the "speed of light"). This speed is 186,000 miles/s, or 300,000,000 m/s, expressed in scientific notation as 3×10^8 m/s.

Example 1.1

What is the wavelength for a frequency of 1 million hertz?

Solution

$$\text{Wavelength} = \frac{c}{1{,}000{,}000/\text{s}} = 3 \times 10^2 \text{ m} = 300 \text{ m}$$

Example 1.2

What is the frequency when the measured wavelength is 6 m?

Solution

Rearranging the equation, we have

$$\text{frequency} = \frac{\text{velocity}}{\text{wavelength}}$$

Therefore, frequency = $c/6$ m = 0.5×10^8 = 50,000,000 Hz.

The propagation velocity is less in any medium other than vacuum. In air, the propagation velocity of electromagnetic energy is about 95 to 98% if the value in vacuum. In wire, it is anywhere from 60 to 85% of c, depending on wire type, construction, and insulation.

For example, a common type of cable used for cable TV has a propagation factor of 75% of c.

Example 1.3

What is the actual velocity of the electromagnetic energy in this cable?

Solution

$$(3 \times 10^8 \text{ m/s}) \times 0.75 = 2.25 \times 10^8 \text{ m/s}$$

Example 1.4

The propagation factor for dry air is 98% of c. What is this, in miles/second?

Solution

$$(186{,}000 \text{ miles/s}) \times 0.98 = 182{,}282 \text{ miles/s}$$

1.2 Electromagnetic Waves and Energy

Example 1.5

What is the signal velocity if the propagation factor is 0.92 of c?

Solution

$$(3 \times 10^8 \text{ m/s}) \times 0.92 = 2.76 \times 10^8 \text{ m/s}$$

Example 1.6

Visible-light frequencies range from 3.9×10^{14} Hz (red) to 7.9×10^{14} Hz (violet). What is the wavelength for a color in the middle of the visible-light band, with a frequency of 6×10^{14} Hz?

Solution

Wavelength = $(3 \times 10^8 \text{ m/s})/(6 \times 10^{14}/\text{s}) = 0.5 \times 10^{-6}$ m, which is about 0.00002 inch! [*Note*: One billionth of a meter, 10^{-9} m, is called a nanometer (nm), so the answer can also be expressed as 500 nm.]

The frequency of a communications system is determined by the transmitter circuitry and is designed to have specific value or range of values. Electronic communications systems make use of values as low as 10,000 Hz to extraordinarily high values in the billions of hertz.

Taken together, the frequency and the velocity of propagation determine the wavelength of the electromagnetic wave. A standard AM broadcast radio signal at 1,000,000 Hz in the vacuum of space has a wavelength of

$$\frac{3 \times 10^8 \text{ m/s}}{1,000,000/\text{s}} = 3 \times 10^2 \text{ m} \quad \text{or} \quad 300 \text{ m}$$

A TV broadcast signal for channel 11 uses a frequency of 200 MHz (1 MHz = 1,000,000 Hz = 1000 kHz). The corresponding wavelength in vacuum is

$$\frac{3 \times 10^8 \text{ m/s}}{200 \times 10^6/\text{s}} = 1.5 \text{ m (about 5 ft)}$$

If the electromagnetic energy is traveling though a substance with a propagation velocity of less than c, the wavelength will be correspondingly less. The channel 11 signal traveling through the cable with a propagation value of 75% of c will have a wavelength of

$$\frac{3 \times 10^8 \times 0.75}{200 \times 10^6} = 1.125 \text{ m}$$

Example 1.7

A signal is normally sent through vacuum, at a frequency of 60 million hertz. However, it is also broadcast through air, with propagation factor = 0.98 of c. What is the wavelength in (a) vacuum and (b) air?

Solution

(a) In vacuum, the wavelength = $(3 \times 10^8 \text{ m/s})/(60 \times 10^6/\text{s}) = 0.05 \times 10^2 = 5$ m.
(b) In air, the wavelength is 5 m $\times$ 0.92 = 4.6 m.

Example 1.8

The same signal is now sent through a cable, with propagation factor of 0.69 of c. What is the velocity? What is the wavelength?

Solution

$$\text{Velocity} = 3 \times 10^8 \text{ m/s} \times 0.69 = 2.07 \times 10^8 \text{ m/s}$$

$$\text{Wavelength} = 5 \text{ m} \times 0.69 = 3.45 \text{ m}$$

The wavelength is important because many aspects of system design and types of electronic components that must be used are determined by the wavelengths in the system. Antenna size, the kinds of wires and cables used, the active components such as amplifiers and transistors, and even the kind of enclosure or chassis are related to the wavelength. As the wavelengths get very short—in the range of centimeters—the electromagnetic energy cannot be easily controlled or confined. Special circuits are needed, together with special components and techniques for design and testing. Nevertheless, modern communications systems use frequencies (and wavelengths) that can be nearly anywhere in the overall wide range of possible values.

Electromagnetic Propagation Characteristics

Electromagnetic energy normally travels in a straight line, but this straight-line travel can be changed by various methods. When the energy is confined because it is traveling in a wire or optical fiber, the energy follows the wire or fiber, of course. But even energy traveling through air or space in a straight line can have its direction changed. A solid metal surface causes energy *reflection*, just as a mirror reflects light, and a nearly solid surface or mesh will have the same effect, depending on the signal wavelength. When the energy waveform travels across the boundary between zones of different characteristics, the travel path will be bent. This is called *refraction*, and you have seen it when a straight stick appears bent in a glass of water. A glass lens also refracts light and bends it—this is what eyeglasses do to change the path and focal point of light reaching the eye. We study these propagation characteristics and their impact in much more detail in Chapter 9.

Review Questions

1. What actually conveys the information in a communications system?
2. What is the relationship among signal velocity, wavelength, and frequency?
3. What is c? Compare its value in vacuum to its value in air and in wire.
4. What is propagation velocity? What is the effect of propagation velocities less than c on the signal wavelength?

5. Why are the frequency and wavelength values important in the design of a system?

1.3 THE ELECTROMAGNETIC SPECTRUM AND ALLOCATIONS

The total span of frequencies and corresponding wavelengths used in communications systems is called the *electromagnetic spectrum*. The overall useful electromagnetic spectrum extends from about 10,000 Hz to several billion hertz. The lowest frequencies are used for special systems that communicate with submarines, because these low-frequency, long-wavelength waves can penetrate several hundred feet of water. As the frequencies increase, the spectrum is used for standard AM broadcast, radio navigation, two-way radio, FM broadcast, TV, radio astronomy, and satellite systems, among many other uses.

Prefixes are used with hertz values, since "millions" or "billions" of hertz is very awkward. The magnitudes of hertz unit prefixes are:

$$1 \text{ kilohertz (kHz)} = 1000 \text{ Hz}$$

$$1 \text{ megahertz (MHz)} = 1{,}000{,}000 \text{ Hz} = 1000 \text{ kHz} = 1 \times 10^6 \text{ Hz}$$

$$1 \text{ gigahertz (GHz)} = 1{,}000{,}000{,}000 \text{ Hz} = 1000 \text{ MHz} = 1 \times 10^9 \text{ Hz}$$

In many cases it is necessary to convert a frequency value from one magnitude multiple to another. For example, a frequency range may extend from 300 to 3000 MHz, which is the same as saying 300 MHz to 3 GHz. Similarly, the frequency of 1.6 MHz may not be immediately recognized, but it is the same as 1600 kHz, the upper end of the standard AM broadcast band.

Example 1.9

Convert 120 kHz to Hz and to MHz.

Solution

$$120 \text{ kHz} = 120{,}000 \text{ Hz} = 0.120 \text{ MHz}$$

Example 1.10

Convert 175.37 MHz to kHz and GHz.

Solution

$$175.37 \text{ MHz} = 175{,}370 \text{ kHz} = 0.17537 \text{ GHz}$$

The total electromagnetic spectrum is shown in Figure 1.5. Note that it is divided into different sections or *bands*. Each band has a name and boundaries, which are just convenient points for dividing the spectrum. The bands used for electronic communication are shown in the figure, with their abbreviations, frequencies, and wavelengths. Also shown are the relative location in the spectrum of visible light and wavelengths near visible light that are used for communications in some systems.

Standard Frequency Allocations

Each band is used for many communications services. Each service, or user, must use an assigned frequency to avoid interfering with other users who may be broadcasting at the same time. Users can share a frequency if they are far enough apart physically that the intended receiver (or receivers) will receive only the desired transmitter signal, without *interference* from other transmitters.

The allocation of the communications services within the overall spectrum is made by general worldwide agreement under the control of the International Telecommunications Union (ITU) and is shown in Appendix A. Countries meet at regular ITU conferences and decide what services should be allowed to use which frequencies. For example, the ITU has allocated the frequencies from 88 to 108 MHz for the FM broadcast band. Within this band, there are assignments for 100 stations, spaced every 200 kHz, at these frequencies: 88.1, 88.3, 88.5, . . . up to 107.9 MHz. The assignment of which station should have specific frequencies is made by the agency in each country which regulates broadcasting in that country. In the United States, the Federal Communications Commission (FCC) does this. The FCC could give a license to one station in New York to use 88.3 MHz, and a license to another New York station at 88.9 MHz. The 88.3 MHz assignment could also be licensed to a station in Los Angeles, for example, because under any reasonable circumstances a signal at this frequency from the New York station could not reach the west coast.

For reasons of historical custom, some band assignments are referred to by their wavelength instead of the frequency. The amateur radio band from 50 through 54 MHz is called the 6-meter band, and the 144 through 148MHz span is known as the 2-meter band.

Many communications systems do not broadcast their signals but, instead,

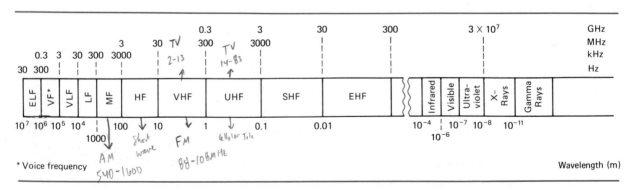

Figure 1.5 Total EM spectrum, dc to gamma rays, with coarse divisions, frequency, and wavelengths, along with detail of radio bands from ELF through EHF.

use a wire or cable to transfer the energy. The wire or cable confines and restricts the electromagnetic energy like a pipe carrying water and prevents it from interfering with user signals in other wires or cables. Therefore, the ITU and FCC allocations do not directly affect the user. Nevertheless, the concept of spectrum is still important since any electromagnetic signal must have some frequency spectrum associated with it, and the circuitry must be able to transmit, receive, and process signals at these frequencies. When several users share the air, wire, or cable, they must also take special steps to prevent interfering with each other.

Overlapping Assignments

It would be much easier if there were enough spectrum so that users did not have to share any frequencies with others. Unfortunately, this is not the case in practice. There are so many users who need to make use of the spectrum and the specific characteristics of the bands of the spectrum that some sharing must be done. In the VHF band, for example, the frequencies from 38 to 39 MHz are used by mobile radios, fixed-base station radios, and radio astronomy (which has no choice because the signals from various galaxies and stars happen to be at these frequencies). Figure 1.6 shows in detail the assignments in the VHF band. Note the many types of users and how tightly they are "sandwiched" into the band.

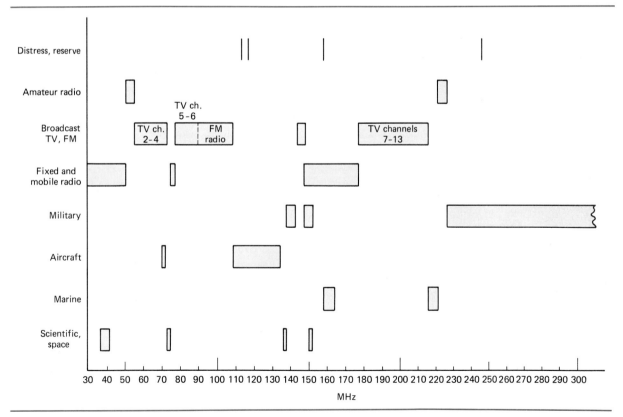

Figure 1.6 Detail of assignments within the VHF band.

Review Questions

1. What is the span of the electromagnetic spectrum?
2. What are the extremely low frequencies used for? Why?
3. What are the applications of the middle bands of frequencies?
4. What is the relationship among hertz, kilohertz, megahertz, and gigahertz?
5. What are standard frequency allocations? How are they assigned? Why is there concern about conflict in allocations?

1.4 BANDWIDTH AND INFORMATION CAPACITY

Bandwidth

It may seem possible to squeeze as many users as desired into a band or part of the spectrum. For example, why not assign FM broadcast stations with 2-kHz spacing, to 88.100, 88.101, 88.102, and so on, MHz, instead of spacing the channel assignments 200 kHz apart? The answer has to do with the need for *bandwidth*. Bandwidth is the span of frequencies within the spectrum occupied by a signal and used by the signal for conveying information.

The function of any communications system is to carry information. This information may be in the form of voice signals, video signals for TV, digital signals from computers, or on/off signals as from Morse code. Regardless of the type of information, there is always one simple fact to remember: carrying information requires bandwidth. To pass information, the system must make use of a specific amount of the spectrum. Music, which is a form of information, uses the range of frequencies from 0 to 20 kHz, and so has a 20-kHz bandwidth. To pass the entire music signal faithfully, the communications system must allocate a minimum bandwidth of at least 20 kHz.

The next issue is how much bandwidth is required to convey the information. The answer depends both on how much information must be transmitted and in how much (or little) time. To send more information in a short time requires more bandwidth. The same quantity of information can be sent in a longer period using less bandwidth. It is the same as transferring water (information) through a pipe (the communications system). To send lots of water quickly requires a larger-diameter pipe (more bandwidth). A smaller pipe has the capability of emptying a huge tank, but it will take a long time, which may be acceptable in some cases and not acceptable in others. If only a little water has to be emptied, a small pipe may be able to do it in the time required.

The same basic ideas apply to bandwidth needs. A communication system that must transfer large amounts of information, such as a regular TV signal that is sending 30 complete pictures per second, requires a large bandwidth. Standard broadcast TV stations use 6-MHz bandwidths. In contrast, a human voice needs much less bandwidth. The total range of frequencies in the voice extends from abut 300 Hz to 8 kHz, but the span from 300 Hz to 3 kHz is all that is really needed to carry the voice information. The bandwidth from 3 to 8 kHz contains the information that identifies who is speaking and makes each voice distinguishable from others, but the message is intelligible with bandwidth to 3 kHz. For this reason, the voice channels of the telephone system provide only about 3 kHz of bandwidth for each user.

The bandwidth of signals from deep-space satellites (those that are beyond our near planets) is selected based on many technical considerations. For a variety of reasons that will become clearer in subsequent chapters, the bandwidth from satellites that are probing the outer reaches of the solar system is very small, sometimes as little as 10 Hz. This means that it can take a very long time to transfer the large quantity of data that the satellite may have acquired. With this small bandwidth, a TV picture can take several hours to transmit, in comparison with almost "instantaneous" transmission using the 6-MHz standard video broadcast bandwidth.

Looking at the total electromagnetic spectrum, it is clear that the bandwidth available in the higher-frequency bands is much greater than that in the lower-frequency bands. The entire HF band has 30 − 3 = 27 MHz of bandwidth, while the UHF band has 3000 − 300 = 2700 MHz, or 100 times as great. Therefore, when large amounts of bandwidth are needed, the communications system must go to higher and higher frequencies. This is true even though the circuitry for these higher frequencies is more complex.

Example 1.11

How many times more bandwidth does the UHF band have than the VHF band?

Solution

The UHF band extends from 300 to 3000 MHz, while the VHF band spans 30 to 300 MHz. Therefore,

$$\frac{\text{UHF bandwidth}}{\text{VHF bandwidth}} = \frac{3000 - 300}{300 - 30} = \frac{2700}{270} = 10$$

Example 1.12

What percentage of the VHF band does a 6-MHz bandwidth TV signal occupy?

Solution

$$\frac{6}{300 - 30} = \frac{6}{270} = 0.022 = 2.2\%$$

Information Capacity

The exact relationship between the rate at which data can be transferred by a communications system for any bandwidth has been analyzed and studied extensively. In 1948, Claude Shannon proved that the information capacity of a communications channel was related to the bandwidth, signal power, and noise in the channel by the equation

$$\text{capacity} = \text{bandwidth} \times \log_2 \left(1 + \frac{\text{signal power}}{\text{noise power}}\right)$$

where capacity is in bits/second, bandwidth is in hertz, and signal and noise powers are measured in the same physical units, such as watts. *Bits* are fundamental units of information. A single bit represents one of two possible states, such as yes/no, on/off, or 0 and 1, and is the smallest element that can convey information. They will be studied more completely in Chapter 10. *Noise* is a major subject in communications systems and is examined in detail in Chapter 3. For now, simply use the fact that any communications system has undesired but unavoidable electrical "contamination" or noise from many sources, and the magnitude of this noise—its power—is an important factor in the performance of the system. [*Note:* If your calculator does not directly provide logs to base 2, use this rule: For any number n, the $\log_2(n) = \log_{10}(n)/\log_{10}(2) = \log_{10}(n)/0.3$. For example; $\log_2(17)$ equals $\log_{10}(17)/0.3 = 1.23/0.3 = 4.10$.]

Some examples show what Shannon's formula indicates. Suppose that a communications link is using a transmitter of 10 W power, and there is 1 W of noise power in the same link. The available bandwidth is 1 kHz. The channel capacity is therefore

$$C = 1000 \times \log_2\left(1 + \frac{10}{1}\right) = 1000 \times \log_2(11) = 3470 \text{ bits/s}$$

which is very low. Increasing the bandwidth to 100 kHz increases the capacity to 347,000 bits/s. The bits/second magnitude indicates how many fundamental information units can be transferred in each second. A typical computer-to-terminal connection requires from 1200 to 19,200 bits/s, for example.

Increasing the ratio of signal power to noise power to 100 increases the capacity to 6681 bits/s:

$$\text{capacity} = 1000 \times \log_2\left(1 + \frac{100}{1}\right) = 1000 \log_2(101) = 6681 \text{ bits/s}$$

From the equation we see that increasing the bandwidth increases the capacity linearly: A doubling of the bandwidth doubles the capacity. In contrast, the effect of signal power and noise power is logarithmic and much less dramatic: Large increases in the power ratio will cause smaller increases in the capacity. This is because the "log" function compresses the ratio and reduces the effect of increases in the signal power/noise power value.

Shannon's formula can also be used to determine what bandwidth is needed to send a desired amount of data through the system. Suppose that 10,000 bits/s must be sent so that the user does not have to wait too long for the critical data, and the system has 100 W of signal power versus 10 W of noise power. By manipulating the formula, the bandwidth needed is

$$\text{bandwidth} = \frac{\text{capacity}}{\log_2(1 + \text{signal power/noise power})}$$

$$= \frac{10{,}000}{\log_2(1 + 100/10)} = \frac{10{,}000}{\log_2(11)}$$

$$= \frac{10{,}000}{3.47} = 2880 \text{ Hz}$$

Example 1.13

What bandwidth is needed, by Shannon's theory, to support a capacity of 20,000 bits/s when the ratio of signal power to noise power is 100?

Solution

$$\text{bandwidth} = \frac{20{,}000}{\log_2(1 + 100)} = \frac{20{,}000}{6.68} = 2994 \text{ Hz}$$

Shannon's analysis provides the mathematical proof for what was approximately, but not exactly, understood since the early days of communications: sending data requires bandwidth, and sending more data at higher rates requires more bandwidth. Therefore, the spacing of the channel assignments in a band must be matched with the data rates or amount of information (per unit of time) that each channel is expected to carry.

Shannon's proof shows the capacity of the channel under ideal conditions and, of course, makes some simplifications. The Shannon formula result is actually the best that can ever be achieved. In the real world, there are other factors that affect how much information the channel can carry. These factors include imperfections in the actual circuit performance versus theory, circuit nonlinearities, temperature effects on stability, and distortion, for a few examples. A practical system can approach the capacity that the equation shows, but can never actually achieve it. Some communications systems come close, whereas others do not reach even half the maximum value that the equation predicts. To a large extent, the more carefully designed the system is, and the better the circuitry that is used, the closer the actual performance will be to the theoretical maximum.

Review Questions

1. What is the relationship between bandwidth and information rate?
2. Does bandwidth limit the total amount of information that can be sent? Explain exactly what it does limit.
3. What is the bandwidth of a voice signal? Of a standard TV signal?
4. What are the higher-frequency bands used when more bandwidth is needed?
5. Explain how capacity increases/decreases with bandwidth. How does it increase/decrease versus the signal-to-noise power ratio?

1.5 SIMPLEX, DUPLEX, AND HALF-DUPLEX SYSTEMS

A communications system can be designed for transmitting information in one or both directions. If the system is capable of sending the electromagnetic energy in one direction only, and has no provision for sending it in the reverse direction, it is called a *simplex* system. Public address systems, broadcast radio, and TV are

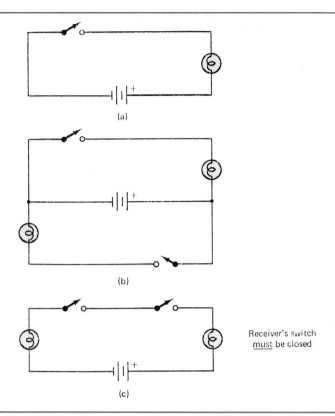

Figure 1.7 (a) Lamp and switch for simplex communications; (b) two lamps and two switches for a full-duplex system; (c) lamps and switches for a half-duplex system.

examples of this. A simplex communications system using a switch as the transmitter and a lamp as the received signal indicator is shown in Figure 1.7a. Only the user at the switch can send information; only the user with the lamp can receive it.

Many communications systems are designed to carry information in both directions at the same time. These are called *full-duplex*, or simply *duplex*, systems. Ordinary telephone systems and many computer systems are examples of duplex systems. The switch and lamp version is shown in Figure 1.7b. It consists of two separate loops, each of which operates independently of the other. Either end can transmit a message, whether or not the other end is sending a message at the same time. The two directions of message travel are independent of each other.

There is another category called the *half-duplex* system. In a half-duplex system, information can go in either direction, but in only one direction at a time. First, the sending end transmits to the intended receiver, and then they reverse roles. What was the sender becomes the receiver, and what was the receiver now becomes the sender. Note that the switch and lamp circuit (Figure 1.7c) requires only as many wires as the simplex version but the same amount of transmitting and receiving circuitry as the full-duplex system. (In the circuit shown, the receiver closes its switch, and then the transmitter switch controls the lamp on/off state.) Half-duplex examples are a two-way radio such as citizens' band, and some computer installations. Half-duplex systems are used instead of full duplex when there is a need for two-way information flow, but the bandwidth, number of wires, or number of signal channels is limited. It takes twice as much bandwidth for full duplex as it does for half duplex. A half-duplex system has each user sharing the same bandwidth or wires but at different times, whereas in a duplex system the users do not have to take turns.

However, a full-duplex system is much more convenient to use. Anyone who has used a half-duplex two-way radio system, where each person must say "over" to let the other side know that it is time to switch, will testify to this. The telephone, a full-duplex system, is much more convenient. Full-duplex operation is also much more efficient in an overall communications system, since the listener can interrupt at any time to say that the message did not make sense, or that there is some really important information needs to be passed. In a half-duplex system, the listener can only inform the sender that the message did not get through properly when the sender/listener turnaround occurs, and the sender cannot interrupt with a critical message until its turn in the talking/listening sequence. Half duplex is often needed even if the main message flow is in one direction only, so that the receiver can at the very least inform the sender whether or not the message was received successfully.

An analogy for simplex, half-duplex, and full-duplex systems is a street for cars. A one-way street is simplex. A street that is so narrow that only one car can pass at a time is half duplex, because cars must wait for their direction's turn before they can go through (often a backup occurs with the waiting cars, and a policeman is needed to direct and control the flow). A wider, two-lane street is full duplex, by comparison, since both directions can proceed without concern for what is happening in the other direction. A wider road is needed (equivalent to bandwidth) and more complicated intersections (similar to circuitry at each end) are often required as well. Developing efficient, reliable communications systems is a challenge and must include whether simplex, duplex, or half-duplex performance is required, together with the technical issues associated with each of these.

Review Questions

1. What is simplex communications? Where is it used?
2. What is a duplex system? Why is it useful? What does it require compared to a simplex system?
3. What is a half-duplex system? How does it compare some of the features and requirements of simplex and duplex?
4. Why is a half-duplex system needed in many applications, even if the information message flows in one direction only? *half bandwidth*

SUMMARY

Electronic communications systems use electromagnetic energy to transfer information from the sender to the receiver. This energy can be in the form of radio waves, signals in a wire, or even light in an optical fiber. All electromagnetic energy follows a simple relationship among wavelength, frequency, and propagation velocity. The greatest possible propagation velocity is through a vacuum, while in air it is a little less, and in a wire or cable it may be much less.

The span of frequencies and wavelengths available ranges from a few hertz to hundreds of millions of hertz—even higher if light waves are included. This total spectrum is divided into subsections, or bands, grouped by some of the general characteristics of the frequencies and wavelengths in each. Users must be assigned specific frequencies so that they do not conflict and interfere with each other. The rate at which information can be sent is determined by the bandwidth available and the ratio of signal power to noise power. Higher information rates require wider bandwidth and higher signal-to-noise power ratios. Bandwidth is a commodity that cannot be wasted in most applications, since only a certain amount is available. Communication systems can be designed for sending information in one direction only (simplex), both directions at the same time (duplex), or both directions but not simultaneously (half duplex). The choice is made by the need of the application, and each requires differing amounts of bandwidth and electronic circuitry to implement.

Summary Questions

1. Give one example each of a short-distance and a long-distance communications system.

2. Explain why a communications system must have similar functions, regardless of size or distance.

3. Why does successful communications require agreement by both the sender and receiver on many technical factors. What are some of these factors?

4. What is the primary relationship among frequency, velocity, and wavelength of an electromagnetic wave?

5. What is the propagation factor? Give typical factors for air and for wire and cable.

6. For a given value of frequency, what happens to wavelength as velocity decreases?

7. Compare the wavelength of visible light with the wavelength of a signal in the LF band.

8. How do the choices of frequency and wavelength affect the overall design of a system? Of components within the system?

9. What are the bands of the electromagnetic spectrum? What are the characteristics of some bands versus others?

10. What is the lowest frequency of the ELF band? The highest frequency of the SHF band?

11. How any hertz are there in a megahertz? How many kilohertz constitute a gigahertz?

12. What is bandwidth? Why is it a critical resource?

13. How does available bandwidth affect the rate at which data can be transferred? How do signal power and noise power affect data rates?

14. Why do the higher-frequency bands of the spectrum offer greater capacity?

15. What is the difference between simplex, duplex, and half-duplex communications systems? Give examples of each.

PRACTICE PROBLEMS

Section 1.2

1. What is the wavelength for a waveform that has a velocity of c and a frequency of 55,000 Hz?
2. Repeat problem 1 when the velocity is $0.8c$.
3. Repeat problem 1 when the frequency is twice the value.
4. What is the frequency when the observed wavelength is 0.35 m and the velocity of propagation is c?
5. What is the wavelength when the velocity is 1.5×10^8 m/s (one-half c) and the frequency is 1 million hertz?
6. What is the signal velocity in a cable with a propagation factor of $0.88c$?
7. Repeat problem 6 for a propagation factor of $0.82c$.
8. What is the wavelength of a 10,000-Hz signal in vacuum? In the case of problem 6?

Section 1.3

1. Convert these frequencies to kilohertz: 25.1 Hz, 15.75 MHz.
2. Convert these frequencies to megahertz: 175 Hz, 125.2 kHz, 1.234 GHz.

Section 1.4

1. What is the bandwidth ratio between the UHF and VLF bands?
2. Compare the bandwidth of the HF and SHF bands. How many times less is the HF band?
3. What percentage of the HF band is occupied by a 3-kHz voice signal? By a standard TV signal?
4. Repeat problem 3 for the UHF band.
5. Determine the channel capacity for signal power of 150 W, noise power of 12 W, and bandwidth of 2 kHz.
6. What is the capacity when the signal-to-noise power ratio is 1000 and the bandwidth is 150 kHz?
7. What is the capacity when the bandwidth is problem 3 is reduced by a factor of 2 but the signal-to-noise power ratio is doubled?
8. What bandwidth is required for a capacity of 15,000 bits/s when the signal-to-noise power ratio is 200?

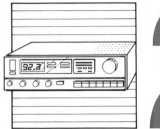

2
Fourier and Spectrum Analysis

CHAPTER OBJECTIVES

When you have completed this chapter, you will understand:

- How a signal may be viewed by its amplitude versus time, or by its amplitude versus frequency, and the pros and cons of each perspective
- How modulation affects the frequency spectrum
- The frequency spectrum of digital signals, and how it varies with pulse width and shape
- How the spectrum of more complex signals is analyzed by the spectrum of its constituent parts

INTRODUCTION

Communications information is represented by a signal that varies with time. The same signal can be represented by a group of specific frequencies, which combine to produce the same result. The time representation and the frequency spectrum representation—or domains—are two equally valid ways of studying any signal. Each domain provides insight into what the communications signal looks like and how it must be handled. It is possible to take the representation in one domain and transform it to the other domain using the theory of Fourier transforms. Instrumentation such as the spectrum analyzer is available to do this easily and conveniently. The spectra of signals before and after modulation, digital signals, and combined signals are also studied in this chapter, together with the power of signals and their frequency components.

2.1 TIME AND FREQUENCY DOMAINS

It is customary to think of a signal as a voltage, current, or power value that changes with time. As time progresses, the signal varies. The most common way of representing this is shown in Figure 2.1a, where the horizontal axis represents time and the vertical axis represents the amplitude, which may be in volts, amperes, or any other unit that is appropriate. This is how an oscilloscope would show a signal on its screen.

Signal amplitude versus time is not the only way of representing a signal precisely. An equally valid way is to use the horizontal axis to show frequency values that are present in the signal, and the vertical axis to show the strength, or amplitude, of each frequency component. An example of this representation is shown in Figure 2.1b. The figure indicates that this particular signal contains a component at 1000 Hz, a much larger component at 1200 Hz, and a small amount at 800 Hz. These three frequency components form the entire signal when they are added together.

The concept of discussing signals in terms of their frequency and frequency components is not a new one. Broadcast TV and radio stations are identified to their viewers and listeners by their frequencies. Identification of an FM station as "104.1 on the dial" means that its frequency components and bandwidth are centered at 104.1 MHz. This gives no indication of the station's signal versus time, but still fully identifies the station signal. Spectrum and bandwidth are also related to the idea of frequency components. Time does not enter directly into the initial discussion of signal spectrum or bandwidth.

A *filter circuit* is essentially a frequency-oriented circuit. It accepts at the input signal (which can have any shape versus time), and then passes all frequency components below the cutoff frequency f_c (for a low-pass filter) or above f_c (for a high-pass filter). A bandpass filter passes only those frequency components that fall within its passband. No element of time is needed to discuss the effect of a filter on an input signal.

The description of a signal versus time is called the *time-domain* representa-

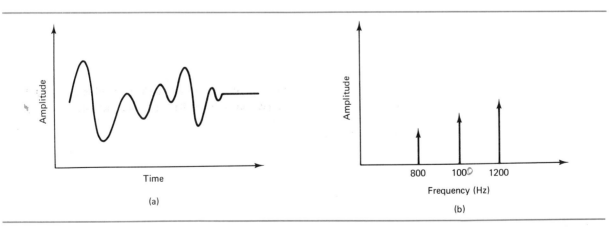

Figure 2.1 (a) Typical signal amplitude versus time; (b) typical signal amplitude versus frequency.

Fourier Analysis

Fourier analysis is named after Jean Baptiste Fourier, who, in 1822, developed some of the basic concepts of mathematics that we now apply to signal analysis applications. Using Fourier analysis, the equation that describes the signal in one domain can, with equal validity, be transformed into the equation that describes the signal in the other domain.

The mathematics can become very complex. The basic equation of Fourier analysis that links a signal $f(t)$ in the time domain with its representation $S(f)$ in the frequency domain is

$$S(f) = \int_{t=-\infty}^{t=+\infty} [f(t)e^{-j2\pi ft}]\, dt$$

This formula shows that the frequency-domain equation $S(f)$ is equal to the integral of the time-domain description $f(t)$, over the range of time from $-\infty$ to $+\infty$, when $f(t)$ is multiplied by the exponential factor. (*Note:* j is the imaginary operator used in signal analysis.) $S(f)$ also shows the phase of each of the component frequencies. There is a similar formula that allows the function $f(t)$ to be obtained from $S(f)$.

---

tion. The description of the signal versus its frequency components is known as the *frequency-domain* representation. Both are equally valid, and neither is better than the other. They are just two different perspectives on the same signal. It is possible to derive the description in one domain from the description in the other, although for different applications the time domain may be more convenient and suitable, and in other applications the frequency domain is a better choice. The theory that links the two domains, and makes it possible to go from one to the other, is called *Fourier analysis*.

There are three ways to obtain the frequency-domain representation from the time-domain representation:

1. By actually performing the mathematical integration if the equation of $f(t)$ is known. This is the theoretical ideal. Even when it is known, however, many functions $f(t)$ may not have known mathematical integrals and cannot be solved "on paper" for $S(f)$.

2. By breaking the time-domain signal into small samples and performing a complex series of calculations on the many pieces. This is called the *fast Fourier transform (FFT)* and can be performed by a computer or microprocessor that is programmed to implement the FFT sequence of calculations or *algorithm*. The algorithm itself requires many thousands of numeric calculations, but this is practical to implement with modern processors.

3. By using an instrument called a *spectrum analyzer* to provide the frequency-domain equivalent, the frequency spectrum. A spectrum analyzer requires a real signal as its input, in contrast to using the equation $f(t)$. This instrument is discussed in the next section.

Review Questions

1. What is the time-domain representation of a signal?
2. What is the frequency-domain representation of a signal? Is it more or less useful than the time domain? Explain.

3. What is the nature of the mathematical equation that relates time and frequency domains? What does it show?

4. Why is it actually often not practical or possible to perform the integration of the equation that relates the two domains?

5. What is the FFT?

2.2 SPECTRUM ANALYZER

The spectrum analyzer is able to take a signal in the time domain and show to the user the signal spectrum in its frequency-domain representation. Spectrum analyzers are available with different ranges and capabilities. Some are designed for handling lower-frequency signals such as those from voice, music, or other audible sources (and therefore called audio signals). Other spectrum analyzers can define the frequency spectrum of signals in the broadcast radio and TV bands, up to about 100 MHz. More advanced analyzers are designed to work with signals at several hundred megahertz and higher frequencies. A typical spectrum analyzer instrument is shown in Figure 2.2.

Construction of a Spectrum Analyzer

One way to build a spectrum analyzer is to use many bandpass filters. Consider a spectrum analyzer designed for use from 0 to 20 kHz, the range of frequencies audible to the human ear. This analyzer can be built with 20 contiguous filters, each with a 1000-Hz passband but with a different center frequency (Figure 2.3). The first filter passes only those frequencies from 0 to 1 kHz, the second from 1 to 2 kHz, the third from 2 to 3 kHz, all the way up the last filter, which has a passband from 19 to 20 kHz. The input time-domain signal is passed through all the filters simultaneously. In operation, the output of each filter is directly proportional to how much of the input signal frequency spectrum is in each band. These

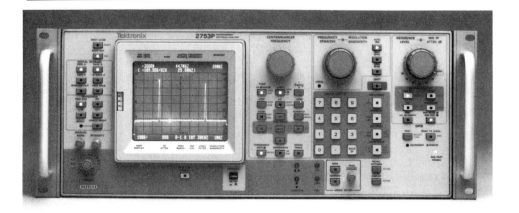

Figure 2.2 Spectrum analyzer (courtesy of Tektronix, Inc.).

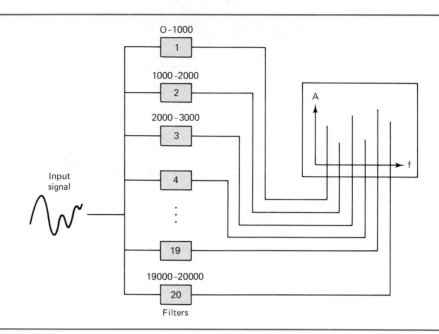

Figure 2.3 Block diagram of a spectrum analyzer built with many individual filters.

output values are then scanned and displayed on the standard cathode-ray-tube (CRT) screen of the analyzer, which shows the frequency values on the horizontal axis and their magnitudes on the vertical. As the input signal changed, the relative shape of the frequency spectrum would also change.

This spectrum analyzer design has a *resolution,* or selectivity, of 1000 Hz. It cannot divide the input signal spectrum into smaller units, because the filters used have a 1-kHz bandwidth. The resolution indicates to what "fineness" the input signal spectrum can be observed. If greater resolution is needed, narrower pass-band filters are used, and more filters are needed.

To see the frequency spectrum with 20-Hz resolution, which is common in audio engineering and voice analysis, 1000 filters would be required. This large number of filters is impractical to build into any instrument. Instead, a single narrow-bandwidth filter that has a tunable frequency is used. This filter can be electronically controlled to have its center at any value within the entire range. The circuitry of the spectrum analyzer makes this filter scan through the entire range of interest, and synchronizes this filter scanning with the sweep of the display across the CRT screen. The user sees the same type of graph as with the previous multiple-filter design.

A single tunable filter is used at higher frequencies because it is extremely difficult to build many narrow-bandwidth filters for VHF and higher frequencies. A single accurate, high-precision tunable filter is used instead. Another advantage of the single-filter approach is that a single filter can be designed with an electronically variable bandwidth. The user selects, via front-panel settings, a selectivity of 1 Hz, 10 Hz, or any other value that the filter in the instrument allows.

Many of the newer spectrum analyzers do not use filters at all. Instead, they use the FFT approach. The input signal is converted into digital computer-compatible format, and then numerically processed with the FFT algorithm by a

2.3 Fourier Analysis Examples

specially designed and programmed microprocessor in the instrument. The results of these FFT calculations are used to drive the CRT in the same way as with the filter design.

Review Questions

1. What is a spectrum analyzer? What does it show?
2. How can a spectrum analyzer be built from many bandpass filters? How does such an analyzer operate?
3. What determines the resolution of a spectrum analyzer? How can the resolution be increased?
4. What alternatives are there to the multiple-filter analyzer? How do they operate? Where are they especially useful?

2.3 FOURIER ANALYSIS EXAMPLES

The formula used in Fourier analysis to express the time waveform $f(t)$ integrates time from $t = -\infty$ to $t = +\infty$. Of course, no real signal in a system spans that time period, but it makes the analysis much easier. For most cases the difference in the result is negligible or can be dealt with separately. Another thing that is done to make the analysis easier is to center the time waveform at $t = 0$. Once again, this is not always the case in the real world, but it is not a problem since any point in the waveform can arbitrarily be called the $t = 0$. If necessary, the final results can be shifted to accommodate the fact that the waveform was not really centered at 0, but this is seldom necessary.

For the frequency spectrum $S(f)$, the analysis also has some aspects that are used to make analysis easier but are simplifications of the real world. The range of frequencies that result from the equation goes from $f = -\infty$ to $+\infty$ hertz. Our normal definition teaches us that frequency must be a positive number, however. The negative frequencies are a mathematical convenience and really represent the same values as the positive frequencies.

The simplest example of the time/frequency-domain pair is the sinusoidal wave. A sine wave of 100 Hz in the time domain looks like Figure 2.4a. In the frequency domain there is a single component (Figure 2.4b), whose amplitude is equal to the amplitude of the sine wave. When the sine-wave amplitude is doubled, the time and frequency graphs are as shown in Figure 2.5a and b. Note that the frequency figure has the single component at the same location, but with double the amplitude.

Example 2.1

The sine wave changes to 200 Hz, but at the original amplitude. What are the time and frequency representations?

Solution

They are shown in Figure 2.6a and b. Note that the frequency component is at the new frequency value, but with a different amplitude.

Chap. 2 Fourier and Spectrum Analysis

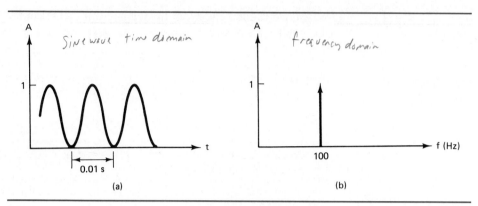

Figure 2.4 Sine wave at 100 Hz: (a) time domain; (b) frequency domain.

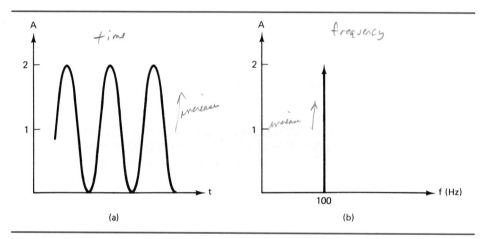

Figure 2.5 Sine wave of Figure 2.4, but with twice the amplitude: (a) time domain; (b) frequency domain.

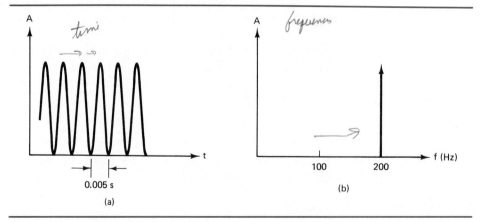

Figure 2.6 Sine wave of Figure 2.4, but at twice the frequency: (a) time domain; (b) frequency domain.

2.3 Fourier Analysis Examples

Fourier Analysis and Phase

Fourier analysis also shows the phase of the frequency components of a waveform. For example, a signal represented by the equation $f(t) = \cos 2\pi ft$ and one represented by $f(t) = \sin 2\pi ft$ have the same shape and the same frequency component. Yet they are not exactly the same waveform, since one is offset in time from the other. The cosine signal has a relative phase of 0°, while the sine wave has a phase of 90°. This *phase difference* will be studied in more detail in Chapter 6.

The frequency spectrum that results from Fourier analysis shows this, usually with a separate graph of the phase of each component frequency (Figure 2.7). A signal with a spectrum component at $\cos 2\pi ft$ is different from one that has the same component but with a different phase, such as $\cos(2\pi ft + \phi)$ where ϕ is the phase shift of the basic cosine waveform. In some applications, the information in a signal is contained primarily in its frequency components; in other situations, the phase of the component is also important.

There are some points to note about the frequency spectrum picture for complex signals: It is much easier to draw, and it provides more useful information than does the time representation. In fact, the ideas of Fourier analysis, spectra, and sine waves are very important for one primary reason: Any signal, no matter how complicated in the time domain, can be fully and perfectly made up from the sum of individual sine waves of the correct frequency, phase, and amplitude. In other words, a frequency-domain picture of any signal can be developed and these frequency components, when added (and subtracted, for some components) together, are identical to what is observed in the time domain. The final time-varying signal can be "built up" using many sine-wave signal generators, each set to the correct frequency and phase and with the right amplitude, then added to and/or subtracted from each other (Figure 2.8).

The frequency domain also shows differences between signals that are very similar in many respects. If both a flute and a violin play the note middle C, they

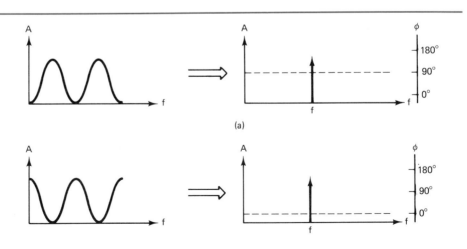

Figure 2.7 Time and frequency domains, with phase, of (a) a sine wave and (b) a cosine wave at frequency f.

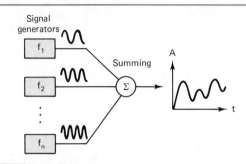

Figure 2.8 Use of separate signal generators to form a final waveform, an analogy to spectrum analysis, and the frequency domain.

are causing oscillations at the same fundamental frequency of 440 Hz, yet they sound very different. This difference is due to the total frequency spectrum of each. Both have the 440-Hz oscillation at the fundamental frequency, and both produce additional oscillations at integral multiples of the fundamental, called *harmonics* (880, 1320, 1760, and so on, hertz). The difference is that each instrument produces different relative amounts of these harmonics (Figure 2.9), which accounts for the different way they sound. Only the frequency spectrum shows this clearly; the time-domain graphs would be very complicated and confusing. A music system needs to have enough bandwidth to pass the fundamental and the significant harmonics if the two instruments, playing the same note, are to be identified by the listener. If the bandwidth is too low, so that only the fundamental and perhaps one harmonic get through, the two instruments will begin to have a dull, flat sound and sound very similar to each other.

Spectra of Basic Waveforms

The spectra of the basic square wave and triangle wave show how a fundamental and harmonics combine in various proportions to create different time waveforms. Figure 2.10a shows a square wave with amplitude $2A$ (from $-A$ to $+A$) and frequency f, together with the equation that describes each of its frequency com-

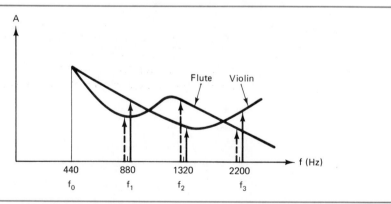

Figure 2.9 Harmonics for note middle C (440 Hz) for a flute as compared to a violin.

2.3 Fourier Analysis Examples

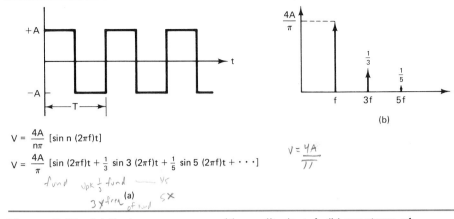

$$V = \frac{4A}{n\pi}[\sin n\,(2\pi f)t]$$

$$V = \frac{4A}{\pi}\left[\sin(2\pi f)t + \frac{1}{3}\sin 3\,(2\pi f)t + \frac{1}{5}\sin 5\,(2\pi f)t + \cdots\right]$$

(a)

Figure 2.10 (a) Basic square wave with amplitude $\pm A$; (b) spectrum of this square wave.

ponents. Note that only odd harmonics are involved (fundamental f, $3f$, $5f$, etc.) because when the harmonic number n is even, the $\sin n\pi/2$ factor is 0. The magnitude of each component is given by the coefficient of the $\sin 2\pi(nf)t$ term and is graphed in Figure 2.10b. The fundamental magnitude is $4A/\pi$, the first harmonic at $3f$ has magnitude $4A/3\pi$, and so on.

Example 2.2

What are the magnitudes of the first three frequency components when $A = 1$ V?

Solution

At the fundamental ($n = 1$) it is $(4 \times 1)/\pi = 1.27$ V; for the next harmonic ($n = 3$), it is $(4 \times 1)/3\pi = 0.424$ V; for the next harmonic ($n = 5$), it is $(4 \times 1)/5\pi = 0.255$ V.

The spectrum of a triangle waveform of the same amplitude and frequency is shown in Figure 2.11a, together with the equation that describes its frequency components. Like the square wave, it contains only odd harmonics of the fundamental frequency, but in different proportions, as shown in Figure 2.11b. Compared to the square wave, the magnitude of these components drops off much more rapidly with increasing frequency. This means that the triangle waveform is made up of a greater proportion of lower-frequency components, with far less of the higher-frequency components added to the total waveform.

Review Questions

1. What assumptions and simplifications are usually made for Fourier analysis in the time domain? In the frequency domain? Why is this usually not a problem?

2. What is the frequency-domain graph of a sine wave?

3. What happens in the frequency domain when the sine-wave amplitude doubles? When the frequency doubles?

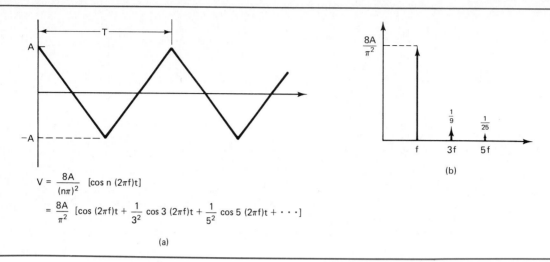

Figure 2.11 (a) Basic triangle wave with amplitude $\pm A$; (b) spectrum of a triangle wave.

4. Why is the frequency domain very useful for representing complex signals?
5. What does the frequency domain show about needed bandwidth? How does it do this?

2.4 MODULATION AND THE FREQUENCY SPECTRUM

Modulation is the process by which two signals are combined, with each signal affecting the other. The resultant signal contains the information of both, but with a different spectrum. There are three ways to combine these two signals—amplitude, frequency, or phase modulation—and the resulting change in spectrum is different with each modulation type used. The signal that is modulated by the information-bearing signal is called the *carrier*.

One of the reasons to modulate is to get the frequency components of the resultant signal into a different part of the frequency spectrum. For example, it would be impossible to have different audio signals with different frequency assignments within the broadcast band without modulation, and all radio stations would be at the same frequency. By modulating, it is possible to shift each user signal to a different frequency so that there is no overlap in frequency components. Modulation can also change the frequency components in the original signals, so that the result of modulation not only moves all the components to new values, but in some cases creates new or additional frequency components. In the time domain, it would be very hard to observe this. In the frequency domain, the changes are clear and obvious.

2.4 Modulation and the Frequency Spectrum

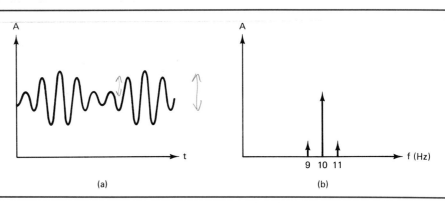

Figure 2.12 1 Hz modulating (AM) 10-Hz carrier: (a) time domain; (b) frequency domain.

An example shows this clearly. In amplitude modulation, studied in detail in Chapter 4, the information signal is used to modulate a carrier signal of much higher frequency. The combined signal is what is transmitted. Figure 2.12a shows a small-amplitude 1-Hz signal modulating a larger 10-Hz carrier in the time domain. The corresponding frequency domain picture is shown in Figure 2.12b.

From the time-domain figure, it is nearly impossible (or at least difficult) to state the important effects of the modulation process. Now look at the frequency-domain drawing. Here it is clear that the original signal, when amplitude modulated, produces new frequency components. One main component is at 10 Hz, and there are two additional frequency clusters on both sides of 10 Hz. These clusters are replicas of the original message signal. By looking at this frequency representation, it is clear that the bandwidth of the result of the modulation is greater than the original signal bandwidth; in fact, it is twice as great. Amplitude modulation is studied in detail in Chapter 4.

The same carrier signal can be *frequency modulated* (Chapter 6) instead of amplitude modulated, which means that the carrier frequency is influenced directly by the modulating signal. In the time domain, the signal that results from the modulation process is confusing (Figure 2.13a), like the result of amplitude modulation. If there are meaningful similarities and differences between the AM and

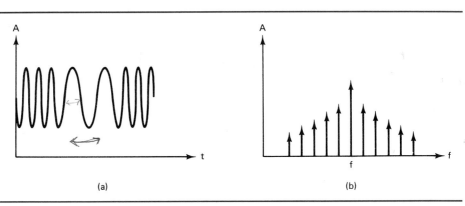

Figure 2.13 1 Hz modulating (FM) 10 Hz: (a) time domain; (b) frequency domain.

FM results, it is difficult to quantify them. However, the frequency spectrum for the modulation shows the differences (Figure 2.13b). The spectral components of the modulation result are spread out over a relatively wide range, and the original signal cannot be located at a single spot on the new spectrum. Nevertheless, the signal is there and in fact can be recovered with the proper demodulation circuitry. The bandwidth required to handle this FM result is much greater than the bandwidth of the original signal alone. (*Note:* The exact bandwidth that results from the FM process depends on how the modulation is implemented. The figure shows one possibility.)

The frequency-domain graph of modulation shows the two effects of a modulation process: The basic spectrum of the signal changes to a new set of values, and the bandwidth of the signal changes as well. Any communications system that is performing modulation or handling modulated signals must be designed accordingly. In addition, the bandwidth of the signal can be seen simply by looking at the spread of frequencies that a signal occupies in the frequency domain. It would be practically impossible to determine the bandwidth simply by looking at the signal versus time display.

Review Questions

1. What are some of the effects of modulation?
2. How is the result of modulation seen in the time domain? In the frequency domain?

2.5 SPECTRA OF DIGITAL SIGNALS

The time representation of an analog signal such as voice or music is very complex, and the frequency spectrum may also be complex. *Digital* signals, which can have one of only a few allowed amplitudes, have much simpler time-domain graphs. It is important to see what the corresponding frequency domain looks like.

Ideal Digital Signal

The simplest digital signal is the *binary* case, where only two signal values can exist. A typical pattern for these bits of binary information is shown in Figure 2.14a, with an amplitude of A, a *pulse width* of w, and a *period* of T seconds (which is the inverse of the *bit repetition rate* $1/T$, corresponding to a frequency of f pulses/second). These are "ideal" signals, because the corners are perfectly sharp right angles, the tops are flat, and the sides are vertical. The frequency domain of this signal extends all the way out to infinity (Figure 2.14b). There is a fundamental component at 0 Hz, and additional components at both plus and minus $1/T, 2/T, 3/T, \ldots$ (corresponding to $f, 2f, 3f, \ldots$). The magnitude of these components is shown by the outer *envelope* (dashed line) of the figure.

Some specific cases are shown in Figure 2.15 for various values of w and T. As the period T becomes larger, the fundamental repetition rate frequency $1/T$ decreases, and more frequency components exist in a given part of the frequency spectrum. At the same time, the spectrum "flattens out" and the lower-frequency

2.5 Spectra of Digital Signals

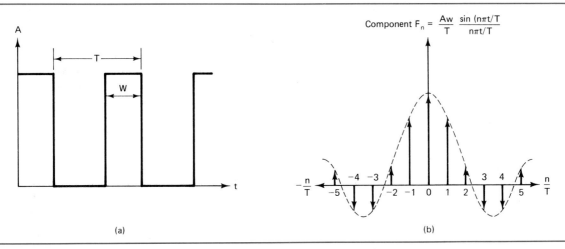

Figure 2.14 Series of digital pulses with width w and period T: (a) time domain; (b) frequency domain.

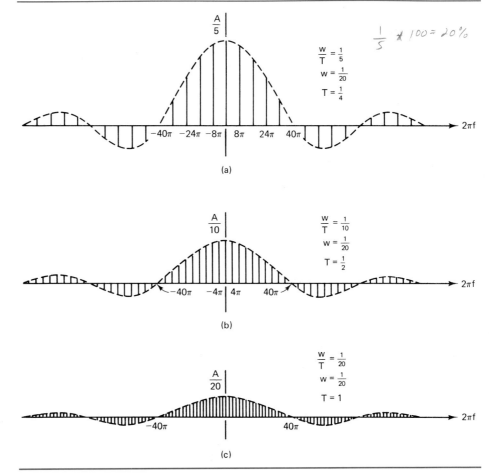

Figure 2.15 Some specific examples of Figure 2.14 for various w and T (from B. P. Latri, *Signals, Systems, and Communications*, copyright John Wiley & Sons, Inc.).

components have smaller magnitudes compared to their previous magnitudes when T was smaller, but there are more of these components. The overall shape of the spectrum envelope does not change when the value of period T changes. Only the number of frequency components and their value changes, within the same envelope.

Compare this to the spectrum of the square wave in Section 2.3. That spectrum had only odd harmonics, whereas this waveform has both odd and even harmonics. The difference is because the basic square wave had a width equal to one-half the repetition rate, whereas this waveform is more general and can have any period.

The total band of frequencies extends to infinity for this ideal case of perfect digital signals. This is because high frequencies contain the energy and information about the sharp corner, and the sudden change in value of the signal at the corner requires an infinite number of the high-frequency components, although in smaller and smaller amounts. Sudden changes in the time domain are the equivalent of these higher frequencies in the frequency domain.

Realistic Digital Signals

A more realistic digital signal is shown in Figure 2.16a. This has some rounding at the corners, so the time-domain change is not so sudden. In the frequency domain (Figure 2.16b) the bandwidth needed is now less wide and drops off more quickly from its peak values. As the signal has more rounding of the corners, the bandwidth needed is less since fewer high-frequency components are needed to convey this slower-changing corner information. Although the bandwidth still extends to infinity, more of the signal energy is concentrated around the peak. In the time domain, a signal with more harmonic content has faster changes and sharper contents, whereas a signal with little harmonic content is slower changing and more rounded.

There is another way to look at the implications of this. If a channel used for communications is limited to a certain amount of bandwidth, that is a constraint on how quickly the input signal can change. For a system intended to carry high data rates—many bits per second—the bandwidth must be correspondingly larger. When the bandwidth is too small, the signal waveform will be rounded off more and more until it no longer resembles the original waveform. Instead, it will

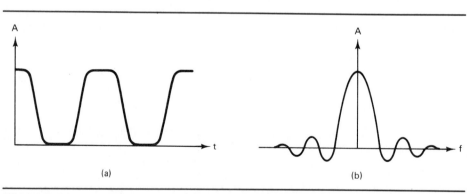

Figure 2.16 Realistic (nonideal) digital signal: (a) time domain; (b) frequency domain.

2.5 Spectra of Digital Signals

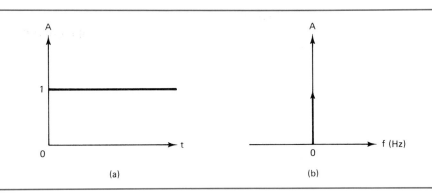

Figure 2.17 DC signal: (a) time domain; (b) frequency domain.

start to look more and more like a sine wave. The original information bits will effectively be lost. To carry the digital signals with low distortion, there must be enough bandwidth.

A great deal of study has been devoted to the amount of bandwidth needed to reproduce digital signals faithfully. The amount depends on how much distortion the receiver can tolerate before it decides incorrectly what was sent. Typically, a bandwidth (in hertz) of five to 10 times the bit rate (bits or pulses/second) is needed. Therefore, to send 10,000 digital bits/second requires a bandwidth of 50 to 100 kHz.

It is also useful to study the time- and frequency-domain graphs of two signal extremes: a steady, unchanging time signal and a sharp "spike" or *impulse*. For the constant signal (Figure 2.17a) the frequency spectrum is very simple and consists of a line at 0 Hz (Figure 2.17b), with an amplitude equal to the amplitude of the signal in the time domain. This is logical because a constant signal is the same as a 0-Hz signal and has only that one frequency component.

Now look at a signal that is an impulse in the time domain: Figure 2.18a. The frequency spectrum extends from 0 Hz to infinity, with a constant amplitude (Figure 2.18b). An realistic impulse that is not infinitely sharp, such as Figure 2.19, has a spectrum that extends out as far, but the magnitudes of the higher-frequency components are less. This type of impulse signal occurs due to atmo-

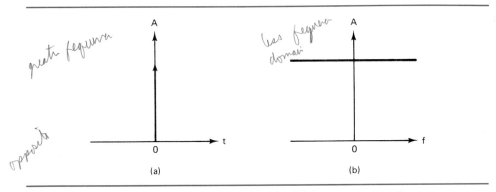

Figure 2.18 Impulse signal: (a) time domain; (b) frequency domain.

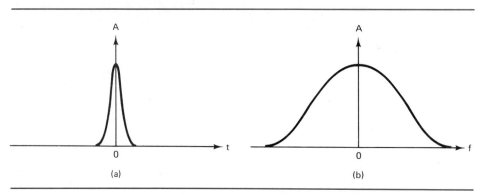

Figure 2.19 Rounded impulse: (a) time domain; (b) frequency domain.

spheric noise, the sudden switching on or off of a circuit, or electrical interference from electrical equipment such as motors and generators.

Note the complementary relationship between a constant signal in one domain and its appearance as a spike in the other domain. The reason for this is that any sudden change in one domain requires a broad spectrum in the other. A signal with a sudden change in amplitude versus time needs a wide spread of frequencies to represent that signal accurately. A slower change (the less sharp impulse or a signal with more rounded corners) requires less bandwidth. Communications systems that carry higher data rates have sharper, faster changes, and must have more bandwidth, since a limited bandwidth signal cannot pass sudden changes in amplitude. Signals that change more slowly require less bandwidth in the communications system.

Review Questions

1. What does an ideal digital signal look like in the time domain? In the frequency domain?
2. What happens in the time and frequency domains as the repetition rate of the digital pulses increases?
3. What is the frequency spectrum corresponding to a sudden change in the time domain? What about the spectrum shape for a slow change in the time domain?
4. What are the implications of the relationship between sudden changes in one domain and the resulting representation in the other domain?
5. What is the relationship between bandwidth and bit rate? Why is this so?

2.6 SUPERPOSITION

Many signals that vary with time are composed of the sum of several individual signals. The spectrum that results is often determined by *superposition*, which means that the spectrum of a signal that is a sum of other signals can be found by adding together the spectra of the individual signals themselves. The equation of

2.6 Superposition

the Fourier transform (Section 2.1) that relates the spectrum $S(f)$ to $f(t)$ is an integral. One of the theorems of integration is that the integral of a sum is equal to the sum of the individual integrals of the parts that make up the sum. In Fourier analysis this is called the superposition principle.

One common case is a sine wave that is offset from 0, shown in Figure 2.20a. This is really the sum of a steady signal (called the *dc signal,* for "direct current," although this is a misnomer) and the sine wave itself (Figure 2.20b). The spectrum is therefore the sum of the spectrum of the dc signal and the sine wave (Figure 2.20c). This clearly shows that this waveform has a component at 0 Hz and one at the frequency of the sine wave. If the offset value doubles, only that component of the spectrum doubles (Figure 2.21). Similarly, if the sine-wave frequency changes (Figure 2.22) only that part of the spectrum changes.

Looking back at the square-wave spectrum discussed in Section 2.3, the first component was at the fundamental frequency f. If this square wave had an amplitude of 0 to $2A$ instead of $-A$ to $+A$, the only difference in the spectrum would be the addition of a 0-Hz term of $+A$ volts to account for the dc offset. The rest of the components are the same, at f, $3f$, $5f$, and so on.

The superposition principle is very useful when studying frequency components in a signal, for several reasons. First, it makes it easier to determine the frequency spectrum of a signal in many cases. Second, it allows someone to look at the spectrum and relate some of the frequency components to those time-domain signals that may be responsible for them. In many systems, certain frequency components cannot be tolerated for technical reasons, and by superposition it is possible to see what time waveform may have to be eliminated or changed.

Such *dc offsets* are sometimes necessary for proper functioning of the circuitry, and in other cases are undesired. Understanding the effect of an offset on the spectrum, by superposition, helps understand the effect of such an offset or its absence. Many systems use transformers between amplifier stages to provide

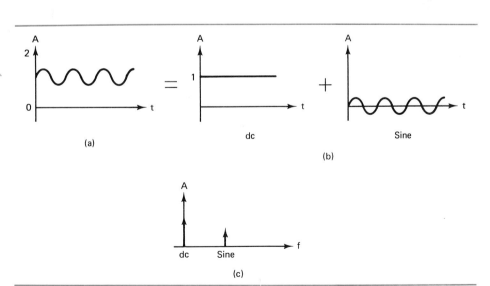

Figure 2.20 (a) Sine + dc offset, as a single signal, time domain; (b) sine wave and dc offset of (a), but as separate signals; (c) frequency domain of (a).

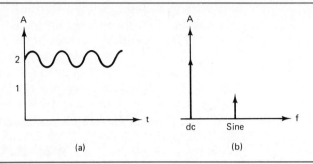

Figure 2.21 Same as Figure 2.20a, but with double the dc offset: (a) time domain; (b) frequency domain.

impedance matching, signal coupling, and isolation. Transformers cannot pass a steady dc waveform, however. In fact, a dc signal may cause the transformer core to *saturate* magnetically and it will not even be able to pass the ac signal. The transformer specification may state "no more than 0.1 V dc can be accepted." Therefore, the dc component must be eliminated before it reaches the transformer.

A capacitor in series between the signal source and the transformer is usually used to eliminate or reduce this component. The Fourier transform shows the effect on the signal after the capacitor is installed and whether it is within specs. The Fourier perspective is especially useful when the signal is much more complex (Figure 2.23a) and it is much harder to see what the dc offset is in the time domain. In the frequency domain it becomes more obvious (Figure 2.23b).

Superposition can be used to derive the time-domain waveform from a frequency spectrum. Consider a system that is sending a 10-Hz signal of 1 V and there is 0.2 V of 60-Hz electrical noise picked up from nearby ac power lines. The frequency spectrum is shown in Figure 2.24a. If this signal is studied on a regular oscilloscope, a technician sees Figure 2.24b, which has the two sine waves added together. The technician can see that a 60-Hz signal is corrupting the desired signal and possibly affecting the performance of the entire system.

For more complicated signals, though, the noise may not be recognizable so easily. Suppose that the signal being transmitted ranged in the frequency spectrum from 1000 to 2000 Hz and looked like Figure 2.25a on the oscilloscope. If there were a source of noise corrupting this signal, it would be very difficult to see it

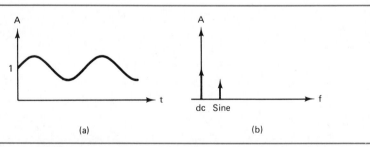

Figure 2.22 Same as Figure 2.20a, but with twice the sine-wave frequency: (a) time domain; (b) frequency domain.

2.6 Superposition

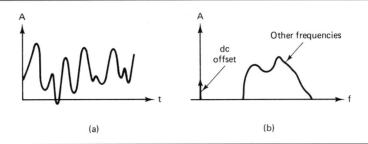

Figure 2.23 Complex signal with embedded dc offset: (a) time domain; (b) frequency domain.

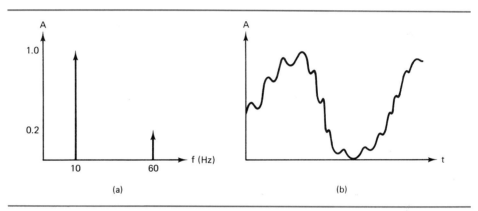

Figure 2.24 A 10-Hz, 1-V ac signal with 0.1-V, 60-Hz noise: (a) frequency domain; (b) time domain.

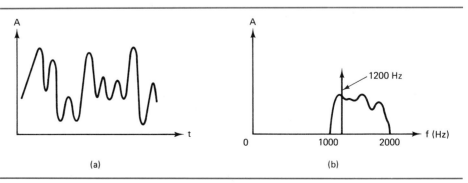

Figure 2.25 Complex signal in 1000 to 2000 Hz, with 1200-Hz noise: (a) time domain; (b) frequency domain.

since the desired signal is so varying that the noise is obscured. However, the frequency spectrum may look like Figure 2.25b with a spread of signal magnitudes in the region 1000 to 2000 Hz, but a sharp, larger value at 1200 Hz. By the superposition rule, the overall spectrum is the result of the sums of the individual signal spectrums. Therefore, this observed signal could be the result of the original signal, which ranted from 1000 to 2000 Hz, plus an additional interfering signal at 1200 Hz that has been coupled in from another part of the system. The observed frequency spectrum shows the many frequencies that make up the observed signal. Since noise is usually an addition to the desired signal, it will show up in the frequency domain quickly as an unexpected or too strong spectral component.

Review Questions

1. What is the superposition principle? Why is it valid?
2. Why is superposition useful?
3. How does using superposition in one domain allow better understanding of a signal and components in the other?
4. How does noise often show up in the time domain? In the frequency domain?

2.7 POWER AND ENERGY SPECTRA

To this point, Fourier analysis has been used to examine and understand the frequency- and time-domain representations of a signal: how they are both valid perspectives, both show characteristics of a signal, and each domain is better than the other for some types of signal study in some applications. The magnitude (and sometimes phase) of the frequency components has been shown as an equivalent way to examine a time-varying signal.

There is another important use for the concept of Fourier analysis and frequency spectrum. This has to do with the energy and power of a signal at the various frequencies. A communications system has the goal of transferring the electromagnetic signal energy from the source (the message sender) to the receiver (the message user). The communications channel—whether it is air, wire, circuitry, or anything else—must allow this energy to pass through it. Therefore, the relationship between the amount of energy that is sent and the amount that is passed through the system and finally received is critical. To understand this, it is often necessary to look at the spectrum of signal energy and power versus frequency, and to understand how it is defined.

In electronics, the definition of power is based on the square of the current or voltage: $P = I^2 \times R = V^2/R$, where current or voltage is used to measure the magnitude of the signal. Power is the rate at which energy is delivered, so the "squaring" applies to the energy value. The Fourier equation that relates time function $f(t)$ and frequency function $S(f)$ is used, except that $f(t)$ is replaced by its square $[f(t)]^2$. Note that the integral of $[f(t)]^2$ is usually not the same as the square of the integral of $f(t)$. The entire integration process must be redone. Once

2.7 Power and Energy Spectra

again, this can be done either mathematically on paper, by a spectrum analyzer with some additional circuitry, or with an FFT algorithm.

Here is an example of how the magnitude spectrum and the energy/power spectrum compare. The frequency components of a series of digital pulses was shown in Figure 2.14b. The power versus frequency spectrum is shown in Figure 2.26. Note how the power spectrum is not just the square of the value in the frequency spectrum shown above. Instead, it is a new graph, with more bumps, or lobes, and a much larger central lobe. The magnitude of the power component at each frequency is a positive value, since power is never negative.

Another waveform that is important is the voice signal. Many communications systems are designed to handle voice signals—the telephone system and the audio portion of a TV signal or radio are examples. Of course, every voice is different, but all voice spectra have a lot in common. In a typical voice energy spectrum, most of the power is concentrated in the region from 300 to 3000 Hz, with much less in the band from 3000 to 8000 Hz. This means that a working communications system can be designed that uses a minimum bandwidth of just 3 kHz, and in fact the telephone system and other voice communications systems often do that.

At the same time, the power above 3 kHz must represent some information, and it does. Research has shown that the higher-frequency power is what allows us to distinguish one person's voice from another. This is why we can understand quite well what someone is saying on the telephone but sometimes cannot recognize the person's voice. The higher frequencies do not carry the message information but do provide information to the listener as to who the speaker is. Of course, where bandwidth must be conserved, the higher-frequency power is not necessary, and we can accept a system that just conveys the message information itself. By studying the frequency and power spectrum of a signal, the system designers can understand what bandwidth is needed, the effects of *filtering* (passing a signal through a filter), and how the limitations of the channel cause problems with the received signal.

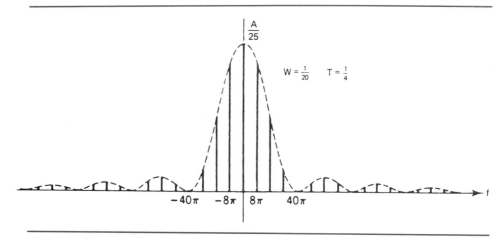

Figure 2.26 Power spectrum of digital pulses of Figure 2.14 (from B. P. Latri, *Signals, Systems, and Communications,* Copyright John Wiley & Sons, Inc.).

Review Questions

1. How are (a) the energy spectrum and (b) the power spectrum of a signal related to the Fourier analysis equation?
2. What is the significance of the power spectrum? Why is it useful to understand it?
3. What is the power spectrum of a series of digital pulses?
4. What are the power spectrum components for a sine wave at frequency f? (No calculations are needed—just think.)
5. What is the spectrum of power versus frequency for voice? What span of frequencies conveys information? What does the energy outside this span convey?

SUMMARY

Signals that carry information can be viewed as changing versus time or as being composed of many unique frequencies added together. Both the time- and frequency-domain representations are useful, depending on the situation. They are two different perspectives on the same signal, and they are linked to each other by the Fourier transform. By measurement with instruments such as the spectrum analyzer, or by mathematical analysis, we can translate a signal from one domain to the other. Fourier and spectrum analysis show that a change in a signal in one domain is seen in the other. A rapid change in the time domain corresponds to a wide band of frequencies, while a slower change versus time requires less bandwidth. Therefore, a higher data rate requires more bandwidth than does a lower rate. This is especially visible for digital signals, where the bandwidth is determined by the sharpness of the corners of the digital waveform and its repetition rate. The spectrum of a signal indicates what frequency components are most critical and where the signal energy is. The Fourier transform of a signal can also be determined using the superposition principle, which shows that the transforms of individual components can be added to form the transform of the sum of the components.

Summary Questions

1. Compare the features of a time- and a frequency-domain representation of a signal. How are they similar? How do they differ? When is one more useful than the other?
2. How are the two domains related? What is the nature of the equation that links them?
3. How can a time-domain signal be converted to the frequency domain? What instrumentation can be used? How does it operate?
4. What is the idea of the fast Fourier transform?
5. Compare a spectrum analyzer with many filters to a single, tunable filter design in terms of complexity and resolution.
6. What does a sine wave look like in both the time and frequency domains?
7. How do the figures of question 6 change when the sine-wave frequency doubles? When the sine-wave amplitude doubles?
8. How can bandwidth requirements be seen or measured in the time domain? In the frequency domain?

9. Compare the spectrum of a square wave and a triangle wave with identical amplitude and frequency, in terms of specific frequency components and their magnitudes.

10. How does modulation affect the time-domain representation of a signal? How about its effect on the frequency domain?

11. What happens to the frequency-domain graph of a digital signal as the corners get sharper? What is the reason for this?

12. Compare a wide bandwidth in the frequency domain to the signal span in the time domain.

13. Compare a slowly changing signal in the time domain to its frequency spectrum.

14. Why does a higher bit or information rate require a wider bandwidth?

15. What is superposition? What makes it useful in signal analysis in both the time and frequency domains?

16. Why is noise often seen in the time domain but analyzed further in the frequency domain?

17. Why is it important to understand the power spectrum of a signal? What does it show?

18. How does the power spectrum vary versus frequency for a digital signal?

19. What is the total power spectrum for a typical voice? How does studying this spectrum allow reduction in the bandwidth required to send a voice signal?

PRACTICE PROBLEMS

Section 2.3

1. Sketch time- and frequency-domain representations of a 5-Hz sine wave with 2.5-V amplitude.

2. Repeat problem 1 for the same amplitude sine wave, but at 10 Hz.

3. Repeat problem 1 for the original 5-Hz signal, but with twice the amplitude.

4. Show the time- and frequency-domain representations when both the frequency and amplitudes are twice their original value.

5. What is the magnitude of the frequency component at seven times the fundamental frequency for a square wave of amplitude 1 V?

6. What are the magnitudes of the first three spectral components of a triangle waveform with a 1-V amplitude?

Section 2.6

1. A time-domain signal is composed of a 1-Hz 1-V signal and a 0.5-V dc component. Sketch the frequency domain of this signal.

2. A signal in the frequency domain consists of a 0.75-V component at 0 Hz and a 0.5-V component at 10 Hz. Sketch the signal versus time.

3

Decibels and Noise

CHAPTER OBJECTIVES

When you have completed this chapter, you will understand:

- The need and use of the logarithmic decibel scale for signals measurement
- The use of decibels for signal gain and loss calculations
- The sources and effects of various types of system noise
- How noise and signal level are measured and compared

INTRODUCTION

Signals in communication systems span a range of extremely wide magnitude. The decibel scale, which uses ratios and logarithms, compresses this wide span into a smaller, easier to manipulate range of numbers. Decibels are used to compare any two signals and can also be used to measure one signal against another signal of defined value. Decibels allow relative signal gain and loss to be measured easily, and the total gain through a multiple stage system to be calculated by simple addition.

Noise—any unwanted, interfering signal—is a serious problem with any communications system. It is often measured in decibels in relation to the desired signal, and the amount of noise that the circuit itself adds to the noise of the received signal must also be measured as part of the overall system performance.

3.1 SIGNAL MAGNITUDES AND RANGES

The magnitudes of signals used in communication systems span a very wide range, with current, voltage, or power values having highest-to-lowest value ra-

3.1 Signal Magnitudes and Ranges

tios of millions or billions to one (1,000,000 : 1 or 1,000,000,000 : 1). A signal at the receiving antenna may be only a few nanowatts (nW; 10^9 nW = 1 W) and is amplified through the stages of the receiver to tens of watts. The microphone that picks up a speaker's voice produces only a few microvolts (μV; 10^6 μV = 1 V) and become several volts as it passes through the system.

It is possible to use these values when measuring signals at various points in the communications system, but there are drawbacks. Any graph or drawing of the signal as it goes through the various stages needs an extremely long scale or will be difficult to read. An oscilloscope display of the signal at the antenna input versus the audio output at the loudspeaker would be impossible—the screen is too small. Second, any calculations or analysis on the signal strength would involve numbers that are too large or too small to deal with easily. Errors would be commonplace as zeros were accidentally dropped or decimal points put in the wrong position (0.00001 versus 0.0000001, for example). Finally, it is very hard to compare two signals, or a signal at one point to its value at another point in the system. Very often, the important fact is not the absolute value of the signal, such as 1.65 V or 0.023 W, but its value relative to another signal, or its change in value within the circuit.

Fortunately, there is a convenient way to overcome these problems. Instead of using the absolute signal values, a *logarithmic* scale is used. The units of this scale are the *decibel* (dB). This equation defines the dB scale for power:

$$\text{dB} = 10 \log \frac{P_1}{P_0}$$

This equation has two aspects. First, it uses the ratio of two power values, P_1 and P_0, instead of a single absolute value such as watts. This makes the dB scale ideal for comparing two signal values. Second, it takes the base 10 logarithm of this ratio. The effect of this is to compress a wide range of signal values into a much smaller range, one that is easier to deal with. Graphing the values is much easier and takes less space as well. Signal changes and differences are easily seen and compared.

Voltages and dB

Signal voltages can also use the dB scale. Since power P is proportional to the square of the voltage, V^2, substituting V^2 for P results in

$$\text{dB (for volts)} = 20 \log \frac{V_1^2}{V_0^2}$$

which by the rule of logarithms is the same as

$$\text{dB (for volts)} = 20 \log \frac{V_1}{V_0}$$

Therefore, by simply using the factor of 20 as the multiplier, the dB scale is used for voltage ratios. Similarly, it can be used for current ratios with the factor of 20, since power is also proportional to the square of the current.

A note of caution: When using the dB scale with voltage or power, the voltage or current must be measured at points with the same equivalent resistance. This is not an issue when comparing signals at the same point in a circuit, but it can be misleading when measuring the signal magnitude at different points and then seeing by how much it has changed.

Review Questions

1. What is the dB scale for signal measurement? Why is it used?
2. How are decibels useful for comparing the relative magnitudes of two signals?
3. What is the dB equation for power measurements? For voltage?
4. How does the dB scale act to compress the range of numbers needed in signal measurement?

3.2 dB CALCULATION EXAMPLES

Some examples will show how the dB scale can be used and how some dB values correspond to common ratios.

Example 3.1

The dB value for a signal at 10 W compared to one at 0.5 W is

$$10 \log \frac{10}{0.5} = 13 \text{ dB}$$

Example 3.2

A signal enters the circuit with a value of 0.1 V and is amplified to 5 V. The input and output resistances are the same. The dB ratio that shows the gain in magnitude is

$$20 \log \frac{5}{0.1} = 34 \text{ dB}$$

Example 3.3

A signal is amplified 100 times in power. The dB gain is

$$10 \log \frac{100}{1} = 20 \text{ dB}$$

The dB scale can also be used when a signal is reduced in value, either by loss in wires, by the effects of filters, by a resistor, or for any other reason. In these cases, the dB result is negative. Consider a signal whose power is reduced by a factor of 100:

$$\text{dB} = 10 \log \frac{1}{100} = -20 \text{ dB}$$

Example 3.4

In the reverse case of Example 2, the signal is attenuated from 5 V to 0.1 V. The dB value is

$$20 \log \frac{0.1}{5} = -34 \text{ dB}$$

This shows that a gain in signal values by some ratio and a loss in signal values by the same ratio have the same dB value but with the opposite sign.

For two signals having the same value the dB equation becomes

$$\text{dB} = \begin{cases} 10 \log \frac{P_1}{P_0} = 0 \text{ dB} & \text{for power since the log of 1 is 0} \\ 20 \log \frac{V_1}{V_0} = 0 \text{ dB} & \text{for voltage} \end{cases}$$

Conclusion: Two signals that have equal power (or voltage) values have a relative value of 0 dB.

Another important case is when the power is doubled ($P_1 = 2 \times P_0$):

$$\text{dB} = 10 \log \left(2 \times \frac{P_0}{P_0} \right) = 10 \log 2 = 3.010 \quad \text{for power}$$

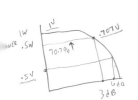

which shows that a 3-dB ratio corresponds to a power factor of 2 (to a very close approximation). Similarly, a −3-dB ratio means a power factor of one-half; that is, the second signal has one-half the power of the first. For voltage, the same calculations show that a voltage ratio of 2 is very close to 6 dB, and −6 dB is a voltage ratio of one-half.

A calculation will also show the relationship between the −3-dB (*half-power*) point and the equivalent voltage ratio. The −3-dB point means that $P_1/P_0 = 0.5$ and that power is proportional to V^2, so the −3-dB power point has $V_1^2/V_0^2 = 0.5$. Take the square root of both sides to solve for the voltage ratio, and $V_1/V_0 = \sqrt{0.5} = 0.707$. At the half-power point, therefore, the voltage (assuming the same resistance) is 0.707 of what it was at the full-power point. A similar exercise shows that the voltage at the 3-dB power ratio is 1.414 times the original value. For example, a signal of 3 V measures 2.121 V when it is at half power, and 4.242 at twice the power.

Figure 3.1 shows the dB values for power and voltage ratios of 10, 100, 1000, and 10,000; a more complete dB chart is given in Appendix B. This chart is used for quickly determining the dB value for many possible power and voltage ratios. Of course, a calculator can also be used, but the chart is often handier.

Ratio	Power (dB)	Voltage (dB)
1	0	0
10	10	20
100	20	40
1000	30	60
10,000	40	80

Figure 3.1 dB values for power and voltage ratios of 10, 100, 1000, and 10,000.

A percentage gain or loss can also be expressed in dB. Suppose that a signal suffers a 10% power loss due to attentuation in the cable. The gain, in dB, is

$$\frac{P_1}{P_0} = \frac{0.9}{1.0} = 0.9$$

$$10 \log(0.9) = -0.46 \text{ dB}$$

also expressed as a 0.46-dB loss.

Example 3.5

A new communications cable is installed and the signal level, in volts, increases by one-half, or 50% (it is now 150% of what it was). What is the increase, in dB?

Solution

$$\frac{V_1}{V_0} = \frac{1 + 0.5}{1.0} = 1.5$$

$$20 \log(1.5) = 3.52 \text{ dB}$$

Review Questions

1. What does a positive dB value indicate? What about a negative one? Does a positive value indicate that the signal is positive, such as +5.7 V?
2. What is the dB value for two signals of the same value?
3. What is the significance of a +3-dB power ratio? Of a −3-dB power ratio?
4. Explain how a dB ratio can be converted to an absolute value.

3.3 dB REFERENCE VALUES

Earlier we showed that the dB value is the ratio of two power or voltage values, not an absolute unit of measurement such as the volt or watt. The convenience of

3.3 dB Reference Values

Determining the Absolute Signal Value from the dB Value

In many instances, the system is specified using dB values whenever possible, but at some point the technician or engineer must calculate the exact value of power or voltage, as an absolute value. The dB formula can be rearranged so that either power value can be determined if the other power value and the dB value are known. For power:

$$P_1 = P_0 \times 10^{dB\ value/10}$$

$$P_0 = \frac{P_1}{10^{dB\ value/10}}$$

For voltage ratios, replace the 10 in the exponent with a 20:

$$V_1 = V_0 \times 10^{dB\ value/20}$$

$$V_0 = \frac{V_1}{10^{dB\ value/20}}$$

For a signal that has been amplified by infinities dB and is measured at infinity × 6, its original value phase

$$P_0 = \frac{1.2}{10^{13/10}} = \frac{1.2}{10^{1.3}} = 0.060\ W$$

If a 0.2-W signal comes into a circuit having a specified gain of 22 dB, its final value will be

$$P_1 = 0.2 \times 10^{22/10} = 31.7\ W$$

A signal of 1.6 V that is amplified by 2.5 dB will become

$$V_1 = 1.6 \times 10^{2.5/20} = 2.13\ V$$

Finally, a signal that comes out the final stage of a 9-dB gain amplifier with a measured value of 15 V must have had an original input value of

$$V_0 = \frac{15}{10^{9/20}} = 5.3\ V$$

using the dB scale, however, has resulted in the use of several common reference values in the electronics and communications industry. These are often used as one of the two values of the ratio. Then, using this reference, all dB values are stated relative to this standard value. Engineers and technicians in industry know from the context the value of the particular reference used, or it is stated somewhere in the technical report if there is any ambiguity.

One common reference is 1 mW (equal to 0.001 W). With this reference, the dB scale is called *dBm*. The letter "m" after "dB" means that the 0-dB point reference is 1 mW. A signal of 200 mW would be expressed in dBm as

$$10 \log \frac{200}{1} = 23\ dBm$$

A very low level signal of one ten-thousandth of a watt would then be

$$10 \log \frac{0.0001}{0.001} = -10\ dBm$$

There are some other standard values that are often used. For higher power circuits, 1 W is commonly used as the reference value. For this case the abbreviation used is *dBW*. An audio amplifier might have an output of 7 dBW. This corresponds to

$$P_1 = 1 \times 10^{7/10} = 5.01\ W$$

Note that 0 dBm = −30 dBW and that 0 dBW = 30 dBm.

Similarly, the voltage ratio dB scale has some standard values. The most common is 1 V, noted by *dBV*. A signal of 8.2 V corresponds to

$$20 \times \log \frac{8.2}{1} = 18.3 \text{ dBV}$$

Example 3.6

What is the dBV value for 25.7 V?

Solution

$$20 \times \log \frac{25.7}{1} = 28.2 \text{ dBV}$$

Example 3.7

Express 0.05 V as dBV.

Solution

$$20 \times \log \frac{0.05}{1} = -26.0 \text{ dBV}$$

Of course, when dB values are expressed using a standard reference, such as dBm, dBW, or dBV, the dB value can be converted back to absolute units (milliwatts, watts, or volts) if needed, as shown in Section 3.2.

The effect of standard reference values is to combine the convenience and advantages of the logarithmic dB scale with the absolute scale, such as volts or watts. The signal magnitude can be expressed in terms of the dB ratio, when needed, and also quickly converted to an absolute signal value since the 0-dB point reference value is indicated by the suffix on the dB. The reference value is often called the 0-dB value, since any signal whose magnitude is the same as the reference will be at 0 dB.

Another dB reference often used is *dBc*, but this does not refer to a specific voltage, current, or power 0 dB value. Instead, it stands for dB referenced to the nominal carrier value. For example, a pure sine wave is applied as the input to a circuit and the circuit output is analyzed with a distortion meter to measure the circuit nonlinearity. For a perfectly linear circuit, the output would be also a sine wave at the same fundamental frequency, but perhaps at a different amplitude, depending on whether the circuit provided gain or loss. This amplitude gain or loss is not of interest, but the appearance of any frequencies besides the fundamental frequency, due to signal distortion caused by the nonlinearity of the circuit, must be determined.

The magnitude of this distortion is often specified as "dB relative to the carrier" or dBc, where the carrier value is the magnitude of the fundamental frequency at the output, whatever that may be. A typical reading for a high-quality, low-distortion circuit is that all outputs—compared output at the original frequency—is −80 dBc, meaning that these distortion results are 80 dB below the undistorted output.

Review Questions

1. What is the idea of standard reference values for the dB scale? Where are these values used?
2. With a standard reference value, how many dB does the reference equal?
3. Explain why dBc does not refer to a specific reference value but is still very useful. Where is it used typically?

3.4 SYSTEM MEASUREMENTS WITH dB

The usefulness of the dB scale is not only that it compresses the wide range of signals into a smaller group of numbers but that it also makes many measurements and types of analysis of system performance much more convenient. There are many cases where the exact, absolute value of the signal in a part of a system is not critical but where the value relative to signals in some other part of the system is critical or must be measured.

A simple example is measuring the response of a filter circuit versus input signal frequency, called the *frequency response*. Here, the parameter of interest is how much the filter reduces the magnitude of each part of the frequency spectrum that passes through the filter. This attenuation factor, called *loss*, is not determined by whether the input is 1 V, 0.12 V, or any other specific input voltage value. The attenuation is an input-to-output ratio, independent of the absolute input signal value.

The filter response is checked using a signal generator which puts out a spectrum of signals across the entire frequency band of interest. The ampitude of this signal is constant at all frequencies, and it forms the reference, or 0-dB, signal, which was called V_0 in the equations. The output of the filter is then compared to the input. This value at each frequency is V_1. The formula for dB is used to get the attenuation, in dB, at each frequency.

The instrument often used for measuring the input and output values is called a *frequency-selective voltmeter*. It measures the voltage value only at the frequency to which it is tuned and ignores voltages at other frequencies. The meter is used to measure the filter output signal value. A quick calculation shows the attenuation of the filter, in dB. There is another advantage to this dB technique. The output of the signal generator does not have to be exactly the same value at all frequencies (called *flat* output). Instead, the meter has two sets of signal leads. One is connected to the filter input and the other is connected to the filter output. The meter reads the input value and the output value, and performs a simple calculation to determine their ratio and the dB of attenuation that results. The effect of this is that the test indicates the true attenuation for any signal, even though the signal generator is not perfect and does not have a flat amplitude versus frequency output. The imperfection in the signal source is automatically compensated for and corrected.

For example, a filter is being tested from 1 to 2 kHz. The input signal is exactly 1 V at all frequencies. The measured filter output over the range of interest is shown in Figure 3.2. Also shown in the dB attenuation at each frequency. Next, the same filter is tested, but with a signal generator that has nonflat output. The

Figure 3.2 Filter (output) response from 1 to 2 kHz for constant input of 1 V.

Frequency (Hz)	Input (V)	Output (V)	dB attenuation
1000	1.00	1.00	0
1100	1.00	1.00	0
1200	1.00	0.89	−1
1300	1.00	0.79	−2
1400	1.00	0.79	−3
1500	1.00	0.708	−3
1600	1.00	0.708	−3
1700	1.00	0.631	−4
1800	1.00	0.56	−5
1900	1.00	0.50	−6
2000	1.00	0.50	−6

filter input values over the band of interest are shown in Figure 3.3. The measured filter outputs are indicated in the next column. The calculated response of the filter is in the last column. Note that the result is the same, even though the absolute values of the inputs were not identical to the first case. Anyone using this chart would know that their input signal at 1500 Hz would be attenuated by 3 dB, regardless of the actual input voltage (as long as the signal was not so large that it overloaded the input and caused the filter output to *saturate*).

The same reasoning applies to specifying amplifier gains. The user of the amplifier wants to know by what factor the input signal will be boosted. The absolute signal value is not immediately critical, or it may vary, such as when the amplifier must reproduce signals ranging from soft-spoken words to loud music. Instead of saying the amplifier takes 1-W signals and boosts them by 7.6 W, to 8.6 W, or takes 0.5-W signals and increases them by 3.8 W, to 4.3 W, the amplifier specification will say that this amplifier has a gain of 9.34 dB.

Frequency (Hz)	Input (V)	Output (V)	dB attenuation
1000	1.00	1.00	0
1100	0.95	0.95	0
1200	0.95	0.847	−1
1300	0.95	0.755	−2
1400	0.90	0.637	−3
1500	0.90	0.637	−3
1600	0.95	0.673	−3
1700	0.95	0.599	−4
1800	1.00	0.562	−5
1900	1.05	0.526	−6
2000	1.10	0.551	−6

Figure 3.3 Same filter as in Figure 3.2, where input is not flat (constant 1 V) for all frequencies. Loss is still the same, as seen from the dB values.

dB and Multiple-Stage Systems

Most practical communications systems consist of many *stages*. A signal begins as the input to one stage, where it may be amplified, modulated, or converted in some way to meet the needs of the system. As it passes from one stage to another, it undergoes some changes in signal level as well as other characteristics. The dB scale is ideal for calculating and specifying the signal level after it has passed through several stages.

An example will show how dB measurement is used for multiple-stage systems. Figure 3.4 shows a signal from a tape cassette as it travels from the playback head of the cassette unit to the antenna of the broadcast station. The signal from the playback head is very low level, only a few μW. There is a first amplifier stage, called the *preamplifier*, which conditions this very weak signal and boosts it by 20 dB to a more practical level. The signal then passes to the main amplifier, where it can be controlled in volume. This amplifier has a gain of from 10 to 20 dB. Finally, the power amplifier and modulator stage takes the signal, modulates the carrier, and amplifies it for the antenna, with a gain of another +15 dB.

The overall gain of the transmitter system is found simply by adding up the gains, in dB, of all the stages. With the main amplifier gain set to a maximum of 20 dB the overall gain is 20 + 20 + 15 = 55 dB. If the volume control were set to the lower value of 10 dB, the gain would be 20 + 10 + 15 = 45 dB. If the actual voltage at the antenna was required, after amplification by all the stages, the input signal value from the playback head would be used in the equation along with the overall dB figure. It is not necessary to calculate the actual signal voltage at each point—simply use the overall gain figure in dB. (Of course, the voltage values must be measured across the identical resistances, or the values must be corrected.)

This could be done without dB. Each stage has a gain that can be expressed as a multiplication factor. The multiplication factors for each stage are all multiplied together, and the result is a larger, harder-to-calculate number. Any change in gain, such as from 20 dB down to 10 dB, involves a large number multiplication instead of simply subtracting 10 dB. The use of dB converts a problem in multiplication of large numbers to a problem of addition of small numbers. This is much easier to do and gives an "intuitive" feeling for the answers more quickly.

The dB scale does this because the mathematics of logarithms shows that the log of a product of two numbers is equal to the sum of their individual logs:

$$\log (A \times B) = \log A + \log B$$

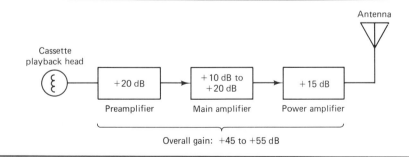

Figure 3.4 Block diagram of signal flow from cassette head to preamp, amplifier, and transmitter.

If A and B are the gains of each stage, expressed as a ratio (i.e., a gain of 150 means that the output is 150 times the magnitude of the input), the overall gain is $A \times B$, and the gain in dB would be related to the log of ($A \times B$). It is equivalent to take the log (or dB) of the gain of each stage and simply add these logs (adding dB values). If the exact signal value must be determined at any point, use the formula to relate the dB value with the ratio of the two power (or voltage values).

Suppose that the signal from the tape head had a value of 3 μW. Its value at the output of the preamplifier is

$$P_1 = 3 \times 10^{20/10} = 300 \ \mu W$$

At the output of the main amplifier, it has a value depending on the gain setting of the amplifier, between 10 and 20 dB. If the amplifier gain is set to 10 dB, the overall gain of the preamplifier and main amplifier is 20 + 10 = 30 dB, so the signal value is

$$P_1 = 3 \times 10^{30/10} = 3000 \ \mu W = 3 \ mW$$

With the amplifier set to 20 dB, the gain is 20 + 20 = 40 dB, and the signal level is

$$P_1 = 3 \times 10^{40/10} = 30{,}000 \ \mu W = 30 \ mW$$

Finally, the signal at the antenna has a gain relative to the tape playback head of between 45 and 55 dB, depending on the setting of the middle stage. The corresponding power values are

$$P_1 = 3 \times 10^{45/10} = 94{,}868 \ \mu W = 94.8 \ mW \text{ for 45 dB gain}$$

$$P_1 = 3 \times 10^{55/10} = 948{,}683 \ \mu W = 949 \ mW \text{ for 55 dB gain}$$

This three-stage example shows the usefulness of the dB scale. Of course, the signal values at each stage can be calculated without them since there are only a few stages in the system, and the gain values are small numbers.

The power of using dB for calculating system performance when there are many stages can be seen from a more complicated but still realistic example. Suppose that the transmitted signal from the cassette is going by satellite link to listeners in England. The overall system looks like Figure 3.5. It consists of the following blocks and signal gains (or losses):

The three stages, as before.

A parabolic dish antenna, to focus the transmitter energy at the satellite in space, gain of 35 dB.

Loss of signal energy because most of the transmitted signal does not reach the satellite. This is a factor of −73 dB.

The satellite receives the signal, amplifies it, and retransmits it back to a receiving station in England. The gain at the satellite is 22 dB.

The loss in the signal path from the satellite to the receiving antenna, −95 dB.

3.4 System Measurements with dB

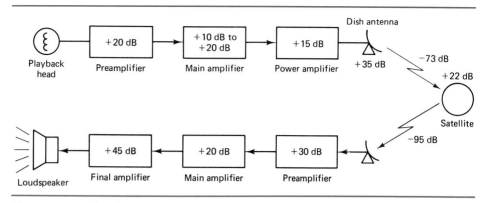

Figure 3.5 Signal source to receiver satellite system, extending Figure 3.5.

The extremely sensitive preamplifier at the receiving dish antenna, which boosts the feeble signal from the satellite by 30 dB.

Another amplifier, similar to the main amplifier of the transmitting side, with a gain of 20 dB.

The final amplifier, which takes the signal and amplifies it for the loudspeaker of the receiving system, with a gain of 45 dB.

The overall system gain is the sum of all the individual components (assume that the amplifier gain of the transmitter side is set to a maximum of 20 dB):

$$20 + 20 + 15 + 35 - 73 + 22 - 95 + 30 + 20 + 45 = 39 \text{ dB}$$

This is easily converted into the actual signal level at the loudspeaker, since the original signal power was 3 μW:

$$P_1 = 3 \times 10^{39/10} = 23{,}829 \ \mu\text{W} = 23.9 \text{ mW}$$

Estimating with dB

The dB scale can be used to estimate very quickly the ratio between the powers or voltages of two signals, without using a calculator. For power, use these two facts: 10 dB is a multiplicative factor of 10, and 3 dB is a factor of 2, while -10 dB is a signal reduction to $\frac{1}{10}$ ($\div 10$) and -3 dB is a reduction by $\frac{1}{2}$ ($\div 2$). Therefore, a gain of 13 dB (10 + 3) is a multiplication of $\times 10$ and $\times 2$, or $\times 20$, and represents a $20\times$ power ratio. Similarly, gain of 36 dB (10 + 10 + 10 + 3 + 3) is the same as a $\times 10$, $\times 10$, $\times 10$, $\times 2$, $\times 2$ combined ratio, or $\times 4000$. A power gain of 7 dB (10 $-$ 3) corresponds to a gain of $\times 10$ and $\div 2$, for an overall $\times 5$ factor.

For voltage, the same concept applies, except that a gain of 20 dB is a $\times 10$ voltage factor and 6 dB is a $\times 2$ factor, while -20 dB divides a signal voltage by 10 and -6 dB divides the voltage by 2. For example, voltage gain of 46 dB (20 + 20+ 6) is the same as $\times 10$, $\times 10$, and $\times 2$, or $\times 200$. Voltage gain of 34 dB (20 + 20 $-$ 6) corresponds to $\times 10$, $\times 10$, $\div 2$, or $\times 50$.

Using these quick factors, you can often estimate the power or voltage ratio that a specific dB value represents and get a good feel for the signal values, without resorting to a calculator. In many applications, a close estimate of the ratio is sufficient for a first level of understanding.

Next, suppose that the receiving antenna preamplifier is replaced by a new design, which increases the gain by 5 dB, to 35 dB. Then the total overall gain is simply increased by 5 dB, and now is 34 dB. The actual signal level at the output is

$$P_1 = 3 \times 10^{44/10} = 75{,}356 \ \mu W = 75.4 \ mW$$

The dB scale and addition/subtraction of dB values at each stage makes it very simple to incorporate changes into the overall result. Any improvement in any stage is simply added, while any deterioration (i.e., the satellite batteries are low, and its gain has been cut in half from 22 dB to 19 dB, a −3-dB change) is subtracted. The engineers and technicians can quickly sense what the overall effect is, by adding and subtracting the gains, losses, and changes as needed.

Review Questions

1. Why is the dB scale useful in system measurements and analysis?
2. Explain why the 0-dB point is often not critical.
3. How are decibels used in multistage systems? Why are they so convenient? What is the principle involved?
4. How does the dB scale accommodate nonflat outputs when checking filter response? How is this useful for the filter user?
5. Explain how a multistage system would use the decibel scale to figure out the system gain through many stages, in both relative gain and absolute units.

3.5 CHARTING dB

In many situations there is a need for an engineer or technician to draw a chart graphing the signal amplitudes in the system, usually versus frequency. The large range of signals in typical systems makes this very difficult and the result is often hard to interpret. Figure 3.6a shows the graph of signals versus frequency in the audio amplifier stage of a receiver, over the range 20 Hz to 20 kHz. Note that the signal ranges in amplitude from 0.01 to 9.3 V. This graph spans three orders of magnitude, as follows:

0.01 V to 0.1 V   0.1 V to 1.0 V   1.0 V to 10.0 V

By looking at this graph, we can see that the frequency response of this amplifier falls off rapidly below 300 Hz and above 8 kHz. This amplifier is intended for amplifying speech, which has a narrower bandwidth than music.

The same information can be presented on a graph that has the vertical axis in dB instead of volts (Figure 3.6b). The dB scale is the logarithm of the linear volts scale, so each major interval of the axis now represents 10 dB. The graph has a different shape, and the information is compressed, yet conveyed more clearly. Often, a regular volts scale is called the *linear scale,* and the dB scale is called the *log scale.* It is easier to read and compare the signal performance on the dB scale.

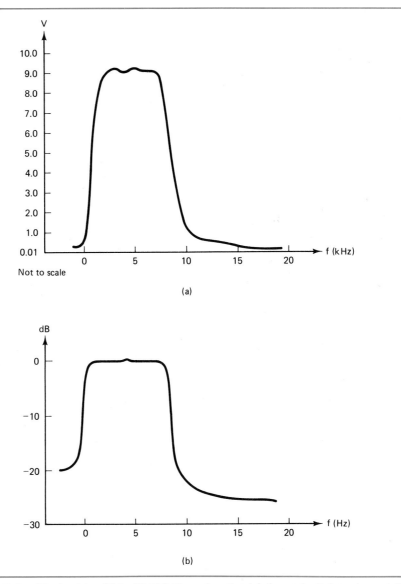

Figure 3.6 dB chart for 20 to 20,000 Hz, using the (a) linear (volts) scale and (b) the log (dB) scale shows how the dB scale is more readable. The linear scale cannot be properly drawn within the space of the page.

Also, the 0-dB point on the scale can be set according to the application. In most cases the maximum amplitude is called 0 dB, and all other signal values are graphed as compared to this 0-dB point. The *log-linear* or *semilog* format is used whenever the signal magnitude spans a wide range or when the dB scale is more convenient for signal calculations.

When the horizontal axis represents a wide span of frequency values, a *log-log* graph is often used. In this graph the frequency axis is also a logarithm scale. This compresses the wide span of frequencies into more readable form and makes each decade of frequency (10 to 100, 100 to 1000, 1000 to 10,000 Hz, and so on)

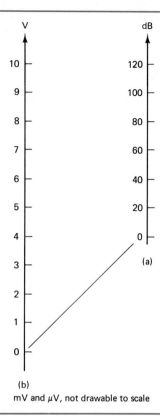

Figure 3.7 (a) Axis spanning 120 dB; (b) V scale for the same span, with 100,000 : 1 range (10 V down to 0.001 V).

occupy the same linear distance of the scale. On the horizontal axis, each major division represents a factor of 10 in frequency, while the vertical axis major divisions are typically 10 or 20 dB.

Most instruments used in communications systems work use the dB scale as much as possible and have a screen that graphically presents the data to the user, allowing the user to select log or linear scales. The log scale, using dB, is most common because of the ease of reading, the usefulness of the information it presents on signal values versus the desired 0-dB point, and the wide range of signals that many communications systems must handle. A total range of 120 dB is common in communications, and a graph using dB would have a vertical scale as shown in Figure 3.7a. In contrast, the linear voltage scale would have to encompass μV, mV, V, and use both very small and very large numbers to span the same range of 100,000 : 1 (Figure 3.7b).

Review Questions

1. What is the log scale for showing signal values? How is the log scale used with decibels?
2. Compare the ability of the linear scale to show a signal with a wide range to the log scale and wide-ranging signals.
3. How is the 0-dB point often set when using the log scale? Why?

3.6
dB AND BANDWIDTH

The concept of bandwidth appears in two related ways in communications systems. First, there is the bandwidth needed to convey the information or data. This is determined by using Shannon's formula or other similar formulas, by looking at the amount of data to be transmitted in a fixed time period, and by seeing how much bandwidth is needed to convey the data signals accurately.

The corresponding aspect of communications systems is to understand what the bandwidth of an amplifier, filter, communications channel, or communications link actually is. How bandwidth is measured and defined is easily understood using dB. Generally, the bandwidth is considered to be the width of the frequency spectrum between the −3-dB power points of the signal amplitude versus frequency graph (often, when talking, this is shortened to "3 dB" and the negative sign is understood). The −3-dB points are the frequency values where the signal power has fallen to one-half its maximum value, the same as a decline by 3 dB (since 3 dB represents a factor of 2 in power). (Remember that at a "−3-dB power level, a signal voltage is reduced to $1/\sqrt{2} = 0.707$ of its previous value, at the same impedance.)

The graph of the frequency response shows this quickly. The maximum signal amplitude, regardless of frequency, is used as the 0-dB reference, and all other amplitudes at the various frequency points are graphed against this (Figure 3.8). The figure also shows the measurement of bandwidth between the frequencies where the amplitude is "*3 dB down.*" Regardless of the general shape of the graph, the bandwidth of a system, or part of a system, is the span between the −3-dB points unless otherwise stated. Figure 3.9 shows some other examples of bandwidth using this definition. Note that in each case the specific curvature of the graph line from the maximum value to its ends varies, but the bandwidth definition ignores this and instead just looks at the points at which the response is −3 dB compared to the maximum.

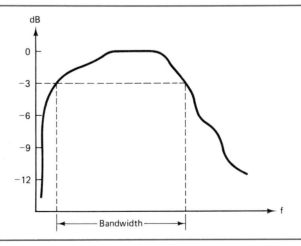

Figure 3.8 Bandwidth defined as the frequency span between −3-dB power points.

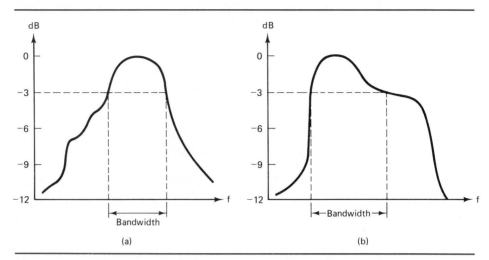

Figure 3.9 Other examples of bandwidth measured at −3-dB points.

The graph of the frequency response of simple filters also shows this. A single resistor–capacitor (*RC*) filter output is shown in Figure 3.10 for a filter with a 10-ms *time constant* (the time for the filter output to reach $1 - 1/e = 63.2\%$ of its final steady-state output magnitude in response to an input that steps from one value to another). The response falls off at the relatively slow rate of 20 dB per decade of frequency. The 3-dB bandwidth of this filter is known from circuit theory to be

$$f_c = \frac{1}{2\pi RC} = \frac{1}{2\pi(0.010)} = 16 \text{ Hz}$$

as shown on the graph.

A more complicated filter is shown in Figure 3.11. This filter has additional components, and the frequency response declines more sharply than in the preceding case. Also, this filter has *ripple* or unevenness of attenuation of about 2 dB in the range of frequencies that it passes with little attenuation. The bandwidth definition ignores this ripple, and instead uses only the −3-dB points. From this, the bandwidth is seen to be 25 Hz.

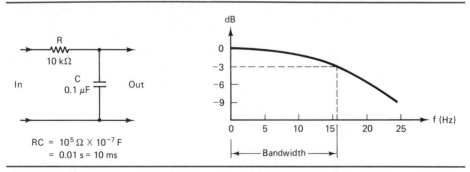

Figure 3.10 *RC* filter schematic and output response versus frequency, showing a −3-dB bandwidth.

3.7 Noise and Its Effects

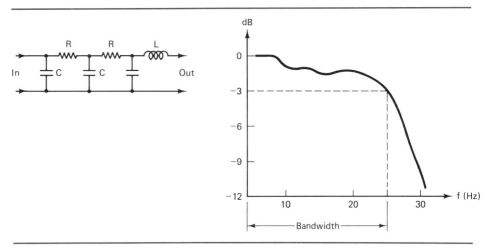

Figure 3.11 More complex filter with a 2-dB passband ripple, showing −3-dB bandwidth points.

Review Questions

1. Which definition of bandwidth is most often used?
2. Why is the dB scale useful in determining bandwidth with this definition?
3. How does this definition accommodate bandwidths where the signal level is not flat across the spectrum, but instead varies within the band?

3.7
NOISE AND ITS EFFECTS

Noise is an important factor in the operation of any communications system. Noise is any unwanted signal that corrupts and distorts the desired signal (or signals) in any way. This corruption can take many forms. It can be added to the communications signal, as shown in Figure 3.12, where a sine wave has a small noise signal combined with it. The receiver of this system may not be able to distinguish the original information of the sine-wave signal from the noise that is added.

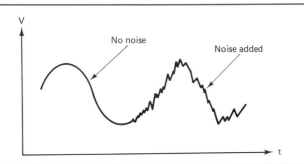

Figure 3.12 Random noise on sine wave.

The noise can cause the original signal to be distorted in shape, increased or decreased in amplitude, delayed slightly in time, or otherwise corrupted and modified. The receiver must attempt to function properly and determine what the original signal was, despite the noise. The success of the receiver at doing this depends on the type of transmitted signal, the type and amount of noise, and the complexity of the receiver circuitry.

Noise is a random signal. This means that it cannot be described simply by an equation of signal value versus time. Instead, the value of the noise signal at any instant is random, like a roll of the dice. The way that noise is described, then, is by its general characteristics, such as its average value (for dice, equal to 7), maximum value (12 for dice), frequency spectrum (how many of 2, 3, 4, . . . , 12), and its effects on the system. If the noise was not random, its value at one instant in time could be predicted by its value at previous instants, since there would be some formula or correlation linking the noise values at different times to each other. Of course, if the value of the noise signal were known, it could be subtracted by special circuitry or otherwise compensated for by the system.

Although noise is random, the more that is known about the noise and its characteristics, the more that can be done by the system designer to minimize its adverse effects. Noise can affect the communications system performance in three areas:

1. Noise can cause the listener to misunderstand the original signal or be unable to understand it at all. The corruption due to the noise can range from slight, which is an annoyance for listening to music, to severe, which makes the signal unintelligible. Of course, any corruption is a problem if the communications signal represents numerical information, since a signal representing a temperature of 107° might be understood as 112°. Here, noise causes errors in the interpretation of the information received.

2. Noise can actually cause the receiving system to malfunction. In many communications systems, the circuitry is designed with some assumptions about the general shape of the received signal. Often, the received signal has some characteristics that the receiver intends to use within its circuitry. The noise may cause this circuitry to function incorrectly, erratically, or improperly.

An example would be a simple case where a stream of digital pulses is being transmitted (Figure 3.13). The receiver must determine when the pulse is a high value representing binary 1 and when it is a low signal for binary 0. To do this, the receiver circuitry must *synchronize* itself (Chapter 11) to the incoming data bits, so that it looks at each pulse once only and does not skip any pulse or look at any pulse twice in quick succession and decide that two 1s (or 0s) exist where there is only one. This synchronization is often done with special timing circuitry that looks at where the pulses cross the 0-V value. Suppose that the original data pulses have a small but varying noise signal, smaller than any pulse, added to their value. If the noise signal occurs at just the right time and with enough magnitude near one of the original zero crossings (Figure 3.14, it might mislead the receiver into thinking that there was another zero crossing or that the real zero-crossing point had moved. In this way, the receiver will lose correct time for synchronization and may misread the data bits themselves.

In this case, noise caused the system to malfunction in its basic operation, regardless of the meaning of the data bits.

3.7 Noise and Its Effects

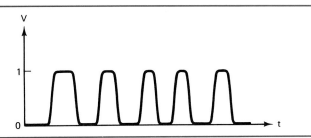

Figure 3.13 Digital pulses without noise.

3. Noise can also result in a less efficient system. The goal of a communications system is to send as much data as possible in the time allowed, using the available bandwidth, power, and channel. If noise affects the system performance, either with errors in the received signal value or with system malfunctions, the system designer and system user will have to compensate for it. This can be done by repeating the information one or more times (since redundancy helps ensure that the correct value does eventually arrive), by using advanced techniques of sending special additional information that allow errors in the received signal to be noted and corrected (Chapter 12), or by devoting extra transmitter power to the original signal, so that it is much larger than the noise signal.

The first two techniques reduce the amount of information sent in the alloted time because some of the signal sent is not new message information, while the third technique requires costly higher-power circuitry. Regardless of the method used, the result is a communications system that is less efficient than it could be, because some communications time or power must be devoted to overcoming the effects of the noise.

The effects of noise must be taken into account whenever a communications system is designed, built, installed, or maintained. There are ways to minimize some of the sources of noise, but some types of noise simply cannot be reduced. Therefore, the design must assume that there will be noise, and take it into account. The effort devoted to handling noise depends on the type of noise expected, the function of the communications system, and the needs of the system

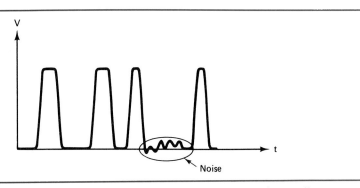

Figure 3.14 Same digital signals as Figure 3.15, but some noise on "zero crossings."

users. Certainly, a system transmitting vital data from one computer to another must be more concerned about noise than standard AM broadcast radio. Noise is an unavoidable fact, and the possible effects of the noise are factored into the system design and operation, in both obvious and subtle ways.

Review Questions

1. What is noise? How does noise affect the original communications signal?
2. What must the system receiver do, despite the noise?
3. What does it mean to say that noise is random? How is noise therefore described?
4. In what three ways can noise affect a system? What is the consequence of each?

3.8 SOURCES AND TYPES OF NOISE

Noise is a constant concern in communication systems. An understanding of the sources and types of noise that may be present will help in determining how the system should be designed, how well it will operate, and which system malfunctions are unavoidable. In some cases, the noise causes system problems that a technician cannot fix; in other cases, the best thing to do is to try to identify the noise source and minimize or remove it, if possible.

Noise can be divided into two main sources: external and internal. *External noise* comes into the communication system at various circuit points, and its exact form depends on the nature of the electrical environment in which the system is used. *Internal noise* is generated by the circuit components themselves. In an ideal world, these components would be perfect and noise-free, but in reality they have noise of their own which adds to the system noise.

External Noise

External noise can be man-made or can arise from natural origins. Virtually all electrical and electronic devices emit electromagnetic waves as a by-product of their operation. Motors, switches, and power lines are all sources of this noise, which is often called *radio-frequency interference* (*RFI*). Every time an ordinary household light switch is turned on or off, the flow of electricity is suddenly started or stopped. This sudden change in voltage and current contains a wide range of frequencies, as we saw in the Fourier analysis of signals. Some of these frequency components then radiate from the switch and house wiring, which act as a miniature transmitter and antenna.

This noise is often called *impulse* noise, because it comes primarily from sudden, on/off events. The spectrum of this noise is relatively wide, and frequencies range up to tens of MHz. The magnitudes of various components is relatively low in most cases, and this type of noise is usually observed in the system to be up to a few hundred millivolts in amplitude, at most. However, in sensitive circuitry handling small signals, such as the initial *front-end* stages of a radio receiver, for example, this is much larger than the desired signal.

3.8 Sources and Types of Noise

Another major source of external noise is the ac power line. Power lines inductively couple 60-Hz signals (50 Hz in some countries, where the line frequency is 50 Hz instead of 60 Hz) into adjacent wires. This interfering 60 Hz then enters the communication system and corrupts the desired signals. In contrast to most types of man-made noise, this noise has a single frequency component at 60 Hz, so it can usually be eliminated within the system by carefully placed filters. If the communications system is carrying low-bandwidth information, such as the signal from a temperature probe, a *low-pass* filter with a cutoff frequency below 60 Hz is used. If the information of the system has a wider bandwidth, such as for voice, a *high-pass* filter or *notch* filter would be needed. The only problem with the filter approach is that the communications system may be carrying information that contains information at 60 Hz, and any filtering at this frequency will also eliminate some of the desired information. This is the case in a system that is transmitting music with a bandwidth of 20 to 20,000 Hz. For these applications, the proper solution is shielding and other techniques to prevent the 60-Hz radiated signal from getting into the system in the first place.

Any transmitter or any signal source can be a source of noise to other systems. A transmitter that is operating perfectly can affect a nearby system when some of the properly transmitted signal enters the circuitry of this nearby system. The system circuitry then adds this "foreign" signal to its own, and the result is an undesired signal corrupting the desired ones. This phenomenon often is called *RF saturation* or *front-end overload,* when a strong signal at another frequency overloads and saturates the front end of a receiver that is designed for very weak signals and providing a large amount of amplification for them. The saturated front end no longer functions as a linear amplifier but instead distorts signals, generates new false signals (*spurious signals*), or even malfunctions completely.

Many sources of external noise are not man-made. They come from natural events in the atmosphere and outer space. Lightning and static discharges in the air cause wideband noise, up to about 20 or 30 MHz, similar to the on/off switch action, but more powerful. Note that just because there is no lightning storm near the communications system does not mean that this noise is not a problem. At almost any time, there is lightning somewhere on the earth, and much smaller atmospheric discharges (not seen by the eye) occur constantly. These travel long distances and cause the crackling and static often heard in the AM radio band.

Another source of noise that affects system performance originates in space. The sun (and all stars) are sources of wideband, low-level noise that begins at approximately 8 MHz and extends out to 1.43 GHz. An antenna pointed at the sun or other stars and galaxies can pick up enough noise to affect sensitive signals. The sun itself has a cycle of generating greater and less amounts of noise, related to the 11-year *sunspot cycle*. In 1957, for example, the sunspot cycle was at its peak and long-distance radio communications were often unusable due to the large noise values.

Internal Noise

Internal noise is also a significant factor in all communications systems and circuits. In an ideal (noise-free) system, any signal that arrives could be amplified or transformed as needed by the receiver, and there will be no loss of signal quality. Therefore, if a very weak signal were received, the preamplifier of the front end could simply amplify it as required to make it useful. Unfortunately, this is not the case.

The movement of the electrons in the circuit generates noise. Different noise mechanisms, related to various aspects of the electron motion, are observed in circuits. The result is that electronic circuitry and components add some noise to any signal as it passes through the circuit. For larger-amplitude signals, this additional noise may be insignificant, but for low-level signals such as those from a satellite or distant transmitter this noise may be a major factor.

The internal noise is generated in any power-dissipating component in the circuit. The simplest type of internal noise is called *thermal noise* (or *Johnson noise*). The amount of noise generated increases with the temperature, since the random motion of electrons is directly related to the absolute temperature. The equation that shows the amount of noise power is

$$P = kT \, \Delta f$$

where P is measured in watts; k is Boltzmann's constant, 1.38×10^{-23} J/K; T is the absolute temperature in kelvin (K = °C + 273); and Δf is the bandwidth, in hertz. This equation shows that the noise power increases with both the temperature and the bandwidth. The noise spectrum extends across all frequencies and has equal amount of power in each part of the frequency spectrum. For this reason this is also referred to as *white noise,* just as white light contains all colors (light frequencies).

Example 3.8

What is the noise power at room temperature, 25°C, when the bandwidth is 1 kHz?

Solution

$$P = kT \, \Delta f = (1.38 \times 10^{-23}) \times (25 + 273) \times (1000) = 4 \times 10^{-18} \text{ W}$$

This is a very small amount of power, compared to the signal power in most applications. Yet there are cases where it is greater than the received signal power, for example, when it is compared to the incredibly miniscule power received from a satellite in space. Note that the noise power is proportional to the temperature and the bandwidth. These extremely low power space and satellite systems minimize the noise power by operating the receiver circuitry at temperatures close to 0 K (using liquid helium) and reducing the bandwidth. Of course, the reduced bandwidth means that the data rate is lower.

Some applications use the specification of *equivalent noise temperature* to easily characterize the noise power. The equation that relates the power to Boltzmann's constant k, temperature T, and bandwidth Δf is used by measuring the noise power and calculating what temperature (in K) would cause this amount of noise at the rated bandwidth. This circuit or component is then rated as having an internal noise of "25 K," for example.

Consider a component with a noise power of 1 pW (picowatt, equal to 10^{-12} W) in a circuit with 1-MHZ bandwidth. The noise temperature is

3.8 Sources and Types of Noise

$$NT = \frac{P}{k\,\Delta f} = \frac{1 \times 10^{-12}}{(1.38 \times 10^{-23})(1 \times 10^6)}$$

$$= 7.24 \times 10^4 \text{ K}$$

When several stages of a system are combined, the equivalent noise temperature is simply the sum of the individual noise temperatures, which makes noise temperatures a useful way to characterize overall noise. A supercooled antenna with NT = 40 K connected to a receiver with NT = 80 K has a noise temperature of 120 K.

Often, it is more convenient and useful to measure the *noise voltage* instead of the power. By Ohm's law, $P = V^2/R$, so the noise power P can be expressed as

$$P = \frac{(V_N/2)^2}{R}$$

where V_N is the noise voltage. [The factor of $\frac{1}{2}$ comes because the maximum amount of noise power (or any power) transfers from the source to the load circuit when the source resistance and the circuit resistance are equal.]

This can be solved for the noise voltage:

$$V_N = \sqrt{4kT\,\Delta f\,R}$$

The value R is called the *equivalent noise resistance*. It is the value of a resistor that would produce the noise value (rms voltage) that was measured. In many situations, the noise of a circuit is summarized by this value. A circuit will be characterized as having a "100-Ω" noise value, meaning that it generates the same amount of noise as a 100-Ω resistor would generate. This is a very convenient

---

Other Internal Noise Sources

Not all sources of circuit and system noise come from the random motion of electrons. Imperfections in the circuit design or in the practical limitations of components can cause effects that are similar to noise. These imperfections cause unwanted and unknown variations in the desired signal. For example, an oscillator may be used in the receiver to provide a local reference frequency, which is then combined with the received signal as part of the *demodulation* process, where the original information signal is recovered. This oscillator must provide a precise frequency value.

In practice, the oscillator may have small but important cycle-to-cycle variations in its frequency and phase, called *phase jitter*. This may be caused by variations in temperature of a fraction of a degree, or by the system power supply not providing an absolutely "rock-steady" voltage without any noise. When measured over several cycles in a longer period, the average oscillator frequency is correct, but the small plus and minus variations in phase appear to the circuit effectively as different frequencies, although only momentarily. This results in inaccuracies and errors in the demodulation process.

Even the layout of the circuitry can cause problems. *Crosstalk* occurs when energy radiates from one wire in the circuit and is coupled to another wire in another part of the circuit. The crosstalk signal may be correct and proper in the circuit it comes from, but it is an interfering noise signal to the second part. It is another form of noise that corrupts the desired signal. The difference between crosstalk and random noise is that the crosstalk is not really random and may have a limited spectrum—instead, it looks like a legitimate signal, only in the wrong part of the system.

shorthand for discussing circuit noise performance. It does not mean that the circuit has 100 Ω of resistance, but that the circuit can be represented (from a noise standpoint) as noiseless but with a 100-Ω resistor in front of it.

Example 3.9

What is the equivalent noise resistance when the measured noise is 300 μV, the temperature is 300 K, and the bandwidth is 3 kHz?

Solution

Rearranging the equation, we have

$$R = \frac{V_N^2}{4kT\,\Delta f} = \frac{(300 \times 10^{-6})^2}{4 \times 1.38 \times 10^{-23} \times 300 \times 3000)}$$

$$= 1.8 \times 10^9 \ \Omega$$

Review Questions

1. Compare the sources of external noise to internal noise.
2. How does nature cause much of the noise?
3. Why do electronic components themselves create noise? How is this noise measured?
4. What is noise voltage? What is equivalent noise resistance?
5. How can minute variations in circuit operation appear as noise?

3.9 NOISE MEASUREMENTS

To work with noise and its implications in communications systems, the noise must be measured in a meaningful way. Noise is a random process and does not have a single, simple value (i.e., 3 V) or an equation to describe it. The *root mean square* (*rms*) value of the noise is the single most important fact about it. When someone says "the noise is 0.5 V," they are really saying "the rms value is 0.5 V."

This rms value is formed by taking the square root of the average of the individual noise voltages, which have been squared. For example, consider a series of 10 noise values measured with a voltmeter as −0.3, 1.0, 0.2, 0.5, 0.6, −0.6, 0.3, 0.1, −0.15, and 0.9 V. The voltages are squared so that the negative values become positive, and then these squared values are averaged. For the values shown, the sum of the squares is 2.9225 and the average is 0.29225. Finally, the square root of this mean value is used, equal to 0.54 V.

3.9 Noise Measurements

Example 3.10

Noise values in millivolts as followed are measured at various times: 10, −100, 35, −57, 90, 26, 26, −10, −15, −20. What is the rms noise value?

Solution

Squaring each value, we have

$$100 + 1000 + 1225 + 3249 + 8100 + 676 + 676 + 100 + 225 + 400 = 15{,}751 \ (mV)^2$$

The average value is 15,751/10 = 1575.1 mV^2. The rms value = 39.7 mV.

If the voltages are simply added, positive and negative noise voltage values partially cancel each other out, and the result of the noise calculation could be a very small value or even 0 V, which is clearly not the case. In practice, special electronic instruments can be used to measure the rms noise value. It is not necessary to measure the noise at many points and then perform the calculation.

Signal-to-Noise Ratio

The value of the noise alone is only one part of the overall situation. The *signal-to-noise ratio* (*SNR*) is extremely important in determining how well the system will operate or how successful the system can recover a weak signal. SNR shows how much stronger (or weaker) the desired signal is as compared to the unwanted noise. This ratio is often expressed with dB:

$$\text{SNR (in dB)} = 10 \log \frac{S}{N}$$

where S is the signal power and N is the noise power.

The importance of SNR is this: Any signal can be amplified to the desired level. But if the signal contains both the desired signal and noise, both are amplified. If the noise value is large enough that the desired information cannot be retrieved, amplifying both by the same factor may not make the situation any different. It is analogous to a small, out-of-focus photograph. The picture can be enlarged (increasing the amplitude), both the words and faces may still not be recognizable if the focus (SNR) is poor. Communications system SNRs range from a low 20 dB to a relatively high 60 to 80 dB, although some deep-space satellite systems operate at SNRs below 0 dB. Of course, higher SNRs make it easier to provide acceptable performance at lower cost and with simpler circuitry.

Noise Figure

The stage at which internal noise affects a received signal is also critical. A simple example shows why. Figure 3.15 shows a gain of 100 (20 dB) preamplifier for very low-level signals, whose output then goes to a second-stage main amplifier, demodulator, or signal converter where the major portion of the signal recovery is performed. Suppose that the input signal is 10 μW (weak but very typical), and that 0.5 μW of noise from internal sources (thermal noise, crosstalk, and jitter) is added before the preamplifier. The the SNR is

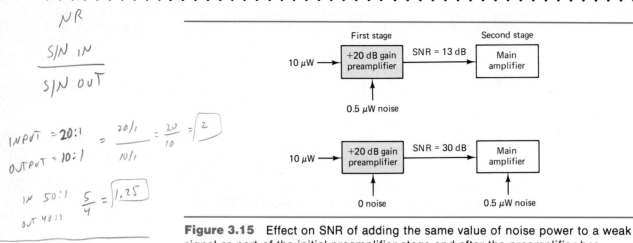

Figure 3.15 Effect on SNR of adding the same value of noise power to a weak signal as part of the initial preamplifier stage and after the preamplifier has boosted the signal.

$$\text{SNR} = 10 \log \frac{10.0}{0.5} = 13 \text{ dB}$$

This relatively low SNR will make recovering the original signal difficult.

Now, suppose the preamplifier has a gain of 100 (20 dB) and is ideal, adding no internal noise. Instead, the second stage adds the same amount of noise to the received signal, but only *after* it has been preamplified. Then the SNR is

$$\text{SNR} = 10 \log \frac{10.0 \times 100}{0.5} = 33 \text{ dB}$$

which is much larger and means that it is more likely that the original signal can be recovered without error. Note that in this example we assumed that the incoming signal itself had no noise. If it did, the adverse impact of the addition of noise before preamplification would be even more severe.

Therefore, an input signal with a high-enough SNR is not the only requirement for proper signal reception. The preamplifier or receiver circuitry must not contribute too much noise of its own, as it amplifies the usually very small input signal. The noise figure measurement shows exactly how much of the noise in the amplified signal is due to the original signal and its noise, compared to noise added by the amplifier itself.

Consider this situation: A signal arrives at the preamplifier of a receiver with a signal power of 1 W and noise power of 0.01 W. The preamplifier has a power gain of 15. The SNR before amplification is

$$10 \log \frac{1}{0.01} = 20 \text{ dB}$$

After amplification the signal is 15 W and the noise is 0.15 W. The SNR remains the same, since the ratio of signal and noise is unchanged, even though the absolute magnitudes have increased. Unfortunately, a real preamplifier adds some of its own noise to the signal, indicated by its *noise figure*. Noise figure is defined

3.9 Noise Measurements

as the dB ratio of the input signal and noise ratio to the output signal and noise ratio:

$$\text{noise figure (NF)} = 10 \log \frac{\text{input signal/input noise}}{\text{output signal/output noise}}$$

(Note that the ratios used are not in dB but are the ratios of input and output magnitudes.) A perfect amplifier would add no noise, so the two ratios would be the same and the NF would be 0 dB. The input and output signal power and noise power ratio is called the *noise ratio NR*, so NF = 10 log NR.

A more realistic case is where the input signal is 1 W, the input noise is 0.01 W, the output signal is 10 W, and the output noise power is 0.3 W. The NF is

$$\text{NF} = 10 \log \frac{1/0.01}{10/0.3} = 48 \text{ dB}$$

(*Note:* The NR is 3.0.) A good low-noise transistor for very low level amplification may have a noise figure as low as 1 dB. Noise figures of 2 to 5 dB are more common, however. Transistors designed for this type of use come with charts that show their NF at various frequencies.

Relationship between Noise Figure and Noise Temperature

The noise temperature of Section 3.8 is related directly to the noise ratio by the equation

$$\text{NT} = T_0(\text{NR} - 1)$$

where $T_0 = 290$ K, a reference temperature expressed in kelvin (approximately 17°C or 63°F).

This relationship allows the system designer to calculate the noise temperature from the noise ratio, and to determine the overall noise ratio and noise figure by adding the individual noise temperatures of the components of the system.

Example 3.11

An amplifier has a noise ratio of 4. What is the noise temperature?

Solution
Using the equation, NT = 290(4 − 1) = 290(3) = 870 K.

Example 3.12

An antenna with NT = 75 K is connected to a receiver input with NT = 300 K. What are the overall noise temperature, noise ratio, and noise figure?

Solution
The noise temperature is 75 + 300 = 375 K. Then, rearranging the equation of noise temperature and noise ratio, we have

$$\text{NR} = \frac{\text{NT}}{T_0} + 1$$

so the noise ratio in this case is

$$\text{NR} = \frac{375}{290} + 1 = 2.29$$

and the noise figure is

$$\text{NF} = 10 \log \text{NR} = 3.6 \text{ dB}$$

Review Questions

1. Why can't noise be measured by a single value?
2. What is the meaning of rms noise value? How is it defined? Why must the noise values be squared?
3. What is SNR? Why is it such an important figure? What is the consequence of low SNR?
4. What is noise figure? What does it indicate?
5. Where is noise figure used? What does it indicate about the noisiness of an amplifier or preamplifier? How is it related to noise temperature?

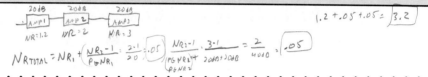

SUMMARY

The decibel scale compares two signals using their ratio and base 10 logarithm. The result is a scale that compresses a wide span of signals into a much smaller one. A voltage span of 1,000,000 to 1 is 120 dB. The dB scale can be used to compare any two signals, to measure their relative size, the gain or loss) of one versus the other, and to measure the performance of specific circuit blocks. Another advantage of the dB scale is that multiple-stage system performance can be calculated simply by adding the dB values for each stage. This allows rapid calculation and study of the effects of gain changes at any point in the system. The dB value can be converted to absolute units, if needed, by knowing the value of one of the signals in the ratio. Several common reference values are used, such as 1 mW and 1 V. Graphs are often drawn with dB as the vertical scale, which is easier to read and shows a wide range of signals in a small space; if the horizontal axis is frequency, it is also drawn on a logarithmic scale so that each factor of 10 frequency decade has the same width. Bandwidth is measured as the span of frequencies between the points where the signal power is −3 dB (half power) below its peak value.

Noise is also a major factor in system applications. It can come from external sources, or be added to the desired signal, or be generated within the system circuitry itself. The noise power of a signal or circuit must be measured to be understood. Noise has no single value, but a special average called the root mean square (rms) value is commonly used. Noise temperature gives the noise that results from the random motion of electrons, and noise resistance shows what value resistor would have that noise value. The SNR shows the relative strength of the signal power relative to the noise power and indicates how easy or difficult it will be to extract the desired information from corrupting noise. The noise figure shows how much noise is added to the received signal by the noise of the circuitry itself.

Summary Questions

1. What is the significance of the dB scale? How is it defined?
2. Compare the dB scale to the traditional linear scale.
3. How do the dB equations for power and voltage differ?
4. How does the dB scale manage to compress a wide range of signals into a smaller range of values?
5. How can dB be used to measure the value of a signal in absolute units such as volts or watts?
6. What does $-n$ dB indicate, compared to $+n$ dB, for any value n?

Practice Problems 75

7. What does 0 dB mean?

8. How are dB useful for testing the characteristics of filters? How does the dB scale compensate for imperfect output levels in the signal source?

9. How are dB used in the analysis of overall system gain and loss through multiple stages?

10. Compare using dB for multiple-stage systems to using gain factors and multiplication factors.

11. Why is the log scale often used for graphing signal levels? Compare its appearance to the traditional linear scale. Why is the horizontal frequency axis also draw on a log scale in most cases?

12. What does 0 dB represent on a graph?

13. How is bandwidth usually defined, in terms of dB?

14. Why is noise difficult to measure and characterize?

15. What are three ways in which noise affects a system?

16. What are the two main sources of noise? How could each be handled?

17. What is the rms value of noise? Why is it used as a measure?

18. Compare impulse noise to narrow-bandwidth noise.

19. Why are equivalent noise resistance and temperature useful ways to talk about system noise?

20. What are SNR and its typical span in a communications system? Why is it a critical factor? What happens to system performance and complexity as SNR decreases?

PRACTICE PROBLEMS

Section 3.2

1. Calculate the dB ratio for these pairs of power values: 3.9 W and 25 mW, 100 mW and 550 mW, 250 mW and 145 mW.

2. What is the dB ratio for these voltages: 2.75 V and 1.93 V, 2.05 mV and 75 mV, 19 V and 10 V, 3 mV and 1 V?

3. Show that doubling a voltage is the same as a 6-dB increase.

4. What is the final power when an amplifier provides a 7-dB gain and the input signal power is 1.5 W?

5. A 100-W signal goes through a circuit with a 30-dB loss. What is the final power value?

6. A voltage signal suffers a 20% loss. What is the loss in dB?

7. Voltage values are amplified by 250% in an integrated circuit. What is the gain, in dB?

8. A signal between a transmitter and a receiver is reduced to just 1% of its original power value because of the distance between antennas. How many dB loss is this?

Section 3.3

1. Prove that 0 dBW = 30 dBm.

2. Convert these mW values to dBm: 150, 1500, 2500, 0.75.

3. Express these power values as dBW: 1 W, 0.73 W, 950 W.

4. A reference value sometimes used in the radio and TV transmitter industry is based on the kilowatt (1000 W), called dBk. Express 1 W in terms of dBk.

5. Express 200 W as dBW and as dBm.

6. Convert 250 mW to dBW and dBm.

7. Convert a signal of 85 dBm to milliwatts and to watts.

8. Convert 12 dBV to volts.

Section 3.4

1. A signal generator is used to generate signals for an amplifier test at a 30-dBm level, from 1 to 10 kHz. The output of the amplifier at each 1-kHz point in this range is, in dBm: 35, 43, 43, 42, 41, 40, 40, 39, 39, and 39, from 1 to 10 kHz. Sketch the amplifier gain versus frequency.

2. For another amplifier being tested in the case of problem 1, the signal generator output is not flat, but is 30 dBm from 1 to 3 kHz, then 32 dBm to 6 kHz, and finally 29 dBm to 10 kHz. The measured output values are the same as in problem 1. Sketch the amplifier gain versus frequency.

3. The signal generator of problem 2, with nonflat output, is used to test the amplifier of problem 1. Using the gain figures from problem 1, what would be the observed output in dBm at each frequency?

4. A filter is used to attenuate unwanted signals outside the voice band from 300 to 3000 Hz. The filter passes the voice band with no loss, but attenuates the power of 0 to 300-Hz signals by 20 dB, 300 to 2500 Hz by 3 dB, 2500 to 5500 Hz by 20 dB, and 5500 to 10,000 Hz by 30 dB. Sketch the filter response.

5. A 0 to 10-kHz signal of +10 dBm is passed through this filter. What is the output signal power at 0, 1, 2, . . . , 10 kHz?

6. A multiple-stage system has three amplifier stages, providing power gains of 3, 7, and 25 dB, respectively. What is the overall gain, in dB? For a signal of 0.05 W, what is its power at the output of the final stage?

7. A deep-space communications system dish antenna receives a signal from a space satellite with −100 dBm of power. The special preamplifier boosts this by 50 dB, and another preamplifier boosts it an additional 40 dB. Further amplifier stages add 20, 25, and 10 dB to the signal. What is the overall gain? What is the signal level, in dBm, after the first two stages (the preamplifiers)? What is the level, in dBm, after all stages? What is the signal level, in mW, after the last stage?

Section 3.8

1. What is the noise power for a bandwidth of 3.5 kHz at 25°C?

2. What is the equivalent noise temperature when the noise power is 1×10^{-18} W and the bandwidth is 300 Hz? By what factor does the noise temperature increase when the bandwidth doubles?

3. What is the noise voltage V_N when the equivalent noise resistance is 500 Ω, the bandwidth is 3000 Hz, and the temperature is 100°C?

4. Repeat problem 3 when the temperature is 4 K.

5. What is the equivalent noise resistance when the noise voltage is 1 nV (1×10^{-9} V) with a bandwidth of 25 kHz and a temperature of 25°C?

Practice Problems

6. By what factor does the noise resistance increase when the measured noise voltage increases by a factor of 3?

Section 3.9

1. What is the rms noise value when the 20 measured noise voltages (in millivolts) have these values: six times, 20 mV; two times, −17 mV; five times, −35 mV; three times, 25 mV; four times, 10 mV?

2. Ten noise voltage readings are made. These readings evenly span from −4 to +4 mV, with two readings at 0 mV. What is the average noise voltage? What is the rms noise voltage?

3. What is the SNR for signal power of 100 W and noise power of 0.1 W?

4. What is the SNR when the signal-to-noise ratio (not in dB) is 1500 : 1?

5. What is the noise figure when the input signal and noise values are 100 and 50 μW, and the output signal and noise values are 1 and 0.75 W?

6. An amplifier takes a signal that has an input signal-to-noise ratio (not in dB) of 1250 : 1, and the output signal power is 25 W with noise power of 0.06 W. What is the noise figure?

7. What is the equivalent noise temperature when the noise ratio is 20?

8. For a noise figure of 16 dB, what are the noise ratio and the equivalent noise temperature?

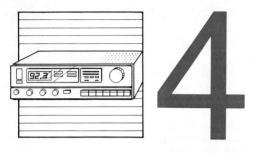

4

Amplitude Modulation

CHAPTER OBJECTIVES

When you have completed this chapter, you will understand:

- The role of and need for modulation
- Amplitude modulation, the most fundamental type of modulation and the simplest to generate and demodulate
- The disadvantages of AM and variations of AM that overcome these disadvantages
- Functional blocks and circuitry used in modulators and transmitters

INTRODUCTION

Modulation is the process by which the communications signal that contains the information is combined with another signal. This results in a signal at frequencies more compatible with the application and in a desired part of the spectrum. The simplest form of modulation is amplitude modulation, where the amplitude of the second signal is varied by the information signal. Amplitude modulation can provide a simple way to send information longer distances, or at frequencies very different than the unmodulated signal spectrum itself. The result of amplitude modulation is both information signal and a carrier, which has power but conveys no information. A modulation index shows the relative strength of these two parts to each other.

Variations of basic amplitude modulation overcome some of the technical weakness of this simple form of modulation. Suppressed carrier modulation minimizes the power of the carrier, while single sideband minimizes both wasted power and bandwidth. The different ways of generating an amplitude-modulated signal, and these variations, are also discussed, along with the two major ways that a complete transmitter can be built.

4.1
NEED FOR MODULATION

The communications signal that represents the information to be sent to the desired receiver is a voltage that varies with time. This signal, without modification, may not be suitable for transmission over the communications system. It may have frequencies that cannot be handled by the circuitry, or require antenna sizes that are impractical (since antenna size is related to wavelength). Instead, the information signal must be used to vary another signal that is more technically appropriate, and has the frequencies, spectrum, and power needed for the application.

This process is called *modulation*. Usually, the signal being modulated is more powerful than the signal that does the modulating. The driver of a car uses pressure on the gas pedal to modulate the amount of gas going to the engine and thus adjust the engine speed to the road situation. Modulation allows a small signal—foot pedal pressure—to control a much larger one, the engine speed. As another example, a person speaking produces a flow of air from the lungs through the vocal passages and mouth. The movement of the vocal cords, mouth, and tongue then modulates this basic flow to produce the sounds which are sent through the air to the listener. The modulated signal is what is transmitted over the communications system by radio, wire, or whatever physical means is being used to carry the signal energy from the sender to the receiver.

At the receiving end, the original information must be recovered. This is done by *demodulation*, and the types of demodulation and the ways it can be done vary depending on the type of modulation and the required system performance. Some forms of demodulation are more complicated but produce better results, more faithful to the original signal or less affected by electrical noise.

The simplest and oldest form of modulation is *amplitude modulation* (AM). In amplitude modulation, the information signal causes the amplitude of a *carrier* signal and information signal to vary with time, in proportion to the instantaneous magnitude of their sum. The carrier can have a frequency that is very different from the frequency of the information itself. Broadcast radio has a band of frequencies from 535 to 1610 kHz, reserved for use by AM radio stations, and is used for voice and music information, which is restricted to a bandwidth of 0 to 5 kHz.

The AM broadcast band example shows the other result of modulation, in addition to transforming a signal into a form and type more suitable for the transmission. Modulation allows many signals which originally had the same frequency spectrum to be moved to different parts of the total available electromagnetic spectrum. By using modulation, the many voice/music radio stations can all be broadcasting simultaneously, without interfering with each other. Without modulation, they would all occupy exactly the same range of spectrum and it would be impossible to choose the desired signal from the resulting jumble.

Modulation serves the same purpose even when a closed communications medium, such as wire, is used instead of broadcast. A wire supports a bandwidth ranging up to hundreds of kilohertz or even megahertz, depending on the wire type and way the wire is made into cable. If several signals were put onto this wire without modulation, the result would be interference with each other. These signals have to be spread out to separate segments of the total available bandwidth of the wire, by using each one to modulate a carrier.

Review Questions

1. What is modulation? What is demodulation?
2. How does the modulation relate to an information-bearing signal?
3. Give two reasons why modulation is needed.
4. How does modulation allow the total frequency spectrum to be used by more than one user at the same time?

4.2 BASICS OF AM

Amplitude modulation begins with a carrier v_c, which is a sinusoidal wave with a frequency f_c and amplitude V_c and described by the equation

$$v_c = V_c \sin 2\pi f_c t$$

Without any modulation, this sinusoidal wave carrier would convey no information, since its value at any time can be calculated perfectly from the previously known values of V_c and f_c.

The AM process modifies the constant value of V_c with the signal which has, for example, voice, music, and computer data information. The result is that the carrier is no longer a simple constant-amplitude sine wave. Instead, the amplitude of the modulated signal varies in proportion to the amplitude of the information signal.

Result of AM   Consider a case where the modulating signal applied to the carrier is a simple tone of constant frequency f_m and amplitude V_m. The equation of the resulting AM

Derivation of AM Equation

The result of the AM process is derived from the definition of AM. Let the carrier be $v_c = V_c \sin 2\pi f_c t$ and the modulating information signal be $v_m = V_m \sin 2\pi f_s t$. The amplitude resulting from modulation is

$$A = V_c + v_m = V_c + V_m \sin 2\pi f_s t = V_c + m V_c \sin 2\pi f_s t$$

where m is defined as V_m/V_c. Collecting the V_c term yields

$$A = V_c (1 + m \sin 2\pi f_s t)$$

The voltage of the resulting AM wave at any instant is

$$v = A \sin 2\pi f_c t = V_c (1 + m \sin 2\pi f_s t) \sin 2\pi f_c t$$

By trigonometric identity, the sin product can be expanded since

$$(\sin a)(\sin b) = \tfrac{1}{2} [\cos(a-b) - \cos(a+b)]$$

with the result that

$$v = V_c \sin 2\pi f_c t + \frac{m}{2} V_c \cos 2\pi (f_c - f_s) t - \frac{m}{2} V_c \cos 2\pi (f_c + f_s) t$$

4.2 Basics of AM

signal voltage v is *contains 3 harmonics*

$$v = V_c \sin 2\pi f_c t + \frac{m}{2} V_c \cos 2\pi (f_c - f_s)t - \frac{m}{2} V_c \cos 2\pi (f_c + f_s)t$$

The factor m has a value between 0 and 1 and is called the *modulation index*. It is the ratio of the unmodulated carrier to the modulated signal magnitudes, V_m/V_c, and is discussed in more detail later.

The important point of the AM process result is that the frequency spectrum of the AM signal is no longer a simple sine wave at the carrier frequency, f_c. Instead, it consists of three parts: a carrier signal component at frequency f_c, plus two new signals at frequencies $f_c + f_s$ and $f_c - f_s$. These two signals are called the *sidebands*. Their frequencies are also known as the *sum* and *difference* signals, because one has a frequency equal to the sum of the original carrier and modulating signal frequency, $f_c + f_s$, and other has a frequency equal to their difference, $f_c - f_s$. The spectrum that results from the AM process for this simple modulating signal is shown in Figure 4.1. Note that the magnitude of the carrier is much larger than that of the sidebands, and that the sidebands are equal in magnitude. Also, the modulating signal no longer exists in its original form. It has been transformed into sidebands of a carrier.

Using the AM radio broadcast band as an example, if the tone had a frequency f_s of 1000 Hz, and the carrier had a frequency of 1500 kHz, the result would be sidebands at 1500 − 1 kHz and 1500 + 1 kHz, or 1499 and 1501 kHz. A carrier at 1250 kHz would produce sidebands at 1250 − 1 kHz and 1250 + 1 kHz after the AM process, equal to 1249 and 1251 kHz. A frequency of 2.5 kHz would yield sidebands at 1497.5 kHz and 1502.5 kHz for the 1500-kHz carrier, and 1247.5 and 1252.5 for the 1250-kHz carrier.

In the time domain, the effect can also be seen (Figure 4.2). The carrier takes on the shape of the modulating signal. This is called the *envelope* of the carrier, and it is this envelope which has the information of the modulating signal. The envelope (the dashed lines) is not a separate or new signal that would be seen on the screen of an oscilloscope. Instead, it is the form and overall amplitude variation that the carrier now has.

Of course, most signals that modulate a carrier are not constant frequencies, but are complicated signals such as voice, music, or data. The result of the AM process with these modulating signals is that the frequency spectrum of the modu-

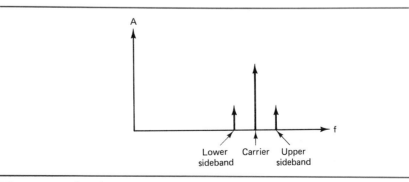

Figure 4.1 Spectrum resulting from amplitude modulation of a carrier by a single frequency.

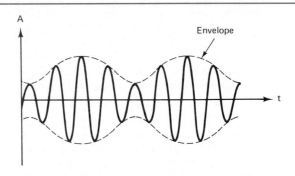

Figure 4.2 Time domain of amplitude modulation, showing information signal forming an envelope on the carrier.

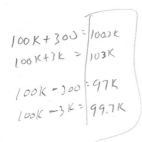

lating signal becomes sidebands of the carrier, just as a single-frequency tone signal did. For example, a voice signal with a spectrum of 300 to 3000 Hz modulates a 100-kHz carrier. The result is a sum sideband from 100.3 to 103 kHz, and a difference sideband at 97 to 99.7 kHz (Figure 4.3). The two sidebands are "mirror images" of each other: the lower end of the voice signal has shifted to 99.7 and 100.3 kHz, while the upper end becomes 97 and 103 kHz.

Amplitude modulation is sometimes said to *translate* frequencies to the frequency of the carrier. This means that the original information signal is moved, or translated, so that it becomes a pair of sidebands of the carrier, regardless of what the carrier frequency is. The bandwidth of the original signal is doubled by amplitude modulation—each sideband is identical in bandwidth to the original signal, and there are two sidebands. This is one of the drawbacks of AM, since bandwidth in the spectrum is a critical commodity and usually must be conserved.

Different Carriers and AM

Due to modulation, many signals that originally had the same spectrum can be translated to new frequencies and thus not interfere with each other. Standard AM broadcast radio is a good example of this. Carrier frequencies in the United States are assigned by the Federal Communications Commission (FCC) to each station,

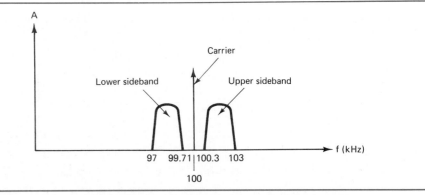

Figure 4.3 The result of AM when the carrier is modulated by a 300- to 3000-Hz signal.

4.2 Basics of AM

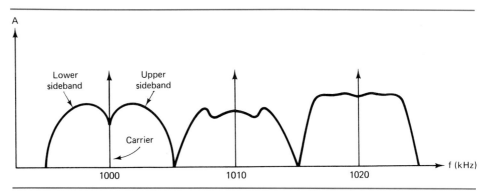

Figure 4.4 Spectrum of a group of carriers, each with its own sidebands.

spaced every 10 kHz in the overall broadcast band. Each station has a voice and music spectrum from 0 to 5 kHz. Without modulation, the information signals from these stations would completely interfere with each other. Instead, each station amplitude modulates its assigned carrier signal, and the result is pairs of sidebands at each carrier frequency. These sidebands do not overlap because the bandwidth of the original modulating signal is limited to less than one-half of the frequency difference between adjacent carriers.

Consider stations that have been assigned to 1000-, 1010-, and 1020-kHz carriers. Each station produces signals with a spectrum band ranging from 0 to 5 kHz on either side of the carrier. The final spectrum looks like Figure 4.4, with each carrier flanked by a pair of sidebands. Note that the sidebands do not overlap. If the carriers were too close to each other, or the bandwidth of the modulating signal too great, these sidebands would overlap and the upper sideband of one station and the lower one of another would interfere with each other. Whenever the spacing between carriers is greater than twice the bandwidth of the modulating signal, there will be no overlap; if the spacing is less than twice the bandwidth, some overlap will occur.

Example 4.1

Carriers are spaced at 20 kHz, beginning at 100 kHz. Each carrier is modulated by a signal with 5-kHz bandwidth. Is there interference from sideband overlap?

Solution

No. The first sideband pair is at 100 ± 5 kHz (95 to 100 kHz and 100 to 105 kHz); the second pair is at 120 ± 5 kHz (115 to 120 kHz and 120 to 125 kHz). Therefore, the upper sideband of one carrier and the lower sideband of the next adjacent one do not overlap.

Example 4.2

The modulating signal is changed to a bandwidth of 15 kHz. Is there sideband overlap and interference?

Solution

Yes. The first sideband pair is at 100 ± 15 kHz (85 to 100 kHz and 100 to 115 kHz), and the next pair is at 120 ± 15 kHz (105 to 120 kHz and 120 to 135 kHz). Therefore, there is overlap between 105 and 115 kHz.

Tuning a radio to select one station consists of using circuitry which is designed to pick out one carrier and its sidebands while ignoring others. Standard radios allow the user to adjust the circuitry to pick out any carrier and its sidebands within the entire band, although some radios are designed for monitoring essential or emergency transmissions and therefore receive only a single station.

Review Questions

1. What is a carrier? How is it defined? Why does it convey no information?
2. What does AM do to the carrier? What is the result of the AM process?
3. What are sidebands? What do they contain?
4. What happens to the frequency spectrum of any signal that goes through the AM process? Compare the bandwidth before and after AM.
5. What is the AM envelope? How is it seen?
6. Where is the information of the original signal after AM?
7. Is any information signal power left at the original frequencies, after AM? Explain.
8. Explain what happens to the frequency spectrum after information signals modulate several adjacent carriers.

4.3 MODULATION INDEX AND SIGNAL POWER

Modulation Index

There is a mathematical and a practical relationship between the amount of power in a carrier before modulation, and the amount of sideband power versus carrier power after modulation. As the magnitude of the information-bearing modulating signal gets larger, it will cause the envelope of the carrier, now modulated, to have a larger high-to-low amplitude span as well. A number called the *modulation index*, m is used to define and measure this span, where

$$m = \frac{\text{modulated peak voltage} - \text{unmodulated carrier voltage}}{\text{unmodulated carrier voltage}}$$

and therefore m can range from 0 to 1 under normal conditions.

Example 4.3

The modulated peak value of a signal is 10 V and the unmodulated carrier value is 8 V. What is the modulation index m?

4.3 Modulation Index and Signal Power

Solution

$$m = \frac{10 - 8}{8} = \frac{2}{8} = 0.25$$

Example 4.4

For an unmodulated carrier of 1000 V and a modulated peak value of 1800 V, what is the value of *m*?

Solution

$$m = \frac{1800 - 1000}{1000} = \frac{800}{1000} = 0.8$$

The effect of varying amounts of modulation and the use of the modulation index is shown in Figure 4.5, with a simple sine wave as the modulating signal for the carrier. The modulation index is often shown graphically:

$$m = \frac{B - A}{B + A}$$

[handwritten: $\frac{4-1}{4+1} = \frac{3}{5} = 60\%$]

[handwritten margin notes: A+B are half = 33% modulation; Decrease A, modulation ↑; Increase A, modulation ↓; A=0 = 100% modulation]

where *B* is the maximum voltage span of the modulated signal and *A* is its minimum voltage span. Multiplying *m* by 100% changes the modulation index from a number ranging between 0 and 1 to a percentage between 0 and 100%, and the modulation index is often expressed as a percent value (% *modulation*). The figure

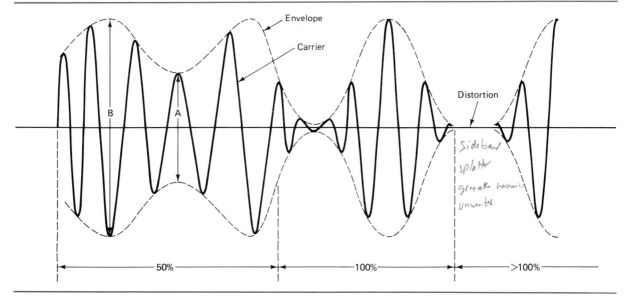

Figure 4.5 Display of modulated carrier with *m* = 50%, 100%, and >100%.

[handwritten: Sideband splatter, greater harmonics, unwanted]

shows one signal with a modulation index of 50% ($m = 0.5$) and another that has an index of 100% ($m = 1$).

Example 4.5

A modulated signal seen on an oscilloscope has a maximum span of 5 V and a minimum of 1 V. What is the modulation index?

Solution

$$m = \frac{5 - 1}{5 + 1} = \frac{4}{6} = 0.67 = 67\%$$

Example 4.6

A signal has a maximum span of 9 V and a minimum of 8 V. What is the modulation index?

Solution

$$m = \frac{9 - 8}{9 + 8} = \frac{1}{17} = 0.059 = 5.9\%$$

Example 4.7

A signal has a maximum span of 8 V and a minimum span of 0 V. Determine the modulation index.

Solution

$$m = \frac{8 - 0}{8 + 0} = \frac{8}{8} = 1.0 = 100\%$$

Modulation Index and Power

The modulation index is a measure of how fully the carrier has been modulated by the information-bearing signal. It is the relative magnitudes of the two signals that causes a carrier to be modulated to a lesser or greater extent. These magnitudes can be measured as the voltage or current value of the signals, but in most cases it is more useful to consider the signal power, and power is proportional to the square of voltage (V^2) or current (I^2).

The reason that power is the most useful measure is that the power available is a critical limitation on any communications systems. Every transmitter—whether a radio station, hand-held portable "walkie-talkie," or circuitry within a computer that sends data to an adjacent part of the computer—is limited in the amount of power it can use. There are limits to the battery size, to the amount of electricity that can be used, or to the physical size of transmitting components (more power means larger devices)—or too much power, causing interference with nearby wires. At the receiver, power is also a concern. The receiving circuitry must extract the original information from the signal power that it receives.

Derivation of Power Equation

The power relationship is derived as follows: Since the upper and lower sidebands each have a value of $m/2(V_c)$ and they are both feeding the same resistance, the power in each side frequency is proportional to $V^2 = [(m/2)V_c]^2 = (m^2/4)V_c^2$, where V_c is the carrier voltage. Power values are *scalars* and add directly (there is no *phase vector* associated with power as there is with voltage), so the sideband powers add up to $2(m^2/4)V_c^2 = (m^2/2)V_c^2$. Total power P_T is proportional to the sum of the carrier power P_c and the sum of the sideband power, so

$$P_T = V_c^2 + \frac{m^2}{2} V_c^2 = V_c^2 \left(1 + \frac{m^2}{2}\right) = P_c \left(1 + \frac{m^2}{2}\right)$$

The greater the received power, the easier it is to recover the desired information, especially if there is significant noise along with the desired signal.

The equation of amplitude modulation and the definition of modulation index can be combined to show the amount of power in a carrier and total signal:

$$\text{total power } P_T = \text{carrier power } P_C \left(1 + \frac{m^2}{2}\right)$$

or

$$m = \sqrt{2} \sqrt{\frac{P_T}{P_C} - 1}$$

This is a very important equation, because it relates total power to wasted carrier power; the carrier conveys no information. Only the sidebands have the desired information. The carrier serves as a vehicle for this information and allows it to be sent and retrieved more easily, but carrier power alone is wasted power. If the modulation index is low, toward 0%, most of the total power in the signal is carrier power, and very little is left for the sidebands. If the carrier is fully modulated (m = 100% or a factor of 1), the total power is 1.5 times the carrier power:

$$\frac{P_T}{P_C} = 1 + \frac{m^2}{2} = 1 + \frac{1}{2} = 1.5$$

This means that if 1.5 W is the total transmitted power, the power division between carrier power and sideband power is 1 W of carrier, and 0.5 W of total sideband power (total power = 1.5 W).

Example 4.8

A carrier of 1000 W is modulated with a resulting modulation index of 0.8. What is the total power?

Solution

$$P_T = 1000 \left(1 + \frac{0.8^2}{2}\right) = 1320 \text{ W}$$

Example 4.9

For a carrier of 250 W, with 90% modulation, what is the total power?

Solution

$$P_T = 250 \left(1 + \frac{0.9^2}{2}\right) = 351.25 \text{ W}$$

Example 4.10

What is total power when the modulation index is 0.75 and the carrier is 500 W?

Solution

$$P_T = 500 \left(1 + \frac{0.75^2}{2}\right) = 640.6 \text{ W}$$

Example 4.11

What is the carrier power if the total power is 1000 W and the modulation index is 0.95?

Solution

$$P_C = \frac{P_T}{1 + m^2/2} = \frac{1000}{1 + 0.45125} = 689 \text{ W}$$

Note that even with maximum modulation of $m = 1$, two-thirds of the power is carrier (1 W out of 1.5 W) and only one-third is information signal (0.5 W out of 1.5 W). As m decreases, the situation gets even worse, as seen by this example:

Example 4.12

What are the carrier power and sideband power with a modulation index of 0.25 (25%) when the transmitter is capable of 100 W maximum total power?

Solution

Using the equation that relates power division and modulation index, we obtain

$$100 \text{ W} = P_C \left(1 + \frac{0.25^2}{2}\right) = P_C (1.03125)$$

Therefore, $P_C = 96.97$ W. The sidebands have the remaining $100 - 96.97$ W = 3.03 W.

4.3 Modulation Index and Signal Power

Determining m If the carrier power and the total power can be measured, the modulation index m can be calculated using the rearranged version of the equation that relates carrier power, total power, and m.

Example 4.13

The total power is 1200 W, and the carrier alone is 850 W. What is the modulation index?

Solution

$$m = 1.414 \sqrt{\frac{1200}{850} - 1} = 0.91 = 91\% \qquad (\text{let } \sqrt{2} = 1.414)$$

Example 4.14

A unmodulated carrier of 10 W is measured at 12 W when modulated. What is m?

Solution

$$m = 1.414 \sqrt{\frac{12}{10} - 1} = 0.63 = 63\%$$

Example 4.15

A 75-W unmodulated carrier has 100 W of power when modulated. What is m?

$$m = 1.414 \sqrt{\frac{100}{75} - 1} = 0.82 = 82\%$$

The relatively low amount of power in the information-bearing sidebands as compared to the total amount of power transmitted, even with 100% modulation, points out one of the weaknesses of AM. A great deal of precious power is wasted in the carrier but conveys no information. In addition, since the total sideband power is shared by the two sidebands, and the sidebands both contain identical information, they are actually redundant, so power is wasted. There is no information in one that is not in the other. Therefore, with 100% modulation, two-thirds of the total power is carrier, one-third is sidebands, and actually only half of the sidebands' one-thirds portion is unique. To put it another way, only one-sixth (approximately 17%) of the total power transmitted contains unique information.

For these reasons, every attempt is made to modulate the carrier as fully as possible and keep the modulation index close to 100%. Many transmitters contain special booster circuitry for doing this and keeping the modulation high. If someone is speaking softly into the microphone, for example, the booster will automatically increase the amplification of the voice signal relative to the carrier, and then adjust it back down when the voice increases in loudness.

Overmodulation

So far, the maximum value of the modulation index has been 100%. It is actually possible for the index to be greater than that, however. This occurs when the magnitude of the modulating signal is too large, compared to the carrier power, and the modulated carrier has a value more than twice its unmodulated value. In Figure 4.5, the value of A is 0 at 100% modulation. A remains at 0 when overmodulation occurs, and the envelope no longer follows the shape of the information signal. Instead, the modulated carrier signal is *cut off* during the overmodulation.

The result of this overmodulation is severe distortion of the modulation envelope. In the time domain, this distortion means that the original information-bearing signal cannot be recovered accurately and is partially corrupted. In the frequency domain, overmodulation results in new frequencies besides the simple sum and difference ones of regular AM. These new frequencies are outside the regular AM spectrum and are called *splatter*. The modulated envelope now contains higher-order harmonics of the original modulating frequency, since the envelope is "clipped" at the zero axis and begins to resemble a square wave rather than the original modulating wave. These harmonics can be widely separated from the carrier frequency and cause interference with adjacent frequencies and broadcast stations. Overmodulation is never desirable and many systems include special circuitry to limit the modulation and ensure that it can never go above 100%, even if the carrier power changes for any reason.

Although the inefficient use (and waste) of power is a drawback of standard AM, this does not mean that AM is not a useful modulation process. In applications where the power is available, such as a standard broadcast station transmitter, AM is a very simple and effective modulation scheme; or where one subsection of a system must send some data to another subsection nearby, AM is very effective.

There is a further aspect to the usefulness of AM versus its inefficiency in the use of power. The fact that a large portion of the power remains in the carrier, even when the modulation index is 100%, makes recovering the information from the received signal very easy. This process of recovering the information is called *demodulation* and is the subject of Chapter 5. The large amount of carrier signal power (recall that the carrier is a sine wave) makes tuning and extracting the sideband information simple. For this reason, AM broadcast receivers can be very simple circuits. Here is the compromise: inefficient use of power, but ease of modulation and demodulation, resulting in low-cost receivers for mass markets.

There are other ways to modulate information onto a carrier that make better use of total power but are more difficult to demodulate and so make recovering the useful information harder. The choice between AM and these other approaches must be made by the communication system designer and engineer based on the requirements of the system, overall power available, and many other factors.

Review Questions

1. What is the modulation index? What does it indicate?
2. Explain how the modulation index can be seen graphically, as on an oscilloscope screen.
3. Why is a higher value of modulation index better? Why is power usage a critical factor in most applications?

4. What is the relationship between total power, unmodulated carrier power, and modulation index?
5. For 100% modulation, what is the fraction of total power in the sidebands? In the carrier?
6. What is overmodulation? Why is it a problem when it occurs?
7. What are the effects of overmodulation in the time domain? In the frequency domain?
8. How does the envelope of the modulated signal compare to the modulating signal when $m > 100\%$?
9. What is the technical trade-off in the amount of sideband power versus carrier power for ease of modulation and signal recovery?

4.4 AM CIRCUITS

The function of an amplitude modulation circuit is to allow the information-bearing signal to vary the amplitude of the carrier. There are many ways to achieve this result. Each method provides technical advantages and disadvantages, and the best method depends on the AM application.

Any nonlinear circuit can act as an amplitude modulator. (A linear circuit is one that reproduces the input signal exactly, but usually with a different amplitude. If the signal has more than one frequency component, they are all reproduced as in the original, and all have the same amplification factor. No new components are introduced, and the existing components retain the same proportions to each as they had in the original.) In a nonlinear circuit, the relationship between the original signals and the new frequencies produced is determined by the types of nonlinearities that the circuit has. For ideal amplitude modulation, the circuit should produce sum and difference frequencies, and not produce harmonics of these sum and difference values (frequencies that are double, triple, and so on, the desired values). A practical AM circuit can use virtually any nonlinear electronic element, such as a diode, to produce the desired signals—as long as the undesired sum and difference harmonics are removed.

A simple AM circuit using a single transistor is shown in Figure 4.6, with the modulation applied to the base. The transistor is biased into a nonlinear operating region, so instead of acting as a linear amplifier, it combines (*mixes*) signals that are applied to its base, in this case, carrier v_c and information signal v_m. The nonlinear mixing produces numerous sum and difference frequencies at the carrier frequency $\pm n$ (information signal frequency), where n is 1, 2, 3, and so on. The *LC* circuit in the collector is tuned to resonate at the carrier frequency and acts as a high impedance at resonance, so frequencies at or near the carrier frequency are available at the output. For frequencies further away from resonance—the undesired sum and difference products—the *LC* circuit appears as a low impedance, and these products are effectively shorted out.

Another, more complete AM circuit, in both tube and transistor configurations, is shown in Figure 4.7. In this plate or collector modulation circuit, the

92   Chap. 4   Amplitude Modulation

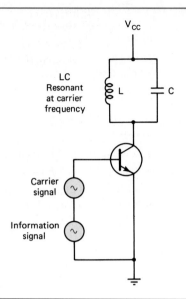

Figure 4.6 Single transistor configured for AM by modulating the base current.

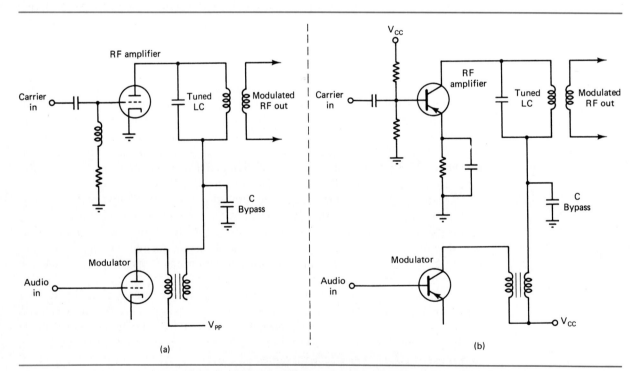

Figure 4.7 More detailed modulator circuits, using either (a) tubes and plate modulation or (b) transistors and collector modulation.

4.4 AM Circuits

modulating signal varies the plate or collector voltage of the RF carrier amplifier. The RF amplifier does not have to be linear in this circuit, and can operate in *class B* or *class C* mode for greater efficiency rather than the linear *class A* mode (all subsequent stages of amplification must be linear, however). The tuned coupled circuit *LC* eliminates the harmonics of the carrier and modulating signal, while capacitor *C* provides the necessary low-impedance path to the carrier currents so that these currents do not flow through the modulation transformer. Capacitor *C* must not bypass the much lower modulating frequencies; if it does, they will not reach the RF amp collector.

Many other circuit designs for AM are also possible. Any simple tube or transistor circuit, however, has some problems with undesired types of nonlinearities and its ability to provide the desired modulation index, as well as stable, repeatable performance with variations in temperature and power supply values. A more modern amplitude modulation circuit makes use of integrated circuits to produce the desired output with high accuracy and few components. An AM circuit based on a *multiplier* which exactly produces the multiplicative product of two signals and the information signal is the central part of this form of modulator, with the carrier multiplied by $[1 + m(\text{information signal})]$.

The Analog Devices model AD534 is an analog IC that uses a clever and sophisticated arrangement of transistors to provide the mathematical function with an output voltage that is the product of two input voltages. Figure 4.8 shows the AD534 specifications and its use as an AM circuit. This particular type of design would be nearly impossible to do with individual transistors, since it makes use of the fact that transistors fabricated on the same IC can have precisely matched currents flows and signal gains, and the changes with current flow with temperature variations (a factor in any circuit performance) will be the same in the various transistors. For individual transistors on a circuit board, this would not be

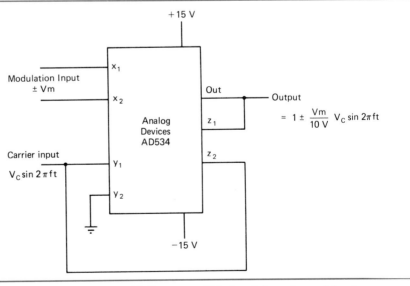

Figure 4.8 Use of an IC multiplier as an AM circuit (courtesy of Analog Devices, Inc.).

> **Vacuum Tubes for AM**
>
> A well-designed transistor circuit can produce AM at levels up to several hundred or thousand watts. For higher powers of thousands of watts and above, vacuum tubes must be used because transistors cannot handle the power levels. Vacuum tubes (their principles of operation are reviewed in Chapter 23) are more cumbersome than transistors and waste more power, but they are the only practical choice when higher power is needed in the AM transmitter. A tube-based AM circuit for screen-grid modulation is shown in Figure 4.9. The RF carrier signal is applied to one grid, while the information signal is applied to a second grid via transformer coupling (grids are roughly analogous to a transistor base). Capacitor C is for RF bypass and has a high reactance at audio frequencies; it prevents undesired coupling between the RF carrier and modulating signal. The AM output is tuned by an LC tank circuit and then inductively coupled to the next stage in the system.

the case—there would be initial differences in performance and different reactions to temperature changes as well.

The AD534 provides the multiplication of any two signals, up to a maximum frequency of 2 MHz. The inputs can range between -10 and $+10$ V, so the output is divided by *scaling factor* of 10 to restrict the output range from -10 to $+10$ V as well (otherwise, the output would have to range from -100 to $+100$ V, which is impractical). To perform AM, the modulating signal V_m is applied as the input voltage between X_1 and X_2, while the carrier $V_c \sin 2\pi f_c t$ is the input voltage between Y_1 and Y_2. The output is the product of these two signals, which consists of the sidebands only (the result of pure multiplication) and which does not include any carrier. The design of the AD534 allows the carrier signal itself to be added to the output, providing the two sidebands (the products) and the desired carrier. By adjusting the magnitude of modulating signal V_m compared to the carrier amplitude V_c, the modulation index can be varied.

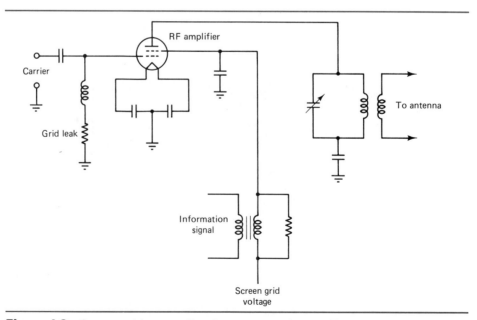

Figure 4.9 Screen grid connection for vacuum-tube amplitude modulator.

Review Questions

1. What is the basic electronic function needed to produce AM?
2. Why is a single-transistor AM circuit not practical?
3. How is an electronic multiplier used for AM? What is the relationship between the input and the output?
4. Why is an integrated-circuit design preferred for the multiplier?
5. How does the AD534 provide AM with carrier?

4.5 SUPPRESSED CARRIERS AND SINGLE SIDEBAND

The amplitude-modulated signal discussed thus far is simple to produce but has two practical drawbacks in application to many real communications systems: the bandwidth of the AM result is twice that of the modulating signal (two identical sidebands), and most of the power is transmitted in the carrier, not in the information-bearing sidebands. To overcome these problems with AM, variations on AM have been developed. These other versions of AM are used in applications where bandwidth must be conserved, or power used more effectively, even if the transmitting and receiver circuitry has to be more complex.

Suppressed Carriers

If the carrier could somehow be removed or reduced, the transmitted signal would consist of two information-bearing sidebands, and the total transmitted power would be information. When the carrier is reduced, this is called *double-sideband suppressed-carrier AM,* or *DSB-SC*. Instead of two-thirds of the power lost in the carrier, nearly all of the available power is used in sidebands. A DSB-SC signal in the frequency domain looks like Figure 4.10. In the time domain, it looks like Figure 4.11.

An example shows the increase in information signal power that results. For a 100-W total available power, compare the power in the sidebands when modula-

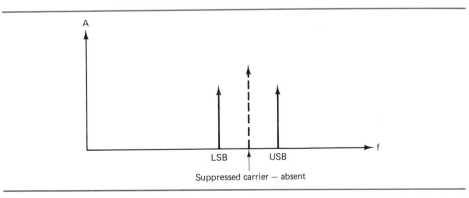

Figure 4.10 DSB-AM in the frequency domain.

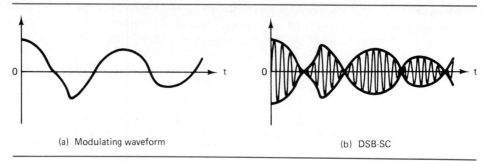

Figure 4.11 DSB-AM in the time domain.

tion is standard AM with a modulation index of 100%, versus a suppressed carrier design where 90% of the carrier power is suppressed. How many times greater is the sideband power in the suppressed carrier case?

For regular AM, we have seen that with 100% modulation, one-third of the power is in the sidebands (in this case, 100/3 = 33.3 W), and the remaining two-thirds (66.7 W) is in the carrier. In this DSB-SC case, 90% of the carrier power equal to 90% × 66.7 = 60.0 W is removed and instead goes to the sidebands. The new sideband power is

previous sideband power + 90% previous carrier = 33.3 + 60 = 93.3 W

The ratio of new sideband power to previous sideband power = 93.3/33.3 = 2.8, or approximately 4.5 dB, which is a significant increase. Note that while in regular AM only 33% of the total transmitter power was sideband, in this DSB-SC case over 93% is sideband power.

Example 4.16

A 500-W DSB-SC system with 100% modulation supresses 50% of the carrier, and the suppressed carrier power goes to the sidebands. How much power is in the sidebands, and how much is in the carrier? By how many dB has the sideband power increased?

Solution

With 100% modulation, the final carrier power is 500 W and the total sideband power is 250 W. If half the carrier power is diverted to sidebands, the new carrier power is 500 − 250 = 250 W and the new sideband power is 250 + 250 = 500 W. The increase in power is 500/250 = 2 = 3dB.

Circuitry for DSB-SC

One approach to generating a DSB-SC signal is to generate a regular AM signal and then filter out the undesired carrier. However, this is usually not a practical approach. First, the carrier signal is very close in frequency to the sidebands, so the filter would have to be very exact, sharp, and stable with time and temperature. It would also have to be designed to work at one frequency, so the carrier frequency could not be changed without changing filter components. Finally, the carrier is larger than the sidebands, and reducing the carrier by filtering might still leave the carrier too strong, relative to these sidebands.

Another approach is to use a circuit that inherently produces sidebands only (no carrier). This would eliminate the need for producing and then eliminating the carrier and would be much more efficient in the use of power: It takes power to produce the carrier, and then this power must be removed as heat from the filter. A preferred approach is to use a *balanced modulator*. For a perfect balanced modulator—and perfect DSB-SC—the output is zero when either input, such as the modulating signal, is zero. This is in contrast, of course, to regular AM, where there is a carrier signal (and power) even if the modulating signal is zero.

A balanced modulator can be built with diodes, such as the *balanced ring modulator* of Figure 4.12. The arrows show the flow of current at one particular instant. The horizontal diodes conduct, and the current flows through the primary of the output transformer are equal and opposite, so the carrier is self-canceled and does not appear at the output. When the carrier goes through one half-cycle, the other two diodes conduct, and the same effect occurs. For the modulating signal alone (no carrier), which appears in the circuit at the secondary of the input transformer, current will flow through either the D_1/D_2 or the D_3/D_4 pair, but not through both and not through the output primary, so the output is zero. But when both the carrier and the modulating signal are present, the diode conduction is determined by the carrier polarity, since the carrier signal is much larger than the modulating signal. The modulating signal either boosts (aids) or bucks (opposes) the signal, and one diode (D_4) will conduct more than another (D_1), or the other way around. The symmetrical current balance into the primary of the output transformer is upset, so the sidebands appear in the output but the carrier is still suppressed.

A typical value for carrier suppression is between 40 and 60 dB, depending on how well the diode characteristics are matched to each other. Balanced ring modulators in IC form are especially attractive because matched component characteristics are easier to achieve than with separate, discrete components. As another alternative, the multiplier of Figure 4.8 can be used as a balanced modulator, since the output contains only the upper and lower sidebands which result from multiplication of the two input signals. An ideal multiplication inherently produces DSB-SC AM.

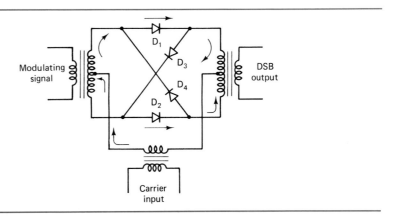

Figure 4.12 Balanced modulator circuit with current flow shown at one instant.

Any real circuit may not be perfect, and may have some output voltage even if an input is zero. This is called *feedthrough*, where the nonzero input signal can "feed through" and appear at the output, even when there should be no output since the other input is zero. The feedthrough produces a small carrier signal, which is undesired for perfect DSB-SC, but may be acceptable if there is to be some carrier allowed. Most multiplier circuits and ICs have provision for a small trimming resistor which can be manually adjusted to compensate for this feedthrough and reduce it to a very low value which will be acceptable.

Example 4.17

A DSB-SC system must suppress the carrier by 50 dB from its original value of 10 W. To what value must the carrier be reduced?

Solution
Using the formula for dB from Chapter 3 gives us

$$P_1 = P_0 \times 10^{dB/10} = 10 \times 10^{-50/10} = 0.0001 \text{ W}$$

Example 4.18

A test specification says that the carrier voltage measured at the output of the DSB-SC modulator must be at least 30 dB below the unsuppressed value of 1.0 V. The technician measures 0.02 V. Does this meet the specification?

Solution
Using the dB formula for voltage (not power) yields

$$dB = 20 \log \frac{V_1}{V_0} = 20 \log \frac{0.02}{1.0} = -34 \text{ dB}$$

which easily meets the specification.

Example 4.19

For Example 18, what is the maximum voltage value of the suppressed carrier that can be accepted?

Solution

$$V_1 = V_0 \times 10^{dB/20} = 1.0 \times 10^{-30/20} = 0.032 \text{ V}$$

Single-Sideband

While DSB-SC can greatly improve the use of transmitter power and result in stronger information-bearing signals being sent and received, the bandwidth of the signal is still twice that of the original modulating signal. Also, these two sidebands are redundant, so there is a waste of power as the transmitter power is

4.5 Suppressed Carriers and Single Sideband

divided between these two sidebands. A solution to both of these problems can be achieved by suppressing the carrier and one of the sidebands. This is called *single-sideband suppressed-carrier*, or simply *single sideband (SSB)*. SSB is the most efficient form of AM, since virtually all the transmitter power is used for a single information-bearing sideband.

Note that it is much more difficult to recover the original information signal from SSB modulation, as we will see in Chapter 5. Simply stated, since there is no carrier present, the receiver has no way to establish its own operating frequency and circuitry at the right frequency separation from the received sideband. In fact, the receiver must generate its own version of the now-missing carrier, and use this to extract the sideband information. The difficulty in doing this properly and reliably was one of the major drawbacks to widespread use of SSB until modern circuitry made it more practical.

SSB Power

Conventional AM power is rated by the power of the carrier signal—a 1000-W transmitter has a carrier of 1000 W of power, and the modulation index is not used to define the power level. With SSB, there is no carrier to use as the measure of signal strength. Instead, power is defined by the peak envelope power (PEP), which is defined as the rms power at the peak of the waveshape (therefore, for SSB, it is the rms power at the sideband peaks).

The time-domain graph (Figure 4.13) of a typical SSB signal does not reveal much about the modulating waveform and bandwidth, but the frequency-domain graph (Figure 4.14) does. The frequency spectrum shows the upper sideband only of a AM spectrum that originally had both a carrier and an upper sideband. The bandwidth is one-half the original AM value, and all the power is in a single information-bearing sideband. Of course, the lower sideband could have been suppressed and the upper sideband used in its place. Most SSB transmitters allow the user to select if the upper sideband or lower sideband should be used for broadcast.

The frequency spectrum no longer used by the suppressed sideband can be used by another transmitter. Suppose that a series of carriers is spaced every 20 kHz, from 100 to 200 kHz, and the modulating signal has a 10-kHz bandwidth. The

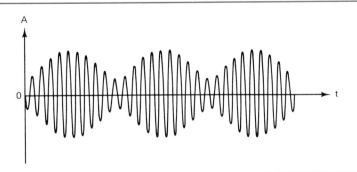

Figure 4.13 Single sideband in the time domain; it is difficult to see how this differs from conventional AM.

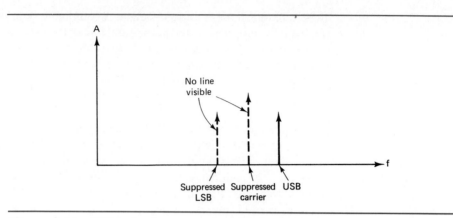

Figure 4.14 SSB is the frequency domain clearly shows difference with AM.

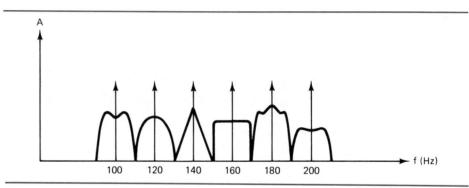

Figure 4.15 DSB AM with a fully occupied spectrum.

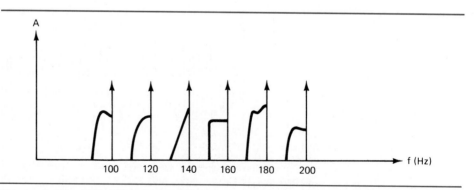

Figure 4.16 SSB AM frequency spectrum with lower sidebands only.

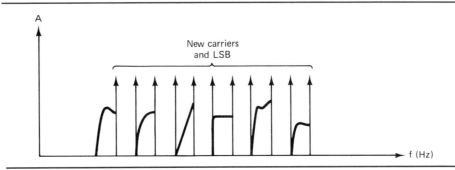

Figure 4.17 Same spectrum as Figure 4.16 with twice as many carriers and lower sidebands only.

frequency spectrum after the AM process would consist of carriers at 100, 120, 140, . . . , 200 kHz, with a 10-kHz lower sideband and a 10-kHz upper sideband on both sides of each carrier (Figure 4.15). This spectrum is fully occupied. If SSB is used with lower sidebands only, the result is shown in Figure 4.16, with half the spectrum available. This newly available spectrum can be used two ways. One approach is to allow each modulating signal to use double its original bandwidth, in this case to 20 kHz. Since the amount of information that can be sent is directly proportional to the bandwidth, this means that each user can send twice as much information in the same amount of time.

Another alternative is to set up new carriers between the original ones, and then use each of these to produce new lower-sideband SSB modulation using a modulating signal bandwidth the same as the original signals (Figure 4.17). In effect, there are now twice as many users (transmitters) allowed in the same spectrum, and each has the same information capacity as in the original AM situation. All of these users—new and old—must agree to use the same sideband (upper or lower). If a transmitter at one carrier (100 kHz, for example) uses the upper sideband, and the user at the next carrier of 120 kHz uses the lower sideband, the two sidebands will occupy the same space in the spectrum and interfere (Figure 4.18).

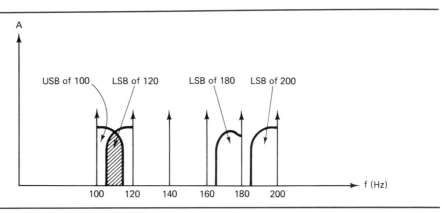

Figure 4.18 All new carriers must use the same sideband to avoid interference.

Making the Choice

In an SSB system, a choice must be made between keeping the same number of users but allowing each of them twice the bandwidth, or adding new users (carriers) and restricting them to the original bandwidth. The decision is made based on the type of system and the information it will carry.

If the communications system is designed for modulating signals that inherently have limited bandwidth, such as voice, there is no purpose to allowing each user to have increased spectrum space available for the result of modulation. It would be better to allow more users, instead. Thus, when the radio links that carry phone calls cross country were upgraded from regular AM to SSB (by modifying the modulation electronics at the transmitting end and the corresponding electronics at the receiving end), the number of phone calls that could be handled at the same time doubled.

For other applications, it is more efficient to allow the user to have the increase in bandwidth for transmitting information. If two computers are passing data back and forth, the rate at which the data can be sent is limited by the bandwidth available in the communications system. By giving each transmitter/receiver pair twice the bandwidth, the computers can start passing these data between them at the much higher rate. The number of computer communication channels remains the same, but the increased bandwidth that is made available by going with SSB modulation is used to increase the channel information capacity.

Another advantage of the reduced bandwidth that SSB needs to convey the same information as DSB is that the amount of channel noise is also reduced. Noise is directly proportional to the bandwidth (Chapter 3); thus by reducing the bandwidth to one-half the DSB value, the received noise power is reduced by 3 dB. This makes recovering the original information easier.

Review Questions

1. What are two technical drawbacks to conventional AM?
2. What percentage of overall power, at best, carries information in AM?
3. What is the filter method of generating DSB-SC? What are the weaknesses of this method?
4. What is a balanced modulator? How can one be constructed?
5. What is the output of a balanced modulator when the information-bearing signal is zero? When it is not zero? Compare this to conventional AM.
6. What is SSB? What are its technical advantages in terms of power use and bandwidth use?
7. What does the frequency-domain graph of an SSB signal look like? Why?
8. Explain how SSB allows a choice of either increasing the bandwidth allocated to a user or allowing more users in the overall spectrum.

4.6 SSB TRANSMITTER CIRCUITS

There are several ways to generate SSB signals. These include carrier and sideband filtering, balanced modulation and filtering, and phase shifting, among others. The first alternative, which generates a regular AM signal and then filters the carrier and one sideband, is very simple but generally impractical. Such a circuit would require precise filters with very sharp *cutoff* (a *high-Q* filter) where the carrier meets the desired sideband, and the carrier frequency could not be changed easily without readjusting the filter. Also, the power used to perform the modulation would be wasted as the filter strips out the carrier and undesired sideband. This filtering method is used most often for a transmitter that is designed to operate at a single, fixed frequency, such as a major transmitter.

Sometimes, the filter specifications are relaxed (a lower-Q filter is used), and some of the unwanted carrier and sideband is allowed to pass through. Called *vestigial sideband (VSB)*, that occupies less bandwidth and makes better use of power than DSB-SC but is not as effective as true SSB. [We will see VSB in the study of television (Chapter 14).]

The balanced modulator provides another approach. Recall that the output of the balanced modulator is a double-sideband signal without carrier. The undesired sideband can be filtered out, leaving only the desired sideband. The requirements for this filter are much less stringent than for a filter that must also remove a relatively large carrier signal between the desired sideband and undesired sideband. It is especially practical when the spectrum of the modulating signal does not begin at 0 Hz, but instead starts somewhat higher, such as 300 Hz for voice signals. Then, after the DSB-SC signal has been generated, there is a gap of ±300 Hz (600 Hz total) in the spectrum. The existence of this gap makes the filter design much easier. In contrast, if the spectrum began at 0 Hz, the filter would have to be ideal to pass all signals of one sideband perfectly, and eliminate the other sideband, without any range of frequencies for the passband to stopband transition to take place.

The requirements for an SSB filter are so severe that in many cases a conventional electronic filter of inductors and capacitors cannot be designed and built to meet the specifications. Instead, a *crystal filter* is used, based on precisely dimensioned quartz crystals whose inherent resonant frequencies correspond to the filter frequency. Crystals are very high-Q LC resonant circuits and are the basis for a filter that is capable of providing stability along with high rejection (>60 dB) of unwanted signals that lie very close to the desired ones. Another type of filter, a nonelectronic filter called a *surface acoustic wave (SAW) device*, may be used and is discussed in Chapter 23.

Figure 4.19 shows a crystal passband filter centered around 5500 kHz. Crystals Y_1, Y_2, and Y_3 are identical, with resonance at 5500 kHz, while Y_4, Y_5, and Y_6 are also identical but with resonance about 1500 Hz different from the first trio. The bandwidth of this fairly sharp filter is 2750 Hz when its response is 6 dB below the center; it has spread to only 3950 kHz when the response is 30 dB down.

Regardless of the type of filter used, it may be difficult or impossible to build a filter with the required characteristics that can eliminate one of two signals separated by only a few hundred Hz at a carrier frequency of tens of hundreds of MHz. Instead, the SSB signal is developed and filtered at a much lower frequency, from several hundred kilohertz to approximately 1 MHz, and then the resulting single sideband is *translated* or shifted up to the desired final frequency (Figure 4.20). This requires additional circuitry, of course. Both upper and lower sidebands can be generated, either by using two filters (one for each sideband) and a single carrier frequency, or by using just one filter and shifting the carrier frequency by an amount equal to the required spacing between sidebands.

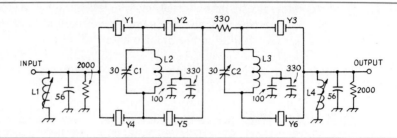

Figure 4.19 A crystal filter provides narrow bandwidth and precise values, here with a 2750-Hz bandwidth at 5500 kHz (from *ARRL Handbook*, American Radio Relay League, 1972, p. 395, Fig. 13-7; courtesy of the American Radio Relay League).

Figure 4.21 shows the schematic of an SSB circuit schematic that operates at 9 MHz. In this circuit, the grounded gate JFET provides impedance matching for maximum power transfer. To the balanced modulator, the JFET looks like the desired 50-Ω load, while the crystal filter appears to be driven by a required 500-Ω source impedance. The *LC* low-*Q* (broad, not sharp) tank circuit is resonant at 9 MHz to shunt any spurious out-of-band signals that may be generated in the modulator to ground, ensuring that they do not propagate into the rest of the system.

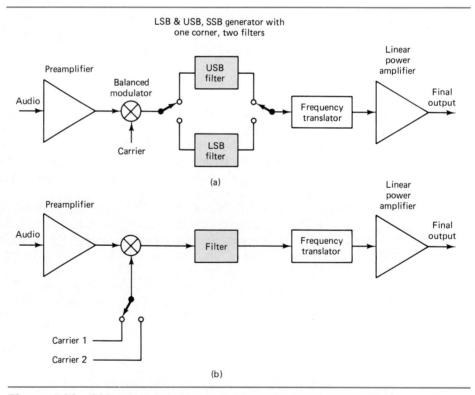

Figure 4.20 SSB generated at a lower frequency and then translated up to the final carrier frequency, using either (a) one carrier and one of two selectable filters, or (b) one of two selectable carriers and a single filter.

4.6 SSB Transmitter Circuits

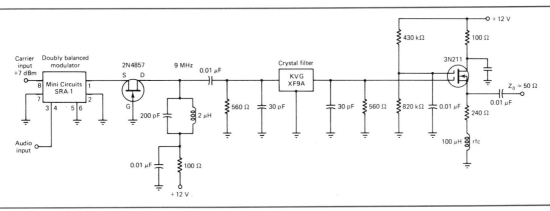

Figure 4.21 SSB generator using standard commercially available balanced modulator and crystal filter modules for operation at 9 MHz (from *ARRL Handbook,* American Radio Relay League, 1988, p. 18-7, Fig. 10-7A; courtesy of the American Radio Relay League).

A much more elegant approach is one that requires no filtering of undesired sidebands. This is the *phase-shift method* of SSB generation (Figure 4.22). The phase-shift method makes use of advanced mathematical analysis of the result of amplitude modulation and balanced modulator operation. This analysis shows that the output of any balanced modulator is a DSB signal with components that have been shifted in frequency and also shifted in phase by 90°. The phase-shift method is based on a simple fact: two signals that differ in phase by 180° (opposites, or "flipped over") cancel each other and produce a zero result if they are added together.

The phase-shift method uses two identical balanced modulators and two 90° phase shifters. The output of the carrier oscillator goes to one balanced modulator, where it is modulated by the information-bearing signal. The output of the carrier oscillator also goes to a 90° phase shifter, where it is modulated in the second balanced modulator by the information-bearing signals, also shifted by 90°. The outputs of the two balanced modulators are added together in a *summing* circuit, and the result of this sum is a perfect SSB signal, one that requires no filtering.

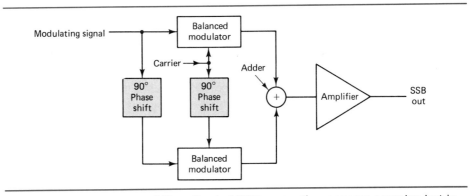

Figure 4.22 The phase-shift method of SSB generation appears complex but is very effective.

Filter Errors and dB Suppression

Analysis shows that sideband suppression is equal to

$$\text{sideband suppression} = 20 \log \left(\text{cotangent } \frac{\phi}{2}\right) \text{ dB}$$

where ϕ is the deviation from a perfect 90° shift. (*Note:* Cotangent $\phi = 1/(\text{tangent } \phi)$. An error of 1° in the phase-shift filter means that the maximum suppression that can occur is

$$20 \log \left(\text{cotangent } \frac{1.0°}{2}\right) = 20 \times \log 114.5 = 41.2 \text{ dB}$$

Although this means that the design of the filter is critical, it is much easier to design and implement a filter that has nearly ideal phase shift over the bandwidth of the modulating signal (typically a low-frequency range, such as for a voice signal) than it is to develop a filter that can eliminate an undesired carrier and sideband which has frequency components very close to the desired sideband.

Example 4.20

What is the maximum suppression for a phase error of 2°?

Solution

$$20 \log \left(\text{cotangent } \frac{2.0°}{2}\right) = 20 \log 57.3 = 35.2 \text{ dB}$$

Example 4.21

If the suppression must be at least 30 dB, what is the maximum phase error that can be accepted?

Solution

Rearranging the equation for suppression and phase error, we have $\phi = 2 \tan^{-1}(1/10^{dB/20})$ ($\tan^{-1}$ is the inverse tangent or "arctangent"). Therefore,

$$\phi = 2 \tan^{-1} \frac{1}{10^{30/20}} = 2 \tan^{-1}(0.0316)$$

$$= 2(1.8°) = 3.6°$$

The advantage of the phase-shift method is that filters for suppressing unwanted frequency components are not needed. In general, such filters are complex, expensive, do not maintain their characteristics with time and temperature changes, waste power, and cannot be tuned easily. In contrast, the 90° phase shift filter is much simpler to design and realize in a practical system. For the carrier, which has only a single frequency component, a nearly ideal phase-shifting circuit is easy to build.

However, this phase-shift filter for the modulating signal is more difficult to achieve. It must provide a precise 90° shift over the entire bandwidth of the modulating signal for complete cancellation of the unwanted sideband. Even a small deviation of a few degrees for some of the frequency components within the bandwidth can leave an undesired partial sideband, which will remain in the final output of the modulation circuit. Nevertheless, this is a very attractive way to implement SSB.

Most new DSB-SC and SSB transmitter designs use multipliers, balanced modulators, and the phase-shift approach. Modern ICs provide near ideal multipliers and balanced modulators at low cost in small packages. These provide better information than discrete components or filters that must supress unwanted signals and consume very little power.

Review Questions

1. How can SSB be generated by a carrier and sideband filter approach? Why is this is a very difficult design?

2. What are the advantages and disadvantages of using a balanced modulator and filter for SSB?

3. What is a crystal filter? Why is it used?

4. What is vestigial sideband? How does it come about?

5. What is the phase-shift method of generating SSB? What is the principle involved?

6. How do the two 90° filters used in the phase-shift method differ from each other? What are the different technical requirements on each?

4.7 BASEBAND AM

There is another specialized subgrouping of amplitude modulation that is important and common in communications. An information signal that is sent directly without modulating any carrier is called *baseband AM*, or simply *baseband*. (*Note:* The term *baseband* is also used in some cases to mean the original modulating signal before actual modulation occurs, or a signal that has been translated to a 0-Hz carrier from another carrier frequency.) In some situations it is not necessary to amplitude modulate a carrier; instead, the signal can be transmitted without modulation of any sort. A common example of baseband modulation is the voice signal from a telephone to the local phone company central office. A single pair of wires connects each phone to the central office. The voice signal on the telephone causes the voltage between these wires to vary, and this voltage variation is what is received at the office. There is no carrier associated with this signal and there are no sidebands.

Baseband modulation is very simple and straightforward. It is really a case of AM where the carrier frequency and amplitude are 0 Hz and 0 V, respectively. Although this may be seen to be a trivial subcategory of AM, it is quite often used when the signal has to travel only a short distance over wires (to prevent interference). There is no additional circuitry needed to perform baseband modulation—the source of the information signal itself is the transmitted signal, such as the phone, or a voltage from a part of a computer circuit. Baseband can be used whenever a single wire or path can be dedicated to one user, since the baseband spectrum begins at 0 Hz and goes on to the bandwidth of the information signal.

In some applications, such as within a single chassis, or in a network linking computers within an office or factory, baseband modulation may be the least complicated technical approach. Some baseband systems, such as the phone wires, have relatively narrow bandwidths. Others have bandwidths that extend to the megahertz range and can handle large amounts of information. Baseband modulation can be considered "no modulation at all," but from a technical and analysis perspective it sometimes must be considered as an AM subgroup.

Review Questions

1. What is baseband AM? Where can it be used?
2. Does baseband AM work only over short distances or for narrow bandwidths? Explain.

4.8 CONTINUOUS-WAVE AM

In the early days of electronic communications, it was not possible to modulate a varying signal, such as voice, onto the carrier. Information was passed by a series of on and off pulses, such as in Morse code, where each letter and number is assigned a specific pattern of dots and dashes. The carrier was turned either fully on or completely off, with the on period corresponding to a dot or dash, depending on the length of time it was on. [There is some irony here in that Morse code and its modulation resembles in many ways modern digital communications (Chapters 10 through 13).]

This form of modulation is really an extreme case of regular AM, since the modulating signal can only have fully on or off values, rather than the many gradations and values that a voice-like signal may have. Since the carrier could be fully modulated for a relatively long period of time, this is called *continuous-wave (CW)* modulation. It really would be better described as an interrupted continuous wave, since a truly continuous, unchanging signal conveys no information. The term CW often is used as a way of describing any Morse-code-like signaling, where the information is carried as complete on/off modulation of the carrier.

In the ideal case, this on and off modulation of the carrier is done with rectangular pulses, corresponding, for example, to when the Morse code key is up or down. (Often, this is called *keying* the transmitter because of this relation to the Morse code operator's key.) As shown in Chapter 2, any sudden change in amplitude results in a very wide spectrum of frequencies. The effect of perfect on/off keying is to generate sidebands with an extremely wide bandwidth. In the early days of radio, this was not a problem: There weren't many other users of the electromagnetic spectrum. Soon, however, the need to assign each user a specific frequency and alloted bandwidth became apparent. The shape of the keying signal was changed from nearly perfect on/off to a more gradual on/off (Figure 4.23) which reduced the bandwidth needed.

Modern CW code systems always have circuitry that prevents the sudden rise or fall but provides a smooth rise (*attack*) and fall (*decay*) to round the sharp corners and reduce the bandwidth of the modulated signal, yet ensures that the carrier is fully modulated for maximum effectiveness. Although many communications systems use voice, or continuously varying signals for AM, Morse code or other CW is still used in some special applications where noise and power problems require reliable communications under difficult conditions. Under difficult signal conditions such as low SNR, it is easier for the listener to distinguish between the presence or absence of a signal (simple *detection*) than to decide what the actual signal value is (more difficult *estimation*). In addition, many communications systems are used for sending signals in digital format, where the information is represented by a group of bits (1s and 0s). In these cases, the most modern

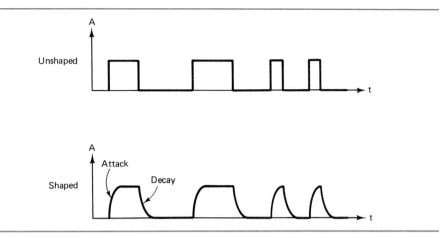

Figure 4.23 CW with both shaped and unshaped keying.

systems are actually making use of a variation of the earliest form of electrical communication: continuous wave with on/off modulation.

Review Questions

1. What is CW? Compare it to modulation by a voice signal.
2. What is the effect on bandwidth and spectrum of perfect on/off keying?
3. Why is it necessary to modify the modulation shape? How?
4. How are the ideas of Morse code applicable for computer- and data-oriented communication systems?

4.9 TRANSMITTER FUNCTIONS AND TESTING

Transmitter Functions

A complete transmitter requires several functions, whether it is designed for regular AM, DSB-SC, or SSB. These include the oscillator, which generates the carrier frequency, modulation circuitry, and amplifiers, which boost the signal to the final power level needed for the application. A typical transmitter for communication over several miles may have a few watts of power, while reliable communication across thousands of miles usually requires several hundreds or thousands of watts (unless extremely complicated antennas and receivers are used which can reduce the power needed, but at great expense).

There are two basic configurations for transmitters. Each has technical advantages and disadvantages in a specific application. In the *low-level modulation* approach (Figure 4.24), the modulating signal is applied to the carrier while both are relatively low power signals. If the application requires more power, the signal that results from modulation is passed to an amplifier or stage of amplifiers which boosts it to the desired final level. In contrast, with the *high-level modulation*

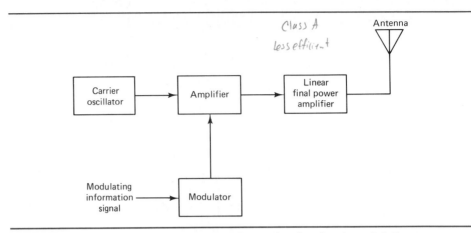

Figure 4.24 Block diagram of a transmitter using low-level modulation.

design approach of Figure 4.25, the carrier is generated and amplified to the final level needed, and then the information-bearing signal is applied to modulate this high-power signal. The amplifier stage between the original carrier or information signal and the subsequent higher-power amplifier (or modulation) stage is often called a *buffer, driver,* or *exciter.*

With low-level modulation, all the stages after modulation must be highly linear, or distortion (and unwanted frequencies) will occur. The linear amplifiers are, unfortunately, not very efficient and convert less than half the applied supply power to useful output—the rest is dissipated as heat. (Some amplifier configurations, such as "push-pull," can reach efficiencies around 70%.) However, since the carrier is at a relatively low power level when modulation occurs, the modulating signal does not have to be very powerful for full modulation.

A complete transmitter for the 2-m amateur radio band (144 to 148 MHz) is shown in Figure 4.26. RF stages include a 12-MHz carrier oscillator, a ×4 frequency multiplier, a ×3 frequency multiplier, and a collector-modulated final power amplifier stage (providing 10 dB of power gain); the audio stages include a

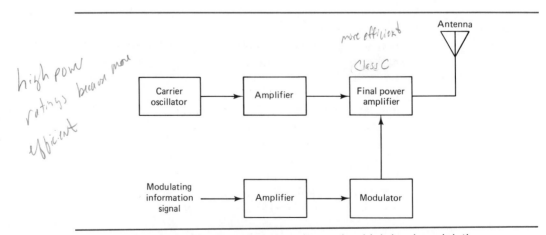

Figure 4.25 Block diagram of a transmitter using high-level modulation.

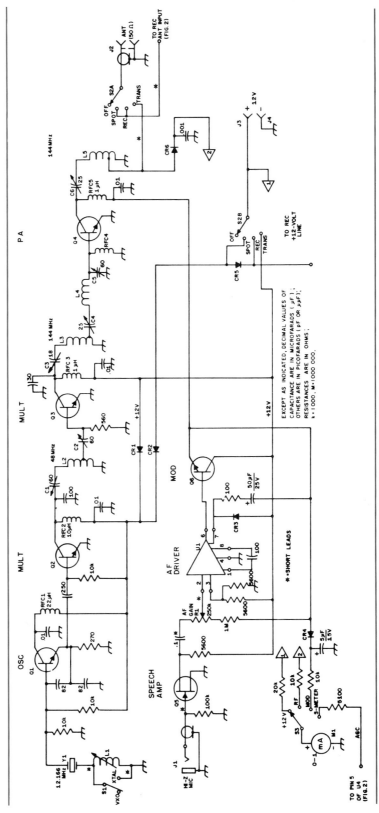

Figure 4.26 Schematic of a complete 2-m AM transmitter providing up to 1 W of output (from *ARRL Handbook*, American Radio Relay League, 1972, p. 388, Fig. 12-21; courtesy of the American Radio Relay League).

speech amplifier (JFET and IC op amp) and a transistor modulation amplifier. The basic crystal oscillator operates at 12.333 MHz, which is multiplied by 12 (3 × 4) to produce the 148-MHz carrier needed. This carrier oscillator covers the entire band since its basic frequency can be adjusted down to 12 MHz by varying the inductance in series with the crystal. The entire circuit requires a single +12-V supply and delivers up to 1 W of output power.

The low-level modulation scheme is reversed for high-level modulation. The information-bearing signal must modulate a very large carrier (hundreds or even thousands of watts is very common) and so must itself be amplified without distortion to a high level for full modulation. However, the carrier oscillator signal can be amplified by more efficient amplifiers with some nonlinearities, and then filtered to produce the single-frequency result. Most low-power transmitters (under 10 W) use low-level modulation, while broadcast stations use high-level modulation.

Modern transmitters mix ICs, transistors, and vacuum tubes in a single overall design. The lower power stages use integrated circuits for preamplifiers, oscillators, filtering, and modulation (if done at low level), while transistor circuits boost signal levels, if needed, above the 10-W (approximate) maximum that ICs normally can supply. This IC or transistor output is then applied to the final amplifier stages, which use vacuum tubes. Although solid-state components offer many advantages, tubes are still the only choice when many thousands of watts of power at radio frequencies must be generated. Mixed designs use both to take advantage of where each is best. The transmitter has meters that show the internal signal levels, power at each stage, modulation index, and antenna conditions. Of course, lower power transmitters—up to about several hundred watts, depending on frequency of operation—are entirely solid state (IC and transistor).

Testing Transmitters

Before testing any transmitter, remember this: **It can be dangerous!** Transmitters with power levels over about 100 W have high voltages, currents, or RF power levels and must be handled carefully to avoid inadvertently damaging components or **serious injury** to the test technician. (Low-power transmitters, such as 5- to 10-W units, *usually* pose no threat to people.) Most transmitters have safety interlocks to force the person working on them to recognize the danger before proceeding. Make sure that you know what you are getting into when testing transmitters: Don't "poke around" to see what signal levels are without a schematic, a parts placement diagram showing where components are located, a troubleshooting manual, and any other relevant documentation. Don't connect test probes from a voltmeter or oscilloscope except to designated test points, and even then, **use extreme care.**

The key to testing transmitters is to recognize that a complete transmitter consists of stages, each of which takes a signal and processes it in some way, thus providing oscillation, amplification, modulation, filtering, or frequency translation. Each stage should be looked on as a complete function, *usually* with a single input signal and a single output signal (although some stages have several inputs or outputs). A good test approach is to trace the flow of signals from the first stage to the final stage and examine the performance of each stage.

The first requirement for successful testing is a signal source. Even if the transmitter is designed for voice, a real or prerecorded voice is a poor source since it varies continuously in amplitude and frequency content. Instead, a single-tone oscillator (an audio signal generator) should be used for basic tests, and a two-tone oscillator for the advanced tests. The frequency or amplitude of this constant source may be varied by the test technician, but this would be a deliberate change, not the randomness of a voice or music signal. A common standard test signal is 1 kHz at 1 mW (0 dBm).

Signal pickup is another issue in transmitters. Because of the high voltages and currents often involved, a test probe cannot be connected directly to the circuit. Instead, a *pickup* that electromagnetically couples some of the signal to the probe, but without physical contact, is used. The pickup provides safety since there is no direct electrical connection between the circuit and the probe, and also reduces the signal levels to a value compatible with the test equipment. Of course, this type of pickup can only be used for medium or high frequencies that couple effectively through coils and electromagnetic fields, and cannot be used at all for dc signal such as power supply voltages. For these cases, test points are usually provided in the transmitter which reduce the high signal levels through resistors or other schemes. Only these designated points should be used!

Power supply levels are measured with a voltmeter (often built into larger transmitters) and should be checked first. Circuit performance should be examined only after the supply voltages are confirmed as correct—if they are not, the cause (either a problem within the supply or a short circuit in the system that the supply is powering) must be located and fixed. The oscilloscope, which shows signal amplitudes versus time, is a key instrument for checking performance of the circuitry. With the recent development of lower-cost, easy-to-use spectrum analyzers, these are also being used increasingly as a test tool, to show performance directly in the frequency domain.

Using the audio oscillator as a signal source (set at a low-enough level not to overload the preamp input) and working through a conventional AM transmitter, here are the areas to check:

1. Make sure that the audio preamplifier stage, which takes the low-level signal from the source, is reproducing the input signal with the correct amplitude. Usually, a first preamplifier stage develops a signal of about 1-V amplitude. This preamplifier may then feed a subsequent stage which further boosts signal power, if needed. The oscilloscope (and spectrum analyzer) display for this stage should be the same from input to output, except that the output has greater magnitude.

2. The carrier oscillator (sometimes called the *master oscillator*) should produce the correct frequency and amplitude sine-wave carrier. In lower-frequency transmitters, the oscillator operates directly at the final frequency. For higher-frequency transmitters, an oscillator cannot be designed to operate at the desired carrier frequency, but instead operates at lower frequency; the oscillator output then passes through a *frequency multiplier* stage which produces the final carrier frequency. The carrier should always be a pure sine wave. This is most easily seen on the spectrum analyzer, appearing as a single steady "spike" at the desired frequency only. There should be no wavering around this frequency (*jitter*), even if the "average" frequency value is correct.

3. In low-level modulation, the carrier and information signal—the audio tone(s)—are combined at the modulation stage. The output of this stage should be a single carrier whose envelope, seen on the oscilloscope, has exactly the same shape as the audio tone(s), indicating no distortion. The modulation index m

should be measured and compared to the specification; it is normally designed to be greater than 75% for a tone. The result of modulation, and any distortion, is seen best on the spectrum analyzer, as the carrier now has two sidebands.

If this is an SSB transmitter using the balanced modulator, the output on the oscilloscope is much harder to judge and interpret. On the spectrum analyzer however, it shows up as two sidebands on either side of the formerly visible carrier. The DSB-SC output goes through a filter, which removes one sideband, producing the final single-sideband signal. This, too, is easily seen and judged on the spectrum analyzer.

4. For low-level modulation, the modulated output is passed through a final amplifier (or several amplifier stages) which produces the required output power. These stages must be linear, and the final output should be a reproduction of their modulated input, but at a higher level. Both the oscilloscope and spectrum analyzer can be used, although with the analyzer it is easier to recognize when distortion—seen as new spectral components—occurs, due to nonlinearities in the power amplifier. It is also easier to compare input and output signals in the analyzer, since new spectral components may be too subtle to see in the time domain, yet are more obvious in the frequency domain.

5. A high-level modulation system can use more efficient nonlinear amplifiers and then filters the nonliniearties. Therefore, the scope or analyzer display will show some distortion (new spectral components) at certain points in the system.

Because the linearity of each stage is important (except for the carrier power amplifier in a high-level modulation design), the single audio tone is often replaced with a two-tone source, where two frequencies (such as 1 kHz and 7 kHz) of the same amplitude act simultaneously as the source. The reason for this is that two independent frequencies pass through a linear amplifier and emerge as the same two frequencies, while a nonlinear amplifier or overmodulated circuit has *intermodulation distortion* and produces new frequencies at the sum and difference of the two input values. (Recall that an amplitude modulator can be made from any nonlinear device, so an amplifier that is not perfectly linear acts as a pseudo-AM circuit.)

With the two-tone source, the sum and difference values are new components and more easily seen on the oscilloscope or spectrum analyzer than is a single tone. Figure 4.27 shows some examples of oscilloscope and spectrum analyzer displays for 50%, 100%, and >100% modulation, using two tones as a source. Note that distortion of the information-modulating signal at $m > 100\%$ results in new spectral components. Similar distortion of the basic audio signal results if any stage in the audio path is not linear (except for the modulator, which "shifts" the baseband spectrum up to the carrier frequency while preserving the shape of the audio signal).

If there is a problem appearing at any stage, do not assume that the cause lies within that stage. Assuming the input to a stage is correct, a subsequent stage may be at fault, loading the previous stage output or even causing instability and oscillation.

4.9 Transmitter Functions and Testing

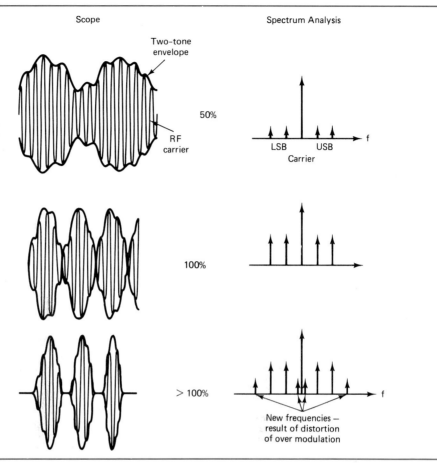

Figure 4.27 Oscilloscope and spectrum analyzer displays for 50%, 100%, and >100% modulation, with a two-tone audio source.

Review Questions

1. What are the functional blocks in a low-level modulation transmitter?
2. How does a high-level modulation transmitter differ functionally from a low-level one?
3. What is amplifier efficiency? How is it defined?
4. What are the basic steps for testing a transmitter? What is checked first? Why must great care be used?
5. How are signals connected to test probes? Why is a simple "touch the probe to the wire" not used?
6. Describe the stage by stage checkout of a transmitter. How can one stage be affected by a problem in the next one?

SUMMARY

In amplitude modulation, a signal containing information is used to vary the amplitude of a continuous high-frequency signal called the carrier. The result of AM is that the information signal is translated into two sidebands of the original carrier. These sidebands contain the information, while the carrier is used for ease of transmission and reception. Some of the weaknesses of AM include wasted power (the carrier) and excess bandwidth (the sidebands occupy twice the spectrum that the original signal required. However, AM is easy to implement. The modulation index indicates how fully the carrier has been modulated and the division of power between carrier and sidebands.

Variations on AM include suppressing the carrier so that more power is available for the sidebands, and single sideband, which eliminates both the carrier and one of the two sidebands. The suppressed carrier and SSB variations of AM can be produced by filtering, by balanced modulators and filtering, or by phase shifting (which requires no filters). Modern ICs make the balanced modulator and phase-shift approach very attractive and precise. Complete transmitters can either modulate the carrier at a low power level and then amplify it, or can amplify the carrier and then perform the modulation.

Summary Questions

1. Why is modulation needed? What roles does it serve?
2. What is the AM process?
3. What does the term *carrier* mean in AM? What about *sidebands*?
4. What is the frequency-domain relationship between the original information signal, the unmodulated carrier, and the sidebands after modulation?
5. What is the largest possible fraction of total power in the sidebands after AM? What is the significance of this?
6. What is the AM envelope? What does it represent?
7. What is the bandwidth of a signal before and after AM?
8. How is modulation index defined in terms of power before and after modulation?
9. How is modulation index seen on an oscilloscope?
10. What is overmodulation? What are its effects in the time and frequency domains? How can it be avoided?
11. Why is some carrier power often desired and even necessary?
12. Compare DSB-SC AM to conventional AM in terms of frequency domain, carrier power, and sideband power.
13. Explain the function of a balanced modulator.
14. Compare SSB to DSB-SC and to regular AM in terms of frequency domain, carrier power, and sideband power.
15. Explain how SSB offers the choice of more users or increased bandwidth.
16. How does SSB make better use of available power?
17. What are two ways of generating an SSB signal? What are the advantages and disadvantages of each?
18. How can balanced multipliers be used with other circuitry to produce an SSB signal without filtering? What are two key factors in the quality of the result of this approach?

Practice Problems   **117**

19. What is baseband AM? Why is it used? Compare it to conventional AM.

20. What is CW? Why is it both the oldest and the most modern form of AM?

21. Compare the block diagram of a high-level and a low-level modulation transmitter. What are the technical points of each?

22. Where are high-level modulation transmitters used? Where are low-level transmitters used?

23. What are the basic steps and procedures for testing a transmitter? What is judged at each stage, using the input signal and output signal?

24. Compare the use of the oscilloscope and the spectrum analyzer in testing of transmitters. Why are two-tone signal sources often used?

PRACTICE PROBLEMS

Section 4.2

1. A carrier frequency of 1 MHz is modulated by a signal at 100 kHz. What are the resulting sideband frequencies?

2. A 10-MHz carrier is modulated by the 100-kHz signal. What sideband frequencies result?

3. A carrier at 100 kHz is amplitude modulated by a signal with a spectrum of 0 to 10 kHz. What is the resulting frequency spectrum?

4. There are five carriers spaced every 100 kHz, beginning at 1 MHz. A 50-kHz signal is used to amplitude modulate each one. What are the resulting frequencies?

5. The same five carriers of problem 4 are modulated by signals with a spectrum of 0 to 5 kHz. What is the spectrum of the sidebands that result?

6. The modulating signal now extends to 60 kHz, with the carriers of Problem 4. Show where sideband overlap, if any, occurs.

7. A carrier at 150 kHz is modulated by a 0- to 25-kHz signal, while an adjacent carrier at 200 kHz is modulated by a 0- to 10-kHz signal. What frequency spectrum is occupied after the AM process?

8. For the situation in problem 7, the second modulating signal bandwidth is increased to 35 kHz. Is there overlap in sidebands? If so, which sidebands and how much overlap?

Section 4.3

1. Determine m for a modulated carrier value of 100 V and an unmodulated value of 85 V. What is m when the modulated value is 125 V?

2. A modulated signal power is measured at 1500 W, while the unmodulated value is 1000 W. What is m? What is the value of m if the unmodulated value is 1100 W?

3. A waveform is seen on an oscilloscope screen. The maximum span is 1 V, while the minimum span is 0.1 V. What is m?

4. What is the total power when the carrier is 700 W and $m = 0.5$?

5. What is the ratio of total power to carrier power when $m = 0.3$? When $m = 0.7$?

6. A 1250-W carrier is modulated with a resulting modulation index of 0.65. What is the total power? What is the sideband power?

7. A 1000-W carrier has sideband power of 300 W. What is m?

8. A 9-kW unmodulated carrier has total power of 10 kW. What is the sideband power? What is the modulation index?

Section 4.5

1. A 1000-W AM carrier signal has a modulation index of 0.5.
   (a) How much power is in the total signal?
   (b) How much power is in the carrier and how much in the sidebands?
   (c) The carrier is suppressed by 20 dB. How much power is taken away from the carrier?
   (d) If this power is instead added to the sidebands, how much power do these sidebands now have?
   (e) By how many dB has the sideband power increased?

2. Repeat problem 1 for a 450-W carrier and modulation index of 0.9.

3. A specification for a DSB-SC system says that the carrier must be 25 dB below the conventional AM value. For a system with total power of 600 W and $m = 0.75$, what is the AM value? What would the suppressed carrier value have to be?

4. A technician is adjusting the feedthrough on a multiplier used for DSB-SC. The specification says that the carrier must be at least 20 dB (in volts) below the AM value of 1.8 V, as seen on the oscilloscope. What must the suppressed carrier voltage be, at most?

5. A technician measures the DSB-SC carrier on an oscilloscope. It has a value of 0.15 V. The specification says that it must be 32 dB (in volts) below the unsuppressed value of 1.0 V. Does it meet the specification?

Section 4.6

1. What is the maximum sideband suppression that can be obtained if the phase-shift filter has an error of 0.25°?

2. Repeat problem 1 for a filter error of 0.75°.

3. A system must suppress the undesired sideband by >40 dB. What is the maximum allowable error in the phase-shift filter?

4. The specification in problem 3 is changed to >60 dB. What is the maximum error in this case?

5

Receivers for AM

CHAPTER OBJECTIVES

When you have completed this chapter, you will understand:

- The basic functions and circuit stages needed to recover information from an AM signal
- Various approaches to implementing these functions and their features
- Techniques for recovering SSB and CW signals
- The importance of these basic receiver functions when receiving other types of modulation, such as frequency or phase modulation or digital signals

INTRODUCTION

The AM receiver has a difficult challenge: to select the desired signal, which is usually very weak, from all the surrounding signals in the band, and recover the original information that modulated the carrier at the transmitter. The receiver requires several stages of amplification and signal processing to do this. Although there are many approaches to receiver design, in this chapter we concentrate on the superheterodyne receiver, which has been the predominant design since its mass use beginning in the 1930s. For comparison, a simpler but less effective design is also examined. Each stage of the receiver is optimized to perform a particular function, and the combination of these acting as a group provides the overall receiver performance. The receiver for AM signals can be adapted, with some changes, to handle other types of modulation, and its fundamental multistage design serves well in virtually all receiver applications.

5.1 ROLE OF THE RECEIVER

Receiving a signal and retrieving the transmitted information from it is a challenging problem. The receiving system must perform many functions in order to extract the original modulating information successfully, and it must do these under electrically difficult conditions. Since AM and AM transmitters were the subjects of Chapter 4, it is logical now to study how the AM process is reversed. This reverse *demodulation* process is only one of the many aspects of recovering information. In this chapter we study receivers and AM demodulation in detail and examine the techniques and technical issues of AM reception.

It may seem that studying AM receivers is a very limiting topic, since there are other types of modulation used, such as frequency modulation (FM) and phase modulation (PM), the subjects of Chapter 6. However, many of the required receiver functions and problems that the receiver must deal with are similar, regardless of the type of modulation used to impress the information-bearing signal onto the carrier. Therefore, the lessons of AM receivers will be directly applicable to receivers for other types of modulation.

The study of receivers is important because the receiver must produce correct results (or correct enough for the application) in order for the communications system to be useful. A perfectly functioning modulator and transmitter is useless without a corresponding receiving system. Between the transmitter and the receiver is the communications channel, or medium, which plays a large role in making the receiver's job more difficult. Let's look at why the receiving system is more difficult and complicated than the transmitter, and some of the issues that a receiving system must deal with.

The transmitter is taking a signal that is usually relatively large (millivolts or greater) with a high signal-to-noise ratio, amplifying it, using it to modulate the carrier, and transmitting it via an antenna or wires. When signal levels and SNR are large, they can be handled by simpler circuitry, without special advanced design techniques. Also, the general type of signal that is being transmitted is known in advance. A standard AM band broadcast transmitter, for example, is designed specifically for signals that are voice and music, with limited bandwidth, and not for computer-type digital signals. The voice or music signal comes into the transmitter circuitry directly, and so has a constant signal level that does not vary inadvertently; the only variation is the change that the voice or music itself produces. The transmitter functional blocks, such as amplifier stages, add only an insignificant amount of noise to the desired information signal. Finally, the transmitter does not have to be concerned about interference: It just performs the transmit function at the assigned carrier frequency.

Compare this with the receiver situation. The signal level that arrives into the antenna is very weak. Its power is measured in nanowatts or microwatts, with a signal level of microvolts. The receiver must be very sensitive to pick up these signals, yet not be overwhelmed by larger signals. There are many competing signals in the spectrum, at frequencies very close to the desired signal, and these may be equal or greater in power. The desired signal must be selected (*tuned*), and the selection circuitry must be stable with temperature variations, power supply variations, and time so that it does not drift. The signal level is not steady, either. Due to the nature of the atmosphere and the way in which the signal propagates through the layers of the atmosphere, the level of the signal will increase and decrease randomly. This *fading* means that the receiver must do more than accommodate a wide range of signal—it must continually adapt to changes in signal level and still perform its function of selecting the desired signal and detecting the information embedded in it.

Noise (Chapter 3) is another severe problem that confronts the receiver. Although the magnitude of the noise may be insignificant compared to the signal power at the transmitter, the noise is very significant when added to the weak signal that reaches the receiver. The SNR may be only 3 to 20 dB for a typical system (although deep-space systems have receivers that use advanced techniques to handle SNRs that are as poor as −100 dB). The internal noise of the receiver affects the performance of the system as well. Although in theory the weak signal that is received can be amplified as much as needed, the reality is that the circuitry of the receiver adds its own noise to this weak signal. This limits the final SNR that can be realized, even if the signal is relatively noise-free.

In summary, the receiver must contend with a very weak signal, corrupted by added noise, and randomly fading and increasing in signal level. It must select (*tune*) this weak signal from within the overall spectrum and not add significant noise to the received signal as it tries to process the signal and extract the information. This is quite a challenge, far more difficult than taking a signal from a microphone or tape deck, boosting its level, then modulating it onto a carrier and transmitting it.

The receiver design is tailored to the type of signal it is expecting. A general-purpose receiver for any AM modulated signal must be very flexible and nonspecific, while a receiver designed for one type of signal only, such as a radar echo or a digital pulse, can make use of that advance knowledge in its design and functional blocks. The more that is known in advance about the signal, the more closely that the receiver can be matched to the type of signal and so do a better job. AM offers both advantages and disadvantages in dealing with these problems; other modulation techniques also offer technical pros and cons—but all must deal with these issues. For these reasons, the study of AM receivers (which are among the simplest type) is instructive and sets a framework of understanding of receiver design for other modulation techniques.

Review Questions

1. What are three reasons why receiving is much more difficult than transmitting?
2. How does external noise make the receiving process more difficult?
3. How does internal circuitry noise affect the ability of the receiver to recover weak signals?
4. What are some typical power and voltage signal levels at the receiver input?
5. Why is the study of AM receivers relevant to receivers for other types of modulation?

5.2 RECEIVER TECHNIQUES AND STAGES

Many system designs can be used to accomplish the required receiver functions, with different degrees of success. In this section we examine the *tuned radio-frequency (TRF)* method, one of the simplest and earliest approaches, and the *superheterodyne* method, which since its refinement in the late 1920s has been the most common and effective receiver design.

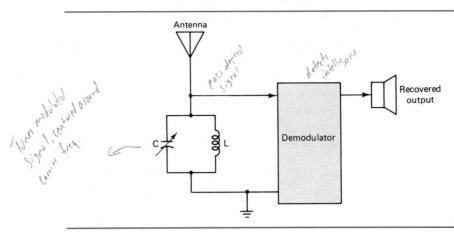

Figure 5.1 Basic TRF receiver block diagram, showing simple structure.

TRF Receivers

The TRF receiver is very straightforward (Figure 5.1). The modulated signal is tuned by an *LC* bandpass circuit that has its passband centered at the carrier frequency. The circuit *selectivity* makes it pass only the desired signal while blocking, or rejecting, all others. The very weak signal, after being tuned, is passed to an amplifier that boosts it to a larger voltage level. The information signal, which was modulated onto the carrier at the transmitter, is then *detected*. This detection, or demodulation process, recovers the original modulating signal. After detection, the signal can be amplified further if necessary, to drive a loudspeaker for example.

Generally, if a simple system can do the job adequately, it is the preferred approach. The TRF design seems to be an example of a simple circuit that can achieve the goal of AM reception. Unfortunately, it has many drawbacks in operation. First, the input *LC* bandpass stage is very critical. It must be stable with temperature and time so that it stays properly tuned; otherwise, the user (a radio listener) will constantly have to retune. More important, its performance varies with the exact frequency to which it is tuned within an overall band. Any bandpass circuit has a selectivity or quality factor called *Q*, which is defined as

$$Q = \frac{\text{reactance (ohms)}}{\text{series resistance (ohms)}}$$

The value of *Q* is determined by the components selected for the filter, and a higher-*Q* filter is sharper with narrower bandwidth than a lower-*Q* one, as shown in Figure 5.2. The bandwidth of the bandpass circuit is

$$\text{bandwidth} = \frac{\text{carrier frequency}}{Q}$$

For example, a bandpass filter with $Q = 100$ at a carrier frequency of 1000 kHz has a bandwidth of

$$\text{bandwidth} = \frac{1000}{100} = 10 \text{ kHz}$$

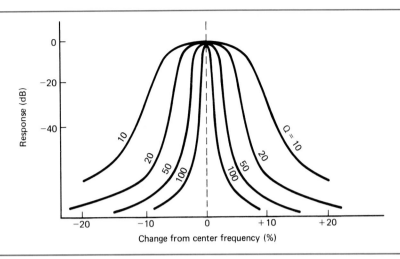

Figure 5.2 Filters with Q values of 10, 20, 50, and 100 can have same center frequency but different bandwidths.

As the TRF receiver is tuned to frequencies other than 1000 kHz, the bandwidth changes while Q remains the same, as a simple calculation will show. Suppose that the circuit with $Q = 200$ is used to tune across the AM broadcast band, 550 to 1600 kHz. We have seen that the bandwidth at 1000 kHz is 5 kHz. At the lower end of the band, 550 kHz, the bandwidth is

$$\text{bandwidth} = \frac{550}{200} = 2.75 \text{ kHz}$$

At the higher end of the band, the bandwidth is

$$\text{bandwidth} = \frac{1600}{200} = 8 \text{ kHz}$$

The result is that the bandwidth varies as the receiver is tuned, which is not good. If the receiver bandwidth is less than the bandwidth of the received information signal, information is lost (which for audio signals means that the signal has distortion and poor fidelity, and perhaps even is not understandable). If the bandwidth is too wide, adjacent signals are also passed, which means interference with the desired signal. Even if there are no adjacent signals, there is always noise, and the wider bandwidth allows more external noise to enter the receiver.

The ideal situation is to have the bandwidth of the input circuit match the bandwidth of the desired signal, and not be wider or narrower. It is very difficult, however, to design a practical filter whose bandwidth remains constant as its center frequency changes. As a result of the practical difficulties of TRF designs, an alternative approach was needed. This does not mean that TRF designs are never used. They can be successful where, for example, a fixed-frequency, nontunable receiver is needed for a specific application. Modern ICs, which provide more stable performance than either vacuum tubes or individual transistors, also enable TRF receivers to operate with very good performance in some cases.

Superheterodyne Receivers

Of all the receiver designs that were tried to overcome the problems of the TRF receiver, one design quickly stood out as superior and has remained superior in most situations. The *superheterodyne (superhet)* receiver was invented by Major E. H. Armstrong, an electronics genius who also invented practical frequency modulation systems (the subject of Chapter 6). Although the circuitry is much more complicated than the TRF, superhets provide excellent performance under many conditions. The superhet divides the overall receiver function into many subfunctions, each of which can be improved as needed or as new technologies are developed, to give even better performance.

Here is the basic concept of the superhet receiver: The receiver tunes to the desired signal and converts the income signal to an *intermediate frequency* via a signal mixing circuit. The intermediate frequency is the same for all received signals, regardless of the original frequency of the signal. Once the signal has been translated to this intermediate frequency (IF), circuitry optimized to perform the critical receiver functions at this single IF value can be used. Instead of requiring that the entire receiver be able to vary its tuning and bandwidth, as in the TRF design, a superhet receiver converts all signals to the IF value and then performs the steps needed to fully recover the modulated information. The key to the superhet design is the practical scheme that performs the frequency conversion as the user tunes the receiver to the desired transmitter frequency.

A superhet receiver is divided into the following main stages (Figure 5.3):

The *radio-frequency (RF) stage*, which takes the signal from the antenna and amplifies it to a level large enough to be used in the following stages.

The *mixer and local oscillator*, which convert the RF signal to the intermediate frequency.

The *IF stage*, which further amplifies the signal and has the bandwidth and passband shaping appropriate for the received signal.

A *detector stage*, which recovers the information signal from the carrier.

The *audio-frequency (AF) stage*, where the received signal is amplified for the loudspeaker, headphones, or for interconnection to the rest of the communications system.

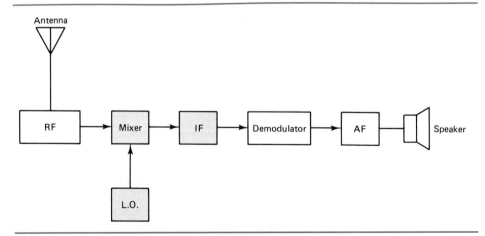

Figure 5.3 Basic stages of a superhet receiver.

5.3 RF Stage

Each stage has a specific function to perform and can be designed to do a superior job, without affecting or being affected in a major way by the design of the preceding or following stages. The operation of each stage is examined in the following sections of this chapter.

Review Questions

1. What are two drawbacks of the TRF design? How does each limit the performance that can be achieved?
2. Why is the input bandwidth value a critical factor in the receiver?
3. What is a superhet receiver? How does it differ in basic concept from the TRF design?
4. What are the stages of the superhet receiver? What is the role of each?
5. Why does the superhet offer the potential for excellent performance?
6. How does the superhet avoid the bandwidth problem of the TRF?

5.3 RF STAGE

The function of the RF stage is to take the weak signals within the frequency band to be received and boost them to practical signal levels. It is sometimes called the *front end* because it is the first stage for the signal coming into the receiver. The typical signal level at the antenna, going into the RF stage, is only a few μV; it must be increased to tens of millivolts or greater to be used in subsequent stages. The RF stage does this with one or more stages of amplification, in sequence.

The first section of the RF stage is the most critical, because the signal is very weak and any noise added to the signal by the amplifier itself reduces the ultimate SNR that can be achieved. As high an SNR as possible is required for good reception and signal recovery in the next stages. Any noise added by the front end makes the signal quality worse and limits the ability of the receiver to extract the information from the signal, regardless of how much amplification is applied in subsequent stages. The major noise factor below 30 MHz is external noise, and relatively little noise is contributed by the receiver. Beyond 30 to 50 MHz, the situation is reversed. The external noise is less than the noise of the circuitry.

For this reason, great care is used in the design of the front end and the selection of components to be used there. Special transistors and other components with very low noise characteristics are selected, although their performance otherwise may not be ideal. In extreme cases, the front-end amplifier is cooled to reduce noise caused by random motion of electrons (this random motion is proportional to the temperature above absolute zero, called 0 kelvin (K) and equal to -273°C), discussed in Chapter 3 together with noise temperature and equivalent noise resistance.

Of course, this cooling is not practical for most receivers, so the components used must have inherently low noise. The RF stage can use several amplifiers in series or *cascade*, with a low-noise amplifier providing the initial gain followed by

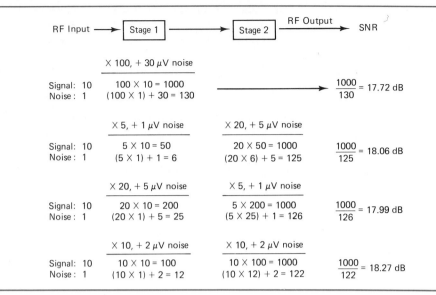

Figure 5.4 Effect of internal noise on performance of multistage RF amplifiers.

Cascade Amplifiers and Noise

The effect of multiple amplifiers in the RF stage is shown by an example and Figure 5.4. A designer can choose among four transistors for the front end. Each can provide some gain (×5, ×10, ×20, and ×100) but also contributes noise. The higher-gain transistors unfortunately are noisier than the lower-gain ones. The received signal is 10 μV with 1 μV of noise, and the signal after the RF stage must be 1000 μV, so a total gain of ×100 is required. The designer can choose to use a single ×100 stage, two ×10 stages, or a ×5 and ×20 stage in either order. Different combinations have different signal gain and noise results. For the ×100 gain transistor, the answer is clear. The incoming signal is increased to the desired 1000 μV, and the 1-μV noise is increased by 100 as well, plus the 30 μV of noise added by the transistor, for a total of 130 μV of noise. The voltage SNR at the output of the RF stage is 1000/130 or 17.72 dB.

The result for the ×5 and ×20 combination requires a few more calculations. After the first amplifier the signal is increased to 50 μV, while the noise is

$$(5 \times 1) + 1 = 6 \, \mu V$$

At the second amplifier, the signal is multiplied by 20 to 1000 μV, and the noise is also multiplied by this factor, then has the internal noise added:

$$(20 \times 6) + 5 = 125 \, \mu V$$

The 1000-μV signal now has 125 μV of noise, so the RF stage output has an SNR of 1000/125 = 18.06 dB.

The figure shows the number for the two other amplifier combinations, ×20 followed by ×5 and ×10 followed by ×10. The best combination consists, in this case, of two ×10 stages, with a final SNR of 18.27 dB. (Note that the incoming signal had an SNR of 10/1 = 20 dB, so the SNR has decreased by nearly 2 dB.)

In this example, the differences between the various combinations were slight. Different multiplication factors and noise factors would produce different results, of course. The specific gain blocks are chosen by the designer after careful calculation of the various possibilities, with the goal of getting the most gain with the least noise and least reduction in SNR. The front-end performance is critical in determining the ultimate achievable performance.

somewhat noisier ones that provide additional gain. The noise of these additional amplifiers is not as critical, once the extremely weak signal is amplified above its original microvolt level. The first amplifier in the RF stage is often called the *preamplifier*, since its role is to amplify the weak received signal with minimum noise, to a level suitable for further amplification by the RF circuitry.

The RF stage bandwidth is usually designed to be much greater than just the desired received signal bandwidth. This RF stage is not designed to select the single signal that the user is trying to receive. In some systems, the RF stage bandwidth is as wide as the entire band that the receiver can tune, if the total band is not too wide. In most cases, it may be technically difficult to design such a wideband amplifier, or the wide bandwidth allows too much noise to enter the receiver system (received noise is proportional to bandwidth). For these situations, the RF stage bandwidth acts as a *preselector*, covering moderately large portions of the overall band and shifting as the receiver is tuned. Since the RF stage is not selecting a single signal, the exact value of the preselector bandwidth is not critical as it is in the TRF design.

The key specifications for the RF stage are its *sensitivity* and its noise figure. The sensitivity specification indicates how small a signal the RF stage can accept and still provide a usable signal output, with a satisfactory SNR. A typical specification is sensitivity of 2 μV, for example, with an SNR of 10 dB. This means that the RF stage takes a perfect 2 μV signal—one that is noise-free—and produces a useful output to the next stage that has an SNR of 10 dB. A more sensitive receiver produces the same 10-dB SNR result with a 1-μV signal. A lower-noise RF stage can take the 2-μV signal and produce a result that has a 20-dB SNR for the next stage. The choice is made by the receiver designer, based on the received signal levels, noise, and intended application. Some applications use the RF stage input power rather than the input voltage as the sensitivity specification, since the antenna is producing power (however small), but the implications are the same. The voltage at the receiver really is the result of this power being applied across the front-end input impedance, where the apparent voltage $V = \sqrt{P \times R}$.

An RF amplifier stage is shown in Figure 5.5. The RCA CA40841 MOSFET (a MOS insulated-gate field-effect transistor with dual gates) receives the signal from the antenna tuning inductor–capacitor (*LC*) preselector, and couples the amplified version into the next stage of the receiver through a similar *LC* circuit. A MOSFET is used because it provides extremely high input impedance and so does not load down the very weak input signal, yet provides the capability to handle a wide span of input levels without overloading (and so developing a lot of spurious frequencies). The circuit also provides, through the second gate, an easy way to set the gain of this stage with a control voltage, as part of the automatic gain control circuit.

Automatic gain control The task of the receiver circuity is complicated by the fact that the received signal level is not constant but varies continually as a result of fading, changes in the atmospheric conditions, movement of the transmitter or receiver in mobile radios or cars, and other factors. This variation is a problem for the user and for the proper functioning of the receiving system. For the user, these variations mean having to constantly readjust the gain or volume to keep the final signal level where it should be—not too loud or too soft. For the receiver, the signal level changes would affect the performance, because the receiver mixer, LO, IF, and AF stage circuits are designed to handle signals only within a small

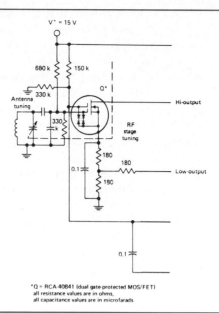

Figure 5.5 CA3088 MOSFET used as an RF amplifier (courtesy of Harris Corp., RCA Div.).

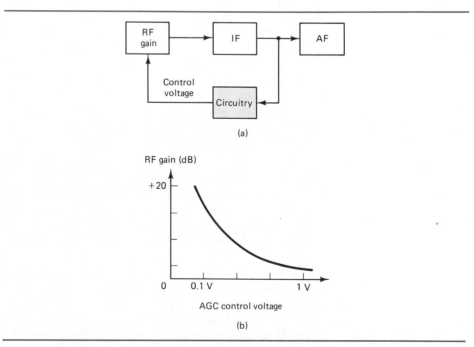

Figure 5.6 AGC function showing (a) block diagram and (b) graph of control voltage versus gain.

range of values. Designing circuits to handle signal strengths that vary over a range of 20 or 30 dB (very common in radio transmissions) is very difficult, with temperamental performance. The mixing effect is like AM, and the LO and RF signals must have proper amplitudes with respect to each other—or the result will be overmodulation or insufficient modulation, instead of the ideal of modulation close to (but not over) 100%.

The solution to this problem is *automatic gain control (AGC)*. The AGC is a very important part of the receiving system, since it must take the received signal and level out the variations. It does this by automatically adjusting the gain of the RF stage as needed to maintain the desired constant output level (Figure 5.6). (In some receivers, the gain of other stages is adjusted as well.) This gain adjustment is done continuously so that it compensates for fluctuations in the received signal level. A typical AGC can reduce variations in signal level from 20 or 30 dB to approximately 1 dB (recall that a 3-dB change is equivalent to doubling the signal power). The AGC must not respond too quickly, however, since it would "level out" the variations of the modulation signal itself if it did, not just the carrier strength. The *attack* and *decay* time of the AGC circuit is designed so that gain changing cuts in at the proper rate and does not release too quickly either.

Review Questions

1. What is the function of the RF stage? Why is it called the front end?
2. Why is the noise figure of the RF stage critical? Why are several amplifiers in series sometimes used?
3. Why is the bandwidth of the RF stage greater than the input signal?
4. What does sensitivity indicate/ Why is it determined by the front end? How is it determined?
5. What is AGC? Why is it needed? Over what range must it operate?

5.4 MIXER AND LOCAL OSCILLATOR

The *mixer* and *local oscillator (LO)* work together to take the signal from the RF stage and convert it to the IF (*intermediate frequency*) of the system. The way this is done is one of the key elements in the superhet receiver and clearly shows where additional circuitry provides significant performance benefits. The principle of mixing two signal frequencies (discussed in Chapter 4) to produce sum and difference frequencies is applied in this stage.

The local oscillator is just what its name indicates. It is an oscillator built into the receiver (hence, local) and designed to produce very stable and pure sine waves. The frequency of the LO output is offset from the frequency of the desired signal by a fixed amount, equal to the intermediate frequency. As the receiver is tuned by the user to pick up a new transmitter signal, the LO frequency is automatically changed so that the difference between the desired signal frequency and the LO frequency is always maintained at this fixed IF amount. Maintaining this constant difference by changing the LO frequency simultaneous with the tuning change is the key to the superhet.

The AM stage is essentially a mixer stage. Recall that the result of the AM process is a new set of frequencies, at the sum and difference values of the carrier and modulating signal frequencies. In the mixer, the RF stage signal can be considered as a carrier, with the LO signal as the modulating signal. Here is how the LO plus mixer works in the superhet receiver:

Suppose that the received signal is an AM signal with a 1-MHz carrier that has been modulated by a 10-kHz tone. The received signal therefore has components at the lower sideband, carrier, and upper sideband frequencies, 990, 1000, and 1010 kHz, respectively. These frequency components go into the mixer, where they are mixed with the LO signal of 0.8 MHz, for example. The result is the sum and difference of the three input frequencies and the LO frequency:

$$990 \pm 800 = 190 \text{ and } 1790 \text{ kHz} \quad \text{(mixing of lower sideband)}$$

$$1000 \pm 800 = 200 \text{ and } 1800 \text{ kHz} \quad \text{(mixing of carrier)}$$

$$1010 \pm 800 = 210 \text{ and } 1810 \text{ kHz} \quad \text{(mixing of upper sideband)}$$

No information has been lost, but the signal has been converted to a new set of frequencies. Now, suppose that the receiver must tune a signal that is 0.2 MHz higher, at 1.2 MHz, that has the same 10 kHz modulation. To track the new input frequency, the LO output frequency is increased by 0.2 MHz, to 1 MHz. The output of the mixer is therefore

$$1190 \pm 1000 = 190 \text{ and } 2190 \text{ kHz} \quad \text{(mixing of lower sideband)}$$

$$1200 \pm 1000 = 200 \text{ and } 2200 \text{ kHz} \quad \text{(mixing of carrier)}$$

$$1210 \pm 1000 = 210 \text{ and } 2210 \text{ kHz} \quad \text{(mixing of upper sideband)}$$

Here is where the key feature of the superhet design becomes clear. Note that the difference output of the mixer (the difference between the received frequency component and the LO output) is the same, regardless of what the tuned frequency is. In this case the incoming signal (carrier and sidebands) is converted to difference frequency components at 190, 200, and 210 kHz, plus sum components at a variety of frequencies. The *difference* between the received carrier and the LO output is the *intermediate frequency (IF)*. Since all tuned signals now have the same IF value, regardless of where they are in the received band, the rest of the receiver is designed to handle signals of this frequency exclusively. This is very different from the TRF design, where the entire receiver has to handle all frequencies that may possibly be tuned.

It may appear that the superhet design has made the problem worse, since a single carrier and its sidebands have now become two sets of carrier and sidebands, with one set at the sum frequency and the other set at the difference frequency. The next section will show why the sum frequency is easily eliminated, while the difference frequency is used in the rest of the receiver stages.

> **Low and High Tracking**
>
> The LO frequency in the previous example was below the designed carrier frequency, called *low tracking* (or *low beat*). The mixer can also use a LO above the carrier at (1 MHz + 200 kHz) = 1200 kHz, called *high tracking* (or *high beat*).
>
> 990 ± 1200 = 210 and 2190 kHz
> (mixing of lower sideband)
>
> 1000 ± 1200 = 200 and 2200 kHz
> (mixing of carrier)
>
> 1010 ± 1200 = 190 and 2210 kHz
> (mixing of upper sideband)
>
> The effective result is similar, with the carrier and sidebands at the same difference frequencies but sum components at different frequencies. We will see in the next section that high tracking is suitable for the IF stage, and sometimes even preferred.

LO and Receiver Alignment

The LO is a critical component in the superhet receiver. It must be very stable with time and temperature, yet be tunable to any frequency within the overall receiver range. Any small variations in LO frequency will appear as variations in the IF output, which will degrade performance through the rest of the receiver. The design of a suitable LO is not too difficult for the medium range of frequencies, from 3 to 30 MHz. At higher frequencies, LO design becomes more difficult. In the higher ranges, such as several hundred megahertz and above, the ability to design an LO that performs properly is sometimes the limiting factor in how well the receiver can perform. (The superhet solution for this problem is a multiple conversion receiver, discussed later in this chapter.)

In a superhet receiver, the tuning of the RF stage and the LO is usually done with mechanically *ganged capacitors*. As the tuning knob is turned by the user to tune a new frequency, two capacitors on the same shaft vary in capacitance simultaneously. One capacitor is for the broader preselection tuning of the RF stage, the other is for the tuning of the LO.

For optimum operation of the superhet receiver, some adjustments are usually needed, especially if a component fails and is replaced. One part of this receiver tune-up is the *alignment* procedure. Alignment ensures that the LO output differs from the received signal by the IF value. Alignment is done by using a signal generator to simulate a received signal at the front end, with precisely known frequency. The tuning capacitor of the LO is then adjusted by small trimming capacitors (in series and parallel) to make sure that its output is at the correct frequency. This can be done by looking at the output of the mixer on an oscilloscope, or by measuring the LO output with a frequency meter.

The LO capacitor (and the mixer capacitor) also must have some trimming done to make them tune uniformly across the entire band. This is needed because the mechanical design of the tuning capacitor cannot be linear across a wide band, with every 1° of rotation providing 1 kHz of frequency change, for example. Any difference in the change in capacitance for one capacitor section versus the other sections on the same shaft as the shaft is rotated will cause problems in the tracking of the ganged capacitors as the shaft is turned to tune new frequencies. The trimming capacitors are used to compensate for these small errors by adding some corrections to the tuning ranges. The adjustment of these trimmers is usually an *iterative* procedure, requiring that the technician go back and forth between sections as the receiver is tuned from one end to the other, to find the best overall combination of trims. This is sometimes a frustrating sequence, but unavoidable with mechanical capacitors and ganging.

Electronic Tuning

The advances in electronic components in recent years have allowed designers of systems to replace mechanical parts of any system—such as its gears, indicators, and adjustments—with electronic circuitry that performs the same function, but more effectively. The same is true for the tuning circuitry in superhets. The mechanical capacitor requires considerable space, can become misaligned, and change value as dirt, oil, or water vapor contaminate the air space between the capacitor plates. It is also relatively costly, with many precise mechanical pieces and a shaft mechanism.

In an oscillator circuit such as the LO, an inductor and capacitor are used to resonate at a frequency determined by the component values. The resonant frequency f is found with the simple equation

$$f = \frac{1}{2\pi(\sqrt{LC})}$$

The inductor is normally made by wrapping many turns of wire around a form with an air or magnetic core, depending on the desired value. The capacitor can be fabricated in many ways, but a variable capacitor usually has parallel metal plates separated by an air gap or other dielectric material. The capacitance is varied by changing the separation between plates or by rotating one plate with respect to the other so more or less of the plate areas overlap.

Many new communications receivers use *electronic tuning*, based on a simple component called the *voltage-variable capacitance* diode, also known as the *varactor* or *varicap* diode, to replace the mechanical capacitor with an electronic one. The capacitance of the varactor diode varies with the dc voltage across it, applied as a *reverse bias* across the diode. For a typical varicap diode, the effective capacitance varies between 20 and 250 pF as the applied voltage spans 20 to 2 V (greater reverse dc bias provides less capacitance). This diode is used in the *LC* circuit of the LO, and similar diodes are used in the mixer and RF stage. Tuning the receiver now involves changing the dc voltage on the varactor, either by a potentiometer connected to the tuning knob, or by a microprocessor controlling the analog output of a *digital-to-analog converter (DAC)*. Either way, the mechanical capacitor is replaced by an all-electronic component with no moving parts (except for the tuning pot, if one is used). This is much simpler and more stable than the mechanical capacitor with parallel plates.

There is one drawback to this approach. The relationship between the voltage applied across the diode and the resulting capacitance is not identical in each system. There is a tolerance of as high as ±20% on the varicap diodes, due to variations in the manufacturing technique. Also, the dc voltage source used on the diode may not be error-free and may change slightly in time or with temperature. Without some way of compensating for this, the frequency indicated on the dial of the receiver could be significantly off. An example shows the effect.

Example 5.1

An oscillator resonates at 1 MHz with a nominal 100-pF capacitor and a 0.00025-H (0.25-mH) inductor, using the formula for resonance. What is the resonant frequency if the actual capacitor value is +20% of the nominal value?

Solution

Using the resonance formula gives

$$f = \frac{1}{2\pi\sqrt{0.00025 \times 120 \times 10^{12}}}$$

$$= 0.91888 \text{ MHz}$$

Example 5.2

What is the change in resonant frequency if the actual varactor capacitance value differs by -5% (0.05) of the nominal value?

Solution

Using algebra, the ratio of the new frequency of f_n to the original frequency f_0 is $f_n/f_0 = 1/(2\pi\sqrt{LC_n})$, divided by $1/(2\pi\sqrt{LC_0})$, where C_n and C_0 are the new and original capacitance values. Since all factors except these capacitances cancel, the result is

$$\frac{f_n}{f_0} = \frac{\frac{1}{\sqrt{C_n}}}{\frac{1}{\sqrt{C_0}}} = \frac{\sqrt{C_0}}{\sqrt{C_n}} = \frac{\sqrt{1.0}}{\sqrt{1.0 - 0.05}} = 1.03$$

Therefore, it is 3% higher. (Note that since we are dealing with ratios and differences, we can arbitrarily set $C_0 = 1.0$ for convenience).

Two solutions to this problem are possible. One is to calibrate each receiver individually for the particular diode installed and voltage source used. This is possible but not practical: If the diode is replaced, the system must be recalibrated. More important, the stability of the dc voltage driving the varactor must be very good, and this is difficult to achieve as the ambient temperature varies.

The other solution is to make use of another function that is now achieved easily and reliably with modern ICs. Instead of trying to calibrate the knob of the receiver by putting markings around the knob, a digital frequency meter and readout is used to directly measure the frequency of the LO. As the LO tuning varies, the digital readout shows the exact frequency of the LO plus the IF for low tracking, and the LO minus the IF for high tracking (the user wants the received signal frequency, not the LO frequency). Even if the varactor or the varactor control voltage is not perfect, or varies with time or temperature, the digital readout shows the true and correct frequency. The accuracy of this method is limited only by the accuracy of the digital frequency meter. With modern electronics, this is relatively inexpensive, and the accuracy and repeatability of the readout are determined almost entirely by the single crystal used for the frequency meter circuit.

Although this may seem a complicated solution to a problem, it is actually simpler and more effective than attempting to design and maintain nearly perfect performance from the varactor and voltage source. It also shows a very different approach to solving a problem. Instead of seeking perfection in the components, accept the imperfection but use a highly accurate scheme of measuring this imperfection and thus showing the true value to the user. It is analogous to a car's gas pedal and speedometer. The gas pedal itself is not calibrated, with a marked number of degrees of angle on the pedal equal to some number of miles per hour. Even if this were done perfectly at the factory, the relationship between gas pedal angle and car speed would change depending on how level the road was, the number of people in the car, and when the engine last had a tune-up. Instead, a speedometer is connected to the driveshaft and shows the true speed at the wheels regardless of these other factors. Electronic tuning and digital readouts provide the same situation as the gas pedal and speedometer, as compared to the mechanical capacitor with factory-calibrated markings around the tuning knob.

Review Questions

1. What does a mixer do? What does the local oscillator do?
2. How do the mixer and LO produce an intermediate-frequency output? Compare low and high tracking for the LO.
3. Why are ganged tuning capacitors needed?
4. How does electronic tuning operate? What does the varicap diode do?
5. How are inaccuracies in the varicap diode and control voltage compensated by the digital frequency readout?
6. What are the benefits of electronic tuning?

5.5 IF Stage

The superhet receiver mixer/LO output is a signal centered around the intermediate frequency. The role of the IF stage is to take this signal, filter and manipulate it as necessary, and boost its voltage up to a level that can be demodulated properly so that the original information can be retrieved. To do this, the IF stage design must take into account the frequencies of the signals that the system is intended to receive.

Choice of IF Frequencies

In the preceding section we showed that the result of the mixing process was a set of sum and difference frequencies. The first task of the IF stage is to eliminate the sum frequencies. Only the difference frequency is desired, because only it has a constant value as the receiver is tuned. The sum frequency will vary and is equal to the received signal plus the LO frequency. The IF is a bandpass filter, centered around the IF value. The ability of the IF stage to reject the undesired sum signal from the mixer depends on two factors: how sharply the bandpass filter *rolls off* from the center frequency, and how close in frequency the sum value is to the difference value.

Standard AM broadcast receivers usually use 455 kHz as the IF frequency. Therefore, the LO is always 455 kHz below or above the received signal. For a signal at the lower end of the band, 550 kHz, the mixer outputs with low tracking are

$$550 - LO = 550 - 95 = 455 \text{ kHz}$$

$$550 + LO = 550 + 95 = 645 \text{ kHz}$$

For a signal at the upper end of the band, at 1600 kHz, the LO is 1145 kHz and the mixer produces

$$1600 - LO = 1600 - 1145 = 455 \text{ kHz}$$

$$1600 + LO = 1600 + 1145 = 2745 \text{ kHz}$$

Therefore, as the receiver is tuned across the entire band, the IF stage filtering must pass the 455-kHz signals while rejecting a sum signal that will range from 645 to 2745 kHz. This bandpass filter performance is not too difficult to achieve, especially if several stages of filtering are used. The bandwidth of the bandpass filter must be slightly wider than the bandwidth of the information signal. For an AM broadcast band signal, this is 10 kHz (recall that the overall bandwidth is twice the bandwidth of the modulating signal). The goal, therefore, of the IF stage is to pass a 10-kHz signal centered on 455 kHz while rejecting signals outside that span.

The discussion of dB in Chapter 3 showed that the bandwidth of a filter is normally considered to be the span at which the filter response is 3 dB down from the peak value. This is not the only factor for characterizing a filter, however. The steepness of the filter rolloff *skirts* determines filter *shape factor* and the ability of the receiver to separate the desired signal from adjacent undesired ones, called *receiver selectivity*.

Shape factor indicates the rate at which the filter response drops and is measured by the ratio of the filter bandwidth when response is 60 dB down from its peak, to the bandwidth at 6 dB down. Figure 5.7 shows this graphically. Both filters have the same 3-dB bandwidth of 5 kHz, but one drops off more quickly beyond that point and has a bandwidth of 10 kHz at −6 dB and 20 kHz at −60 dB. Its shape factor is 20/10 = 2. The broader filter bandwidths are 12 kHz at −6 dB and 40 kHz at −60 dB, for a shape factor of 40/12 = 3.3.

The filter's ability to select the desired signal while rejecting others is specified by the amount of rejection the filter provides at various frequencies from the center. A typical specification for a filter is as follows: a 6 dB bandwidth of 10 kHz, and signal suppression of 20 dB at 20 kHz, and a 30-kHz bandwidth at −60 dB, for a shape factor of 30/10 = 3. A less selective 10-kHz filter might have a suppression of only 10 dB at 20 kHz and a bandwidth of 40 kHz at −60 dB (shape factor = 4); a more selective filter provides rejection of 40 dB at 20 kHz with a bandwidth of just 25 kHz at −60 dB (shape factor −2.5). (This is often called the filter Q, defined in the same way as the Q of a tuned circuit in the TRF receiver.)

Sharper filters are generally preferable, but they are more complex and

Chap. 5 Receivers for AM

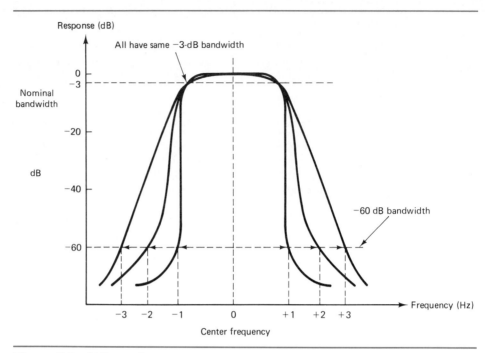

Figure 5.7 Different filter rolloffs and shape factors, all with the same −3-dB bandwidth.

costly than less sharp filters. Also, they may unavoidably cause phase shift in the frequency components of the received signal, which results in distortion of the recovered signal. This is not generally a problem for voice and music since the ear is not very sensitive to phase changes, but it may cause digital signals for computers to look very different than they originally did, which results in errors in the interpretation of the data bits. The shape factor must take into account the type of received signal expected.

Images and Other IF Frequencies

In the AM broadcast band situation, the IF used a 455-kHz IF value. The IF stage had to reject signal that varied from 645 to 2745 kHz (with low tracking) while passing the 455-kHz signal. Suppose that the desired received signal is at 25

Variable-Bandwidth IF

The same receiving system can be used for signals from different sources, such as a low-bandwidth Morse code signal, a medium-bandwidth voice signal, or a wider-bandwidth high-fidelity music signal. Each has different bandwidths, yet in each case the ideal is to allow only the necessary bandwidth to pass while rejecting all signals outside this bandwidth and reducing adjacent signal interference and noise.

To adapt the receiver to a variety of needs, advanced communications receivers sometimes offer adjustable bandwidth, where the user can, via a front-panel control, select one of several 3-dB bandwidths. Typical choices are 0.3, 1, 3, 5, and 10 kHz, although the specific values may differ depending on the intended receiver applications. Of course, if the receiver is intended only for one type of signal, such as standard AM broadcast reception, there is no need for adjustable bandwidth.

MHz—a much higher frequency—and the same IF—455 kHz—is used. There is also an adjacent, undesired signal from another transmitter at 24.090 MHz. Both pass through the RF stage preselector and into the mixer, which is fed by a LO signal at 25,000 − 455 = 24,545 kHz. The mixer output contains the difference between each of the signals (ignore the sum frequency mixer output):

$$25{,}000 - 24{,}545 = 455 \text{ kHz from the desired signal}$$

$$24{,}545 - 24{,}090 = 455 \text{ kHz from the undesired signal}$$

In other words, an undesired signal on the other side of the LO output will have the same difference frequency and pass into the IF amplifier. This undesired signal is called the *image frequency*. Note that since it is so close to the desired signal (25,000 kHz versus 24,090 kHz) it is very difficult to filter out in the RF front end.

The image signal is not a problem when a 455-kHz IF is used for the AM broadcast band since it is relatively far away from the desired signal and more easily filtered before the mixer. For example, if the desired AM broadcast signal is at 1 MHz, the LO is at 1000 − 455 = 545 kHz. The image frequency that when mixed with 545 kHz also produces 455 kHz is 545 − 455 = 90 kHz, and a 90-kHz signal is easily separated from a 1-MHz signal simply by filtering at the RF stage.

The 455-kHz IF frequency has problems in rejecting an unwanted signal that is on the "opposite side" of the LO as the receiver frequency increases. This is because the relative frequency spacing between the desired signal and the image becomes less, as the desired signal has a higher frequency. Fortunately, a solution exists: Increase the IF frequency significantly. The most common IF used for these cases is 10.7 MHz.

For example, repeat the case of the desired 25-MHz signal using an IF of 10.7 MHz and low tracking. The LO frequency must be 25 − 10.7 = 14.3 MHz. The image frequency is the frequency which when mixed with 14.3 MHz also has a 10.7-MHz difference. The answer is 14.3 − 10.7 = 3.6 MHz, which is easily filtered at the RF stage from the desired signal in the RF stage.

High Tracking

In many cases, high tracking for the LO is used to provide better rejection of the undesired mixer output since this places the sum output frequency farther away from the IF. Repeating the AM band with 455 kHz IF shows this. For a signal at the lower end of the band, 550 kHz, the LO is 550 + 455 = 1005 kHz and the mixer outputs with high tracking are

$$\text{LO} - 550 = 1005 - 550 = 455 \text{ kHz}$$

$$\text{LO} + 550 = 1005 + 550 = 1555 \text{ kHz}$$

For a signal at the upper end of the band, at 1600 kHz, the LO is 1600 + 455 = 2055 kHz and the mixer produces

$$\text{LO} - 1600 = 2055 - 1600 = 455 \text{ kHz}$$

$$\text{LO} + 1600 = 2055 + 1600 = 3655 \text{ kHz}$$

Therefore, as the receiver is tuned across the entire band, the IF stage filtering must pass the 455-kHz signals while rejecting a sum signal that will range from 1555 to 3655 kHz. This is more easily filtered than the 645- to 2745-kHz signals that the mixer produces with low tracking.

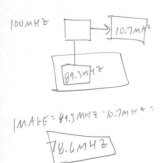

Example 5.3

A receiver for a signal at 100 MHz uses a 10.7-MHz IF and low tracking. What is the LO frequency? What is the image frequency?

Solution

The LO frequency is 100 − 10.7 = 89.3 MHz; the image frequency is 89.3 − 10.7 = 78.6 MHz. Here, the importance of the higher IF frequency is seen again. The 78.6-MHz image signal is removed by a relatively broad filter ahead of the mixer stage.

IF-Stage Circuitry

Once the sum and difference signal are produced by the mixer, the IF stage must filter this signal and provide the necessary selectivity. Most IF stages consist of a number of sections of bandpass filters made of tuned circuits, with inductor/capacitor values chosen to put the bandpass filter center at the IF frequency (Figure 5.8). Multiple stages, each with a slightly different frequency offset from the IF value (*stagger tuning*), combine to provide the required shape factor, with a wider overall response and yet sharp roll-off. The LC components of the IF stages are used to couple the signal from one stage to the other through *IF transformers*. This provides the signal coupling along with the interstage isolation needed.

Between the IF stages there is usually an amplifier, which serves two purposes. First, it provides gain which makes up for the loss through the filter (all real filters have some loss, even at their center frequency). Second, the AGC control voltage is used, in some receiver designs, to vary the gain through the IF stage; this provides additional AGC range beyond that of which the RF stage is capable and helps keep the final signal within the desired range with tighter tolerance. The IF transformer inductances have an adjustment for alignment, to make sure that each stage is centered exactly at the IF frequency, or alternatively staggered properly around the IF center frequency, depending on the design. This alignment is part of the overall superhet receiver alignment process. Many newer receivers use ceramic or crystal filters as precise filters that require no alignment and do not drift with time or temperature.

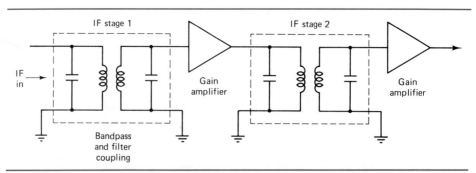

Figure 5.8 Schematic of an IF stage, showing coupling, filters, and transformers.

Double-Conversion Superhets

Although increasing the IF frequency improves the image frequency rejection tremendously for higher-frequency signals, there can still be problems in rejecting the image frequency while providing the necessary selectivity. Remember that the receiving system may be trying to provide a selectivity of a few kilohertz for an input of tens or even hundreds of megahertz. While the higher IF frequency makes image rejection easier, it makes the design of a narrow bandwidth IF filter more difficult, especially if the IF bandwidth must be adjustable to accommodate different received signal bandwidths.

The solution is to perform the mixer/LO/IF process twice, in what is called a *double-conversion* receiver (Figure 5.9). Here, the signal from the RF stage is first converted to the higher of two IF frequencies, such as 10.7 MHz, with the LO tracking the input frequency. It passes through an IF stage and is then converted again to a lower IF frequency, such as 455 kHz. There are several points to note about this design. First, because of the 10.7-MHz IF frequency, it is easier to reject images. Second, the LO frequency for the second mixer does not have to be variable. Since the signal coming into the second mixer has a fixed frequency equal to the IF frequency (10.7 MHz in this case), the second LO can be a fixed frequency oscillator at $10{,}700 - 455 = 10{,}245$ kHz (low tracking) or $10{,}700 + 455 = 11{,}155$ (high tracking). This simplifies the design. Third, the final selectivity setting can be done at the second IF frequency of 455 kHz, where it is easier to provide bandpass filtering with sharp roll-offs and adjustable bandwidth.

Some receivers for extremely high frequencies, designed to tune into hundreds of megahertz or gigahertz ranges, use *triple conversion* with a first IF stage at a very high frequency. This allows image rejection that cannot be achieved even with a 10.7-MHz IF. The first IF stage may be at 50 or 100 MHz, then a fixed LO frequency develops the 10.7-MHz IF, followed by another fixed LO resulting in the 455-kHz IF.

Review Questions

1. What functions must the IF stage perform?
2. How does the LO vary as a receiver is tuned across the band to maintain a constant IF value?
3. What is selectivity? How does the roll-off of the IF filtering determine the selectivity? Why is a high amount of selectivity often desired?
4. What is shape factor? How does it show the sharpness of the filter roll-off?
5. What is an image frequency? How does it occur? How is it eliminated?
6. Explain why higher-frequency signals need higher IF values.

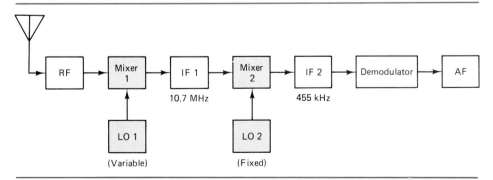

Figure 5.9 Double-conversion superhet block diagram.

7. What is a double-conversion superhet? Why is it needed? What benefits does it provide? What about a triple-conversion superhet?

8. What are high and low tracking for the LO? What is the advantage of high tracking?

5.6 AM DEMODULATION AND AUDIO STAGES

The output signal from the final IF stage is at a relatively high level (several hundred millivolts and greater) and contains only the modulation due to the desired original modulating signal. All adjacent signals that could interfere with it have been rejected by the selectivity of the IF stage. The final stages of the receiver are the *demodulation stage,* followed by the *audio-frequency stage* (*AF*), which reproduces the audio frequencies that were transmitted as the signal source. The function of the demodulation stage is to recover (*detect*) the information that was modulated onto the carrier. The AF stage amplifies this detected signal so that it is large enough to drive a loudspeaker (if the receiver is to be used as a radio or the audio channel of a TV) or to make the signal voltages compatible with whatever the receiver output feeds in the overall system, such as a computer input.

A detector for AM can be very simple. This is one of the virtues of AM and one of the reasons that AM was historically the first type of modulation: It is relatively easy to do amplitude modulation and demodulation. A *diode detector* can be used as the demodulation circuit (Figure 5.10). To understand how such a simple circuit can demodulate an AM signal, recall the AM waveform and envelope. The information that must be demodulated is this envelope itself. The diode rectifies the ac shape of the AM waveform, since it passes only signals with a positive polarity. When the signal is negative, it is blocked by the diode. The output of the diode is used to charge the capacitor, which "holds" the voltage

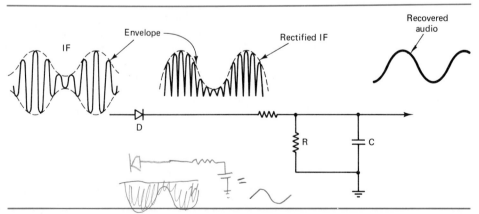

Figure 5.10 Basic diode detector and key voltage waveforms.

value of the rectified AM signal. The diode prevents the capacitor from discharging when the input signal goes negative with respect to the voltage stored on the capacitor.

In this way, the diode and capacitor act as a *peak detection and following* circuit, which follows the shape of the envelope and ignores the high-frequency IF fluctuations embedded within the envelope. The resistor allows the voltage stored on the capacitor to discharge, or "bleed off." The values of the capacitor and resistor are selected to allow the capacitor to follow faithfully the increases in the envelope, while the resistor allows the capacitor to discharge quickly enough to track properly the decreases in the envelope.

If the resistor value is too large, the capacitor will follow the positive, upward-direction travel of the envelope faithfully, but be unable to follow the reverse direction, as the envelope voltage decreases. (In fact, a circuit with an infinitely high resistance value is a *peak detector,* since it follows peaks and ignores all decreases in signal value.) Conversely, if the resistor value is too low, the capacitor will be unable to hold the envelope shape and will begin to follow the IF cycles instead. Fortunately, the modulation frequencies and the IF frequencies are spread widely apart, so that the selection of the *RC* time constant is not critical at all.

The diode detector performs this demodulation function without any "knowledge" of the original carrier frequency or phase, another advantage of such a simple approach. The output of the diode detector is the voltage across the capacitor, and this voltage can then be amplified by regular audio amplifiers to the level required by the user.

AGC Circuitry

The AGC continuously changes the gain of preceding stages to maintain a constant carrier + sideband signal level and consists of two parts. First, there is an RF amplifier with gain set by a control voltage; there may also be a similar IF amplifier stage. This *variable-gain amplifier* (*VGA*) has a precise relationship between the gain it provides and the applied control voltage. A typical value is that gain changes of +20 dB to 0 dB as the control voltage changes from 100 mV to 1 V. Note that the gain is greatest when the control voltage is smaller, and there is minimum gain when the control voltage is at its maximum. In the MOSFET RF amplifier of Figure 5.5, the AGC voltage goes to one of the two gates to control the gain.

The AGC control voltage is developed by a circuit very similar to the diode detector. The *RC* values are selected for a greater time constant than for the demodulator version of the detector. This makes the signal after the rectifying diode equal to the average value of the signal from the IF rather than its envelope. The larger capacitor charges slowly and discharges slowly through the larger resistor, ignoring the fast variations of the envelope and the much faster variations of the carrier. (*Note:* Since many circuits require a control voltage that is just below 0 V for maximum gain and much more negative for minimum gain, the diode of the detector circuit is reversed so that the circuit follows the negative direction of the envelope rather than the positive, but the operating principle is the same.)

The circuitry that develops the AGC voltage is essentially a low-pass filter, with time constants selected to ignore the faster variations that result from the modulated information, yet respond to the slower variations in the overall signal

> **Crystal Radios**
>
> Historically, the first mass-market radios were *crystal radios*, which simply have a tuned RF stage to select one of many signals from the broadcast band, followed by a crystal (a piece of quartz or similar crystalline material acting as a semiconductor, just like a modern silicon or germanium diode) and R-C pair to demodulate the signal (Figure 5.11). Headphones are connected to the detector directly. No battery is needed since the received signal provides its own power for the headphones. Of course, this type of receiver has poor selectivity and sensitivity (a long antenna, typically 100 ft or more is needed) and very low output volume, but it brought AM radio at very low cost to millions of early radio enthusiasts!

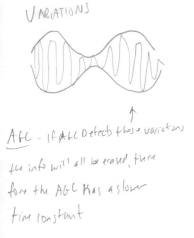

level. Some receiving systems allow the AGC time constants to be set by the user to match the conditions under which the receiver is used. These AGC *attack and decay* time constants determine how quickly the AGC responds to changes and how long it retains the variations in signal level. Recall that the input signal has the desired variations which are the actual information that has been modulated onto the carrier. If these variations are used for setting the gain level, the AGC and VGA interaction will have the effect of erasing the desired information as it quickly changes the VGA gain, which is certainly not desirable. When the AGC responds too quickly, the gain will be trying to chase a fast-moving target and will not have a chance to settle to the desired value—it would be like tracking every tiny curve in the road but always a little too late for a smooth ride. If the AGC responds too slowly, the undesired fluctuations in signal level will be missed and performance will suffer. The selection of these attack and decay times is based on the type of application for the receiver, the rate of anticipated fluctuations, and the modulated information bandwidth.

Importance of AM Receivers

At the beginning of this chapter we said that the study of AM receivers is important as a general introduction to the study of receivers for any type of modulation. Looking back at the structure of the superhet, we see that the only place where the nature of the modulation actually was a factor was the demodulation stage.

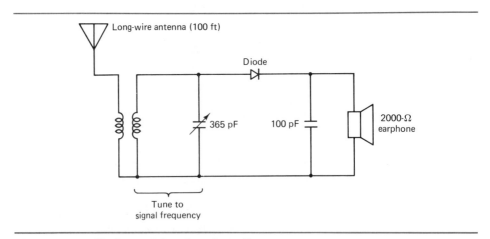

Figure 5.11 Basic crystal radio schematic.

The RF and IF stages, along with the AGC, mixer, and LO, really didn't have any conditions placed on them just because the received signal was the result of AM. How the carrier and the various associated sidebands were originally developed did not affect the operation of these stages. All that mattered was that signals with certain frequency components had to be received, amplified, converted to an IF, and passed to the demodulation stage. A receiver for non-AM signals can use essentially the same superhet design. The only difference will be the bandwidths needed (a function of the type of modulation), the kind of signal amplification required, and the demodulator.

In the next chapter we show that a receiver for frequency modulation is similar to the one for AM, except for the addition of a *limiter* and a change in the demodulator to one that can extract the original modulating signal from a carrier that has been frequency, rather than amplitude, modulated. Many sophisticated communications receivers have a front-panel switch so that the user can select AM, SSB (either upper or lower sideband), or FM. By setting this switch, a few components that set bandwidths and time constants in the superhet functional blocks are changed and the appropriate demodulator is switched in. Otherwise, everything remains the same.

Review Questions

1. Why is the final stage of the superhet receiver called the audio stage? What are the two functions of this stage?
2. What is a diode detector for AM? How does it work? What are the functions of the resistor, capacitor, and diode? How does the resistor value affect performance?
3. How is AGC implemented with a VGA and control voltage? What circuitry is often used to derive the AGC control voltage?
4. Why must the AGC operate not too slowly or quickly? How is this done?
5. Why can the basic superhet design be used for non-AM signals? What must be changed?

5.7 SSB AND CW DEMODULATION

The diode detector requires the presence of the carrier to determine the signal envelope and so demodulate the transmitted signal, but for an SSB signal there is no carrier. Demodulating an SSB signal requires that a carrier of the same frequency as the original somehow be restored and reinserted at the receiver, just before demodulation. The discussion of SSB in Chapter 4 showed that one of the drawbacks of SSB is that it requires much more complex transmitter circuitry, and this applies to the receiving circuitry as well. Among the receiving methods used, two important ones are the *beat frequency oscillator* (*BFO*) and the *product detector*. Despite these difficulties, the advantages of SSB are enough in many applications to justify its use.

The BFO method basically performs a re-modulation process by reinserting

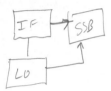

a replacement carrier for the one that was suppressed. After the IF stage the SSB signal is mixed with a signal from a local oscillator, which makes the SSB signal into a conventional double-sideband AM signal. This signal is then demodulated by a diode detector. The frequency and phase of this local oscillator must be precisely the same as the original carrier. If it is not, the demodulated signal is shifted in frequency by the difference between the original signal and the local BFO signal. For a voice signal, it sounds like a "Donald Duck" distortion and is unpleasant to listen to (for small differences) and totally unintelligible (for larger differences).

Getting the BFO output to match the original carrier requires care, and any BFO drift with temperature or time means that the user has to retune the BFO manually. The term "beat frequency" derives from the fact that matching two frequencies is often done by listening to them, or *beating them,* together in a primitive mixer: the nonlinear ear! If there is no difference in frequency, this beat output is zero. The beat output frequency is equal to the difference between the two signals. (*Note:* This frequency-matching method is still used today in electronics, and in many applications such as tuning musical instruments. A violin playing middle C at 440 Hz and one at 460 Hz will result in an audible 20-Hz sound, their beat frequency.)

Some SSB systems use a BFO combined with an *automatic frequency control.* Instead of transmitting no carrier, a *reduced carrier* or *pilot carrier* is sent in the SSB process. At the receiver, the BFO is designed to lock onto this carrier and track it automatically. This eliminates the need for manually tuning the BFO but can be used only when the transmitter is sending this modified SSB. It cannot be used for standard "pure" SSB systems.

As an alternative to the BFO, the balanced multiplier of SSB generation (Chapter 4) is used with slight changes as a *product detector.* In this configuration it has two inputs: the locally generated oscillator signal (which replicates the original but now absent carrier) and the received RF signal. The product detector multiplies these two signals and the output is the demodulated version SSB signal, thus recovering the original modulating signal.

For CW modulation, the reinserted carrier does not have to be exactly on frequency. Any difference between the reinserted carrier and the original carrier will sound like a slightly different or distorted pitch to the listeners, but the result is still quite useful. This is because the information in a CW signal is determined by the presence or absence of tone; in speech, distortion of the original tones makes it hard or even impossible to understand the demodulated signal.

BFO and Morse Code

SSB signals are more complex to generate than regular AM and require a reinserted carrier to be recovered. Ironically, CW type of modulation, which is one of the simplest (carrier either on or off), also requires a BFO. The frequency difference between the BFO and the carrier is the audible output tone of the Morse-like code, although the exact frequency (pitch) of this beat difference is not critical. In other words, both simple CW and complex SSB need a BFO (or equivalent scheme) to demodulate properly. Without a BFO, the CW signal will produce no sound when the key is up (no carrier) and no sound when the key is down (carrier present, but no low-frequency sound modulating it). The BFO mixes with the carrier/no carrier and the difference is a tone or no tone, corresponding to carrier on and off.

One of the factors that makes SSB reception and demodulation difficult is that the receiving systems discussed so far are designed to tune to any transmitted frequency in the entire band. This makes it difficult to achieve the precise tuning needed to successfully restore the carrier and eliminate distortion in SSB demodulation. In practice, however, most transmitters are intended to operate at only preset frequencies or channels. For example, a system for airport tower to pilot communication uses only a specific set of frequencies in the allocated band, and both the pilot and the air traffic controller know these in advance. For these applications, the receiver may use *crystals* instead of a continuously tunable receiver front end.

The crystal sets the oscillation frequency of the LO and the BFO, and so allows precise tuning without any drift. The crystal—also used as part of the crystal filter in SSB transmitter filters—functions as a precise *LC* tuned or resonant circuit to set the LO operating frequency and so tune the receiver. Tuning the receiver to a different channel is done simply by switching in a different crystal. This is precise, but it becomes space consuming and costly for many channels. A modern alternative, called frequency synthesis, is discussed in Chapter 15.

Review Questions

1. Why can't SSB be demodulated by a diode detector?
2. What is the BFO method of SSB demodulation? How does it work?
3. Why is the BFO frequency critical? What does the receiver output sound like for voice and music if the BFO output isn't exactly right? Why does it have to be readjusted in use?
4. How is automatic frequency control done with a BFO? Why can't it always be done?
5. How is on/off modulation (such as in Morse code) demodulated? Compare this demodulation to demodulation of voice signals via SSB.
6. When can crystals be used to improve receiver tuning performance, especially useful for SSB? How are they used? What are the advantages and drawbacks of crystals?

5.8 COMPLETE RECEIVERS

The design of the low-cost AM superhet receiver with very good performance, for use by the average consumer, was refined during the 1940s and 1950s. Of course, this vacuum-tube design was soon replaced by the transistor receivers, due to the size, power, and weight, and cost advantages that transistors offer over tubes.

A basic yet complete receiver for AM broadcast band signals is relatively simple and can be built with five transistors and one diode as shown in Figure 5.12. The signal from the antenna is tuned by L_1–C_1 and coupled into the base of transistor Q_1 via L_2, which is wound on the same ferrite rod as L_1. This 9-V radio uses transistor Q_1 as a combined LO and mixer, with the LO frequency set by L_5–

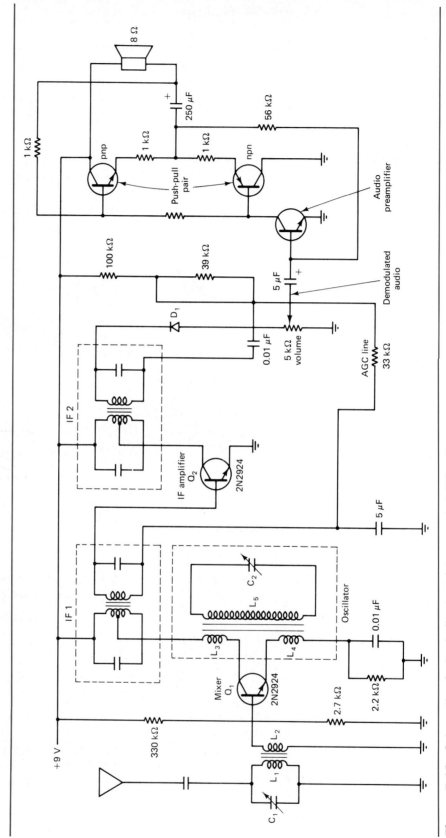

Figure 5.12 Schematic of a basic five-transistor superhet AM radio.

C_2 (C_2 is mechanically linked to front-end bandpass filter capacitor C_1). The LO signal is inductively coupled to Q_1 via L_3 and L_4.

There are two IF transformers, plus an IF gain amp. The mixer output passes through IF transformer T_1, amplified by Q_2, and goes to the second IF transformer, T_2. The signal is demodulated by diode D_1, and the demodulated output is passed to the audio amplifier stage through the potentiometer (which sets the signal level and thus the final volume) and 5-μF coupling capacitor. Additional IF stages and amplifiers would provide higher IF signal levels and improved filtering, if needed, for low signal levels or crowded bands and possible interference from adjacent signals.

The audio amplifier stage uses three transistors, with an initial single-transistor preamplifier followed by an NPN–PNP power amplifier pair in a "push-pull" configuration; one transistor amplifies the positive part of the signal while the other transistor in this pair is cut off; the complementary transistor amplifies the negative part of the signal and the first transistor is cut off. This *class B* operation provides good power gain with high efficiency (low current drain from the power supply) and fairly low distortion. The output of the push-pull pair drives the low-impedance speaker directly. (Note that for lower distortion, but at higher current drain, the class B power amplifier can be replaced by a class A design. This choice in the audio amplifier is independent of the earlier stages of the superhet receiver.)

However, such a design with many transistors and discrete components requires many individual components and is relatively costly to build and physi-

Vacuum-Tube AM Superhet Radio

Millions of low-cost vacuum-tube superhet AM receivers were produced using advances in mass production and effective but simple design, to bring broadcast radio to the public. This is the first example of a sophisticated electronic design transformed into a simple-to-use, reliable, and affordable product for the general public. Many manufacturers offered designs that were fairly similar, using five vacuum tubes: a LO plus mixer, an IF amp, two for the audio amplifier stage, and one acting as power rectifier (to convert ac line power to dc power). These AM receivers were sensitive enough for general use, and they did not need an RF amplifier stage. The antenna was a simple long wire external to the radio or many turns of wire wrapped around a ferrite core and placed within the radio itself (see Chapter 9); later designs used the ac power line itself as an antenna with capacitors to separate the RF signal from the ac power and pass this RF to the front end while isolating the ac power. Earlier sets used a power transformer between the ac line and the rectifier, plus an output transformer to convert the relatively high output impedance of the audio amplifier tube to the lower impedance (8 Ω) of the speaker; later designs eliminated the power transformer.

Variations on the design allowed battery operation, using relatively large batteries for the filament power (6.3 or 12.6 V from the A battery) and plate (60 to 90 V from the B battery). (Incidentally, this is why our more common batteries today are designated as AA, AAA, C, or D, to avoid confusion with the tube-radio traditional A and B battery designations.)

This type of radio was so popular that it was dubbed "the All-American Five" design. Replacement sets of the five vacuum tubes were sold in a single kit, so the average user could repair a dead set by swapping tubes (the tube is the device most likely to fail). **Warning:** If you should come across a tube radio, limit your troubleshooting activities to replacing tubes (if you can find replacements!). The voltages in a vacuum tube radio receiver are dangerous and lethal! Although the signal levels are low—microvolts to volts—the vacuum tubes themselves require high voltages for operation. In addition, the transformerless sets are tricky to work with since they have no isolation between ac line and their internal electronics, so misconnection of a test probe can end up grounding the power line through the test technician!

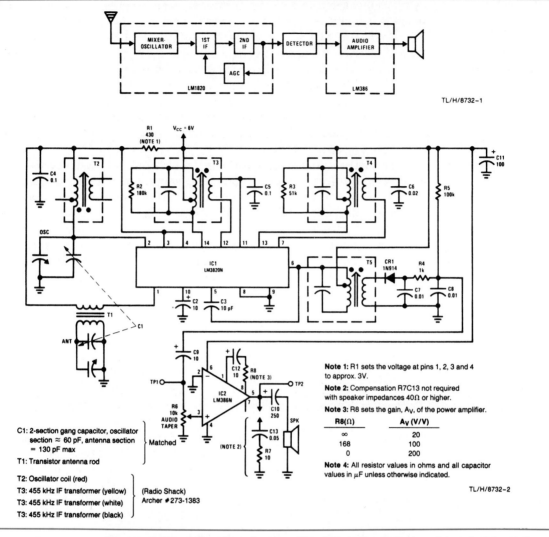

Figure 5.13 AM radio using two ICs; CR_1 is the detector: (a) radio block diagram; (b) radio schematic (courtesy of National Semiconductor Corp.).

cally large, compared to the equivalent function in an IC-based design. Most AM radios and radio functions are now built using ICs and a few discrete components. The circuit of Figure 5.13 shows a receiver that uses one IC, the National Semiconductor LM3820, for virtually all functions except the RF amplifier and audio amplifier, along with a second IC (National LM386) as the audio amplifier driving a speaker. The LM3820, IC1, provides an RF amplifier, mixer, local oscillator, and AGC functions. Specifications for the LM3820 include input RF sensitivity of 35 μV for 10 mV of output, and a SNR of 28 dB with a 100-μV RF input. The antenna signal is preselected with C_1 (130 pF) and coupled into the IC via transformer T_1. T_2 is the coil for the LO with the ganged half of C_1 (60 pF) setting the LO frequency, while IF transformers T_3 and T_4 filter and couple the signal for the two IF stages. IF output is coupled to the diode detector CR_1 through T_5. An AGC signal is developed internally using capacitor C_2; it sets the gain of the first IF stage. If the AGC voltage is needed for any other function, it can be picked up at this point. The output of the IC is demodulated with a diode detector and then fed to IC_2, the LM386 audio amplifier, which can drive the 8-Ω speaker with up to $\frac{1}{4}$ W of power from a 6-V supply.

A *communications receiver* is a general-purpose receiver that can tune across a very wide range of frequencies and handle AM, SSB, and CW as well as other forms of modulation. A block diagram of such a receiver is shown in Figure 5.14. This complete design provides several additional features that increase the quality of the received signal or the usefulness of the receiver to the listener. These features include *noise limiting, squelch,* and *signal strength meters.* A specialized receiver, such as one that is designed for receiving one type of signal or intended for use as part of an overall system such as a radar system, may not have these features. However, they do indicate the type of improvements in performance that advanced circuitry can provide.

Noise Limiters

A *noise limiter* circuit is designed to minimize the effects of noise that occurs within the demodulated signal band. The filtering of the RF and IF stages eliminates a large part of noise outside the band, but noise within the band cannot be filtered without eliminating some of the desired signal at the same time. This noise, from sources such as nearby ignition systems or atmospheric lightning, can range in amplitude from less than the received signal to much greater—as much as several hundred times greater. While the recovered audio may be understandable with a little noise, it is painful to listen to or useless with significant noise added. In addition, the noise may temporarily overload the amplifier, and most circuits require time—up to several milliseconds—to recover from the overload (or may even be damaged by the suddenly applied, sharp noise pulse).

A *noise limiter* is a diode-based circuit that automatically *limits,* or *clips,* the amplitudes of signals. Signals below the clipping level are unaffected, while signals above it are limited to the maximum value that the diodes allow. A simple noise-limiting circuit is shown in Figure 5.15a. The diodes are placed across the signal path, with one diode in each direction (since the noise peaks may be positive or negative). For signals below the conduction voltage of the diode (0.7 V for

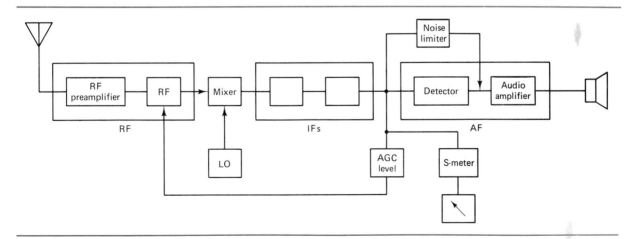

Figure 5.14 Communications receiver block diagram and schematic.

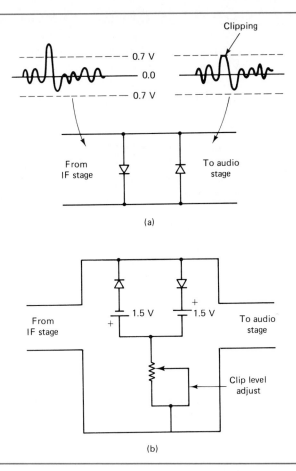

Figure 5.15 (a) Simple diode noise limiter; (b) noise limiter with adjustable limits.

silicon) the diodes look like open circuits; for signals above this voltage, they clamp the signal to 0.7 V and prevent a signal greater than 0.7 V from reaching the amplifier stages that follow.

A *noise blanker* is even more effective but more complicated. The blanker actually shorts the input to the next stage when the noise is too great, instead of limiting to the diode voltage. Since the noise is large but of a very short duration, the user does not hear that the output has gone to zero for a few milliseconds. The ear and brain tend to bridge over this short time gap in voice or music.

The simple noise limiter with shunt diodes assumes that the noise should be clamped at the 0.7-V diode voltage, but this is not usually the case. The noise level is related to the signal level. For weak signals, the noise level that is acceptable will be smaller, while for higher-level signals it will be larger. Also, the listener may need to adjust the noise level cutoff depending on how much noise there is—if there is a lot, too much of the desired signal may be clipped out. The solution is a noise limiter with an adjustable threshold (Figure 5.15b). Here, the user can set the level at which the noise-limiting action begins, for the best combination of minimizing the noise while retaining the most of the received signal.

Squelch

Noise limiters take care of large-value noise impulses that come into the receiving system with the desired signal. Another type of noise that can be very annoying occurs when there is no received signal. This is the internal receiver noise that every electronic circuit produces, but is normally small compared to the desired signal. Consider a two-way radio used between a police car or taxi and the central office. The receivers at both ends must always be on, tuned to the channels that the police or taxis have assigned. Much of the time, however, there is no signal (or only very faint, undesired signals) to be received. The system AGC, sensing no signal, turns the receiver gain to the maximum, since the function of the AGC is to adjust gain to accommodate the different signal strengths. The result is that the receiver is set to maximum gain, and the internal noise of the receiver is amplified by this maximum gain value. The listener hears a constant, loud, and very irritating hiss from the loudspeaker when there is no signal. If the user turns the volume down to reduce the hiss level, then when a real signal comes in, it will be at the low volume.

This dilemma is neatly solved by a _squelch_ function (Figure 5.16). The comparator continuously compares the AGC voltage to the squelch threshold, which is adjustable. The output of the comparator controls an analog switch, which is an electronically controllable on/off switch which passes any signals without distortion within a range of -2 to $+2$ V. The squelch circuit cuts off the input to the AF amplifier when there is no received signal level. Only when the received signal level is greater than some specific value is the AF stage input "released" to act in the normal fashion. The result is nearly absolute silence at the receiver output, except when a signal comes through the RF and IF stages.

The level for the squelch is derived from the AGC signal, which is near zero (corresponding to maximum gain) for no or low input, and increases with the received signal level to reduce the receiver gain. The squelch circuit has an adjustment that allows the user to set the value where the squelch allows signal to

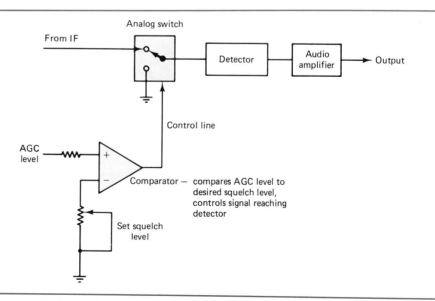

Figure 5.16 Squelch circuit compares AGC level to desired squelch level, thus controlling passage of IF signal to the detector and final amplifier.

reach the AF stage. The user sets this depending on the acceptable noise level and the level of the faint, useless signals that are causing noise but convey no useful information. The squelch can also be disabled in most receivers. This is needed where the listener is searching for a very weak signal such as from a low-power transmitter or from a transmitter that is located thousands of miles away, and the signal may be so faint that the squelch would cut off the receiver amplifier. However, for systems where the listener is always monitoring specific channels but there are only sporadic transmissions, the squelch circuit is very useful.

Signal Strength Indications The AGC signal is also used for indicating the strength of the received signal to the user. The action of the AGC keeps the signal level relatively even, which is a good result. However, the listener then has no way of judging how well the signal is being received: Is the new antenna bringing in a stronger signal? Is this location better than the previous one? Are atmospheric conditions causing signal fading? The AGC signal is used to drive an *S-meter,* which indicates signal strength. This meter essentially is calibrated so that a near-zero AGC voltage indicates a weak signal, while a higher AGC voltage indicates a stronger received signal. The AGC signal for the S-meter is derived using essentially the same circuit as the diode detector, except that the time constant of the *RC* after the diode is much longer. In this way, S-meter voltage ignores the envelope of the signal, but indicates its average value instead. The S-meter can also serve as a tuning indicator, to show if the receiver is tuned to the center of the desired signal bandwidth. The signal strength will be at a peak when the tuning is "right on" and be less if the receiver is tuned off the optimum center point.

Clearly, the general-purpose communications receiver is complex and provides many features and options. Many communications systems do not need a receiver with all these functions. A receiver for receiving radar signals, for example, is intended for one specific type of modulation and for only a limited range of frequencies. A TV signal receiver needs to work just on specific channels and does not need an S-meter or squelch. However, the study of the general communications receiver is an overall guide to the more specialized receivers tailored to specific applications, which are in many ways subgroups of the general case.

Review Questions

1. Explain how an AM broadcast receiver can be built with just two ICs. What functions does each IC provide?
2. What is the role of each of the five transformers in the two-IC receiver? Where and how is demodulation done?
3. What is a noise limiter? What are two reasons why it is needed? Why is the level value adjustable?
4. What is the difference between a noise limiter and a noise blanker?
5. What is the AGC value when the input signal is low or zero? Why is the audio output noisy in this case?
6. How does the squelch circuit operate? Why is the squelch value adjustable?

5.9 AMPLITUDE MODULATION FEATURES AND DRAWBACKS

In Chapter 4 we studied AM signal generation, while this chapter has covered AM receivers. Occasionally, there has been a mention of other forms of modulation, especially in relation to the overall functions of a communications receiver. Before studying these other types, frequency modulation (FM) and the closely related phase modulation (PM), it is important to understand the characteristics of AM and why other forms of modulation may be needed.

The major problem with AM is that the AM signal is greatly affected by noise from any source; whatever its cause, it gets added to the amplitude of the AM signal. At the receiver, there is no way to separate this added noise from the original modulation. No amount of filtering can reduce noise that is in the frequency band of the desired signal. The added noise becomes a part of the signal, and the demodulated output consists of the transmitted signal plus the added noise. The only way for the receiving system to eliminate this noise is to have some idea of what the modulating signal looked like, which is not the case. If the modulating waveform was known, there would be no need to send the signal at all in the first place!

The noise from sources internal to the receiver may be reduced by careful design techniques, but any external source noise that is added by nearby ignitions, lightning, and static cannot be reduced. This noise becomes a limiting factor in the quality of the received signal. The noise may range from annoying, making conversations hard to understand, to disturbing, ruining the fidelity of music, to excessive, making the received signal useless and full of errors. For signals that are inherently weak, such as those from space satellites or transmitters thousands of miles away, the noise may be stronger than the received signal (SNR < 0 dB!) and make signal recovery very difficult or even impossible.

The second major problem with AM is that the received level of signal and modulation actually conveys the information about the amplitude of the modulating signal. As the received signal level varies for any reason, it is impossible to absolutely determine the original signal level. For example, suppose that the modulation at the transmitter indicated the temperature, from some sort of temperature-to-voltage probe. Certainly, as the temperature increased, the modulating voltage would increase, and the demodulated signal at the receiver would be stronger. But assigning an exact value to this (0° equal 100 mV, while 100° is 1 V) in a direct, one-to-one relationship is not possible with AM, because absolute signal levels cannot be maintained from the transmitter to the receiver output (of course, noise aggravates this problem, but this problem exists even without any noise).

Finally, conventional AM is not efficient in the use of transmitter power, as was seen in Chapter 4. Even with 100% modulation, only one-third of the power is information-bearing sideband power. The use of SSB improves this situation but requires more complex transmitter and receiver circuitry and requires that circuitry be very stable as power supply voltages vary even slightly or the operating temperature changes.

This does not mean that AM is not useful—hundreds of millions of AM receivers in operation prove it is. AM is useful where a simple, low-cost receiver and detector is desired, such as in the mass-market standard broadcast band. The transmitters used are large enough to supply sufficient power so that the numerous users can receive the signals with simple, low-cost receivers. The noise is either not a problem or can be tolerated in the application. The precise signal levels and signal fidelity is not an issue, as when listening to news, weather, or traffic reports.

The weaknesses of AM were known from the earlier days of radio. The solutions that were proposed were theoretically good, but the circuitry to implement these ideas had not been invented. However, by the 1930s the alternative to AM, known as frequency modulation, was put into use in practical systems and is the subject of the next chapter.

Review Questions

1. Why is noise a major and unavoidable problem with AM?
2. What is the resulting signal when noise combines with the transmitted signal?
3. What are the effects of noise, ranging from minor to severe?
4. Why can't AM be used for indicating absolute signal levels? Is noise a factor in this, too?
5. Where is AM useful and practical? Why?

5.10 AM RECEIVER TESTING

Just as receiver design is more sophisticated than transmitter design (due to noise, tuning unknowns, low signal levels and their variations, and related factors, Section 5.1), the testing of a receiver is more difficult. For this reason, many technicians and engineers prefer to test receivers starting with the final output stage and then working backward, since signal levels and frequencies are larger and easier to measure in the later stages. The stage that does not produce a good output for a known-OK input is assumed to be the source of the problem.

Of course, the first thing to check is the power supply output and to make sure that power is reaching all stages. This is checked with a standard dc voltmeter. As noted earlier, do not troubleshoot a tube receiver unless you know exactly what you are doing: **lethal voltages** are present throughout the receiver. In contrast, a battery-powered transistor or IC receiver uses safe, low voltages (such as 9 V, 12 V, or ±15 V). If the receiver is line-powered, there is a single power supply section that is responsible for converting line ac to the low-voltage dc levels. This section is clearly marked, and as long as you stay away from this, there is no physical danger. However, there is still the opportunity to damage good components by careless probing and short circuits! Also switch off any special function the receiver has, such as squelch or noise limiting.

After checking power, disconnect the antenna and short the antenna terminal together (if they are accessible) so that no external signal reaches the front end. Instead, the input to each stage is replaced by an injected signal from a test instrument to replicate the signal from the previous stage. Beginning with the final audio amplifier stage, a relatively high-level audio signal (400 to 1000 Hz, 10 to 100 mV typical) is injected to represent the demodulated signal from the IF stage. This signal should appear, amplified, at the final output of the receiver. If it does not, the problem is with this final stage.

Working back to the previous stage, a modulated signal at the IF frequency (usually 455 kHz) is injected to represent the mixer output. This should appear amplified and demodulated at the output of the IF stage. If it does not, the IF amplifier or its filters are bad or misaligned. The IF rejection of unwanted mixer frequencies is analyzed by injecting frequencies other than the nominal IF value.

The LO output is measured with a voltmeter. Its frequency should be checked with a frequency meter and examined on a scope or spectrum analyzer for low distortion (a clean sine wave). The LO frequency should be equal to the tuned frequency $\pm$ the IF, depending if high or low tracking is used.

Troubleshooting a receiver at the RF stage requires more than the standard voltmeter and oscilloscope, although these are essential. A signal source that can generate AM signals at varying levels and frequencies is also needed, since it is difficult to use an unreliable, not always consistent "off the air" signal. The signal generator acts in place of the received signal.

A signal generator for AM receiver testing (and also FM receivers) must generate RF waveforms with the RF level set from a few microvolts or greater. The signal can be an unmodulated carrier, or can be modulated anywhere from 0 to 100% by an internal oscillator at a user-selectable frequency. The signal generator is a signal source that can be tuned across the band, to check the receiver dial calibration and allow the RF stage and LO/mixer to be aligned properly by the technician. It also allows the technician to check front-end sensitivity, as the generator output is set lower and lower. The technician can measure the signal with the scope at the output of the mixer to see how the signal has been mixed down to the IF frequency, as well as check the level there. (Be sure to "un-short" the antenna input when the signal generator output is applied there.)

To check the IF stage, the signal generator can be used once again. This time it is used to inject a signal that replaces the mixer output. This signal is stronger than the RF signal and does not have to be tuned, since the modulation is now at the fixed IF frequency. The AGC is checked by watching the AGC voltage with the scope or voltmeter (the AGC signal is relatively slow to change) as the injected signal level changes. Some signal sources can be set to step automatically through various carrier levels, so that the AGC action can be tracked easily and observed without the technician constantly readjusting the level manually.

A single source of RF signals may not be able to provide the wide range of frequencies that are used in communications systems. Signal generators with built-in modulation capability span various ranges. For work at relatively low frequencies, a 0.01- to 30-MHz generator might be used. For the ranges at VHF and above, more expensive generators are required. Usually, these are used only to check out the front-end circuitry (which requires precisely calibrated RF signal levels and tuning over a wide bandwidth), while less complicated sources are used for the IF stage and AF stage, to verify performance of the IF filtering, demodulator, and audio amplifiers.

If a signal source is not available, a known broadcast station signal can be used, although the signal level will be uncalibrated and may vary during testing. When troubleshooting with a received station, it is best to begin with the receiver input and work toward the output, after checking the power supply and LO output. Then trace the received signal to the mixer, out of the mixer, out of the IF stage, out of the detector, and then as the final amplifier output.

Review Questions

1. Why is it impractical to use a regular broadcast signal for receiver testing?
2. What type of signal should the signal generator provide? What should be variable?
3. What is checked at each stage? How?
4. Why are several signal generators usually needed for receivers that operate at the highest frequencies?
5. Why does receiver testing usually start at the final stage, whereas transmitter testing begins at the first stage?

SUMMARY

In this chapter we have seen the complexity needed at the receiver to recover the original modulating signal, under the realistic, yet typically difficult conditions of noise, weak signals, interference, and varying carrier frequencies. The simple tuned-radio-frequency design unfortunately has many performance limitations. The more complicated superhet provides far superior performance, because it separates the needed functions of high sensitivity, tuning, selectability, and signal demodulation into separate stages, each of which can be optimized for a specific task. The central concept of the superhet, translating all received signals to a single IF frequency through a local oscillator and mixer, is the key to the way the superhet operates. An automatic gain control circuit eliminates the effects on the IF and AF stages of unavoidable changes in received signal level so that the entire receiver can be designed to operate on signals of known values rather than accommodate frequent, wide-ranging changes.

The superhet receiver was originally designed for AM. Unfortunately, AM suffers from drawbacks in noise resistance and the ability to transmit absolute signal levels. Other forms of modulation can overcome this, and still use the basic principles of the superhet receiver with some changes. Testing and adjusting receivers requires injecting modulated, precise RF and IF signals in the appropriate points of the receiver circuit, in order to check the functions of each block and adjust the alignment of the RF, local oscillator, and IF stages, as well as verify the performance of the demodulator itself.

Summary Questions

1. Why is receiving much more difficult than transmitting?
2. Compare the effects of internal noise and external noise on the ability of the receiver to recover a signal.
3. What elements of AM receivers are common to receivers for other types of modulation? Why?
4. How does the superhet overcome many receiver problems?
5. What are the major blocks of a superhet receiver? What does each do?
6. Why is the performance of the RF stage critical to overall receiver sensitivity, especially for frequencies above 30 MHz?
7. Why is the RF stage often comprised of several amplifiers in series?
8. How does the AGC function? What specifically does it need and do?
9. How do the LO and mixer combine to produce the IF? What happens to the sum frequency? What are low and high tracking, and their differences?

10. What is the advantage of having a fixed IF?

11. What are the advantages of electronic tuning? How can variations in varactor capacitance and driving voltage be compensated?

12. What is an image frequency? What is the cause? How can it be eliminated? Why is it a problem when the IF frequency is much less than the tuned frequency?

13. How does a double-conversion receiver reduce image problems for higher frequencies? Where is triple conversion needed?

14. Explain how a diode detector works for AM.

15. How is SSB demodulated? Why is a simple diode detector insufficient? What happens if the BFO frequency for SSB is off, or drifts?

16. What advantage do crystals have in receiver tuning? What drawbacks?

17. What is the function and need, in a general communications receiver, for noise limiting, squelch, and an S-meter? How do each of these work?

18. Why is external noise a problem for AM? Why can't it simply be filtered out? Why is AM still very useful, despite noise problems?

19. How is an AM receiver tested and aligned? What does the signal generator do?

20. Compare AM receiver testing to transmitter testing, in terms of procedure and test equipment used.

PRACTICE PROBLEMS

Section 5.2

1. What is the Q of a tuned circuit with a center frequency of 10 MHz and a bandwidth of 30 kHz?

2. A tuned circuit has to pass a 50-kHz bandwidth signal at a frequency of 7 MHz. What Q is required?

3. A tuned circuit with a Q of 250 is used to tune from 5 MHz to 8 MHz. What is the bandwidth at both ends and the center of this band?

4. A TRF receiver is used for receiving AM signals in a band from 20 to 25 MHz. These AM signals are voice, with a bandwidth of 8 kHz. What Q is required at the lower end of the band? At the upper end of the band?

Section 5.3

1. An RF stage uses a single amplifier with $\times 1000$ gain and $+50$ μV noise, for a received input signal of 20 μV + 2 μV noise. What is in original SNR? What are the output signal level and the output noise level? What is the output SNR?

2. The amplifier of problem 1 is used for a signal of 20 μV + 10 μV noise. What are the input SNR and the output SNR?

3. Instead of the single RF amplifier and signal of problem 1, two amplifiers are used in sequence. The first provides gain of $\times 10$ plus 10 μV noise, and the second provides gain of $\times 100$ plus 50 μV noise. What are the signal level and noise level after the first amplifier? After the second amplifier? What is the final SNR?

4. Repeat problem 3, but use the $\times 100$ amplifier (and its noise) first, followed by the $\times 10$ amplifier.

Section 5.4

1. What LO frequencies could be used to tune a 7.7-MHz signal if the desired IF is 0.8 MHz?

2. A receiver is tuning from 10 to 15 MHz, and the mixer produces an IF of 1.5 MHz. What are the frequency spans of the LO for low tracking and for high tracking? What are the frequencies of the mixer outputs for both cases?

3. An AM signal with a carrier of 5 MHz and modulating sidebands at 5 kHZ on both sides of the carrier is mixed with a LO of 3.5 MHz. What are all the resulting output frequencies? What is the IF?

4. The AM signal of problem 3 must have an IF of 455 kHz to use standard IF components and circuitry. What low-tracking LO frequencies could be used? What would all the mixer outputs be?

5. A resonant LC circuit is tuning a signal at 5 MHz using a varactor with a nominal capacitance of 100 pF, but the actual value is 10% low: 90 pF. What is the resonant frequency that results? What is the resonant frequency if the varactor is 110 pF (10% high)?

6. The resonant LC frequency in a receiver must be within 1% of nominal or the user will have a great deal of difficulty tuning properly. What percentage of tolerance can be accepted in the varactor capacitance? (*Hint:* Rearrange the formula slightly, then solve for C_n/C_0.)

Section 5.5

1. What is the image frequency when the received signal is at 9 MHz and the IF is 455 kHz, with the LO using low tracking?

2. For a signal at 30.5 MHz, with a 455-kHz IF, what is the image frequency with high tracking? What is the image frequency when the IF is changed to 10.7 MHz?

3. A double-conversion receiver uses a first IF of 10.7 MHz and a second IF of 455 kHz, both with high tracking. A signal is received at 50 MHz.
   (a) What are the LO frequencies?
   (b) What are the outputs of the first mixer?
   (c) What are the signals going into the second mixer, and what signals come out of the second mixer?

4. A receiver intended for the aircraft to control tower signals at 110 MHz uses single conversion with a 455-kHz IF. What is the low-tracking image frequency? What is the image frequency if the IF is changed to 10.7 MHz?

5. The receiver of problem 4 is changed to double conversion, with 10.7-MHz and 455-kHz IFs and low tracking.
   (a) What are the LO frequencies?
   (b) What are the frequencies after the first mixer?
   (c) What frequencies are the input of the second mixer?
   (d) What are the outputs of the second mixer?

6

Frequency and Phase Modulation

CHAPTER OBJECTIVES

When you have completed this chapter, you will understand:

- The nature of frequency (and phase) modulation and its spectrum
- How an FM (or PM) signal is generated
- How an FM (or PM) signal is demodulated in the receiver
- The benefits and weaknesses of FM/PM compared to each other and to AM
- Test techniques unique to FM and PM

INTRODUCTION

Frequency modulation, together with phase modulation, represents the alternative to AM for impressing an information signal onto a carrier. FM and PM offer many advantages in noise reduction, signal fidelity, and use of power, but more complex circuitry is required at the transmitter and receiver. In this chapter we examine the modulation equations for FM and PM, along with the bandwidth and sidebands that result from modulation. Some types of FM transmitters and receiver designs are shown, along with circuits and ICs that actually implement the required functions. The role of the limiter, which is the FM equivalent to AGC, and the effects of limiting, are also studied, as is narrowband FM, which requires less bandwidth than conventional FM. A very powerful circuit function called the phase-locked loop provides excellent performance in many FM and PM applications, as well as in other aspects of communications systems. Commercial stereo FM broadcasting also involves two audio channels multiplexed onto a single carrier. We will look at how this stereo encoding is done at the transmitter and how the receiver decodes the two original channels. Finally, the key features and benefits of AM, FM, and PM are compared and contrasted, and some complete FM receiver schematics using high-performance ICs are discussed.

6.1 THE CONCEPT OF FREQUENCY MODULATION

The same person who developed the superheterodyne receiver was also responsible for the development of an alternative to amplitude modulation. Major E. H. Armstrong did much of the theoretical work and practical circuit design on frequency modulation during the 1930s in order to produce a noise-resistant, static-free type of radio transmission. Armstrong's dedication to FM included being slung in a bosun's chair to adjust his first full-scale FM antenna on its tower in New Jersey, 1200 ft above the Hudson River.

You are probably familiar with the standard FM broadcast band which spans 88 to 108 MHz. Although AM broadcast was the dominant type of broadcast for many years, FM has been the most popular since the 1970s because it provides much clearer signals, much lower distortion, less noise and static, along with the capability of stereo broadcast (which is not due to the use of FM rather than AM, but rather that the higher music fidelity of the standard FM band makes stereo more meaningful). This use of FM in standard commercial broadcasting has been a major incentive to the popular use of this type of modulation and to the development of FM specific ICs by various manufacturers.

This does not mean that the use of FM is limited to broadcasting music with great fidelity and clarity to home and car radios. FM is used extensively in radio links where signal quality is critical, such as between a radio studio and transmitter, in two-way mobile radios, in sending signals from space satellites, and in computer communications links and circuitry such as modems (Chapter 17). For all these and many other applications, the ability of frequency modulation to allow the original signal to be received at the other end of the link without error makes FM very useful and necessary. Compared to AM, FM usually requires more bandwidth to transmit a given amount of information, and its circuitry tends to be more complex. Nevertheless, in many cases the advantages of FM are absolutely necessary for successful communications systems.

FM is resistant to noise and static due to a simple fact about most noise and noise effects on signals: At each instant the noise adds to, or subtracts from, the amplitude of the desired signal by some unknown, random amount. If a signal of 1.1 V is affected by noise, the noise appears as a change in the amplitude of the signal, perhaps to 1.05, 1.13, 1.7, or any other voltage. More noise causes a greater change in the voltage of the desired signal, and this noise cannot be separated from the original signal value in many cases. The noise becomes an indelible part of the signal. If the information is transmitted by AM, the amplitude of the signal is in fact carrying the information, and any change in amplitude results in a change in the perceived information of the signal.

In FM, a different approach is used. The frequency of the carrier is modulated by the information signal. The resulting signal is demodulated by looking at its frequency and changes in frequency, and ignoring the amplitude. Changes in amplitude, caused by noise, are of little consequence. The receiver is designed to ignore these changes but responds only to changes in frequency. There are certain types of noise and channel problems that can affect frequency, but these are usually relatively minor compared to the ease with which noise affects amplitude.

Another type of modulation that is closely related to FM is *phase modula-*

tion (PM). In PM, the phase of the carrier is directly modulated instead of the frequency. The mathematics that explain FM also explain PM, since phase and frequency are closely related to each other (mathematically, frequency is the rate at which the phase changes, so it is also called the time derivative of phase). For this chapter, with few exceptions, the statements that are made about FM also apply to PM.

There is an important difference between demodulating an AM signal and demodulating an FM or PM one. While the receiver front end has to be tuned to the carrier for an AM signal, extracting the modulated information at the detection circuitry does not require a specific knowledge of the carrier frequency or phase. The detector takes the IF signal and simply recovers the envelope, which represents the information part of the total signal. For FM and PM, the demodulator itself must have some indication of the absolute frequency or phase of the carrier, since deviations from this frequency or phase represent the modulated information. This means that the receiver circuitry must somehow be able to extract a signal that represents the frequency or phase of the carrier, in order to perform the demodulation properly. Depending on the system and the application, there are several ways to do this, with different complexity and various degrees of performance.

Review Questions

1. In basic terms, how does FM differ from AM? How is the information at the source encoded onto the carrier?
2. Why is FM much more noise and static resistant than AM?
3. Where is FM used? Why?
4. Why does demodulating an FM or PM signal require a different approach at the detector than demodulating an AM signal, beyond just a different detector circuit?

6.2 FM SPECTRUM AND BANDWIDTH

Frequency modulation uses the modulating signal to vary the frequency of the carrier. The variation has two key elements: the *amplitude* of the modulating signal determines the amount of carrier variation or *deviation* from the original value—which yields the amplitude information at the receiver—while the *rate* at which the modulating signal varies determines the rate of the carrier deviation, yielding the frequency information in the received signal. Let's examine the effect of this on the spectrum of the carrier and on the spectrum and bandwidth of the overall signal to be transmitted.

Intuitively, it may seem that the result of causing the carrier to deviate from the nominal value is simply a new spectrum component at a frequency related to the frequency of the modulating signal, similar to AM. (Recall in AM that the modulation process caused sidebands at the carrier frequency ± the modulating frequency.) This is not the case, however. The spectrum of an FM signal is complex. The modulating signal does not simply cause the carrier to jump to a new frequency. The FM process creates a whole range of new sidebands. The analysis of an FM signal is based on the description of the signal voltage $V(t)$ as a function of time:

$$V(t) = A \sin(2\pi f_c t + m \sin 2\pi f_m t)$$

where A = amplitude of the carrier
f_c = carrier frequency
f_m = modulating signal frequency
m = *modulation index*, which is defined as the maximum carrier frequency shift (deviation or "delta," Δ)/the frequency of the modulation signal f_m
$= \Delta/f_m$

Example 6.1

What is m for a deviation of 100 kHz and a modulating frequency of 15 kHz? What happens to m if the deviation triples?

Solution

Since $m = \Delta/f_m$, then $m = 100/15 = 6.66$ in the first case, and $m = 300/15 = 20$ in the second case.

Before studying the equation of FM and its implications, note that the voltage $V(t)$ requires taking the sine of a sine. This is a complex operation and one that does not result in a simply expressed result as in the AM case. The analysis requires the use of *Bessel functions*, named after the Prussian mathematician and astronomer who developed them in the first part of the nineteenth century. The result of FM is a carrier with sidebands at the carrier frequency ± all multiples of the modulating signal frequency: $f_c \pm nf_m$, where n is an integer from 0 to ∞. This spectrum is infinitely wide! The time and frequency domains of a typical FM signal are shown in Figure 6.1. Note the differences in both domains as compared to AM: The time-domain amplitude is constant, while the frequency domain is spread widely rather than consisting of just two sidebands.

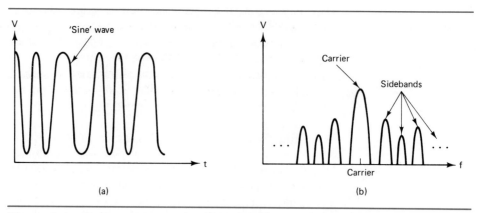

Figure 6.1 Typical FM signal: (a) time domain; (b) frequency domain.

6.2 FM Spectrum and Bandwidth

Modulation index	Carrier J_0	Sidebands										
		J_1	J_2	J_3	J_4	J_5	J_6	J_7	J_8	J_9	J_{10}
0.0	1.00	—	—	—	—	—	—	—	—	—	—
0.25	0.98	0.12	—	—	—	—	—	—	—	—	—
0.5	0.94	0.24	0.03	—	—	—	—	—	—	—	—
1.0	0.77	0.44	0.11	0.02	—	—	—	—	—	—	—
1.5	0.51	0.56	0.23	0.06	0.01	—	—	—	—	—	—
2.0	0.22	0.58	0.35	0.13	0.03	—	—	—	—	—	—
2.5	−0.05	0.50	0.45	0.22	0.07	0.02	—	—	—	—	—
3.0	−0.26	0.34	0.49	0.31	0.13	0.04	0.01	—	—	—	—
4.0	−0.40	−0.07	0.36	0.43	0.28	0.13	0.05	0.02	—	—	—
5.0	−0.18	−0.33	0.05	0.36	0.39	0.26	0.13	0.06	0.02	—	—
6.0	0.15	−0.28	−0.24	0.11	0.36	0.36	0.25	0.13	0.06	0.02	—
7.0	0.30	0.00	−0.30	−0.17	0.16	0.35	0.34	0.23	0.13	0.06	0.02
8.0	0.17	0.23	−0.11	−0.29	0.10	0.19	0.34	0.32	0.22	0.13	0.06
15.0											

Figure 6.2 Table of Bessel functions.

It may seem very impractical to transmit an FM signal since infinite bandwidth is not available as this would allow no other transmitters in the spectrum. Bessel functions can show the answer. The magnitude of each spectral component varies as the sideband number n increases. A table of Bessel functions (Figure 6.2) shows the magnitude of each successive sideband for different values of m. In practical terms, the FM system is designed to allow almost all, but not all, of the sidebands and their energy to pass. A practical limit is to allow 98% of the total modulated signal energy to be used and ignore the remaining 2%. The Bessel function table can be used to add up the energy of each sideband and see how many sidebands—and therefore what bandwidth—is needed. The effect of limiting the number of sidebands and bandwidth is that there will be some distortion in the transmitted and recovered signal. If more sidebands are eliminated, the distortion is greater; using more sidebands and a wider spectrum reduces this distortion.

The amount of power in each sideband relative to other sidebands does not follow any simple pattern. Depending on the modulation index, the overall shape of the envelope formed by the carrier and FM sidebands may be roughly triangular (with the carrier as the peak and all other sidebands decreasing evenly with distance from the carrier) or may have hills and valleys. A few examples using the table of Bessel functions illustrate this.

Example 6.2

What is the relative amplitude of each sideband when the modulation index is 1.0?

Solution

Using the table, for the horizontal line at $m = 1.0$, we have the following:

Carrier: $J_0 = 0.77$

First sideband pairs: J_1 (carrier ± modulating frequency) = 0.44

Second sideband pairs: J_2 (carrier ± twice the modulating frequency) = 0.11

Third sideband pairs: J_3 (carrier ± three times the modulating frequency) = 0.02

This is shown in Figure 6.3a. The rest of the modulation sidebands are close to 0 and negligible.

Example 6.3

What are the relative amplitudes of the frequency components for a 1-MHz carrier, modulated by a 10-kHz signal, with $m = 2.0$?

Solution

Using Table 6.2 once again, we see what is graphed in Figure 6.3b:

J_0 (the carrier, at 1 MHz) = 0.22

J_1 (1 MHz ± 10 kHz, or 990 kHz and 10,010 kHz) = 0.58

J_2 (1 MHz ± 20 kHz, or 980 kHz and 10,020 kHz) = 0.35

J_3 (1 MHz ± 30 kHz, or 970 kHz and 10,030 kHz) = 0.13

J_4 (1 MHz ± 40 kHz, or 960 kHz and 10,040 kHz) = 0.03

Although the relative amplitudes from one sideband to the next do not follow a fixed pattern, the general trend is that sidebands farther away from the carrier have decreasing amplitudes and are therefore less significant. Watch out, though, since there can be some strange bumps along the way, as the table of Bessel

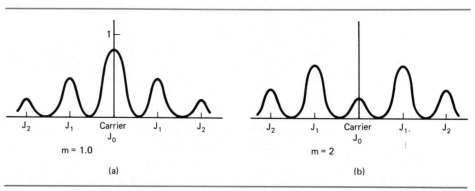

Figure 6.3 FM and sidebands for (a) $m = 1.0$; (b) $m = 2.0$.

Carson's Rule

Instead of using the tables, *Carson's rule* can be used to estimate the 98% level of the Bessel functions. This rule approximates the relationship of the 98% energy bandwidth to the deviation Δ and the modulating frequency:

$$\text{bandwidth} = 2(\Delta + f_m)$$

Example 6.4

A system uses a deviation of 100 kHz and a modulating frequency of 15 kHz. What is the approximate bandwidth needed?

Solution

By Carson's rule, the bandwidth = 2(100 + 15) = 230 kHz.

Example 6.5

A system has 150 kHz of bandwidth available for a 10-kHz modulating signal. What is the approximate deviation to be used?

Solution

Rearranging the equation, we have

$$\Delta = \frac{\text{bandwidth}}{2} - f_m = \frac{150}{2} - 10 = 65 \text{ kHz}$$

---

functions shows. In the first example above, the magnitude of the sidebands steadily decreased; in the second example, the carrier was weaker than the first sideband. Note that there are values of the modulation index m for which the carrier amplitude is 0, such as just under 2.5, between 5 and 6, and close to 9 (if the carrier amplitude makes a transition from positive to negative, or vice versa, it must have passed through zero along the way).

The interplay of modulating frequency, deviation Δ, modulation index m, and bandwidth can get confusing. Keep these points in mind: The modulating frequency f_m is determined by the spectrum of the information source, such as voice, music, or data. The deviation Δ is determined by the design of the FM transmitter, which has a relationship between the modulating frequency and the resultant frequency shift of the carrier. Modulation index m is the ratio of this deviation (change in frequency) to the modulating signal frequency that caused it. Finally, bandwidth is calculated from the modulation index using Bessel functions, or approximated using Carson's rule.

If intuition tells you that causing a deviation of 10 kHz from a nominal carrier of 1 MHz should result in a spectrum at 1 MHz + 10 kHz, your intuition is correct in only one situation. If the modulating signal causes the carrier to shift to the new frequency and you look at the spectrum *after* this shift, you would see only the single spectral component. But the infinitely wide spectrum of FM results from the act of *sweeping* from the original nominal carrier frequency to the new one, and this sweeping contains information as well. Recall that the frequency of the modulating signal determines the rate of deviation, not the final value, so information is contained in the sidebands, which the modulation generates, as it causes the deviation to occur.

FM and Power

In AM, modulating the carrier requires power. The modulating circuitry has to apply power to the carrier, in order to change the amplitude of the carrier and produce AM. The power in the carrier and the sidebands are determined using equations that use the AM modulation index (very different from the FM modulation index). The corresponding situation for FM power before and after modulation is different than that for AM.

In FM, the amplitude of the signal after modulation is the same as before modulation: since power is proportional to the square of the amplitude, the power is unchanged. What happens in FM is that the power in the carrier is distributed over the various FM sidebands that result from the modulation. The modulation circuitry does not have to add power, so it can be an ordinary low-power circuit rather than the high-power circuit that AM required for high levels of modulation. Unlike AM, which has a carrier that contains no information but uses power, FM uses all the power for information. This power is contained at the various frequency spectrum components of the FM result, in amounts determined by the modulation index m and the corresponding Bessel functions. Higher values of modulation index for a fixed modulating signal f_m (which results from a greater deviation Δ for a given f_m) spread the power out more evenly (the spectral "mountain" is less steep), and a wider bandwidth is needed to encompass a fixed fraction (such as 98%) of the total power.

Broadcast FM The standard FM radio broadcast band spans 88 to 108 MHz. Stations are assigned carrier frequencies spaced every 200 kHz, beginning at 88.1 MHz and going up to 107.9 MHz. The modulating signal is 30 Hz to 15 kHz, and each station and its sidebands can spread ±75 kHz of the carrier, with a 25-kHz *guard band* at both ends (25 + 75 + 75 + 25 kHz = 200 kHz), shown in Figure 6.4. The purpose of the guard band is to provide an extra measure of protection against signal energy from one transmitter overlapping that of another transmitter. This may be due to inadvertent drift in the carrier or improper filtering at the transmitter, both of which are normally controlled to very tight limits. The modulating signal that represents voice or music can have a bandwidth from 30 Hz to 15 kHz. The maximum allowable shift Δ is 75 kHz, so the modulation index m varies from 75 kHz/30 Hz = 2500 down to 75 kHz/15 kHz = 5. Note that the modulation index m varies inversely with the modulating frequency: a higher value of m means that a smaller modulation frequency is causing a relatively larger swing in frequency.

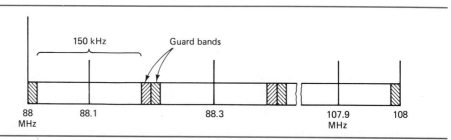

Figure 6.4 Commercial FM broadcast band allocations and sidebands.

Narrowband FM

Broadcast FM allows high fidelity and relatively low distortion, but requires considerable spectrum bandwidth for each station. Bandwidth is always a resource that must be conserved, and many applications that can benefit from the noise resistance of FM do not need the signal fidelity. For example, police, fire, and taxi radios need minimum static to make sure that the message gets through—but the voice must only be understandable, not necessarily recognizable.

This can be achieved with a form of FM called *narrowband FM (NBFM)*, developed for these applications. NBFM uses low modulation index values, with a much smaller range of modulation index across all values of the modulating signal. An NBFM system restricts the modulating signal to the minimum acceptable value; which is 300 Hz to 3 kHz for intelligible voice (but the voice may not be recognizable, which is acceptable in the application). The user is allocated anywhere from 10 to 15 kHz of spectrum (sometimes less), depending on the frequency band.

Example 6.6

What is the modulation index for NBFM when the modulating signal varies from 300 to 3000 Hz and the user has 12 kHz of spectrum ($\pm$ 6 kHz deviation)?

Solution

At 300 Hz, $m = 6{,}000/300 = 20$; at 3000 Hz, $m = 6{,}000/3000 = 2$.

The *deviation ratio*—the ratio of the maximum carrier frequency deviation to the highest modulating frequency used—is 10 kHz/3 kHz = 4 for this NBFM case. For standard commercial broadcast FM, the deviation ratio is 75 kHz/15 kHz = 5. *greatest BW*

Review Questions

1. How does the equation of FM differ from the equation of AM?
2. What is the modulation index? What ratio does it represent?
3. Why is the FM equation difficult to analyze, requiring use of Bessel functions?
4. What is the true bandwidth of an FM signal? Why?
5. What is a practical bandwidth for an FM signal? How can it be approximated?
6. What happens to the power in the carrier after frequency modulation?
7. Explain the relationship among modulating frequency, Δ, m, and bandwidth.
8. What is the difference between standard FM and NBFM? Where is each used? Why?

6.3 TRANSMITTERS

There are many ways to design an FM transmitter that will provide the necessary frequency swing in relation to the modulating signal. Each of these transmitters can provide different combinations of performance in various applications. In some situations, absolute stability of the carrier frequency is not critical, and a simple oscillator with variable frequency can work well. In other cases, the basic carrier frequency must be absolutely accurate and stable. This requires a crystal-controlled oscillator—yet the goal of frequency modulation is to vary this "rock-solid" carrier, which is in conflict with the stability of the crystal itself. We can now look at a few types of FM transmitters.

Direct FM

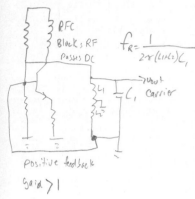

One straightforward way to vary the frequency of a carrier is to use an oscillator whose resonant frequency is determined by components that can be electronically varied. The varactor diode used as a tuning circuit component in Chapter 5 can be controlled by a modulating signal. The variations of the modulating signal are applied to the diode bias, which results in capacitance changes. These, in turn, vary the resonant frequency of the oscillator. By choosing the right combination of voltage change on the varactor, capacitance range of the varactor, and LC combination ratio, the oscillator can provide the necessary modulation index for the application.

Another direct approach is the *variable reactance* method. An active component, such as a transistor or MOSFET, is employed in such a way that it causes an apparent change in reactance (the frequency-determining aspect of the oscillator) in a circuit. The modulating signal is applied to cause the value of desired transistor characteristic to change, which in turn affects the reactance as seen by the circuit (Figure 6.5). The L_1–C_1 tank circuit combination determines the oscillator frequency. Normally, RF current in the drain circuit of the modulator MOSFET is in phase with the grid voltage, and 90° lagging with respect to the current through C_3 (the effective input capacitance of the MOSFET) and the tank voltage. This lagging current is drawn through the oscillator tank, yielding the same effect as if an inductance were connected across L_1–C_1. A modulating voltage applied to the MOSFET gate varies the transconductance of the transistor and so causes the RF drain current and apparent inductance across the tank to vary. The oscillator frequency is thus changed by the modulating signal amplitude.

Indirect FM

The relatively simple design of the direct FM transmitter has a major drawback. The exact carrier frequency is determined by resistors, capacitors, inductors, and transistors. These components have inherent variations in manufactured value, and so cannot provide a carrier frequency that is exact, without first being calibrated. This may not be a major problem where the components can be calibrated and adjusted once by skilled technicians, if needed, to get the precise carrier

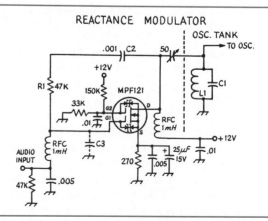

Figure 6.5 Reactance modulator using MOSFET (from *ARRL Handbook*, American Radio Relay League, 1972, Fig. 14-5a; courtesy of the American Radio Relay League).

6.3 Transmitters

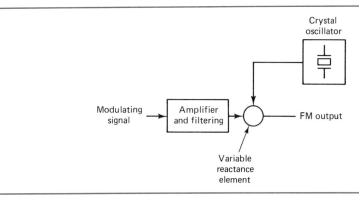

Figure 6.6 Indirect FM transmitter using crystal for stability and accuracy of carrier.

value. It is a problem for a transmitter that has to change carrier frequencies while in use, such as in a mobile radio used by police, fire, and taxis.

Even if the transmitter frequency is fixed and can be calibrated once, the unavoidable drift in the carrier due to temperature changes, component aging, and voltage variations would be unacceptable. For example, the Federal Communications Commission (FCC) rules for commercial FM broadcast stations require that the carrier, which ranges between 88 and 108 MHz, be within ±2 kHz of the correct value. This is certainly a very small allowable drift, yet this stringent limit is necessary to prevent interference in tightly packed parts of the spectrum.

The solution to this dilemma is *indirect FM*, where the basic carrier frequency is determined by a crystal oscillator circuit which is inherently more accurate and stable than any combination of passive and active components. The challenge, then, is to make use of the fundamental stability of the crystal, yet still vary the carrier frequency as required by FM. One indirect FM transmitter circuit is shown in Figure 6.6. The phase of the crystal oscillator is varied by changing the phase angle of an *RC* network, comprised of a capacitance and the resistance of the FET. Although the crystal oscillation frequency is unchanged, its phase variations can be related to frequency modulation, since phase and frequency are closely tied together (as discussed later in this chapter).

The amount of frequency shifting that can be obtained from various indirect schemes is relatively low, since the crystal is a very sharply tuned circuit element. A shift of up to 50 to 100 Hz per megahertz of nominal crystal frequency is the maximum practical amount, and crystals cannot be made to oscillate reliably and without drift above approximately 10 to 20 MHz. Fortunately, FM transmitters can use a variety of schemes to provide both the carrier frequency and deviation needed. *Frequency multiplication* is the key. A frequency multiplier is an amplifier circuit with an output that is the same waveform shape as the input, but at the original frequency times 2, 3, 4, and so on. Multipliers with factors of ×2 and ×3 are most practical. All frequencies in the carrier and sidebands are multiplied by the same amount. For example, a 1-MHz carrier with sidebands spanning ±5 kHz will become a 2 MHz ± 10 kHz signal after passing through a ×2 multiplier.

A multiplier can be built in several ways. One way is to use a very nonlinear amplifier (such as class C), one whose output distorts the original input signal and thus contains many harmonics. This output is filtered and the desired harmonic (second or third, corresponding to ×2 or ×3) is passed; all other frequency components are severely attenuated.

An IC multiplier can also be used. If both inputs to the IC multiplier are the same signal (cos ϕ), the output is the input value squared (cos^2 ϕ). By trigonometry, for any value of angle ϕ,

$$\cos^2 \phi = 0.5(1 + \cos 2\phi)$$

so squaring produces a dc component (which is easily filtered) plus a spectral component at twice the original frequency.

The basic FM process is done at a relatively low frequency, where circuit component values are more manageable and the performance is more stable. Then one or more stages of multiplication are used to scale the carrier and deviations up to the desired final values. For example, a 2-MHz carrier with ±100 Hz deviation can be repeated by multiplying and filtering with ×2, ×3, and ×3 factors to produce a carrier at 36 MHz ± 1800 Hz, which is 18 times the original value.

Multiplication affects both the carrier and the deviation by equal factors, but the resultant deviation is still fairly small. Large values such as the 75 kHz required for commercial broadcast FM can be obtained with some clever schemes. The multiplied carrier plus deviation side frequencies are mixed, just as in AM, with an oscillator frequency close to the "multiplied up" carrier. The result of mixing is, of course, sum and difference frequencies. By going through the calculations, we can see that the result is that the difference frequency has the carrier shifted down to a lower frequency, but the deviation value is unchanged. The new, lower-frequency carrier and the deviation are then multiplied again by the necessary ×2 and ×3 factors, to produce a final carrier and deviation of the desired values.

To summarize the steps for indirect modulation:

Modulate a low-frequency carrier, resulting in a small amount of deviation around the carrier value.

Multiply this carrier plus small deviation by ×2 and ×3 factors, to increase both by the multiplication factor.

Bring the multiplied carrier back down in frequency by AM mixing and low-pass filtering, while the deviation remains at the multiplied value.

Multiply the carrier and its deviation again by ×2 and ×3 factors, to bring both the carrier and deviation up to the final values.

Example 6.7

A 1-MHz carrier is modulated with a resulting 100-Hz deviation. It undergoes ×36 multiplication, followed by mixing with a 34.5-MHz signal and remultiplication by 72 (Figure 6.7). What is the final carrier and its deviation?

Solution

After ×36 multiplication, the carrier is at 36 MHz with 3600 Hz deviation. The output of the mixing with 34.5 MHz is a sum signal at 34.5 + 36 = 70.5 MHz and sidebands with a deviation of ±3600 Hz; the difference signal of 36 − 34.5 = 1.5

6.3 Transmitters

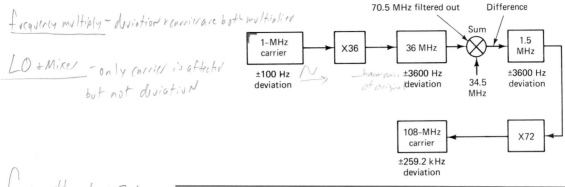

Figure 6.7 Use of multipliers for FM generation attempts to overcome limitations of crystals as frequency control element.

MHz, also with ±3600-Hz sidebands. The difference signal is multiplied by 72 and the sum signal is removed with a low-pass filter. The ×72 multiplication yields a final carrier of 1.5 × 72 = 108 MHz with a deviation of 72 × 3600 = 259.2 kHz.

Multipliers can be used for both direct modulation or indirect modulation. Figure 6.8a shows the stages of an FM transmitter with direct modulation, while Figure 6.8b shows the stages in indirect modulation with a crystal-controlled carrier. The complete series of stages for producing the FM signal with the desired carrier and deviation is the *exciter*. The output of the exciter is the desired signal, but usually at lower power than needed in the application. The output of the exciter is fed to a power amplifier, which simply boosts the FM signal to the necessary level.

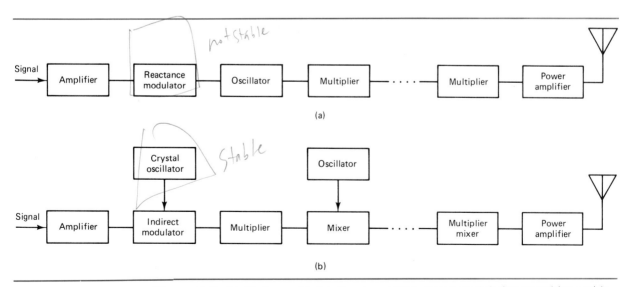

Figure 6.8 Multipliers with (a) direct and (b) indirect modulation to achieve wide deviations and higher carrier frequencies.

This separation of exciter function and final power amplifier shows a key difference between AM and FM transmitters. In AM, there is a choice to be made between low-level and high-level modulation. In high-level modulation, to produce full modulation the modulating signal has to be comparable in power to the final carrier power. However, it is not desirable or practical to do frequency multiplication, signal filtering, and other forms of signal processing at these high power levels, which may be tens, thousands, or even millions of watts.

In contrast, all FM exciter functions are done at the lower power levels of ordinary circuitry. The modulation (whether direct or indirect), the frequency multiplication, and the filtering are all done with low voltages and standard electronic components. This makes the design easier and less costly, and means that large amounts of power do not have to be supplied or excess heat removed. Only the final amplifier stage (or stages) which takes the exciter output and boosts it to the required level is a high-power circuit. In modular transmitter designs, the same exciter stage is used for various final output power levels. Only the power amplifier is changed to suit the application. This makes the design and maintenance of an entire series of transmitters relatively simple.

Stereo FM and Transmitters

The ability of FM systems to provide low-noise, high-fidelity music broadcasts soon led to the next phase of audio enhancement: *stereophonic* or *stereo FM*. In stereo systems, two signals which together contain information for the left and right speakers are sent out simultaneously. The idea of stereo is to provide a sound wavefront in the listener's room, which replicates the depth and realism of an actual, in-person performance. Two separate broadcast transmitters could be used, but this would occupy twice the bandwidth, be incompatible with single-channel systems (*monophonic*), require special receivers with independent tuners, and present various practical operating problems if the two transmitters were not located in the same physical place. Instead of this simple but impractical approach, stereo FM uses a much more effective scheme (Figure 6.9).

The total signal broadcast consists of a left channel (L) and a right channel (R) from microphone, disk or tape player, or any other source, each with bandwidth from 30 Hz to 15 kHz. These L and R signals are combined through simple circuitry to form a *sum signal* $L + R$ and a *difference signal* $L - R$. This is done by using a *matrix* circuit with two adders which combines two analog signals: $L + R$ is the sum of L and R, while $L - R$ is the sum of L and $-R$, which requires simply inverting R before summing. The $L + R$ signal and $L - R$ signal are *encoded* versions of the basic L and R information. Note that the result of the $L + R$ and $L - R$ encoding contains new amplitudes but no new frequency components.

The transmitter next must modulate the carrier. The $L + R$ signal is used to modulate the carrier just as a nonstereo signal does. The $L - R$ signal, however, is used differently. Its frequency is shifted by a balanced modulator using a special carrier (usually called a *subcarrier* to distinguish it from the broadcast frequency carrier) at 38 kHz. The output of the balanced modulator is the $L - R$ signal, DSB with suppressed carrier, with a lower sideband spanning $(38 - 15) = 23$ kHz (from 15 kHz up to 30 Hz below 38 kHz) and an upper sideband from 30 Hz above 38 kHz, up to $38 + 15 = 53$ kHz. This operation *frequency multiplexes* the information, allowing the basic bandwidth to contain several signals instead of just one, without increasing the required bandwidth. For this reason, stereo FM is also

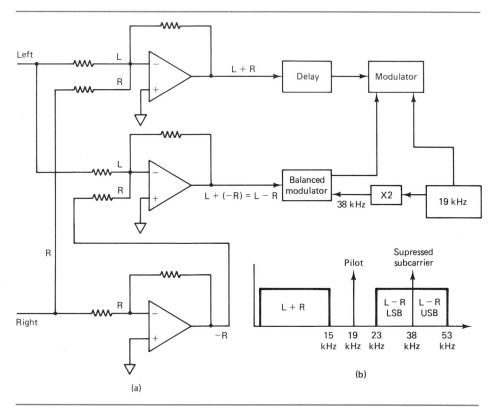

Figure 6.9 Stereo FM transmitter: (a) block diagram; (b) resulting spectrum.

known as *MPX* or *multiplex*. Of course, this multiplexing can be done only if the available bandwidth is not being used to its capacity by the first signal.

The 38-kHz subcarrier is derived by frequency doubling a 19-kHz *pilot signal* generated at the exciter. A 38-kHz value puts the $L - R$ signal, after multiplexing, into an unused portion of the available channel bandwidth. The final signal to the modulator has three components which are added together: $(L + R)$, the DSB-SC modulated $(L - R)$, and a small amount of the 19-kHz pilot tone (needed by the receiver to demodulate the $L - R$ signal properly) with an overall spectrum as shown in Figure 6.9.

Before going to the modulator, the $L + R$ signal is delayed by an amount equal to the very small but still significant delay of the $L - R$ signal through the balanced modulator circuitry. This preserves the precise time and phase relationship that the original $L + R$ and $L - R$ signals had with each other. The transmitted pilot tone energy is about 10% of the total, so it reduces slightly the total available power for actual information.

This all seems fairly complicated, compared to simply sending out two independent signals, one for L and one for R. There are several advantages, though, which make it a very good method. First, it is compatible with mono systems. A mono receiver is designed to demodulate FM signals with the information bandwidth of 30 Hz to 15 kHz. If an ordinary mono receiver demodulates a stereo signal, it will see the $L + R$ sum information, which represents the complete signal. Only receivers specifically designed for stereo (discussed in the next sections) need to have circuitry to also demodulate $L - R$ and combine it with the $L + R$ to recreate and *decode* the separate L and R signals. The 19-kHz pilot tone is above the 15-kHz bandwidth of $L + R$, and so can be filtered out easily. Second, the design for stereo requires no more bandwidth than the mono design since it occupies otherwise unused bandwidth. Finally, a transmitter being changed from mono to stereo only has to upgrade its exciter, with no changes to its basic carrier oscillator or power amplifier. These are all major advantages.

[Note that stereo broadcast is not inherently limited to FM. Several schemes are proposed to broadcast mono-compatible stereo signals in the standard AM band, using various matrixing approaches. The FCC has authorized some forms of stereo AM broadcasts, and some U.S. stations are trying it. However, the relatively high noise (AM is noisier than FM) and low fidelity (the narrow AM bandwidth limits true sound quality) mean that the improvements offered by AM stereo may not be worth the effort. So far, the public has shown little interest in AM stereo.]

FM Transmitter ICs

Most FM transmitters now use ICs for the majority of their required functions, except for some passive components and the power amplifier. The Motorola MC1376 monolithic IC (Figure 6.10) is a complete FM modulator in an 8-pin dual-inline package (DIP) which is intended for applications such as cordless telephones. The carrier frequency can be set between 1.4 and 14 MHz, and an auxiliary transistor which is fabricated on the IC die can be connected to boost the signal level to the antenna up to 600 mW from a 12-V supply. The heart of the modulator is a voltage-controlled oscillator (VCO), which is an oscillator whose output frequency is controlled by the applied voltage. The relationship between the VCO output and the modulating input is shown in the figure and is fairly linear over the range of 2 to 4 V. For a nominal carrier of 1.76 MHz, the frequency shifts down to 1.55 MHz with 2 V applied, and up to 1.9 MHz with 4 V applied. The nominal carrier frequency is set by the value of L_1 acting with the 6.0-pF internal capacitance and any circuit board stray capacitance.

Review Questions

1. What is direct FM? Indirect FM? What is the principal difference between them?

2. How does indirect FM resolve the conflict between a stable carrier frequency and the need to vary the carrier as part of the FM process?

3. Why are frequency multipliers needed? What role do they serve?

4. How is mixing used to generate increased deviations that multiplication alone cannot provide?

5. Explain how multiplication changes both the carrier and deviations by the same factor, while mixing translates the carrier but leaves the deviation unchanged.

6. What is the exciter? What is different about the signal generation and modulation for FM in the exciter compared to AM?

7. Why is FM used for stereo? Why aren't two transmitter channels used, one for left and the other for right?

8. What are three advantages of using the encoding/decoding of L and R, with respect to receivers and compatibility?

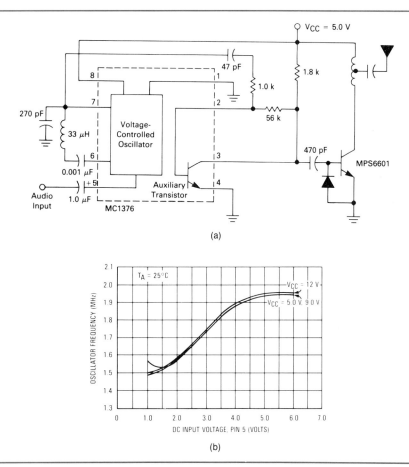

Figure 6.10 (a) MC1376 FM transmitter and (b) VCO frequency versus modulating input voltage (courtesy of Motorola, Inc.).

6.4 RECEIVER FUNCTIONS

There are many similarities as well as key differences between the AM and FM superhet receiver. In this section we concentrate on the areas where the receiver functions must differ since the information is carried as variations in frequency, not amplitude.

The RF stage, mixer, and LO of the FM receiver are basically the same as for the AM receiver. The function of these stages is to produce an IF signal, which can then be amplified and demodulated. The LO frequency, of course, is determined the same way that it is for an AM system since the type of modulation has no effect on how the IF is produced. For standard broadcast band FM, an IF of 10.7 MHz is most common, and it provides good image rejection and performance even in a single-conversion superhet. AGC is used to try to keep the signal levels approximately constant in the RF and mixer. While it may seem that since amplitude variations are not a concern in FM, the circuitry of the mixer stage and subsequent IF stage will perform better if the signals stay within a predefined range. This increases the SNR (always desirable), minimizes the effects of noise within the receiver, and allows the receiver designer to optimize various component values and voltages, based on the typical signal voltage values expected.

Limiters

The IF signal needs to be filtered and amplified to reject undesired adjacent signals, just as in the AM case. However, while the information is contained in the frequency variations, the circuitry that demodulates these frequency variations is unavoidably sensitive to amplitude variations. These variations cause less than ideal performance of the demodulator and distortion in the demodulated signal. The function of the *limiter* is to eliminate these amplitude variations.

A limiter acts as a small-scale AGC circuit. The gain of the limiter stage automatically adjusts, based on the input amplitude, to make the average output level approximately constant. If the input increases by 10%, the limiter gain decreases by 10%; if the input decreases to half, the limiter gain doubles. The input-to-output transfer function of the limiter and the result of this action are shown in Figure 6.11, where the FM signal contains both the desired frequency variations along with the unavoidable amplitude variations due to static, fading, and noise. Since these AM variations contain no information, but can obscure the desired variations and make recovery more difficult, it is far better to remove them. The limiter action smoothly and continuously reduces the gain as the input signal increases, and increases the gain as the input signal decreases in amplitude. This *soft limiting* produces an output of nearly uniform average amplitude.

The limiting action of a real limiter circuit does not cut in exactly at 0 V. Typically, limiting begins at some small value, such as 10 or 20 mV. The onset of limiting action causes the FM receiver suddenly to go *quiet* and background noise disappears, as limiting takes place and the receiver begins "normal" operation. This defines the sensitivity of the receiver. The amount of RF input from the antenna needed for full limiting and consequent *quieting* is used to express the receiver sensitivity. A typical specification is 3 μV input for 25-dB quieting. The

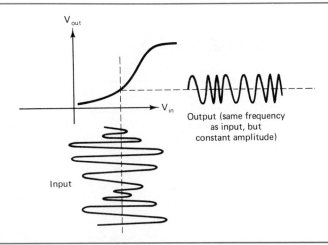

Figure 6.11 Input to output transfer function of a limiter and the effect on an FM signal with undesired amplitude variations.

Automatic Frequency Control

One interesting aspect of FM reception that is quite different from AM is the need for the extremely good frequency stability in the LO. Any deviations from the correct carrier frequency will result in highly distorted demodulation. In standard broadcast AM the drift problem is less severe for two reasons: the operating carrier frequency is much lower than the FM band, and it is easier to maintain good stability at lower frequencies; also, mistuning in the AM band results in less severe and annoying distortion than for FM. While the mistuned output in AM is usually understandable, in FM it is not.

To overcome this, the LO has circuitry designed to "pull" its operating frequency so that the signal is centered properly in the IF stage. This *automatic frequency control (AFC)* is more common on older FM receivers, which have greater drift than the newer, IC-based systems. While it may seem that AFC is always useful and therefore should be included on all receivers, it does have a drawback: A weak station carrier near a stronger carrier will be "crowded out" as the AFC pulls the LO toward the stronger carrier (*capture effect*). For this reason, many receivers with AFC have a manual switch that allows the AFC to be enabled or disabled (and note that the capture effect occurs to some extent even without any AFC, but as a result of the action of the limiter). Receivers that always have the AFC engaged are primarily useful for tuning strong, local stations, but not for receiving situations where signals, from many different transmitters, have received signal levels ranging from relatively weak to strong. Receivers that use a phase-locked-loop detector (discussed later) do not need AFC since the phase-locked-loop design copes with LO drift inherently.

quieting specification is used to show receiver sensitivity in terms of the amount of front-end input signal required to produce a specific amount of quieting, usually 20 or 25 dB. A smaller receiver input magnitude for quieting indicates a more sensitive receiver front end with greater amplification in subsequent stages before the limiter.

Example 6.8

A receiver limiter requires a 30-mV signal for quieting operation. The receiver is specified to have full quieting when the input at the antenna terminals is 5 μV. How much voltage gain is there between the RF input and the limiter, assuming equal resistance values?

Solution
The total gain must be 30 mV/5 μV = 6000 = 75.6 dB.

Review Questions

1. Why is AGC needed in FM, even though amplitude variations do not convey information?
2. What is a limiter? How does it work? Why is it needed? Where is it located in the overall FM receiver block diagram?
3. Explain the meaning of limiting action. What is a typical limiting specification? How does it indicate the sensitivity of the receiver?
4. What is AFC? Why is it needed in FM more than AM?
5. Why can AFC be disabled? Why do many newer receivers not have AFC?

6.5
FM DEMODULATORS

The demodulator, or detector, for an FM signal must extract the frequency variations from the IF signal. The FM detector is often called a *discriminator*, because it discriminates (selects) the changes in frequency from the nominal carrier frequency. There are many circuit designs that will act as an FM detector. We will examine a few of these—the slope detector, the Foster–Seeley and ratio detector, the quadrature detector, and the phase-locked loop-detector—to see how they operate and their strengths and weakness.

Slope Detectors

Slope detection is the simplest FM detection method. It makes use of the shape of the IF filter frequency response roll-off versus frequency. Recall from Chapter 5 that the IF filter could have different bandwidths (−3 dB) as well as shape factors and roll-off rates which indicated how sharp the filter was. Suppose that the IF signal center frequency is centered on one of the side slopes of the filter. Then, as the frequency varies, the filter output amplitude varies as well.

Figure 6.12 shows this as the transfer function (input-to-output relationship) for a filter. The input signal is shown as the input to the filter stage, and its changes in frequency are transformed into amplitude variations (corresponding to the original variations of the modulating information signal) by the output response versus frequency of the filter. The slope detector output is calculated by the roll-off of the filter versus the input frequency.

Example 6.9

A filter with a roll-off of 6 dB per kilohertz is used as a slope detector. The input signal varies with ±3 kHz deviation from center carrier frequency. How many dB down is the output at full deviation? At half deviation? For an input of 100 mV, what is the slope detector output at these two deviations?

Solution
With 6 dB/kHz of roll-off, the output is down 3×6 dB = 18 dB at full deviation and $\frac{3}{2} \times 6$ dB = 9 dB at half deviation. For a 100-mV signal, −18 dB corresponds to 12.6 mV, while −9 dB is 35.5 mV.

This simple scheme seems to do the job, but in reality it can give only fair performance, for several reasons. First, the limiting action must be very good, since any amplitude variations in the input will result in amplitude variations in the output, regardless of the frequency variations. The slope detector assumes that the input amplitudes are all the same, yet for any filter, the filter output amplitude (in absolute volts) depends on both the input amplitude and the input frequency.

Second, the relationship between the input frequency changes and the output amplitude that has been demodulated is only as linear as the straightness of the filter slope. If the filter slope has any deviation from a straight line, the same amount of frequency variation will cause differing amounts of amplitude change, depending on where on the slope these changes occur. This nonlinearity means

6.5 FM Demodulators

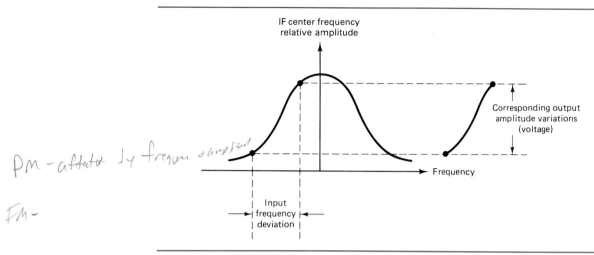

Figure 6.12 Transfer function for FM demodulation using slope detection.

that the recovered signal will have distortion, which negates the high-fidelity capability of FM. To avoid this problem, the filter must have slopes that are perfectly straight lines, which is extremely difficult to design and implement, and may conflict with other objectives of the filter.

Finally, the signal range over which the slope detector can operate is very limited. The allowable frequency changes can span only the frequency range spanned by the slope itself, which is usually relatively small for a sharp filter. Therefore, the slope detector is suitable only for narrow deviations in frequency.

The slope detector is used, for these reasons, where distortion and fidelity are not critical, such as in FM used for voice in police and fire radios. Modern ICs can provide more advanced detectors with better performance and at about the same cost, so the slope detector is no longer as common as it once was. It is a scheme sometimes used to convert an AM receiver into an FM receiver in an emergency: The last IF filter is "detuned" slightly so that it is no longer centered on the IF frequency. This last stage of the IF then becomes an IF amplifier + slope detector, although it is capable of only poor to fair performance.

Foster–Seeley and Ratio Detectors

These detectors are similar in operation, except that the ratio detector is much less sensitive to variations in amplitude than the Foster–Seeley (sometimes known as the *phase discriminator*). Both require transformers to provide a balanced signal path; the inductances of both of the transformer windings also form part of two LC tuned circuits (Figure 6.13). The analysis of the operation of these detectors is complex, but the general idea is this. At the resonant frequency (equal to the nominal IF frequency) the primary and secondary voltages are exactly 90° out of phase. When the input signal is greater or less than the resonant frequency, the voltages are less than or more than 90° out of phase with the input. The phase differences cause the *RC* circuits to charge more or less, depending on the phase differences. The diodes allow signal to flow only in one direction, and cut off the capacitors from discharging when the diodes are back biased. When the input frequency is equal to the IF, the net output voltage is 0 V, whereas it is negative when the input frequency is below the nominal IF frequency and positive above it.

180   Chap. 6   Frequency and Phase Modulation

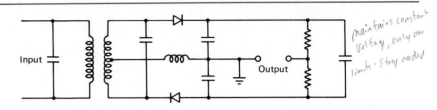

Figure 6.13   Ratio detector for FM demodulation.

The Foster–Seeley and ratio discriminators can give good results, especially if the Foster–Seeley is preceded by a good limiter stage. The major drawback to these two designs is that they require transformers for coupling and balance. Modern designs prefer to use ICs for lower cost, higher performance, and greater reliability, yet a transformer cannot be fabricated into an integrated circuit chip design. This has led to increased use of the quadrature detector, discussed next, and phase-locked loop.

Quadrature Detectors

Quadrature detectors require no transformer and can be built as part of the overall FM receiver IC, if desired. The idea of a quadrature detector is to compare the input signal to be demodulated with its phase-shifted version. The phase shifting is done by an *LC* combination which is resonant at the nominal IF frequency and so provides a shift of precisely 90° via the phase-shift capacitor, at that frequency only (Figure 6.14). The amount of phase shift will vary from exactly 90° as the received signal frequency differs from the *LC* resonant frequency. The term *quadrature* refers to the use of signals whose phases are at right angles (90°) to each other. The abbreviations *I* and *Q* are often used in literature for the original in-phase signal and the quadrature, 90° phase-shifted version.

In operation, the quadrature detector uses the modulated IF signal and its phase-shifted version to control the flow of current through two transistor switches. When the two signals are in perfect quadrature (exact 90° phase differ-

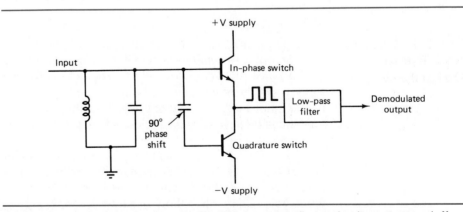

Figure 6.14   Quadrature detector for FM demodulation varies its output on/off ratio with input frequency deviation from nominal value, and is then filtered.

6.5 FM Demodulators

ence) because the signal to be demodulated is at the carrier frequency, the on/off ratio of the switching action is 1:1, and a series of square-wave pulses is produced. The average value of these square waves is 50% of full scale. As the received signal frequency differs from the nominal value, the amount of phase shifting varies above and below 90°. The switching action now produces pulses that are not square waves but have on/off ratios of less than 1:1 or greater than 1:1. The average value of the pulses then goes below half scale, or above it, in direct proportion to the phase shift that resulted from the frequency variations. The conversion from the on/off ratios to the average value of the pulses is made by simple low-pass filtering (RC) of the pulse train output.

A complete quadrature detector circuit requires some additional components to give performance that is linear and consistent over the desired range, with output voltage proportional to frequency deviation, but this is not a problem with ICs. In IC design, it is relatively simple and usually desirable to add extra transistors and related components to compensate for nonlinearities, temperature effects, and other imperfections, instead of requiring costly external components such as transformers plus manual trimming and adjustment. For these reasons, the quadrature detector is often used in modern IC FM demodulators.

The CA3089 FM IF System IC

One very popular IC used in FM receivers is the CA3089 (Figure 6.15a). This IC provides IF amplifier, quadrature detector, an audio-frequency preamplifier, and associated circuitry for AGC, AFC, squelch, and a tuning meter in a single 16-pin package that operates from a single +8.5- to +18-V supply. It requires only a few external components (resistors, capacitors, inductors) and has excellent limiter sensitivity of 12 μV for −3-dB quieting (measured at the input to this IC—a preamplifier ahead of it in an RF stage will result in greater sensitivity, such as only 2 μV needed for −3-dB quieting). The IF amplifier has three stages, while the quadrature detector uses a balanced modulator design for low distortion in the recovered signal, typically less than 0.1%. Figure 6.15b shows how this FM IF system is used in a typical FM receiver. Note that many of the features, such as the tuning meter and the squelch, do not have to be used in the basic FM receiver.

Review Questions

1. What is slope detection? How does it work? What are its good and bad points? Where is it used?

2. What are the prime characteristics of the Foster–Seeley and ratio detectors? What is the main difference between them?

3. Describe the general operation of the Foster–Seeley and ratio detectors.

4. Why is the quadrature detector now favored over Foster–Seeley and ratio types? How does the quadrature detector operate?

5. How does the quadrature detector convert frequency differences to a series of pulses and then to a corresponding amplitude variation?

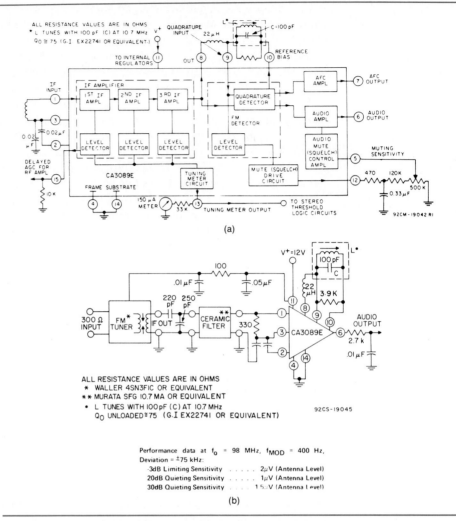

Figure 6.15 CA3089 FM IF: (a) IC block diagram (note its internal complexity); (b) as used in receiver (note the final simplicity) (courtesy of Harris Corp., RCA Div.).

6.6
THE PHASE-LOCKED LOOP AND STEREO DEMODULATION

One of the most effective circuits for demodulating an FM signal is the *phase-locked loop* (PLL). The PLL is capable of outstanding performance under many difficult conditions. It is used not only in demodulation, but in LOs that track received carrier drift (sometimes unavoidable), in demodulation of signals with very low (or even negative) SNRs, in frequency synthesizers, in tracking filters, and in many aspects of communications system design. We can study the PLL principle and operation in FM demodulation for both mono and stereo to gain some appreciation of the PLL versatility.

The concept of the PLL has been studied extensively for many years, and there are several hundred books and technical articles available that analyze its operation in great detail, with many possible variations of some of the PLL specifics. As with other electronic functions that were understood for many years, the IC has made the PLL function very practical, with high-performance PLLs available in small, inexpensive, low-power, high-performance packages. The basic operation of the PLL involves the phase of a signal. This may seem unrelated to frequency and FM, but as we shall see in the next section, phase and frequency are closely related. For now, we can work with the statement that if the frequency of a signal is known, the phase is easily determined. The reverse holds true as well: Knowing the phase of a signal is just as good as knowing its frequency, since the phase can easily be converted to the frequency.

A basic PLL is shown in Figure 6.16. It consists of the following elements:

A *phase detector*, whose role is to allow the relative phase relationship of two inputs to be compared with each other. A larger phase difference produces a larger magnitude output. The phase detector is often implemented with a multiplier circuit, and the output is the product of the input signals.

A *low-pass filter*, to filter out the high-frequency components of the multiplication of the phase detector. The output of the filter is a dc voltage corresponding to the phase difference of the two phase detector inputs. This dc voltage acts as the control voltage to the next element, the voltage-controlled oscillator.

A *voltage-controlled oscillator (VCO)*, which has a natural, free-running frequency when the applied control voltage is 0 V (or some other nominal idle value), and whose frequency varies up and down around this free-running frequency in proportion to the control voltage.

These three basic elements combine to form a powerful system function in the PLL operation as follows: The input signal and the VCO signal are compared in the phase detector (multiplier), and the result is low-pass filtered to produce a dc signal proportional to this phase difference. The difference between them represents the error between the two signals. The error signal controls the VCO frequency. If there is no error (the two signals are matched in phase), the VCO control voltage is 0 (or at the nominal value) and the VCO frequency stays unchanged. If there is a difference, the error voltage is nonzero (or different than the nominal) and acts to cause the VCO output to change its phase (and frequency) so that it matches the input signal. The net effect is that this is a *closed-loop, negative feedback* system that serves to compare the VCO constantly against the input

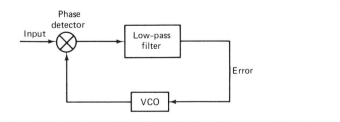

Figure 6.16 Block diagram of PLL shows its three major functional blocks.

signal and automatically correct the VCO phase for any difference between them.

[*Note*: In an actual implementation of a PLL, the VCO and the input are considered matched when they actually are at quadrature, with a 90° phase difference between them. At quadrature, the control voltage is at the nominal value. As the phase difference shifts from 90° toward 0° the control voltage increases from nominal; as the phase difference shifts toward 180° the control voltage decreases from nominal toward its minimum value. This practical implementation does not change the concept of the preceding discussion.)

The error voltage is the key to the usefulness of the PLL. It represents the difference between what the input signal frequency (and phase) is, as compared to the VCO's last estimate of the input. By using the error signal in different ways, the input signal can be demodulated and manipulated. Note that the amplitude of the input versus the VCO amplitude is not involved. As long as the input voltage is within the range of the phase detector capability, the PLL will function properly. This is another advantage of the PLL: It is relatively insensitive to variations in signal level. Only the phase and frequency really affect the overall operation.

To examine a PLL used as an FM demodulator, consider a carrier at 1 MHz that has been frequency modulated by a 3-kHz voice signal, with a maximum deviation of 30 kHz. The VCO is set to free-run at the carrier frequency, while the error voltage from the phase detector and low-pass filter can span ±0.5 V and cause the VCO to swing ±30 kHz around the 1-MHz point. As the modulated signal is received, the closed-loop action of the PLL will cause the error voltage to constantly correct the VCO to track the input frequency at any instant. The amplitude of the error voltage, therefore, is exactly proportional to the frequency modulation of the carrier. As the received frequency changes due to the FM, the phase detector determines the amount of difference, an error voltage is formed, and the error forces the VCO to correct itself (Figure 6.17).

There are some critical elements in a good PLL circuit. First, the phase detector (multiplier) must produce a linear output (multiplication result compared to actual phase difference) over the range of both inputs. The low-pass filter is

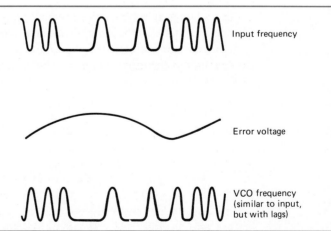

Figure 6.17 Input/output/error action of PLL shows how an error signal from a phase detector forces the VCO to follow the PLL input.

6.6 The Phase-Locked Loop and Stereo Demodulation 185

> **PLL as Filter**
>
> The PLL also offers another key advantage: as a precise, yet tunable filter. Using regular components, a traditional filter cannot achieve narrow bandwidths at high center frequencies: A 20-Hz bandwidth around 50 MHz is not realizable in practice. With a PLL, the center frequency is set by the VCO, while the bandwidth is set by the low-pass filter. These two are completely independent of each other, and a PLL can easily be designed for 20 Hz around 50 MHz (assuming that the VCO can run at 50 MHz). For this reason the PLL is often used in recovering data from space probes, where extremely narrow bandwidths are used to minimize received noise, although the carrier may be at a high frequency. Finally, a change in one parameter (either desired center frequency or bandwidth) is accomplished by changing only the VCO free-running frequency, or the low-pass filter bandwidth, without affecting the other parameter.

critical. It determines the bandwidth and stability of the PLL and its ability to lock quickly onto the input and follow it. A PLL with the wrong filter parameters will be sluggish in following the input changes, possibly overshoot the correct error value, or cause the entire PLL to hunt constantly but never settle to the proper value. A narrow bandwidth is preferred to minimize noise, while a wider bandwidth is needed for faster response and higher data rates. Fortunately, the filter can be analyzed "on paper" very thoroughly and matched to the anticipated signal. Virtually all IC PLLs use external components which are not part of the IC for the filter so that the designer can tailor the filter to the intended application.

Finally, the VCO must be linear and produce a constant change in frequency per volt of control signal (a typical spec is 1000 Hz/mV of control signal). Different PLLs offer a variety of VCO ranges. One model, for lower-frequency operation, has a VCO that can operate from 0 to 300 kHz (LM565), while the LM568 VCO operates up to 150 MHz and provides highly linear operation for deviations up to ±10% around a 70 MHz center frequency.

From this discussion, the full power of the PLL is not yet apparent. Consider this: What the PLL does is try to find some underlying pattern, or *coherence*, in the input signal, by the inherent action of the closed loop formed by the phase detector, low-pass filter, and VCO. PLLs are sometimes called *coherent detectors* for this reason, since they attempt to extract the coherence that may exist in the noisy, modulated received signal. A properly designed PLL can actually find a signal whose power is less than the noise (SNR < 0 dB) and track changes in the signal.

PLL Capture and Lock

When the receiving system is turned on, or a new signal first received, the PLL VCO may be at a frequency significantly different from the input. The range over which the VCO can lock onto a new signal is called the PLL *acquire* or *capture range*. Once a signal has been captured and the PLL is tracking it, the PLL is said to be in *locked* condition. The span of frequencies over which the PLL can remain locked and track a signal is called the *tracking* or *lock range*. The lock range is always greater than the capture range, since the error signal is better able to control the VCO once the VCO is already close to the input signal and needs only small adjustments to continue to follow the input. The tracking bandwidth is typically 10 to 100% greater than the acquisition bandwidth, depending on the PLL design and the filter used.

Example 6.10

A PLL with a VCO free-running frequency of 100 kHz has a 10% capture range and a 20% lock range. Over what frequencies will the PLL be able to capture and subsequently maintain lock?

Solution

100 kHz × 10% = 10 kHz, so the capture range is 100 ± 10 kHz, or 90 to 110 kHz.
100 kHz × 20% = 20 kHz, so the lock range is 100 ± 20 kHz, or 80 to 120 kHz.

Example 6.11

A PLL is used for tracking signals using a VCO that has a center frequency of 20 MHz. The capture range of this PLL and filter is 12%, while the lock range is 18%. What is the PLL operation with a 17.5-MHz input?

Solution

(20 − 17.5)/20 = 12.5%; therefore, the PLL will not be able to capture this signal. If it could, however, it would be able to track it afterward.

PLL Applications

The PLL is used in demodulating *telemetry* signals (data sent by radio), recovering data bits that have been frequency modulated onto a carrier, and decoding the *L* and *R* signals from stereo FM. In telemetry, standard channels and modulations have been set by groups such as the Inter-Range Instrumentation Group (IRIG). IRIG channel 13, for example, has a center frequency of 14.5 kHz and a maximum deviation of ±7.5%, with a bandwidth of 220 Hz, corresponding to a deviation ratio of 5. Figure 6.18 shows a demodulator for IRIG channel 13 with the LM565 PLL at the heart of this demodulator. The VCO of the LM565 is set to free-run at 14.5 kHz by the resistor connected to pin 8 and the capacitor to pin 9, while the low-pass filter bandwidth is set at 1890 Hz by components at pin 10. The output of the PLL is the error signal (an amplifier within the LM565 increases its magnitude for convenience) and the amplified error signal is directly proportional to the signal that modulated the carrier. This signal is applied to an LM107 op amp to increase the gain further, and shift its level to values compatible with the rest of the system. This particular implementation can recover information even with SNRs as low as −8.4 dB.

The PLL is also useful when the data is digital and has only two values. In *frequency shift keying*, a carrier is modulated between one of two frequencies, with one frequency representing a logic 1 and the other a logic 0. In one standard set of frequencies commonly used, 2025 Hz is a 1 and 2225 Hz is a 0. The PLL VCO is set to a nominal frequency midway between these, 2125 Hz. The loop filter is selected to handle the bandwidth that results from the rate at which the data is *keyed* between 1 and 0: lower keying rates require less bandwidth and can

6.6 The Phase-Locked Loop and Stereo Demodulation

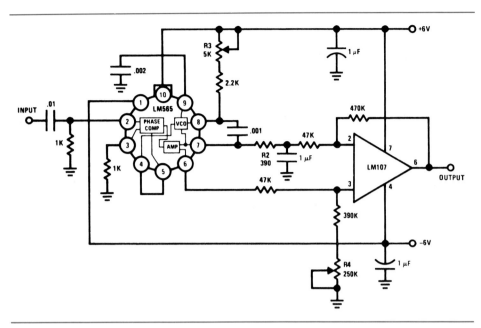

Figure 6.18 PLL demodulator for IRIG-13 (from National app. note, Fig. 17; courtesy of National Semiconductor Corp.).

use a lower-bandwidth filter. When the PLL receives the FM signal, the error signal changes as the received frequency shifts between 2025 and 2225 Hz (Figure 6.19). When the VCO is at 2025 Hz and the input shifts to 2225 Hz, the error goes positive to drive the VCO toward 2225 Hz and hold it there, offset by +100 Hz from the free-running frequency of 2125 Hz. If the input shifts to 2025 Hz, the VCO is shifted by a negative error signal to have an offset of −100 Hz from the

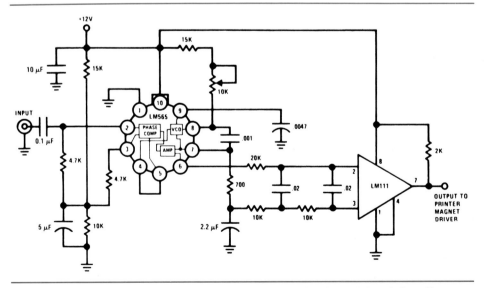

Figure 6.19 PLL used for FSK: the error signal is the demodulated data (from National app. note, Fig. 19; courtesy of National Semiconductor Corp.).

nominal 2125-Hz value. The result is that the error signal goes between a large positive voltage value for binary 0 and a large negative voltage value for binary 1. The original information signal has been recovered by the PLL demodulation process.

Stereo FM Demodulator

To demodulate the stereo FM signal properly, the 19-kHz pilot tone of the transmitter must be used as a reference so that the $L-R$ signal is precisely recovered. A PLL-based design is ideal for this, because the PLL can be used to filter the pilot tone from the received signal, and remove any noise that may be present. At the same time, the PLL will automatically track and synchronize to the nominal 19-kHz pilot tone, which may differ from a perfect 19 kHz due to the unavoidable imperfections and drift in any oscillator. In contrast, if the 19-kHz tone for the demodulator were generated at the receiver, the small differences between it and the signal used at the transmitter would result in poor demodulation and distortion. Significantly, the VCO output itself is what is used in this application. This is in contrast to use of the PLL for demodulation of the FM information itself, where the error signal and not the VCO output contained the desired information.

This problem of demodulating with an imperfect pilot signal is not unique to stereo FM. Whenever a receiver must demodulate based on the precise frequency and phase of a transmitted signal such as a carrier, any differences and drift between the transmitter signal and the receiver's version of it can cause problems (think about this and FM reception in general). The ability of the PLL to reconstruct the reference signal from the received signal itself is an important reason for the popularity of the PLL. Instead of simply creating a hopefully close copy of a critical frequency, a PLL allows the system to extract the one that was actually used and regenerate a stronger, noise-free version of it.

The LM1800 is an IC used for the stereo FM demodulation process (Figure 6.20). It contains the circuitry needed to recover the 19-kHz pilot tone as well as demodulating the received signal into $L + R$ and $L - R$. It then performs the reverse operation of the matrix that created these left and right sum and difference signals, with a final output of L and R, just as were used to modulate the transmitter. As added benefits, the LM1800 provides as outputs both a signal for a stereo indicator lamp, to show that the received signal has a 19-kHz pilot tone and is therefore stereo, and the reconstructed 19-kHz pilot, useful in test and receiver adjustments. Internal circuitry in the LM1800 checks the level of the received pilot signal. If it is too low to recover properly, the LM1800 automatically switches to mono mode and demodulates only the $L + R$ signal, which does not require use of the pilot tone.

The block diagram of the internal functions of the LM1800 reveals some of the complexity that modern ICs allow. The received signal, already demodulated by a PLL or ratio detector as a 30-Hz to 53-kHz signal, is applied to pin 1. This signal goes to both the PLL, which recovers the 19-kHz pilot, and to the decoder, where the actual demodulation and recreation of L and R takes place. The PLL operates by locking onto 76 kHz, the fourth harmonic of 19 kHz. (The reason for using 76 kHz instead of 19 kHz directly is that a much better VCO can be designed in the IC at the higher frequency, and the required low-pass filter is easier to design using smaller-value components.) Filtering is provided by 0.22-μF and 0.47-μF capacitors and a single 3.3-kΩ resistor. The VCO frequency is set by an RC pair, which is less expensive and easier to adjust than an LC combination, since a potentiometer can fine-adjust the free-running frequency.

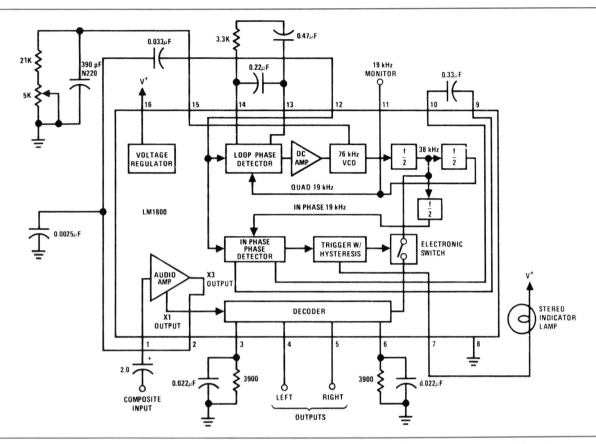

Figure 6.20 LM1800 stereo FM demodulator IC incorporates considerable amount of circuitry (from National data sheet, Fig. 2; courtesy of National Semiconductor Corp.).

The output of the VCO is a relatively large and noise-free signal that is passed through two "divide by 2" circuits, recreating a 19-kHz signal of the exact frequency and phase as the original pilot tone at the transmitter. This signal is used to demodulate the input, using a variation on the quadrature detector, to yield $L + R$ and $L - R$ signals. Adding the $L + R$ and $L - R$ signals together produces $2L$, which is the left-channel signal but with twice the amplitude. The two signals are also subtracted, $(L + R) - (L - R)$, with the result of $2R$. Of course, the final $\times 2$ demodulated amplitude factor is no problem, since the signal has to be amplified considerably anyway before going to the speakers; the original modulating information has been recovered, and that is what counts.

In this section we have examined the basic operation of the PLL and how it is used for FM demodulation. In later chapters we will see how the PLL is used to demodulate AM signals with carriers that are unstable, and to build frequency synthesizers that can create and digitally select any one of many frequencies from a single master oscillator. It may seem that an excessive amount of time has been devoted to the PLL, but the reality is that the PLL is one of the most useful functions in many communications systems because of its versatility and performance. Modern technology provides many types of PLLs, with a variety of VCO ranges and sensitivities, in easy-to-apply ICs.

Review Questions

1. What are the advantages in using a PLL versus traditional FM detectors?
2. In the PLL, what does the phase detector do? What about the low-pass filter and the VCO?
3. Explain the operation of the PLL as a closed-loop system. How does it produce a signal that tracks the input signal frequency and phase?
4. Why is a PLL relatively insensitive to signal amplitude?
5. Why is the design of the low-pass filter critical? What does it determine?
6. How can a PLL extract information from signals with low SNR?
7. How does a PLL allow tracking of a signal whose carrier drifts? How does the PLL allow separate setting of a filter center frequency and bandwidth?
8. What is the PLL capture range? Compare it to the PLL lock range.
9. Explain how a PLL is used to demodulate a carrier that has been shifted to one of two frequencies.
10. How is a PLL used in recovering the exact pilot tone frequency for stereo FM demodulation?

6.7 PHASE MODULATION

The frequency and phase of any signal are closely related parameters. Frequency, in fact, is the rate at which the phase changes. Mathematically, it is the derivative of the phase with respect to time. For this reason, frequency modulation and *phase modulation* have many similarities. They are often discussed together under the heading *angle modulation* because both involve changing the phase angle of the carrier, not its amplitude as in amplitude modulation.

In Section 6.2 we saw that an FM waveform is created by a frequency deviation that is proportional to the voltage of the modulating signal at a given instant. The modulation index m is the maximum carrier shift Δ divided by the frequency of the modulating signal, and the modulation index increases as the modulating frequency decreases while the modulating voltage amplitude remains constant. For PM, the modulation index m is defined differently: it is proportional only to the amplitude of the modulation signal, and is independent of the frequency of the modulating signal. The result of PM is that the frequency deviation of the carrier is not constant as it is in FM. Instead, the deviation is proportional to the amplitude and frequency of the modulating signal.

If the carrier is modulated by a single, constant-frequency signal, the difference between FM and PM is not seen. This is because the receiver will see a carrier that is now offset from its original frequency. Whether the cause of the offset is FM or PM cannot be determined by the receiver. However, any real system must transmit information by varying the modulating signal—after all, a constant, never-changing signal carries no information. As soon as the modulation waveform varies, the difference between FM and PM becomes visible: the FM modulation index will increase as the modulation frequency decreases, while the PM modulation index will remain the same.

To reiterate: For FM, the modulation index increases as the modulating frequency decreases; the deviation of the carrier is constant with frequency. For PM, the modulation index is constant with modulating frequency; the deviation is proportional to the amplitude and frequency of the modulating signal.

This points out another interesting and useful relationship between FM and PM. Suppose that the modulating signal has energy across a spectrum from 300 to 3000 Hz and is used in an FM transmitter, but the receiver is designed for PM. The lower frequencies then would have much greater phase deviation than they would have received from a PM transmitter, due to the different definition of m for FM versus PM. For a PM receiver, the demodulated output is directly proportional to the phase deviation (same as modulation index), so lower frequencies will have relatively greater amplitude than they had at the signal source. This is called *bass boost*. By the same reasoning, a signal that has been phase modulated at the transmitter will have reduced bass amplitude when received with an FM system.

This fact is often exploited to build an FM system using a PM transmitter, by first shaping the energy spectrum of the modulating signal before it is applied to the carrier. Increasing the bass signals prior to PM in the exact mirror image of how they will be reduced by the FM receiver results in a demodulated output that correctly reproduces the spectrum of the unmodified modulated signal. The original indirect FM transmitter used by Armstrong used this principle of *equalization*. The audio to be modulated was first bass-boosted, then used to phase modulate the carrier. The reason that PM was used at the transmitter, rather than FM, was that a stable, crystal-based carrier is relatively hard to frequency modulate but much easier to phase modulate.

PLL and FM/PM   In the discussion of the PLL, we used a phase detector to compare the phase and frequency of the received signal to the VCO output. A PLL can be used for demodulation of either FM or PM. Intuitively, the situation is this in a PLL: The goal of the closed loop of the PLL (phase comparator, low-pass filter, VCO) is to match the phase of the VCO output to the phase of the input. But inherently, if two signals are matched in phase, they are also identical in frequency. Therefore, the very act of tracking phase and phase changes also results in tracking the frequency and frequency changes. The PLL, which is essentially a phase-tracking circuit, also provides frequency tracking and can be used in phase- and frequency-dependent applications.

Review Questions

1. What is the relationship between phase and frequency? Why are both sometimes grouped under the term *angle modulation*?
2. What is the definition of modulation index m for PM? Compare to the definition for FM.
3. Explain the proportionality of the frequency deviation of a carrier under FM and PM, and the proportionality of the modulation index under FM and PM.
4. How can a PM transmitter be used for FM broadcasting and reception?
5. Why is a PLL suitable for either FM or PM demodulation?

6.8 COMPARISON OF AM, FM, AND PM

Amplitude modulation and angle (frequency/phase) modulation have technical advantages and drawbacks. In FM/PM, the modulation can be done at low signal levels and high-efficiency class C amplifiers used for power amplification. For AM, low-level modulation requires low-distortion, class A final amplifiers, but these are inefficient and waste power. As an alternative, the modulation of an AM carrier can be done after the carrier has been amplified to its final level by class C amplifiers, but then the modulator must be high power as well.

AM is much more noise prone than FM/PM, since the noise adds to the signal and thus corrupts the modulation itself which carries the information. In FM/PM, the amplitude variations are not relevant, and are in fact removed by the limiter in the receiver. Only larger amounts of noise begin to affect the phase or frequency of the received signal. In FM, the effect of noise can be reduced further by increasing the deviation, which spreads the information energy into a wider spectrum. Then, noise at any single frequency affects a much smaller portion of the total energy space and has a smaller relative effect. In effect, FM can overcome noise by using more bandwidth: In contrast, AM must overcome noise by increasing power while the AM bandwidth remains the same.

The FM/PM signal makes good use of the transmitted power, since all of the signal spectrum conveys information. This is unlike AM, where the major portion of the power is in the carrier, which carries no modulation information at all.

Unfortunately, the FM signal also requires much wider bandwidth than AM. The AM bandwidth is exactly twice the bandwidth of the modulating signal, while the FM/PM bandwidth is much wider than that of the modulating signal—typically, 10 times as wide. Of course, NBFM requires much less bandwidth than conventional FM, but the received signal has much greater distortion and would not be acceptable for many types of signals, such as music. Due to the wide bandwidth needs of FM/PM, the available frequencies for carriers are much higher in the electromagnetic spectrum, where these bandwidths are a smaller percentage of the spectrum. Note that the broadcast AM band is from 550 to 1600 kHz, while the broadcast FM band spans 88 to 108 MHz. This means that more sophisticated RF circuitry is needed and the physical layout of the circuitry design is more critical.

Another difference between AM and FM/PM is that AM is much easier to analyze and interpret, since the equations and instrument displays of AM are much simpler than those for FM and PM. The performance of systems with the different types of modulation for various circuit, signal, and noise conditions, especially as these change with time, can become very involved and may not be solvable on paper. Even the basic equations of FM and PM took many years to fully develop, whereas AM was fully understood early in the history of electronics.

Finally, the circuitry needed to modulate and demodulate with AM is much simpler than the circuitry for FM/PM. This means that lower-cost AM systems can be built (receivers and power transmitters), although modern ICs have made the difference less than it was in the past. At higher transmitter power levels, the major complexity and expense are in the final amplifier stages, which are similar for AM and FM/PM, not in the exciter, so there is little difference.

FM/PM and Noise

Up to this point, we have emphasized that FM (and PM) is inherently noise resistant. Amplitude variations caused by noise corrupt the AM signal, but while they also corrupt the FM signal it does not matter since the frequency and phase, not the amplitude, convey the information. This does not mean that FM is absolutely immune to noise. The effects of small and larger amounts of noise in an FM system are worth examining.

Noise can obviously cause problems if it is large enough, so that when added to the modulated waveform, it causes false cycles to appear (Figure 6.21a). The receiver assumes that the frequency cycle—the waveform crossing the zero or midway point in amplitude—is from the modulating signal. Small amounts of noise have a less dramatic effect. At the beginning of this chapter, we said that noise adds to or subtracts from the amplitude of the desired signal. Looking at Figure 6.21b, you can see how this addition/subtraction of noise can cause an advance or delay in the shape of the modulated waveform. The effect of smaller noise magnitudes is to cause phase shifts in the modulated signal. The total number of cycles in the signal is unchanged, but the phase of the signal has changed. We can say that noise appears to "phase modulate" the carrier. A full analysis of the effect of noise involves complex mathematics, with some differences for FM versus PM.

One interesting result of the analysis is that noise in FM affects higher modulating frequencies more severely than it affects lower frequencies. At the same time, most modulating waveforms contain less energy in the higher frequencies. The result is that the higher-frequency components—which are already weakest—are most seriously corrupted and have the poorest SNR, and overall performance falls off rapidly with increasing frequency.

Fortunately, there is a solution to the varying effect on noise on the different parts of the spectrum. The higher frequencies are boosted in amplitude before modulation occurs, and then reduced after demodulation. This boost is known as *preemphasis* and the corresponding reduction is *deemphasis*. As long as the pre-

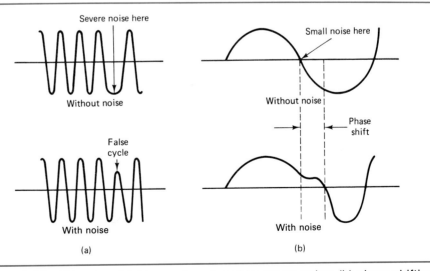

Figure 6.21 FM noise: (a) false cycles caused by severe noise; (b) phase shifting and phase error caused by small amounts of noise.

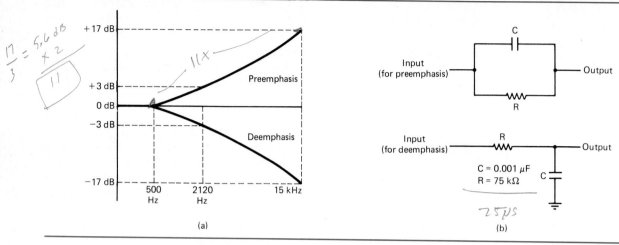

Figure 6.22 Standard pre- and deemphasis curves; (b) RC circuits for pre- and deemphasis.

emphasis and deemphasis are exact mirror images, the original signal will be recovered with its correct amplitude versus frequency. All standard broadcast FM stations use this scheme to increase the effective SNR at the higher frequencies.

The pre- and deemphasis time consistant is standardized in the United States to give a response curve as shown in Figure 6.22a and has a -3-dB point at 2120 Hz. It is called *75 μs preemphasis* because the *RC* time constant of the filter is 75 μs (recall that the 3-dB point is at $f = 1/2\pi RC$). The circuit that provides preemphasis is shown in Figure 6.22b together with the corresponding deemphasis circuit. This scheme is not used to boost effective SNR in AM because the effect of noise on AM is uniform across the spectrum of the modulating signal. Therefore, the advantage gained would be much less.

Review Questions

1. Compare where modulation is done for FM/PM versus AM. How do the approaches differ? What are the advantages and drawbacks of each?

2. How can an FM/PM system be made more noise resistant? What about AM?

3. Explain the use of transmitter power and signal spectrum in an FM/PM system. Does the same hold true for AM? Explain.

4. How does NBFM overcome the need for large bandwidth in conventional FM? What is the problem with this technique?

5. How does noise cause undesired phase shifting in FM and PM? At what part of the modulating signal spectrum is it more significant?

6. What are preemphasis and deemphasis? Where and why are they used? Why aren't they used for AM?

6.9 FM RECEIVER SYSTEMS

In this section we look at two receiver systems built using IC technology to minimize the number of components while providing performance beyond what can be achieved with more complicated circuits made up of dozens of discrete devices.

A complete single-conversion FM receiver using the MC3367 is shown in Figure 6.23. This IC is designed for use in low-power NBFM transmitter/receivers, called *transceivers* or *walkie-talkies*, which are usually battery powered. The IC contains a mixer, LO, two IF amplifiers, limiters, and a quadrature detector, plus an audio amplifier for the final output to headphones or a smaller speaker. It also contains a data buffer and data comparator, which can be used to properly filter and amplify data (not voice) for applications where the modulating signal is in binary form (1s and 0s, Chapters 10 through 13) representing letters and numbers that are received and then displayed to the user (in a message beeper such as used by doctors and emergency personnel, for example). The RF frequency to the mixer can be up to 75 MHz, and only one inductor, a few capacitors and resistors, and two 455-kHz IF filters are needed external to the IC. The mixer may be preceded by an RF amplifier, if necessary, although the sensitivity of 0.2 μV for -3-dB limiting is sufficient for many applications.

For maximum flexibility, the LO frequency can be set by either a lower-cost LC circuit or by a more stable crystal. The mixer output IF is 455 kHz, which means that this IC should be used only where image signals are not a problem. If they may be a problem, a dual-conversion design should be used, with a first conversion stage ahead of this stage at a higher IF, such as 10.7 MHz, which then feeds into the MC3367. There is also a low-voltage detector circuit built in, which monitors the battery voltage and signals if the voltage drops below the level needed for proper operation of the rest of the IC. The entire single-conversion FM receiver function requires only 1.1 to 3.0 V at a current of less than 3 mA (not counting the current driving the audio output). This makes the MC3367 ideal for radios that will be on continuously and monitoring relatively quiet channels for transmissions, such as police and public-safety walkie-talkies.

A complete AM/FM/stereo receiver can be built with only five ICs, shown in the schematic of Figure 6.24. The receiver uses a preassembled tuner to avoid the fairly complex design and layout of the high-frequency, low-noise front end, and includes an input and amplifier for signals from a tape deck or other audio source not part of the main receiver function. Tone controls (bass and treble) are also included. The three communication ICs include an LM3089 FM IF system (identical to the CA3089 of Section 6.5 but from another manufacturer), an LM1800 stereo demodulator, an LM3820 AM receiver. Two op amps provide necessary audio amplification: the LM382 dual preamplifier is used for low-level inputs from local audio sources (a tape player, for example) while the LM378 dual power amplifier supplies up to 3 W of audio power to each of the two speakers. If more power is needed, the LM378 can drive discrete transistor power amplifiers.

The performance specifications of this receiver include a sensitivity of 2.5 μV for 30-dB quieting (FM section), FM frequency response from 5 to 15 kHz, and distortion of less than 0.3%; for AM signals, the sensitivity is 20 μV for 20 dB SNR, and distortion below 2%. A signal strength meter with a 0- to 200-μA

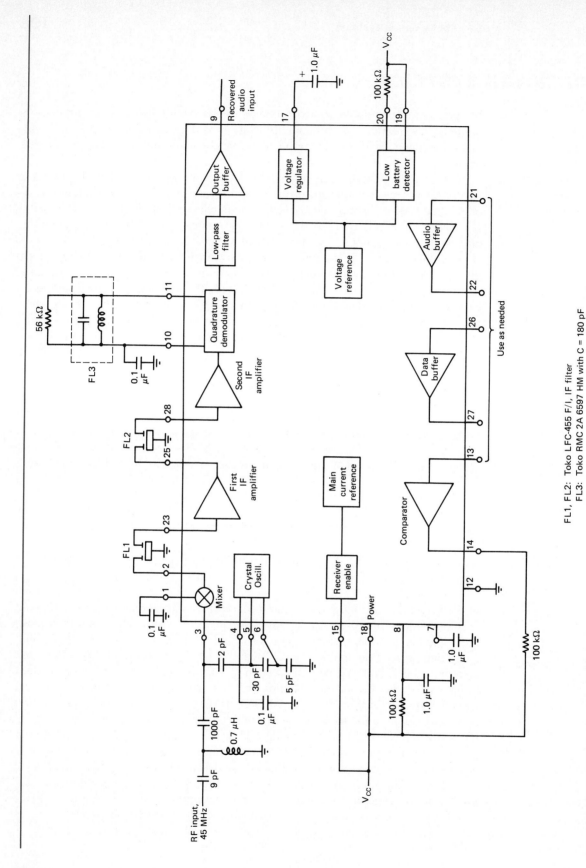

Figure 6.23 MC3367 IC is a nearly complete NBFM receiver (from Motorola data sheet; courtesy of Motorola, Inc.).

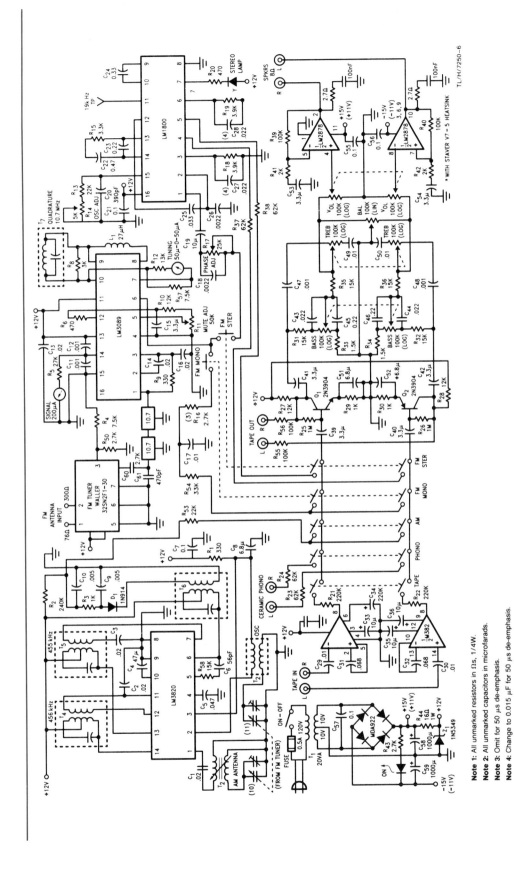

Figure 6.24 Complete AM/FM stereo receiver using five ICs (from National app. note, Fig. 6; courtesy of National Semiconductor Corp.).

movement is driven by an output of the LM3089, while the only critical adjustments are two trimming resistors for the LM1800 which set the VCO oscillator nominal frequency and correct for some unavoidable phase shifts within the IC.

Review Questions

1. Why is a low-power IC desired for the NBFM application of the MC3367?

2. What is the mixer output of the MC3367? When would a double-conversion receiver based on this IC be needed?

3. What ICs perform the real "communications" functions in this five-IC receiver? If the complete AM/FM/stereo receiver of this section were to be used only as a broadcast signal receiver and not for any other signal inputs, which IC could be eliminated?

4. If this receiver did not have to drive speakers but only had to power headphones, what IC would be deleted or replaced with a lower power version? Why?

6.10 FM TESTING AND EQUIPMENT

Testing FM systems requires equipment for checking the transmitter and receiver separately. The transmitter is checked with a spectrum analyzer that can display the frequency components of the signal. This is used to determine the modulation patterns, the deviation, and the modulation index. The table of Bessel functions is used in conjunction with the observed spectrum or may be built into a computer program that combines the spectrum analyzer results with some numerical processing of the data.

The receiver is checked with a test instrument which functions as a simulated transmitter that allows many key parameters to be set by a test technician. One such instrument is the Hewlett-Packard 8640B (Figure 6.25). This signal source can provide carrier frequencies from 500 kHz to 512 MHz, along with simultaneous internal AM and FM capability (used for checking the performance of the FM receiver limiter at eliminating amplitude variations of the FM signal). The FM deviation is selected by the user and can be up to 1% of the lowest frequency of the particular range that the 8640B is set on. For the 2 to 4 MHz range, the maximum deviation is 20 kHz; for the 64 to 128 MHz range (which covers broadcast FM) the maximum deviation is 640 kHz. The modulating frequencies are fixed at 400 Hz and 1 kHz, although the user externally may supply any frequency to the unit from another signal source.

The output level of the generator is calibrated and can be set to any value between +19 and −145 dBm, in 1-dB steps. The wide range of output allows the generator output signal to act as the received RF input at the antenna (very low level) or the signal in the mixer or IF stages (a much higher level). A built-in frequency counter can be used to measure the frequency of either the signal from the 8640B or from within the receiver under test.

It is important that the basic oscillator of the signal source have extreme purity, meaning it should be a nearly perfect sine-wave source. Any unintentional

Summary

Figure 6.25 HP8640B test instrument (courtesy of Hewlett-Packard Co.).

variations in the frequency or phase of this oscillator will corrupt the measurement of the receiver. In the Hewlett-Packard unit, the harmonics of the oscillator at any frequency are at least 30 dB below the fundamental, and the *phase noise* (minute variations around the correct phase, although the overall average frequency is correct) is better than 100 dB below the carrier value.

The particular unit is designed for general coverage AM and FM testing over a very wide range of frequencies and conditions. Other instruments are available which are designed specifically for complete testing and alignment of standard commercial 88- to 108-MHz FM broadcast systems. To make testing the standard transmitter or receiver easier, these units contain circuitry to generate mono and stereo signals along with variable pilot tone level (for testing stereo decoding capability), and for aligning the 10.7-MHz and 455-kHz IFs used in standard FM receivers. Although limited to the frequency range of 88 to 108 MHz, this is all that is required for the application. The general signal source, such as the HP8640B, has a wider range of frequency and signal level output than these more limited units, but it is also more complicated to set up and use just for testing of broadcast band stereo.

SUMMARY

In this chapter we have shown the theory and practical aspects of frequency and phase modulation, which offer noise-resistant signal transmission. The bandwidth of FM and PM is infinite, but practically can be limited and still preserve the modulating signal and produce minimal distortion. FM and PM transmitters are more complex than AM transmitters, but all signal manipulation can be done at low levels, in contrast to AM with high-level modulation. Direct exciters use a varactor or other reactance as the frequency-determining element, while indirect FM (or PM) uses a crystal as the heart of the oscillator and thus has a more stable and predictable carrier.

FM and PM receivers are also more complex than their AM counterparts. The FM receiver uses a limiter, similar to an AGC, in the IF stage to eliminate variations in signal level. The FM detector, or discriminator, can be a simple filter-based slope detector, a ratio or Foster–Seeley detector, or a quadrature detector that uses both the in-phase signal and its 90° phase-shifted version. The phase-locked loop is a powerful circuit function used for FM demodulation and signal tracking. It is also used for recovering the pilot tone of FM stereo, which multiplexes the left and right audio channels into a single broadcast using a 38-kHz carrier. Pre- and deemphasis are used to minimize the effects of noise on the higher modulating frequencies of standard broadcast FM.

Each form of modulation—AM, FM, and PM—has distinct technical features. AM is simplest and requires the smallest bandwidth but is easily corrupted by noise. FM and PM are closely related and differ in how they define the modulation index. They provide greater noise immunity than AM but require more bandwidth (except for narrowband FM, which uses less bandwidth but distorts the modulating signal). FM and PM circuit functions are often used together, depending on which is technically easier, since the results of one can be derived from the other with simple circuitry.

Summary Questions

1. Compare the modulation equations and the noise resistance of AM, FM, and PM.
2. Explain in general terms, how an FM signal is demodulated. Compare to AM?
3. What is the meaning of modulation index for FM, for PM, and for AM?
4. What is the bandwidth of FM? Why? What do Bessel functions show?
5. How is carrier power distributed in the spectrum after FM? Compare to AM.
6. Define *modulating frequency, deviation,* and *modulation index for FM.*
7. What is NBFM? Where is it used? Why isn't it always used?
8. Compare direct and indirect FM transmitters. What is an exciter?
9. Compare high- and low-level AM transmitters to FM transmitters in terms of power levels and modulation signal power.
10. How are multipliers and mixers used to obtain the desired FM carrier and deviation in indirect modulation? What are the effects of each on the carrier and deviation?
11. How is stereo broadcast on commercial FM? Compare this to using two separate channels. How is it demodulated?
12. What is the role of the limiter? Compare it to AGC. What is quieting?
13. What is AFC? Why is it sometimes needed? What is the drawback?
14. Explain the operation of the PLL. What is the role of the filter?
15. How is a PLL used for FM demodulation?
16. Why can a PLL provide excellent performance at low SNRs?
17. What controls the PLL center frequency when acting as a precise filter? What sets the effective PLL bandwidth?
18. How are signal phase and frequency closely related?
19. What is the relationship of Δ and the carrier frequency for FM and for PM? What about the relationship of m?
20. What are pre- and deemphasis? Why are they used in broadcast FM?
21. How can a PM transmitter be used for FM results? Why does this work?

PRACTICE PROBLEMS

Section 6.2

1. What is m for Δ of 60 kHz and f_m of 10 kHz?
2. What is Δ when m equals 9 and f_m is 3 kHz?
3. What f_m caused a deviation of 150 kHz with a modulation index of 8?
4. What are the relative magnitudes of the first six sidebands for $m = 3.0$?
5. What is the approximate bandwidth, using Carson's rule, for deviation of 125 kHz and f_m of 10 kHz?
6. Repeat problem 5 when f_m is 15 kHz.
7. What are the minimum and maximum values of m for a voice signal that spans 300 to 5000 Hz and produces a deviation of 50 kHz?
8. If the voice of problem 7 is used in an NBFM system with deviation of only 8 kHz, what are the extreme values of m?

Section 6.3

1. A crystal is capable of being pulled up to 65 Hz/MHz of resonant frequency. If the crystal is operated at 3.5 MHz, what is the maximum deviation?
2. The crystal of problem 1 is to be used in a transmitter at a carrier frequency 18 times higher than its resonant frequency. What series of multipliers should be used? What is the deviation at the new carrier frequency?
3. An FM transmitter for the standard broadcast band must operate at 101.7 MHz. The crystal used in the transmitter can only be used at between 0.9 and 1.1 MHz. Show what $\times 2$ and $\times 3$ multipliers could be used and what the nominal crystal frequency should be.
4. For the system in problem 3, the frequency deviation of the crystal is 100 Hz/MHz. What is the deviation after multiplication?
5. An indirect FM generation transmitter generates a crystal-controlled carrier at 1.2 MHz and causes it to deviate by 75 Hz. The signal is multiplied by a total of 36 factors, then mixed with an oscillator at 43 MHz, and then remultiplied by $\times 81$ factors. At each stage, what are the carrier and the deviation? What is the undesired mixing result that is filtered out?
6. A very stable crystal at 0.5 MHz can only be deviated by 15 Hz. The result is multiplied by 81, mixed with a 40-MHz signal, and then multiplied (after discarding the sum signal) by 81 again. What are the final carrier frequency and deviation?

Section 6.5

1. A slope detector has a filter with straight-line response in "dB versus frequency" roll-off slopes. The output of the filter at these slopes ranges from 0 dB at nominal center to -20 dB at 5000 Hz above or below the center frequency. If the 0-dB signal is 1 V, what is the slope detector output in volts at 2500 Hz off center? At 5000 Hz off from the center frequency?
2. Repeat problem 1 for a slope detector with output that ranges from 0 to -40 dB between center frequency and 20 kHz from center.
3. A receiver has a quieting specification of 2 μV for full quieting, using a limiter that requires a 20-mV signal minimum. What is the total gain between the RF input and the limiter?

Section 6.6

1. A VCO has a specification of frequency of 1500 Hz/mV control signal. What error voltage does it need to swing ±10 kHz around the free-running frequency?

2. A VCO running at 50 MHz is designed to capture and track signals from 45 to 55 MHz. The control voltage is ±0.5 V. What is the necessary change in VCO frequency versus control voltage, in MHz/mV?

3. The input to a VCO operating at 10 MHz is a signal that may range from 8.5 to 11.5 MHz. Can a PLL with 10% capture range and 20% lock range be used?

4. The VCO of a PLL is specified with 17% lock and 23% capture ranges. For a free-running frequency of 5.8 MHz, what are the minimum and maximum capture and lock frequencies?

Section 6.8

1. The pre- and deemphasis used in other parts of the world are not necessarily 75 μs. Suppose that a 50-μs time constant is used. What is the necessary −3-dB frequency? What resistor value can be used if the capacitor of the 75-μs preemphasis circuit in the system is retained?

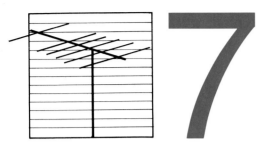

7

Wire and Cable Media

CHAPTER OBJECTIVES

When you have completed this chapter, you will understand:

- The electrical appearance of wire and cable at low and high frequencies
- How cables are physically built and the characteristics of each type
- The use of balanced and unbalanced cables, together with line drivers and receivers
- Key cable driving parameters such as slew rate and common-mode voltage
- Use of time-domain reflectometry to locate cable faults working from one end only

INTRODUCTION

The basics of using wire and cable for carrying the modulated signal of a communications system are examined in this chapter. Signal loss, capacitance, useful signal bandwidth, noise resistance, and ease of connection are issues that help decide what type and size of cable to use. Two common choices for cable are twisted pair and coaxial types, which offer many trade-offs in performance. Single-ended electrical connections use the fewest wires but are susceptible to noise, especially when the ground wire is not ideal. Balanced, differential signals offer an alternative but require more signal wires. The effect of common-mode voltages must be minimized, as measured by the common-mode rejection ratio. Special circuits called line drivers and receivers are optimized for driving and receiving signals over cable, and have different characteristics for analog and digital signals. Time-domain reflectometry allows the location and type of cable fault to be determined with high accuracy, with a simple connection to one end of the cable.

7.1 WIRE AND CABLE PARAMETERS

Wire and cable seem to be very simple parts of any communication system. A *wire* is a strand of copper that carries current at some voltage. A *cable* is the complete wire assembly, which includes insulation, connectors, and any support needed for mechanical strength. Despite this apparent simplicity, there are many critical technical aspects to wire and cable in systems. The terms *wire* and *cable* are often used interchangeably, so we will just use one or the other from here on, unless there is a specific reason not to do so.

At dc or low frequencies, the wire is primarily a simple resistive conduit for the current, and its resistance depends on the thickness of the wire. For No. 22 AWG (*American Wire Gauge*) this resistance is 16.5 Ω per 1000 ft, so a 10-ft cable between two systems has a resistance of 0.165 Ω. Thicker wires (lower AWG numbers) have less resistance and can carry more current without overheating; of course, the opposite applies to thinner wires. Wires thinner than No. 28 (higher AWG numbers) are very weak and require strong support in the final cable. As the current flows through the wire, some of the voltage is lost as characterized by Ohm's law: $V_{drop} = I \times R$, where R is the resistance of the wire and I is the current flowing through the wire. This is usually called *IR drop* and is significant in some applications, where a large fraction of the voltage applied to the wire is dropped across the wire itself and never appears at the far end.

As the frequency of the signals impressed on the wire increases, the reality of the true nature of the wire and its characteristics becomes more critical. An accurate model of a cable shows resistance, inductance, and capacitance, all distributed along the entire length of the cable (Figure 7.1). The specific values of these R, L, and C elements depend on the wire diameter, the way the wire is made into the final cable, spacing between wires, and the type of insulation used.

At low frequencies, the effect of the L and C factors is negligible (and the cable is nearly lossless), but at higher frequencies they actually are the major factors in determining the bandwidth and signal capacity of the final cable. Instead of looking like a simple resistance, the cable now has a more complex *characteristic impedance*. This impedance is not due to a physical resistor across the wires of the cable—it is due to the voltage/current transfer ratio looking into the cable and is determined by the L and C values. The inductance and capacitance combine to

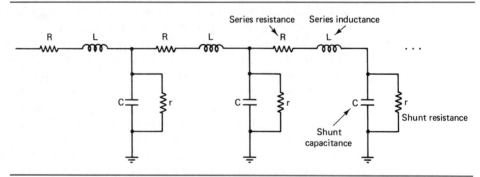

Figure 7.1 A realistic model of wire, showing resistive, capacitive, and inductive lumped elements.

make a crude low-pass filter which affects the amplitudes and phases of the various frequency spectrum components that comprise the modulated signal being transmitted on the wire. Even the system ground, or zero potential point, is severely affected by the nonideal reality.

Another important factor in the overall cable design is the ability of the cable to resist external electrical noise. The wire within the cable is a crude antenna and is sensitive to electromagnetic energy around it. The amount of induced noise depends on the cable construction and the circuitry that the cable is connected to at either end. Some cables have *shielding* to minimize the noise pickup. The shield is a separate conductor around the main signal-carrying wire and is usually connected to ground. The function of the shield is to prevent electromagnetic energy from reaching the signal carrying conductor, by establishing a grounded enclosure around the conductor. Different types of shielding are needed as the frequency of the noise increases.

Review Questions

1. What is *IR* drop? How is it calculated?
2. What is the simple, low-frequency view of a cable?
3. What is a more accurate model of a cable, necessary at higher frequencies? What does it show?
4. What is shielding? Why is it needed? How is it implemented?

7.2 BALANCED AND UNBALANCED LINES

In many situations, the cable carrying the signal uses one wire for the signal itself together with a second wire, or the shield, as the ground wire. The ground is shared by the rest of the electronic system circuitry, and many signal wires may share this ground. Therefore, each signal being transmitted requires just one wire plus one ground (or common) wire for the entire system (Figure 7.2). This scheme is called *single-ended* transmission, using *unbalanced* lines.

In theory, this should work well, and in many cases it does. Unfortunately, the simple ground wire in the system is not perfect, and is often far from perfect. A perfect ground has exactly the same potential anywhere along its length, with no potential difference between any two points. A real-world ground has resistance, inductance, and capacitance that result in some potential difference from one point to another. Therefore, the ground is no longer a perfect reference point, but really another wire with voltage and current. This is a problem because an imperfect wire can pick up noise, currents can be induced into this wire, and voltage potentials can be imposed on it.

Equally important, the value of the transmitted signal now depends on where it is measured. Remember that a voltage value exists only as the *potential difference* between two points. When we say "6 volts," for example, we mean that the signal was measured at 6 V with respect to some specific point. If the

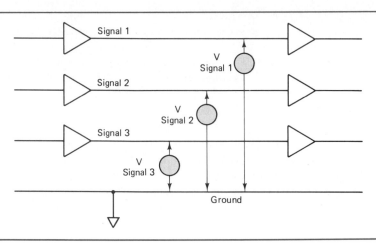

Figure 7.2 A single-ended system has several signal wires sharing a common ground wire.

ground is not always at the same potential everywhere, any signal measurement made with respect to ground will depend on the exact reference location used. As a result, a signal of 4 V at the transmitter may be measured at 3.8 V elsewhere, and 4.2 V still elsewhere. Line losses caused by *IR* drop have a similar effect, since voltage will be lost as current flows through the ground wire resistance.

The alternative to assuming a perfect ground and living with the consequences of this incorrect assumption is not to use the ground as the reference for the transmitted signal. This is done by using two wires for each signal, in a method called *differential* or *balanced* signal transmission. The transmitted signal is measured everywhere as the difference between the two wires and is seen at the receiver as this difference value (Figure 7.3). This scheme works to eliminate most of the problems of an imperfect wire or ground because it does not assume that either of the two wires is ideal. Instead, it assumes that the two wires have very similar characteristics and are thus affected equally by noise, voltage drops, and induced signals. Since they are affected identically, the difference between them will remain the same everywhere. One wire is usually called the *HI* signal, and the other is the *LO* signal.

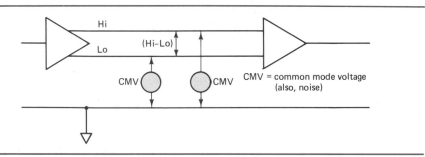

Figure 7.3 A differential system requires two wires per signal; the CMV is measured between each signal wire and ground.

7.2 Balanced and Unbalanced Lines

Example 7.1

Two wires are used to transmit a signal. The HI wire is at 10 V with respect to the local ground, while the LO signal is at 2 V. Along the way, both signals lose 1 V due to *IR* drop. What is the differential value at each end of the wires? Compare this to using the single-ended method.

Solution

At the transmitter, the difference is $10 - 2 = 8$ V. At the receiving end, the signals each are now $10 - 1 = 9$V and $2 - 1 = 1$ V. The difference is therefore $9 - 1 = 8$ V, the same as at the transmitting end. In a single-ended design, the measured signal value is equal to the original value minus the *IR* loss, or $10 - 1 = 9$ V.

The differential method is used because only the voltage difference between the wires matters, not the actual measured value with respect to ground. If the ground wire has voltage drops or noise, the difference between the two signal wires should remain unchanged. This is *balanced* because the two signal wires used are measured the same way, with respect to the local ground. This differential method is used when noise pickup is a problem, when the ground wire imperfections are significant and unavoidable, or when very low signal levels are involved (such as in low-level audio signals from a tape player in a broadcast studio). The drawback to balanced designs is that two wires are needed for every signal. In contrast, single-ended systems need only one wire for each signal, plus one ground that can be shared by all.

Common-Mode Voltage and Rejection

The circuitry that receives the two wires of a differential signal must look at the difference only, and convert this differential signal back to a single-ended, ground-referenced signal for use within the circuitry. The two differential signals are, of course, corrupted by noise. This corruption is called the *common-mode voltage (CMV)* signal, because it is common to both signal wires. The receiving circuitry ignores the common-mode signal while responding only to the difference. The ability of the receiving circuitry to do this is called *common-mode rejection*, and it is usually measured as the *common-mode rejection ratio (CMRR)*, often expressed in dB. The CMRR is the ratio of the received common mode voltage to its voltage value after passing through the differential receiving circuitry. As the CMRR increases, the undesired corrupting CMV is reduced to less significance versus the desired differential signal. Ideally, the CMV will cancel out completely, since the differential receiver will produce the difference:

$$(HI + CMV) - (LO + CMV) = HI - LO$$

The CMRR shows how good the receiver is at providing complete cancellation of the CMV.

Example 7.2

A 10-V differential signal is received with 5 V of CMV. After the receiving circuitry, this CMV is reduced to 0.01 V. What is the CMRR?

Solution

$$\text{CMRR} = \frac{5}{0.01} = 500 = 54 \text{ dB}$$

Example 7.3

A 5-V differential signal, transmitted as a 6-V and a 1-V signal pair, is received with 3 V of CMV. The receiver provides a CMRR of 40 dB. (a) What are the received voltages? (b) To what value is the CMV reduced by the CMRR? (c) What is the actual difference, after receiver rejection of the common-mode signal by 40 dB?

Solution

(a) The received voltages are $6 + 3 = 9$ V and $1 + 3 = 4$ V. (b) The 3-V CMV is reduced by 40 dB; using the standard dB formula, the resultant voltage is 0.03 V. (c) The actual difference is $5 + 0.03 = 5.03$ V after the common mode has been reduced by 40 dB.

Manufacturers provide ICs for both single-ended and differential applications. These ICs are usually used in pairs, as complementary transmitters and receivers. Some examples are shown in the next section. Technicians must be careful when checking differential circuits, for two reasons. First, the CMV may be relatively high, and in some cases, dangerous. A 5-V differential signal may have CMV of tens or hundreds of volts, and the circuitry may be designed to handle it—but a careless technician is not.

Even if the CMV is low and harmless, thoughtless use of the voltmeter or scope probes can defeat the differential circuitry and perhaps damage components. If a grounded instrument is connected across the differential pair to measure the difference voltage, one of the two difference wires will be grounded through the probes (Figure 7.4). This completely upsets the balance with respect to ground and makes the measurements meaningless. One solution is to measure the differential voltages separately, each with respect to the system ground, and then calculate the difference; with an oscilloscope, connect channel A to the one differential wire and channel B to the other differential wire (some scopes have a

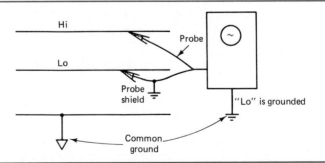

Figure 7.4 Differential circuitry can be affected and even defeated by grounded probes that are improperly used.

7.3 Line Drivers and Receivers

switch to display A − B for this reason). Another choice is to use a *floating*, ungrounded instrument such as a battery-powered unit, or one that is electrically isolated from the ac line ground. Most test equipment circuitry is not connected to ac line ground except for the power supply and enclosure, and can be optionally disconnected from ground for differential measurements by the user.

Review Questions

1. What is single-ended transmission? How is it implemented in a system?
2. Contrast single-ended and differential signals as to wiring used and potential performance.
3. Why is the system ground not perfect? What are the effects of this imperfect ground on system performance?
4. How does differential wiring overcome problems with grounding and *IR* drop?
5. What is common-mode voltage? What is CMRR? How is it measured?
6. Why must differential circuits be tested with care for correct results? How should the instrumentation be used?

7.3 LINE DRIVERS AND RECEIVERS

The circuitry used for putting signals onto single-ended or differential cables must be able to drive the electrical load of the cable. This is not as simple as it may seem. The typical cable has a relatively large amount of capacitance (from 10 to 1000 pF/ft), and this capacitance affects the ability of the driving circuitry to transfer the electrical signals faithfully from circuit to cable. The capacitance can be up to several thousands of picofarads, depending on cable type and length.

The special circuits that drive cables are called by various names, including *line driver, transmitter,* and *buffer.* The receiving end of the cable has corresponding circuitry, known as the *line receiver* or *buffer.* Some specialized line drivers and receivers are designed to drive the *bus* that interconnects various circuit boards in a single chassis; these are called *bus drivers* and *receivers* to distinguish them from the drivers and receivers for longer signal cables. Besides driving cables, these circuits usually have various types of protection as part of their design to prevent damage from inevitable accidental connections to the power supply, other high-voltage sources, or short circuits to ground or other powered wires even if they themselves are not powered.

The drivers and receivers for analog signals are very different than those for digital signals. These differences are shown with some examples in this section.

Slew Rate

The function of the driver is to take the modulated signal and impose this signal voltage (and frequency) onto the wire. To do this, the driver must be able to cause the voltage to swing (*slew*) from one value to another value at the frequency of the modulated signal. The voltages used are determined by the system design. For example, many audio systems use a 1 V maximum value at full modulation.

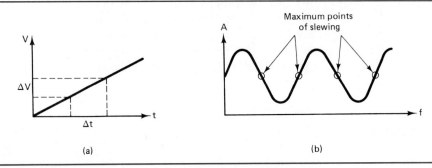

Figure 7.5 (a) Slew rate for a simple ramping waveform; (b) the slew rate for a sine wave is maximum at the sine-wave amplitude midpoint.

Slew rate is the maximum voltage span that the signal passes through in a specific amount of time (Figure 7.5a). It is defined as the (change in voltage)/(time period), measured in V/s (or V/ms or V/μs, depending on range). A higher slew rate means either that the voltage span is greater or that the time period is shorter. A shorter time period corresponds to a faster signal rate or higher frequency. The driver circuitry must be capable of slewing the voltage at the required rate, or the full signal voltage will not be reached on the wire.

Example 7.4

A signal must slew from -10 to $+10$ V in 50 μs. What is the slew rate?

Solution

$$\text{SR} = \frac{10 - (-10)}{50} = \frac{20 \text{ V}}{50 \text{ μs}} = 0.4 \text{ V/μs}]$$

For a signal that is increasing linearly with time (a ramp), the slew rate is simple to determine: It is the rate of rise in the signal following the definition above. But how does this relate to the sine-wave type of signal from a modulated waveform? It turns out that it is easy to determine the slew rate of a sine wave. The maximum slew rate occurs when the sine wave crosses its midpoint (Figure 7.5b). For a sine wave of amplitude A and frequency f, the maximum slew rate is

$$\text{SR}_{\text{max}} = A(2\pi)f$$

Example 7.5

What is the maximum slew rate of a 10-kHz sine wave of 2 V amplitude?

Solution

$$\text{SR}_{\text{max}} = 2(2\pi)(10{,}000) = 125{,}663 \text{ V/s} = 125 \text{ V/ms}$$

If the driver is unable to achieve this slew rate with the cable and capacitance, the waveform on the cable will be distorted with resultant errors or poor signal quality. Two solutions to insufficient slew rates are used. One is to reduce the amplitude of voltage through which the signal must pass. This reduces the required slew rate but results in a smaller signal that is more easily corrupted by noise. The other solution is to reduce the frequency of the signal. This means that a lower bandwidth is being used and less information can pass through the system in a fixed amount of time. If the signal frequency cannot be reduced, as is often the case, this choice is not practical.

The slew rate that is needed is determined by the voltage levels that must be used to pass the signal with minimal noise and by the system bandwidth. A more powerful drive circuit is the other way to achieve the slew rate. To understand what more powerful means, consider the cable as a capacitance that must have its voltage changed. From basic electronics, $I = C(\Delta V/\Delta t)$ where I is the current to be supplied, C is the capacitance, and $\Delta V/\Delta T$ is the required rate of voltage change (the slew rate). This means that a drive circuit that can produce more current in the same time period is what is needed to slew the signal voltage at the required rate. For this reason, more powerful drivers have larger transistors, lower internal impedances, lower internal capacitances, and sometimes higher power supply voltages, to provide the necessary current quickly.

Example 7.6

A cable of 1000 pF total capacitance must be driven with a 10-V swing in 10 μs. What is the current needed?

Solution

$$I = C\frac{\Delta V}{\Delta t} = (1000 \times 10^{-12})(10/10 \times 10^{-6}) = 0.001 \text{ A} = 1 \text{ mA}$$

Analog Drivers   A driver for an analog signal, regardless of the type of modulation used, must drive the cable with a low distortion signal. The role of the driver is primarily that of an amplifier that is capable of providing more current to a load, especially a load with capacitance, than the circuitry within the transmitter (or exciter) itself can provide. Although it is an amplifier, it does not necessarily increase the voltage of the signal. Instead, it may simply take a source that can supply a 1-V signal at up to 10 mA and provide the 1-V signal at up to 500 mA. The slew rate of this amplifier must also be suitable for the application—even a high-power amplifier is useless unless it can dynamically provide the current at the rate needed.

One such driver is the National LH0033 high-speed op amp IC. It is designed to be used as a *voltage follower* with an output voltage that follows the input signal, at the same amplitude, but with much greater drive capability. The LH0033 as shown in the complete circuit of Figure 7.6a can slew at up to 1500 V/μs with a bandwidth of 100 MHz and supply up to $\pm$100 mA into the cable load. The 100-Ω resistors limit the current that the IC draws to prevent burnout and self-destruction if the output is short-circuited. As the capacitance of the load increases, the slew rate decreases as shown in Figure 7.6b.

Digital Drivers   For digital signals, which have only a fixed group of voltage values, there are some similarities and major differences compared to analog drivers. The slew rate (and

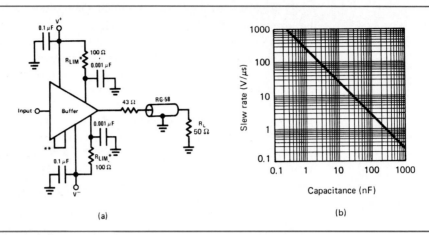

Figure 7.6 (a) LH0033 line driver circuit for driving a coaxial cable; (b) the achievable slew rate versus load capacitance (courtesy of National Semiconductor Corp.).

sufficient current to achieve this slew rate) must still provide the transitions from one value to another at the rate of the signal from the data source. The nature of the digital signal, however, changes some of the requirements.

Absolute fidelity to the original signal and low distortion is not as critical. If the driver signal has a small amount of *overshoot* (going past the desired final value, then settling to the correct value) or *ringing* (damped oscillations around the final value), the receiver will still be able to interpret the voltages correctly. This is one of the advantages of digital systems versus analog, as we will see in subsequent chapters. Equally important, small values of noise or line *IR* drop are much less of a problem for digital signals. The receiver is designed to recreate the received signal at the closest correct digital value in a process called *regeneration*.

Take a simple case of a binary digital signal that has only two values, called logic 0 and logic 1. This system may use a voltage of -1 V for logic 0 and $+1$ for the logic 1. If any noise corrupts the signal, the received value will be different from the transmitted value. The line receiver is designed to decide which value the received signal is closest to, and then produce an output at this nearest value. If the received signal is 0.8 V, the receiver will produce 1 V; if the received value is -0.1 V, the receiver produces -1 V. In this case, the receiver uses 0 V as a threshold of decision: all signals below 0 are converted to -1 V, while all signals above 0 V are converted to $+1$ V. The effect is to recreate the original signals without noise or error (Figure 7.7a). The system can then determine if a logic 1 or a logic 0 was received. An error will occur only if the noise is greater than ± 1 V, which will mislead the receiver into classifying the received signal into the wrong logic category.

For applications where there are more than two digital levels, the receiver establishes thresholds between each level and bases its decision on these. For a four-level digital system with transmitted values of -3, -1, $+1$, and $+3$ V (each spaced 2 V apart), the receiver uses thresholds at -2, 0, and $+2$ V (Figure 7.7b). Only a noise value greater than the difference between the ideal value and the threshold value will cause an error. In this case, as long as the received signal was within ± 1 V of the correct value, it will be regenerated at the correct value by the receiver.

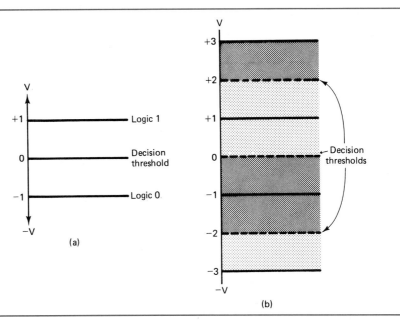

Figure 7.7 (a) Decision threshold for binary signals; (b) decision thresholds for four-value digital signals.

Digital Line Driver/Receiver ICs

There are many types of lines drivers and receivers available for digital signals. One driver type is the Texas Instruments SN75109/110/112 series, all identical except for their output drive current capability. Each of these drivers operates from a +5-V and a −5-V supply and can provide differential outputs at up to 6, 12, and 27 mA, respectively. The common-mode output imposed on the driven line can range from −3 to +10 V, and these ICs will still function properly. A single IC package contains two independent drivers, and each driver may be switched "off" via an "enable/inhibit" line controlled by the system processor. The data signal to the driver can be standard digital logic TTL levels of 0 and 5 V (nominal) values.

The corresponding line receiver is the Texas Instruments SN75107 or SN75108 (which differ only in their interface to the subsequent electronic circuitry. These line receivers accept differential signals (HI minus LO) up to ±5 V and decide what the input value was based on this differential input voltage. When it is more positive than +25 mV, a logic 1 output results; when the difference is more negative than −25 mV, the output is logic 0. If the difference is between −25 and +25 mV, the output is invalid because the input signal difference is too small to be meaningful and error-free. These ICs can operate with a common-mode voltage (the average of the two input voltages) between −3 and +3 V, and can be deactivated ("turned off") by built-in digital control lines.

Example 7.7

The SN75107 input voltages are +1 V (LO) and −2 V (HI), measured with respect to the receiving system ground. Does this exceed the CMV specification? If not, what is the output level?

Solution

The common-mode value is the average of the inputs, $[1 + (-2)]/2 = -0.5$ V, which is within the acceptable ± 3-V CMV boundary. The input difference is $-2 - (+1) = -3$ V, so the output is logic 0.

Example 7.8

The same IC as in Example 7 receives voltages of $+5$ V (LO) and $+3$ V (HI). What is the output result?

Solution

The average of these two inputs is $(5 + 3)/2 = 4$ V, which exceeds the allowable CMV specification of $+3$ V. The IC output is therefore not valid and cannot be judged.

Review Questions

1. Why is driving a wire or cable electrically difficult? What is the effect of line capacitance?
2. What is slew rate? How is it measured? What factors determine the maximum slew rate for a sine wave? What is the formula?
3. How is the current required of the driver determined by capacitance and slew rate?
4. What is the function of an analog signal driver? Compare this to the functions of a digital signal driver.
5. How does a digital signal line receiver regenerate the original signal values? Compare this to the effect of noise in an analog situation.

7.4 TWISTED-PAIR AND COAXIAL CABLES

Many types of cable are used in communications systems, such as between computers, from exciters to power amplifiers, and in networks linking individual terminals to a central computer. Two of the most common are twisted-pair cable and coaxial cable. Each provides a different combination of electrical and mechanical characteristics, as well as installation convenience and cost.

Twisted-Pair Cable

Twisted-pair cable is made from two insulated wires which are continuously braided. Typical wire gauges are No. 20, 22, and 24 AWG, although heavier wire is sometimes used for longer runs to reduce *IR* loss, and thinner wire is used for short-distance installations, such as between subsystems of a single chassis. The cable may be enclosed in another outer insulating jacket for mechanical protection, especially if it is going to be pulled through various ducts and conduits during installation. Twisted-pair cable is relatively low in cost and easier to connect: The wire ends are stripped and then screwed to the equipment terminations or soldered to connectors.

Typical capacitance of twisted pair is low, about 5 to 15 pF/ft. This results in a large useful bandwidth, which can range from 1 to 20 MHz, depending on the wire gauge, insulation, and signal levels. The two wires of the twisted pair are identical and symmetrical with respect to the system ground, so twisted pair is often used for balanced signal (differential) installations.

Unfortunately, twisted-pair cabling is not suitable for all applications. Because it is not shielded, it will pick up electrical noise. The use of differential signals minimizes the effect of this noise, but there is a practical limit to how much common-mode noise the circuitry of the system can reject, as shown by the CMRR. Finally, the twisted pair can act as a signal-energy transmitting antenna, as the frequencies on the cable increase. Instead of the signal energy reaching the receiver, it radiates from the cable, does not reach the receiver at the desired strength, and causes interference and noise to nearby wires and systems. When twisted pair (or any unshielded wire) passes through or near metallic surfaces, the signal will see impedance discontinuities (bumps) which cause some signal reflection and waveform distortion.

Shielded twisted-pair cable is available and often used. The shield is connected to ground, and this reduces the noise pickup considerably. It also reduces any unwanted radiation from the cable. The shielding, however, adds to the cable and connector cost and also increases the effective capacitance of the wire, which reduces the achievable slew rate and bandwidth.

Coaxial Cable

One alternative to twisted pair is *coaxial cable* (Figure 7.8). This consists of a central wire running through an enclosing conducting cylinder on the same axis (hence *coaxial*). The enclosing cylinder is usually made of fine wire woven into a braid, to allow flexibility of the overall cable, and usually grounded to act as an electrical shield. The space between the center conductor and the shield can be filled with any one of several types of insulators (air, Teflon, polystyrene, for example) which keep the conductors apart and also affect cable capacitance, signal propagation, and power-handling capability. The overall cable is enclosed in another insulator for electrical and mechanical protection.

Many different standard coaxial cables are manufactured, most with diame-

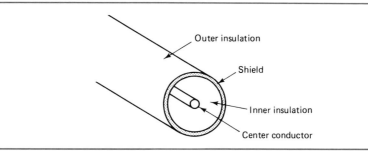

Figure 7.8 Coaxial cable construction.

Chap. 7 Wire and Cable Media

Coax type	Impedance (Ω)	Velocity of propagation	pF/ft	50 MHz attenuation, (dB/100 ft)	Diameter (in.)
RG58	53	66	29	3.1	0.195
RG58, polyfoam insulation	50	79	25	2.2	0.195
RG59	73	66	21	2.4	0.242
RG8	52	66	30	1.4	0.4

Figure 7.9 Key specifications for industry standard coaxial cables and twisted-pair gauges.

ters from 0.2 to 0.75 in. Each provides a different combination of electrical and mechanical characteristics, such as impedance, dc resistance, power capacity, flexibility, and bandwidth. The table in Figure 7.9 shows some of the more common coaxial cables used in communications systems, with industry standard "RG" numbering prefixes. While coaxial cable can support bandwidths in the range of several hundred megahertz, it is much more costly to buy and install than twisted pair. Coaxial cable can be installed near or through metallic surfaces without impedance discontinuities, since the entire electromagnetic field is within the shielded cable and the metallic surfaces are therefore not seen by the signal.

Special connectors (Figure 7.10) such as BNC, J-type, and PL series are needed to connect both the inner conductor and shield properly, and it requires care to attach the connector without special tools to strip the outer and inner insulations cleanly, yet not nick the wire. A careless connector installation on a cable often results in a single hair-thin strand of the shield braid shorting to the inner conductor, which will cause many intermittent system problems and aggravation for users and technicians.

Review Questions

1. What are the capacitance, bandwidth, and connectability of twisted pair?
2. Why is twisted-pair cable often used in differential designs? Why isn't twisted pair used for all communications cables?
3. How does shielded twisted pair overcome some drawbacks? What are its weaknesses?
4. What is coaxial cable? How does it compare to twisted pair for bandwidth, noise resistance, and connectors?
5. What is the effect of insulation between the coax center wire and shield?

7.5 TIME-DOMAIN REFLECTOMETRY

Just as with any other component in a system, cables can develop problems. Yet unlike equipment that can be examined, pulled apart, or replaced for repair at a

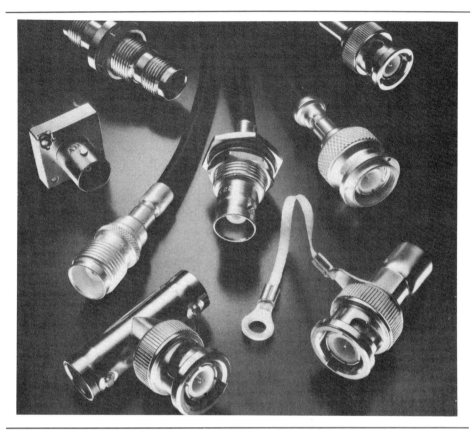

Figure 7.10 Some connectors used in industry (courtesy of AMP, Inc.).

single physical location, cables often stretch for miles. It is usually difficult, time consuming, and costly to locate a cable problem. Imagine pulling out miles of wire or digging up cables that are buried underground to locate a break somewhere in the cable!

Fortunately, there is a way to identify the location of a cable problem working from one end of the cable only, without needless exploration. This technique is called *time-domain reflectometry (TDR)*. The principle behind TDR has been known for many years: any discontinuity in a cable—a short, an open, or a high resistance between conductors acts as a partial mirror to some of the energy sent through the cable. Some portion of the transmitted energy will be reflected back to the transmitter by the discontinuity, rather than continue on to the intended receiver at the far end (Figure 7.11). By carefully measuring the time between sending a signal and seeing the reflected energy, the distance to the fault can be calculated, since the propagation factor of a given type of cable is known. It is the same idea as using an echo to locate the distance to a target, using the echo time and the speed of sound. The time between the transmitted pulse and the received echo is divided in half to find the one-way time, then multiplied by the propagation velocity to calculate the distance to the fault.

While the principle of TDR has been known for many years, there were several technical barriers to practical TDR equipment. A very short duration, fast-

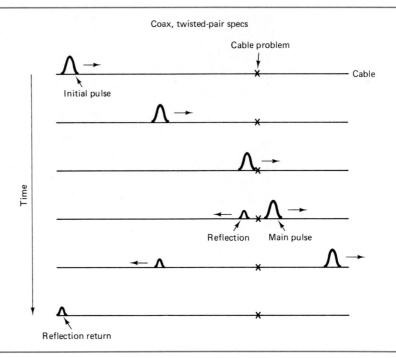

Figure 7.11 TDR basic concept of traveling wave and reflection from discontinuity.

rise time pulse must be generated by the TDR instrument, which must also receive the much weaker echo. A highly sensitive, low-noise receiver is required to sense the weak reflection, yet this receiver input must not be overloaded by the much stronger transmitted pulse. The time between the pulse and the echo has to be measured with great precision, since the speed of signal propagation is approximately 1 ns/ft for signals traveling at c; it is proportionally less for signals with a lower velocity of propagation.

Example 7.9

A pulse is sent through a cable with a propagation factor of $0.9c$. The echo is received 1 μs later. What is the distance to the fault? (Use $c = 3 \times 10^8$ m/s.)

Solution

$$\text{Distance} = \frac{\text{time}}{2} \times \text{velocity in cable}$$

$$= \frac{(1 \times 10^{-6}) \times 0.9 \times (3 \times 10^8)}{2} = 270 \text{ m} \quad \text{or} \quad 884 \text{ ft}$$

Example 7.10

A fault is located 1 mile (5280 ft) away from the cable end, with a cable propagation factor of 83% c. What is the time between the initial pulse and the reflection? (Use $c = 186{,}000$ miles/s.)

Solution

By rearranging the equation and doubling the distance (since we need the round-trip time), we have

$$t = \frac{\text{distance}}{\text{velocity}} \quad \text{where the actual velocity is } 0.83c$$

$$= \frac{2 \times 1}{0.83 \times 186{,}000} = 0.000013 \; s = 13.0 \; \mu s$$

A typical TDR instrument is shown in Figure 7.12a. This instrument looks like an oscilloscope, but combines a pulse source and receiver in the unit. The purpose of the screen is to allow the technician to look at the amplitude and shape of the received reflection in case there are multiple faults (several reflections), and to study the type of fault (which takes some experience and judgment). The time between the pulse and the reflection is determined by multiplying the CRT horizontal sweep rate by the distance on the screen (Figure 7.12b).

There is a technical trade-off between the maximum detectable fault distance and the precision of fault location. Since the reflected signal is very weak compared to the initial pulse (from 40 to 80 dB less), a longer pulse is used to put more energy onto the cable, thus increasing the strength of the reflection. A stronger reflection is easier to see, of course. However, a longer pulse spreads the reflection out in time so that the precision with which the fault distance can be determined is less. TDR instrumentation can locate faults to within 0.6 in. on short cables (several feet long) and to 3% of the distance for lengths to 50,000 ft.

Most TDR equipment lets the operator adjust the pulse width to get the best compromise for the application. In many cases, the fault distance does not have to be known to a few inches, but only to a segment or area of the cable (such as

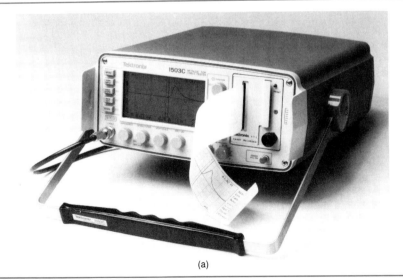

(a)

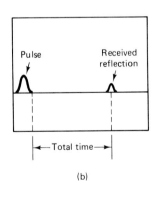

(b)

Figure 7.12 (a) TDR test instrument (courtesy of Tektronix Inc.); (b) TDR screen shows reflection as vertical pulse along horizontal time/distance axis.

between which two telephone poles or access points the fault lies). Although a fault at a greater distance usually produces a weaker reflection than a fault closer to the TDR equipment end (and so can be located with less precision), TDR is a very powerful and effective technique for locating most faults and minimizing needless digging or cable removal.

Review Questions

1. What is the principle of TDR? Why is effective TDR instrumentation only a recent development, although the principle has been known for many years?
2. How does pulse width determine the maximum distance at which faults can be located?
3. How does pulse width also determine the resolution and precision of the distance measurement?
4. Why do most TDR instruments have a CRT screen for the operator? What can the technician adjust? Why is this adjustment available?

SUMMARY

Communications systems often transmit the modulated signal from sender to receiver using wire and cable. Twisted-pair wire and coaxial cable are among the different types of wire and cable commonly used. The resistance, capacitance, and inductance of the wire at various frequencies play a major role in the performance that can be achieved. All cables have maximum useful signal bandwidth, different amounts of resistance to external noise, and practical considerations such as flexibility and ease of connection.

The technical issues of balanced and unbalanced signal lines, as well as common-mode voltage and rejection, show how the use of any wire does not ensure ideal performance, unless the application is matched to the cable performance. Line drivers and receivers act as the signal interface between the modulated signal and the cable, and must be able to drive the cable and its capacitance with a voltage (and current) that slews from one value to another at a rate corresponding to the frequencies in the signal. Cable and wire problems anywhere in a cable can be located using time-domain reflectometry, working from one end of the cable only. This technique is very powerful and avoids the need to dig, cut, or waste effort looking for a fault in a long length of cable.

Summary Questions

1. Explain how *IR* drop can be significant in long lengths of cable or when the current carried is relatively large.
2. Why is the simple view of a ground as a perfect conductor with the same potential everywhere not correct? What is it really like?
3. Compare single-ended and differential interconnections in terms of number of wires needed, resistance to noise and CMV, resistance to *IR* drop, and connection of test instruments.
4. Explain the effect of high CMRR on received CMV.
5. What are the roles of the line driver and receiver?
6. What is slew rate? Where does the maximum slew rate for a sine wave occur?

What formula links the sine-wave amplitude and frequency to the maximum slew rate?

7. How does cable capacitance affect achievable slew rate? Why must the driver provide current at a high enough rate?

8. Compare an analog signal driver/receiver with a digital pair. What is the primary difference in how they must function?

9. Compare coax cable and twisted pair for ease of connection, noise resistance, and variety of choices.

10. What is the result of shielding twisted pair compared to unshielded?

11. What is the concept of TDR? Why must time be measured precisely?

12. Why is TDR for long distances difficult to use to locate faults to very small areas of the cable? Why is this usually not a problem?

13. What determines the maximum distance at which TDR will operate? What determines the ability to locate the fault with great precision?

PRACTICE PROBLEMS

Section 7.2

1. A system is using HI and LO voltage of 10 V and 5 V. What is the differential signal voltage?

2. A CMV of -7 V is added to the HI and LO values of problem 1. What are the received signal voltages?

3. A 5-V differential signal is received with 10 V CMV, which then is reduced to 0.15 V. What is the CMRR?

4. The system of problem 3 now has a CMV of 8 V, with CMRR unchanged. To what voltage value is the CMV reduced?

5. A system uses -5-V and 5-V signals and has a CMRR of 80 dB.
   (a) What is the differential voltage in use?
   (b) To what value is 15 V of CMV reduced?
   (c) What is the final differential voltage after the receiver?

Section 7.3

1. What is the slew rate of a signal that is going from -3 to $+6$ V in 25 μs?

2. A signal must go from -4 to $+4$ V in 10 ns. Can a driver that provides a slew rate of 1000 V/μs be used?

3. What is the slew rate (maximum) of a 5-V 1-MHz sine wave?

4. A 10.7-MHz IF signal of 1 V amplitude must be transferred via a driver/receiver to another stage in the system. What is the slew rate required?

5. How much current is needed to drive the signal of problem 1 into a cable with 175 pF of capacitance?

6. Can a current drive capability of 100 mA drive the signal of problem 4 into a 25-pF load?

7. An SN75107 line receiver sees -2 V (HI input) and $+1.5$ V (LO input). What is the logical output level?

8. The same line receiver as in problem 7 has received signals of −5 V (LO) and −3 V (HI). What is the output state?

Section 7.5

1. What is the time between the initial TDR pulse and the received reflection when the fault is 5 m away for a cable with propagation of $0.85c$?

2. A TDR reflection is received 125 ns after the initial pulse, with a cable with propagation at 67% of c. How far away is the fault?

3. A TDR unit can measure reflection time with 2 ns precision. Within what uncertainty of distance can it locate faults in cables with a propagation of $0.95c$?

4. A TDR instrument for finding faults in short cables must locate these faults to 1 cm (0.01 m). To what precision must the TDR unit measure the reflection time for signals traveling at the speed of light?

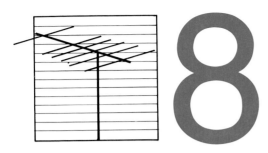

8

Transmission Lines

CHAPTER OBJECTIVES

When you have completed this chapter, you will understand:

- Concepts of transmission lines and line impedance at high frequencies
- Use of microstrip transmission lines on circuit boards
- Waveguides for carrying higher frequencies and power levels
- Effects of impedance mismatch between line and load, and techniques for correcting this mismatch (including the Smith chart)
- *S*-parameters for studying line/load networks and their relationship to the Smith chart

INTRODUCTION

The study of transmission lines—which carry signal energy at relatively high frequencies—requires understanding line impedance, termination, signal reflection, and matching the line and load impedances. The coaxial line and open wire parallel line are two choices for transmission line. Microstrip lines, which use carefully dimensioned traces on printed circuit boards, and waveguides, which carry signal energy in a confining conducting enclosure, are important alternatives. Line and load matching require understanding signal reflections on a line, how they are measured, and the effects of mismatch. Many systems are now characterized by *S* parameters, which show how signal voltage divides and reflects. These *S* parameters, in turn, can be applied to the Smith chart, a very useful tool for analyzing line/load matching and determining how to correct for mismatch at various frequencies.

8.1 IMPEDANCE AND LINE FUNDAMENTALS

The goal of a transmission line is simple: to transfer the energy of a signal from a source at one point to a load at another point, with very little loss. The application can be a transmitter connected to an antenna, an antenna feeding a weak received signal to a receiver front end, or signal transfer between two subsections of a larger system chassis. Since this energy contains high frequencies (not just low frequencies or direct current), the wire used for carrying the energy cannot be considered as a simple resistance. Instead, the characteristics of the wire at the signal frequencies must be considered and the simple wire is now considered as a transmission line.

The most important specification of a transmission line is its *characteristic impedance*, Z_0. This impedance and its value relative to other impedances at the source and the receiver determines how well the signal energy will propagate through the line. As we saw in the discussion of TDR, any discontinuity in the line will reflect energy back to the source instead of to the intended receiver. Although the most severe discontinuities are shorts or opens, any impedance differences (*mismatch*) will cause energy to be wasted. Maximum power is transferred from the source to the line, or the line to the receiver load, when the impedances are *complex conjugates* with the same real (resistive) values and opposite imaginary (reactive) values, such as $(6 + 3j)$ and $(6 - 3j)$. Properly matching these system impedances is the key to making sure that transmitted energy reaches the receiver with minimal loss.

Coaxial cable, discussed in Chapter 7, is a very common transmission line. It is used, in various sizes, at frequencies up to approximately 20 GHz (20,000 MHz). Above 1 GHz, however, coax may not provide acceptable performance, and other types of transmission lines such as waveguides (Section 8.3) are used. A model of coax that includes its frequency-dependent elements is shown in Figure 8.1. (This figure is a more realistic model than the simpler wire and cable model of Chapter 7.) The *L*, *C*, *R*, and *G* elements represent a much more correct view than simply a copper wire surrounded by a shield. The *admittance G* is the inverse of the resistance of the *dielectric* (air, polyethylene, Teflon, or other material) between the shield and the center conductor, and is the result of the small but measurable leakage current that flows between them.

Although the schematic shows these elements as individual discrete components, they are not physically lumped clusters of resistance, capacitance, or in-

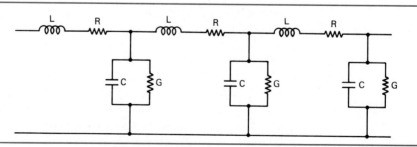

Figure 8.1 Electrical model of coaxial cable as a transmission line.

ductance to which you can point. Instead, they exist as continuously distributed values along the length of the coax and are specified in electrical units per foot or meter of cable length.

If the impedance of an infinitely long piece of transmission line is measured, the value indicated will be Z_0. Of course, infinitely long lines do not exist. However, a line of finite length will have the same value of Z_0 if the far end is terminated in a resistance equal to Z_0. An analysis of the impedance of a coax line shows that

$$Z_0 = \sqrt{\frac{R + j2\pi f L}{G + j2\pi f C}}$$

which is a complex impedance. Fortunately, for most practical purposes the dc resistance R and the leakage admittance G are very low compared to the reactances of L and C, so they can be ignored. The result is

$$Z_0 = \sqrt{\frac{j2\pi f L}{j2\pi f C}} = \sqrt{\frac{L}{C}}$$

which is not a complex impedance but a simple resistance. (Note that this is not a tangible, physical resistor that you can point to, but is the V/I transfer ratio.) The impedance can be constant versus frequency, or may vary depending on the behavior of the L and C values as the frequency changes.

Measuring Z_0 involves measuring the inductance and capacitance, per unit length, of the coax.

Example 8.1

A cable has inductance of 15 nH/m and 85 pF/m at a specified frequency. What is the impedance?

Solution

$$Z_0 = \sqrt{\frac{15 \times 10^{-9} \text{ H}}{85 \times 10^{-12} \text{ C}}} = 0.177 \times 10^3 = 177 \; \Omega$$

The physical dimensions of the coax can also be used. Analysis of the electrical and magnetic fields of coax cable shows this relationship between physical dimensions and characteristic impedance:

$$Z_0 = \frac{138}{\sqrt{\epsilon}} \log \frac{D}{d}$$

where D is the outer coax diameter, d is the inner conductor diameter, and ϵ is the *dielectric constant* of the material between the center conductor and the outer shield, equal to 1.00 for vacuum, 1.0006 for air, and up to about 2.8 for various other materials (2.3 for polyethylene, 2.55 for Teflon).

With this equation, the range of impedances available is determined. For example, coax with an outer diameter of 0.35 in., center conductor diameter of 0.05 in., and vacuum dielectric will have

$$Z_0 = \frac{138}{\sqrt{1.00}} \log \frac{0.35}{0.05} = 117 \, \Omega$$

Example 8.2

The coax with inner and outer diameters specified above now uses polyethylene dielectric ($\epsilon = 2.3$) to keep the inner conductor and shield separated. What is the new value of Z_0?

Solution

$$Z_0 = \text{previous } \frac{Z_0}{\sqrt{\epsilon}} = \frac{117}{\sqrt{2.3}} = 76.9 \, \Omega$$

Open Wire Transmission Line

An alternative to coaxial cable is the *open wire transmission line*, which consists of two wires running parallel to each other. These wires are commonly separated only by air and are held apart by spacers every few inches. Open wire line is useful in balanced systems where neither transmission wire should be grounded (in most designs, the coax shield must be grounded—or special nongrounded, insulated

Losses

No real transmission line can be perfect, passing all the energy that enters it without loss. There are three sources of loss for coax: resistive heating, radiation, and dielectric heating. The resistance of the conductor and shield means that any current carried, regardless of frequency, is subject to *resistive heating* I^2R loss. At lower frequencies, this loss is reduced simply by using thicker wire with less resistance. However, as signal frequencies increase, more and more of the signal energy stays closer to the outer surface, or *skin*, of the conductor, so loss increases with frequency due to the *skin effect*. Increasing the conductor size to reduce IR loss helps but not for the usual reason, since the increased thickness does not affect the way that the signal energy stays on the conductor skin. Larger-diameter conductors are used to reduce loss since they offer more circumferential "skin" perimeter, not because they have less dc resistance. Hollow conductors are sometimes used to save weight and cost, with no change in effective resistance at high signal frequencies (since nearly all of the electromagnetic energy is concentrated on the surface).

There is also *radiation* loss of signal energy when the coax cannot restrain all of the signal energy to stay within the shield. The shield is typically made of braided wire, foil, or metal tubing. The braiding is the easiest to use and least costly—but it is not 100% perfect. Some energy will escape through the weave of the braid, typically 1 to 20%. Since higher frequencies have shorter wavelengths and can pass through the braid weave more readily, coax for these higher frequencies uses solid foil shielding (in low-power applications) or tubing (which is very hard to handle).

Dielectric heating comes from the *leakage* current that flows through the dielectric of the coax. Vacuum and air have the lowest dielectric loss but are often impractical since some sort of spacer must be used to keep the center conductor away from the outer shield. Polyethylene is often used instead, and some special "foam-like" plastics are also common. Overall, coax losses range from 1.5 dB to over 10 dB per 100 m, depending on the signal frequency, coax size and construction, and dielectric used.

8.1 Impedance and Line Fundamentals

connectors must be used). Compared to coax, open wire provides lower loss, but installation must be away from nearby metal and the lines must be fairly straight, with gentle curves and no sharp bends. This makes it impractical in many situations. Coax, by contrast, can be bent and routed as needed and can be run near metal or even through metal conduit since it is shielded.

The impedance of open wire line can be determined by analysis, just as it is for coax. It is

$$Z_0 = \frac{276}{\sqrt{\epsilon}} \log \frac{2s}{d}$$

where s is the conductor center-to-center spacing, and d is the conductor diameter. For air as the dielectric, the ϵ factor does not differ significantly from 1; other dielectrics produce different values of k and so affect Z_0.

Example 8.3

What is the value of Z_0 for open wire line made up of standard No. 18 AWG wire (0.040 in. in diameter) spaced 0.4 in. apart?

Solution

$$Z_0 = 276 \log \frac{2(0.4)}{0.040} = 276 \log 20 = 359 \, \Omega$$

Unlike coax, open wire transmission line can easily be constructed as needed for nonstandard Z_0 values. For example, the load may be a transistor input with 112-Ω Z_0, for which there is no available standard coax. Special coax could be fabricated, but at great cost—which may be impractical if only a small amount is needed. The open wire line with desired 112-Ω impedance, however, can be built as needed for the application with some wire and spacers.

One common form of parallel-wire transmission line is standard *twinlead* used between a TV antenna and the TV, with conductors held apart along their entire length by plastic insulation. This 300-Ω line is very low cost, easy to terminate, and available with different protective jackets for indoor or outdoor use. (The impedance of low-cost, low-quality twinlead changes and shunt resistance decreases as the insulation characteristics are affected by pollution and by changes in chemical characteristics of the plastic insulation with temperature and time.) Another form of parallel wire is *multiple-conductor flat cable*, which can have between 10 and 50 wires running in parallel. For both twinlead and flat cable, the basic Z_0 formula is used with the value of ϵ corresponding to the type of plastic insulation (ϵ approximately 2 to 2.5) that occupies the space between the conductors.

Review Questions

1. What is the function of a transmission line? Why must the wire be viewed as much more than a simple resistance?

2. What is the characteristic impedance of a line? Compare Z_0 for an infinitely long line to its value when a finite-length line is terminated in Z_0.
3. What is the practical frequency range of coaxial cable? What is the true model of coax? Explain what it means to say that the R, L, C, and G elements are not lumped but are distributed along the line.
4. What is Z_0 for coax, determined from the L and C values? What is the value from its physical dimensions alone?
5. What is the role of the dielectric constant? What is the range of k?
6. What are line losses? Explain the three sources. What is the skin effect?
7. Compare the features of open wire line versus coax with respect to line load and ease of installation.

8.2 MICROSTRIP LINES AND STRIPLINES

Coaxial cable, parallel open wire, and other forms of transmission line are practical when the distance between the signal source and load is more than a few inches. At shorter distances, the connectors and terminations may take up too much space. Modern circuitry makes extensive use of *printed wiring boards* [also called *printed circuit (PC)* boards] with many components in a small space. The coax or open wire to connect a transistor or IC output to the input of the next stage would be relatively large, difficult to install in tightly packed circuitry, and costly in the final application.

Fortunately, an alternative exists for short-distance transmission lines on PC boards. *Microstrip lines* use the traces (tracks) of the PC board itself, in various configurations. These traces can be designed and etched just as any other traces on the board, and therefore not add to the manufacturing process. (When the lines are only on the surface of the circuit board, they are called *striplines;* when they are in a middle layer of a *multilayer* circuit board they are known as microstrip lines. We will use the microstrip lines term for both, for convenience.)

Using microstrip line is like getting the transmission line for free, as part of the board layout. The specific dimensions of the traces are calculated using various formulas, to give the desired characteristic impedance. This is another advantage of microstrip lines: a wide range of Z_0 values can be designed by simply varying the line dimensions. The practical range of microstrip impedances is between 50 and 200 Ω, typically, but the values can be outside this range by careful choice of key dimensions.

Several different microstrip configurations are used. Figure 8.2 shows cross sections of three versions, along with their formulas for Z_0. The simple microstrip line of Figure 8.2a uses a single track on the PC board top side and a large, grounded section of copper on the bottom. This microstrip line is grounded, like a coax line, and the ground plane must be wide compared to the top conductor (at least 10 times wider), so it appears like a nearly infinitely wide *ground plane* with

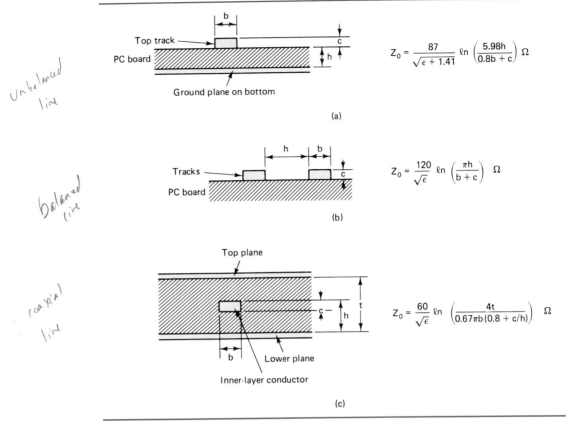

Figure 8.2 Three examples of microstrip lines and impedance, viewed from the edge of the PC board.

only very small electric field fringes at its edges. The second configuration (Figure 8.2b) is similar to parallel open wire transmission line. Neither track is grounded, so this can be used for balanced signals.

Example 8.4

What is the value of Z_0 for the single 0.1-in.-wide, 0.005-in.-thick track plus ground plane microstrip line? Assume that the PC board is 0.075 in. thick and that the dielectric constant of the board is 2.0.

Solution

$$Z_0 = \frac{87}{\sqrt{2.0 + 1.41}} \ln \frac{5.98(0.075)}{0.8(0.1) + 0.005}$$

$$= (47.1) \ln 5.27 = 78.4 \; \Omega$$

(Note: ln is the natural logarithm, to the base e = 2.718. . .)

Example 8.5

Use the figures of the previous examples to find the impedance of microstrip lines used in parallel open wire transmission. Assume that the spacing between conductors is 0.075 in.

Solution

$$Z_0 = (120 \sqrt{2.0}) \ln \frac{\pi (0.075)}{0.1 + 0.005}$$

$$= (84.85) \ln 2.24 = 68.6 \ \Omega$$

The third version (Figure 8.2c) is a true microstrip line and uses a *multilayer* PC board. The inner layer has the signal-carrying conductor, while the top and bottom of the PC board have ground planes similar to the single plane in the first microstrip design. This microstrip design offers better performance than that of the simpler two-sided PC board version, with less signal loss through radiation. It does, however, require a more expensive multilayer board. Also, since the inner conductor is inside the PC board, it cannot be physically trimmed to change its dimensions—it must be designed and manufactured correctly. Many microstrip lines need some manual fine adjustment to compensate for nearby components, manufacturing tolerance in dimensions, or changes in required Z_0.

Microstrip lines are used extensively with PC boards used for communications at microwave and shorter wavelengths. Maintaining the desired Z_0 requires that the track width and spacing dimensions be held to tight tolerances and that these dimensions and the dielectric constant of the PC board stay constant with changes in temperature and humidity. For this reason, special PC board material, such as Teflon and fiberglass, is often used instead of the more common fiberglass–epoxy material.

Review Questions

1. Why are coax and parallel open wire transmission lines often not suitable for short distances?
2. What is a microstrip line? How is it implemented? What determines Z_0?
3. Compare the microstrip line for a grounded design with a balanced line. What is the ground plane? Why must it be large compared to the signal conductor?
4. Compare the grounded microstrip line on a simple two-sided PC board with a multilayer board stripline. What are the advantages and disadvantages of each?
5. Why must the microstrip line traces often be manually "trimmed"? What factors must be controlled to achieve the intended value of Z_0?

8.3 WAVEGUIDES

At frequencies of approximately 1 GHz and above, the ability of cable (coax or other type) to effectively carry energy decreases quickly, and above 18 to 20 GHz, coax is generally not usable except for very short distances (a few centimeters or inches). The high losses and attentuation in the cable from skin effect and radiation result in little of the initial energy reaching the load, even if the line-to-load match is perfect. Also, it is difficult to transfer large amounts of power, because the voltages that occur at high power levels will break down the dielectric barrier between conductors (so very wide spacing is required).

An alternative to cable is the *waveguide* (Figure 8.3). A waveguide is a conducting tube through which the energy is transmitted, in the form of electromagnetic waves. This waveguide tube is not carrying a current in the same way that a regular cable does. Instead, it acts as a boundary or enclosure for the space through which the electromagnetic wave propagates. The complete enclosure and skin effect prevent any energy from radiating outside the waveguide, so there are virtually no losses due to radiation. The signal energy is *injected* at one end of the waveguide by a signal launcher and received at the other end, where it is removed by a signal absorber. The waveguide confines the propagating energy by reflecting it off the conducting walls. Waveguides can be joined together using their flanged ends to form longer lengths, if needed.

Warning: *Never look into a waveguide or stand in front of the open end* unless you know that the other end is disconnected, or the associated electronics system is off and cannot be switched on by someone else. Waveguides can carry large amounts of high-frequency/short-wavelength electromagnetic energy, and this energy radiation can cause eye damage, genetic damage, and other temporary

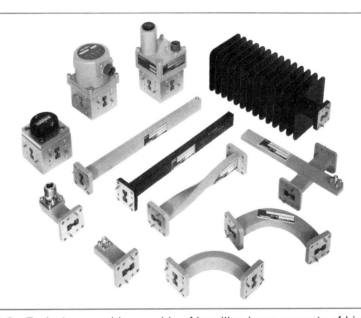

Figure 8.3 Typical waveguide, capable of handling large amounts of high frequency and microwave power with little loss (courtesy of Waveline, Inc.).

or long-term injury. Be careful, double check, keep safety in mind. Just because you don't see any wires, do not think that the situation is necessarily a safe one.

The electrical and magnetic fields patterns in the waveguide have varying complexity. Many different distributions of field energy are possible, with a typical case shown in Figure 8.4. The electric field intensity is greatest along the center of the x dimension (the waveguide width), decreasing sinusoidally to zero at the walls. The magnetic field consists of closed loops in the plane that is normal (at right angles) to the electric field, and is therefore parallel to the top and bottom of the waveguide and is the same in any plane from the top and bottom.

Modes of Propagation

Figure 8.4 illustrated a relatively simple distribution of the fields through the waveguide. Each possible distribution of energy is called a *mode*, and these modes can be separated into two major groups. The *transverse magnetic (TM)* mode has the magnetic field at right angles to (transverse) the direction of propagation along the guide, and the electric field in the direction of propagation. The other mode, *transverse electric (TE)*, has the electric field traverse the direction of propagation while the magnetic field is along the propagation direction.

The specific mode of transmission is shown by two subscript numbers added to the TE or TM designation. The most common mode is the $TE_{1,0}$ mode, called the *dominant mode*, shown in Figure 8.6. The number of potential modes increases with the frequency being used, for a given size of guide. As the frequency increases, the waveguide propagation shifts to the next mode and has more complex fields. The dominant mode is the lowest mode that can be used for the frequency of interest, and is usually the mode used in practical applications. The electric and magnetic field pattern for modes above the dominant mode is more

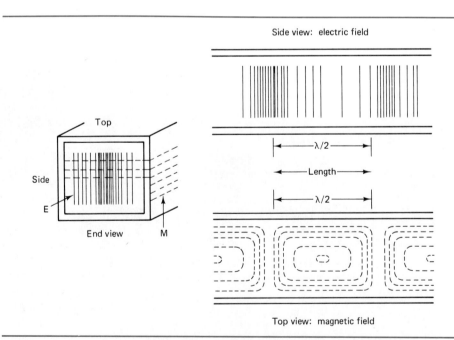

Figure 8.4 Electric and magnetic field patterns for dominant mode of waveguide operation, as viewed from the top, side, and end.

8.3 Waveguides

complex, with a larger number of tightly spaced, smaller loops and circles. Higher-order modes offer additional flexibility in some applications, but are more difficult to control properly.

Waveguides have rectangular or circular cross sections. When energy must be coupled from a source to a load and both are fixed in place, rectangular guides are used because they are smaller than circular waveguides for a given wavelength. The dimensions of a waveguide are very critical. If the wavelength of the frequency in use is too long, the waveguide will not develop a mode of operation and no signal will pass. The lowest frequency that a waveguide will handle is its *cutoff frequency;* only frequencies above this value—corresponding to shorter wavelengths—will propagate. For a rectangular waveguide, the formula linking the width x and this longest wavelength (the *cutoff wavelength*) is

$$\lambda = \frac{2x}{m}$$

where m is the mode number (1, 2, 3, . . .). Simply stated, the x dimension must be more than one-half the wavelength desired for mode 1, the dominant mode. A 1-cm wave, 30 GHz, requires an x dimension of greater than 0.5 cm. The y dimension (waveguide height) can vary over a wide range, but it is usually made equal to about half the x value, to inhibit modes other than the dominant one.

Example 8.6

What is the longest wavelength that a 2.5-cm-wide waveguide will support in the dominant mode ($m = 1$)? How about for the next mode ($m = 2$)?

Solution

$$\lambda = \begin{cases} \dfrac{2(2.5)}{1} = 5 \text{ cm (6 GHz)} & \text{for the dominant mode} \\ \dfrac{2(2.5)}{2} = 2.5 \text{ cm (12 GHz)} & \text{for the next mode} \end{cases}$$

From this we see that for a fixed-size waveguide, only higher modes can be supported at higher frequencies. Conversely, a fixed-size waveguide can support lower frequencies if the lower modes are used.

Some simple formulas link the cutoff wavelength, the longest wavelength that can be transmitted without attenuation, and the shortest wavelength that can be transmitted before the next mode above the dominant mode becomes possible. These are given in Figure 8.5 for rectangular and circular waveguides.

Example 8.7

What is the minimum x dimension needed to support a 0.7-cm signal in the dominant in a rectangular waveguide?

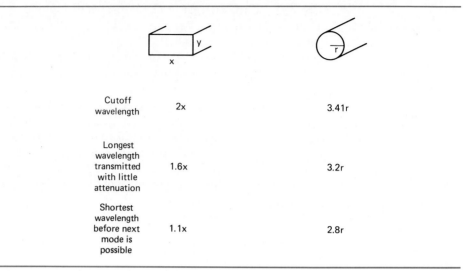

Figure 8.5 Key wavelength formulas for rectangular and circular waveguides.

Solution

$$x = \frac{\lambda}{2} = \frac{0.7}{2} = 0.35 \text{ cm}$$

Example 8.8

At what wavelength does the waveguide of Example 8.7 become capable of supporting mode 2?

Solution

$$\text{Next } \lambda = 1.1x = 1.1(0.35) = 0.385 \text{ cm}$$

Therefore, to ensure dominant-mode operation, the rectangular waveguide should be used only for wavelengths between 0.7 and 0.385 cm. Note that a waveguide for a 100-MHz signal (λ = 30 cm) would require an x dimension of 15 cm, or about 1 ft. This is impractical in most situations, but it has been done on occasions in specialized applications such as radar transmitters.

The waveguide is a sharp high-pass filter with slightly increasing attenuation versus frequency, unlike most circuits, which pass lower frequencies relatively well but have increasing attentuation with increasing frequency. Below the cutoff frequency, no signal is passed. Above the cutoff value, signals can propagate but with some steadily increasing attenuation as the frequency increases. At some frequency value the operation shifts to the next mode, which is still effective but has more complex field patterns.

Circular Waveguides

Rectangular waveguides are convenient to use, mount, and install. However, they are not symmetrical around an axis down the center. For rotating systems such as a radar antenna turning to scan the horizon, this asymmetry will cause physical problems (a rectangular joint cannot mechanically rotate) as well as electrical problems. The energy pattern and mode will change as the antenna rotates and so presents a different perspective to the rectangle. A circular waveguide, of course, looks the same from any angle around its centerline, and will be used instead. A special transition rectangular to circular waveguide can be used to allow the use of both in a system, if needed.

For a circular waveguide with radius r, the longest wavelength is

$$\lambda = \frac{2\pi r}{k}$$

where k is 1.84 for the dominant mode; therefore, $\lambda = 3.41r$ for the dominant mode. The value of k for each mode is determined by a complex Bessel function equation, unlike the simple $m = 1, 2, 3$ factors for the rectangular waveguide.

Example 8.9

What is the minimum radius r needed to support a 0.7-cm signal in the dominant mode in a circular waveguide?

Solution

$$r = \frac{\lambda}{3.41} = \frac{0.7}{3.41} = 0.205 \text{ cm}$$

Compare this result to the $x = 0.35$ cm dimension in a rectangular waveguide at this wavelength.

Waveguide Couplers

Energy must be *coupled* into the waveguide from the signal source. Obviously, this is not done by soldering the cable to the guide! The energy transfer can be done using either the waveguide electric field or its magnetic field (hence the interest in the field direction, indicated by the TE or TM designation). Note that the techniques for coupling energy into a waveguide also apply to extracting energy from the waveguide.

Coaxial cable can be used if oriented properly. The inner conductor of the coax is extended into the waveguide as a probe, shown in Figure 8.6a. When the probe is parallel to the electric field lines of force, the energy will be coupled by the electric field. If the probe is arranged in a loop so that it encloses the magnetic lines of force, magnetic coupling is achieved (Figure 8.6b). The same probe or loop can also be used to extract energy from the waveguide.

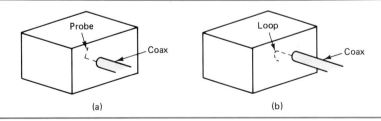

Figure 8.6 (a) Coax E field coupler for waveguide; (b) coax M field coupler loop for waveguide. Each must be inserted at the correct physical point in the waveguide to intercept the desired field.

The amount of energy transferred by the coupling is affected by two factors. The coupling is maximum when the coupling device is inserted into the most intense part of the electric or magnetic field. Second, the amount of coupling at a given location in a field can be varied by rotating or tilting the probe or loop from 0 to 90°, relative to the field. Coupling for the probe in an electric field is proportional to cos ϕ, where ϕ is the angle between the probe and the field lines. It is a minimum when the probe is perpendicular to the field and is maximum when parallel to the field. For magnetic coupling, the minimum occurs when the loop is parallel to the magnetic lines, and the maximum, when the loop is perpendicular. It is proportional to sin ϕ, where ϕ is the angle between the loop and the magnetic field.

Example 8.10

A probe for coupling into the electric field has a 20° tilt with respect to the electric field. Compare the coupling to the maximum value.

Solution
Coupling is proportional to cos ϕ and is maximum at 0°, where cos ϕ = 1. At 20°, cos ϕ is 0.94, so the relative coupling is 94% of maximum.

Example 8.11

Repeat Example 10 for a 20° rotation from parallel for a magnetic loop coupling.

Solution
Coupling is proportional to sin ϕ, with the maximum at 90°, where sin ϕ = 1. At 20°, the coupling is sin 20° = 0.34 of maximum.

Note that coupling signal energy into a waveguide is very different from transferring energy into a cable. For a cable, the connection is made at the end by physical (metallic) contact, with voltage, current, and impedance the main parameters of interest. Waveguide coupling is associated mainly with induced electric and magnetic fields, and the coupling location and the angle are important. Rigorous analysis with electromagnetic field theory shows that these are two perspectives on the same general case of the propagation of electromagnetic energy.

Slot Coupling
Another method used to couple signal energy into or out of the waveguide is to cut a *slot*, or *aperture*, in the guide wall. This slot allows some of the electric or magnetic field to "escape" the confinement of the waveguide. The field now extends into territory outside the guide, and signal energy can be coupled through these field lines.

The location of the slot is critical and depends on the frequency in use. At specific points and orientations in the waveguide wall, the electric field will be a maximum, while the magnetic field is a maximum at other locations and orientations. These maximum locations are preferred, so there is maximum field exposure and coupling. Two waveguides with slots can be used to couple energy from one waveguide to the other if their slots are aligned next to each other. The slot also acts as an energy radiator and can be used as a basic antenna in some applications, or is used as a pickoff point for monitoring the energy in the waveguide. Higher-order modes, above the dominant mode, are more difficult to couple into because the electric and magnetic field patterns are more complex.

Waveguide Losses

Waveguides are capable of transferring enormous amounts of power with little loss. A waveguide as small as 0.5 × 1.0 in. can propagate several hundred thousand watts of power. The main sources of energy loss for waveguides are from mechanical misalignment between joined sections (which may be done deliberately to attenuate the signal using a special section with built-in adjustable obstacles), losses due to energy flowing in the waveguide walls, and losses in the waveguide dielectric (similar to I^2R and dielectric loss in cables).

Loss figures for typical air-filled waveguides, per 100 m of length, are 5 dB at 5 GHz, 12 dB at 10 GHz, and 20 dB at 35 GHz. Note that 100 m is much longer than most waveguides in application, so the actual losses are much less. The 100-m figure is used as a standard of convenience, and the loss per meter is 1/100 the loss per 100 m. The material of the waveguide affects the loss factor, and some waveguides are plated on the inside with silver or gold to reduce losses. Air, the most common dielectric, has very low dielectric loss; any other dielectric within the guide will have higher loss.

Review Questions

1. What is a waveguide? Compare it to cable in terms of frequency span, effectiveness, losses, radiation, and method of propagating signals.
2. What are the modes of waveguide operation? What is the dominant mode? What is the cutoff wavelength?
3. Why is it necessary to know the general configuration of the electric and magnetic fields in a waveguide?
4. Compare physical and electrical characteristics of rectangular versus circular waveguides.
5. How is signal energy coupled into and out of a waveguide electric field? How about for a magnetic field?
6. What is a slot coupling? How is it used? What are some critical factors?
7. What are the sources of waveguide signal loss? What are some typical loss figures, in dB/100 m? What reduces this loss?

8.4 LINE AND LOAD MATCHING

Electrical theory shows that maximum power is transferred from a source to a load when the source impedance is the *complex conjugate* of the load impedance (such as $6 + j10\ \Omega$ and $6 - j10\ \Omega$). The signal source is usually referred to as the *generator* but does not have to be the actual starting point of the signal. A trans-

Electrical Length

An important concept in transmission lines and matching is *electrical length,* which indicates the length of a cable in terms of the number of wavelengths of signal. Electrical length is critical in line/load matching applications when lines need to be specified in signal wavelengths, not in ordinary length units of feet or meters. It differs from the measured physical length because it is a function of the frequency of the signal and the propagation velocity of the signal in that type of cable. A 10-m cable has an electrical length of $10/300 = 0.033\lambda$ for a 1-MHz signal (300-m wavelength), and 3.33λ for a 100-MHz (3-m) signal, assuming propagation at c. Electrical length is expressed as number of wavelengths:

$$\text{electrical length} = \frac{\text{physical length}}{\text{signal wavelength}}$$

$$= \frac{\text{physical length}}{\text{propagation velocity/frequency}}$$

Example 8.12

What is the electrical length at 150 MHz for a 2.5-m cable with a propagation velocity of $0.85c$?

Solution

$$\text{Electrical length} = \frac{2.5}{(3 \times 10^8 \times 0.85)/(150 \times 10^6)}$$

$$= 1.47\lambda$$

Example 8.13

What physical length corresponds to $\frac{1}{4}\lambda$ at 500 MHz for cable with a propagation velocity of $0.68c$?

Solution

Rearranging the equation, we have

$$\text{physical length} = \text{electrical length} \times \frac{\text{propagation velocity}}{\text{frequency}}$$

$$= 0.25 \times (3 \times 10^8) \times \frac{0.68}{500 \times 10^6}$$

$$= 0.102 \text{ m}$$

Note that electrical length is a concept associated only with transmission lines—it never occurs in discussions of cables or wire that supply power to circuitry.

---

mission line connected to a broadcast antenna is the signal source for the antenna acting as load. A receiving antenna can be the source for the front-end preamp input circuitry (the load). We will use the simple line/load terminology, while recognizing that the "line" being matched to the "load" can represent any pair of elements with signal energy passing from one to the other.

A transmission line that is terminated with its characteristic impedance is a *nonresonant* or *flat* line, since energy put into the line from the source end will not oscillate back and forth—it will all be transferred to the load and not seen again at the source end. The effects of mismatch between line and load, and ways to measure and correct for this mismatch, have resulted in some simple but effective techniques.

We can better understand the line and load mismatch by looking at two extreme cases: when the load is an open circuit, and when it is a short circuit. Consider an alternating waveform that is impressed on one end of a line which is open at the load end. The voltage, when it reaches the load, has nowhere to go. It reflects back toward the source due to the mismatch, in phase with the way the signal appears at the load end. Note that since the termination is an open circuit, the voltage across the load end is equal to the incident voltage (ignoring any losses

8.4 Line and Load Matching 239

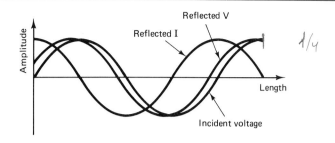

Figure 8.7 Incident signal voltage and reflected voltage, and current on transmission line with open circuit at far end shows relative phase angle between reflected current and voltage.

in the line). Also, since there is no path for electrons at the load end, the current is zero.

Compared to the signal incident at the load, the reflected voltage is in phase while the current is 90° out of phase. A drawing of the reflected voltage and current shows that the voltage repeats periodically, at the frequency of the incident signal, and the current also repeats periodically, but with $\frac{1}{4}$ cycle (90°) phase shift (Figure 8.7). The points along the transmission line where the voltage, or current, is a minimum (ideally 0) are *nodes,* while the points where they are at their maximum are *antinodes.*

The situation for a short at the load is similar, but with some important differences. The incident signal voltage reflects its mirror image and is therefore 180° out of phase with the signal as it reaches the load (Figure 8.8). Since the end is shorted, the voltage across the load is zero while the current is a maximum. As in the case of an open at the load, there will be a 90° phase difference between the current and the voltage everywhere along the line, but the phase shift is in the opposite direction of the open-circuit case. The nodes and antinodes still exist, as in the case of the open termination.

The definition of electrical impedance says that it is the ratio of voltage to current. The reflections from the load end of the line cause periodic repetition of voltage and current cycles but 90° ($\frac{1}{4}$ wavelength) out of phase with each other. Therefore, the ratio of the voltage to the current is also periodic and depends on

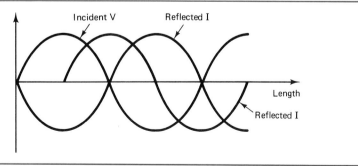

Figure 8.8 Incident signal voltage and reflected voltage and current on transmission line with short circuit at far end, also showing relative phase angle between reflected current and voltage.

the point on the transmission line where it is measured, as seen by looking carefully at the figures. The impedance of the line and its load that the user sees is not constant. Instead, the impedance repetitively cycles from inductive to capacitive, and then back to inductive, as the user connection point moves along the transmission line.

When the transmission line is terminated in its characteristic impedance, the impedance seen at any point on the line is the same everywhere. But when the line and load are mismatched, the situation is counter to intuition (and your observations at low frequencies): The impedance changes radically (capacitive to inductive to capacitive) with position, even though the physical components are unchanged. Put another way, although the transmission line and the load are unchanging physical pieces connected together, the impedance result of the line and load connection does in fact vary with position along the line and line length.

The input impedance of a section of transmission line that is exactly $\frac{1}{4}$ wavelength long (equivalent to 90°) and short circuited at the far end looks like an open circuit, while an open-circuit termination looks like a short circuit at this distance. If the load impedance is purely capacitive, then at $\frac{1}{4}$ wavelength distance it will look purely inductive, and vice versa, due to the phase difference between voltage and current. Over a $\frac{1}{2}$-wavelength distance, the impedance changes from inductive to capacitive (or the reverse, depending on your direction of travel). This complete cycle of change in impedance repeats every $\frac{1}{2}$ wavelength along the transmission line.

Figure 8.9 shows the current and voltage measured along a transmission line with different loads (impedance is the vector ratio of voltage to current). Traces (a) and (b) show the most extreme situations, for a shorted load and an infinite load (open-circuit load). Note the $\lambda/2$ periodicity in voltage and current and that the voltage minimum value is 0. In trace (c), the line is terminated in its characteristic impedance with $Z_L = Z_0$, so the current and voltage are constant and nonzero everywhere. A purely capacitive load and a purely inductive load are shown in (d) and (e), respectively. Here the voltage and current values repeat periodically but with different phase relationships for the capacitive and inductive cases, and the minimum value is 0. Trace (f) shows a most general termination with $Z_L = R + jX$. Here the voltage and current repeat, with their exact phase difference determined by the relative proportion of reactive to resistive load and Z_0. Since the line has real resistance, the current and voltage do not go down to zero.

The figure also shows the significance of electrical length. The impedance characteristics of the mismatched line and load repeat every $\frac{1}{2}$ wavelength. Therefore, in transmission-line issues related to line and load matching, what matters is not the line length in exact number of electrical wavelengths, but the fractional length with the integral number of wavelengths disregarded. For example, a line that is $6\frac{3}{8}$ wavelengths long can be considered as a $\frac{3}{8}$-wavelength long line; one that is $10\frac{3}{4}$ wavelengths long is considered as a $\frac{3}{4}$-wavelength line. However, note that for other issues such as line loss, it is the actual length, not just the fractional part, that is critical.

Standing Waves

The nodes and antinodes of the reflected signal do not change position along the transmission line. Their location is determined entirely by the electrical length of the line and the frequency of the signal. If you look at the voltage (or current) everywhere along the line at the same time, you will see the voltage (or current) varying in amplitude, with the nodes and antinodes in unchanging locations as

8.4 Line and Load Matching   241

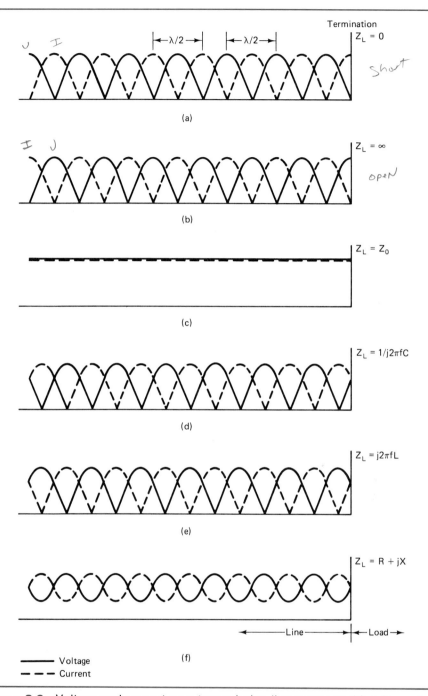

Figure 8.9 Voltage and current on a transmission line versus position for various line and load combinations. Note that the values repeat every half-wavelength.

standing waves. It is very similar to stretching a rubber band (or tight string) and then plucking it. The band will oscillate and will have nodes and antinodes. As the energy of the band dissipates, the amplitude of the oscillations will decrease, but the nodal points stand in place and do not change.

Between the two extreme cases of an open or shorted termination, not all the signal voltage is reflected. Some of the incident signal voltage is accepted by the load, and some is reflected. The standing waves that result from the reflections and subsequent nodes and antinodes can be characterized by a *voltage standing wave ratio (VSWR)*, equal to the ratio of the maximum voltage to the minimum voltage on the line:

$$\text{VSWR} = \frac{V_{max}}{V_{min}}$$

The VSWR is often referred to as the *SWR*, with the "voltage" prefix word eliminated. The SWR is infinite for total signal reflection, as from an open or short termination, since $V_{min} = 0$, as it is for a purely reactive load. A perfect match (a flat line) has SWR = 1 since $V_{max} = V_{min}$.

Example 8.14

A line has a voltage node of 4.5 V and an antinode value of 6 V. What is the VSWR?

Solution

$$\text{VSWR} = \frac{6}{4.5} = 1.333$$

When the load is resistive, the SWR is easily measured as the ratio of characteristic impedance to load resistance:

$$\text{SWR} = \frac{Z_0}{R_L}$$

Example 8.15

A line with 75 Ω impedance is terminated with a 30-Ω load.

Solution
The SWR is 75/30 = 2.5.

Here we see that the VSWR can be determined from just the knowledge of the line characteristic and actual load impedances, which are both relatively easy to measure—and unlike measurement of line voltage minimum and maximum, this does not require that the line or load have any power applied. Since you do not known in advance if Z_0 is less than or greater than R_L, the VSWR may range from 0 to infinity when measured this way (depending on whether $R_L > Z_0$ or $Z_0 > R_L$). It is customary to invert the SWR result if it is less than 1, saying that the SWR is 4 : 1 or simply 4, rather than 0.25 : 1 or 0.25.

Example 8.16

A 50-Ω line is terminated by a 300-Ω antenna. What is the SWR?

8.4 Line and Load Matching 243

Solution

The SWR is 50/300 = 1/6; however, we invert this to say that the SWR = 6.

Reflection Coefficient

Another useful indicator of the line and load match, and the effects of mismatch, is a *reflection coefficient* Γ. This complex number indicated how much voltage (and power) supplied by the signal source—the line—is reflected back to the source as a result of mismatch at the line/load:

$$\Gamma = \frac{V_{\text{reflected}}}{V_{\text{applied}}} = \frac{Z_L - Z_0}{Z_L + Z_0}$$

Example 8.17

What is Γ for a 100-Ω characteristic line and a 300-Ω load?

Solution

$$\Gamma = \frac{300 - 100}{300 + 100} = \frac{200}{400} = 0.5$$

The real power that is reflected back—and thus not delivered to the load as useful power—is the magnitude $|\Gamma|$ of complex number Γ. In the example above, the imaginary part of Γ is 0, so $|\Gamma|$ also equals 0.5. this means that one-half the power incident on the load will be reflected back to the line. When Γ is complex, complex arithmetic must be used.

Example 8.18

For a 75-Ω line and capacitive load (a very common situation) with $Z_L = 50 - j25$, what is Γ? What is $|\Gamma|$?

Solution

$$\Gamma = \frac{(50 - j25) - 75}{(50 - j25) + 75} = \frac{-25 - j25}{125 - j25}$$

$$= -0.154 - j0.231$$

$|\Gamma|$ = real part of Γ, so convert Γ from rectangular to polar (magnitude/angle) form:

$$-0.154 - j0.231 = 0.277\underline{/-123°} \quad \text{so} \quad |\Gamma| = 0.277$$

For a perfectly terminated line with $Z_L = Z_0$, the numerator is 0, so no incident signal power is reflected back to the source. Note that only a resistive load can absorb any power at all ($|\Gamma|$ between 0 and 1), and only a matched load

Effects of SWR > 1

The importance of maintaining a value of SWR near 1 is seen by looking at the effect of higher SWR values (higher reflection coefficient Γ). In general, SWR between 1 and 1.5 is considered an acceptable or moderate value; values above 2 are high and difficult to work with reliably. There are several problems associated with high SWR.

First, the transmitter generates signal power at the RF frequencies, but only a fraction of it is actually delivered to the load. This wastes battery power in portable units, reduces the effective power actually accepted by the load, and requires larger transmitter components to achieve a specific value of useful power delivered to the load (often an antenna).

The high SWR also affects the loss in the line due to I^2R heating of the transmission-line conductors. Even more critically, the presence of a voltage across the conductors of the line means that the dielectric between these conductors can *break down* and *flash over*, effectively shorting the line. The spacing of the lines and the dielectric must be designed to withstand any SWR voltages that can occur, even when the termination is accidentally disconnected or shorted. Active components in the transmitter must also be designed to withstand the SWR level, and this is often difficult or costly.

Example 8.21

The VSWR is 2.5, and the node voltage is 100 V. What is the peak voltage across the dielectric?

Solution

$$VSWR = \frac{V_{max}}{V_{min}}$$

so

$$V_{max} = VSWR \text{ times } V_{min} = 2.5(100) = 250 \text{ V}$$

Another problem is that the signal waveform is corrupted and distorted by the reflections and standing waves. This is a problem even with systems where the signal is simply going between two circuit boards—not just a concern in conventional transmission line-to-antenna connections. Consider a clean, sharp edge digital signal that is impressed on a system backplane PC board traces, with impedance determined by microstrip analysis techniques. The signal goes into an input buffer acting as load but with impedance mismatch. Now the digital signal has *ringing* (damped oscillations around the desired value) as the signal energy resonates on the line, which can cause false binary 1s and 0s to appear. As computers and digital circuits operate at higher and higher rates (frequencies), the circuit board traces that carry the digital signals from IC to IC and board to board must be analyzed as transmission lines and cannot be assumed as simple, ideal conductors.

absorbs all the incident power ($|\Gamma| = 0$). A shorted load, open load, or purely reactive load (no resistive component) absorbs no power at all ($|\Gamma| = 1$).

Another perspective of applied power and reflected power relates the SWR and the $|\Gamma|$:

$$SWR = \frac{1 - |\Gamma|}{1 + |\Gamma|}$$

Example 8.19

What are Γ and SWR for a 75-Ω line terminated with a 100-Ω load? What proportion of the incident power is reflected back from the load?

Solution:

$$\Gamma = \frac{100 - 75}{100 + 75} = \frac{25}{175} = 0.1428$$

which also equals $|\Gamma|$

$$\text{SWR} = \frac{1 - 0.1428}{1 + 0.1428} = 0.75 \quad \text{or} \quad \frac{1}{0.75} = 1.333$$

(inverting SWR to be >1, as customary). Using impedances directly, SWR = $Z_0/R_L = 75/100 = 0.75$, or 1.333. 0.1428 of the incident power is reflected back and not delivered to the load. If the line is applying 100 W of power from the transmitter to the load, 14.28 W will reflect back and the load will absorb only $(100 - 14.28) = 85.72$ W.

Example 8.20

For the 75-Ω line and 30-Ω load, find Γ and then SWR.

Solution

$$\Gamma = \frac{30 - 75}{30 + 75} = -0.4286$$

Therefore, $|\Gamma| = 0.4286$.

$$\text{SWR} = \frac{1 - 0.4286}{1 + 0.4286} = 0.4$$

so we invert, and SWR = 2.5. Note here that nearly half of the incident power is reflected, and only slightly over half is actually absorbed by the load where it can be useful.

Line/Load Matching Techniques

Several matching network methods are commonly used to match the line and load impedances for maximum power transfer and minimal SWR. The most obvious way to match the line and load is to change the load impedance so that it has the same value as the line, but this is often not practical. The load impedance is sometimes determined by the physical dimensions of the load (such as an antenna, cable, or preamp transistor base–emitter geometry) and cannot be changed without completely changing the size and shape of the load. Or, if the match must be performed between the output stage of the amplifier and the load (the transmission line), there is nothing that can be done to change the signal source impedance—it is determined by the amplifier components.

The goal of matching is to make the signal source see the desired load value, even if the physical load impedance actually has a different value. When the load is resistive, one way to do this is to use a *quarter-wavelength matching transformer*. This is a special intermediate section of line between the existing line and load (Figure 8.10). It makes use of the principle that a section of line that is $\frac{1}{4}$ wavelength long can act to transform impedances, with this formula to relate the new impedance value, called Z_n, and the existing Z_0 and R_L:

$$Z_n = \sqrt{Z_0 R_L}$$

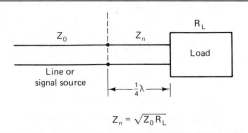

Figure 8.10 Use of $\frac{1}{4}\lambda$ matching section to match line and load impedances for minimal reflection.

Therefore, in order to match Z_0 and R_L, a special matching section must be built that is $\frac{1}{4}$ electrical wavelength long with impedance Z_n. This can be a special piece of coax, open wire parallel cable, or microstrip line. Often, a section of standard coax of another type can be used, so custom coax does not have to be fabricated.

Example 8.22

What characteristic impedance is needed to match a 50-Ω line to a 300-Ω load?

Solution

$$Z_n = \sqrt{(50)(300)} = 122 \ \Omega$$

The $\frac{1}{4}$-wavelength method works because the mismatches between the line and the matching section and between the matching section and the load produce two reflections. These reflections are separated in phase by $\lambda/4$. The reflection from the matching section to load interface therefore travels $\lambda/2$ farther than the line to matching section reflection. The reflections are 180° out of phase due to the $\lambda/2$ distance difference, so they add together and cancel each other out.

This scheme has the advantage of being very simple and reliable. Unfortunately, it is only effective at a single frequency value since the special section must be exactly $\frac{1}{4}$-wavelength long. If the signals passing through the line and load span a wide bandwidth, there will be significant error in matching because the actual electrical wavelengths in use are different than the one for which the section was calculated.

Another technique used to match line and load involves the *stub,* a relatively short length of cable with its end open or shorted (shorted ends are preferred because they radiate less energy than open ends do). This stub acts as a reactance that is placed in parallel with transmission line. By varying the position and length of the stub, the stub takes on the full range of impedance reactances because the impedance of the stub on the unmatching line varies with position due to the phase difference between current and voltage on the line. The stub can be made of any characteristic impedance cable, but usually is made from the same type as the transmission line for convenience.

In stub matching, the parameters that must be selected are the stub length, position, whether to open or short the end, and the stub Z_0. The most common tool used to do this is the *Smith chart,* discussed in the next section.

Review Questions

1. Why is line and load impedance matching important? What is electrical length? What three things determine the electrical length of a wire?

2. What is the nature of voltage and current reflection from an open termination? From a short-circuit termination? Compare and contrast.

3. Why does the impedance of a resonant line vary with position along the line? What does the line impedance appear as when you move along it? How does it repeat itself periodically?

4. What are standing waves? How do they come about? Why are they undesirable? What is the meaning and range of VSWR? What is the VSWR of a flat line?

5. What is reflection coefficient in terms of line and characteristic impedances? What does the reflection coefficient represent? What does its magnitude represent? What does low Γ indicate in terms of power actually absorbed by the load?

6. When is $\Gamma = 0$? What must the load impedance be for Γ greater than 0 and less than 1? Under what conditions is $\Gamma = 1$?

7. What is VSWR in terms of Z_L and Z_0? What are the problems that occur in systems with relatively higher values of VSWR?

8. What is $\frac{1}{4}$-wavelength matching? Why does it work?

8.5 S PARAMETERS AND THE SMITH CHART

To characterize a transmission line or its load properly, its key parameters must be defined and measured. A transmission line, or a load, is a *two-port network* with a signal able to enter or leave only from one port or the other. Traditionally, an *n-port network*—one with exactly *n* points for signal entry and exit—is characterized by *y, h,* or *Z parameters*, which are defined in terms of total voltage and current (these parameters are often used to characterize transistors). Measuring these parameters is increasingly difficult as higher frequencies are used. The standard parameters require that the network's terminals be opened or shorted, which cannot be done easily at high frequencies. To achieve an open or short termination on a transmission line, a short is used and the line length is varied until a short or open is reflected back. The line length has to be varied as the frequency is changed. Also, many active devices such as transistors are not stable or will self-destruct with a short as the termination.

S Parameters

An alternative to *y, h,* or *Z* parameters is *S parameters* (for scattering), which are defined in terms of the signal that is applied to a port (incident power) and reflected back to the port. These factors are much easier to measure at high frequencies than are voltage and current alone. *S* parameters require only that the line (or network) be terminated in its characteristic impedance (generally standardized at

50 Ω in the industry). They describe the network completely, (including magnitude and phase information), and can, with additional calculation, be converted to *y*, *h*, or *Z* parameters. (Note that although we are discussing transmission lines, which are passive lengths of cable, all of this discussion applies to any network, including active circuits and their devices, such as transistors and associated capacitors, inductors, and resistors.)

Transmission-line theory shows that there will be a reflection of any incident signal from the end of a line unless the line is properly terminated in its characteristic impedance, and this reflection sets up standing waves in the line. The incident and reflection waves are used to derive the *S* parameters. The letter *a* is used for the incident power and *b* is used for the reflected power. Define four simple equations:

$$a_1 = \frac{E_{i1}}{\sqrt{Z_0}} \qquad a_2 = \frac{E_{i2}}{\sqrt{Z_0}}$$

$$b_1 = \frac{E_{r1}}{\sqrt{Z_0}} \qquad b_2 = \frac{E_{r2}}{\sqrt{Z_0}}$$

where E_{in} represents the voltage wave incident on port *n*, and E_{rn} is the wave reflected from port *n*. Variables a_1 and b_1 are the incident signals, and a_2 and b_2 are the reflected signals. The signals a_n and b_n have magnitude and phase; and the magnitude of a_n or b_n squared is the power incident or reflected at the respective port.

Figure 8.11 shows a two-port network and the signal paths. A signal a_1 applied at port 1 will divide, by some proportion, to a path to port 2 and also back to port 1. Similarly, the signal a_2 applied at port 2 can go on to port 1, with some of it diverted back to port 2. The total signal seen at port 1, called b_1, is the sum of the fractional signals from a_1 and a_2; the signal b_2 seen at port 2 is the sum of the same signals but with the different paths and proportions.

These can be defined to form the *S*-parameter equations:

$$b_1 = S_{11}a_1 + S_{12}a_2 \qquad b_2 = S_{21}a_1 + S_{22}a_2$$

S_{11} is the *input reflection coefficient* and shows what fraction of the incident signal at port 1—the input port—is reflected back to the source. It is thus directly

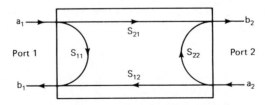

Figure 8.11 Basic two-port network and definition of *S* parameters (courtesy of Hewlett-Packard Co.).

related to the input impedance. S_{21} is the *forward transmission coefficient* and is the fraction of the incident signal at port 1 that is transmitted through the network to port 2. It is also called the *forward power gain*. The *output reflection coefficient* S_{22} is the fraction of signal incident at port 2—the output port—reflected back to port 2. S_{12} is the *reverse isolation coefficient* (or *reverse power leakage*), which shows the fraction of signal applied to port 2 that reaches port 1. The S parameters are complex numbers and are expressed in rectangular form, $A + jB$ or polar form, $M\ \underline{/\phi}$. They are most commonly plotted in polar form. The plot will then show how the S parameters vary with frequency, as they do for many real networks.

If either port is terminated in the characteristic impedance, the reflection term for a signal sent to that port becomes 0. Thus if port 2 is terminated in Z_0 and an incident signal applied to port 1 (which means that no signal is applied to port 2, $a_2 = 0$), then

$$S_{11} = \frac{b_1}{a_1} \qquad S_{21} = \frac{b_2}{a_1}$$

Similarly, if port 2 is driven by a signal (a_1 set to 0) and port 1 is terminated with Z_0,

$$S_{12} = \frac{b_1}{a_2} \qquad S_{22} = \frac{b_2}{a_2}$$

Note that for an open-circuit termination, the entire signal is reflected back in phase (assuming that the cable itself has no losses) and $S_{11} = 1$. In contrast, a short circuit reflects back the incident signal, but with opposite phase, so the value of S_{11} is -1.

By examining S parameters S_{11} and S_{22}, the exact values of complex input and output impedances are determined and then used to match the line and load for maximum power transfer and minimum reflection.

Example 8.23

A lossless network output is terminated in Z_0 and a 1.3-V signal applied to port 1. The signal measured at b_1 is 0.15 V. What are parameters S_{11} and S_{21}?

Solution

$$S_{11} = \frac{b_1}{a_1} = \frac{0.15}{1.3} = 0.115$$

The signal at b_2 is the difference between the incident signal and the signal measured at b_1, since this is a lossless network. Therefore, b_2 is $1.3 - 0.15 = 1.15$.

$$S_{21} = \frac{b_2}{a_1} = 0.885$$

Note that since this is a lossless network, $S_{11} + S_{21} = 1.00$, since all of the incident signal either reflects back to the input or appears at the output.

Example 8.24

A lossless network with $S_{11} = 0.87$ and $S_{21} = 0.13$ is properly terminated and then a 5.0-V signal is applied. What are the values of b_1 and b_2?

Solution

Using the formulas shown above, we have

$$b_1 = S_{11}a_1 = 0.87(5.0) = 4.35 \quad \text{and} \quad b_2 = S_{21}a_1 = 0.13(5.0) = 0.65$$

In both of these examples, the S parameters were real, for simplicity. In most practical systems, they are often complex coefficients. If the networks has some losses, some of the signal energy is dissipated as heat, so the incident signal energy does not equal the output signal energy.

Smith Chart

To match the line properly with the load requires great care, especially when the load impedance Z is not simply resistive ($Z = R + jX$). A load impedance that varies with frequency—which is the case for many loads, such as line receivers, buffer amplifiers, and even antennas—must be examined and understood before any impedance matching is done. The *Smith chart,* developed and refined in the mid-1930s by P. H. Smith, is a very useful tool for understanding impedances and to match line and load. Because it is a graphical tool, it allows the user to see and estimate the range of possibilities in line/load impedance applications, and to judge trade-offs and compromises in a way that a table or even computer program cannot easily do.

A simplified version of the Smith chart is shown in Figure 8.12. The chart has two basic sets of curves. The *circles* of increasing diameter which all touch at

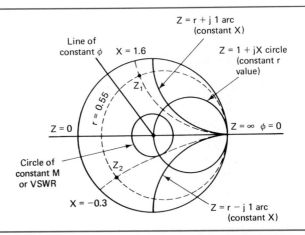

Figure 8.12 Simplified Smith chart, showing arcs of constant reactance and circles of constant resistance using $z = R + jX$ (with $R = X = 0$ at right) and circles of constant VSWR and lines of constant phase using $z = M \underline{/\phi}$ (polar form, with $M = \phi = 0$ at center). Also shown are some sample values on the chart (courtesy of McGraw Hill, Inc.).

8.5 S Parameters and the Smith Chart

the right-hand side represent increasing values of R. Any point on the same circle will have the same value of resistance R as any other point on that circle. The larger circles indicate larger values of R. The group of *arcs* of increasing size coming from the right-hand point indicate constant reactance X values. A point on any arc has the same X value as all other points on the same arc, with larger values of X represented by the larger arcs. The upper half of the chart represents inductive reactance ($+jX$) and the lower half indicates the capacitive ($-jX$) reactance. The horizontal line across the middle has no reactive component, only pure resistance, with the left end a short circuit ($R = 0$) and the right end an open circuit ($R = \infty$). Two examples of Z values are plotted on the chart, for $z_1 = 0.55 + j1.6$ and $z_2 = 0.55 - j0.3$. Note the effect on changing the real or complex coefficient on the plotted location, but not both: The plotted point moves along the arc or circle.

The chart can also be used with the center of the circle as the starting point. A circle centered on the center point represents a constant VSWR anywhere on its circumference, and larger circles have greater SWR. Any point on a straight line beginning at the center and radiating outward will have a given phase angle ϕ corresponding to the polar plot angle. The distance from the origin at the center to the plotted point represents the magnitude M of polar $M \underline{/\phi}$ form.

With the detailed Smith chart (Figure 8.13) any impedance can be plotted, its SWR studied, the effects of matching seen, and the admittance (inverse of the impedance, useful for stub tuning) read easily. Using the Smith chart to perform these functions requires practice, because of the unusual orientation of the curves and the nonlinear scale markings. We will not show how to fully use it, but will look at how an impedance is plotted and what the Smith chart reveals, as well as how the S parameters can be used with the chart. Many books and handbooks discuss use and applications of the Smith chart in detail.

Normalized values The Smith chart is labeled in *normalized* impedance values. This allows a single blank chart form to be used for any characteristic impedance and keeps all the numbers down to more manageable values. A normalized impedance z is simply the actual value Z, divided by the characteristic value Z_0: $z = Z/Z_0$.

Example 8.25

For a complex impedance $Z = 150 + j50 \, \Omega$ with Z_0 of $50 \, \Omega$, what is the normalized value of z?

Solution
Dividing Z by Z_0, the result is simply $3 + j1$.

To convert back to actual impedance values (if needed), simply multiply the normalized value by the characteristic impedance. Since z is dimensionless, not in ohms, the Smith chart can also be used to graph other dimensionless parameters, such as S values.

Example 8.26

Normalize an impedance of $120 + j90 \, \Omega$ to $50 \, \Omega$.

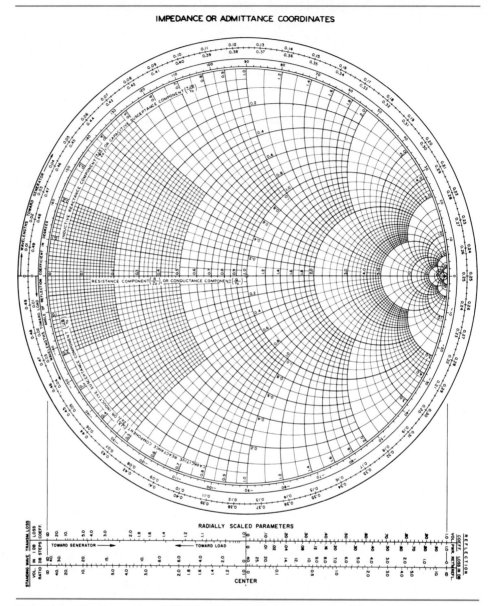

Figure 8.13 Complete Smith chart is a useful tool for graphing and studying impedances, transmission lines, and matching.

Solution

The normalized value is $120/50 + j90/50 = 2.4 + j1.8$.

Example 8.27

Convert a normalized value (to 50 Ω) of $1.8 - j.3$ to the actual impedance value.

8.5 S Parameters and the Smith Chart

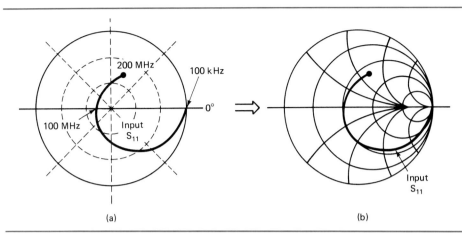

Figure 8.14 (a) S parameters of op amp plotted in polar graph form; (b) the same data plotted on a Smith chart that replaces the polar graph (courtesy of RF Design magazine).

Solution

The actual value is $1.8(50) - j0.3(50) = 90 - j15 \, \Omega$.

The Smith chart is used to chart any resistance $z = r + jx$ (all normalized) and instantly shows four separate parameters: r, x, the magnitude of the reflection coefficient Γ (indicated by $|\Gamma|$), and the phase angle ϕ of the reflection coefficient.

The Smith Chart and S-parameters

The S-parameter values can be transferred to the Smith chart to show the changes in impedance with frequency. S parameters S_{11} (the input port reflectance) and S_{22} (output port reflectance) are really the same as the reflectance Γ, in that they show how much incident power is reflected back to the source.

An example is shown in Figure 8.14 for an S-parameter test of an op amp input over the frequency range from 100 kHz to over 100 MHz. The magnitude and phase of S_{11} is plotted in polar form, and then the Smith chart is overlaid. The result is the plot of z (normalized to 50 Ω). Note that the input impedance of this op amp is not constant, but the resistance and reactance vary with frequency. For this op amp, the input impedance at 100 kHz is primarily resistive and becomes increasingly capacitive as frequency increases. As the frequency passes 100 MHz, the capacitance decreases and the input impedance becomes inductive, due to the reactance of the op amp lead wires and package.

This plot of z versus frequency can be used to match the line to the load at the frequency of interest. It also shows how the VSWR changes for a fixed matching circuit as the frequency varies. This is the connection between S parameters—which are easier to measure at higher frequencies and reveal a great deal about the network—and the Smith chart, which shows the system impedance and helps match impedance and study VSWR. The S parameters are measured and plotted on a polar chart, then the polar chart is overlaid with a Smith chart to indicate impedances at various frequencies. Put another way, you can draw the impedance data on a Smith chart and then replace the chart markings with a simple polar chart with the same diameter and same center, and you will have an S-parameter chart; or you can go the other way, and measure and plot S parameters and then replace the polar chart with a Smith chart and read off impedance information. Depending on the circumstances, one measurement may be easier to make, in practice, than the other.

For example, the point $z = 0.55 + j1.6$ is shown on the simplified Smith chart. The value of $|\Gamma|$ is simply the distance from the center of the chart to the $r = 0.55$, $x = 0.1.6$ point. Angle ø is measured from the horizontal to the line drawn from the chart center to the r, x point.

If all the Smith chart did was allow an impedance to be drawn so that the relationship among the four parameters could easily be seen, it would be a handy accessory but not really very useful beyond that. The chart shows much more, however, and serves as a quick calculation aid to determining the best length and location of stub impedance matching sections, finding the result of added impedances, and determining the SWR that results from mismatch, among other applications. An auxiliary scale, not shown in the simplified version, allows the Smith chart to be used with realistic transmission lines that have loss instead of ideal, lossless lines.

Review Questions

1. What is a two-port network? Why is a transmission line or load such a network?
2. What are the four S parameters? What does each represent?
3. How can S parameters be measured? Why is this very convenient?
4. What is the meaning of the horizontal line across the middle of the Smith chart? What are the values of the leftmost and rightmost points? What do the circles of the Smith chart, which originate from the right-hand side, represent?
5. What do the arcs emanating from the right-hand side represent? What do arcs in the upper half indicate compared to those in the lower half? What is common to any point plotted on the horizontal axis?
6. When used as a polar plot, what do circles centered on the origin mean?
7. What four items can be read off the Smith chart from a plotted point? What are normalized values? Why are they used?
8. What is the relationship between the Smith chart and S parameters? Which S parameters correspond to line/load reflectance?

8.6 TEST EQUIPMENT

Transmission-line testing begins with two basic pieces of equipment: the power meter and the dummy load. Unlike a circuit, a transmission line cannot be probed or have its "internal points" measured: all measurements consist of signal in, signal out, and the relationship between these factors.

Power Meters A *power meter* is a version of an RF energy voltmeter that measures the voltage at any point in the line it is inserted. The voltmeter markings for power levels indicate power at a specific impedance, since $P = V^2/R$. Figure 8.15a shows the schematic of a basic voltmeter/power meter. A fraction of signal passing through

8.6 Test Equipment   255

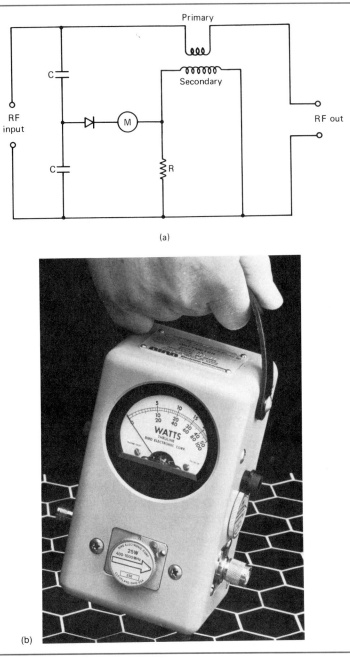

Figure 8.15 (a) Schematic of basic power meter, for indicating power in forward direction; (b) power meter for commercial use (courtesy of Bird Electronic Corp.).

the instrument is inductively coupled to the indicator meter itself via the transformer; the exact fractional value that is directed to the meter is determined by the ratio of primary to secondary turns and how the primary and secondary are coupled together. The secondary signal is rectified to drive the meter itself. The arrangement and symmetry of the coupling windings in the transformer are designed so that only signal passing from the left side to the right is measured by the circuit (forward power). Reversing the leads on the secondary of the transformer sets the meter for measuring the reverse power only. A switch controlled by the user can implement this reversal, so the meter can indicate either forward or reverse power, and then SWR calculations can be done. (Of course, the user can also disconnect the meter and reconnect it the other way around to read reverse power, but this is a major inconvenience.)

A commercial power meter is shown in Figure 8.15b. This meter incorporates both forward and reverse direction circuitry, so that the user does not manually have to switch the direction of reading. Such a meter is especially useful for semipermanent or permanent installations, where it is left in-line as a monitor on the operation of the transmitter, transmission line, and antenna system.

The power meter is used to measure the power output of a transmitter or the input into a transmission line. It is also used to measure the output of the transmission line, which when combined with the line input power reading, shows line loss. The forward and reverse readings are used to determine VSWR.

Dummy Loads

A transmission line can connect two pieces of equipment or sections of circuitry, or connect a transmitter to an antenna (or an antenna to a receiver). For testing the transmitter-to-antenna combination it is often impractical, or even dangerous, to use an actual antenna. First, the antenna may not be installed or available. Second, the system may be operating improperly, so it will broadcast at incorrect frequencies and with poor modulation performance, causing interference to others. Third, it may be physically difficult to have the equipment near the antenna for test purposes.

All of these problems are avoided with a *dummy load,* which is basically a resistor of correct value that is installed in place of the antenna. The transmitter and transmission line see the same terminating impedance as the antenna presents (discussed in the next chapter), so electrically it is identical to the real and final connection to an antenna. However, the dummy load is designed to be nonradiating, so all of the RF energy is dissipated in the resistor rather than transmitted into the air. The dummy load makes testing and measuring convenient, safe, and much more practical.

The key technical requirements for a dummy load are that it have the proper impedance, that it be nonreactive (the load should be purely resistive), and that it be capable of dissipating the applied power without burning up. The physical size and construction of the dummy load depend on how much power it must handle. For a small, relatively low power transmitter (1 to 5 W) a power resistor about 2 in. long and $\frac{1}{2}$ in diameter is used. At higher power levels, this resistor can be cooled with an oil bath in a 1-gallon can and handle hundreds of watts (one popular home-made design uses a resistor in a paint can). Often, many smaller resistors are combined in parallel and series combinations to provide both the required ohmic value and power rating. This must be done with care in the overall assembly to minimize unwanted lead inductances which make the final dummy load look reactive instead of resistive. Finally, for testing transmitters at thousands and hundreds of thousands of watts, advanced cooling and dissipation schemes are used with coolant circulated around the load to keep its temperature down while it dissipates the RF power.

Although it may seem that a dummy load is more trouble than it is worth,

> **Dummy Loads for Audio**
>
> Dummy loads are not used just for RF signals. Technicians who test and repair high-fidelity (stereo) amplifiers use them in place of the actual loudspeakers. In this case, the impedance is usually 4 or 8 Ω and the power levels range from about 10 W to several hundred watts (but can be higher in systems for concerts). The need for this dummy load is simple: The audio level from even a small (20-W) amplifier running at full volume can be deafening. It would be impossible for the technician (or co-workers) to do the job with this loud music blasting only a few feet away, and would cause serious ear injuries over the long term. Also, the loudspeakers used would have to be able to handle the power applied, so the test shop would need a variety of small through large speakers. (Using the actual speakers is not a practical option, since they may be in fixed installations or weigh hundreds of pounds for the higher power levels.)
>
> The dummy loads for audio applications, like those for RF use, are designed to dissipate the applied power and be resistive. Unlike RF dummy loads, this resistive requirement is much easier to meet at audio frequencies (20 to 20,000 Hz), where a few microhenries of inductance produce insignificant reactance, in contrast to their reactance at megahertz frequencies.

especially for high-power loads, it is a very useful and often absolutely necessary part of transmission line testing. It is much easier to test, measure, and adjust the transmitter and transmission line with a dummy load than with an actual antenna (which may be located outside, making it hard to access). It is also against Federal Communications Commission regulations to use a transmitter that may be operating improperly and thus causing interference to others.

Instrumentation for S Parameters and Smith Charts

Both S parameters and the Smith chart are useful for additional testing and adjustment of transmission lines and even active circuits. With the proper equipment, S parameters are easy to measure, while the Smith chart is an extremely useful tool for matching impedances, determining SWR, and seeing change in parameters over a frequency range of interest.

To perform the S-parameter measurements, the network under test must be terminated at one end while a signal is injected at the other end, and this signal must sweep over the frequency range of interest. Then the situation must be reversed, with the termination and signal source swapped. This could be done manually, but modern test equipment does the entire operation automatically. The Hewlett-Packard HP3577A network analyzer with HP35677A S-parameter test set (Figure 8.16) requires only that the two ports of the network under test be connected to two connectors on the front panel. The test operator sets some controls for the desired range of frequencies, signal levels, and some other test conditions.

The instrument combination runs the complete S-parameter test and displays the S parameters on a polar chart on the CRT screen. The screen markings are drawn by special circuitry in the HP3577A, and the operator can select either the S-parameter polar chart for the screen or select that the Smith chart markings, instead, be drawn on the screen by the circuitry. The actual lines of the data measurements made by the instrument do not change; just the graph used to interpret these markings as either S parameters or Smith chart impedances is switched as needed. This is equivalent to the operator drawing the results on paper and then overlaying a clear Smith chart or polar chart on the data.

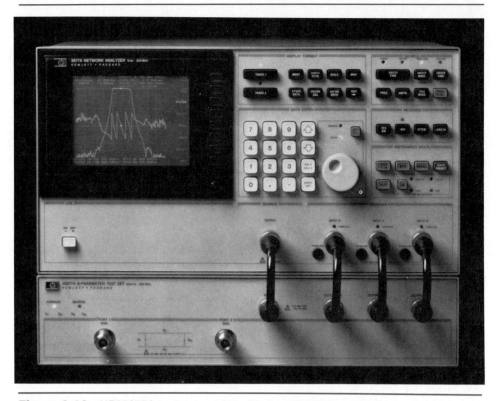

Figure 8.16 HP35677A test set works with the HP3577A network analyzer to measure *S* parameters of a two-port network from 5 Hz to 200 MHz and plots them using a polar graph or Smith chart (courtesy of Hewlett-Packard Co.).

Review Questions

1. What is a power meter? Why is it really a specially marked RF voltage meter?
2. How does the power meter measure forward signal strength only? In what two ways can it be changed to measure reflected power?
3. How is the power meter used to measure power levels, line loss, and SWR?
4. What is a dummy load, electrically? What are its advantages versus the actual load? Why is the dummy load both a convenience and a necessity?
5. How does a modern network analyzer make measurement of a network in terms of *S* parameters and the Smith chart relatively easy?

SUMMARY

The transmission line that connects a signal source to a load is a complex element at the frequencies used in communications. The most common transmission lines are coaxial cable and open wire parallel line, each providing a variety of technical specifications. The impedances of these lines are essentially nonreactive, determined by the wire size and spacing used, as well as any dielectric between the conductors. Microstrip lines are etched into a circuit board and are miniature versions of coax and parallel wire transmission lines. The microstrip line is especially useful where the signal has to travel a short distance on a single board, or a special impedance value is needed.

At frequencies beyond the several gigahertz capability of coax, open wire, and microstrip lines, waveguides are used to propagate signal energy. A waveguide is very different from a wire in that it acts as an envelope for electromagnetic energy carried as electrical and magnetic fields, instead of voltage and current. Waveguide dimensions are critical for proper operation and the dominant mode of operation is most often used. Signal energy is coupled to a waveguide using a simple probe into the electric field, a loop for magnetic fields, or slots which allow some of the waveguide field to leave the guide. Waveguides can carry large amounts of signal power with low loss, at frequencies well above regular transmission-line capability.

For maximum power transfer, the impedance of the signal source must match the impedance of the load it is driving, whether the source is an amplifier, transmission line, or antenna. A mismatch will result in wasted power, standing waves, and high voltages across the conductors, as seen by the standing wave ratio. Proper matching requires understanding the impedance of the line and load as a function of frequency and use of special matching lines such as the quarter-wavelength section. The length of a transmission line is often measured in electrical wavelengths as compared to physical length. S parameters offer a way to study the signal reflection and transmission characteristics of a two-port network (such as a transmission line) at high frequencies. They are relatively easy to measure and can then be transferred to a Smith chart. The Smith chart is a very useful tool for studying the magnitude, phase, resistive, and reactive parts of impedance, as well as seeing SWR quickly and matching line and load to reach desired SWR values.

Summary Questions

1. What electrical factors determine Z_0 for coax? Why can the resistance and admittance of the model be ignored? How does dielectric constant affect Z_0?

2. What are distributed parameters? Compare them to lumped ones.

3. What are three sources of line loss?

4. What are the physical appearances and geometries of three common microstrip configurations? What parameters affect microstrip Z_0?

5. How does signal propagation in a waveguide differ from that in a cable?

6. What is dominant mode? What is the relationship between waveguide width or radius and cutoff for dominant mode?

7. Why is the electric and magnetic field shape important for waveguide signal coupling? What about coupler angle relative to field?

8. Compare waveguide losses to coax losses.

9. What is electrical wavelength? Why is it important?

10. Explain how Z varies dramatically versus measured position on a resonant line. How is this possible?

11. What does SWR indicate? Why is high SWR undesirable?

12. How can SWR be calculated by voltage maximum and minimum values? By impedances?

13. What is the reflection coefficient? What does it indicate about power absorbed for useful work by the load? How is it used to indicate the fraction of applied power reflected, and actually absorbed at the load? What loads make $\Gamma = 0$, and which make it greater than 0 but less than 1? What loads make $\Gamma = 1$?

14. How does a ¼-wavelength matching section operate?

15. What do S parameters indicate? Why are they more useful than traditional parameters for communication systems, in many cases?

16. What is a Smith chart? How does it show resistance and reactance values? How does it indicate impedance magnitude and angles and SWR?

17. What is the relationship between S parameters and Smith charts? Why is this very convenient?

18. Explain how an RF voltmeter/power meter is used in transmitter and transmission-line testing. Compare its use for measuring power levels, forward and reflected power, and SWR.

19. How is the dummy load used? Why is it very helpful and often essential and important in many aspects of transmitter and transmission-line testing?

20. How is a network analyzer used to determine S parameters and Smith chart data? Why is the changeover from showing S-parameter results to showing Smith chart impedance plots relatively simple for advanced instruments?

PRACTICE PROBLEMS

Section 8.1

1. What is Z_0 for coax wire with an inductance of 100 nH/ft and a capacitance of 150 pF/ft?

2. Repeat problem 1 for a twofold greater inductance but the same capacitance.

3. A system requires a characteristic impedance of about 120 Ω. The available wire has an inductance of 55 nH/m. What is the necessary capacitance of the wire to achieve this Z_0?

4. What is the characteristic impedance of coax with an inner diameter of 0.025 in., outer diameter of 0.3 in., and air dielectric?

5. The air is replaced with polypropylene, $\varepsilon = 2.3$, in problem 5. What is the new Z_0?

6. An open wire transmission line is formed with two pieces of No. 18 wire spaced 0.25 in. apart in air. What is Z_0?

7. A value of Z_0 of 200 Ω is needed in an application, using No. 18 wire and air dielectric. What is the spacing required?

Section 8.2

1. A microstrip line is formed using a 0.095-in.-thick PC board (dielectric constant = 1.8), with a bottom ground plane and a single 0.15-in.-wide, 0.008-in.-thick track on the top. What is Z_0?

2. The circuit board of problem 1 is now used, with the same values, to fabricate an open wire transmission line, with 0.2-in. spacing. What is the characteristic impedance?

3. A stripline is formed using a multilayer board ($\varepsilon = 2$). The center track is 0.15 in. wide and 0.005 in. thick, and the PC board first-layer thickness is 0.05 in. thick, with an overall board thickness of twice the single layer. What is Z_0?

4. The track in problem 1 is now doubled in thickness. By what percent does Z_0 change?

Practice Problems 261

5. The open wire microstrip equivalent of problem 2 is changed so that the track spacing is cut in half and the dielectric constant of the PC board is 2.3. What is the new value of Z_0?

Section 8.3

1. What is the cutoff wavelength for a rectangular waveguide with $x = 0.85$ cm in the dominant mode? What is the wavelength in mode 3?

2. Use the figures of problem 1 for the case of a circular waveguide with a radius of 0.85 cm in the dominant mode.

3. What is the shortest dominant-mode wavelength that a 0.55-cm rectangular waveguide will support while cutting off higher modes? What about a circular waveguide with the same radius?

4. What is the span of mode 1 wavelengths that a 0.45-cm-radius circular waveguide will support without the possibility of higher modes?

5. A waveguide with $x = 0.75$ cm will support which modes for a wavelength of 0.075 cm?

6. A simple electric field probe is tilted at 30° to the field. What is the relative coupling factor?

7. The loop coupler for a magnetic field is rotated at 35° to the field. What percent of maximum is the coupling?

Section 8.4

1. Find the electrical length of a 5-m cable that propagates the signal at $0.75c$ at 125 MHz.

2. What is the electrical wavelength of a 20-m cable (propagation factor of 0.90) for a 12-MHz signal? For a 108-MHz signal?

3. A $\frac{1}{4}$-wavelength matching section is needed at the center of the FM band (88 to 108 MHZ) using coax with propagation at 80%c. What is the correct length?

4. What is the SWR when the line impedance is 75 Ω and the load impedance is 125 Ω?

5. Calculate Γ and the VSWR for a line with $Z_0 = 100$ Ω feeding a load of 300 Ω impedance. How much power, as a percent, is reflected at the line/load interface? What percent is actually absorbed by the load?

6. Calculate the VSWR with a V_{max} of 100 V and a V_{min} of 20 V.

7. A line has a characteristic impedance of 50 Ω feeding a load with a complex impedance of $150 + j100$ Ω. Calculate the reflection coefficient as a complex number, and finds its magnitude.

8. A system uses 65-Ω coax to feed a 300-Ω antenna. What is the SWR? How much power is actually absorbed at the load compared to the incident power?

9. A coax cable is feeding a mismatched load with VSWR of 3.2. The coax has a dielectric breakdown value of 1200 V. Will the coax flash over if the node voltage is 300 V?

10. What is the matching impedance needed for a $\frac{1}{4}$-wavelength section between a 50-Ω line and a 150-Ω load?

Section 8.5

1. An S-parameter test is run on a properly terminated lossless network. The voltage applied to the input port is 1.7 V and the output is 0.7 V. Calculate the two applicable S parameters.

2. The same network has 1.7 V applied to the output port, and the signal at the input port is 0.8 V. Which two S parameters can be determined, and what are their values?

3. A network has $S_{11} = 0.3$, $S_{12} = 0.7$, $S_{21} = 0.25$, and $S_{22} = 0.9$. Determine the voltage seen at the input when 1.7 V is applied to the input port.

4. Using the figures of problem 3, what is the voltage at the output port?

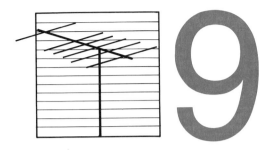

9

Propagation and Antennas

CHAPTER OBJECTIVES

When you have completed this chapter, you will understand:

- How an antenna radiates or captures electromagnetic energy
- The various ways that a radiated signal propagates through the atmosphere and space
- How antenna performance is characterized, and the importance of the various aspects of antenna performance
- The specifics of simple, fundamental antenna designs
- The need for, and design of, advanced antenna systems

INTRODUCTION

Antennas for transmitting are designed to radiate signals into the surrounding air or vacuum; antennas for receiving must capture as much electromagnetic energy as possible to be effective. Beyond these simple statements is the complex world of antenna design, function, trade-offs in antenna performance, and the application: how and where the antenna is to be used. Part of antenna selection and design is determined by its intended application and location, which depends on how the signal will travel from the transmitter to receiver: by a direct wave along the ground, by refracting off layers of the atmosphere, or by other modes of propagation. Atmospheric conditions play a major role in this propagation.

Antenna performance is characterized by the shape of the transmitted signal field, the antenna's ability to reject signals that are to the side of its main line of strength, and its bandwidth capabilities. Simple antennas, such as the long wire or dipole, are effective in some situations; other applications require more complicated antennas to achieve the system performance needed. The more complicated antennas include additional elements that provide more shape to the radiated field than a simple antenna design can, which results in the high gain needed for transmitting long distances or receiving weak signals with good SNR. Antenna size is very closely related to frequency of operation, and this affects practical designs, as well.

Combinations of multiple identical antennas can also be useful. The phased array design uses hundreds of small antennas combined into a large array, with a computer controlling the signal flow to each antenna. The result is an electronically shaped and steered antenna, without moving parts.

9.1 PROPAGATION AND THE FUNCTION OF ANTENNAS

In Chapter 8 we saw that the function of a transmission line was to carry signal energy, with minimal loss, between a generator and an antenna (for transmitters), or between an antenna and an RF input stage (for receivers). The antenna performs the opposite function: It must *radiate* as much energy as possible into the surroundings (for transmitting) or capture as much energy as possible from the surroundings (for receiving). A good transmission line is a very poor antenna, while a transmission line that loses energy via radiation is actually acting as an antenna!

There is another important difference between a transmission line and an antenna. The transmission line uses a conductor to carry the signal energy in the form of voltage and current. Yet a radio signal traveling through air or vacuum is carrying energy in an insulating medium. What an antenna does is act as an energy *transducer* (converter) by transforming the energy as the voltage and current on a transmission line to energy as an electric and magnetic field in space. In a similar way, a loudspeaker is a transducer that takes electrical energy from an amplifier and converts it to sound energy for air.

To understand how an antenna transforms the energy mode, electromagnetic field theory and equations must be used to show how voltages and currents in the antenna conductors radiate energy and how the frequency of the signal affects the radiation. Intuitively, here is what happens. The voltage and current variations in the transmission line produce an electromagnetic field around these conducting elements, which expands and contracts at the same frequency as the variations. The field returns the energy to the source—the conductors—as it collapses. Therefore, all the energy of the circuit or line remains within the system. In contrast, an antenna is designed to prevent most of the field energy from returning. The physical dimensions of the antenna and the electromagnetic wavelengths it is designed for are selected to cause the field to radiate out a distance of several wavelengths before the cycle reversal allows the field to collapse and energy to be returned.

An antenna produces a three-dimensional electromagnetic field. The radiated wave consists of an electric field and a magnetic field at right angles to each other. The direction of wave energy propagation is at a right angle to both of these fields.

The same electromagnetic theory applies, at the most basic level, to all radiated signals, electromagnetic fields, and antennas. However, the practical

> **Antenna Reciprocity**
>
> There is no difference, in theory, between an antenna intended for transmitting and one designed for receiving. All the discussion about radiation fields, patterns, antenna gain, and similar issues applies equally to both. An antenna designed for transmitting can be used for receiving, which is called the *reciprocity* principle. The difference is one of practical implementation, since a transmitting antenna handles relatively higher power signals (tens of watts through megawatts) while a receiving antenna is a milliwatt and microwatt device. The transmitting antenna uses larger conductors, heavier support structures, and higher power connectors as a result.

implementation of antennas is very frequency dependent. First, antenna dimensions are related to frequency, and a complex, multiple-element antenna cannot easily be made many meters or even miles long (although some have been made). Second, a great deal of the antenna performance is related to the way the radiated signal actually travels in the environment formed by the earth's surface and curvature, the layers of the atmosphere, and the ability of these surfaces and layers to distort and change the propagation of various frequencies by different amounts. Any study of antennas must take the signal propagation into account in the real-world surroundings, which will be studied in the next section.

For the purposes of this chapter, we will usually talk about antennas from the perspective of a transmitter, for convenience and simplicity. Keep in mind, however, that whatever is said about a transmitting antenna applies equally to a receiving antenna.

Review Questions

1. What is the function of an antenna? Contrast this to a transmission line.
2. How does an antenna allow signal radiation to occur?
3. What are two major factors that determine the antenna design used?
4. Explain the reciprocity principle. How does it affect the antenna design, in theory and in practice? What are the similarities and differences between transmitter and receiver antennas?

9.2 PROPAGATION MODES

The signal, or wave, that radiates from the antenna can take several paths. It can follow the contour of the earth's surface as a *ground wave,* it can travel in a direct "line of sight" to the receiver as a *space wave,* or it can radiate as a *sky wave* toward the various atmospheric layers and be bent back toward the earth. Finally, it can travel from the transmitting antenna into space beyond the earth's surface and atmosphere, in which case it is a *satellite wave.* (Perhaps "space wave" would be a clearer name, but "space wave" was already claimed for signal radiation before communications from earth to space satellites or the moon was even thought about!)

Propagation refers to the specifics of how these various waves travel under different circumstances. Unlike simple signal radiation in a vacuum, the travel of a radiated wave is very much affected by the type of wave selected, the earth itself, the atmospheric conditions, the time of day, and the season. The wave travel used is determined by the antenna design, its angle of signal radiation, and the frequency. The sun's activity has major effects on propagation: *Sunspots* are linked to electromagnetic waves and particles generated by the sun, which can severely hinder communications by disrupting the usual atmospheric propagation patterns and via interference and noise. Sunspots wax and wane in a roughly 11-year cycle.

An important factor in communication by electromagnetic waves is that their direction is affected as they pass through layers of the atmosphere. These layers cause *refraction,* or bending as the wave passes through the boundary from one layer to the next. The effect is the same as the refraction of light, where a straw in a glass of water appears to be bent. Of course, the straw is still straight, but the electromagnetic information—the light—from the part of the straw that is below the water line bends as it crosses from the water to air layer. Refraction bending can become actual *reflection,* depending on the angle of the light passage and the *index of refraction* of each atmospheric layer compared to its adjacent ones.

Ground or Surface Waves

Ground or surface waves (Figure 9.1) hug the contour of the earth and are affected by the terrain. Mountains get in their way and cause attenuation. In contrast, open water, with its high conductivity (especially for salt water) forms a low-loss path for surface waves. The ground wave provides a reliable 24-hour/day communications capability for frequencies up to about 2 to 3 MHz at distances of several hundred miles maximum. The antenna for a ground wave must project the signal at a very small *radiation angle* (essentially parallel to the earth surface) so that the radiated energy is not transmitted toward the atmosphere instead of along the ground.

The limitation of ground waves is that they can provide only relatively short distance communications, not the worldwide contact that is needed for some applications. Also, they cannot be used for the higher frequencies because of high signal attenuation, or provide the wide bandwidth that higher frequencies make possible.

Space Waves

Space waves travel in a straight line from the transmitting antenna to the receiving antenna (Figure 9.2a). The potential distance depends on the height of the antennas, since the curve of the earth limits the line of sight. In practice, the distance that a space wave can reach for a given antenna height is about $\frac{4}{3}$ greater than the distance that strict geometry analysis shows. This is because of *diffraction* effects that occur as a wave reaches the horizon at a low angle, a sort of "glancing along."

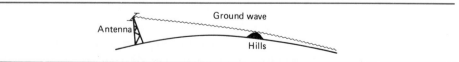

Figure 9.1 Ground wave going along earth's surface.

9.2 Propagation Modes

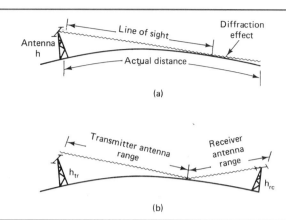

Figure 9.2 Space wave: (a) showing effect of antenna height and how diffraction extends distance; (b) showing additive effect on distance of two antennas.

The distance achieved due to the height of the transmitting antenna ($height_{tr}$) and the receiving antenna ($height_{rc}$) are independent of each other and are added together to find the overall distance (Figure 9.2b). The empirical formula that relates the distance, in miles, to the height of each antenna, in feet, is

$$\text{total distance} = \sqrt{2 height_{tr}} + \sqrt{2 height_{rc}}$$

For maximum signal strength at the intended receiver, the angle of signal radiation from the transmitter antenna should be toward the horizon (a low radiation angle).

Example 9.1

A transmitting antenna is on a 50-ft tower and the receiving antenna is on an identical tower. How far apart is the potential distance between them?

Solution
$$\text{distance} = \sqrt{2(50)} + \sqrt{2(50)} = 20 \text{ miles}$$

Example 9.2

For the 50-ft tower, what is the distance if the receiving antenna is at ground level? What if it is on a 100-ft tower?

Solution
For the ground-level case,
$$\text{distance} = \sqrt{2(50)} + 0 = 10 \text{ miles}$$

For the 100-ft tower,
$$\text{distance} = \sqrt{2(50)} + \sqrt{2(100)} = 10 + 14 = 24 \text{ miles}$$

Note that doubling the height of the tower from 50 to 100 ft increased the distance associated with that tower from 10 to 14 miles (which is not doubling the distance).

Unlike ground waves, space waves are used at any frequency, as long as the antenna can be fabricated in a size compatible with the frequency in use. The space wave is used for commercial FM and TV stations, mobile communications, and microwave telecommunications system links.

Sky Waves

The sky wave is aimed not at the intended receiver but at the sky, unlike the ground wave, which is aimed low to the horizon and travels along the surface of the earth, or the space wave, which is aimed directly at the receiver.

The sky wave takes advantage of the *ionosphere* that surrounds the earth to provide worldwide communications with reasonably good quality, reliability, and moderate power. The basic idea of a sky wave is to radiate the signal toward the ionospheric layers and have it refract and return to earth a substantial distance away (Figure 9.3). Some of the signal passes through the layers and out into space, but enough returns to earth to be picked up by a sensitive receiver. Since these layers are relatively high, from 20 to 250 miles up, they can provide tremendous distance in a single refractive *bounce*. Additional distance is possible when the signal reflects from the earth and goes back up to the ionospheric layers for another *hop*. These multiple hops are what provide the capability for globe spanning communications.

The angle at which the transmitted signal is critical to the distance. If it is too high, the refracted return (bounce) will come down too close, or too much of the signal will pass through the layers rather than refract from the layers as it should. If the angle is shallow (too low to the horizon), the return of the bounced signal will be much farther away. In this case, nearer listeners may receive nothing while distant ones will have a good signal. The area between the transmitter and the return of the usable signal is the *skip zone*. Careful adjustment of the antenna angle of signal radiation allows the transmitter operator to direct the energy to reach the desired listeners, and avoid others. For example, broadcasts from the United States to Europe via short-wave radio (to about 30 MHz) would be aimed to skip over the ocean, where there are few listeners. Similarly, Radio Moscow's

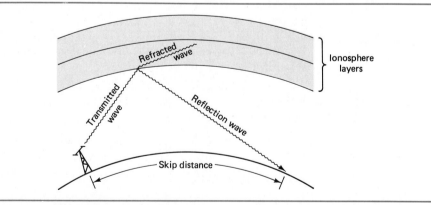

Figure 9.3 Sky wave refracting back to earth extends distance far beyond ground or space wave.

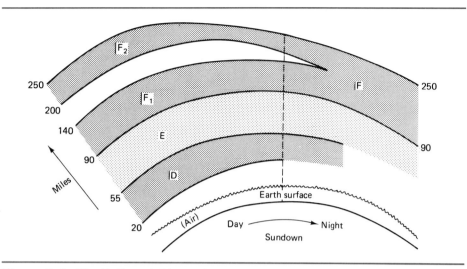

Figure 9.4 The D, E, and F ionospheric layers during the day and night.

English language broadcasts could be targeted to go to the United States but skip Europe.

While sky waves and refraction provide effective worldwide communications, successful application requires understanding of the specific factors that determine what frequencies will skip, by what distance, and at what times. These are the key factors of reliable and predictable propagation by sky waves. First, we will look at the ionospheric layers that are responsible for "refraction and skip" (the shorthand term for propagation by refraction from atmospheric layers).

The atmospheric layers that produce the refractive effect on radiated signals consist of free ions and electrons in the upper atmosphere region (60 miles and higher) called the ionosphere. The ionosphere is the layer of partially ionized gas that is above the oxygen-rich layer we live in, with the *troposphere* above that, and the *stratosphere* above both of them. This ionization is caused by ultraviolet radiation from the sun. The ionosphere consists of a series of layers of varying ion density at different heights. Each of these layers has a central region of relatively dense ionization, with this ionization tapering off above and below it. The amount of ionization depends on many factors: amount of sunlight, season of the year, sunspots, weather conditions, and local terrain.

The ionosphere is divided into three major layers, Figure 9.4:

The D layer, 20 to 55 miles above the earth's surface, exists during the daytime only; it ceases to exist after sundown. It can be used for signals up to several megahertz.

An E layer, at roughly 55 to 90 miles, is most useful at the sun's noon peak. It can be used until a little after sundown and at frequencies up to about 20 MHz. By midnight it is gone completely.

The F layer, which actually has two sublayers, F_1 and F_2, at 90 to 250 miles height. This is the long-distance layer, used for frequencies up to 30 MHz and available around the full 24 hours. The two F layers are sepa-

270   Chap. 9   Propagation and Antennas

rate during daylight hours, at about 140 and 200 miles, but merge after sunset.

Researchers publish prediction of *skip conditions* for use by broadcasters based on theory, past experience, and known sunspot and seasonal factors. The broadcasters then try to use the layers and associated skip opportunities to their best advantage for low noise communication with sufficient quality. For frequencies above 30 MHz, communication by sky wave is possible but much harder to predict and less reliable. Some regular skip communication is done up to 50 MHz, and there are many recorded skips covering thousands of miles, at frequencies above 100 MHz. But these are the exception and are not used in regular commercial communication systems because of their unpredictability and poor consistency.

Critical and Maximum Usable Frequency: Critical Angle

A wave of low frequency that is sent vertically toward the ionosphere will be reflected back to the transmitter. As the frequency of the signal is increased, eventually a frequency will be reached that does not reflect. This is the *critical frequency* for the layer being studied and is an indication of the highest frequency, called the *maximum usable frequency* (*MUF*) that can be used to transmit via skip. Since absorption of signal energy by the ionosphere (and therefore signal loss) is much less at higher frequencies, best results occur when the MUF is used rather than lower frequencies.

The angle that the transmitted wave energy makes with the tangent to the earth at the transmitter is the *angle of radiation*. A wave leaving the earth at a smaller angle (more parallel to the earth surface) requires less ionospheric refraction to bring it back to the earth. Sometimes the bending is not enough to return the wave unless the angle of radiation is less than what is called the *critical angle*. The critical angle is therefore one factor in determining the minimum skip distance, along with the height of the ionospheric layers that will cause the bounce. Higher layers give longer skip distances for a fixed angle of radiation. If the critical angle is very small, only small, shallow angles of radiation can be used and the skip zone will be long. The desired listeners may be too close in to receive the signal but more distant listeners will be satisfied.

When the single transmitted signal wave takes slightly different paths to the destination, or the transmitted signal travels in more than one propagation mode (ground, space, sky), there will be a main signal and the delayed signal (which sounds like an echo or appears like a "ghost" on a TV picture) at the receiver. This is because the time to reach the receiver depends on the path length. The space wave is a straight line and so has the shortest distance; the ground wave follows the earth and has a longer distance; and the sky wave must travel to the earth's atmospheric layers and back, so it has the longest distance. As propagation conditions change, one or more of the signals can increase or decrease in strength, so the delayed signal intensity changes. The delay time can also change as the reflecting layer moves or the point of maximum layer reflection changes, too. The phrase *multipath* is also used to describe the problems that occur when a primary signal and a slightly delayed version are received.

Satellite Waves

Satellite waves are intended to pass through the earth's ionosphere and into space, or travel from a space-based transmitter to a receiver on the ground. Like

9.2 Propagation Modes

the space wave, this is line-of-sight communications, except that the curvature of the earth and the horizon play no direct role. Unlike a space wave, the distances involved are very large: A satellite orbiting the earth synchronously (turning as the earth turns, once every 24 hours) is approximately 23,000 miles above the earth. Space probes and interplanetary satellites are, of course, much farther away.

Satellite wave systems use frequencies which are much higher than the critical frequency, high enough to penetrate the ionosphere without refracting back to the transmitter. These higher frequencies also provide the bandwidth that many space vehicles need. The major problem in using satellite waves is the high *path loss* caused by the large distances. The electromagnetic energy spreads (disperses) with distance, and relatively little reaches the receiver, just as the light beam of a flashlight spreads out and the light intensity is much less at 20 ft than it is at 5 ft. A sharply focused radiation pattern, which aims a larger proportion of the energy at the receiver, is a major help in increasing the signal strength at the receiving antenna, as we will see in the next few sections.

This loss of energy as the signal travels through space unimpeded and spreads out is the *free-wave path loss*. The power P_r at the receiving point will be far less than the power P_t at the transmitting point, by the formula

$$\frac{P_t}{P_r} = \left(\frac{4\pi f d}{c}\right)^2$$

where c is the speed of light, d is the distance in kilometers, and f is the frequency in megahertz. By substituting the numerical values and converting this to a dB ratio, we have the path loss in dB:

$$\text{path loss (in dB)} = 20 \log \frac{4\pi f d}{c}$$

$$= 20 \log \frac{4\pi}{c} + 20 \log f + 20 \log d$$

$$= 20 \log \frac{4\pi (10^6)(10^3)}{3 \times 10^8} + 20 \log f \text{ (in MHz)} + 20 \log d \text{ (in km)}$$

$$= 32.4 + 20 \log f + 20 \log d$$

This equation makes clear the extremely high path loss over the large distances of space communication, as some examples show.

Example 9.3

What is the path loss in dB at 300 MHz between the earth station and a satellite at 37,000 km?

Solution

$$\text{Path loss} = 32.4 + 49.5 + 91.4 = 173.3 \text{ dB}$$

Example 9.4

What is the dB path loss at 100 MHz when tracking an interplanetary satellite in space at 100 km distance? (Note that this is only about 63,000 miles, so this satellite is just starting on its journey.)

Solution
$$\text{Path loss} = 32.4 + 40 + 40 = 112.4 \text{ dB}$$

The path loss equation can be used to determine the signal levels over the distance and the effect of the antennas being used. Any antenna has *gain*, studies more fully in the next section. This gain factor indicates how well the antenna projects energy in the desired direction of interest, and shows how an antenna can magnify the effective transmitter power or increase the effective signal level at the receiver. The overall signal loss, including transmitting antenna gain G_t and receiving antenna gain G_r (both in dB), is

$$\text{total loss} = G_t + G_r - \text{path loss}$$

The antennas increase the apparent strength of the signal sent or received, while the path loss reduces the signal. The actual signal strength at the receiver is the transmitted strength minus the total loss (in dB).

Example 9.5

A 1000-W transmitter power amplifier for 150 MHz is to be used with a broadcast antenna that has a gain of +70 dB. The receiving antenna on the spacecraft 1,000,000 km away has a smaller gain, only +10 dB. What is the signal strength at the receiver front end?

Solution
$$\text{Total loss} = 70 + 10 - (32.4 + 43.5 + 120) = 115.9 \text{ dB}$$

$$\text{Receiver signal} = 1000 \text{ W reduced by } 115.9 \text{ dB} = 2.5 \times 10^{-9} \text{ W!}$$

Note that this is not an unrealistic example. There are many applications with even greater loss, which emphasizes the need for high-gain antennas and a low noise, sensitive front end in the receiver. The situation is even more difficult in radar, where the transmitter signal is returned to a receiver at the transmitter location. In radar, a small fraction of the transmitted signal is reflected, the overall distance is twice the distance to the target, and there is significant loss in the reflection itself, in contrast to gain from antenna.

Review Questions

1. What factors make signal propagation a complex subject? How does the ionosphere affect propagation? What causes refraction and reflection?

2. What is a ground wave? What is the maximum distance? What limits the distance?

> 3. What is a space wave? Compare it to a ground wave.
>
> 4. What are sky waves? Why are they used? How do the heights of the refracting layers affect sky-wave distance?
>
> 5. What is the skip zone? What is the effect of signal angle on skip zone?
>
> 6. What are the major characteristics of the D, E, and F layers? What are propagation predictions? Why are they done?
>
> 7. What frequencies are used for satellite waves?
>
> 8. What factors enter the equation that links power transmitted and power received, for satellite waves?
>
> 9. What is the equation for path loss? What does it clearly show for long-distance communications? Why are high-gain antennas needed, along with low-noise receivers?

9.3 ANTENNA CHARACTERIZATION

The electromagnetic field pattern that an antenna develops can be described by equations, but these mathematical descriptions are usually complex and not very useful for actually using the antenna effectively. Although the equations are exact in theory, there are many practical antennas that cannot be described, modeled, and analyzed mathematically with sufficient precision, especially when the effect of nearby structures (walls, ship hulls) is considered. For this reason, many antenna designs are studied and verified by use of small-scale models (described later) in electromagnetically isolated rooms. Fortunately, the key characteristics of the antenna can be described with some relatively straightforward terms. In this section we examine how the performance of any antenna, regardless of its design or size, is characterized and interpreted.

One of the general tools used to show the antenna performance is a *polar plot* of the relative signal strength of the radiated field versus distance and angle from the antenna. Some typical radiation patterns are shown in Figure 9.5. The 360° of the total circle represents a complete sweep of the horizon, in all directions. The plot shows how the antenna shapes the field strength versus distance as it radiates signal energy, and which angles from the antenna have more or less field strength as a result. While the field strength can be plotted in absolute numbers versus distance, most polar plots are done in a relative scale. The location of maximum signal strength is the arbitrary 0-dB reference, and each graduated ring closer to the center of the plot represents reduced signal strength, typically by −3 or −10 dB. The angle of the signal strength information of the polar plot generally has its 0° orientation aligned with the major physical axis of the antenna.

The shaping of the radiated field pattern is analogous to using a lens to shape light to the desired pattern. The lens takes the light from a bulb and directs it toward a target. The type of direction and focus required is determined by the application. A spotlight for an entire cast on stage would be focused broadly, while the same light would be more directed, and be apparently brighter, when focused onto the face of a single performer.

274 Chap. 9 Propagation and Antennas

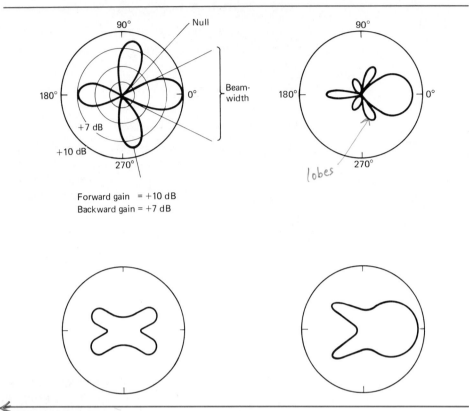

Figure 9.5 Some typical radiation patterns, along with definitions of gain, beamwidth, and front/back ratio.

By looking at the polar plot, key antenna factors such as *radiation field pattern, gain, lobes, beamwidth,* and *directivity* are seen. Polar plots are very useful for comparing different antennas. The use of the relative scale for signal intensity, rather than absolute numbers, is normally not a problem. This is because the absolute value is related to the power applied to the antenna at the transmitter (or received by the antenna from the surrounding space) and may therefore change. The relative pattern, however, is determined primarily by the antenna design.

The antenna field pattern is the general shape of the signal intensity, in all directions, measured relatively far from the antenna (many wavelengths) and called a *far-field* measurement. The far-field measurement is used so that variations caused by the detailed physical construction of the antenna, visible only when very close (called *near-field*), do not confuse the overall "big picture." It is very similar to studying a picture printed in a newspaper or book with many tiny dots. The picture looks fine and clear at a distance of a few inches, but if you take a close-up look (or use a magnifying glass) you will see the many dots of varying intensity and not the overall, recognizable scene of the picture.

Measuring the Pattern

In practice, the field pattern is measured with a *field strength meter*. This is basically a sensitive, calibrated receiver which is tuned to the frequency of interest and indicates the amplitude of the received carrier signal. The field strength

> **Scale Antennas**
>
> Sometimes it is impractical to build and measure an actual antenna, especially if the antenna design is complex, or the antenna will be used near structures that can affect its pattern (ships, aircraft bodies, even buildings). The alternative to an outside field test is to build a precise scale model of the antenna, at $\frac{1}{10}$ or $\frac{1}{20}$ the final size. Frequency of operation must be scaled as well, to much higher frequencies proportional to the decrease in size. This is because antenna dimensions are closely related to wavelength, and smaller antenna elements require shorter wavelengths to maintain the relationship between the scale model and the full-size antenna. The interfering structures can be scale-modeled as well.
>
> A special room called an *anechoic chamber* is used for the measurements. This room has two main features: It is shielded from the outside world, so no electromagnetic radiation enters to interfere with the tests and corrupt the results; it also has inside walls that absorb nearly all radiated energy, so that signals from the test antenna do not echo within the room and cause false readings. The field strength meter tests are then run in this room with the miniature versions of the antenna and its operating environment, and completely valid results are obtained. Of course, it is also easier to make changes on the small model than on a full-scale antenna.

meter is calibrated for its own receiving antenna, and the relationship between field strength it indicates (in μV/meter) and its received power (in μW) is known in advance. Therefore, the meter's received power can directly indicate the electromagnetic field strength.

Conceptually, the overall pattern of the antenna under test is determined by using some nominal field strength value, such as 100 μV/m, going around the antenna for 360° and noting the distance from the antenna where the indicated field strength has this value. The connection of all these points of equal field strength forms the radiation polar plot pattern. In practice, the distance between the meter antenna and the antenna under test is maintained at a constant value. The antenna under test is then rotated, while a plot of the field strength produced by the antenna under test, versus angle, is made.

There are several reasons for shaping the antenna field pattern: First, this produces higher effective power in the direction of the intended receiver, by concentrating power in the correct direction and not wasting some transmitted power where there are no listeners. For receiving antennas, the field is shaped to concentrate the energy-collecting ability of the antenna in the direction of the transmitter and make the antenna more sensitive in that direction—there is no point in being sensitive in the wrong direction, especially since it reduces the ability of the antenna to look in the right one.

Second, there may be receivers that the transmitter does not want to reach. This is especially important in military communications, but also when several stations are using similar frequencies. A carefully directed signal can allow two or more transmitter/receiver pairs to be located near each other, yet not interfere. A shaped antenna pattern means that the receiver will pick up only the desired signal, while minimizing signals from transmitters that can overload the front end and cause malfunctions.

Finally, the shaped pattern can be used to locate targets. A radar system works by transmitting a signal and seeing how much is reflected by a target. A narrow field pattern means that the specific target direction can be identified with greater precision. A direction-finding antenna can also locate the direction of unknown transmitters if the receiver antenna is tightly focused and its pattern known.

Not all antennas need to shape the field pattern. An antenna for a standard broadcast station, or for a police/fire mobile radio, needs to have a circular pattern (called *omnidirectional in azimuth*) to cover the overall territory of listeners. The modifications to the field pattern would be done only if there were large areas of the 360° pattern which had few potential listeners—such as broadcasting to open ocean water where the only listeners would be boaters.

The *gain* of an antenna is specified relative to some reference antenna. This reference antenna is often *isotropic,* ideally radiating power equally in all directions in a perfectly spherical pattern. The gain is then called *gain over isotropic*. The polar pattern of an isotropic antenna is a circle, a slice through the sphere. The gain in dB (dBi is often used to indicate when an isotropic antenna is the reference) shows the field strength of the antenna compared to the field strength of the reference antenna, when both are fed with the same signal power and measured at the same distance and location.

An isotropic antenna is an ideal concept but cannot be realized in practice. For this reason the gain of an antenna is often measured with respect to the *half-wave dipole* as a reference, which is an easy-to-build antenna with a simple field pattern (and usually indicated simply as dB without any suffix). The half-wave dipole is studied in the next section.

Note that the word "gain" is somewhat misleading. An antenna does not produce gain as an amplifier does, since the total signal power output of the antenna is the same as the input power from the transmitter (less any losses in the transmission line). What the antenna does by its design specifics is to shape the radiated field pattern so that the power measured at a specific location is greater than (or less than) it would be for an isotropic antenna or half-wave dipole. The total power is unchanged, but the apparent power is different in a given direction.

Once again, the light and lens analogy is appropriate. A lens provides a gain (or loss) in intensity in certain direction but does not increase the actual power of the lamp. To the receiver (eye) seeing and using the light, however, the increased intensity has the same impact as using a brighter lamp. For example, an antenna with a gain of +6 dB in one direction over isotropic shapes the radiated field so that a receiver sees the same amount of power at that location, as if the transmitter power had been increased by 6 dB while using an isotropic antenna—equivalent to quadrupling the transmitter power. A receiver antenna that provides gain of 3 dB is equivalent to using a receiver with a front end that is twice as sensitive.

Gain can also be negative: A gain of −3 dB means that the power at that location is the same as would be measured with a transmitter providing just half the output power, using an isotropic antenna. Therefore, using gain at the transmitting antenna is the same as increasing transmitter amplifier power; having antenna gain at the receiver is equivalent to a more sensitive front end. Achieving this increased power or improved sensitivity through use of an antenna and its gain is usually less costly than increasing transmitter power or receiver sensitivity, and is often the only technically practical way. More power may not be available, while receiver sensitivity cannot be increased to arbitrarily large values since front-end noise determines the maximum sensitivity available.

Antenna patterns have *lobes,* which look like flower petals. The radiated field pattern does not have to be a smooth, continuous curve. As a consequence of the antenna design for high gain, the overall pattern will have segments of very high signal strength (lobes) and very low signal strength (*nulls*). A receiver in the null area sees very little field intensity, while a receiver located at a lobe sees greater strength. Some field patterns have a main lobe as well as smaller, secondary lobes.

For applications where antenna gain is needed, but the direction of this gain must change, antennas are designed to rotate. Some applications require repetitive rotation over a complete circle or part of the circle, while others must rotate only in specific circumstances to try, for example to receive a new signal from another transmitter. Rotation can be done by hand, or with the special motor drives, which turn the antenna support mast as needed. Of course, this complicates the mechanical design of the antenna and the transmission line between the communication system and the antenna.

The *beamwidth* of the antenna is the total angle at which the relative signal power is 3 dB below the peak value of the main lobe. Beamwidths can be as small as 1°. They can be as wide as 360° for an antenna that radiates nearly equally in all directions, but many antennas provide some focusing of the pattern, with a 30–60° beamwidth; special antennas produce extremely narrow beamwidths on the order of 1°.

An antenna that is designed to radiate primarily in one direction has *directivity*. This is a measurement of the signal strength in the desired forward direction compared to the backward direction. For example, an antenna with gain of +3 dB in the forward direction and −3 dB in the backward direction has a *front-to-back (F/B) ratio* of 6 dB. The F/B ratio is a quick indication of how directive an antenna is. It does not show the ''sharpness'' of the pattern focus—the lobes and the beamwidth indicate that factor. It shows the contrast between the field strength in the main radiation direction versus the opposite direction. An antenna with a sharp forward lobe gain of +10 dB and equal backward lobe gain has a F/B ratio of 0 dB, yet is quite different from an antenna with a perfectly circular pattern, which also provides a 0-dB F/B ratio.

There are other antenna characteristics that do not use the polar radiation pattern. The *radiation angle* of the main lobe measured with reference to the horizon (Figure 9.6) is important in determining the way the majority of the sky wave power is projected. This, in turn, affects the skip pattern or the space-wave angle. Radiation angles between a few degrees and 60° are commonly used. If the angle is too high, the signal will not reflect properly from the atmospheric layers and instead will go right out into space—which may be good for satellite communications, but not suitable for medium-wave or short-wave communications around the globe.

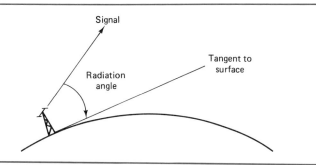

Figure 9.6 Measurement of radiation angle with respect to the horizon.

Antennas also have *bandwidth*. The physical design and dimensions of the antenna are optimum at a specific frequency, called the *center frequency*. Some antennas perform nearly as well at frequencies far removed from the center value, while the performance of others falls off rapidly away from the center. Just as with filters, the bandwidth can be defined by the 3-dB points of the antenna performance. An antenna intended for 100-MHz operation, with a 10-MHz bandwidth, is only one-half as effective at radiating energy (or at receiving it) at 95 and 100 MHz. Bandwidth is expressed as a percentage of the center frequency, or as a ratio between the highest and lowest frequencies at the "3-dB down" frequencies.

Example 9.6

What is the bandwidth, in percent, for a 220-MHz center-frequency antenna that has 3-dB points at 190 and 240 MHz? What is the bandwidth ratio?

Solution

As a percent,

$$\text{lower bandwidth} = \frac{220 - 190}{220} = -13.6\%$$

$$\text{upper bandwidth} = \frac{240 - 220}{220} = 9.1\%$$

As a ratio, maximum to minimum is 240/290 = 1.3 to 1.

The amount of bandwidth needed depends on the application, and having more (or less) bandwidth is not inherently a good (or bad) thing. An antenna expressly designed for transmitting a broadcast radio station signal is designed for the frequency and bandwidth of that station. In contrast, an antenna for receiving one of many signals within a larger band must have wider bandwidth, or received signal performance will suffer at the ends of the band. A standard FM broadcast receiver antenna must have an 88- to 108-MHz bandwidth, while the broadcast station antenna need only have a 200-kHz bandwidth.

It may seem that it is always good to have a wide-bandwidth antenna. However, wider-bandwidth antennas tend to have less gain than narrow-bandwidth ones, so some of the desired antenna directivity may not be achievable. Second, a narrow bandwidth reduces interference from adjacent signals (which can overload the front end) and reduces the received noise power as well. Both of these are factors that limit how well a receiver can function for very low level signals. The antenna designer must look at the many goals of the application: directivity and beamwidth, acceptable lobes, maximum gain, radiation angle, and bandwidth, and trade these various goals and constraints against each other.

Antenna Dimensions

As we will see in the next sections, the antenna operating frequency (and corresponding signal wavelength) determines, in large part, the physical size of its various elements. For this reason, dimensions are often specified in terms of wavelength λ. An element is said to have a "1.5λ length" or to have "0.25λ

Polarization

The electromagnetic field that defines the radiated signal power has a specific orientation of its electric field and magnetic field. The direction of *polarization* is determined by the antenna design and its physical orientation. Fields are classified as horizontally, vertically, or circularly polarized, based on the orientation of the electric field. A *vertically polarized* signal has the electric field in the vertical direction; a *horizontally polarized* one has its electric field parallel to the horizon.

The polarization of the transmitting antenna and the receiving antenna must be the same for maximum signal energy to be induced in the receiving antenna. If the polarizations are not the same, the electric field of the radiated signal will be trying to induce an electric field into a wire that is at right angles to the correct orientation, and in theory there will be no induced voltage (in reality, there will be a little). The corresponding situation holds for the magnetic field if it tries to induce a field into an incorrectly oriented wire.

There are situations where the orientation of one or both of the antennas is not fixed, but varies. Aircraft and satellites are examples of this. To overcome the possible mismatch in polarization that would occur, *circular polarization* is used. In circular polarization the electric field orientation is not fixed horizontally or vertically, but is constantly rotating. It is used because it is compatible with any polarization angle, from horizontal to vertical. The drawback to circular polarization is that its maximum gain is 3 dB less than a correctly oriented horizontally or vertically polarized antenna, but this is still much better than using a vertically polarized antenna for a horizontally polarized signal.

---

spacing.'' The formulas used for design often use λ as the variable, and then the actual wavelength of operation is substituted to provide real dimensions in meters (most commonly) or feet.

Testing and adjusting a transmitter for proper modulation, amplifier operation, frequency accuracy, and other operating characteristics requires the electrical load of antenna. Using a real antenna for these tune-ups may cause interference with other users of the spectrum or cause interference with test equipment nearby. The solution is a *dummy antenna* (discussed in Chapter 8), which looks like an antenna electrically and has the impedance of a real antenna. The dummy load does not radiate any signal but instead dissipates the transmitter signal as heat.

Although the dummy load tells you nothing about the performance of an antenna itself, it allows the transmitter to be adjusted as if it had the antenna. Then the real antenna can be connected and if there is a problem, the technician knows that the problem lies with the antenna and not the transmitter or transmission line, since the transmitter has been checked out separately and independently.

For receiver antenna testing, the problem is not preventing actual signal radiation and possible interference as it is with transmitters. First, the transmission line and receiver must be checked out, by testing with a signal of the correct frequency and power. An RF signal generator is connected in place of the antenna to generate a signal of known amplitude and power. After the receiver has been adjusted or repaired with this single signal, the performance of the receiver must be checked in the presence of more than one receiver signal. Special signal generators are available that produce simultaneously several independently adjustable signals, which represent the RF energy that a real antenna sees and sends into the receiver front end. Once the receiver operation is verified and assured, the real antenna is installed in place of the signal generator and its performance checked.

Review Questions

1. What is the radiation pattern polar plot? How is it oriented? Why does it show relative dB rather than absolute numbers?
2. How is the radiation pattern measured? Why is it measured at a significant distance?
3. Explain how scale models can be used to measure antenna characteristics. What are the advantages of the scale model method?
4. Why is the antenna pattern shaped? What are the goals of this?
5. What is antenna gain? How is it defined? Why is the term *gain* a misnomer? What does the antenna really do to provide gain?
6. What is gain compared to an isotropic antenna? What is the isotropic radiation pattern?
7. What are radiation pattern lobes and nulls? What is antenna beamwidth? How is it measured?
8. What is the front-to-back ratio? Compare it to antenna lobes. What does the F/B ratio indicate, and what does it not indicate?
9. What is antenna radiation angle? Where is it critical?
10. What factors affect antenna bandwidth? How is this bandwidth defined and expressed? What situations require greater bandwidth? Why? Which applications need less bandwidth, and for what reasons?
11. What is antenna polarization? How is it defined? Why should both antennas have the same polarization?

9.4 ANTENNA FUNDAMENTALS

We can understand a lot about antenna functions by looking at one of the earliest and simplest practical antennas, the half-wavelength dipole. Despite its simplicity, this antenna is used in many applications, and a large proportion of more advanced antennas are basically refinements and enhancements of it.

The *half-wavelength dipole* is formed by spreading the conductors of the two-wire transmission line out into a straight line (Figure 9.7). Each of the conductors forms an arm that is $\lambda/4$ long, for a total span of $\lambda/2$, with a small gap between the arms at the center. The voltage and current in this antenna have distinct patterns. The current is zero at the ends and a maximum in the middle where the antenna is driven by the transmission line. The voltage of this antenna is zero in the middle, at the gap, with a positive maximum at one end and a negative maximum at the other, as shown in the figure. The impedance of this antenna is purely resistive at 73 Ω, which is relatively easy to match to many transmission lines.

The electric and magnetic fields of this antenna are used to determine the overall radiation pattern. The electric field goes through the ends of the antenna, shown in Figure 9.8a, spreading out in ever increasing loops, while the magnetic field looks like a series of concentric circles centered at the gap of the two arms.

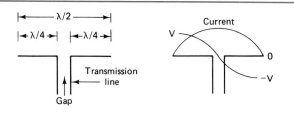

Figure 9.7 Basic half-wavelength dipole antenna design.

The resultant three-dimensional field pattern has a doughnut shape, with the antenna as a shaft through the center of the doughnut. To determine the radiation pattern, take a slice through the doughnut: When the antenna is horizontal, a slice reveals a "figure 8" pattern (Figure 9.8b). Maximum radiation is broadside to the antenna arms.

A half-wave dipole in free space, unaffected by any surrounding structures or the earth's surface, has a beamwidth of 78° and a maximum gain of 2.1 dB relative to an isotropic antenna (whose polar radiation pattern is a circle). This number is important, because many practical antennas are compared to the dipole, not to the isotropic antenna (which cannot be built). The tests are run by feeding the same signal power to a half-wavelength dipole and to the antenna under test, and comparing the field strength in all directions.

The actual construction length of a half-wave dipole differs slightly from the theoretical values for two reasons. First, the propagation velocity in the wire is less than in air or vacuum, and this must be factored in. Second, the fields are not perfect at the ends of the elements that form the antenna arms, but have "fringe" effects and are affected by the actual capacitance of the antenna elements (which is a function of the diameter of the wire or tubing used to construct these elements). A good first estimate to compensate for these factors is to make the real length of the elements 5% less than ideal. For example, a half-wavelength dipole using ideal figures has an impedance of $73 + j45 \, \Omega$, while one that uses 5% shorter dimensions is nearly perfectly resistive.

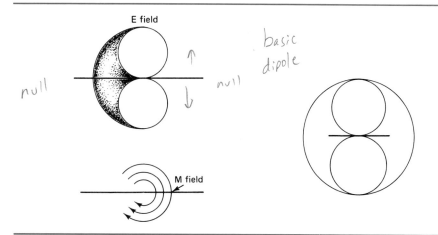

Figure 9.8 (a) Electromagnetic fields of a basic dipole; (b) dipole polar radiation pattern is a figure 8.

Example 9.7

What is the span of a $\frac{1}{2}\lambda$ dipole for 50 MHz, in theory, and when is it compensated for antenna element diameter?

Solution

$\lambda = c/f = (3 \times 10^8 \text{ m/s})/(50 \times 10^6 \text{ Hz}) = 6$ m, so the dipole size is 3 m, in theory. In practice, it is 5% less: $3 - 0.05(3) = 2.85$ m.

The bandwidth of a $\frac{1}{2}\lambda$ dipole is between 5 and 15% of the center frequency. The major factor for determining bandwidth is the diameter of the conductors used—smaller conductors result in a narrower bandwidth than from larger-diameter conductors. This fact is used to design antennas for differing bandwidth needs, within limits.

Multiband Dipole Antennas

In many applications it is impractical to use more than one antenna, although several widely separated frequency bands are in use, such as in international short-wave broadcasts. For example, amateur radio (ham radio) stations use 80 m (3.75 MHz), 40 m (7.5 MHz), 20 m (14 MHz), 15 m (21 MHz), and 10 m (29 MHz) bands, in the HF part of the spectrum. Individual antennas would be large, unwieldy, costly, and difficult to install and maintain.

A solution to using a single physical antenna structure on these widely separated bands is with *traps*. These traps (Figure 9.9) are inductor–capacitor combinations inserted in the dipole arms to make a single dipole appear to have different physical and electrical lengths at different frequencies. As the frequency of antenna operation increases, the traps present an increasingly higher impedance to the signal, and thus electrically sever the remainder of the dipole arm from the center parts. Using traps, a single folded dipole can serve several bands.

There are some practical and electrical issues with traps. For outdoor use, they must be in weather-resistant housings. Transmitter antenna traps must use

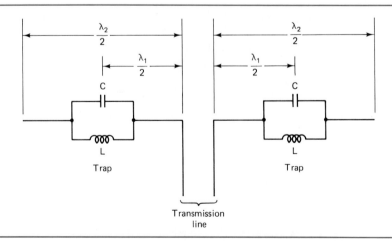

Figure 9.9 Traps allow multiple bands to use the same dipole structure.

9.4 Antenna Fundamentals

Transmit/Receive Switches

The use of a single antenna for both transmitting and receiving is very convenient and is some cases a practical necessity. This is clear from the principle of reciprocity, which shows that a transmitting antenna performs identically to a receiving one. For mobile radio, radar, and half- or full-duplex communications systems, the same antenna means that field patterns will match perfectly.

There is one problem with using the same antenna, however. The relatively high power signal from the transmitter amplifier can overload and even damage the sensitive, high-gain RF front end of the receiver. To overcome this problem, a *transmit/receive (T/R) switch* is used between the power amplifier, receiver front end, and the transmission line (Figure 9.10). It operates in conjunction with the transmit/receive mode switch of the system. The T/R switch prevents most of the transmitted energy from reaching the receiver via a variety of isolation schemes, including electromechanical relays. The figure of merit for a T/R switch indicates, in dB, how well it prevents this energy feedthrough effect. Typical T/R isolation is 80 to 100 dB. At 100-dB isolation, a 10-W signal to the transmission line appears as a 10^{-9}-W (10-nW) signal to the receiver front end, which is well below the limit that a front end can handle without overload.

Many T/R switches are specified with their effective isolation versus frequency of operation. In general, it is technically difficult to provide isolation at higher frequencies with mechanical switching contacts. The miniature contacts of the T/R relay can act as tiny antennas and allow some transmitter signal to be radiated and then received within the relay enclosure, reducing the achievable isolation figure of merit. At microwave frequencies, entirely different techniques are required (Chapter 23).

inductors and capacitors that can withstand the full output power and voltage. Secondary effects of traps affect the calculation of dipole arm length, so an antenna with traps usually must be manually adjusted after it is built. The overall performance (gain, beamwidth, radiation angle) of an antenna with traps is not the same as a nontrap antenna, and the user must accommodate the differences. Finally, traps cannot be used at the higher frequencies (above approximately 30 MHz) since their detrimental impact on performance is more severe, and the relatively small inductance and capacitance of the trap wires themselves render the trap scheme ineffective.

Traps are not limited to use in half-wave dipoles. The Marconi (whip) antenna discussed in the next section also can handle several bands using traps.

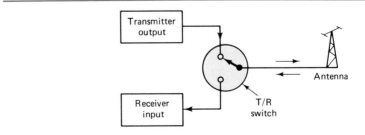

Figure 9.10 Location of transmit/receive switch which prevents transmitted signal from overloading or damaging the receiver front end.

Review Questions

1. What is the half-wavelength dipole? How is it formed? What are the voltage and current pattern for the half-wavelength dipole?
2. For the dipole, what are the radiation pattern and F/B ratio? What are its beamwidth and bandwidth?
3. Why does the actual length of the arms of the dipole differ from the theoretical value? How much is the typical difference?
4. What are traps? Why are they used? What is the benefit of using them?
5. What is a transmit/receive switch? How is it used? Why is it needed? How is its performance measured?

9.5 ELEMENTARY ANTENNAS

The design of practical antennas spans the range from a simple long wire to complex units with many elements of precise length. Each of these antennas serves a need in some application, and no single antenna design is suitable for all applications. Historically, of course, the simpler antennas were developed first. As understanding of antenna theory and implementation increased, and as the uses of electronic communications expanded in different areas, enhanced antennas were put into use. The simpler antennas are still often used, serving many needs at low cost and with great flexibility. In this section we examine some of these relatively simple antennas, and in the next section, the more advanced designs that have been created.

Long Wire Antennas

A piece of wire that is several wavelengths long can be an effective, wide-bandwidth antenna. The *long wire* antenna is driven at one end by the nongrounded output of the transmitter, and the transmitter electrical circuit ground is connected to earth ground. The far end of the long wire is terminated by a resistor connected to ground with the resistor value equal to the Z_0 of the antenna, typically several hundred ohms.

In operation, the long wire antenna acts as a lossy transmission line. The terminating resistor prevents standing waves that would reflect back to the transmitter. The polar radiation pattern of the long wire antenna has two main lobes, on either side of the antenna, and these lobes are pointed forward, in the direction of the antenna termination (Figure 9.11). There are also smaller lobes on both sides of the antenna in forward and backward directions. The radiation angle for the long wire antenna is approximately 45°, depending on height above ground. This makes the antenna useful for skywave applications.

One major feature of the long wire antenna is that it is very broadband, and can be used for wavelengths from one-half to one-tenth the antenna length, a 5 : 1 frequency span. This is an advantage for general-purpose short-wave receivers which often cover 500-kHz to 30-MHz bandwidths in several bands. It is a drawback for systems intended to operate over a limited band, since avoidable interfering signals and noise will be received. The long wire antenna is not efficient: Only

9.5 Elementary Antennas

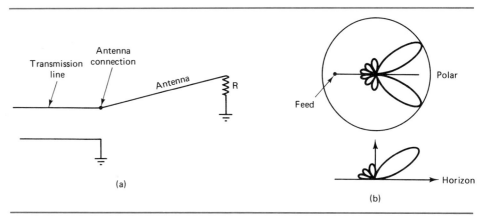

Figure 9.11 (a) Long-wire antenna schematic and (b) its radiation patterns (polar and angle).

half the power it receives from the transmitter is radiated, and half is dissipated in the termination resistor. The same situation applies when it is used for receiving, since only half of the electromagnetic energy it captures is delivered to the receiver front end. Despite these drawbacks, the long wire antenna is effectively used in broadband applications and where the antenna cannot be installed precisely, such as in an attic or on a roof.

Folded Dipole Antennas

The basic dipole of Section 9.4 can be $\frac{1}{2}\lambda$ long but folded around so that it forms a complete circuit. This *folded dipole* (Figure 9.12) is the heart of many advanced antenna designs because it is mechanically more rugged than a simple dipole. The bandwidth of the folded dipole is about 10% greater than the open dipole, while the radiation pattern is very similar.

The folded dipole impedance is 292 Ω, four times the 73-Ω impedance of a half-wavelength dipole and a close match to standard 300-Ω twinlead parallel-wire transmission line. The folded dipole does not have to use wire or tubing of the same gauge or diameter throughout. Using different diameters for the top part versus the lower arms produces different impedance values, so the folded dipole can be fabricated to match a wide range of impedances, if needed.

Loop Antennas

When a length of wire is bent into a circular or nearly square shape, a *loop antenna* is formed (Figure 9.13a). When used with a receiver, the maximum

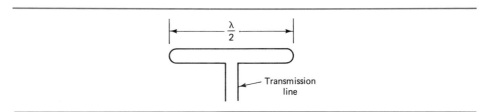

Figure 9.12 The folded dipole antenna is a variation of the basic $\frac{1}{2}\lambda$ dipole.

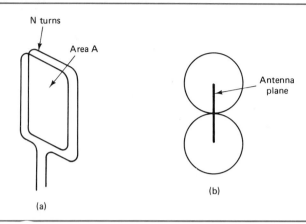

Figure 9.13 (a) Basic loop antenna and parameters; (b) loop antenna radiation pattern, viewed from above, is greatest along the plane of the loop.

voltage induced in this antenna by an electromagnetic field is a function of the magnetic field strength flux B, the loop area A, and the number of turns N in the loop, as well as the frequency of the signal, by the formula

$$V = k(2\pi f)BAN$$

where k is a physical proportionality factor. Note that signal strength is proportional to the factors f, B, A, and N and will increase (or decrease) as they increase (or decrease).

The radiation pattern of the loop antenna is maximum perpendicular to the center axis through the loop and very low broadside to the loop (Figure 9.13b). For this reason, the loop antenna is often used in direction finding (Chapter 19). The loop is rotated until a signal minimum (*null*) is observed, and therefore the transmitter is either on one side or the other of the loop. A second reading is taken at another location, another null is observed, and a second bearing taken. The two null lines meet, by simple triangulation, at the transmitter location.

The loop antenna is broadband and has been used from the early days of AM broadcast radio, where a 500- to 1600-kHz bandwidth must be received. Most portable AM radios still use the loop, but with a variation so that the radio is not dwarfed by the antenna. Many turns of wire are used (ranging from tens to hundreds) and these loops are wound on a magnetic ferrite core. The ferrite concentrates the magnetic flux so that more flux lines are cut by the loop. As a result, the induced voltage is much higher than if a nonmagnetic material such as air, plastic, or wood formed the core.

Marconi (Grounded Vertical Antennas)

A very simple antenna which has uniform radiation in all directions is formed by a single vertical element insulated from ground (Figure 9.14a). This antenna is called the *Marconi* antenna (after early radio pioneer Guglielmo Marconi, who developed and used it extensively) or the *grounded vertical* antenna. The Marconi antenna is commonly used at frequencies up to 2 MHz (such as standard AM broadcast) and also used as a whip antenna for mobile applications (police, taxi) because of its simplicity and omnidirectional pattern.

9.5 Elementary Antennas

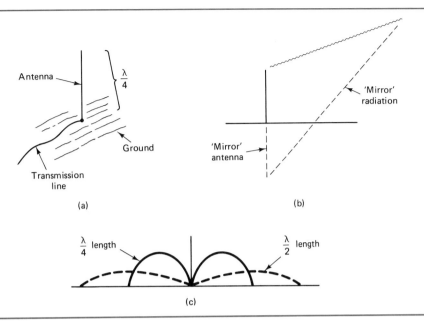

Figure 9.14 (a) Marconi antenna design; (b) Marconi "mirror" effect due to ground; (c) the shape of the Marconi vertical radiation pattern depends on its length.

The Marconi antenna makes use of the fact that the earth, as ground, is a *reflecting plane* for radiated signals. A mirror image of the antenna signal from the metal element appears to be coming from a phantom image below the surface (Figure 9.14b). If the antenna elements is $\frac{1}{4}\lambda$ long and is physically close to ground, the Marconi antenna appears to be a $\frac{1}{2}\lambda$ dipole, oriented vertically. The coaxial transmission line from the transmitter is connected with the center connected to the base of Marconi element, and the coaxial line ground is connected to ground right under the element. The impedance of a Marconi antenna is 37 Ω, one-half that of a true dipole.

The vertical radiation pattern of a Marconi antenna (Figure 9.14c) with a $\lambda/4$ element (total effective length $\lambda/2$) has significant energy in both ground- and sky-wave directions. If the element is shorter than $\lambda/4$, the pattern is still along the ground but with higher lobes. The field strength at a given distance will be less than it is for a $\lambda/4$ element, so the antenna will be less useful for long-distance ground waves. An antenna length of $\lambda/2$ results in more energy along the ground and less toward the sky. The radiation angle also varies with the height of the antenna height.

The ground under the vertical must have good conductivity for the Marconi antenna to be effective. Wet soil, marshes, or wetlands are ideal, and the coax ground can be connected directly to a rod sunk into the earth. Where the soil has poor conductivity (dry sand or rock, for example) an enhanced grounding scheme must be used. Many deep spikes are put into the ground all around the vertical element. These are connected together to form a ground plane or mat using radial wires which originate under the vertical element.

The $\lambda/4$ vertical element length may be too long at lower frequencies to be practical. Shorter elements can be used effectively if they are matched to the

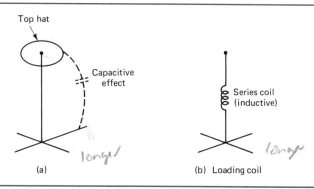

Figure 9.15 (a) Capacitive tophat loading of Marconi antenna to provide a better match; (b) loading coil at the base of the Marconi antenna also provides a better match.

transmission line to minimize SWR. For example, a $\lambda/8$ Marconi antenna (a typical car-mounted whip antenna size) has an impedance of about $8 - j500\ \Omega$ looking into the base, where the transmission line is connected. Two impedance-matching techniques are commonly used. In one method the vertical element has a *top hat* that loads the top of the element with additional shunt capacitance to ground

Antenna Couplers

When the impedance that the transmitter output would like to drive and the impedance of the antenna are different, an *antenna coupler* is used to match these impedances to maximize power transfer and minimize VSWR. Most transmitter outputs have impedances of 50 to 75 Ω, while antenna systems have a wide range of impedances due to the inherent physical design, or because the impedance varies over the wide bandwidth being used and the antenna is no longer operating at its design frequency. Of course, the antenna coupler is also useful for a receiving antenna, which must deliver maximum power to a receiver front end. The coupler also tunes out any reactive component of the antenna impedance, so the antenna appears as a resistance and maximum power is transferred.

Figure 9.16 shows the schematic of a very flexible coupler that performs several functions and match the impedances over a wide range of values, for frequencies from 3 to 30 MHz. This coupler also matches a transmitter output that is referenced to chassis and ground (as most are) with both grounded, unbalanced antennas such as the long wire, and ungrounded, balanced antennas such as the dipole. Without this sort of balance/unbalance transformation (called a *balun*) the circuit and antenna combination will not function properly, since a grounded transmitter output must see what looks to it like a grounded load (and a grounded receiver input needs to be fed by what appears to be a grounded source). The antenna coupler achieves the impedance matching by using inductors and capacitors in series and parallel with the antenna to transform the reactive load impedance into a resistive 50- to 77-Ω load.

To use the coupler, the transmitter output coaxial (grounded cable) is connected at J_1. A coax line to the antenna is connected at J_2; an end-fed wire is connected at J_3, which is electrically the same point as J_2 but has a simple binding post for the wire instead of a shielded coax-type connector. Transform T_1 is a balun that transforms the unbalanced signal at C_2 into a balanced signal available at J_4 and J_5, with the jumper from C_2 installed. It is removed for driving unbalanced loads, disconnecting T_1.

All components between J_1 and C_1 are part of a power monitoring circuit, where L_1 and L_2 inductively sense the current in the line between J_1 and C_1 without physical contact and sensed current is indicated on meter M_1. Diodes D_1 and D_2, with switch S_1, let the meter monitor forward and reverse current, indicating incident and reflected power and thus SWR. The capacitances and inductances are adjusted to minimize SWR, indicated by a minimum value of reflected power on the meter.

9.5 Elementary Antennas

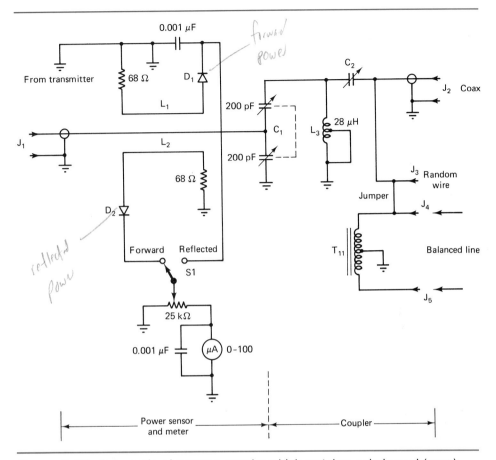

Figure 9.16 Schematic of antenna coupler which matches unbalanced (coax) transmission line with various balanced and unbalanced antennas and includes a forward/reflected power meter (from *ARRL Handbook,* courtesy of the American Radio Relay League).

(Figure 9.15a). This reduces the antenna capacitive reactance while reducing the radiating effectiveness only slightly. A second technique uses a series inductance called a *loading coil* near the base (Figure 9.15b), which acts to tune out the capacitive reactance and makes the antenna appear resonant at the transmitter frequency. The loading coil wastes power through I^2R loss but is mechanically more rugged, especially in mobile applications, since it can be wound and encapsulated at the vertical element base.

Although the Marconi antenna is inherently omnidirectional, it can be used in directional applications where the goal is to direct (or receive) power in one direction more than others. This is done by feeding an array of several antennas in parallel, but with precisely calculated lengths of cable between the antennas. The function of these cable lengths is to delay the signals in time and thus phase shift the radiated signals with respect to each other. The phases of the radiated waveforms from these antenna combinations then reinforce by adding, or interfere by subtracting with each other, depending on the distance and angle from the antenna array. This *phased array* can be used to distort the pattern from the original omnidirectional pattern into one more appropriate to the application. Advanced versions of this phased array antenna are discussed in the next section.

Review Questions

1. What are the radiation pattern and bandwidth of the long wire antenna?
2. Compare the folded dipole to the open dipole in terms of size and impedance. How can the impedance be changed?
3. What factors determine the loop antenna induced signal? How can a physically small loop be made useful?
4. What is the radiation pattern of the loop antenna? Why is it used for locating transmitters?
5. How does a Marconi antenna operate? What are its radiation pattern and impedance?
6. What is the radiation angle for the Marconi? How can it be changed? Where is it useful?
7. What is an antenna coupler? What are its goals in use? How does it achieve these?

9.6 ADVANCED MULTIPLE ELEMENT ANTENNAS

The antennas studied to this point have been relatively simple physical and electrical structures, are used extensively in many applications, and provide acceptable to outstanding levels of performance. Nevertheless, there are many other antenna designs in use which are more complicated and are used because they provide specific characteristics that are necessary for the communications system.

There are so many of these advanced antenna designs that it is impossible to study even a large fraction of them. The next two sections examine some of the more useful and common of these designs:

Antenna arrays, with multiple and parasitic elements, including the Yagi antenna, the log periodic antenna, and phased arrays

Slotted antennas, used for microwave and higher-frequency applications

Helical antennas, which provide circular polarization

Dish antennas, which provide the highest gain and directivity

Antenna Arrays

Antenna arrays form a large group of antenna designs with a variety of capabilities. The goal of the *array* is to shape the radiation pattern of the antenna field to focus it more sharply and thus increase the apparent power in a specific direction. The array may also be used to provide a higher front-to-back ratio or to reduce sidelobes so that interference from physically adjacent transmitters is reduced.

The array consists of three basic elements: the *active, powered,* or *driven* element, plus the *director*, and the *reflector*. The transmission line is connected only to the driven element. The director and the reflector are *passive* and serve to deliberately and carefully distort the radiated field pattern. As *parasitic* elements,

9.6 Advanced Multiple Element Antennas

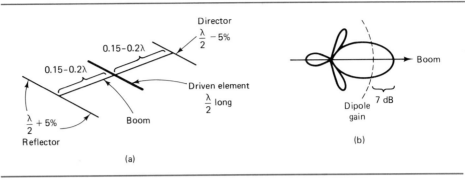

Figure 9.17 (a) Basic three-element Yagi design; (b) Yagi radiation pattern.

they do this via the natural coupling that occurs whenever conducting elements are in an electromagnetic field. Their close physical proximity to the driven element (presence in the near field) means that significant distortions can be formed. The parasitic elements are not normally connected to the driven element and may be insulated from the support mast.

The location and length of the director and reflector elements are critical, and there can be more than one of each. In practice, adding more reflectors has little benefit, so nearly all arrays have only one reflector element. The *Yagi–Uda array* (named after the two Japanese scientists who developed it) has a single driven element, usually a $\frac{1}{2}\lambda$ dipole or folded dipole (Figure 9.17a). The reflector is 5% longer than the driven element and is spaced between 0.15 to 0.20λ behind the driven element. A single director element, 5% shorter than the driven element, is spaced the same distance on the other side to complete the basic Yagi antenna. Due to the relatively large physical dimensions of the elements and their spacing, the Yagi is rarely used for frequencies below the VHF band.

The typical performance of a three-element Yagi antenna provides a forward gain of 5 to 7 dB compared to the dipole alone. Equally important, the pattern has a large main lobe and small minor lobes (Figure 9.17b). Typical front-to-back ratios are 15 to 20 dB. This shows the directivity of the Yagi. The forward gain, front-to-back ratio, and the beamwidth (and minor lobes) can be adjusted and traded off against each other by varying the element spacing.

Example 9.8

Dimension the elements of three-element Yagi for 100-MHz operation using 0.2λ interelement spacing.

Solution

Driven element: at 100 MHz, $\lambda = 3$ m; therefore, the driven element is $\frac{\lambda}{2} = 1.5$ m.

Reflector: 1.5 m + 5% = 1.575 m

Director: 1.5 m − 5% = 1.425 m

Spacing is 0.2λ = 0.2(3) = 0.6 m between elements, so the overall support length is twice that distance (reflector to driven element, and driven element to director).

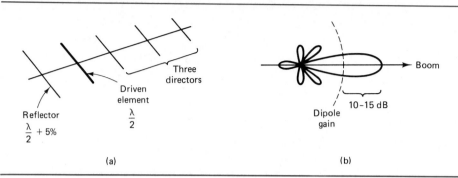

Figure 9.18 (a) Multiple-element Yagi design; (b) radiation pattern.

Note that the calculated lengths of the elements are ideal, and actual lengths will be 5% shorter to compensate for end fringing and related effects.

The Yagi is not restricted to a single director and reflector. Additional directors, each 5% shorter than the preceding one, can be added to increase the gain and sharpen the main lobe. Each additional director is initially spaced slightly closer than the previous one. A Yagi with three director elements has a gain of about 10 to 15 dB compared to the dipole, while a Yagi with eight director elements has gain of about 13 to 18 dB (Figure 9.18).

Example 9.9

What are the dimensions of the elements and mast for a 100-MHz Yagi with three directors, and one with eight directors (using 0.2λ spacing)?

Solution

Begin with the results of the single-director problem (Example 8). The first director is 1.425 m. Director 2 is 5% shorter, or 1.354 m; director 3 is 5% shorter than director 2, or 1.286 m, and director 3 is 5% shorter than 2, or 1.22 m. As there is a total of five elements, the support boom length is $0.2\lambda(4) = 2.4$ m, which is over 7 ft.

For the eight-director Yagi, use the results of the three-director design. Director 4 is 1.16 m, and the remaining director dimensions are 1.1, 1.05, 0.995, and 0.945 m. The boom length is $0.2\lambda(9) = 5.4$ m, nearly 17 ft. This is impractical unless heavy gauge tubing and additional supports are used, and the *wind load* (and possibility of bending) of this large antenna is high.

The range of gain occurs because the interelement spacing does not have to be the same between each element—in fact, one advantage of a multiple-director Yagi is that the spacing can be adjusted for a variety of different combinations of gain, front-to-back ratio, and beamwidth, compared to the simple three-element (one reflector plus one director) Yagi. Yagis have been built with extremely high gains, up to 30 dB, but mechanically these can become very difficult to construct and maintain, especially in outdoor environments.

Despite all this capability and flexibility, the Yagi is not ideal for all applications. The narrow bandwidth of the Yagi—anywhere from 1 to 10% of the nominal center frequency depending on number of elements and spacing (greater directivity from additional elements reduces the bandwidth)—can be a problem. For some applications, this limited bandwidth is a major advantage, since it minimizes adjacent channel interference. However, a Yagi would be a poor choice for the total broadcast FM band, which has a bandwidth of about 20% of the center frequency, or in a system that had to change frequencies depending on radio conditions. There are methods of developing broadband Yagis by using thicker antenna elements, but forward gain and the front/back ratio suffer considerably.

Log-Periodic Antennas

The *log-periodic antenna* of Figure 9.19, developed in 1957, is a multiple-element antenna that uses an entirely different principle than that of the Yagi antenna, with its single driven element and parasitic reflectors and directors. It provides good gain and wide bandwidth, although the achievable gain is not as high as a Yagi can provide for the same size. The log-periodic antenna consists of an array of dipoles extended along a horizontal axis. The length of each dipole is shorter than the dipole preceding it. All the dipoles are electrically connected, and the transmission line is connected to the shortest dipole. Maximum radiation is from the small end.

The length and spacing relationship are governed by a formula that incorporates a *design ratio factor r*, along with dipole length l and spacing d: $r = l_2/l_1 = l_3/l_2 = l_4/l_3$; $l_1 = \lambda/2$ is the length of the longest element $r = d_2/d_1 = d_3/d_2 = d_4/d_3$; d_1 is the distance between the first (longest) and second elements. Looking at the formula, we see that beginning at the short end, each successive element is longer and spaced farther apart than the previous one. The longest dipole (number 1) is cut for the lowest frequency (longest wavelength) over the band of interest. The ratio factor is between 0.7 and 0.98, and d_1 is between 0.07 and 0.09λ for the lowest frequency. Input impedance is determined by the physical dimensions of the elements, with nominal values between 300 and 600 Ω for typical dimensions. This impedance is fairly constant but has a small periodic variation around the

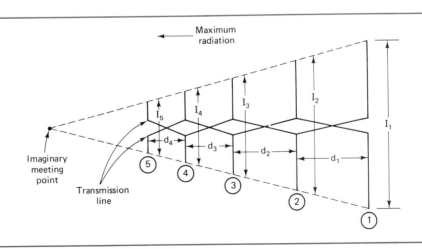

Figure 9.19 Log-periodic antenna design.

typical value, and this impedance variation is a function of the logarithm of the frequency (hence the name: log-periodic).

The design factor r determines the maximum gain that the log-periodic provides, which can range from 7 to 12 dB. It would seem that the largest gain is desirable, but this is not the case—there are trade-offs, as always. Higher gain requires larger elements and spacing, which may be impractical to build or maintain.

The log-periodic provides wide bandwidth, especially when compared to the Yagi, because it is really the sum of many separate dipoles, each tuned to a slightly different frequency. Dipoles adjacent to the one that is resonant at a given frequency act as reflectors and directors, as in a Yagi. As the frequency of interest shifts, the specific dipole that is acting as the "driven" element changes and the ones that are directing and reflecting change as well.

The log-periodic antenna can operate over a 4 to 1 or greater frequency ratio, meaning that the highest usable frequency is four times the lowest. This is in contrast to the Yagi, which can barely cover the 88- to 108-MHz FM band at constant gain. Over the desired frequency range, the log-periodic antenna must have an element at the highest and lowest values, as well as the ones in between as determined by the formula.

The design of a log-periodic antenna begins based on the lowest frequency of the operating range and the required gain, which are both applied to another formula (and engineering judgment) to determine the design ratio value which provides this gain at this frequency. Next, the design ratio is used to calculate the lengths and spacings of the remaining elements (beginning with the longest element and proceeding to the shortest one for the high end of the band), which fixes the number of elements and the overall size.

Example 9.10

Design a log-periodic antenna for the FM broadcast band (88 to 108 MHz) using a ratio factor of 0.95 and d_1 equal to 0.08λ.

Solution

Begin by determining the wavelength at each end of the band. At 88 MHz, $\lambda = 3.4$ m, so a half-wavelength dipole is 1.7 m (this is l_1). At 108 MHz, $\lambda = 2.78$ m, so the half-wavelength dipole is 1.39 m. From the formula

$$l_1 = \lambda/2 = 1.7 \text{ m}$$

$$l_2 = rl_1 = 0.95(1.7) = 1.615 \text{ m}$$

$$l_3 = rl_2 = 0.95(1.615) = 1.53 \text{ m}$$

$$l_4 = rl_3 = 0.95(1.53) = 1.46 \text{ m}$$

$$l_5 = rl_4 = 0.95(1.46) = 1.38 \text{ m}$$

which is shorter than the length for the upper end of the band.

$$d_1 = 0.08\lambda \text{ at the lower end of the band} = 0.08(3.4) = 0.272 \text{ m}$$

$$d_2 = rd_1 = 0.95(0.272) = 0.26 \text{ m between elements 1 and 2}$$

$$d_3 = rd_2 = 0.95(0.262) = 0.25 \text{ m between elements 2 and 3}$$

$$d_4 = rd_3 = 0.95(0.25) = 0.233 \text{ m, between elements 4 and 5}$$

Overall length of the antenna support boom is the sum of d_1 through $d_4 = 1.015$ m, which is just over 3 ft long.

Phased Array Antennas

Phased array antennas represent a unique approach to making an antenna that provides a desired radiation pattern and direction that can be varied as needed. The term *phased array antenna* does not refer to a specific type of driven antenna element. Instead, it is a method of combining many identical elements electronically and then controlling the signals going to these elements.

The phased array antenna is an *electronically steerable* antenna. Consider the problem of an antenna that must provide high gain, high front-to-back ratio, and low sidelobe pattern. This highly focused antenna provides good performance in one direction. If the application required aiming this antenna at any point (360° sweep) such as in radar at an airport control tower, or receiving weak but critical signals from transmitters located anywhere on earth, the antenna would have to be mechanically rotated. This would require mechanical supports, which add expense and reliability problems and require special connections to bring the weak electrical signal from the antenna to the receiver while allowing rotation. These mechanical problems would be more severe if the antenna is large and outdoors, with wind and weather factors.

The phased array uses electronics to steer the antenna pattern and has no moving parts at all. The principle of the phased array antenna is based on the *interference* of electromagnetic waves, which can be both *constructive* and *destructive*. Suppose that an antenna is built of many identical driven elements, arranged in a line (Figure 9.20). The transmitted signal is applied to all these

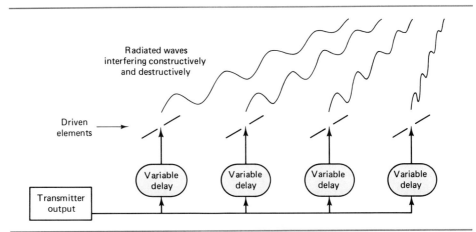

Figure 9.20 Identical antenna elements, each with adjustable delay, form a phased array antenna.

elements, but with electronically varied phase from one element with respect to another. The signals will then radiate from each element with some difference in phase. At some points in the far-field radiation pattern, the radiated waves from the various elements will add together constructively, so the overall signal strength will be greater. At other points, the waves will actually interfere destructively and reduce the signal strength at those points. Careful control of the phase to each element determines the position of the constructive and destructive interference, so the overall pattern can be changed.

In operation, a phased array antenna uses a single row of tens or hundreds of identical antenna elements (or there may be a two-dimensional antenna array to allow shaping of the pattern in the up/down direction as well). The antenna does not look like a traditional antenna, with many wires and rods of different diameters, or dish reflectors in varied configurations. Instead, the entire front of the phased antenna is fairly flat and easily covered with an electrically transparent cover, to protect the antenna elements. Without the cover, the phased array antenna looks like a mosaic, a repeated row of identical units.

The key to operation of the phased array antenna is the complex electronics needed to drive it. Although the principle of wave interference and selective phase shifting has been known since the eighteenth century and was well understood in the early days of electronic communication, the practical components needed to control a phased array did not exist. There are two elements to the phased array control.

First, some sort of electrically controllable time-delay line is used between the transmitter and the actual driven element. This delay line passes the RF signal without distortion, providing a precise amount of time delay (which is equivalent to a phase shift, Chapter 2). A controlling signal determines the time-delay value, and the delay line responds quickly to changes in this signal to implement new delays values.

Second, a computer is needed to precisely generate the control signals to all the delay lines. The delay that is associated with each element determines the radiation pattern, so the computer is constantly performing complex calculations of what delay time is necessary at each element to provide the desired final pattern. Each delay value is continuously reevaluated, so that new patterns can be realized.

In operation, the computer is programmed for the desired pattern and then varies all the delays to change the specific pattern, or its direction, as required. For example, a phased array radar antenna designed to scan one-fourth of the sky (90°) with a sharp radiation pattern (1°) is set up for the delays to each element that produces that pattern. Then the delays are changed so that the overall pattern is turned by 1° on the next radar transmit/receive cycle. To an outside observer looking down onto the beam radiated from this radar antenna, it would seem as if the transmitted energy was being sent by a rotating antenna, but the reality is that the beam is being steered electronically.

Besides its mechanical advantage of not requiring physical rotation to scan the sky, the phrased array antenna performance is flexible. The radiation pattern can be instantly changed from high gain/narrow beamwidth to a lower gain/wider beamwidth, simply by changing the delay pattern. This cannot be done with mechanical elements that are driven directly by a transmitter. Another feature is that the scanning sequence can be varied to meet changes in the application. A mechanically rotating antenna can usually have only a fixed rotation speed and direction. It cannot go backward to recheck some signal—the mechanical drives simply cannot do that. In contrast, a phased array antenna can be directed by the control computer to go back, reexamine some area of the sky again, or change the search pattern based on some possible signal of interest.

The results of analysis of the received signal are used by the controlling computer to vary the radiation pattern and scan cycle, based on some rules that have been programmed into the computer. Thus if a radar return indicates a possible aircraft in one direction, the antenna can rescan that area immediately. Thus the phased array antenna is dynamically controlled to change as the needs of the application change.

Slot Antennas

The slot that was used to couple energy into and out of the waveguide in Chapter 8 can also be used as an effective antenna, especially when multiple slots are combined into a phased array. A properly designed slot is efficient, radiating 80 to 90% of the power in the waveguide and with low sidelobes if designed properly. Slot array antennas have relatively narrow bandwidth of about ±2% of center frequency for an array designed for sidelobes that are 25 to 30 dB below the main lobe. In general, narrow bandwidths and greatly reduced sidelobes come in the same design; thus a wideband slot array would have significant sidelobes. Of course, narrow bandwidth is not necessarily a disadvantage and is often very desirable in systems designed for operation only at specific frequencies.

A key advantage of the slot antenna and slot array is that it can blend into the skin of an airplane or ship. It is not practical to have wires, tubing, or any protrusions from a airplane, especially one traveling at supersonic speeds, since the protrusion will cause air drag and probably get ripped off unless it is very rugged. In contrast, the slot array can be built to fit into the curve of the surface metal and then covered with a protective skin that is transparent to RF but makes the overall surface smooth and seamless.

Review Questions

1. How does an antenna array, with directors and reflectors, function to shape the radiation pattern? What is the objective of this shaping? What are the basic elements of the array and their functions?

2. What is the typical spacing for a Yagi antenna? How is the final size related to the signal wavelength?

3. What is the effect of adding more reflectors to the Yagi? What about more directors?

4. What is the operating principle of the log-periodic antenna? What is its gain?

5. Why is the bandwidth of the log-periodic antenna much greater than a Yagi? What are typical bandwidth ranges?

6. What formulas are used to design a log-periodic antenna? What is the basic approach to the design? What are some of the design choices?

7. What principle does a phased array antenna use to shape the radiation pattern? How is this pattern varied in use? What do the delay lines do? What are the benefits of phased array antennas?

8. What is a slot antenna? Why does it often use an array of slots? What are some bandwidth and sidelobe values? For what reasons is a slot antenna a good choice in aircraft?

9.7 ADVANCED SINGLE-ELEMENT ANTENNAS

Not all antennas have many elements to achieve required performance. Two antennas, the helical antenna and the parabolic dish antenna, show what can be achieved with designs that use one major element.

Helical Antennas

For applications where circular rather than horizontal or vertical polarization is needed, the *helical antenna* is used. The driven element is wound like a screw thread (helix), and the direction of radiation is along the axis of the helix (Figure 9.21). The helical antenna is mounted onto a ground plane made of solid metal or open-wire mesh with a minimum side length of about one wavelength. Helical antenna gain is a function of the helix diameter D, number of turns N of the active element, and spacing S (pitch) between turns, as well as the wavelength:

$$\text{power gain} = 15 \left(\frac{\pi D}{\lambda}\right)^2 \frac{NS}{\lambda} \quad \text{[for dB, calculate 10 log(power gain)]}$$

Typical helical antennas have at least 3 to 4 turns, up to a maximum of 20, and provide gains from 14 to 20 dB. The centerline axis length of the helical element is approximately equal to the pitch times the number of turns. Beamwidth of a helical antenna is relatively broad (usually 20 to 45° but can be up to 90°), with the 3-dB angle, in degrees, given by

$$\text{beamwidth} = \frac{52}{(\pi D/\lambda)(\sqrt{NS/\lambda})}$$

As expected, for a given diameter and spacing, the gain increases and the beamwidth decreases as the number of turns increases.

Example 9.11

What are the gain, in dB, and beamwidth of a helical antenna with diameter $= \lambda/3$, 10 turns at pitch of $\lambda/4$, used at 100 MHz ($\lambda = 3$ m)?

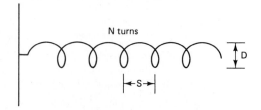

Figure 9.21 Basic helical design and key performance variables.

9.7 Advanced Single-Element Antennas

Solution

From these figures, $D = 1$ m, $N = 10$, and $S = 0.75$ m. Using the formula yields

$$\text{power gain} = 15\left[\pi\left(\frac{1}{3}\right)\right]^2\left[10\left(\frac{0.75}{3}\right)\right] = 15(1.097)(2.5) = 41.1 = 16 \text{ dB}$$

$$\text{beamwidth} = \frac{52}{\pi(1/3)\sqrt{10(0.75/3)}} = \frac{52}{(1.047)(1.58)} = 31.4°$$

Note that these equations could also be solved in terms of λ alone, without reference to frequency. The frequency value is needed only for actual physical dimensions.

Example 9.12

What is the change in gain and beamwidth that occurs when the diameter D is doubled?

Solution

Looking at the formula, gain increases with the square of D, so a new diameter equal to $2D$ will have gain $2^2 = 4$ compared to diameter D. The increase in gain is therefore 6 dB. Similarly, the beamwidth varies with the inverse of D, so the new D causes beamwidth to decrease to one-half its previous value.

Example 9.13

A helical antenna is operated at twice its original frequency, so the new λ is $\frac{1}{2}$ the previous λ. What is the change in gain?

Solution

The formula for gain shows that it is proportional to $1/(\text{wavelength})^3$. A new λ at $\frac{1}{2}$ the previous value will therefore increase the gain by $1/(\frac{1}{2}^3) = 8$ times.

Helical antennas are often *stacked* horizontally, vertically, or as a 2 by 2 array to increase gain and decrease the beamwidth further. Unlike the parabolic dish discussed at the end of this section, with its critical shape and unique focal point, the helical antenna has fairly tolerant dimensions and is easily aligned—the location of the helix axis on the ground plane only has to be far enough from the ground plane edge to minimize fringe effects. The helical antenna provides bandwidth from a relatively narrow ±20% of center frequency, up to a 2 : 1 span from maximum to minimum frequency. This bandwidth is determined primarily by the diameter of the helix and the diameter of the wire or tubing used for the helix.

Dish Antennas For applications where very high gain and very narrow beamwidth are needed, the *parabolic dish antenna* of Figure 9.22 is used. This antenna provides the required gain and beamwidth with a principle that is different than the Yagi or phased

Figure 9.22 Parabolic dish antenna [courtesy of NASA (National Aeronautics and Space Administration)].

array. Instead of directing the energy from the antenna using an array of passive reflectors and directors to shape the radiated field, the dish antenna uses simple reflection. Just as a mirror can reflect light and a curved mirror can reflect and focus light at a single point, the dish reflects and focuses radio waves. This is the same principle and shape that is used as a reflector in a flashlight or headlight behind the bulb.

Dish antennas are used for systems that transmit and receive as well as receive only. Smaller ones are commonly seen with satellite TV receivers, which pick up the signal from a satellite that is relaying the signal from a central broadcast point to many users. Larger dish antennas are used for the uplink/downlink to satellites in fixed orbit locations, used for major communications links. The most powerful (and sensitive) dish antennas are used to communicate with space vehicles and deep-space probes, as well as receive signals from natural galactic sources, where signal distances are very large and received signal power is very small.

A dish antenna system consists of two distinct parts shown in Figure 9.23: the *feed*, which is the active element, and the dish itself. The dish reflects received energy to its *focal point*, and the feed is usually placed at this point to actually collect the signal. When transmitting, the signal is goes to the feed element, radiates to the dish surface, and then is reflected outward by the dish.

9.7 Advanced Single-Element Antennas

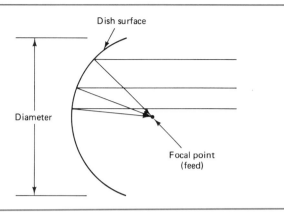

Figure 9.23 Parabolic dish design with feed at dish focus.

The dish is in the shape of a parabola. The reason for this is basic geometry: A parabola reflects any signals that come in parallel to its main axis (as from a single point far away) toward a single focal point; conversely, signals that begin at the focal point are reflected by the dish surface as a parallel stream (toward a single target point).

The required diameter of the dish is partially related to the signal wavelength and also to the desired gain. For better gain than an array can provide, the dish needs to be at least 1λ in diameter—although even at lower gains, it provides sharper beamwidth and higher front-to-back ratio, which are also important when trying to receive a weak signal. Dishes with larger diameters can provide higher gain and lower beamwidth but are much more complex mechanically and more expensive. The narrow beamwidth of these larger antennas means that the dish must be precisely aimed (not an easy task) and maintain that exact aim despite winds and changes in antenna dimensions due to temperature variations. Larger dishes, used for frequencies of several hundred MHz and higher, are 30 to 60 m in diameter (some are larger), providing up to 100 dB of gain. Typical dish beamwidths are from $\frac{1}{2}$ to 3°, depending primarily on the diameter of the antenna and the wavelength of operation. For protection against the weather and for ease of maintenance, dishes are often housed in *radomes* (RF-transparent shelters).

The surface of the dish does not have to be solid metal to reflect the signal. If the openings are less than 0.1λ, nearly as much of the energy will be reflected as would be from a solid surface. Openings larger than 0.1λ can also provide good results, with the advantage that the dish is lighter (easier to steer) and deflected less by wind. A large-diameter dish with larger than 0.1λ mesh may provide better overall results in terms of physical position accuracy and stability than a smaller, nearly solid surface dish which reflects more energy.

The ideal power gain of a dish antenna with respect to an isotropic source is given by this formula, where D is the dish diameter:

$$\text{dish gain} = \frac{6D^2}{\lambda^2} \quad \text{for gain in dB, take 10 log (dish gain)}$$

The approximate -3-dB beamwidth for a dish antenna, in degrees, is

$$\text{beamwidth} = \frac{70\lambda}{D}$$

Example 9.14

What are the gain and beamwidth for a dish with $D = 20\lambda$?

Solution

$$\text{Gain} = \frac{10 \log 6(20\lambda)^2}{\lambda^2} = 10 \log [6(400)] = 33.8 \text{ dB}$$

$$\text{Beamwidth} = \frac{70\lambda}{20\lambda} = 3.5°$$

Note that for signals at 100 MHz, a 20λ diameter is 60 m, nearly 200 ft!

Example 9.15

How large a dish diameter, in wavelengths and meters, is needed for 60 dB gain at 300 MHz?

Solution

Solving for D, we have $D = \sqrt{G/6}\lambda$, where G is the power gain expressed as a simple ratio. A gain of 60 dB corresponds to a gain G of 1,000,000. Therefore, $D = \sqrt{1,000,000/6}\lambda = 408\lambda$. In this example, $\lambda = 1$ m, so the diameter must be 408 m, about 1338 ft.

This is larger than can be practically built. The very largest dishes in use are about 100 m, or a little over 300 ft. (There is a larger dish built into a valley in Areicibo, Puerto Rico, but this is an exception.) At a given performance level, dish size decreases with increased frequency (decreased λ), which is another reason why satellite communications systems use frequencies in the multigigahertz range. A less costly, more practical dish size is less than 10 m (30 ft), which makes gigahertz frequencies essential. Home satellite receiving dishes as small as 1 to 2 m are now used, combined with high gain and low-noise receiver front ends.

A real dish will not have as large a gain or as small a beamwidth as the equation indicates. There are several reasons for this. First, the reflectance from the dish surface is not perfect, due to unavoidable absorption at the surface and dimensional imperfections, so usually only 50 to 75% of the energy is actually reflected. There is fringe area *spillover* or leakage around the edges of the dish, as some of the energy near the edge does not reflect back at all but goes past the dish edge due to diffraction. Finally, a real feed is not a true point, so cannot be located entirely at the single focal point of the parabola. Finally, the feed of the dish actually shadows and obscures part of the dish center, and the shadowed area is thus ineffective for gathering and focusing energy, from either the feed to the outside (for transmitting) or the outside to the feed (for receiving).

The shadow and imperfections of the feed have other effects. They distort the radiation pattern of the antenna and cause extra off-center lobes to be generated. These lobes affect the dish and can result in wasted energy not directed at the target, as well as interference with nearby systems. For receivers, these lobes mean that nearby signals of no interest may be picked up, which makes receiving the desired signal more difficult.

9.7 Advanced Single-Element Antennas

The *feed system* or *feedhorn* is critical to overall system performance. Ideally, a feedhorn would have no shadow and direct all the energy to the dish surface but not beyond the edges. A real feedhorn is a compromise. Different feed designs are used for accomplishing the system requirements. A Yagi is sometimes used, although mechanically it is difficult to mount and aim accurately. Also, the Yagi pattern is not symmetrical for all 360° perpendicular to its axis, so some parts of the dish would get more signal than others. Other designs include horn-shaped antennas which have conical radiation patterns, but a larger shadow (although several variations of this design have smaller shadows).

A *Cassegrain feed system* (Figure 9.24) is another choice. Here the active element is placed at the exact center of the dish surface (not to be confused with the parabola focal point) and directs its signal to a reflector at the dish focal point, which then reflects the signal to the entire dish surface. This design means that the complex active antenna element is located in a place that is easier to install and adjust mechanically, and the rest of the feed system is passive and reliable. However, the Cassegrain feed does use one more reflective surface than an active feed at the focal point and casts a larger shadow, both of which cause additional loss. The antenna designer must calculate the best choice by examining the benefits and drawbacks of each feed system.

Most parabolic dish receivers use an active preamplifier as close to the receiving element as possible. This preamp is designed to boost the weak received signal level prior to sending the signal to the receiver through a long cable that unavoidably adds noise while causing signal loss. The goal is to maximize SNR, and the best way to do this is to increase signal strength before any additional noise or loss occurs in the system. Special designs are used for the front-end preamp to provide the needed signal gain while minimizing noise added by this preamp itself. Deep-space and astronomy dish preamps usually are cooled to nearly absolute zero (0K −273°C or −459°F) to minimize the thermally induced random motion of molecules and the noise that this random motion generates. Smaller dish antennas designed to point-to-point microwave links or home TV satellite reception use simpler high-gain, low-noise preamps, of course, since

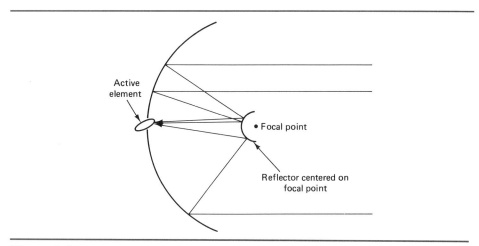

Figure 9.24 A parabolic dish with Cassegrain feed system is mechanically more rugged but has additional losses compared to feed at focus.

signal levels are much greater and the distance between the feed element and the receiver is much shorter than for a huge deep-space dish.

Review Questions

1. Where is a helical antenna used? What are the structural details? What are typical gain, beamwidth, and bandwidth values for the helical antenna? What physical details determine these?

2. Where is a parabolic dish antenna used? Why is the parabola shape used? What is the principle of operation?

3. Contrast the dish antenna to the Yagi and log-periodic in terms of gain, bandwidth, and beamwidth. What is the effect of antenna diameter on antenna gain?

4. What are some of the mechanical issues for dish antennas? Explain why a solid dish surface may not be necessary and an open mesh may be a better overall choice.

5. What factors make dish performance less than the theoretical idea?

6. What feed designs are used? Compare the Cassegrain feed to a Yagi or similar feed system.

7. Why do large receiving dish antennas have a preamp right at the feed or focal point? What preamp is used for space-oriented dish antennas? Why?

SUMMARY

The antenna is the last link in the transmitter and the first in the receiver. Antenna design is heavily influenced by the operating frequency, signal gain needed, and desired radiation field pattern, as well as the propagation conditions of operation. Antenna design is often done in dimensions of signal wavelength, to make calculations and ratios easier. Actual dimensions are then substituted for the frequency of interest. Simple antennas like the long wire, the half-wavelength dipole, and the Marconi vertical are useful in many applications and are often used. However, their performance cannot be tailored to specific needs.

More complex antennas, with multiple elements, are designed to provide the specific electromagnetic field patterns of gain, beamwidth, lobes and nulls, directivity (F/B ratio), radiation angle, and bandwidth that are often needed. Every antenna design involves compromise among all these elements, as well as practical size limitations. Multiple-element Yagi and log-periodic designs provide high gain, wide bandwidth, and high F/B ratios in different combinations, but with performance and complexity trade-offs.

Other designs are used to build antennas that can be flush with an airplane's surface (the slot array) or electronically shaped and steered, as in the phased array. The highest performance gain is available in the parabolic dish design, which concentrates energy at its focal point and provides gains up to 100 dB and beamwidths of less than 1°. The dish antenna is physically large compared to the signal wavelength.

Signals from an antenna may travel allow the ground, in a straight line of sight, up to the earth's ionosphere and be reflected back, or pass through the ionosphere and into space. Each mode results in different signal strength and distance as a function of frequency. The ionosphere and its ability to reflect, at different heights, various frequencies results in different propagation characteristics. This is used to allow reliable long-distance communication or to direct signals beyond undesired listeners.

Summary Questions

1. How is an antenna like a transmission line, and how is it different?
2. What is reciprocity? Why is it an important principle?
3. Why not simply analyze antennas mathematically? Why complicate the theory?
4. What is the role of the ionosphere in propagation? How do refraction and reflection occur?
5. Compare ground, space, sky, and satellite waves in terms of distance, frequencies, reliability, and applications.
6. How does increasing antenna height increase line-of-sight distance? By what proportionality factor?
7. What is radiation angle? What role does it play in skip distance?
8. Why are critical angles and MUF important factors?
9. How does an antenna produce gain? What is the meaning of the word *gain* in this context? How is gain defined?
10. What does a radiation pattern show? How is it measured in the field? What are: lobes, nulls, beamwidth, and F/B ratio?
11. How and why are scale models used in antenna measurements? What must be scaled up and down? What is special about the room used in tests?
12. What are the pros and cons of more versus less antenna bandwidth?
13. What is an isotropic antenna and its pattern? Why is it often not used for purposes of comparison?
14. Why is the $\frac{1}{2}\lambda$ dipole studied and used extensively? What are its impedance, gain, radiation pattern, and beamwidth?
15. What is the function of the T/R switch? What would happen without it?
16. Compare long wire and loop antennas with respect to construction, size, and radiation pattern. What are key features of each?
17. What is the Marconi antenna? Where is it used? What is its normal size? Why is the ground conduction critical?
18. How does the Yagi antenna shape the field pattern? What are typical gains and bandwidths? Repeat the question for the log-periodic antenna.
19. Compare and contrast the steps in designing a Yagi versus a log-periodic antenna.
20. What is a phased array antenna and its operating principle? What can it do that no fixed antenna can?
21. What are the features of the parabolic dish antenna performance? What affects gain and beamwidth? What are some highest achievable values?
22. Give three reasons why real dish antenna performance is less than theory indicates.
23. What is the function of the antenna feed? How is it physically done? What technical concerns are involved in feed design and position?

PRACTICE PROBLEMS

Section 9.2

1. A TV station has a 1000-ft tower. How far can its signal reach if the viewers have antennas on the TV which may be at ground level? How much additional distance is gained if a TV viewer puts antenna on the roof, 30 ft off the ground?

2. A police/fire radio link has an antenna on a 100-ft tower at headquarters to reach vehicle-mounted antennas at a height of 10 ft. What is the maximum coverage distance?

3. What is the path loss, in dB, over 10,000 km, at 50 MHz? At 500 MHz?

4. A system can tolerate a path loss of 120 dB, but no more, at the distance of 50,000 km to a satellite. What is the highest frequency that can be used?

5. What is the total loss when the system of problem 4 has a transmitting antenna with gain of 20 dB and receiving antenna gain of 10 dB?

6. What is the received signal power when the system of problem 5 is driven by a 100-W signal?

7. What must the transmitter power be if the receiver front end in problem 5 must see a power of 10^{-6} W for satisfactory operation?

Section 9.3

1. What is the front-to-back ratio for an antenna with a forward gain of +20 dB and a backward gain of +15 dB? Compare this to an antenna with a forward gain of 14 dB and a backward gain of 9 dB.

2. Which antenna has greater F/B ratio: one with a forward gain of +6 dB and a backward gain of +3 dB, or one with forward and backward gain figures of +10 and +8 dB?

3. What is the bandwidth, as a percentage and as a ratio, for a 175-MHz antenna with 3-dB points at 200 and 150 MHz?

4. An antenna for the FM broadcast band must span from 88 to 108 MHz. What is the required bandwidth ratio? Can an antenna designed for 100 MHz with ±10% bandwidth suffice?

Section 9.6

1. What are the theoretical elements lengths and overall boom length for a three-element Yagi designed for 220-MHz operation with element spacing of 0.18λ?

2. Design a 500-MHz Yagi with eight directors. What are the ideal element lengths and the boom length required if spacing is 0.15λ?

3. Broadcast TV channels 2 through 6 span 54 to 88 MHz, called the lower VHF group. How many elements, and of what length, are needed for a log-periodic antenna with $r = 0.9$ and $d_1 = 0.08\lambda$? How long a boom is used?

4. The upper VHF TV channels, channels 7 through 13, broadcast from 174 to 216 MHz. Dimension a log-periodic antenna with $r = 0.95$ and $d_1 = 0.075\lambda$.

Section 9.7

1. A helical antenna has 20 turns of diameter $\lambda/4$, pitch of $\lambda/4$. What are the gain and beamwidth? If the frequency of operation is 250 MHz, what are the actual dimensions of turn diameter, pitch, and length of helix axis?

Practice Problems

2. The antenna of problem 1 has the number of turns reduced to 15. What are the resulting gain and beamwidth values?

3. A helix antenna has its pitch halved while everything else remains unchanged. What is the change in gain, in dB? In beamwidth? In axis length? What happens to gain and beamwidth if the number of turns is doubled when the pitch is halved?

4. A helical antenna is designed for one frequency, but then is used at half that frequency. What is the change in gain (in dB) and beamwidth?

5. A parabolic dish has a diameter of 10λ. What is the gain, in dB? What is the beamwidth? If the dish is built for operation at 200 MHz, what would be the actual diameter?

6. To increase gain at a given frequency, a dish is built with a diameter three times the previous model. What is the increase in gain, in dB, and the change in beamwidth?

7. A dish designed for operation at 150 MHz is operated at twice that frequency. By what factors do gain and beamwidth change?

8. A system requires gain of +18 dB for a 150-MHz signal. What diameter dish is required, in wavelengths? What is the physical dimension? What is the resultant beamwidth?

9. To minimize interference, a 500-MHz dish needs to have a 1° beamwidth. What diameter dish is required, in wavelengths and meters? What is the corresponding gain, in dB?

10
Digital Information

CHAPTER OBJECTIVES

When you have completed this chapter, you will understand:

- The relationship between analog signals and their digital representation in various digital formats
- Why digital representation offers major benefits in modern communication systems
- Key specifications in analog-to-digital conversion, including accuracy, resolution, sampling rate, bandwidth, and bit rate
- Why traditional analog test equipment is inappropriate for testing digital signals and systems, and what types of test equipment are used instead.

INTRODUCTION

An effective communication system requires an understanding of the electronics of the transmitter, amplifiers, antennas, receiver, and associated hardware, along with the information that the system is designed to carry. In digital communications, the signals that represent information are restricted to a specific, limited group of values. However, this set of values can be used to represent, completely and faithfully, any analog information signal which is not limited to only a specific set of values. The advantages of digital signals is their ability to provide virtually error-free results and the potential for signal improvement through additional processing that they offer.

The bridge that links the analog signal and the digital equivalents is the conversion process, which considers the required accuracy and rate at which the analog signal is converted to its digital equivalents. These factors are combined with the bandwidth of the analog signal, the bandwidth of the system, the number of digital symbols available to convey the information, and the length of time available for transmission.

10.1 DIGITAL INFORMATION IN COMMUNICATIONS

Up to this point, we have looked at many elements of a communications system, but not at the information that the system is carrying. The nature of this information, along with the various forms it may take, is a very important element of the overall communications system. Subsections of the communications system are specifically designed or optimized depending on the specific nature of the information itself.

The information signals can be divided into two major categories: analog and digital. Increasingly, digital formats are used in communications systems because of many benefits they can provide. However, analog signals are often required because of some of the limitations of digital signals or some inherent characteristics of the communications system. To understand the roles of analog and digital signals, we can look at the sources of these signals, their features, and the conversion from one form to another.

In an *analog* system (Figure 10.1a) the signal can have any value within the total amplitude, phase, or frequency range that is available. For example, this range can be 0 to 1 V, 0 to 10 V, −1 to +1 V, −45 to +45°, or 1000 to 2000 Hz, but many other spans are used. The analog signal represents a varying physical quan-

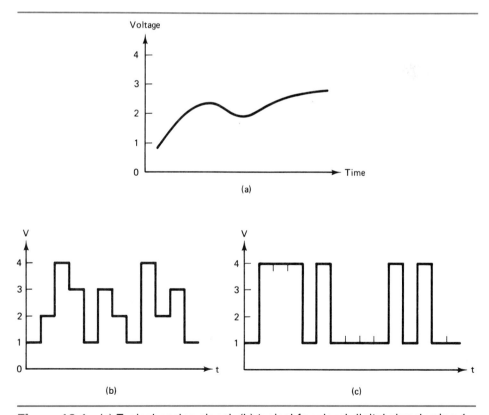

Figure 10.1 (a) Typical analog signal; (b) typical four-level digital signal using 1, 2, 3, and 4 V; (c) binary signal.

tity, such as a voltage, voice, temperature, pressure, or video intensity. The physical quantity being measured can be 1.135 V, 87.74°, 25 lumens, or 167 or 167.3 pounds, for example. The analog signal does not impose a set of "allowed" values within the range—any value within the range is legitimate and meaningful.

In contrast, the *digital* signal is allowed to have only one value from a set of previously defined specific values. For example, voltages of +1, +2, +3, and +4 V may be used (Figure 10.1b). These are the *discrete* signal values, and they are the actual values that digital information must use to represent the information it is trying to convey. The number of discrete values in the set of allowed values varies depending on the system design and many technical trade-offs and compromises. Most commonly, the set has 2, 4, 8, or 16 (2^1, 2^2, 2^3, or 2^4) allowed values. The discrete values can be represented by amplitude (AM), phase (PM), or frequency (FM) variations. When there are only two discrete values, the digital system is reduced to a special and important subgroup called the *binary* system (Figure 10.1c) and the discrete values are called *bits* (a contraction of "binary digit").

A shorthand notation is used to represent these values, since it is impractical to write and refer to +1 V, −0.5 V, or the other discrete values being used. In the two-value binary system, this shorthand uses 1s and 0s. A four-value digital system can use 0, 1, 2, and 3, or the letters A, B, C, and D. Of course, all the people involved in the system design, analysis, and repair must know the "translation" between the shorthand representation and the actual signal values. Interestingly, users of the communications system do not have to know what these values are: whether a binary 1 equal to 5 V or 2 V is irrelevant to the system user (but not to those who design and repair systems).

Of course, there are many cases where the information is inherently digital. Financial numbers have discrete intervals of 1 cent, for example, and under ordinary circumstances do not show values such as $7.87657. Similarly, typed text needs only discrete values for each letter, along with numbers and punctuation. There are no values "between" the letters—the set of digital values consists only of these letter symbols, and no others. Although this chapter deals primarily with

Transparency

The fact that the exact voltages or signal values used to represent digital signal are not seen by the system user is part of a theme that we will see more of as we progress through a complete communications system. The many layers of a communications system isolate the user (someone on the phone, a computer talking to a terminal or other computer, someone watching TV) from many of the technical details. In fact, as long as users at both ends of the link agree in advance to certain standards of communication, many of the intermediate steps of the overall communications process will be *transparent* to them.

Transparency means that each step of the signal processing and transformation does not try to interpret the signal, but carefully follows some predefined rules on how it will handle the signal before passing it to the next layer or stage in the system. Since each stage does not need to interpret, each stage can be optimally designed to perform its specific operation, following specific rules, regardless of the actual meaning of the signals it is processing. The signals can represent different information and messages at different times, yet the stages respond the same way in their operation.

The situation is analogous to writing and mailing a letter. The operation which actually transports the letter has only to read the address on the envelope and route it accordingly. It does not have to look at the contents of the letter or understand them—the message—in order to perform the routing to destination function.

the conversion from analog to digital format, the benefits of digital format apply regardless of the inherent nature of the information. For these reasons, understanding digital concepts in communications is important for both analog data which has been converted, as well as inherently digital information.

Comparison of Analog and Digital Performance

Traditionally, most communications systems have used information that was in analog, continuously varying format. A radio station that is transmitting voice and music is providing information that is analog by its nature. The modulation index for this signal can be anywhere from 0 to 100% under normal operating conditions. Ironically, the first electrical communications system—the telegraph—was a binary digital system using *Morse* code. The letters of the alphabet in telegraphy are represented by dots and dashes. Each dot or dash is a "fully on" signal, and when there is no dot or dash the signal is off (a pattern of 1s and 0s). The receiver must determine what letter was sent by the dot/dash pattern, where dashes are approximately three times as long as dots.

In recent years, the extensive use of computers, communications between computers, and computer (and digital) circuitry has been part of a large migration to digital communications. There are two major reasons for this: improved performance with respect to noise resistance (and consequently, fewer errors), and the potential ability for the receiver to process and improve the quality of the received information.

Noise is a constant and unavoidable concern in communications. Even when the signal is relatively strong, noise affects the accuracy of the signal. Was the received 100-mV signal really 100 mV, was it a 99-mV signal plus 1 mV of noise, or was it actually 101 mV with −1 mV of noise? The signal cannot be corrected, and the noise cannot be compensated for, since the exact value of the noise at any instant is unknown. Even knowing the SNR does not help, since the noise value in the SNR equations refers to the average (or sometimes root-mean-square, rms) noise value, not its exact value at any instant.

A digital signal, in contrast, provides a large amount of immunity to the effects of noise because of the relatively wide gaps between the allowable values. For example, if the set of allowable signals consists of 1-, 2-, 3-, and 4-V signals, a signal at 1.2 V was most likely really a 1-V signal with +0.2 V of noise. Similarly, a 1.78-V signal is probably a 2-V signal with noise of −0.22 V. In this way, the corrupting effect of the noise on digital signals is lessened considerably and minimized. The original signal value is restored despite some corruption by noise.

The process of restoring a noise-corrupted signal to its original value is called *regeneration*. Digital signals allow for nearly perfect regeneration, as long as the noise is not so large that one digital value is misinterpreted as another digital value.

Here is how regeneration works, with the example of binary digital signals of 1 and 5 V (Figure 10.2): a digital signal is sent at its correct 1-V value, but along the way is corrupted by +1 V of noise that appears at the receiver as +2 V. The receiver has a special input *buffer* circuit which is designed to force all signals below 3 V back to 1 V, and all signals above 3 V back to 5 V, before passing the signals on to the next part of the system. The received signal (and its noise) is therefore regenerated at the original 1 V value. Similarly, a 5 V signal that has noise of −1.5 V is received as 3.5 V, but the regenerative buffer converts this to 5 V.

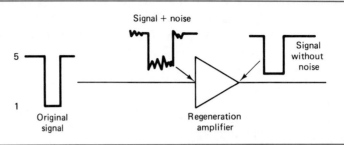

Figure 10.2 Regeneration of a digital signal operating with 1- and 5-V levels.

As a result of this regeneration process, digital signals free of noise (and subsequent errors) are restored and recreated, even after they have been corrupted by noise. In comparison, analog signals can only be corrupted, but not improved. If the signal can take on any value in the range 0 to 5 V, there is no way to distinguish between the original signal and the original signal plus added noise. An input amplifier can only increase the amplitude of the received signal (plus its noise) but cannot remove the noise itself. In fact, the amplifier actually can make the situation worse, since it adds its own noise to the signal as it amplifies it.

An example with audio recordings shows the difference between the ability of a digital signal to be repeatedly regenerated, versus the additive noise effect of analog signals. Suppose that an analog tape recording of a critical music performance is to be sent, using cable-based transmission channels, from the studio on the east coast to a film maker in California (3000 miles away), to be used as a film music score. The audio signal is recorded expertly and has a high SNR of 70 dB. As the signal passes from the studio to the film maker, it passes through amplifier stages spaced every 300 miles, which boost the signal level to compensate for signal loss in the wires. Over each 300-mile stretch (Figure 10.3), the signal voltage decreases and the SNR deteriorates by 4 dB due to additive noise. The amplifiers restore the original signal level, but add their own noise of 0.5 dB. What is the resulting SNR at the receiver?

At the first amplifier stage input, the SNR has declined to 66 dB, and at the amplifier output the SNR is 65.5 dB (although the signal voltage is restored to its original value). In fact, over each 300-mile link and amplifier stage, the SNR decreases by 4.5 dB. At the end of the 10 stages, the 70 dB SNR is reduced by $10 \times 4.5 = 45$ dB and is only $70 - 45 = 25$ dB. This is unacceptable in most

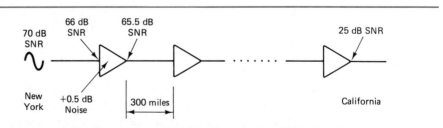

Figure 10.3 Analog signal amplifier stages add noise at each point while increasing signal amplitude, thus reducing final SNR.

applications. The solutions are to use more costly, lower noise links, or to mail the tape of the recording to film makers.

In comparison, the digital signal uses special *regenerative amplifiers* and arrives with essentially the same SNR as it had at the originating point. The noise added at each stage is relatively small compared to the signal level, so at each stage the digital signals are recreated and reestablished at their original voltage values. Any noise from the signal link is eliminated and there is no repeated summing of the noise from each stage. The ability of digital signals to be passed through many stages of necessary amplification, yet have the quality they had at the starting point, is a major advantage of digital versus analog signal representation.

Noise margin or *noise immunity* is the amount of noise that can be tolerated before the signal (plus noise) is confused with another signal value and regenerated at the wrong signal value. The boundary between digital values is typically set midway between the adjacent values, so the margin is one-half the difference between these adjacent values. For signals spaced 1 V apart, the noise margin is $\frac{1}{2}$ V. A 1-V signal with 0.49 V of noise will be understood as a 1-V signal, but when it has 0.51 V of noise it is interpreted as a 2-V signal instead.

Greater spacing (called signal *distance*) between the allowed digital states results in larger noise margins. However, these larger distances mean that fewer discrete digital values can be fit in the overall available signal range of the system. This is the technical trade-off that the system designer must make: use fewer, more widely spaced values with greater noise margin, or provide more discrete values but with reduced noise margin.

Besides noise immunity and error resistance, there are other reasons for using signals in digital formats. After all, a little noise on a telephone signal is often not a problem, and if the noise is so severe that the voice is unintelligible, it is probably greater than the noise margin; a digital system would not be much better than an analog one. There is another major reason, however, why digital signals are increasingly used: They provide the potential for signals to be processed at both the transmitting and receiving ends for better results than can be obtained even with a nearly perfect, almost noiseless analog system. When signals are available in a digital format, they can be stored in computer memory and extensively manipulated. Stored signals can be processed before transmission (*pre-processed*) to enhance the information and eliminate unneeded, redundant portions. They can be processed after reception (*post-processed*) to improve communications performance and eliminate any errors. Computer calculations and analysis of the data represented by the digital format make this possible.

Pre- and post-processing are not limited to digital signals. Analog signals can be processed to enhance their usefulness and efficiency: for example, a transmitted voice signal often has its modulation index boosted, and received signals are nearly always filtered to reduce noise. However, the kinds of processing that can be done with analog signals is much more limited than what can be done with digital signals that can be stored, manipulated, and computed with as many times as needed.

In contrast, a signal in analog format cannot be stored, since there is no way to store analog voltages indefinitely. While the single voltage value at any instant can be stored for a short time on a capacitor, this is cumbersome and the voltage eventually falls as the capacitor discharges. One capacitor would be required for each voltage value, so a lengthy signal would need hundreds of capacitors arranged in sequence. In contrast, digital signal storage simply requires digital memory, which is inexpensive and flexible.

Two examples show the potential benefit of pre-processing and post-processing. In video signal transmission, the image being transmitted may have low *contrast,* which means that there is not much span in intensity between the light areas and the dark ones (perhaps the picture was taken under very low light levels). *Image enhancement* is used to increase the contrast, by converting the intensity levels of the picture to a digital format, using special computer *algorithms* to process the numbers that represent the intensities, and expanding the range of intensity levels to improve the contrast. This would be impossible to do using the original analog signal levels alone.

As another example, consider a stream of digital values being transmitted where the noise is large enough that errors may still occur. However, before these data bits are transmitted, they are put through a special calculation that creates additional bits, called *check* bits, that are derived from the digital values themselves. The check bits are sent along with the data bits. At the receiver, the data bits are used to recreate the check bits, which are then compared to the received check bits. If both sets of check bits agree, there is probably no error. However, if the check bits determined at the transmitter differ from those recalculated at the receiver, then an error has occurred. Depending on the number of check bits and the complexity of the calculation algorithm, the check bits can be used simply to indicate an error or even to correct the error. The result is virtually error-free communication despite errors in individual parts of the overall system. This scheme for error correction is discussed in detail in Chapter 12.

In summary, the two basic advantages of digital formats compared to analog are:

The greater noise and error immunity that results from noise margin

The potential of digital signals (which are inherently compatible with computers) to be stored, recalled, processed, and manipulated for signal enhancement and improved performance, which simply cannot be done with analog formats

Review Questions

1. What is the basic difference between analog and digital representation of a signal?
2. What is meant by the discrete, allowed values of a digital system? How are these discrete values represented?
3. What is the relationship of binary to digital representation?
4. What are the advantages of digital versus analog representation with respect to noise? Signal pre- and post-processing? Signal storage? Why are these significant?
5. How does a digital system regenerate a signal after noise has corrupted it, with no degradation in SNR? How does this regeneration process prevent the accumulation of degrading noise in a multistage system?
6. What is the noise margin? How does it determine how much noise can be tolerated before the digital signal is incorrectly received?

Algorithms

An *algorithm* is the plan or set of instructions that is followed to reach a specific goal. In communications, it is the specific type of signal processing done to get the desired results. You use algorithms in daily activity. For example, suppose that you are given a list of names to alphabetize. Using one algorithm, you go through the list and find all the words beginning with "A," then those beginning with "B," and so on, down to "Z." If there is more than one word beginning with any letter, you repeat the alphabetization algorithm, but this time on the second letter of the word.

Of course, there is more than one valid algorithm that accomplishes a task, just as there is more than one way to alphabetize. Some people prefer to alphabetize by starting with the first word on the list, comparing it to the next. If the second word comes before the first alphabetically, then switch the two words. Repeat this procedure for the second and third words, all the way to the bottom of the list. Then start again from the top. After enough passes through the entire list, all the words will be alphabetized (the number of passes needed is not random, but can be calculated in advance by studying the problem and the algorithm in more detail).

Algorithms range from very simple to extremely complex. A simple algorithm is one that tries to minimize the effects of noise by taking the average value of a signal at 10 points in time. The 10 values are summed, and this sum is then divided by 10. The theory of this particular algorithm is that noise is random and is both plus and minus values with respect to the desired signal, so the average of the signals plus noise will result in many of the noise values canceling each other out.

Many communications algorithms are more involved. The signal is processed and manipulated in a variety of ways to extract the desired signal while minimizing the effect of noise, distortion, or errors. Algorithms can be implemented by circuitry alone—a *hardware* implementation—but many sophisticated algorithms are actually implemented by *software* instructions which tell a microprocessor how to perform specific calculations on the digital values that represent the signal. The software approach has the advantage that the algorithm can be changed or adjusted to meet new situations, simply by changing the software program; however, a hardware implementation dedicated to performing this one algorithm is usually faster. Finally, the algorithm actually used may seem awkward and inefficient to a person, but may be relatively easier to perform with hardware and software than one that seems simpler to a human being!

The use of algorithms is not limited to general-purpose computers that sit as separate units in a room or on a desk. Most new communications systems incorporate *microcontrollers*, which are microprocessors specifically programmed and optimized to do one task only. For example, in a receiver that uses frequency synthesizer (Chapter 15), a microcomputer is connected to the frequency-entry keyboard, the display readout for the user, and the synthesizer counters. Using a variety of algorithms, the microcomputer checks to make sure that one and only one key is pressed at a time, interprets these keys, issues the appropriate instructions to the synthesizer counters, and puts messages for the user on the display.

---

7. What are the advantages and disadvantages of wider spacing?
8. What is an algorithm? Why are they used in communications?

10.2 DIGITAL SPECIFICATIONS

The relationship between analog signals and their digital equivalents is important to understanding the overall performance of a system and some of the choices that are made in the system design. There are many factors that determine the relationship between the analog and corresponding digital value. The two most important are accuracy and resolution.

Simply stated, *accuracy* is how perfect and correct the digital equivalent of the original analog value is when this equivalent is compared to some "better" standard of absolute correctness (such as the official meter defined by the U.S. Government National Institute for Science and Technology, formerly called the National Bureau of Standards). *Resolution* shows how fine the gradations of the digital value are, or into how many distinct values the overall signal span has been divided. Although both specifications are important, they are separate measures. Some systems can accept poorer accuracy but require higher resolution; others need the opposite situation.

Accuracy

Suppose that you were using a ruler to measure the length of a line. The ruler is made of a poor material and actually has stretched by 5%, so that when it said 1.0 in., the real value (compared to a better, standard ruler) is 0.95 in.; when it said 10.0 in., it is really 9.5. This is not an accurate ruler—it has 5% error—but it may be accurate enough for the application, such as cutting a piece of paper to fit a frame. However, the accuracy would not be good enough if the measurement was used to place an IC in the precise position on a circuit board—the IC would miss the solder pads!

If the analog signal-to-digital equivalent converter accuracy is not perfect, what would be the effect? If the original signal was a voice or music signal, the inaccuracy means that the signal volume might be a little too loud or soft, yet this is not a major problem. The listener could easily adjust the volume to the preferred level. However, if the analog signal represented the temperature of a laboratory experiment, the exact value is critical and the accuracy of the analog-to-digital conversion important. Otherwise, a true temperature of 100° would be recorded as 105°, using the 5% error factor.

In many cases, some inaccuracy is acceptable. A special calibration procedure can be performed by the system that compares the accuracy of the conversion process to a more correct standard. The difference can then be factored back into the system as a known inaccuracy, and taken into account. If you know that a ruler is off everywhere by 1 in. but is otherwise perfect (stable with time and temperature, legible, and rugged) you simply have to take this 1-in. error into all calculations based on the ruler's readings.

Resolution

Resolution is an indication of how many divisions the analog-to-digital conversion process uses. It is equivalent to the fractional inch markings on the ruler. A ruler can have each inch divided with markings for sixteenths, or thirty-seconds, or sixty-fourths of an inch. The finer markings mean that the item being measured—the analog signal—can be described to a finer degree. Note that an extremely accurate ruler could have divisions only to sixteenths, while a very inaccurate one may have divisions to sixty-fourths of an inch. The accuracy and resolution are different indications of the performance.

We can see the effect of differing amounts of resolution by looking at a 1-in-4 part and a 1-in-8 part analog-to-digital conversion. Figure 10.4a shows an analog waveform that spans 0 to +1 V. With four part resolution, the span 0 to 1 V is divided into four equal *bins, zones,* or *subgroups* (Figure 10.4b). With resolution to 1 of 8 parts (Figure 10.4c), there are eight bins. The digital equivalent of the analog value at any instant is the name of the subgroup into which it falls. The

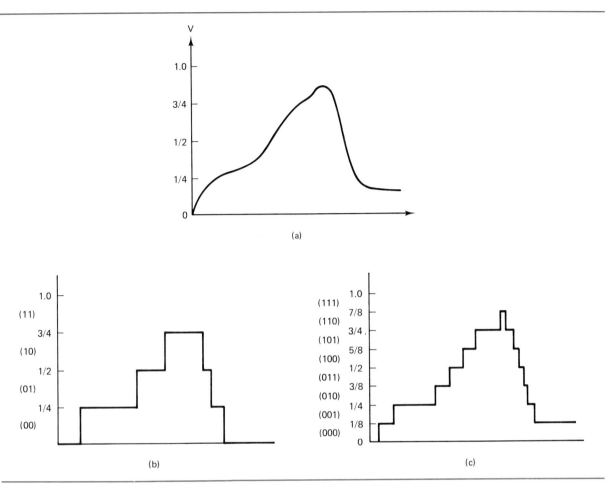

Figure 10.4 (a) A 0- to 1-V analog signal; (b) the same signal with a resolution of 1 in 4 parts (two bits); (c) with 8-part (three-bit) resolution.

resolution is sometimes called the *step size*.

Note one very important point about these digital subgroups: Differences between analog signals that fall into the same subgroup are lost forever. This is the consequence of the digital resolution. With four bins, a 0.15-V signal and a 0.20-V signal both have the same digital equivalent. The 0.05 V difference is information that cannot be recovered later, since the same bin (and value) represents 0.000 to 0.249 V. The term *quantization error* (or *quantization noise*) is used to describe the fact that a digital value corresponds to a distinct zone of analog signals. Any analog signal within the zone will have the same digital value and therefore look identical to all others after the analog-to-digital process. To reduce the quantization error, more bins are used to represent the analog value.

The resolution of an analog-to-digital conversion is related to the number of *bits* used to represent the number: 8 bits results in $2^8 = 256$ bins, 10 bits is $2^{10} = 1024$, and so on. (The number of bits positions is analogous to the number of columns used for decimal numbers.) Another way of expressing resolution is as a percent of the full-scale span. Ten bits (1024) means that the overall span is divided into 1024 parts, so a single bit is equal to 1/1024 of the span, very close to 0.1%; 12 bits divides the span into 4096 parts, or nearly 0.025%. The resolution of the conversion process provides fineness of reading to the percentage shown, but does not say anything about the absolute accuracy of the result.

A two-bit representation (00 through 11) of an analog signal provides $2^2 = 4$ distinct subgroups. For a 0 to 1-V signal, each subgroup zone is $\frac{1}{4}$ V. When three bits (000 through 111) are assigned to the same span, each digital value now represents $\frac{1}{8}$ V (compared to $\frac{1}{4}$ V when two bits are used), so the uncertainty surrounding the original analog value is cut in half. Binary 000 is 0.000 to 0.125 V, while 0.125 to 0.250 is 001. Additional bits increase the resolution and reduce the uncertainty, but at a cost: More bits require more communications time to transmit, wider bandwidth, and even more complex analog-to-digital conversion schemes. This is part of the technical trade-off that the system designer must consider when deciding how many bits are to be used for the conversion from analog format to digital format.

The resolution is determined by the inherent design of the analog-to-digital converter, which will be studied more thoroughly in Chapter 11. Most communications systems use 8, 10, or 12 bits for audio and video signals. Signals that represent scientific data such as temperature usually require resolutions of 12 through 16 bits. Some advanced systems use 18 bits.

Recall that digitization and limits of resolution represent an irretrievable loss of some information in the analog signal. Once the analog signal has been converted to digital format, the blurring of values within a subzone cannot be undone. There is no way to look at a digital pattern and determine what analog value it came from with any better fineness than the number of bits represented. In the two-bit example, all we can say is that binary 11 means the original analog signal was between 0.75 and 1.00 V, while three-bit binary value 111 indicates a signal between 0.875 and 1.00 V.

Once the analog value has been converted into the equivalent digital value, the bit pattern can be used in many ways to transfer the information. This is part of the overall encoding process between the basic analog signal and the way it is physically conveyed in digital format. The details of some encoding schemes are the subject of Chapter 12.

Dynamic Range

Another way of expressing the effect of using more bits when representing an analog signal in digital format is through *dynamic range*. This represents the ratio of the largest signal value that can be expressed compared to the smallest signal value (as represented by a single bit).

The relationship between achievable dynamic range and the number of bits is very simple. In binary numbering, each column to the left has double the weight of the previous column: the column weights for a four-bit number are 8, 4, 2, and 1. (Compare this to the traditional decimal numbering system, where each column represents 10 times the column to the right: 1, 10, 100, 1000, and so on.) In the earlier study of the decibel, we saw that doubling the voltage value represents a 6-dB increase. Therefore, each new bit position corresponds to a 6-dB increase in range (it is actually 6.02, but 6 is close enough for nearly all purposes). This leads to the very simple relationship

$$\text{total dynamic range (in dB)} = 6 \times (\text{number of bits})$$

10.2 Digital Specifications

Example 10.1

A 10-bit converter has 60 dB of dynamic range, while a 12-bit converter has 72 dB of range.

The dynamic range specification means that an analog signal of specified maximum to minimum span can be fairly represented, with a single bit count corresponding to the required resolution (in mV, V, or any other units) and all 1s representing the largest signal value. If the signal span is greater than the dynamic range, one of two things happens: Either the analog signal is greater than the range maximum so that the binary value "tops out" and overranges at 1111 . . . 111 (equivalent to 9999 . . . 9 in a decimal system), or the smallest binary value of 000 . . . 001 is still too coarse for the smallest analog value that must be resolved.

Example 10.2

A 10-V signal must be resolved to 1 mV. What is the dynamic range? How many bits are needed to represent it?

Solution

$$dB = 20 \log \frac{10}{0.001} = 80 \text{ dB}$$

Using the 6-dB/bit formula, we need 80/6 = 13.3 bits, which means that the conversion process must use at least 14 bits. The answer can also be determined from basic binary principles: This converter needs resolution of 1 part in 10 V/0.001 V = 10,000. Using 13 bits provides 2^{13} = 8192 parts; 2^{14} = 16,384 parts.

Example 10.3

What is the dynamic range of an 8-bit conversion? What is the maximum signal value when resolution of 10 mV is required?

Solution

$$\text{Dynamic range} = 6 \text{ dB/bit} \times 8 \text{ bits} = 48 \text{ dB}$$

The maximum signal is 48 dB greater than 10 mV, or approximately 2.5 V.

Binary versus Multilevel Digital Signals

At the beginning of the chapter we saw that "digital" means that only a specific number of discrete values are available. *Binary* is a special case of digital, with only two allowed signal values to be transmitted. Since binary uses only two values, it is at the opposite extreme from analog, where any value is allowed (so there are an infinite number of possible values). In multiple-level digital, the signal is represented by more than just the two symbols of binary.

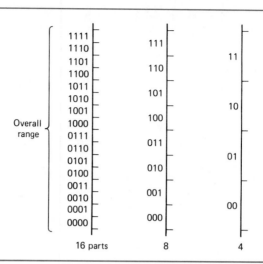

Figure 10.5 The number of bits needed to represent a signal span depends on the resolution. Shown are resolutions of 1 part in 16, 8, and 4, corresponding to 4, 3, and 2 bits, along with the names of the bins.

It is important to distinguish between the process of representing an analog signal by a digital equivalent, and the actual set of *symbols*—voltages, frequencies, or phases—used to transmit the digital version. The first step in converting an analog signal to its digital equivalent is to divide the overall analog signal span into subzones, or bins. Each of these bins represents the resolution to which the analog signal will be characterized. This is the heart of the analog-to-digital conversion process. More (and therefore smaller) bins provide better resolution and reduce the uncertainty that quantization error can cause.

The next step is to provide names to each bin. These names are the symbols used to refer to the digitized values. If there are two allowed symbols as in the binary case (bits called 1 and 0), four bits (also called *elements,* positions, or columns in the number) are required for 16 bins, and the bins are named 0000 through 1111 (Figure 10.5). When more symbols are available, the situation changes somewhat. For example, when four voltage values are available in the set of symbols, the same 16 bins can be uniquely identified by just two positions or elements.

Suppose that voltages of 0, 1, 2, and 3 V are used. The 16 bins are uniquely named using X, Y format, where X or Y can be 0, 1, 2, or 3 V. The first bin is 0, 0, followed by 0, 1, then 0, 2, while the last bin is 3, 3. (Note that the symbols do not have to be of the same type. In one common format for digital communications, two voltage values and two phase values are used for the four symbols: symbol 1 is 0 V, 0°, symbol 2 is 0 V, 180°, symbol 3 is 1 V, 0°; and symbol 4 is 1 V, 180°.)

The English alphabet is really a digital system with 26 distinct allowable symbols. Because of the relatively large number of symbols, a single element—a letter in a word—conveys a lot of information, and a few elements can form many possible words. For a five-element word, there are $26^5 = 11,881,376$ unique five-letter combinations.

Virtually all digital electronic circuitry uses binary format internally, but the multilevel digital is common for the transmitted signal format. An *encoding* portion of the modulator converts the binary to multilevel digital; the reverse process occurs at the demodulator and *decoder* of the receiver. The reason for switching to multilevel digital signals is that they make much better use of the bandwidth available. However, multilevel digital provides less noise immunity since the various symbols are "closer" to each other than they would be for binary and have less noise margin.

It takes less noise to cause an error in a multilevel digital system, as compared to a binary scheme where the two symbols are more widely separated in phase, frequency, or voltage values. A binary system with 0 and 1 represented by 0 and 9 V has more noise immunity than a four-level digital system with symbols 0, 3, 6, and 9 V. However, since fewer elements are needed to represent any bin number in a multilevel digital system, the time required to send the digital value is reduced, as the number of digital levels increases (if the elements are sent at a fixed rate). This is because each element in a multilevel digital system conveys more information than an element in a binary system. The more levels used, the greater the potential information content of each element: a binary system conveys only one of two possibilities with each element, while a 16-level digital system points to 1 of 16 unique pieces of information with each element.

Example 10.4

A system must convert an analog signal to digital format with resolution of 1 part in 16. The digital version is then transmitted using voltage that ranges between 0 and 15 V. Compare the use of binary digital format versus 4- and 16-symbol format in terms of noise margin and number of symbol elements that must be sent (which therefore affects transmission time).

Solution

In binary, the 16 bins are called 0000 through 1111; with 4 symbols these bins are called 00 through 33; with 16 symbols they are called 0 through F (0, 1, 2, . . . , 9, A, B, C, . . . , F); see Figure 10.6. Sending the bin numbers in binary format requires four elements, but values of 0 and 15 V (for 0 and 1, respectively) are used, so the distance between symbols is 15 V and the noise margin is very wide. With four symbols, only two elements are needed, which takes one-half of the time needed to send four elements. However, the voltages used would be 0, 5, 10, and 15 V (for symbols 0, 1, 2, and 3, respectively), so the distance and noise margin are reduced from 15 V to 5 V. Finally, for a 16-symbol scheme, only one element is needed to describe each bin uniquely, so transmission takes one-fourth of the time it did for binary. The voltages used are 0, 1, 2, 3, . . . , 13, 14, and 15 V, which have the least distance and noise margin. It is thus more error prone.

The trade-off is clear: Binary offers the greatest noise immunity but requires that more elements be sent compared to multiple-level digital symbols, where noise immunity is reduced but the number of elements needed to represent information is also reduced. Reducing the number of elements needed means that at a given rate of sending symbols—which implies use of bandwidth—the multilevel system conveys more information; conversely, a multilevel system makes better use of the bandwidth and sends the same total amount of information in less time than a binary system.

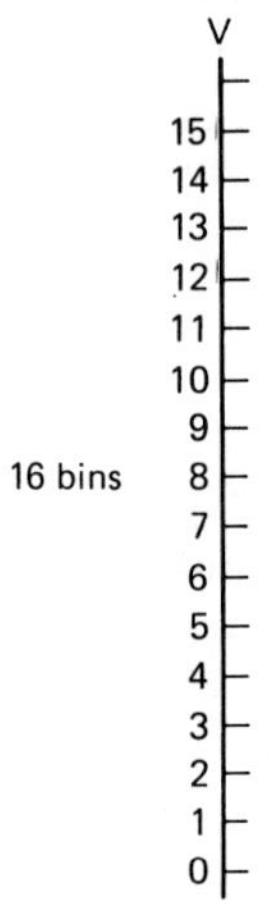

```
                    V
                    |
                 15 ┤   1111      33      F
                 14 ┤   1110      32      E
                 13 ┤   1101      31      D
                 12 ┤   1100      30      C
                 11 ┤   1011      23      B
                 10 ┤   1010      22      A
                  9 ┤   1001      21      9
       16 bins   8 ┤   1000      20      8
                  7 ┤   0111      13      7
                  6 ┤   0110      12      6
                  5 ┤   0101      11      5
                  4 ┤   0100      10      4
                  3 ┤   0011      03      3
                  2 ┤   0010      02      2
                  1 ┤   0001      01      1
                  0 ┤   0000      00      0
                       Binary   4 symbols  16 symbols
```

Figure 10.6 Signal resolved to 1 part in 16, with the 16 discrete values represented by 2-symbol (binary), 4-symbol, and 16-symbol format.

In Chapter 13 we will study some digital modulation schemes in more detail. One popular method used by the telephone system employs four amplitude and four phase-shift values for a total of 16 combinations. Extensive engineering analysis showed that this provided the best compromise between bandwidth and immunity to noise and errors in this particular application.

Review Questions

1. Compare and contrast accuracy and resolution. Where is one needed, and where is the other more critical?

2. What information is lost, irretrievably, when representing an analog signal in digital format? Why?

3. Explain quantization error. How does it decrease when resolution is increased?

4. What is the relationship between the number of bits of the digital representation and the resolution?

5. What is dynamic range? Why is it important? How is the dynamic range determined as a function of the number of bits?

6. Distinguish between the binary case and the multiple-level digital representation. How does the number of digital symbols affect how many symbols are needed to represent the digital value?

7. Why does multiple-level digital representation decrease the noise resistance but reduce the bandwidth required?

10.3 SAMPLING, BANDWIDTH, AND BIT RATES

We have seen that a continuously varying analog signal can be converted into a digital signal with accuracy and resolution determined by the conversion process and the application requirements. However, this is not all there is to making the transition from analog to digital signals. There is another very critical aspect of completely conveying the information that was originally in analog format via a digital format. This is the *sampling rate*, the number of times per second that the conversion process is performed (Figure 10.7).

In 1928, H. Nyquist showed that an analog signal could be perfectly reconstructed, without any loss of its information, if the sampling rate is at least twice the bandwidth of the signal. This means that a signal with a bandwidth of 1000 Hz must be sampled at 2000 samples/s or greater, for example. The *Nyquist sampling theorem* is the most important link between the analog world and the equivalent digital world. Note that the sampling theorem says nothing about the accuracy or the resolution of the conversion. It simply states that if sampling is done at the Nyquist rate or above, the samples can capture all the information inherent in the analog signal. Samples of the variations in the analog signal between the sampled instants are not needed, because the sampled instants are sufficiently frequent that they convey all the information that the signal (and its bandwidth) can hold.

Note that the sampling frequency, bandwidth, and filters are often set up so that only the required bandwidth of the overall signal is used and the remainder deliberately discarded. For example, suppose that a voice signal with bandwidth

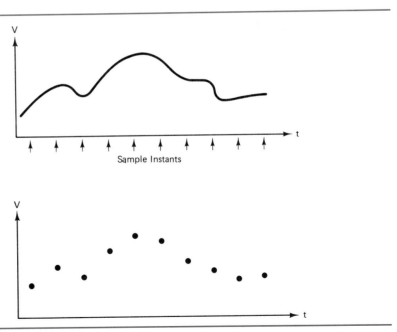

Figure 10.7 Sampling of a signal occurs at specific instants of time, and these sample values are all the system has to use.

Aliasing

If the actual sampling rate is below the Nyquist rate, the condition leads to samples that do not faithfully represent the original analog signal. It is not as if the samples capture only part or a portion of the entire signal when sampling below the Nyquist rate. Instead, *all* samples taken are virtually meaningless. The result of sampling too infrequently (known as *undersampling*) is called *aliasing*, since the samples *appear* to represent the analog signal but in reality do not.

Aliasing is a serious problem and therefore firm action must be taken to prevent it from occurring. First, the designer must use a sampling rate at twice the bandwidth of the signal of interest. Second, the signal being sampled must be strictly *bandwidth limited* before it goes to the analog-to-digital conversion circuitry to eliminate any frequency spectrum components beyond the Nyquist-specified bandwidth, through the use of low-pass filters with a sharp cutoff at one-half the sampling frequency or less. These *antialiasing* filters are essential to meaningful sampling and analog-to-digital conversion.

---

from 0 to 15 kHz is to be converted to digital format. This requires a sampling rate of at least 30 *kilosamples/s* (*ksps*) (also referred to as 30-kHz sampling, although use of the Hz units is a misnomer). However, the useful information in the voice signal exists primarily up to 3 kHz, and while the frequencies above 3 kHz serve to make different voices recognizable, they do not convey actual information or increase the intelligibility. The designer sets the sampling rate at 6 kHz and uses a low-pass antialiasing filter with a sharp cutoff frequency of 3 kHz to satisfy Nyquist's theorem.

It may seem simpler to sample the voice signal at 30 kHz, twice the original bandwidth of 15 kHz. However, there are reasons to resist doing this. First, conversion circuitry that operates at higher rates is more complex and expensive than slower-speed circuitry. More important, the higher sampling rate means that more data are being generated, and more bits/s will have to be transmitted. This requires more bandwidth, but once again, bandwidth is a resource that must be used carefully and conserved when it is not really needed. Also, even though the conversion circuitry can provide the higher rates, the remainder of the communication system must be faster, with more capacity to handle more data.

An example shows how digital signals quickly increase the bandwidth and information capacity required to convey a specific amount of information. Consider a voice signal with only a 3-kHz bandwidth, which is being converted to digital format with 8 bits of resolution. Using the Nyquist theorem, this signal must be sampled at 6 ksps (or higher). Thus 6000 ksps × 8 bits = 48,000 bits/s (48 kilobits/s or kbps) are generated from a signal that has only 3 kHz of bandwidth in the analog world. As we saw in Chapter 2, a bandwidth of approximately five times the bit rate is needed to transmit digital bits without excessive distortion and error, so the overall bandwidth required for the 48-kbps rate is 240 kHz.

Here is the digital format dilemma: Digital format provides the potential for much lower error rates and more reliable communications, but requires a much higher bandwidth than analog signals for the same amount of information. This is because the information content (*efficiency*) of digital signals is low compared to analog signals. Digital signals waste a great deal of the available bandwidth, since they use only a few discrete values to convey information. Binary is least efficient, while multilevel digital is more efficient at using the available bandwidth. In contrast, analog signals are allowed to have any value within the total range, and so make best use of the available bandwidth.

There are several ways to reduce the bandwidth needed. First, the sampling rate (and signal bandwidth) are set as low as possible. If the necessary information is contained in the band 0 to 5 kHz, then bandwidth, sampling rates, and data from beyond 5 kHz serves no purpose except to increase the data rate. For this reason, the system designer must look carefully at the signal being converted to digital format and set the sampling rate and antialiasing filter bandwidth to the lowest value that encompasses the required signal and satisfies the Nyquist theorem.

Second, the resolution of the conversion that occurs at each sample must be compatible with the system requirements. There is no point in performing 16-bit conversions (1 part in 65,536) on a signal that is measured to only 1% (1 part in 100).

Third, multilevel digital symbols are used where possible. Binary (two symbols) is replaced by 4, 8, or even 16 levels of digital representation, so that each transmitted element represents and conveys more information. The trade-off is the reduction in noise immunity. Of course, a binary set of symbols could be used but at a higher rate to send the information in the same amount of time as a multilevel symbol set, but the higher rate requires more bandwidth than the lower rate.

Review Questions

1. What is the Nyquist sampling theorem? Why is it critical in going from an analog signal to its digital representation?

2. What is aliasing? Can the results of undersampling approximately represent the original signal?

3. What are the two steps that are taken to prevent aliasing? What is the role of the antialiasing filter?

4. Why not simply sample at rates much higher than Nyquist indicates?

5. Why do digital signals require much more bandwidth than the original analog signal, to convey the same information? What factors are involved?

10.4 DIGITAL TESTING

Signals that are in digital form require test equipment and methods that are different from those for analog signals. The voltmeter and oscilloscope are of limited usefulness when checking the performance of digital systems. Instead, tools such as *logic analyzers* and *error rate* testers are required, along with other specialized equipment.

Logic Probes and Analyzers

Beyond the voltmeter and scope, the logic probe and logic analyzer (Figure 10.9) and some of its specialized variations are essential. The *logic probe* looks like a fat pencil and draws power from the circuit under test. It shows any digital activity on a single line, by a flashing light. If the light is dark, the line is low; a bright light shows that the line is high. The probe also generates a visible pulse of light if any

> **Voltmeters and Oscilloscopes**
>
> Initial system troubleshooting uses the voltmeter to check power supply values to see if there are any open wires (especially grounds and ground connecting straps) and verify proper dc supplies at the various ICs in the system. Communications systems typically use one or more supplies of both positive and negative 5, 12, and 15 V, depending on the ICs in the design and the speed of operation. High-power transmitters, of course, have higher voltage supplies to develop the required output power, but many of the analog-to-digital circuits and modulation functions operate from lower voltages for convenience and simplicity.
>
> The oscilloscope is used to look at the shape of the digital waveforms. The ideal digital waveform with absolutely vertical sides and a perfectly flat top does not exist in practice, due to waveform *slew rate* limitations (the signal does not instantaneously go from one level to the next but makes its transitions at some maximum rate of V/s), circuit capacitance and inductance, *ringing* and *overshoot* (the signal goes past the desired final value, then homes in and finally settles at the correct value), and other real-world circuit performance constraints (Figure 10.8). The real signals can look nearly ideal, and the scope is used to judge if the actual signal is good enough for proper operation. Excessive noise can ruin digital signals, and this noise can be observed with the scope.
>
> Beyond this, the scope is not very useful for examining digital signals and systems. In analog systems, the signal *shape* conveys the information, and it is the accuracy of this shape—along with the transformations the signal undergoes at it passes through the system—that carry the information from its source through modulation, amplification, transmission, reception, demodulation, to the final user. In digital signals, however, the shape is not critical as long as it is within certain acceptable and relatively broad boundaries. Instead, it is the *presence, position,* and *timing* of signals that are critical. These cannot easily be studied on the scope screen for comparison to what they should be like.

signal pulse occurs, even if the signal pulse itself is much too brief to be seen by the eye. The logic probe is very useful for quickly checking activity of key circuit wires: Are lines that should be showing activity stuck at a high (1) or low (0) state? Do lines that should be high or low actually have pulse activity, due to failures of associated components, noise, or unusual circumstances?

The *logic analyzer* is designed to display the signal state of many lines simultaneously. Although some analyzers can show the waveforms themselves,

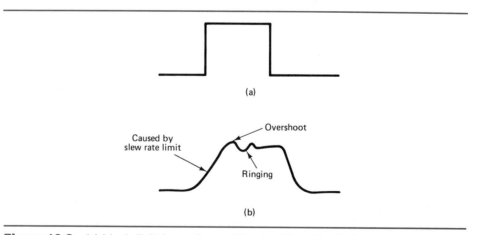

Figure 10.8 (a) Ideal digital waveform; (b) actual (nonideal) digital waveform with slewing limitations, ringing, and overshoot. Too many such distortions can cause errors in the interpretation of the signal.

10.4 Digital Testing

Figure 10.9 Simple logic probe, together with a logic analyzer capable of displaying various screen formats (courtesy of Hewlett-Packard Co.).

this is not useful for most applications. What is useful is to see the logic state of each line with respect to the others and with respect to the passage of time. The logic analyzer shows these as groups of 1s and 0s. Most analyzers also allow the 1s and 0s to be displayed to the user in different formats, such as eight-level (*octal*) and 16-level (*hexadecimal*) for easier reading.

One key signal for the logic analyzer is the *clock input*. Nearly all digital systems are *synchronous* systems, with a single master clock that acts to pace all the digital circuitry operation. The master clock ensures that all operations occur at the proper instant. Without the master clock, the various *propagation delays* of the signals traveling through the circuitry wires, along with delays within the ICs themselves, would mean that the same information would reach different parts of the circuit at widely varying times. This would lead to many circuit malfunctions. The clock paces system operation and indicates when digital signals are valid so that the circuitry can perform the next function.

The clock makes troubleshooting easier, since the logic analyzer has to look at the digital signal only when this clock signal indicates that the digital signal is now valid to use. Logic analyzers have a clock wire lead that is connected to the clock of the system under test. Any minute variations in the system clock or offsets from the theoretical ideal are not a problem since the analyzer is now operating in synchronization with the clock of the system.

Network Analyzers

In most cases the digital information is not solely a *stream* or *string* of 1s and 0s that represent just the original analog signal. Instead, various communications-related information is appended to these data bits as part of the overall communications structure. These added bits are elements of the system *format* and *protocol* and will be studied in detail in subsequent chapters. One example of a format is to send the bits in clusters, with each cluster containing from 8 to 256 bits. Before each cluster, a *header* block is sent that indicates the following cluster length. This clustered scheme makes it easier to recover and regroup after any error is detected, especially if the receiver has to have an opportunity to respond. (Without the clustering, the receiver might have no chance to answer, just like trying to interrupt a nonstop speaker who monopolizes the microphone.)

A *network* or *protocol analyzer* is a specialized logic analyzer designed with knowledge of the rules of the clustering. It separates out the various parts of the overall run of data bits and extracts the actual information bits from the bits that are part of the scheme to manage the communications. In this way the network analyzer is much more complex, yet much more useful than a simple scope. It makes use of communications specific knowledge to sort out the various parts of the signal, which is something a scope cannot do.

The technician or engineer who is using the analyzer selects to view only the data bits, or just the bits that are part of the management of the communications flow, or both. Different parts of the troubleshooting will require looking at and understanding different parts of the message. In some testing procedures, the actual data bits are not of interest, but the interaction of the data bits and the format is of concern. The logic analyzer is able to highlight the required portion of the bit stream.

Not all communications systems have a separate clock line for synchronization and circuit timing. For example, a radio that is transmitting signals in digital format is sending out a single signal. While there are ways to combine the clock signal and the actual information, this is usually not a good technical solution since it uses bandwidth and requires a more complex transmitter modulation scheme. Fortunately, there is an attractive alternative. Whenever digital signals are generated using a synchronizing clock, remnants of the clock frequency and phase become embedded within the digital signals themselves (to be studied in detail in the next chapters). The network or protocol analyzer designed for such applications contains special circuitry, such as a filter or phase-locked loop, to extract and retrieve the embedded clock signal. The logic or network analyzer then uses this recovered clock to synchronize to the digital signals.

Review Questions

1. Compare the function and display of the logic analyzer to these features on the scope. What does the analyzer show that the scope does not? What does the scope show that the analyzer does not? Compare a logic probe to a logic analyzer.
2. Why do many digital systems use a system clock? How does the analyzer use this clock to synchronize its operation to the circuitry?
3. How does a communications system without a separate clock signal still convey clocking information? How can this clock be recovered or derived?
4. What is a network analyzer? How does it differ from the logic analyzer?
5. What additional system "knowledge" does the network analyzer have as part of its design? How is this helpful to the engineer or test technician?

SUMMARY

Digital communications systems offer the ability to deliver information that is not corrupted by the noise, multiple stages of attenuation and amplification, and similar problems that affect every system's performance. In addition, a signal in digital form can be stored in a computer memory indefinitely, and processed to produce even better quality information or to reduce the amount of information that must be transmitted.

The first step in producing the digital equivalent of an analog signal is to divide the overall signal range into a distinct number of subzones. The analog signal at any instant is matched to the closest bin or subzone it falls into. If there are more bins, the quantization error that results is smaller than when there are fewer, larger bins. The names of the bins are the set of distinct values—symbols—that the digital system is restricted to using. If there are more names, fewer of these symbol elements are transmitted, but the signal distance between the symbols is less, so there is more of a chance that noise will corrupt a signal symbol enough that it is confused with another symbol. The extreme case is the binary—two-value—system, which has simply two widely spaced signals. This provides the greatest difference between signals but requires the maximum number of signal elements of any digital system to convey the same amount of information.

Digital systems are noise resistant because the original signal value can be regenerated by the amplifier, which only has to produce an output at the original value. When the original signal, plus noise, reaches the amplifier, it simply has to be recreated at its original value. The amplifier can perform this regeneration accurately because the noise is small compared to the distance between digital symbols.

To completely represent an analog signal, it must be sampled at twice its bandwidth. This is called the Nyquist sampling rate. These samples are sufficient to convey all the information in the signal. If the sampling is done at less than the Nyquist rate, then aliasing and false samples occur. Low-pass filters are used to limit the signal bandwidth to prevent aliasing.

Logic analyzers and network analyzers are much more useful than voltmeters and oscilloscopes for analyzing digital systems. The scope shows signal shape and amplitude, but in the digital world the precise shape and value are of less importance than the relationship between multiple signal lines, and the digital values compared to the system clock, which synchronizes operations within the circuitry. The logic analyzer can display data in multiple formats. The more advanced network analyzer is designed for a specific format and protocol and can actually separate the many digital elements into various categories, such as overhead information or actual data bits.

Summary Questions

1. Compare analog and digital representation in terms of potential for error-free transmission, required bandwidth, and efficient use of bandwidth.

2. What can digital representation accomplish that analog data transmission, storage, and manipulation cannot? How can algorithms for signal processing be implemented on analog signals and on digital signals?

3. What is the difference between resolution and accuracy? What determines the value of each?

4. Give an example of where high resolution is important but accuracy is not critical. Give an example of the reverse situation.

5. How much resolution does an n-bit conversion provide in binary digital format?

6. What are the advantages and disadvantages of using more (or fewer) symbols to represent a signal value digitally in terms of error resistance and number of elements that must be sent? Why are these important factors?

7. What is dynamic range? What is the simple relationship between the number of bits used and the achievable dynamic range?

8. What is the sampling theory? How does it link the analog and the digital worlds?

9. Why must aliasing be prevented? Why are bandwidth limitation and antialiasing filters critical?

10. Besides preventing aliasing, why is analog bandwidth limiting often necessary when converting an analog signal to digital format?

11. Outline the sequence of steps for taking an analog signal and presenting it as a series of digital signals as the analog signal changes.

12. Explain the difference between sorting an analog signal into its corresponding digital bin and the use of digital symbols to represent the bin number.

13. What is quantization error? How is it reduced? At what cost?

14. Explain how the number of digital symbol elements per second required is determined by the signal bandwidth, required resolution, and the number of digital symbols used.

PRACTICE PROBLEMS

Section 10.2

1. An analog-to-digital converter has inaccuracy of 2%. What is the possible range of values it will determine when the analog signal value is 4 V?

2. The converter of problem 1 provides resolution to 8 bits over a span of −5 to +5 V. What is the size of each digital subzone? Compare this resolution to the 2% accuracy of the converter.

3. Into how many steps does a 14-bit conversion divide an analog signal? What about an 18-bit conversion?

4. A signal that ranges from 0 to +3 V must be converted to digital format with a quantization error of less than 1 mV. How many bins must be provided? How many bits would a digital converter have to provide to meet this goal?

5. What is the dynamic range of the signal in problem 4? What is the dynamic range if the signal span is 0 to +6 V? How many bits would be needed in this case?

6. A 0- to 1.6-V analog signal is digitized to 16 steps. What is the voltage span that the first three steps each represent? The last three steps?

7. For the signal of problem 6, binary format is used to identify each bin. What are the binary values of the first three and last three steps?

8. The digital results from the signal of problem 6 are now represented by 16-symbol format, using symbols 0 through F, instead of binary format. What are the representations of the first and last three steps? How many elements must be conveyed to identify the signal value uniquely?

Section 10.3

1. A signal is bandwidth limited to 12 kHz. What is the Nyquist sampling rate required?

2. A system can sample up to 6000 Hz, although the signal bandwidth extends to 5 kHz. At what cutoff frequency should the antialiasing filter be set?

3. A signal has a bandwidth of 3300 Hz. What is the Nyquist sampling rate? If the signal is digitized to 10-bit resolution, how many bits are generated per second? What is the digital bandwidth required?

4. A signal is being digitized at 25 kHz. What is the maximum bandwidth of the signal that avoids aliasing? If the samples are digited to 8 bits, how many bits/s must the system handle?

5. A 25-kHz bandwidth signal is digitized at the Nyquist rate to 8 bits and the digital representation is done with binary signals. How many samples/s must be transmitted? How many elements in each sample? How many elements/s are transmitted?

6. The digital samples in problem 5 are now represented by four symbol levels. How many elements in each sample? What are they? How many elements/s are transmitted?

7. The digital signals in problems 5 and 6 can have any value from 0 to 7 V. What would be a good choice of symbol values for the binary case? For the four-level representation? What is the difference in noise immunity in each case?

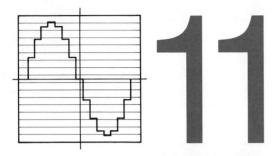

11
Digital Communication Fundamentals

CHAPTER OBJECTIVES

When you have completed this chapter, you will understand:

- How analog signals are converted to digital format, and vice versa
- How a signal value is represented as a multibit digital word
- The need for various types of synchronization of bit and word clocks, and how this is achieved
- Delta modulation, a unique form of digitization that eliminates the need for synchronization and shows signal changes rather than absolute signal values
- Basic approaches to testing of systems that convey analog information in digital form, along with synchronization information

INTRODUCTION

In this chapter the many functions required for communicating analog information in digital format are combined to form a complete system. The simplest approach to this requires only an analog-to-digital converter at the information source, and a corresponding digital-to-analog converter at the receiver, and transmits the data as a parallel group of bits with one signal line or path for each bit. This method is often modified with additional circuitry to send all the bits in sequence over a single line using a single serial data link.

The role of a clock for timing of data bits and synchronization is studied in both parallel and serial versions. A complete system for pulse code modulation (sending analog information in digital form) must make sure that the bits are received with the proper timing and that the receiver can determine where the overall digital byte or word representing the analog value actually begins. There are various techniques used to pass both bit and byte or word timing information to the receiver or to derive it at the receiver, so that the receiver is properly synchronized with the activity of the transmitter.

In the last part of this chapter we study delta modulation, an alternative approach to communication of analog signals in digital form. In this scheme the bits represent the change in analog signal value instead of its absolute value. The circuitry for delta modulation is very simple, at both ends of the system, so it has wide use in some applications such as digitized voice from telephone units.

11.1 ANALOG-TO-DIGITAL AND DIGITAL-TO-ANALOG CONVERTERS

In Chapter 10 we saw how any analog signal, such as a voice, could be represented faithfully by its digital equivalent value, using the Nyquist theorem, sampling, and analog-to-digital conversion. We will now look at some of parts of the complete digital communications system before putting all the pieces together.

The *analog-to-digital converter* (also known as an *A-to-D converter, A/D converter,* or simply *ADC*) is the circuitry responsible for taking an analog input and converting it to binary digital format. Most modern ADCs are single integrated circuits, although some specialized ones require complete circuit boards with many individual components for the necessary speed, resolution, or accuracy in performance. A typical ADC is the AD673 (Figure 11.1), available from Analog Devices, Inc. This ADC takes analog signals of 0 to +10 V or −5 to +5 V range and provides a complete eight-bit conversion every 30 μs. For signals with maximum values less than the range of the converter, an amplifier is used between the signal source and the converter input to boost the signal as needed.

The process begins when a *convert* control line to the ADC goes low (to binary 0) to initiate a conversion cycle. The end of the conversion cycle is indicated by a *data ready* or *conversion complete* signal that the ADC generates for use by the rest of the circuitry. At this point the eight-bit representation of the analog input is available at the converter output and must be passed on to the rest of the communications system before the next A/D conversion is initiated.

The ADC, although the heart of the digital system, is only one-half of the conversion process. At the receiving end, the binary signal must be converted back to analog form to be useful. This is done by a *digital-to-analog converter* (*D/A converter* or *DAC*) to recreate the original analog signal.

One such D/A converter is the Analog Devices AD558 (Figure 11.2), which takes an eight-bit digital byte and produces the equivalent analog output spanning either the 0 to 2.56 V or 0 to +10 V range. If this output range does not match the receiver application requirements or the original signal value, a *buffer* amplifier at the output amplifies or attenuates the voltage to the required value. The time for conversion from digital representation to analog value is very fast: The AD558 settles to the correct analog value in 1 μs. This is typical of ADCs and DACs: the

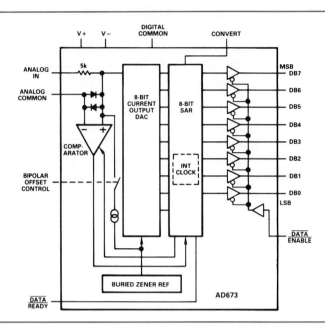

Figure 11.1 AD673 analog-to-digital converter (courtesy of Analog Devices, Inc.).

conversion time from analog to digital format is longer than for the reverse and is due to the inherent design architectures of ADCs and DACs.

In a simple system design, the digital outputs of the ADC can be connected to the inputs of the DAC (Figure 11.3), producing a complete digital communications system for transferring information that was originally in analog form. This relatively crude system is the beginning nucleus of a complete data communications system. One major problem is that this design is not efficient in use of signal lines: it transfers data bits in *parallel* requiring eight signal lines or carrier frequencies—one for each bit—to carry the data between the ADC and DAC. A more effective system is discussed in the next section of this chapter.

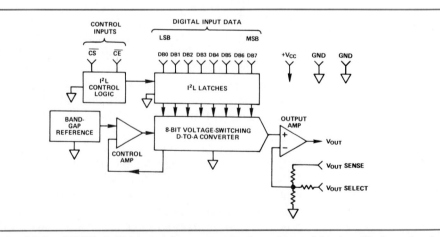

Figure 11.2 AD558 digital-to-analog converter (courtesy of Analog Devices, Inc.).

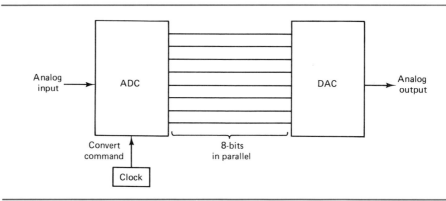

Figure 11.3 Eight-bit ADC-to-DAC interconnection and communication with parallel connections.

Review Questions

1. What initiates an analog-to-digital conversion? What indicates that it is completed?
2. Compare the conversion time for an ADC to the time for the reverse process.
3. Why is the ADC parallel to DAC parallel input an inefficient scheme?

11.2 PULSE CODE MODULATION

The very simple ADC + DAC approach leads to a system that can transmit and receive analog information in digital format. To make the system complete and effective, some additional functions are needed, shown in Figure 11.4. These include:

An *analog input filter,* to prevent aliasing (the result of digitizing analog signals at less than twice their bandwidth)

A *clock,* to initiate A/D conversions and provide timing signals

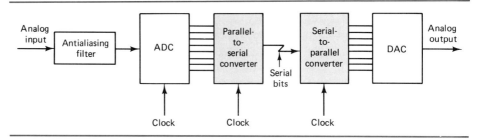

Figure 11.4 A complete communications system for data in serial format requires additional components than simpler parallel communication.

Serial-to-parallel and parallel-to-serial converters, to transform the multiple bits from the ADC into a single *serial stream,* and perform the reverse operation

Let's look at the role of each of these in a *pulse code modulation* (*PCM*) system in more detail. The term "pulse code modulation" has historically been used when an analog signal is converted to a digital (pulse code) format, then used to modulate a carrier. There are many specific ways of implementing PCM.

Analog Input Filter

The particular ADC chosen here can perform a conversion every 30 μs, which is equivalent to conversions at a 33.3-kHz rate. By the Nyquist theorem, the input signal bandwidth must be less than $33.3/2 = 16.6$ kHz to prevent aliasing. A sharp cutoff *low-pass filter* is used between the analog signal and the ADC to ensure that no frequency components beyond this bandwidth reach the converter. These undesired components can be part of the signal, such as the higher frequencies of voice and music (with a bandwidth up to 20 kHz), but they can also be wideband noise that is corrupting the signal. Due to the noise, a low-pass antialiasing filter is essential even when the analog signal bandwidth is known to be less than half the Nyquist frequency.

Clock

The system clock paces the conversion process and related operations. The clock, running in this case at 33.3 kHz, acts as a signal to the ADC to begin a new conversion. If the clock rate is faster than the ADC conversion rate, the converter will ignore the clock pulse that occurs during conversion, or will stop the conversion cycle in midoperation and begin another cycle (depending on the particular converter design). Either way, this would interfere with proper operation of the conversion and result in sampling at too slow a rate. Since the bandwidth of the input signal is set by the filter at 16.6 kHz, any sampling below the 33.3-kHz rate results in aliasing, just as when the 33.3-kHz rate is used for bandwidths beyond 16.6 kHz.

The role of the clock at the receiving side is less obvious. For the case where the data bits are transferred as eight parallel bits, the DAC simply takes the bits and produces the equivalent analog output. In concept, a clock is not needed to tell the DAC to perform this operation. However, not having a clock can cause problems when the parallel data bits do not reach the DAC in exactly perfect synchronization, as is often the case. Slight differences in the physical length among the eight paths, or momentary shifts (called *jitter*) in circuitry or path propagation delays, can cause some bits to reach the DAC after others. The result at the DAC end can be severe. The DAC will respond to the digital pattern at any instant, yet at some instants the pattern will be wrong, as bits from a previous conversion and bits from a new conversion combine to form a false new pattern. These results are *glitches,* which are very short false outputs of the DAC.

To see how this erroneous analog output value can occur, let's look at an example. Suppose that one byte is 01111111 (just below half-scale), followed by 10000000 (exactly half-scale). These two bytes differ only in their *least significant bit* (LSB, the rightmost bit) and will normally cause the DAC output to make the smallest possible incremental change. However, if there are propagation delays among the various bits and the MSB of the second byte arrives at the DAC slightly before the rest of the bits, the DAC input will be the combination of the *most significant bit* (MSB, the leftmost bit) of the second byte and the seven LSBs of the first byte: 11111111. This is full scale, quite a difference from the correct half-scale values.

To prevent this problem, the DAC uses a clock signal to indicate when all the bits are valid as a group. The clock signal is the same one as the ADC used but is delayed by a small amount to guarantee that all the bits of a conversion have arrived.

Example 11.1

An eight-bit ADC is used over the span 0 to 2.56 V. (a) How many volts does a single LSB represent? (b) What is the binary representation of 1.00 V and of 1.63 V? (c) What is the binary number and the equivalent analog value if the three MSBs of the representation of 1.00 V and the five LSBs of 0.35 V momentarily combine to become the DAC input?

Solution

(a) One bit = 2.56/256 = 0.01 V. (b) 1.00 V/0.01 V = 100 (decimal) counts = 01100100 (binary); 1.63 V = 1.63/0.01 = 163 counts = 10100011. (c) combined value is 01100011 = 99 counts at 0.01 V/count = 0.99 V.

Parallel-to-Serial and Serial-to-Parallel Conversion

There is a major practical limitation in the system outlined above: eight parallel data lines, plus one clock line, are required to carry the data. Using this amount of circuitry and communication channels is unacceptable in most applications, except within a single chassis where distances are very short. It would be much better if a single wire (or carrier frequency) could serve for all the data bits and clocking information, with the bits traveling in *serial* form, one immediately after the other. *Parallel-to-serial* and *serial-to-parallel* converters make this possible. Note that these converters are very different from A/D or D/A converters. They do not change the data bits or signal representation, but they do change the way the data bits are presented within the system.

In operation, the eight parallel bits from the ADC are *loaded* or *clocked* into the parallel-to-serial converter, which is a *digital shift register* that accepts data bits in parallel and then clocks these out in serial format. The parallel-to-serial converter requires a clock line that operates at a higher rate than the ADC. For an eight-bit conversion, the clock must be eight times faster, so that the eight bits can be clocked out in the same time that a single analog-to-digital conversion occurs.

This is shown in Figure 11.6. The system master clock is now at eight times the ADC clock, or 8 × 33.3 = 266.66 kHz and goes to the parallel-to-serial shift register. At each clock pulse, another bit from the shift register is output onto the single output line. The ADC clock is derived from the master clock by a digital "divide by 8" circuit. There is one A/D converter clock pulse and cycle for every eight bits.

At the receiving end, the complementary operation is performed by a serial-to-parallel converter. As the serial stream of bits is received, each bit is clocked and loaded into the shift register successively. After the eighth bit, an entire byte is ready to be passed to the DAC as a single entity. As new bits come in, they

Shift Registers

The *shift register* is one of the most important and versatile elements in digital communications. With a shift register, many goals can be accomplished: Data can be temporarily stored and later retrieved, data bits can be accepted in either serial (one bit at a time) or parallel (a cluster of bits at the same time) and then recalled later in either serial or parallel form, and data can be accepted at one rate and then retrieved at another rate.

A basic shift register for eight bits is shown in Figure 11.5 and consists of eight flip-flops connected to each other. (A flip-flop is an interconnection of digital gates that stores a single bit of data; it accepts data at an input, passes it out through an output, and is controlled by a clock signal that tells it to accept a bit present at the input point and hold it, while also making it available at the output point.) The output of one flip-flop is the input to the next. Also, the inputs and outputs of the eight flip-flops are available directly to outside circuitry, without going through adjacent flip-flops. A shift register can be built into a special IC or can be a part of a larger, more complex IC.

In operation, data bits can enter serially into the first flip-flop. With each clock pulse (cycle), the next bit is accepted at the first flip-flop and the previous bits all shift over, to the next flip-flop to the right. After eight clock cycles, all eight bits have been accepted serially, shifted in, and now are stored in the shift register. They can be read in parallel at the outputs of all eight flip-flops, if needed (serial input, parallel output). If clocking continues, the eight bits start to exit via the rightmost flip-flop, one bit per clock cycle (serial output). Finally, all eight flip-flops can be loaded simultaneously (parallel input) by presenting the eight bits to the flip-flop inputs: On the next clock cycle, all the data bits will be stored into their respective flip-flops and can then be retrieved either in parallel, or serially with a succession of clock cycles (parallel input, serial output).

One interesting use for a shift register is to allow data to be accepted serially at one rate from a source and then transmitted at another rate. For example, suppose that the data bits are coming at a high rate as serial inputs but are going to a system or communications line that operates at slower rates. The clock for this portion of operation operates at the higher rate. When all eight bits are loaded, the clock is switched or changed to the lower rate. The bits now exit the shift register at this new, slower rate but with the same digital values and in the same sequence as they arrived.

begin to fill the serial-to-parallel converter again and overwrite the previous bits, thus forming the new serial-to-parallel transformation of bits.

The receiving end must also have a clock, but not because of a potential glitch problem. The receiving clock is now needed to make sure that the serial-to-parallel converter looks at the input signal line at the right time instants and then clocks the next bit into the register. If there were no receiving clock, the serial-to-parallel converter might load input bits too quickly or too slowly compared to their actual rate of arrival, and so make errors. As shown in Figure 11.7, the bit rate at which the bits are clocked in must agree exactly with the rate at which they were sent. An example shows the significance of this:

Example 11.2

Two clocks operate nominally at 1 MHz and are initially identical, but one clock differs from the other by 0.01% over time. After how many bits do the clocks differ by a full bit period?

Solution

One bit period is 1/1 MHz = 1 μs; 1 μs is 0.01% of 10,000 μs. Therefore, after 10,000 μs = 10,000 bits, the clocks will differ by one complete bit period.

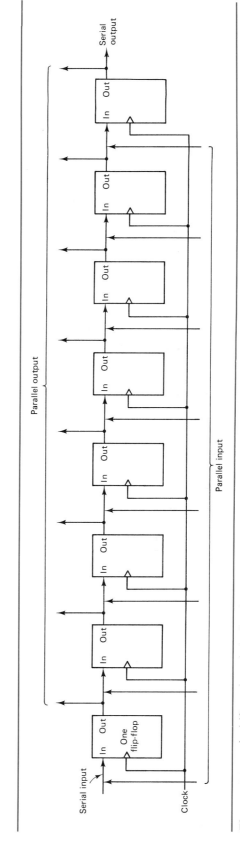

Figure 11.5 A shift register is built of an interconnected series of flip-flops, accepting and transmitting data as series or parallel bits.

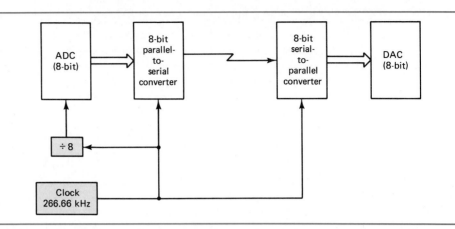

Figure 11.6 Serial communications at a desired rate of samples/s (with eight-bit/sample resolution) requires use of system clock at 8× ADC rate.

By using a parallel-to-serial converter at the transmitting end of the PCM system, along with a serial-to-parallel converter at the receiving end, the eight parallel lines have been reduced to a single data bit line plus a separate clock line. There are techniques for eliminating even this separate clock line, and these will be examined as part of the larger issue of synchronization in the next section. The technical "cost" for reducing the number of lines is increased transmit and receive circuit complexity, along with the need for a single line that can carry data at eight times the rate of a single line. A single higher bandwidth line is needed to replace many lower bandwidth lines to achieve the same number of eight-bit digital signals each second.

From another perspective, the maximum number of digitized values that can be transferred serially is reduced when the bandwidth of a line is limited to a fixed value, since this line must support a greater number of bits/s compared to the parallel case. For this reason, where the data rate requirements are very high, such as between circuit boards within a computer, the choice is usually made to stay with parallel lines. Where there is a need to minimize the number of communications lines or carrier frequencies used, the data are usually converted to serial format.

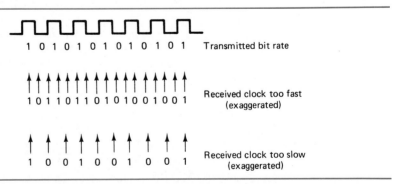

Figure 11.7 Bit rates must be precisely equal at both transmitter and receiver for error-free results, or else the receiver will check bits too frequently or not often enough.

The discussion of PCM has shown how an analog signal can be converted to digital representation, then converted from parallel to serial format and transmitted and recreated at the receiving end. This is the heart of any system that transmits analog information using digital circuitry. In many applications, the data are inherently in digital format—such as from a computer keyboard or memory—but the concepts of a PCM system are still valid. The computer memory bytes or keyboard stroke information are converted to serial format, transmitted at a known rate, and then reformatted into the set of parallel bits at the receiver. This requires the use of shift registers as parallel to serial/serial to parallel converters, clocking, and synchronization, just as in a system designed for signals that were originally in analog form.

In telecommunications, the A/D circuitry, which takes an analog signal and produces the final digital bit stream, is often referred to as an *encoder*. The D/A circuitry, which performs the reverse operation at the receiving end, is known as a *decoder*. The two functions often exist as a pair called a *codec*, short for coder–decoder. The codec function is often implemented by a single IC, with the same IC able to function as both an encoder and a decoder since there is considerable overlap in functional blocks between an encoder and a decoder. A combination codec can use many portions of its circuitry for either and has only to have some internal blocks switched in or out when going from encoder to decoder mode. Many codecs also have special circuitry for communications functions such as *companding*, discussed next, built in.

Companding

Greater resolution results from using more bits to represent the analog signal, but sending these additional bits takes longer at a given clock rate, or requires more bandwidth and faster clocks. Yet if the number of bits is too low, the resolution is poor and the ± 1 LSB quantization error inherent in any analog-to-digital conversion is large: the signal sounds "coarse" or a video image appears "grainy" instead of varying smoothly from one value to the next.

In telecommunications, a solution that is often used is to *compand* (compress/expand) the analog input signal. This scheme makes use of an understanding of the technical nature of the typical voice or video signal. Higher resolution is needed when the signal is at its lower amplitudes, but at higher signal amplitudes a coarser resolution is sufficient. It is the same as measuring distance with a ruler. For distances between 0 and 1 in. you may want to measure to $\frac{1}{32}$ of an inch; from 1 in. to 1 ft, measure to $\frac{1}{4}$ in., and from 1 to 10 ft, measurements with resolution of $\frac{1}{2}$ in. would be fine. Instead of resolution being a fixed value, the resolution is now a percentage of the reading, such as "resolution to 0.1% of actual reading."

To implement companding, the normal straight-line *linear* relationship between analog input and corresponding digital output value is deliberately curved (Figure 11.8a). A digital bin at lower signal amplitudes corresponds to a smaller analog signal span, while the bins at larger amplitudes represent a wider analog span. The equation of companding determines the curve, which then shows how the bin sizes change along the overall analog signal range. Companding is done by passing the analog signal through a circuit that *compresses* the signal range according to a predetermined equation, before the signal is digitized (Figure 11.8b). At the receiver the compressed analog signal passes through the DAC and then expands to recreate the original analog signal that went into the compressor.

The industry has standardized on several companding equations, such as the

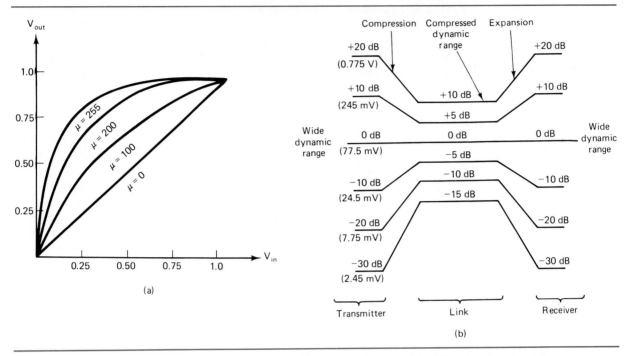

Figure 11.8 (a) Companding curve, for $\mu = 0$ to 255, showing how the relationship between input and output is deliberately made nonlinear; (b) companding showing how overall dynamic range is compressed, the signal is transmitted, and then the received signal dynamic range is expanded back to the wider value, all for one specific companding ratio. (*Note*: Other companding ratios are also used.)

μ-*law* (used in the United States) and the *A-law* (Europe). The μ-law equation relates the companded output voltage V_{out}, the actual analog input V_{in}, and the full-scale voltage V_{max} as follows:

$$V_{out} = V_{max}\left[\ln\left(1 + \frac{\mu V_{in}}{V_{max}}\right)\right] \div \ln(1 + \mu)$$

where the *companding constant* μ indicates the amount of companding. Figure 11.8a shows the amount of compression for differing values of μ. When $\mu = 0$, there is no compression; compression is a maximum at $\mu = 255$. A value of $\mu = 100$ is most commonly used.

Example 11.3

For $\mu = 100$, what is the companded output voltage when the input is 0.5 V, for $V_{max} = 1$ V? What is the companded voltage for $V_{in} = 0.8$ V?

Solution
For $V_{in} = 0.5$,

$$V_{out} = 1\left\{\ln\left[1 + 100\left(\frac{0.5}{1}\right)\right]\right\} \div \ln(101) = 0.85 \text{ V}$$

for $V_{in} = 0.8$,

$$V_{out} = 1\left\{\ln\left[1 + 100\left(\frac{0.8}{1}\right)\right]\right\} \div \ln(101) = 0.95 \text{ V}$$

Companding has been used for many years, even before digital communications systems became common. It provides other advantages in improving overall signal-to-noise ratio for lower-amplitude signals, while reducing the signal voltage span—dynamic range—that is required. The circuitry that implements companding consists primarily of a *variable gain amplifier* (*VGA*) (Figure 11.9). The amount of gain of the amplifier determines the relationship of the input (the original signal) and the output (the compressed signal). The control voltage of the VGA that determines gain is derived using nonlinear circuitry which looks at the magnitude of the uncompressed signal. The control voltage varies via the companding equation to change the gain of the amplifier at all signal levels within the total signal range.

With increasing use of digital circuitry, companding is often done with the digital signal itself. To compress an analog signal with 72 dB (12 bits) of dynamic range to 48 dB (8 bits), it is first converted to digital format with a 12-bit ADC. The output of the ADC goes to a *look-up table* stored in *read-only memory* (*ROM*) where each 12-bit ROM input produces an 8-bit ROM output. The relationship between the 12-bit value and the 8-bit companded value is calculated as part of the engineering design; several of the $2^{12} = 4096$ possible input values correspond or *map* to identical values among the $2^8 = 256$ possible outputs. At the receiving end, the 8-bit value is passed to an 8-bit DAC, converted to analog format, and then expanded by analog circuitry to restore the original dynamic range.

Companding may seem like getting "something for nothing," but this is not the case. Some resolution is lost for larger signals compared to smaller signals. SNR usually suffers (for lower-amplitude signals) when the signal is expanded since the noise of the receiver is expanded along with the signal (noise of a given value will appear much larger after expansion). Typical companding applications reduce the dynamic range from 72 dB (equivalent to 12 bits) to 48 dB (8 bits). This eases the requirements on the analog and digital circuitry, including amplifiers, buffers, A/D and D/A converters, and shift registers.

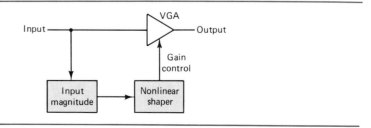

Figure 11.9 Analog companding with VGA, using actual signal level to set gain.

Companding is not simple *attentuation*. When a signal is attentuated, its amplitude is reduced by a fixed factor, unrelated to the signal's amplitude: for example, a ÷10 attenuation reduces 1-V signals to 0.1 V and 0.01-V signals to 0.001 V. The dynamic range—the ratio between the maximum value and the minimum value that must be resolved—remains unchanged; in this example, the range is 1/0.01 (unattenuated) and 0.1/0.001 (attenuated), both with dynamic range of 100 : 1. A compressed signal varies the attenuation (or gain, depending on the final signal range needed) in inverse proportion to the signal amplitude, so that a 1-V signal is reduced by a much larger factor than the tiny 0.1-V signal. In this way the initial dynamic range is represented by a smaller range, and thus by a reduced number of bits.

Incidentally, the Dolby system of noise reduction used on audio tapes is an advanced companding system. The major feature of the Dolby approach is that the companding is frequency selective: the audio band is divided into subbands, and each subband is companded by a different amount. The result of Dolby companding is that high-frequency noise and hiss on tapes (normally greater than the low-frequency hiss and noise and also more annoying) is reduced substantially.

Review Questions

1. What is the role of the low-pass filter? Why is it vital even for bandwidth-limited signals?
2. Why is a clock needed in a parallel system? How does it function as a glitch remover? What might happen without the clock?
3. What happens if the receiver clock is not at precisely the same rate as the transmitter clock?
4. What is the function of a codec? Why is a single circuit or IC often designed to be both a coder and decoder?
5. What are the advantages and disadvantages of companding? What is a typical reduction in the number of bits needed to represent a signal when companding?
6. How does a shift register accept data in either serial or parallel form and pass it on in either form? How is a shift register used to accept serial data at one rate and then transmit it at another rate?

11.3 SYNCHRONIZATION

Synchronization, at its simplest, is the process that allows the clock of the receiver in a digital system to operate at the same frequency (and phase) as the clock used to transmit the bits. Of course, there are many aspects to this simple statement. As we saw in the preceding section, if the clocks are not at "exactly" the same frequency, there will be bit errors even if the data signal itself is received perfectly.

Suppose that a clock at the transmitter and the one at the receiver are both designed to operate at the same frequency. Let's look at why synchronization is still necessary, how "exact" to each other the clocks have to be, and how these two clocks are used in a communications system.

Two separate clocks set to the same frequency can be used for a low-performance, relatively low-data-rate system, within some limits of performance and synchronization. One example is the *RS-232 format* for communications, which operates up to 19,200 bits/s and is studied in detail in Chapter 17. The limitations of this relatively unsynchronized system are due to the fact that no two clocks can provide exactly the same frequency, for a variety of reasons. Operating temperature variations cause frequency drift even if the clocks are initially at the same frequency, components age and change in time, and in reality the initial settings cannot be precisely identical. There may be a small variation, as low as 0.001% difference, but this is still significant at high data rates. In addition, even if the clocks were perfect, the small but significant variations in data travel time from the transmitter to the receiver will cause errors, since the receiver clock is a constant frequency whereas the signal as received has jitter and delay.

The result is that a nonsynchronized system can be used reliably only where a small number or *cluster* of bits is sent at a time, up to about 15 to 25 bits in a row. Beyond that number, even slight differences in clocks will cause bit errors as the receiver tries to determine what bits were sent but at the wrong time instants. For lower error rates, the bit data rates must be relatively low so that these clock variations are a small percentage of a bit time period. Typically, this means that the bit rates must be limited to less than 100 kilobits/s, which is low by modern rates but still sufficient in many situations.

Synchronization Choices

For higher rates, or sending longer streams of bits than a nonsynchronized system allows, some sort of scheme must be used so that the receiver clock is derived from the actual clock as used at the transmitter. Three methods can be used:

1. Send the clock signal along as a separate signal.
2. Derive the clock timing from the received data bits (clock recovery).
3. Use special bits as part of the data bit stream to reestablish synchronization and timing at the receiver.

The concept of using a separate line was discussed earlier in this chapter. It is an effective method when the communications channel consists of wire or cable, since no additional carrier frequencies are needed in the spectrum. It is also practical when distances are relatively short, so that wire costs are low. A separate clock line is most common when the data is being sent in parallel with one line for each bit of the data word, such as between circuit boards within a single computer.

A more sophisticated approach is to derive the clock from the received bits themselves. Rigorous mathematical Fourier analysis of the spectrum of the data bits shows that the clock frequency and phase itself are spectral components of these bits. As the clock frequency varies for any reason, its spectral frequency component varies accordingly. If there were some way to extract this component at the receiver, then the clock for bit timing at the receiver would also be automatically correct, since really it is the transmitter clock itself that is being used for recovering the data bits at the receiver.

The most common technique uses the *phase-locked loop* (*PLL*), studied for use in FM receivers in Chapter 6. Since the nominal frequency of the transmitted clock is known by the receiver, the design of the PLL is fairly easy. The VCO free-running frequency is set to this nominal clock rate. The PLL low-pass filter needs only a very narrow bandwidth, since the variations around the nominal clock rate are relatively small. The PLL is connected to receive the incoming stream of bits (Figure 11.10) and by its nature locks onto the clock rate that is embedded in the bit stream. As the transmitted clock jitters, whether due to propagation changes or variations at the transmitter, the VCO output tracks this, and the result is that the VCO output is a nearly perfect recreation of the transmitter clock and its variations.

The PLL method is very effective and used where highest performance is needed, even when the data rates are high or the clock performance is hard to predict precisely. One common application is receiving signals from space satellites, which is probably the most challenging of all communications design problems. The transmitter clock frequency varies with the temperature of the satellite (not constant but affected by the angle to the sun) and with the vagaries of signal propagation through the atmosphere.

To complicate space satellite reception, even an absolutely perfect clock at the satellite transmitter has a different frequency as seen by an earth-based receiver, due to *Doppler shift* (we will see more of the Doppler shift in Chapter 21). This shift is the apparent change in frequency that a receiver sees when there is a large relative velocity between a transmitter and receiver. It is roughly the same situation as the change in pitch that a person hears when a car or train passes by at high speed while blaring its horn. The amount of Doppler shift relative to a carrier of 2450 MHz is about 7 Hz per mi/hr speed, or 385 Hz at 55 mi/hr (small but measurable with modern electronics, so it is a way of measuring the speed of a car with a *radar gun*). The PLL is well suited for tracking and adjusting to this change and then generating a clock signal that is correct for the data bits.

Although the PLL method is very powerful and practical with modern ICs, there is still a problem with which it must deal. The PLL VCO is driven by tracking the signal transitions of the received bit pattern. The 1 and 0 pattern of the received bits are the variations that the PLL, or any synchronizing circuit, must look at to derive the embedded clock. If the bit stream contains a long string of 1s, without 0s (or vice versa), the received signal has no variations, but instead is a steady signal. In this case there is nothing for the PLL to try to lock onto. After all, a string of 100 continuous 0s looks like a dc signal. The PLL will lose lock and wander randomly. Yet there is no assurance that after it has been digi-

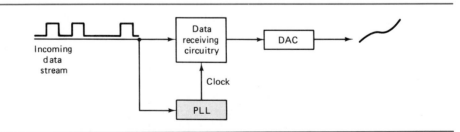

Figure 11.10 A PLL is used to derive the transmitter bit clock from received data; this recovered clock is used to determine if a 1 or 0 was actually sent.

tized, the signal (or even an inherently digital signal) is not all 1s or all 0s for many bits in a row. The analog signal could be a zero or full scale, for example. In practice, more than 8 to 12 unchanged bits in a row can cause problems for most clock recovery systems.

The solution involves some clever circuitry at the transmitter, with corresponding circuitry at the receiver, and a technique called *bit stuffing*. There are many variations of bit stuffing, but the idea is this: Whenever more than a previously specified number of 1s (or 0s) occur in a row, the circuitry automatically stuffs a 0 (or 1) after that number of bits. If the maximum number of allowed 1s in a row is 8, the ninth bit is automatically made a 0 whenever the 8 in a row pattern occurs. This gives the PLL a small toehold with which to maintain lock. Of course, this is not as simple as it seems. The receiver, which is looking at the bits to determine what numbers they represent, must separate out a real 0 from one that was stuffed in. Various sophisticated schemes are used to indicate which bits are "real" and which are "stuffed," which adds to the complexity at both ends of the system.

Special Bits

Another method is to use special sequence of nondata bits as part of the overall bit stream. The method works as follows: The receiver clock runs at the nominal value of the transmitter clock with only small variations expected. It is designed to run at this nominal frequency and resynchronize itself precisely at regular intervals using some special sync bits that come with the data bits. Typically, there will be one synchronizing bit pattern for every 8 to 256 data bits. The exact number of bits that can be transmitted between sync patterns is determined, in advance, by how close the clock frequencies will be and by the bit rate.

There is a technical trade-off here. Tighter synchronization can be assured by using the sync bits more frequently, such as sending a 1010 sync pattern every 16 data bits. However, this wastes 4 out of every (4 + 16) = 20 bits with sync information rather than new data. At the other extreme, a 1010 sync pattern for 256 data bits is relatively efficient, "wasting" only 4 out of every (4 + 256 bits), approximately 1.5%. However, since the sync bit pattern occurs relatively infrequently, there is a longer time between re-synchs and a greater chance of clock differences and data errors.

Many systems use a combination of both separate sync bits and deriving timing from the bits themselves. The reasons for this are discussed in the next part of this section.

Frame Synchronization

So far we have looked at the problem of maintaining synchronization between the clock of the transmitted bits and the clock that the receiver uses to process the bits as received. This is called *bit synchronization*. It is a part, but not all, of the general synchronization problem for digital signals.

Looking back at the example of an A/D converter and parallel-to-serial converter, along with the corresponding circuitry at the receiving end, there was one point that we have ignored until now. The receiving end of the link needs to know where the first bit (the MSB) of the eight bits is in the overall time sequence. Even if the receiver clock is perfect, it still could be out of sync with the bits overall. The serial-to-parallel converter at the receiver might be offset by one bit to the right, so that the real MSB is in the second MSB position while the real LSB is in the MSB location. The results would be dramatic: Every bit would be correct in itself and received without error, but the overall number represented by the bits would be completely wrong. A 10000000 would be interpreted as 00000001, which is a very different value.

The grouping of eight bits constitutes a *frame*. The problem of ensuring that receiver knows which bit represents the MSB, which represents the second MSB, and so on to the LSB, is called *frame synchronization*. To establish frame sync, there must be some way of indicating where the frame starts. This is done by a special bit (or bits) called the *framing pattern*. The framing pattern is a unique code that the circuitry of the receiving is designed to recognize as indicating "start here." In simple systems that are sending small frames, the framing may be indicated by a simple 0,1 bit pair. For longer frames, a longer framing pattern of 4, 8, or even 16 bits is used.

Note that the framing pattern can also serve to help synchronize the received clock for the bit sync operation. For this reason, many systems that could use the PLL approach to recover the bit timing also make use of the frame pattern for bit sync. Since the framing bits are going to be transmitted anyway, they can be put to use for bit sync. Depending on the system design and situation, the receiver may actually be able to eliminate circuitry for deriving timing from the received bits themselves, and instead use the framing bit pattern for both frame sync and bit sync.

Review Questions

1. What is synchronization? Why is sync a necessary function?
2. Why is it usually insufficient to set both the transmitting and receiving clocks to the same nominal frequency for proper communication at higher data rates?
3. When is a clock sent as a separate signal? What are the pros and cons of this technique?
4. Why can the clock be derived from the received data bits themselves? What circuitry is used?
5. How do special bits provide sync? What is the drawback of this technique?
6. Compare frame and bit sync. Why is frame sync needed? How is it done?

11.4 DELTA MODULATION

The PCM system studied in this chapter requires several functional blocks at both the transmitter and the receiver to provide digital format communication for analog signals. There is another design approach that is often used in communication systems, which requires much less circuitry and is much simpler and lower in cost. This technique is called *delta* modulation.

The diagram of a delta modulation system (Figure 11.11) shows its simplicity. The analog input signal is compared to the last "estimate" of the signal value, and the *comparator* output indicates if this last estimate was too high or too low. A binary 1 (positive voltage) output indicates "too low, should be higher," while a

11.4 Delta Modulation

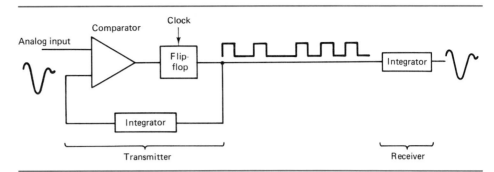

Figure 11.11 Basic delta modulation system passes analog input through a comparator whose output is clocked through a flip-flop for transmission and also fed back via an integrator to the comparator.

binary 0 (negative voltage) means "too high, should be lower." This revised estimate is then added in the *integrator* to the previous estimate to form a new estimate. The delta modulator combines ADC and clock function and inherently puts out a serial stream of bits, eliminating the need for a parallel-to-serial converter.

Let's look at the operation of the delta modulator in more detail. The clock for the modulation process drives a *flip-flop*, a basic digital circuit which simply passes the present comparator output state that it sees at its input to its output—the communications line—when the clock signal makes a transition from high to low. This simple digital function is responsible for taking the comparison function (which is continuously producing an output based on the difference between its two input signals) and making it into a spaced, clocked signal composed of a series of bits. The key waveforms of a delta modulator are shown in Figure 11.12.

The *integrator* of the delta modulator is an op amp with capacitor in a standard analog circuit, where the op amp output is equal to the integral of the input signal. As the comparator and flip-flop output have more 1s (positive value), this integral increases, while any 0s (negative value) will cause it to decrease. The output of the integrator is thus the integral over time of the comparator 1 or 0

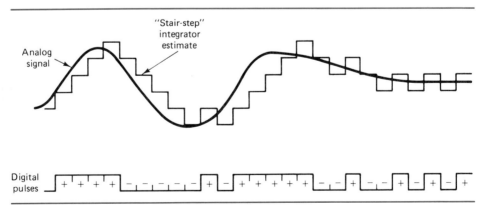

Figure 11.12 Delta modulation waveforms show how the 1,0 pattern causes the integrator output to track the analog as it varies, in a series of up and down stair-steps.

outputs, and looks like small stair steps that approximate the original input signal. Some of the steps are above the analog input, and some are below. The *closed-loop* scheme where the integrator output (best guess) is used to set the new comparison level for the comparator means that the overall circuit serves to correct itself automatically. When the integrator output is above the analog signal (estimate too high), the next comparator output is a 0, which corrects for the error in the estimate. The reverse is also true: If the integrator output is below the analog signal, the estimate is too low, so the next comparator output should be a 1 to increase the estimate. The delta modulator is a *tracking* ADC since it follows the contours of the input and indicates the changes in the input rather than reporting the exact value at any instant.

What the delta modulator does is produce a running string of bits that continually track the changes in the analog input signal. For reconverting the digital bit stream into the analog signal at the receiver, the delta modulator is even simpler. All that is needed is an integrator, identical to the one in the transmitter. This integrator sees the same digital sequence as the integrator at the transmitter and generates its estimate of the original analog signal based on the stream of 1s and 0s. A low-pass filter at the output of the integrator smooths the staircase of the steps into a continuous waveform. The only difference between the two integrators is that the one at the transmitter is used to drive a comparator and so develop a new estimate. Note that the receiver has no clock and no sync issues to deal with, which is a major advantage.

Unlike the ADC and DAC schemes, the delta modulator does not produce an equivalent complete digital representation of the sampled analog signal value at each conversion. You cannot look at the bits and say that the original signal was really 1.04 V, for example. What you can say is that the present value of the signal is larger or smaller than it was at the preceding clock period. Therefore, delta modulation is useful where the changes in the analog signal are important, but the specific value is not. This is typical of voice signals, since the necessary amplitude can be set by a simple volume adjustment. (*Note:* It is possible to add additional circuitry to the delta modulator and produce digital output words that correspond to the analog signal voltage. However, when this is done, the converter appears to the user just like any other conventional A/D converter and requires bit and frame sync.)

Delta Modulator Performance

Although the circuitry of the delta modulator is extremely simple, the choices of clock rate and step size are critical. If the clock rate is low (desirable for a low data rate and narrow bandwidth), the time between new estimates is large. In that time period, the input may change considerably, so many steps will be needed to correct the estimate (Figure 11.13a). A fairly large error between estimate of the signal value and its actual value may occur and exist for a significant time period. A higher clock rate means more frequent updating of the estimate, so errors are smaller, but the data clock rate is higher.

The step size is also critical. If the steps are large, the delta modulator will quickly be able to follow large, sudden changes in signal level (Figure 11.13b). However, for a fixed signal level the delta modulator will have a large quantization error since the step size will determine the resolution of the system. Here again, a technical compromise is necessary.

In summary, although the delta modulator can provide good performance

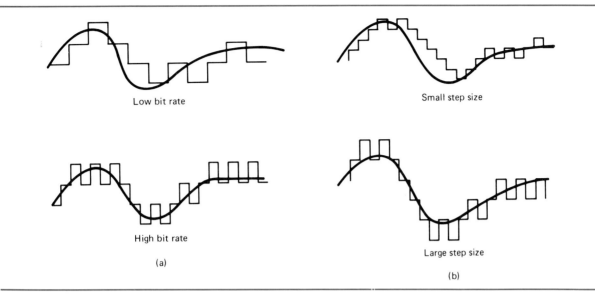

Figure 11.13 Delta modulation waveforms at integrator output for (a) low and high bit rates and (b) smaller and larger step sizes.

with very simple circuitry, it is very susceptible to *slope overload,* which occurs when the input signal changes rapidly and has a large slope (rate of signal amplitude change per unit of time). If there are small steps, it will take a long time to catch up to the actual signal value. If the steps are larger, the resolution will be poor. Higher clock rates can help, but these require more bandwidth.

CVSD

The simplicity of delta modulation makes it attractive for digital communication, where low cost and ease of use (no synchronization) are needed and the exact magnitude of the analog signal is not critical. One example is a telephone that transmits voice digitally, with the ADC function built right into the phone itself. The Motorola MC34115 (Figure 11.14) is an IC specifically designed for telecommunications delta modulation. It also incorporates an improvement algorithm called *continuously variable slope delta modulation* (*CVSD*) to overcome the slope overload problem.

Delta Modulation Bit Rate

Since delta modulation is so simple and effective, it may seem that it should be used nearly all the time in place of a convential ADC and DAC. The reason it is not is that the bit rate required for acceptably low distortion and quantization noise in delta modulation (and the corresponding bandwidth) is generally higher than the bit rate of the conventional ADC/DAC pair, by a factor of 2 to 5. When delta modulation bit rates are too low for the application, the quantization noise and slope overload cause distortion that makes the signal unpleasant to listen to and even difficult to understand. Techniques such as CVSD improve the overall performance of the delta modulation system in some ways, but there are still distortion and general fidelity shortcomings. However, for applications where the bandwidth is available—such as a single digitized voice phone connected to a

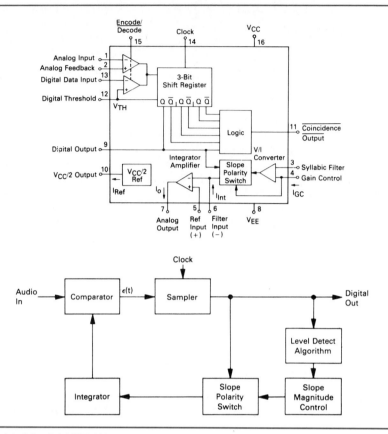

Figure 11.14 (a) MC34115 delta modulation IC pinout and design. (b) The functional block diagram shows basic elements within the IC (courtesy of Motorola, Inc.).

phone company central office—delta modulation is a very useful approach and is often used. The absence of clock and sync makes it very attractive in these applications.

Review Questions

1. What is delta modulation? What is the overall concept? Why is delta modulation used in some applications?

2. How does the delta modulator track the analog input signal? Compare this tracking to the conventional ADC in terms of absolute signal value and the meaning of the resultant bits.

3. What does a delta modulation receiver (DAC) look like? How does it operate? What clock is needed? What synch? Why is a low-pass filter used at the receiver integrator output?

4. How does the clock rate affect the ability of the delta modulator to track the changes in input?

5. What is the effect of larger and smaller step sizes? What is the trade-off? What is quantization error caused by in delta modulation?

Continuously Variable Slope Delta Modulation

The CVSD modulation algorithm is an *adaptive* scheme, where the actual signal level and modulator 1,0 pattern are monitored and used dynamically to adjust a key parameter in the modulator itself. For CVSD, this parameter is the gain of the integrator, which is determined by both the amplification factor of the integrator and/or the amplitude of the pulses being integrated.

The IC looks at the stream of pulses and, by looking at the sequence of pulses, automatically decides if the output is lagging the input by too much. Ideally, when the delta modulator is tracking the signal closely, the bit stream will be alternating 1s and 0s. When there is a longer run of 1s, the modulator and its integrator are trying to run up to catch the input signal; when there is a string of 0s, they are trying to run down to catch the signal. In either case, the delta modulator estimate would catch up more quickly if the pulses it was integrating were larger.

This is the concept of CVSD: to adjust the step size (by varying the pulse amplitude) in accordance with how well the modulator is doing at tracking the input (Figure 11.15). When a large change in input occurs, the step sizes increase to accelerate the integrator output more quickly to a better estimate. As the estimate gets closer to the correct value, the step sizes are reduced. The specific algorithm used by the MC34115 is simple: A run of three 1s (or three 0s) means that the step size should be increased, since the modulator is trying to catch up to the input by going in the same direction for several bits in a row. When the modulator output bit stream does not have three identical bits in a row, it indicates that the estimated output from the integrator is hovering around the correct value, so the gain is at a correct level. If three identical bits in a row occur after the step size is increased, the step size is increased yet again, until there are no "three in a row" bit runs.

For the receiving end, the same IC is used. Recall that one of the virtues of delta modulation is that the receiving circuitry is a subset of the transmitter circuitry. The MC34115 can be used for converting a delta modulation signal back to analog via a single control line that sets some internal switches and circuitry to transmitting or receiving mode. The same CVSD algorithm is used by converting the pulse stream back to analog: When three pulses in a row have the same binary sense, the integrator gain is increased.

The effect of the CVSD algorithm is similar to companding. The normally straight-line relationship between input level and output representation is deliberately distorted so that larger signals are handled more effectively. Unlike companding, however, which is a fixed algorithm that is independent of time or the signal's past values, the CVSD algorithm depends on the past and adapts to it.

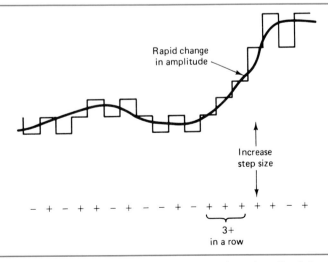

Figure 11.15 The CVSD concept increases step size when the integrator output lags the actual signal for too long, as seen by a run of three 1s or 0s in a row at the flip-flop output.

> 6. How does CVSD compensate partially for slope overload? Explain why it is an adaptive scheme.

11.5 TROUBLESHOOTING

Troubleshooting A/D and D/A converters requires examining two aspects: Is the device functioning, and is it functioning properly? Basic checks begin by measuring with a digital multimeter that the proper power supply voltages are available at the converter, usually some combination of +5, −5, +12, −12, +15, and −15 V. The ground(s) must also be checked for dc continuity.

Next, the presence of the correct clocking signals is verified. The converter clock should be observed and measured on an oscilloscope. For an A/D converter, the *convert* signal that initiates conversion should be judged on the oscilloscope. The *data ready* or *conversion complete* signal from the A/D converter which indicates to the system that it is done should be examined on the second channel of the scope. The time between these two signals should be estimated from the scope screen and should agree with the specified conversion time for this type of A/D converter. Typical conversion times range from 10 μs to 1 ms, although much faster and slower ones are used as well.

If the A/D converter seems to be converting, the next step is to make sure that it is capable of presenting its digitized result to the rest of the system. The digital lines between the converter and the system (usually a microprocessor bus) should be checked for proper activity. The pattern of signals in most systems usually begins with the A/D converter signaling on the bus that it has a new *conversion complete* which appears to the processor as an *interrupt,* and the bus typically responds within 10 to 100 ns with a control signal telling the converter to put these results in the bus. If this sequence does not occur, the converter may not be generating the conversion complete, or the microprocessor bus may not be recognizing this interrupt and responding due to bus, microprocessor, or microprocessor software problems.

If all this seems to work yet the A/D converter gives incorrect results, the problem may be between the point where the analog signal first enters the system and the various amplifiers between this point and the input pin of the converter, or amplifiers within the converter. This is checked by putting a known signal on the analog input of the system and tracking it through to the A/D converter input. If the value at the A/D input is correct but the converted results are wrong, the problem is probably within the analog circuitry of the converter.

It may also be that one of the many digital data lines between the converter and the bus is defective and unable to *drive* the bus properly with the 1 or 0 voltage values, so these must be measured as well. A standard voltmeter is needed for checking the analog side of the system, but a logic analyzer is very useful for the digital side, since it allows all data bus lines to be examined simultaneously.

Digital-to-analog converters are simpler to check than A/D converters. Besides checking for correct power and ground, the quality of the *clock* signal, if any, should be checked with the scope. The interface between the system bus and the D/A converter should be checked to make sure that the proper data are being presented to the converter. Unlike most A/D converters, there is usually no signal back from the D/A converter signaling that conversion is complete, so it must be assumed that data loaded into the D/A converter are actually being acted upon.

The analog output of the D/A converter should be measured with a voltmeter. If it is incorrect, one of three causes is likely: The data bits did not reach the converter; the converter has a problem with its internal switches and amplifier, so that it cannot produce the correct value; or any amplifiers external to the converter output are faulty. Be sure to check the analog output right at the converter, before any buffer amplifiers, to help narrow down the problem.

Finally, note that most converters either pass their results to a processor bus or accept data from the bus. Without access to this bus, the test technician will have a difficult time to check the bus data and compare it with the desired results and will not be able to generate special data patterns to produce desired analog outputs from the D/A converter (a logic analyzer can observe data but cannot write new data). Most systems have a test mode that allows the technician to write to the converters directly with known patterns and then check if the actual result is the correct one.

SUMMARY

A practical system for transmitting analog information in digital format can be built simply using an ADC and a DAC, with antialiasing filters. This system would require one signal path for each bit and be useful only for short distances. Parallel-to-serial and serial-to-parallel converters reduce the number of signal lines to one but increase the system bit rate by the number of bits in a conversion. Compression and expansion of the analog signal is often used to reduce the dynamic range required, and consequently, reduce the number of bits/sample that must be sent.

The receiver and its DAC have to know the correct timing of the clock used at the transmitter to make sure that they are properly synchronized. Otherwise, errors would occur even in perfectly received data bits as the receiver clock checks the incoming bit stream a little too often or not often enough. Bit timing can be transmitted on its own signal line, derived from the bit stream by a PLL, or derived from special sync bits inserted into the data stream. Besides bit timing, the receiver must establish frame sync to indicate the beginning of the digital word. This is usually done with special frame sync bits sent along with the data. Synchronization is a major issue in digital communications and often determines the limits of achievable performance.

An alternative to the traditional ADC and DAC design for PCM (with supporting circuitry) is delta modulation. In delta modulation, very simple circuitry, consisting of a clock, a comparator, and an integrator, compares the last estimate of signal value to the signal's present value. The bit stream indicates if the estimate at any time is too high or too low. The receiver for delta modulation is very simple, with just an integrator identical to the one at the transmitter. Delta modulation has the advantage of requiring no sync or timing at the receiver, but it is somewhat noisy and prone to distortion when the analog signal slews rapidly. An adaptive form of delta modulation reduces this distortion but increases complexity. Delta modulation is often used for low-cost digitization of voice signals, such as providing digital voice from a telephone unit to a phone central office.

Summary Questions

1. Compare and contrast ADC versus DAC operation in speed and control signals.
2. Compare the speed, number of signal lines, and other technical considerations in serial versus parallel ADC-to-DAC connections.

3. What is the role of the clock in a parallel system and in a serial system?

4. Why is a low-pass filter almost always essential ahead of an ADC?

5. Discuss the construction of a shift register from flip-flops. Explain the operation of a shift register as a parallel-to-serial and serial-to-parallel bit format converter.

6. Why is clock sync important? What happens when the receiver clock is slow or fast compared to the transmitter clock?

7. What is the function of a codec? Why can a single circuit or IC often be used for other mode?

8. Why is companding useful? What are its drawbacks?

9. Why do digital systems almost always require sync?

10. Compare and contrast frame versus bit sync.

11. What are three approaches to maintaining sync? What are the characteristics of each method?

12. What is the principle of delta modulation? Why is simple circuitry sufficient?

13. How does the meaning of the bits in delta modulation differ fundamentally from their meaning with a convential ADC?

14. Why does the delta modulation receiver require no clock?

15. What does the delta modulation clock rate determine in terms of performance and required bandwidth?

16. What happens to signal fidelity as the step size or clock rate changes? Why not simply increase the clock rate and decrease the step size?

17. What are the first things checked when testing converters?

18. How is the analog side of an A/D converter checked? What about the digital side? What about for the D/A converter?

19. What can cause an A/D to appear to convert but still give incorrect results? What causes the same symptoms in a D/A converter?

20. What can a logic analyzer do in testing converters? Why are special test programs often for complete operation verification?

PRACTICE PROBLEMS

Section 11.2

1. An eight-bit system has some propagation differences between bits, so that bits 7, 5, 3, and 1 are combined for an instant with bits 6, 4, 2, and 0 of the previous byte (bit 0 is LSB). What is the DAC input during the transition when the existing byte is 01010101 and the new byte is 1000000? What is the apparent DAC output if full scale is 5.12 V? What should the correct new output voltage be?

2. Repeat problem 1 if the four MSBs of the first byte combine with the four LSBs of the new byte; also, for the reverse situation where the four LSBs of the first byte combine with the four MSBs of the new byte.

3. After how much time will two clocks that differ by 0.3% have a 5-μs difference? How many bits is this if the clock and bit rate is 100 kHz?

4. A system requires that the clock timing be accurate to $\frac{1}{10}$ the bit period to ensure that the bit value is checked near its center, not near its edges or transitions. The system clocks differ by 1%, and the bit rate is 250 kHz. After what elapsed period of time is the error greater than $\frac{1}{10}$ bit period? How many bits can be transmitted under these conditions?

5. For $\mu = 100$ and $V_{max} = 10$ V, what are the companded output values for every 1-V input increment from 0 to 10 V?

6. Repeat problem 5 with $\mu = 255$. Compare the change in output voltage as the input changes from 0 to 1 V and 9 to 10 V for both values of μ.

Section 11.3

1. What is the Doppler shift for a space satellite moving at 25,000 mi/h for a nominal frequency of 2450 MHz? What percentage of the nominal frequency is this? (Use the Doppler shift numbers given in this section.)

2. A PLL can track frequency changes up to 1% of the nominal carrier frequency. Up to what transmitter velocity can the PLL be used for the nominal 2450-MHz transmitter frequency?

3. Some low-performance systems are one sync bit for every eight data bits. What percentage of the bits carry no information? Compare to a system with four sync bits for every 64 data bits.

4. A system is designed to send 90% of its bits as information. Data are sent in clusters of 256 bits. Up to how many sync framing bits can be used?

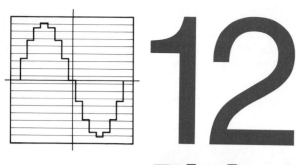

12
Digital Communication Systems

CHAPTER OBJECTIVES

When you have completed this chapter, you will understand:

- The need and function of key elements of a digital communication system, including coding, format, and protocol
- The types of electrical signals used on the physical interface to represent digital signals
- The rate at which data can be effectively transferred and how this overall rate is affected by coding, format, and protocol in both synchronous and asynchronous systems
- The use of state transition diagrams to map out the possible states of the communication system along with the events that shift the system from one state to another
- The use of special bits, algorithms, and circuitry for both detecting errors, as well as detecting and correcting them

INTRODUCTION

A complete system for communication of signals in digital form has many aspects besides simply sending signals that correspond to the binary 1s and 0s. Multiple levels of coding, formating, and protocol precede the actual connection of the data bit signal to the physical channel at the transmitter; the receiver reverses these operations to recover the original information. The advantages of this apparently unneeded complexity are that the system becomes reliable, flexible, and can automatically handle many problems without operator intervention. These problems include smooth startup, recovery from equipment or power failure, and the detection (and even correction) of errors in individual bits.

ICs are available that implement many of the levels of the system with relatively little effort by the user. The system performance is judged by the bit error rate versus SNR and the effective throughput of information in bit/s or characters/s.

12.1 COMPLEXITY OF DIGITAL COMMUNICATIONS

A primitive digital communications system simply sends a voltage or current to represent the data bits. If the analog voltage value, in binary, is 10010011, a +5-V signal might be sent for binary 1 and a 0-V signal might represent binary 0. Data bits are sent one after the other, in serial format. This sort of simple system provides only a low level of performance in a real system. Practical digital systems perform many complex operations on the bit pattern before the signal is transmitted and the receiver performs the opposite, complementary operations to recover the original information.

The reasons for this added complexity fall into two main categories. First, a complete communications system must take into its design all the possible operating modes and conditions. Issues such as how the system begins communication, detects and deals with errors, and recovers after a power, channel, or link failure must be accommodated in the design. Second, digital communication inherently provides the potential for these additional operations and enhancements of the data, which analog systems do not. It is very difficult to process analog information and transform it into other formats, without losing signal accuracy, fidelity, or meaning. In contrast, the digital signal can be put through a series of "recipe" steps to make it more usable in a communications system, and then can easily be recovered by reversing the recipe steps.

All communications systems require agreement between the sender and the receiver on many aspects of the communications and signal. If the two ends do not agree, or at least have some set of rules for resolving disagreements, the information that is transmitted will be meaningless to the receiver—even if the message is received perfectly. This applies to analog as well as digital systems. In digital systems, there are many more things that can be changed and agreed upon, and that contribute to better system performance. In contrast, an analog system has relatively few things that can be changed, beyond adjusting the signal level, filtering, and selecting the modulation parameters.

Actually, the concept of agreement applies to nonelectronic communications. Consider writing a letter to someone in another country. This simple activity actually implies many levels of agreement. Both the letter writer and the letter reader must use the same alphabet as a first step. They must both use the same language, too: A letter written in English uses essentially the same character set as one done in Spanish, but clearly the languages must agree (or else a translation scheme must be used). The way that symbols are used must also be the same: In the United States, a decimal point is used in numbers (6.32), whereas in Europe the custom is to use a comma (6,32). The symbols (punctuation) used for indicating the beginning and end of the message sentence must agree. The overall form and addressing the envelope that carries the letter, regardless of its contents, should also be the same. For example, the house number can precede the street name (100 Smith Avenue, in the United States) or follow it (Smith Avenue 100, in Europe). Finally, the words themselves used must be understood with the same meaning, even if all other points agree (American "elevator" versus British "lift").

Communications is a *multiple-level* or *multiple-layer* activity (Figure 12.1). In communications between people, there are rules to follow to ensure that each person has a chance to speak, to interrupt, to finish. This is called the *protocol* of communications. Communications systems also have protocols that specifically define how the communications is to start, finish, recover from problems (such as due to noise or equipment failure), for the receiver to indicate if a message was received properly and without error, and to define what to do if an error is detected. The next-to-highest level is the *coding,* which defines how the initial data to be sent and transformed into the symbols that represent it, before being actually transmitted with these specific signal values. One level below is the *format*, which is responsible for adding additional information about the message, such as who it is for, how long it is, and where it ends. Format also provides framing and additional information that helps the receiver determine if the message, as received, contains any errors. At the lowest level are the specific voltages (or currents or frequencies) used in modulation to represent the digital information.

Transparency

All these layers add apparent complexity to the communications system. At the same time, they allow the construction of systems that can operate reliably, nearly free of error, and automatically recover from problems (not requiring a person to intervene to restart operation). Although the layers add more complexity, they are designed to be *transparent* to each other, which is an important concept in data communications.

What transparency does is to make the action and processing within each layer—the algorithm of each layer—independent of the activities of the layer above it and below it (Figure 12.2). Each layer simply has a set of rules to follow, taking the data bits it receives from the layer above (at the sender), processing these bits following a set of rules, and then passing the results to the layer below. At the receiver, the reverse operation is performed: The data are passed from a

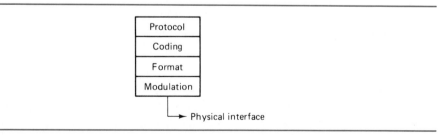

Figure 12.1 Multiple layers of communications system show the structure from the physical interface to the channel up to the part of the system that formulates messages and handles the rules of "conversation."

12.1 Complexity of Digital Communications

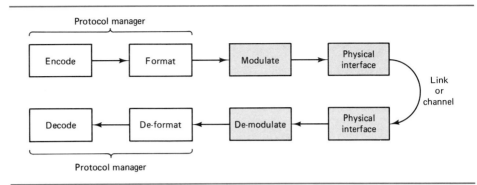

Figure 12.2 Transparency between layers; each layer processes what it receives according to preset rules, without trying to understand what the bits actually mean.

lower layer to a higher one. The same transparent concept applies, and each layer follows the rules. The operation at each layer does not have to study or understand the operation that was performed at the higher and lower layers. Any change in rules at one layer of the sender is transparent to the next layer, and the overall system will still function as long as the same rule change is implemented at

The Significance of Transparency

Transparency is important because it allows the overall complexity of the system to be broken down into individual functions—*layers*—that operate essentially by themselves. A change in the format implemented by the sender, such as using a colon (:) to separate phrases instead of a slash (/) does not affect the coding function. Instead, the coding just operates on the results that the new format procedure provides. Data are transformed within each layer and passed from layer to layer without having to worry about its impact on the overall system. In addition, transparency means that the operation at each layer does not have to be concerned with the actual meaning of the bits themselves. The layer must only follow some predetermined rules that specify how to operate on bits it receives. It does not have to worry that some of the bits may represent information one way or another way. For example, the protocol does not care if the data bits represent information that originally was an analog voltage from 0 to 10.0 V, or a series of bits from a keyboard. The meaning of the data is irrelevant to the layer, and data pass through the layer transparently.

Once again, an analogy with letter writing shows the effect and impact of transparency between layers. To be delivered properly, a letter written to an English-speaking friend in France needs only to be addressed in the style that the French postal system can accept. The postal employees do not need to understand the contents of the letter, because the contents are handled transparently by the system, which looks only at the address. The same letter could be sent to a friend in another country simply by changing the addressing (format) or could be sent by the French postal system by a different route (modulation) without any changes in the body of the letter itself.

Not all systems use transparency and multiple layers, since there is a cost in additional system hardware, software, complexity, and potential for problems when each message requires so much processing. Communications systems that are *dedicated* to one specific, unchanging, very well defined and characterized application may use fewer layers or combine the layers together and not have transparency between layers. These systems are less common in industry, because they lack flexibility; any desired change often requires redoing large parts of the system since small changes quickly affect large portions of the system. In contrast, a layered system with transparency allows changes or improvements to be implemented with minimal confusion or problems. If there is a problem, it can be isolated to the appropriate layer.

the corresponding layer in the receiver. (There is additional discussion of layers in Chapter 18.)

Review Questions

1. What are the two reasons that digital communications systems have more complexity than simply sending 1s and 0s corresponding to the binary information? Why don't analog systems have this complexity?
2. Define and compare modulation, coding, format, and protocol. Why are they called layers?
3. What is transparency? Why is it useful? What is the drawback to multiple layers and transparency?
4. What is the impact on the system when the "rules" of one layer are changed in a transparent system?

12.2 CODING

The first step in preparing data for communication with a digital system is *coding*. Coding takes the data bits from their source and converts them into a standardized form. Both sender and receiver must agree on what specific digital patterns will represent the information to be transmitted. Many different codes are commonly used, and each can provide advantages in specific applications. We will concentrate on the most common code but also look at some alternatives. To make the examples realistic, we use two types of information: the numerical results of an A/D conversion (from 0.00 to 2.55 V), and the data that come from an inherently digital source such as a keyboard of a computer—any letter, symbol, or punctuation mark).

The code most often used to represent information is called *ASCII* (pronounced "ask-key"), an abbreviation for "American Standard Code for Information Interchange." The ASCII code of Appendix C assigns a binary field of 7 bits to represent each character, so a total of $2^7 = 128$ unique characters can be represented. Some systems use an expanded version of the ASCII code with 8 bits (256 characters). To represent a typed word, such as HELLO, the ASCII patterns for each letter are used (the most significant bit is on the left):

```
H    1 0 0 1 0 0 0
E    1 0 0 0 1 0 1
L    1 0 0 1 1 0 0
L    1 0 0 1 1 0 0
O    1 0 0 1 1 1 1
```

so the overall bit pattern (shown with spaces for clarity) is

12.2 Coding

H	E	L	L	O
1001000	1000101	1001100	1001100	1001111

Note that in the convention of ASCII format, the least significant bit of each character is actually transmitted first. The bit pattern observed on the communications channel is

| 0001001 | 1010001 | 0011001 | 0011001 | 1111001 |

True spaces between words or numbers in the message (not those added for clarity in printed text) are represented by the ASCII code for the space character (0100000). There is even a code called *null* (all 0s) which, by definition, causes no action but only "kills time"; it is needed for test purposes and sometimes to allow the receiving system to catch up with the transmitted information. ASCII codes can be used for numbers or for mixed number/letter combinations: The message 25F in ASCII is represented by the three symbols for 2, 5, and F.

Nonprinting ASCII Characters

The ASCII codes for letters, numbers, and punctuation symbols use more than half of the 128 character code set and are called *printable* codes. This is because they will actually print out on a printer or be visible on a computer terminal screen. The remainder of the code set are called *nonprinting* codes, or *control* codes. These are ASCII symbols that cause specific actions but do not print a tangible character on the screen or paper. The nonprinting codes indicate things like line feed, form feed (feeding a new sheet of paper), shifting, or that the end-of-line or warning bell should be sounded. They are used to control the activities of the equipment and are very important to a properly operating system.

There is one problem with nonprinting codes: how to see them when troubleshooting, since only the printing codes produce a character on the terminal. A standard terminal will show nothing when it receives a nonprinting ASCII code, although it may take some action, such as returning to the beginning of a line when it sees "carriage return." For this problem, special troubleshooting terminals are used which actually display unique symbols for each nonprinting character. A tiny "CR" will be printed when carriage return is received, and a tiny "LF" is printed for line feed.

Numerical Information

When the data are entirely numerical, it is not necessary to use ASCII code. The output of the ADC can be used directly as the encoded bit pattern. The eight-bit ADC output for half-scale (1.27 V) is sent as 01111111, for example, while full scale of 2.55 V is 11111111. Numbers can still be transmitted in ASCII code, either by sending the ASCII value of each bit (0110000 for 0, 0110001 for 1) or more commonly, the ASCII value for each number in the decimal equivalent. The advantage of using the binary value directly is that fewer bits are required to send a number compared to ASCII, as seen in the following example.

Example 12.1

An eight-bit converter output is transmitted as the direct digital output and also as the ASCII equivalent of the output decimal value. What is the bit pattern for half-scale with both of the methods? How many bits does it take in each? What percentage of bits does the more efficient scheme take compared to the less efficient one?

Solution

As an eight-bit value from the ADC, half-scale is this eight-bit pattern: 01111111. As an ASCII value for 1.27-V half-scale value, there are four symbols: one each for the three digits and one for the decimal point:

1	.	2	7
0110001	0101110	0110010	0110111

This is a total of 28 bits, compared to 8 for the direct binary approach. The binary approach requires only 8/28 = 28.5% as many bits as the ASCII code.

For this reason, some systems that transmit numerical data use the binary code directly. However, this efficiency has a drawback. The system cannot send any nonnumerical messages, since they would be interpreted as numbers. This makes it very difficult to set up, test, and maintain the communications system. There are many times when letters, punctuation, and words must be sent as a part of the overall message or group of messages, although the bulk of the data is numerical.

Another problem with using the binary value directly is that it makes it difficult to change the number format. Suppose that the ADC resolution is increased to 10 bits from the previous eight. Instead of an eight-bit group for the ADC output, the system must now handle 10. This is not a problem when there is a single, isolated value to transmit. However, in most cases, the system is transmitting a series of numbers. The receiver needs to know that the eight-bit result has been changed to 10 bits. All the receiver sees is a long series of bits, and it must have some way of knowing where to separate the series of bits into 10-bit groups. Any boundary markers between groups, called *delimiters,* would be confused with bits representing numbers.

In contrast, the ASCII scheme is more transparent to the resolution of the converter, since the ASCII code works on the decimal value associated with the ADC output. Even when the increased ADC resolution changes the decimal resolution and adds an extra decimal place after the "0.01" column, the ASCII code simply converts this new number. The receiver now sees five characters, with four representing digits and one for the decimal point. The ASCII coding on a "one code per digit" basis is much easier to adapt to changes. For added clarity, the letter V for volts can be added after each number as a boundary delimiter since the V cannot be confused with a numerical value. Although this costs an extra character, it does make interpreting the series of bits much more convenient.

Communications systems usually do not use the direct binary value. Instead, they use the ASCII, or similar code, to transmit numerical information and accept the fact that it takes more bits to send the number as the trade-off for a system that can also send letters, punctuation, and nonprinting control codes if necessary. The main application for the binary value approach is where data rates and effi-

Non-ASCII Codes

Besides ASCII, several other codes are in common use. The *EBCDIC* (Extended Binary Code for Data Interchange) is an IBM-originated code pattern that uses eight bits to represent each character. It is similar in concept to ASCII, although the specific bit patterns are different. Integrated circuits are available that translate ASCII to EBCDIC (or the reverse) so that two otherwise incompatible computers or terminals can be connected to each other.

For information that is primarily numerical, a modified four-bit binary code set is sometimes used. Binary 0000 to 1010 represent digits 0 to 9, and the unused codes 1011 to 1111 can stand for symbols such as decimal points and delimiters. The value 1.27 would be

$$0001 \quad 1011 \quad 0010 \quad 0111$$

where 1011 represents the decimal point. (Compare the number of bits in this code to the number of bits required in seven-bit ASCII code.)

Other codes are used in different applications. For transmitting special symbols not in the ASCII code set, such as integrals ($\int$), some substitutions can be used to replace the standard symbols. Of course, both sender and receiver must agree, or else what the transmitter intends to mean $\int$ will be interpreted differently. In systems where only alphabet letters along with a few punctuation and control symbols are needed, a five-bit code called *Baudot* is available. The advantage of Baudot coding is that only five bits are needed per character, which saves transmission time.

Ironically, Baudot code was the first digital (on/off) code to be compatible with modern electronics, although it was invented in 1874 for early electromechanical teletype machines. The original telegraph *Morse code* uses a series of dashes (a long "on" signal), dots (a shorter "on" signal), and spaces (shorter "off" signal) to represent each character. Each of the two code elements—the dot and the dash—has different lengths, with a dash approximately three times longer than a dot. To complicate things, some characters are represented by a single element, while others need up to four elements (E is a single dot, while O is three dashes ---). The differing length of elements and varying number of elements per character in Morse code are identifiable by the ear and brain, but it is relatively difficult to build circuitry to interpret this coding.

Electromechanical machines and digital electronics, in contrast, work much better with codes where each element has exactly the same time period and number of elements per character (which the ear does not like!). Baudot code uses five bits of identical length to represent $2^5 = 32$ characters directly (and this can be extended with a special *shift* character to represent an additional 32 characters). Digital systems are much better at recognizing characters when they are represented by a series of bits of equal length, spacing, and number.

ciency are critical, such as in space satellites that are sending data back at low bit rates, or where the nature of the data is fixed in advance and not likely to change, such as digitized voice sent by the telephone company over long-distance links.

Most standard computer terminals are designed to accept ASCII character representation. The terminal receives the seven-bit pattern and displays the equivalent character on the screen. The conversion between the ASCII data and the pattern of screen dots needed to draw it is made by circuitry within the terminal and is completely transparent to the communications system. For data originating at the terminal, the terminal translates a key stroke by the user's finger into the ASCII pattern that represents that character, and then sends that pattern out.

Review Questions

1. What is the ASCII code? How does it represent a character? How many characters can be represented?

2. What is the difference between seven- and eight-bit ASCII? What are the nonprinting (control) codes? Where are they needed?

3. How are the bits of an ASCII character actually sent?
4. What is the efficiency of ASCII versus straight binary code for transmitting numbers? What are the drawbacks of straight binary?
5. What is a delimiter? What role do they serve?
6. What are some non-ASCII code variations? What purpose do they serve?

12.3 FORMAT

The *format* of the communications message defines what additional information the message needs in order to be effectively transmitted and understood. As in most other aspects of communications systems, formats range from very simple to fairly complex. Each format meets the needs of different applications, and there is no single format that is best for all situations.

The simplest format uses the characters themselves, with little else. This format is used between a keyboard and a computer terminal, for example, where the rate at which characters are generated is relatively low. A message consists of the ASCII code for each character, and a line on the screen is terminated by the ASCII pattern for carriage return (referred to as CR or Return), so the message HELLO has five ASCII character patterns, followed by a sixth for the terminating carriage return. The keyboard transmits the message to the terminal one character at a time, on a *character-by-character* basis, with each character transferred as soon as it is typed. Other systems wait until the entire message line has been entered, followed by the terminating return, and then send the entire line of

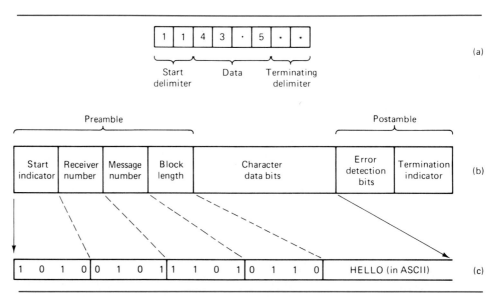

Figure 12.3 (a) Simple format for temperature reading; (b) more complex format with preamble, postamble, and their details; (c) bit patterns for the example in the text.

characters (called *line-by-line* transmission). This scheme is more efficient in use of the communications channel and other components of the system.

A simple format to transmit a temperature reading from a sensing device to a computer is to use delimiters to mark the beginning and end of the message. Each temperature reading begins with a starting ASCII character, or characters, that cannot be confused with the temperature, such as two slashes (//). The temperature value is transmitted as two digits, decimal point, another digit, and *terminating* characters (**). A typical reading is //43.5** (Figure 12.3a).

The drawback to this extremely simple format is that it does not handle large amounts of data well, nor does it provide for additional information that is needed in a system for best overall performance. A more advanced format adds a *preamble field* (a field is a related group of bits in an assigned position) or *header* of bits (information at the beginning) and a *postamble* field (at the end of the message) (Figure 12.3b). The preamble and postamble fields can be divided into other, smaller fields that contain specific information about the message.

The fields of the preamble typically indicate information such as the number of characters in the message (the *block length*), the message number, and the specific number (address or ID) of the intended receiver, together with some type of *start of message* indicator. The preamble fields can use ASCII format or any other that is appropriate. Each field in the preamble is fixed at a specific number of bits in length, and the receiver is designed to interpret the fields and their contents properly.

A typical preamble has a start pattern consisting of four bits in a 1010 pattern. After this is a four-bit message number field, so message numbers from 0000 to 1111 can be used (0 through 15 in decimal). Message numbers let the receiver know the proper sequence of messages in case one does not get transmitted completely. They also let the receiver indicate to the transmitter which specific message was received with some sort of error due to noise, timing problems, or any other cause. (The methods used to detect errors are discussed in Section 12.7.) The transmitter then follows system protocol rules for handling messages that it determines were not received properly.

The receiver number in the format is not normally needed when a single transmitter is sending to a single receiver. However, more complex *networks*, which have many receivers and transmitters on the same wire or frequency, need to have some method to identify the intended receiver unit. A five-bit field allows up to 32 receiver IDs, while eight bits can accommodate 256 receivers. Often, several receiver numbers are reserved for test messages and cannot be assigned to a real receiver.

The field in the preamble which indicates the block length—the number of actual message characters to follow—is essential to some systems and not used in others. *Variable-block-length* systems allow the sender to transmit a block of characters with length ranging from 1 up to a large maximum value of characters in the data block. This maximum number of characters is determined by the number of bits in the block length field and is equal to 2^n, where n is the field size. Other systems are designed for a *fixed block length,* so they do not need a message block length indicator.

The advantage of fixed block length is that it requires less circuitry and processing on both ends of the system, and is usually easier to handle in the receiver since the receiver does not have to count characters and compare the count to the specified block length. However, the fixed length can be inefficient if the data source generates new characters slowly, and therefore the block fills up slowly. The block cannot be sent until it is filled. As a result, the first characters will not be transmitted until all the characters have been generated, which may take a significant amount of time and make the first character arrive too late to be useful.

Example 12.2

The preamble for the sixth message, HELLO, for receiver 13, using four-bit fields in the preamble (Figure 12.3c):

Start field	Message number	Receiver	Block count	Message (in ASCII)
1010	0110	1101	0101	HELLO

The postamble is the closing of the message. The receiver knows that this is the postamble because it is either designed for a fixed block length or has been counting arrived characters. The postamble contains a field of bits that are part of the error detection scheme (discussed later in this chapter), together with some sort of terminating pattern such as 0101, in this case the opposite of the start pattern. The terminator can also consist of a particular string of ASCII characters that will normally never appear in the message, such as three colons in a row (: : :). The terminating pattern is not strictly needed, since the receiver knows where the end should be, but it is very useful for test purposes. It allows the technician to observe and note the end of the block of transmitted characters, without having to know what the fixed block length is or interpreting the block length field of the preamble and then counting the number of characters. The entire message, with preamble and postamble, is referred to as a *frame*. The start field and the terminating field are often used by the receiver circuitry as framing bits to aid in synchronization (Chapter 11).

Once the initial information has been encoded, either into ASCII or some other code, and the proper format information placed in with the character bits, the message is ready to be actually placed on the signal channel link. This is the function of the physical interface level, discussed in the next section.

Review Questions

1. What is the role of the format? Give an example of a simple format and where it is used.
2. What are the limitations of a simple format? What are the parts of a more complex format?
3. What information is contained in the preamble and postamble fields?
4. Compare features of the fixed-block-length and variable-block-length format. What is a frame?
5. Why are the preamble and postamble not always needed?

T-1 Format

A relatively simple format used in communications is called the *T-1 format* (Figure 12.4). It was developed by the Bell Telephone System to handle the data that result from digitization of voice signals to seven bits of resolution (companding is used to extend the dynamic range) with sampling at 8000 samples/s. This format is very simple, since it is dedicated and optimized for one specific purpose. The T-1 format consists of the seven-bit digital field (corresponding to the analog value of the voice) plus an eighth bit which indicates the users' telephone line status to the receiving office. The 7 + 1 information bits from 24 separate users are sent in sequence, for a total of 192 bits. Each group of 192 bits is preceded by a single framing bit, used for synchronization, for a total of 193 bits per complete T-1 frame. Since the voice under the T-1 format is sampled 8000 times per second, the overall bit rate is 8000 × 193 bits = 1.544 Mbits/s.

Although it lacks the preamble and postamble fields, this is not a major problem. In practical systems, a single bit error in a digital voice value will sound like a small click on the phone, which is not considered a major problem. This format is not intended for general use but is optimized to transmit a high percentage of data bits with minimum "waste" due to preamble and postamble. Note that (193 bits − 1 frame bit − 24 line status bits) = 168 bits out of the 193 bits represent actual voice information.

12.4 PHYSICAL INTERFACE AND THROUGHPUT

Regardless of the number of levels in the communications system or the processing performed at each level, the signal must be electrically connected to the link of the system. This connection to the wire, optical fiber, or antenna is called the *physical* layer. It is the most tangible layer of the system, since the signal may be easily observed and measured at this point. It is also very critical to overall system performance, since the signal quality must be maintained and even enhanced so that the receiver will be able to recover the information bits without error. Unlike the activities of the higher layers, which are completely internal to the system and consist of the manipulation and transformation of data bits into other data bits by hardware circuitry and software, this layer is the interface between the internal system and external channel.

The circuitry to launch the transmitted signal into the link, or to retrieve the signal from the link at the receiver, is of course matched to the type of link itself

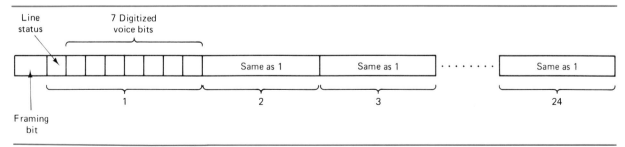

Figure 12.4 The Telco T-1 frame format is relatively simple, with 24 voice signals each digitized to seven bits plus a line status bit and a single framing bit.

by this lowest layer of the communications system. To match the signal and the link properly, this layer performs the modulation that makes the signal compatible with the link itself. Higher layers are isolated from directly seeing the physical link by the action of this lowest layer. This maintains the transparency and means that higher levels do not have to be concerned with the actual physical path used.

In fact, in some advanced systems the actual path is continuously changing, yet the higher layers are unaffected. For example, the telephone system uses wire, fiber optic, and satellite link for long-distance communication. A phone call from New York to Los Angeles may be traveling exclusively by wire, while phone company equipment is monitoring the performance of all circuits. If the wire circuit has excessive noise problems or a signal loss, the computers that manage the phone system links will automatically switch to an alternative route that may use fiber optics. The information source—the phone in New York—does not see this switchover and is unaffected by it, since the physical interface level of the system automatically changes the signal voltages and modulation to make it compatible with optical fibers instead of copper wire.

Voltage Levels

The simplest physical interface uses two voltage values to represent data bits. For example, 0 V can represent binary 0, while +5 V can represent binary 1, as is the standard in TTL (transistor–transistor logic) digital circuitry. This method is effective and useful in some limited situations, such as where the overall distance between sender and receiver is short—several feet—and both sender and receiver share the same chassis and ground wiring. The advantage of this voltage approach is that very little additional circuitry is needed to put the digital signal onto the wire, or retrieve the signal and bring it into the receiver.

The standard digital logic gate is not a good choice for the physical interface circuitry. Its output circuitry does not have the ability to drive a cable (and its capacitance) at the slew rates that the data rate requires. Even at lower rates, the gate output is not protected against the kinds of electrical problems that occur when circuitry is connected to other circuitry: connection while power is on, shorts to power supplies, or miswiring, for example. To overcome this problem, special digital circuits called *line drivers*, and corresponding *line receivers*, are used as the physical interfaces between the digital circuitry and the cable (Chapter 7). The line driver drives a signal into capacitive loads while withstanding overvoltages and shorts. The line receiver input has the same "withstand" capabilities as the driver output, and it is the interface between the harsher external world and the much less hostile digital signal environment within the communications circuitry.

Although the line driver and receiver take voltage signals and make them more suitable for transmission, a 0 is still represented by 0 V while a 1 is represented by +5 V. This is a problem for testing and troubleshooting because a system that is transmitting a string of 0s will present a steady 0-V signal, which may look like a grounded or shorted line. The *unipolar* scheme (meaning that one data bit voltage is the same as system ground, 0 V) is therefore not often used in practical systems. Instead, a *bipolar* pair of voltages is used instead, with one voltage below ground and the other above it. Typically, these are +5 and −5 V or +15 and −15 V (Figure 12.5), but some systems use other values. The line driver/receiver IC therefore serves two purposes: first, it translates the unipolar voltages of the digital circuitry within the system to bipolar values; and second, it provides drive capability and protection that is needed when interfacing to cables.

12.4 Physical Interface and Throughput

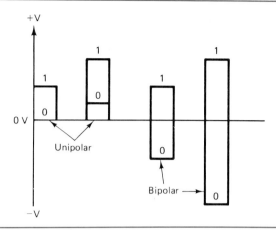

Figure 12.5 Bipolar signal voltages are symmetrical around 0 V; in contrast, unipolar signals use 0 V and another voltage to represent the two binary values.

Data DC Levels

Whether unipolar or bipolar signals are used, on average, there are an equal number of 0s and 1s. Therefore, the average voltage will be the average of the voltage values used for binary 0 and 1. For a 0- and +5-V unipolar system, this is 2.5 V, while a bipolar system will have an average value of 0 V. This average voltage is called the *dc voltage* or *dc component* of the data.

Many communications links cannot tolerate a nonzero average value. This is a further incentive for using bipolar rather than unipolar voltages. The reasons for needing an average voltage of 0 V is that the communications circuitry may have elements that block dc voltages. There is no guarantee that the physical link is a solid piece of wire, for example, that will simply pass through any voltage impressed across one end. Capacitors in series with the line, often needed to isolate various stages of amplifiers from each other, will block the dc component of the

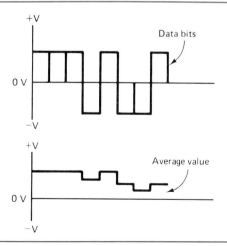

Figure 12.6 Dc drift with data bits occurs when the average value of the bit pattern is not zero.

> **Computing Average Value**
>
> The average value is computed like any other arithmetic average: Take the sum of the preceding bit values and divide by the number of bits to that point.
>
> **Example 12.3**
>
> A series of bits is transmitted using bipolar -5 and $+5$ V to represent 0s and 1s. (a) What is the average value after each bit if the bit pattern alternates 010101, etc.? (b) What is the average if there are eight 0s followed by eight 1s?
>
> **Solution**
>
> After the first bit, the average value settles to 0 V since every $+5$-V signal bit is balanced by a -5 V-bit. For the series of eight 0s, the average is constant at -5 V.
>
> When the ninth bit, a 1, is sent, the average begins to change. After the ninth bit, it is
>
> $$\frac{(-5 \times 8) + (5 \times 1)}{9} = -3.88 \text{ V}$$
>
> while after the tenth bit it is
>
> $$\frac{(-5 \times 8) + (5 \times 2)}{10} = -3.0 \text{ V}$$
>
> For the remaining bits, the averages are -2.27 (11th bit), -1.66 (12th bit), -1.15 (13th bit), -0.71 (14th bit), -0.33 (15th bit), and 0.00 (16th bit). Note how the average value has changed with each succeeding bit in the series.

voltage. This is a problem, since the average over many bits will be 0 V, but there will be long streams of only 0s (or only 1s) and the average will drift away from 0 V (Figure 12.6). Any blocking of the dc component will seriously distort the resulting bits at the receiver, since the receiver will see the signal minus the dc value.

There is another problem associated with a long run of 0s or 1s of the digital bits: the loss of clock information that lets the receiver synchronize to the transmitter clock. This was discussed in Chapter 11, and some of the techniques used to overcome it include adding special bits for synchronization or using circuitry to make sure that there is never a continuous run of more than a few 0s or 1s. The conventional digital format is *nonreturn to zero (NRZ)*, where the digital level remains at the last value until a new bit is generated (Figure 12.7a).

A technique used to guarantee that bit transitions between 1 and 0 are always present is to perform a *return to zero (RZ)* encoding process based on the data bits rather than use the simple data bit-to-voltage relationship. The most common technique for this is *Manchester encoding*. In Manchester encoding, the basic data bits are used to drive a special, yet simple circuit. The time period for each bit is divided in half. Suppose that $+5$ and -5 V are used for communication. For a binary 1, the first half of the period is set to $+5$ V, and the second half is -5 V (Figure 12.7b). The situation is reversed for a binary 0, with the first half of the time period at -5 V and the remaining half at $+5$ V. The result is that every data bit has a level transition, and there cannot be a longer time period (or run of several bits) where the voltage remains at either $+5$ or -5 V.

The data bits therefore also contain their own clocking information, and it is relatively easy for the receiver to synchronize to the transmitter clock. In addition, the average voltage will always be 0 V, regardless of the number of 1s versus 0s at any time, since the average of each bit itself is 0 V. Circuitry for Manchester encoding at the transmitter, and decoding at the receiver, is built from digital logic gates (Figure 12.8). Note that the encoded bit rate is twice the unencoded bit rate: The output has two transitions for every one data bit.

12.4 Physical Interface and Throughput

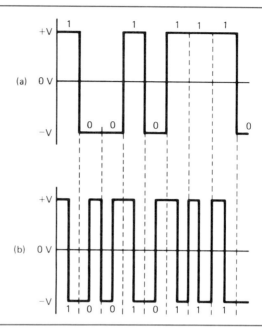

Figure 12.7 (a) Traditional NRZ bits stay at their last value unless the next bit is different; (b) Manchester RZ bits alternate between two voltages regardless of the bit value.

Schemes such as Manchester encoding are performed at the physical interface level, just before the data bits are transmitted. Since this type of encoding solves both the sync and dc average drift problems, it may seem that this completes the story of the physical interface, especially since the circuitry needed to implement Manchester encoding is so simple. Unfortunately, the drawback to Manchester encoding is severe: Twice as many transitions are needed to transmit the data bits, so the channel bandwidth must double or the bit rate must be cut in half if the channel bandwidth has a fixed value.

Of course, many communications systems do not have a direct wire link between transmitter and receiver. For cases where a radio link is used, some type of modulation in required so that the basic data bits can be transmitted at the desired carrier frequency. AM, FM, or PM, or combinations of these types of modulation are used. Modulation of a carrier by digital signals is the subject of Chapter 13.

Data Rates

Data rates in use range from a relatively low 300 bits/s to hundreds of megabits/s. The rate that can be achieved depends on many factors: how much information is to be sent, the type of communications channel, the complexity of the circuitry, and the acceptable rate of errors. Signals transmitted at lower data rates are easier to send, more tolerant of timing problems and noise, and require less bandwidth. They also require more time to send a given amount of information, and time is a valuable commodity in a system. The overall technical trend is toward higher rates, even though higher data rates require more advanced circuitry and more complicated manipulation of the data bits to achieve low errors.

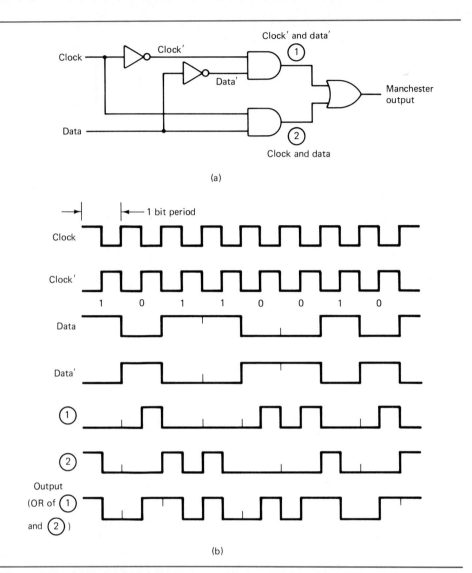

Figure 12.8 Manchester circuitry produces an RZ bit pattern from a conventional NRZ pattern. (a) Schematic; (b) waveforms at key points.

Data rates are measured in *baud,* named after Maurice Emile Baudot, who in 1874 devised one of the earliest encoding schemes for transmitting data, using only two signal values (Section 12.2). One baud is generally the same as 1 bit/s. In some forms of modulation, however, each value that is transmitted represents more than one bit, so 1 baud is not 1 bit/s (more on this in Chapter 13). However, common industry terminology uses the terms *baud* and *bits/s* interchangeably unless it is specifically noted that they are not equal.

The slower industry standard values include 300, 600, 1200, 2400, 4800, 9600, and 19,200 baud. The 300 through 1200 baud values are easily achieved on ordinary voice bandwidth telephone lines, while the 2400 and 4800 baud values require special communications techniques of modulation. Present technology

12.4 Physical Interface and Throughput

achieves rates greater than 9600 baud on phone lines, but error rates may be high and information must often be retransmitted, unless sophisticated formats and circuitry are used (techniques for detecting and compensating for errors are discussed later in the chapter).

Of course, many data communications systems often use signal transmission lines and channels specifically designed and installed for the application, and do not use the readily available (but low-bandwidth) voice telephone lines. With these data channels, much higher data rates are easily obtained. Coaxial cable, twisted-pair, or wideband radio channels can provide the wide-bandwidth, low-noise environment needed for megabit rates. The telephone system T-1 standard between central switching offices runs on coaxial cable, microwave, or fiber optic links at 1.544 megabits/s, for example, and this rate has been adopted by other nontelephone system applications as well. Special *dedicated* channels for use between high-speed computers located near each other can run at 15 to 50 megabits/s, but require more expensive, faster circuitry at both the transmitting and receiving ends. Optical fiber (Chapter 24) provides the highest data rates, easily reaching over 100 megabits/s. Of course, the other benefits of optical fibers (immunity to electrical noise, isolation, electrical safety) are often needed by lower-rate users, so fiber is sometimes used at the much slower 300 through 19,600 baud rates, with correspondingly simpler driving circuitry and receiving circuitry.

When bits are sent continuously, the system is producing maximum *throughput*, which refers to the number of useful data characters sent, received, and processed per second. The relationship between character throughput and baud rate depends on the coding and format scheme used. A more complex format uses many bits for the preamble and postamble, which in themselves do not contain any message information. Therefore, the actual throughput is less than the bit rate implies at first inspection.

For example, a message that contains 10 characters in seven-bit ASCII format, and also has five four-bit preamble fields and two four-bit postamble fields, transmits 70 character bits plus 28 format preamble and postamble bits, for a total of 98 bits in the frame. When sent at 300 baud ($\frac{1}{300}$ s per bit), the complete frame requires 0.33 s, and the effective throughput is 10 characters in that time, or 30 characters/s. Longer blocks are more efficient, since the same "overhead cost" of the preamble and postamble bits are used for many more character bits. Longer frames have greater efficiency, regardless of baud rate, but may have error and sync problems: This is the system design performance trade-off.

Example 12.4

(a) What is the throughput in the previous case when the character field is expanded to its maximum length of 15 characters? (b) What is it when the character count field is expanded to six bits (31 characters)? (c) What is the throughput in cases (a) and (b) when the baud rate is quadrupled to 1200 baud?

Solution

(a) Total number of bits sent is 28 format bits, plus $15 \times 7 = 105$ character bits, for a total of 133 bits. At 300 baud, this requires $133/300 = 0.44$ s, so the throughput is 15 characters in 0.44 s, or 34 characters/s. (b) The total number of bits is 30 format bits (the 4-bit count field was expanded to 6) plus $31 \times 7 = 217$ character bits, for a total of 247 bits. This requires $247/300 = 0.823$ s at 300 baud; therefore, the 31 characters are sent in 0.823 s, for a throughput of 31/0.823 = 37.6 characters/s. (c) The bit rate—and thus the number of bits/s and complete frames—increases by the same factor as the baud rate. The throughput therefore increases in all cases by the same factor of 4.

To the extent practical and compatible with synchronization, framing, and acceptable error rates, communications systems use the maximum baud rate and the longest blocks they can, in order to achieve the greatest throughput at the baud rate.

Review Questions

1. What is the physical layer? How does it differ from other layers in the system?
2. Why is transparency important at the physical layer?
3. Where are regular digital gates used at this layer? Why are they not the best choice? What are the limits to their capability?
4. What do line drivers and receivers do?
5. Why can a nonzero "dc average" value cause problems?
6. What is a NRZ signal? Contrast it to an RZ signal.
7. What is Manchester encoding? What is its benefit and its drawback?

12.5 PROTOCOL AND STATE DIAGRAMS

A *protocol* defines the rules of conversation. In a communications system, this means that the protocol specifies how a transmission is initiated, how it ends, and most important, how to handle all the special situations that occur. These situations include normal startup, recovering from an unplanned shutdown, and dealing with noise on message bits or with garbled messages where message, preamble, or postamble bits are in error. The protocol defines what steps must be taken by the transmitter and the receiver when any message is to be transmitted or received.

The range of protocols in use spans from simple to extremely complex. The choice of the proper protocol is made by looking at the type of application and the data that will be sent, together with understanding how critical the data are and how important it is that the data eventually be received without error. Simpler protocols are used where an error is acceptable, and there is no need to retransmit the data. For example, bits representing digitized voice can be accepted with an occasional error, since the listener will hear only a momentary click when the DAC reconstructs the data into analog. (In most cases, the person won't hear anything wrong when one sample out of thousands is wrong, and the person will certainly be able to understand the message anyway.) In contrast, when the data bits represent numbers from a bank account, every single one must be correct, since a single bit error results in a completely incorrect number.

Protocols are needed even when the flow of information is primarily in one direction. The receiver must have some way to indicate to the transmitter that the information was received and how well it was received. Communications systems therefore use *half-duplex* or *full-duplex* channels (Chapter 1) even in these applications. *Simplex* systems, where there is no communications channel in the reverse direction, usually do not have a protocol (or at most, a very simple one) since the transmitter can only send the information and hope that it is received.

A simplex example is a space satellite transmitting data from the far distances of our solar system. Because of the distances involved, the channel from the earth back to the satellite is often not useful—it takes hours for the signal from earth to reach the satellite and any message from the earth to the satellite indicating a signal problem would be received long after the problem began. Communications systems prefer full-duplex or half-duplex channels whenever possible, so an effective protocol can be established and used. The reverse channel is important, and it can be a lower-speed channel if that is all the system can provide, since it is carrying relatively little information.

The most straightforward protocols are for *point-to-point* systems where there is a single transmitter and a single receiver. The protocol is more complex when there are multiple receivers each of which get different messages, such as when a computer is broadcasting data to one of several other computers. The most complex protocols are reserved for *networks* (Chapter 18), where there are many devices interconnected and any device may originate a message for one or more other devices in the network.

Protocol Activity

Protocols are divided into *stop and go* and *continuous* types. Half-duplex channels must use stop and go, while a full-duplex channel can use either stop and go or the continuous type. The receiver knows the format and uses the character count in the preamble and the postamble to determine where the message block ends.

When the end of the message is received, the receiver decides if it received the message with or without error, using techniques described in the next section. If there is no error, it sends back an *acknowledgment*—an *ACK* message—indicating message received OK, as shown in the timing diagram of Figure 12.9. However, if the received message had some sort of error, or was too garbled for the receiver to determine anything about it, the message that is sent back to the

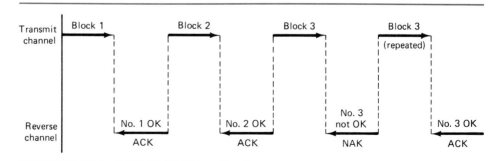

Figure 12.9 Stop-and-go protocol timing shows how, after each message, the sender stops and waits for a response from the receiver.

transmitter is a *negative acknowledgment*, or *NAK*. When the ACK is received back at the transmitter, it sees that all is well and then proceeds to send the next message block. For a NAK, the transmitter repeats the last message.

This simple protocol of a message answered by an ACK or NAK is effective but very inefficient in use of time and channel capacity. After each message block, the transmission of new information stops while the ACK or NAK is returned. This is more of a problem than may appear at first. For a half-duplex channel, the electronics at the normally sending end must be switched from transmit mode to receive mode and the normally receiving end switched to transmit mode. The switchover time can be significant, especially if long-distance links are involved with internal time delays. The problem is most severe on satellite links, where propagation delay is significant and adds to the unusable, wasted time during switchover.

As an example of how protocols must be designed to handle all conceivable situations, look at what happens if the ACK or NAK itself is corrupted or garbled. The protocol at the transmitter must have a special message it sends to the receiver to indicate that the last ACK or NAK was not received properly. Then the receiver must resend the ACK or NAK itself.

Also, look at the situation where the message itself, or the ACK or NAK, are never received. To accommodate this, *timeouts* are used. These are predetermined time periods that each end of the link waits after sending its message. If the meaningful response is not received within the time period, the protocol "times out" and takes a special action based on the assumption that the message was never received, either because of a major channel problem or severe noise which made the message almost incomprehensible. After a timeout, the transmitter or receiver may decide to resend the message, to see if it can elicit some sort of understandable response.

Full-duplex channels have potential for much greater efficiency and overall throughput of data (the total amount of information data communicated over a period of time) than half-duplex, because a continuous protocol can be used, with timing as shown in Figure 12.10. Instead of having to stop and wait after each message block, the receiver continuously sends back an ACK or NAK on its reverse direction channel. The receiver sends back the ACK or NAK block for the last block it received while simultaneously receiving the next block of bits. When the receiver detects an error, it sends back a NAK along with the number of

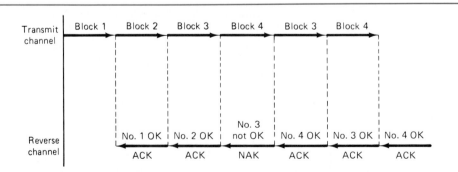

Figure 12.10 Continuous protocol timing allows the sender to accept responses to previous messages while it continues to transmit new ones, without waiting.

State Transition Diagrams

State transition diagrams are very useful pictorial representations that clearly show the protocol operation. Protocols are confusing to express in words or conventional drawings.

We have seen how even a simple stop-and-go protocol can become relatively complex, with ACKs, NAKs, and timeouts. A *state transition diagram* (or simply *state diagram*) (Figure 12.11) is a tool to describe the protocol and to design the electronics that implements the protocol and troubleshoot communications problems. In a state diagram all the possible activity states of the system are shown in *nodes*. At each state node, the system must respond to some event occurring and then proceed to the allowed next state. For example, in the stop-and-go case, one state node is "send new message and stop." There are three possible events that occur next: ACK received, NAK received, or nothing received and timeout occurs.

The diagram shows, at each node, what action occurs after each one of these events. A received ACK leads back to the same "send new message and stop" node. In contrast, a received NAK leads to another node called "resend last message and stop." Unlike a conventional programming *flowchart*, which shows the diagram of program steps of software, the state diagram shows what situations exist in the system and what responses the protocol must provide as well as the paths between the situations. When a timeout occurs, the paths lead to a timeout node where special procedures are implemented. These include sending a special message, and if there is no response, to keep trying a specific number of times. If there is eventually a valid answer (the communications link is restored), the action is for the system to go to the "send next message" state.

If some changes to the protocol are made, the state diagram is easily updated to show these. A state diagram is a valuable tool for understanding the operation of a communications system and its protocol in both design and troubleshooting. Many software programmers use the state diagram as a guide to show what the program they are developing must accomplish. The state diagram allows a test technician to see if the system is sensing events and responding to them properly. The state diagram also helps show a test technician where unanticipated events occur, ones that were not incorporated into the original design. These events—which may normally be unallowed (such as two senders trying to seize control of a channel simultaneously)—can cause the communication system to fail if the state diagram and subsequent design of the system make no provision for dealing with them.

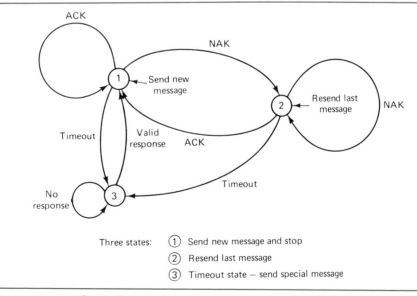

Figure 12.11 State diagram for stop-and-go protocol shows all the possible states of the system and the events that cause a transition from one state to another along various paths.

the last block correctly received. The transmitter, on getting a NAK and block number, completes the current transmission. It then backs up and begins transmitting again, starting at the block that was received with an error (the one after the last one received correctly).

The continuous protocol provides a nearly constant stream of message blocks. The only interruptions occur when an error is detected at the receiver. Since these are relatively infrequent (ranging from one bit in 10^6 for a poor-quality system channel to one bit in 10^{12} for a high-quality, low-noise channel) there are very few cases where the transmission must be stopped and begun again from a previous block. Contrast this to the stop-and-go approach, where the system must stop after each block whether or not an error occurred.

Review Questions

1. Where are simpler protocols adequate? Where are more complex ones needed?
2. Compare the complexity of protocols for simplex, half-duplex, and full-duplex systems.
3. What is a stop-and-go protocol? How does it operate?
4. What are ACK and NAK messages? What is their role in the protocol?
5. What is a continuous protocol? How is it different from stop and go?
6. Why are full-duplex systems more efficient than half-duplex?
7. What is a state diagram? What does it show? Why is it useful?

12.6 ASYNCHRONOUS AND SYNCHRONOUS SYSTEMS AND EFFECTIVE THROUGHPUT

All data communications systems can be divided into two groups: *asynchronous* and *synchronous*. They differ considerably in achievable data rates, throughput, complexity, and the nature of the data that must be transmitted. An *asynchronous* system is designed to send characters of the message with the specified bit rate but without any fixed or specified timing relationship from one character to the next. In an asynchronous system, the bit rate says nothing about the time period between characters. Asynchronous is very useful where the actual data source is generating characters at sporadic intervals, without precise timing (Figure 12.12a). Once again, a computer keyboard is a good example (although asynchronous is used for many other types of data sources). The person typing at the keyboard is hitting the keys at some speed, but the speed may vary or be erratic, or there are long gaps where no keys are pressed. An asynchronous system takes each new character as it is generated, performs the required encoding, and then transmits the character. *Parity*—a special bit added after the data bits for error detection, detailed in Section 12.7—is often used to provide a first level of error detection for each character.

In general, because of the sporadic nature of the asynchronous characters,

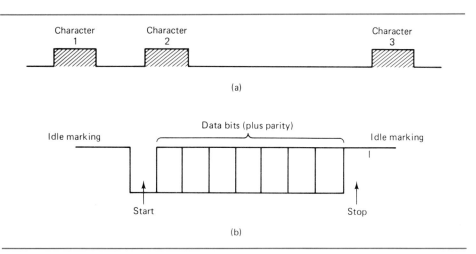

Figure 12.12 (a) Asynchronous character spacing allows any time period between characters; (b) start and stop bits mark the beginning and end of the asynchronous character.

there is much less protocol used. The message that is being typed, for example, could be a few characters or dozens in length, so it is more difficult to formulate the protocol blocks with character counts and related features. This simpler protocol is not a bad thing in itself, since this is all that many applications need. However, it does limit the usefulness of asynchronous in high-speed transmission.

Since the beginning of a new character can come at any time in an asynchronous system, the transmitter must somehow indicate a new character to the receiver. It does this by using a special *start* bit. When no characters are being sent and the channel is idle, the communications line is marking time and at a constant signal level (Figure 12.12b). (Note that because of this long time period without change, it is impossible to derive the bit timing from the data signal.) When the bits of a new character are to be sent, they are preceded by a single bit transition to the opposite state. The receiver sees this transition and gets set to capture the actual data bits as they arrive.

At the end of the data bits of the character (typically, seven or eight bits, plus a parity bit in some cases) a *stop* bit is also transmitted which indicates to the receiver that the character is complete. The stop bit has a transition that is the opposite of the start bit. Of course, the receiver knows in advance how many bits it should expect for each character, so the stop bit serves as a sort of "double-check" and sets the signal line to the ready for the next start bit. (*Note:* Some older systems use $1\frac{1}{2}$ or 2 stop bits, but these are less common.) When viewed on a scope, the start and stop bits look just like data bits, although they have a completely different meaning.

Here we see one weakness of asynchronous: Each character requires two extra bits to delimit the character bits and indicate the start and end points. These bits convey no information, so they waste channel time. However, many asynchronous applications have extra time between characters, so this may not be a problem.

The second issue is how the receiver recognizes the data bits themselves. The receiver clock is preset to the correct baud rate and begins counting time from the leading edge of the start bit. The receiver then checks each bit to read if it is a 1 or a 0, in what it thinks is the center of each bit. At lower speeds, up to a maximum of about 50 to 100 kilobits/s, this can be done reliably. Beyond that rate, timing jitter and propagation problems will cause errors in the receiver clock timing, as discussed in Chapter 11.

In summary, asynchronous provides a simple, convenient way to send data at lower rates without a lot of protocol. The maximum throughput occurs when there is no idle time between characters and each character stop bit is immediately followed by the start bit of the next character. In these cases the channel reaches its maximum theoretical throughput. A quick "estimate" of this value for ASCII coding is that the maximum possible throughput, in characters/s, is $\frac{1}{10}$ of the baud value. This is based on the simple fact that a single character requires seven or eight ASCII bits, plus an optional parity bit, plus a start and a stop bit, for a total of 10 or 11 bits/character. Transmission in ASCII at 1200 baud corresponds, at best, to about 120 characters/s.

In practice, the actual throughput achieved is less than the maximum for a variety of reasons. The receiver often cannot accept characters at this continuous maximum rate for a lengthy period, since it must process each character. Also, the channel may have noise or other problems that prevent it from operating at the full potential. However, since many asynchronous applications do not generate data at high rates, the difference between theory and what is actually achieved may not be a concern.

An Asynchronous IC

Many manufacturers provide ICs that implement the start, stop, and parity bits needed in asynchronous systems. Parity bits are optional bits used for error detection and are discussed in detail in Section 12.7. One such component is the COM81C17, *Universal Asynchronous Receiver/Transmitter (UART)*, from Standard Microsystems Corp. This 20-pin IC is the interface between a microprocessor bus and the line drivers and receivers (Figure 12.13a).

For transmitting, it is designed to accept character bits from the processor as a parallel group, determine and add a parity bit (if desired), and then generate a start bit, serially send the character bits, add the optional parity bit, and end the transmission with one or two stop bits. It next signals the processor via *interrupt* (a special signal line into the microprocessor that gets immediate attention from the software program that the microprocessor is executing) that it can accept the next character for transmission.

When receiving, the COM81C17 recognizes the start bit, accepts the character bits and optional parity bit, and recognizes the stop bits. After the stop bit (or bits), the IC indicates to the processor via an interrupt that a new character is received and available. The processor then reads the new received character and transfers it into its own memory for further processing. As an added feature, the UART checks the parity bit versus the received bits to see if there is an error in the data; if there is, it indicates this to the processor.

The internal block diagram of the COM81C17 (Figure 12.13b) shows the complexity of circuitry that is required, now available on a single IC. The system microprocessor reads from and writes to this IC just as it does to any memory address. Two memory addresses are occupied by the UART: one address (write only) for new characters to be transmitted and one write-only address for controlling the IC; and the same two addresses, but read-only, for retrieving received characters and for checking the status of the IC.

12.6 Asynchronous and Synchronous Systems and Effective Throughput

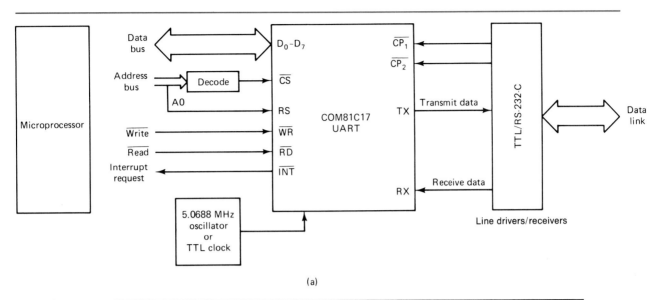

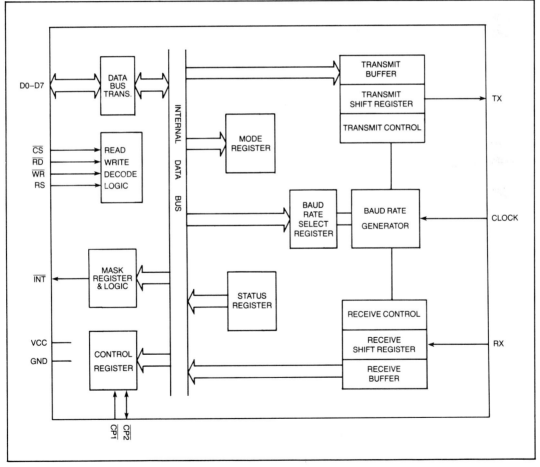

Figure 12.13 (a) A UART is the link between the microprocessor bus and serial data channel; (b) internal block diagram shows the complexity and flexibility of the UART (courtesy of Standard Microsystems Corp.).

The *command register* at the write address allows the microprocessor to set the desired baud rate. The communications baud rates are all derived by special dividing circuitry in the UART driven by a 5.0688-MHz *master clock*. The microprocessor also selects the desired number of stop bits, parity, and some other modes of operation. The *status register,* located at the same address, indicates when the character *buffer* (a collection of flip-flops that store bits) within the UART is full with a new character from the microprocessor but not yet transmitted, or the received character buffer in the UART has a newly received character which needs to be passed to the microprocessor. When the new character is transmitted, or the received character passed back to the microprocessor, an interrupt is generated and the microprocessor checks the status register for details on which situation caused the interrupt. The status register also indicates if a parity or other signaling error was detected, which also initiates the interrupt.

The serial output of the UART is not designed to interface directly at the physical level to the communications link. Line drives and receivers appropriate to the application are needed. Regular TTL ICs or special ICs designed for the physical layer are required. The UART provides one signal line for transmitted data (Tx) and one for received data (Rx) so it can be used in full-duplex as well as half-duplex operation. It also provides control lines CP_1 and CP_2 (*handshake lines*) which indicate to the physical interface and the other end of the link if the UART (and its microprocessor) can accept a new character, or has a new character to send. The function and operation of these handshake lines are studied in more detail in Chapter 17, discussing the RS-232 Interface.

Synchronous Systems

For much higher performance in terms of data rates and protocols, a *synchronous* system is used. A synchronous system takes many characters and sends them as a continuous block, governed by the protocol. Block lengths typically begin at 64 characters (equal to 512 bits if each character has an eight-bit code) plus preamble and postamble, and can be as long as several hundred characters. The goal of the synchronous design is to keep transmitting data at high rates without any dead time or breaks between characters. The efficiency of the system is determined by the ratio of the number of character bits to the total number of bits. As the block lengths get longer, the preamble and postamble are a smaller and smaller part of the overall communications, so the efficiency is high. (We saw this same effect when studying formats earlier in this chapter.)

Compare the efficiency of a synchronous system to an asynchronous one. Each character in an asynchronous system requires start and stop bits, regardless of the number of characters. In contrast, a synchronous system can use the same preamble and postamble for larger and larger blocks. The efficiency increases as the block length increases, since the "penalty" of the preamble and postamble account for a smaller percentage of the bits sent.

The receiver in a synchronous system derives its clock timing from this long stream of continuous bits, so there is much less of a clock recovery or timing problem at high bit rates or with long bit blocks. For internal consistency, some synchronous systems use a fixed block length with the same number of characters for each transmission. If there are no new characters to send, the protocol simply fills in the unused character spaces with "nulls" that occupy time with a pattern but which convey no information. As more data are generated, the next block is sent with fewer nulls and more real characters.

12.6 Asynchronous and Synchronous Systems and Effective Throughput

In a synchronous system there is no idle time between the characters where nothing at all is transmitted. There is time between blocks, and this time between blocks varies with the system design. A higher-performance system has very short time gaps between complete blocks, since any protocol responses from the receiver (such as "error in block 22") is done on a full-duplex basis. In some cases, the channel is half-duplex, so there is time between blocks where the transmitter stops, the channel direction is reversed, and the protocol responses are sent. Even so, longer blocks contribute to higher throughputs.

Protocols are usually more advanced with synchronous systems. They are capable of sending more data, and to maintain the data flow, the protocol must smoothly handle any disruptions such as errors or noise. The circuitry for the synchronous protocol is more complex, but standard ICs are available that contain the protocol rules and states and implement them automatically. All the user has to do is provide the IC with the characters to send.

Although the initial stages of the data transmission may be asynchronous, this does not mean that the characters will be sent this way to the final receiver. Many computer systems use asynchronous mode to collect the characters from various terminals or sources. The characters are then combined into longer blocks

---

Implementation of Protocols

The transmitting and receiving ends of the communications system must be designed to implement the protocol and incorporate the activities and events described by the state diagram. Either a software or a hardware approach can be used.

In the *software* method, a microprocessor is designed into the circuitry of the transmitter and programmed to take the required actions. This method provides a great deal of flexibility, since a change of program is all that is needed to support a different protocol. However, there are two drawbacks to this method. First, it is relatively slow, since there are dozens or hundreds of software instructions that must be executed by the microprocessor in order to determine exactly what is happening, what the present state is, and what action should be taken to reach the correct next state. As a result, the data rate that a software-based protocol can handle is relatively low, since the microprocessor and its time to execute the complex software become limiting factors on how fast new messages can be sent and the protocol executed.

The second weakness of the software method is that there is a great deal of opportunity for programming errors (*bugs*) to exist. Since the protocol is implemented by a programmer working from a state diagram and related descriptions, any error by the programmer will result in an improper protocol. These problems often do not surface immediately, but occur only in the more subtle, hard-to-test aspects of the protocol, often when the equipment is in field use rather than in development at the manufacturer. This makes life for the technician frustrating: a system that generally seems fine but fails only in rare, often unrepeatable circumstances is hard to diagnose and fix.

The hardware approach uses a *dedicated* protocol IC. This IC contains internal digital logic which precisely implements the protocol. As a digital circuit, it is much faster than a microprocessor executing software code. There is no chance that the program steps of the protocol will be lost due to power failure. Most important, these ICs are produced by companies that have specialists who understand the protocols and have the facilities to test the ICs thoroughly after they are designed. (Of course, sometimes even these have problems.) By using one of these protocol ICs, the communications system designer is essentially assured that the protocol will be implemented properly and does not have to spend time programming and verifying the software of the protocol.

The use of a hardware protocol IC relieves the system designer of producing a major and complex function. To overcome the fact that a protocol IC is not as flexible as the software-provided protocol, many of these ICs provide several variations of a standard protocol. A different variation is selected by loading the ICs mode of operation control registers with codes to indicate which protocol is desired. Although this does not provide the complete flexibility of software, it provides enough for most applications.

of data and sent by high-speed synchronous channels to the receiving system. This provides a good combination of simple asynchronous mode where it is best, and the more complex but efficient synchronous mode where the system can take advantage of the benefits of transmitting longer data blocks.

A Protocol IC

The Intel 8273 programmable protocol controller is a 40-pin IC designed to implement synchronous protocols. It supports two similar protocols called *High-Level Data Link Control (HDLC)* and *Synchronous Data Link Control (SDLC)* in full- and half-duplex modes. An internal phase-locked loop synchronizes the IC clock to the data rate at rates up to 64 kilobaud.

An HDLC/SDLC frame format (Figure 12.14) consists of three eight-bit fields within the preamble, the actual message bits, and two fields in the postamble. The three preamble fields are the opening delimiter *flag* 01111110, an address field, and a control field. The information field contains the actual bits to be transmitted, and can be from 0 to 2^{16} bits in length. This is a *bit-oriented* protocol, since the data do not have to consist of ASCII code (or any other code) but are considered and handled as a stream of individual, unrelated bits which may represent ASCII characters, binary numbers, or anything. The postamble field has a 16-bit frame check sequence, which is used for error detection on the message bits (Section 12.7), followed by a closing flag terminator 01111110.

Basic operation of the 8273 consists of three phases. In the first phase, the many command parameters needed are *set up* by writing to the IC. These specify the length of the information field, mode of operation (several regular modes, plus special test modes), and several other operating characteristics. Next comes the actual *execution* phase, where the data bits are transmitted. The message bits are transferred from the processor memory to the 8273, which places them in the frame, transmits them, and adds the error-checking bits needed. When receiving data, the received data bits are passed to the processor memory while an error-related *frame check* operation is performed. In the *final* phase, the results of the communication activity are available to the processor via status bits. The processor can see if the communication was completed successfully. If the result is unsuccessful, the reason, as determined by the protocol controller IC, is reported to the processor.

The operation of the 8273 protocol controller is designed to minimize direct involvement by the system microprocessor. This frees the microprocessor to form up the data bits of the message or interpret the received bits. Command and status registers within the protocol controller appear to the processor as memory addresses, just as in the COM81C17 asynchronous UART. The many details of managing the protocol and format themselves are handled entirely by the 8273 once it is set up by the processor.

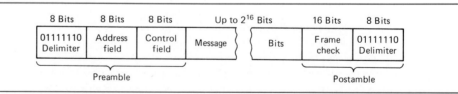

Figure 12.14 Frame format of one type of synchronous system (courtesy of Intel).

Review Questions

1. Explain why, for an asynchronous system, the bit rate and character rate are unrelated.
2. What is the function of the start and stop bits in an asynchronous system? Why can't the first bit be a character data bit?
3. What is the effect on efficiency of the start and stop bits? What is a quick rule for the maximum throughput, in characters/s, for an asynchronous ASCII system?
4. What is the structure of a synchronous system? What is the objective of the transmitter?
5. Why is the efficiency and throughput of a synchronous system higher than an asynchronous system? What happens to efficiency as the block length increases in each?
6. Explain the functions that a UART provides. What does it look like to the microprocessor? To the physical layer of the link?
7. What does the UART do with received characters? How does it indicate errors? How does it get the next character to be transmitted? What does it add to this character before transmitting it?
8. How does the microprocessor set the UART up to the desired conditions of operation? How does it determine what the UART's internal status is? What does the UART indicate when it cannot receive a character from the link or cannot accept the next character for transmission from the microprocessor?
9. What does the HDLC/SDLC protocol frame look like? What information is contained in the preamble and postamble fields? How many message bits can be sent in one frame?

12.7 ERROR DETECTION AND CORRECTION

As we have seen to this point, a large part of the communication system protocol relates to dealing with the inevitable errors that occur in communications systems. Errors can be divided into two groups: the large, gross errors and the much smaller-scale errors that affect just one or a few bits. In the first case, the system has a severe and very obvious problem. Messages are not completed, or not responded to, or the number of characters received does not match the received character count information. These gross errors are the easiest to detect and deal with, since the system has a problem and knows it.

The smaller bit errors are another situation. In this case a single bit within a character may have been corrupted by noise and changed to its opposite sense. Ordinarily, there is no way that the receiving part of the system can detect this error, since in all other ways the received message is perfect. The format is correct, the character count agrees with the number of characters actually received, and the postamble and preamble make complete sense. But a single bit error can have a severe impact when a character is changed or a number is corrupted. The message passes transparently through the layers of the system to be interpreted by the end user, and the layers do not interpret the meaning of the data; in fact, there is no way for the various layers to know that what was received as 6.75 should have been 7.75.

To overcome the problem of a few bits in error and the consequences of these small but significant errors, techniques of detecting and even correcting errors have been developed. These techniques once again show the power of digital signal communications compared to analog signal format. For an analog signal, there is no way for the receiver to convey that the signal that was measured as 1.27 V really was a 1.1-V signal corrupted by 0.17 V of noise. For a digital system it is possible to detect single-bit errors, multiple-bit errors, and even to correct these detected errors from the corrupted bits themselves without requiring a retransmission. In this section we study two methods of *error detection*—called *parity* and *checksums*—along with a more advanced technique that detects and corrects errors automatically.

The theory that shows the effectiveness of checksum error detection, and error detection and correction, is extremely complex and based on a branch of math called algebra theory, which is quite different from what you know as traditional algebra. The fortunate result of this theory is that the actual circuitry to implement its conclusions is very simple and requires only a small number of digital logic gates for its circuitry.

The basic idea with all error detection (and correction) techniques is to use all the message bits at the transmitter to generate one or more special bits used for error detection (and correction) purposes (Figure 12.15). The error detection (and correction) bits are then transmitted along with the actual data bits as part of the final system format. At the receiver, the procedure used to generate these special bits is repeated on the received message bits. The results of this second procedure are compared to the value of the error-related bits sent along with the message. If the bits as received agree with the bits as recalculated by the receiver, it is assumed that there is no error. If they differ, there is some difference between the message as sent and its received version.

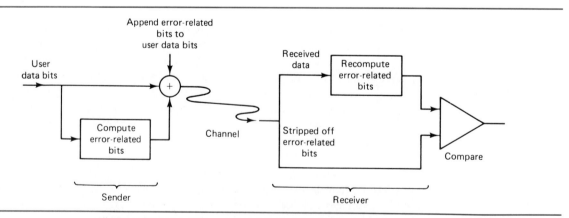

Figure 12.15 All error detection schemes calculate and add special bits to the transmitted bits, then recalculate these bits at the receiver to see if the special bits (as received) agree with the recalculated version.

Parity Bits

The simplest approach to detecting an error in a group of bits is to use a single extra bit called a parity bit. Typically, there is one parity bit for each seven or eight bits of the message. This is especially convenient for seven-bit ASCII coding, where seven data bits represent the character code and the eight bit is the parity bit: The data bits plus parity bit occupy exactly one byte. Parity can be either odd or even, and both odd and even offer the same amount of error detection potential. The use of either odd or even parity in an application does not provide any technical advantage and is mostly a matter of historical custom in the application.

Parity is calculated very simply. For odd parity, the total number of binary 1s in the character and its parity bit must be odd. For even parity, this total number of 1s is even. The parity bit is calculated by looking at the character data bits and seeing whether the parity bit should be a 1 or 0 in order to make the total number of 1s odd (for odd parity) or even (for even parity). Here are some examples:

Character bits	Parity bit for odd parity	Parity bit for even parity
01010101	1	0
11111111	1	0
0010011	0	1
0000000	1	0

In operation, the parity bit is sent after the data bits. The receiver calculates its own version of the parity bit based on the data bits and compares this to the received parity bit. If the two versions of the parity bit agree, all is apparently OK, but if the two parity bits differ, there has been an error:

Transmitted bits	Odd parity bit	Received bits	Recalculated odd parity
01101000	0	01111000	1
10010000	1	10010001	0

The simple parity scheme seems to take care of detecting all errors, but it does not. Unfortunately, with its simplicity comes a problem: A parity error (as it is called) indicates only that an odd number of bits are in error (regardless of whether odd or even parity is used). When one, three, five, or seven bits are in error, the parity bit will show it, while if two, four, six, or eight bits are in error, the parity bit will remain unchanged:

Transmitted bits	Odd parity bit	Received bits	Recalculated odd parity
01101000	0	01111000	1 (one error)
01101000	0	01111100	0 (two errors)
01101000	0	00111100	1 (three errors)
01101000	0	10111100	0 (four errors)
10010000	1	10010001	0 (one error)
10010000	1	00010001	1 (two errors)
10010000	1	00010011	0 (three errors)

As a result of this relatively weak ability to detect errors, parity is used as the error detection scheme only when the overall conditions are relatively good and very few errors are expected. The assumption used when parity is chosen for error detection is that only one error will occur in any group of bits over which the parity bit is calculated (in practice, rarely more than eight bits in group, although sometimes up to 16 bits are involved). Typical applications for parity are in the communications link between a keyboard and computer terminal, where the distance is small and the noise is low. Parity is well suited for asynchronous communications, where the characters occur at random intervals, because the parity bit is associated directly with each character as it is generated.

Circuitry for parity bit generation A few digital logic gates can be used to generate odd or even parity. Figure 12.16 shows a circuit that generates a parity bit for eight bits of data. A single control line selects odd or even parity. The data bits are presented in parallel to the circuit, which computes the parity bit. The communications electronics at the transmitter takes this bit and appends it to the stream of transmitted bits. The receiver develops its own version of the parity bit and then compares its version to the received value. The receiver may use a different circuit configuration, but that does not matter as long as both circuits implement the same definition of parity.

Checksums

To overcome the weakness of parity bits, another approach is used. Instead of associating a single parity bit with a field or group of data bits, the idea of the checksum is to develop a group of bits that acts as the nearly unique "fingerprint" of the data bits. The checksum fingerprint is a condensed version or summary of the data bits. The receiver develops its summary checksum using the same rules as the receiver used and then compares results. If they agree, the odds are extremely high that there has been no data bit error; if they differ, there has almost certainly been an error. As an added benefit, a single checksum of just 16 bits length can act as the summary of several hundred or thousand data bits, so the number of extra bits that have to be sent as part of the data stream is relatively small. Unlike parity, which is a good match for asynchronous data, checksums are better suited to the longer data streams of synchronous systems since the checksum can be calculated only after the final data bit is generated.

The circuitry to produce the checksum varies with the particular type of

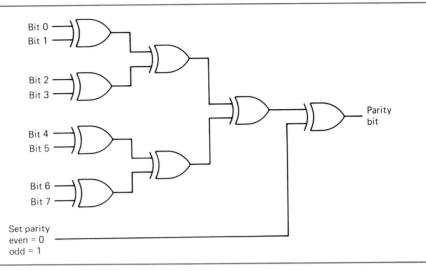

Figure 12.16 Circuitry for parity generation uses standard digital logic gates.

checksum used. For the most common type of checksum, called a cyclic redundancy code (CRC) checksum, the circuitry consists of a 16-bit shift register with feedback taps and exclusive-OR gates wired as shown in Figure 12.17. The shift register is initially set to all 0s. As the data bits are transmitted, they also pass through the shift register. The bits combine and recombine because of the feedback and exclusive OR functions that connect selected bit locations within the shift register with new, incoming data bits. When all the data bits have been passed through the shift register, the *remainder* in the register is the checksum, which is then transmitted. Figure 12.18 shows the contents of the shift register as each bit is sequentially shifted in. Note how the register content is not simply the data bit stream but is greatly modified by the feedback and XOR gates.

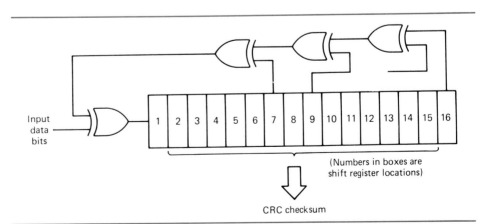

Figure 12.17 Shift register and feedback taps through XOR gates used to generate CRC checksums from a stream of input bits.

CRC for data 0000 0000 0000 00001[a]

Bit number	Input bit value	Register contents														
		1	2	3	4	5	6	7	8	9	10	11	12	13	14	15	16
1	1	0	0	0	0	0	0	0	0	0	0	0	0	0	0	0	0
2	0	0	0	0	0	0	0	0	0	0	0	0	0	0	0	0	1
3	0	0	0	0	0	0	0	0	0	0	0	0	0	0	0	1	0
4	0	0	0	0	0	0	0	0	0	0	0	0	0	0	1	0	0
5	0	0	0	0	0	0	0	0	0	0	0	0	0	1	0	0	0
6	0	0	0	0	0	0	0	0	0	0	0	0	1	0	0	0	0
7	0	0	0	0	0	0	0	0	0	0	0	1	0	0	0	0	0
8	0	0	0	0	0	0	0	0	0	0	1	0	0	0	0	0	0
9	0	0	0	0	0	0	0	0	0	1	0	0	0	0	0	0	1
10	0	0	0	0	0	0	0	0	1	0	0	0	0	0	0	1	0
11	0	0	0	0	0	0	0	1	0	0	0	0	0	0	1	0	1
12	0	0	0	0	0	0	1	0	0	0	0	0	0	1	0	1	0
13	0	0	0	0	0	1	0	0	0	0	0	0	1	0	1	0	0
14	0	0	0	0	1	0	0	0	0	0	0	1	0	1	0	0	1
15	0	0	0	1	0	0	0	0	0	0	1	0	1	0	0	1	0
16	0	0	1	0	0	0	0	0	0	1	0	1	0	0	1	0	1

—————— Remainder ——————

CRC for data 0000 0000 0000 0011

Bit number	Input bit value	Register contents														
		1	2	3	4	5	6	7	8	9	10	11	12	13	14	15	16
1	1	0	0	0	0	0	0	0	0	0	0	0	0	0	0	0	0
2	1	0	0	0	0	0	0	0	0	0	0	0	0	0	0	0	1
3	0	0	0	0	0	0	0	0	0	0	0	0	0	0	0	1	1
4	0	0	0	0	0	0	0	0	0	0	0	0	0	0	1	1	0
5	0	0	0	0	0	0	0	0	0	0	0	0	0	1	1	0	0
6	0	0	0	0	0	0	0	0	0	0	0	0	1	1	0	0	0
7	0	0	0	0	0	0	0	0	0	0	0	1	1	0	0	0	0
8	0	0	0	0	0	0	0	0	0	0	1	1	0	0	0	0	0
9	0	0	0	0	0	0	0	0	0	1	1	0	0	0	0	0	1
10	0	0	0	0	0	0	0	0	1	1	0	0	0	0	0	1	1
11	0	0	0	0	0	0	0	1	1	0	0	0	0	0	1	1	1
12	0	0	0	0	0	0	1	1	0	0	0	0	0	1	1	1	1
13	0	0	0	0	0	1	1	0	0	0	0	0	1	1	1	1	0
14	0	0	0	0	1	1	0	0	0	0	0	1	1	1	1	0	1
15	0	0	0	1	1	0	0	0	0	0	1	1	1	1	0	1	1
16	0	0	1	1	0	0	0	0	0	1	1	1	1	0	1	1	1

—————— Remainder ——————

[a] Rightmost bit goes into register first.

Figure 12.18 Two examples of data bits shifted into the register and the remainder (the CRC result to be sent with the data).

The receiver contains the identical shift register circuit, also initialized to all 0s. As data bits are received, they pass through the shift register and eventually produce a remainder. If this remainder agrees with the received checksum (the transmitter's version of the remainder) there is no bit error. A different checksum means that there has been a bit error.

It is possible, mathematically, to show that there are some error patterns that will produce the same checksum as the original, correct data bits. However, the choice of feedback taps locations, and a 16-bit checksum produces a certainty that typically 99.99% of errors will be detected by this CRC scheme. Longer checksums using 32 or 64 bits (longer shift registers and more feedback taps) increase the percentage of errors that will be caught. Compared to parity, which requires one bit for every seven or eight data bits (14% or 12.5% wasted), checksums are very efficient: a single checksum field of 16, 32, or 64 bits performs error detection on thousands of bits.

Another way of looking at the meaning of the CRC checksum is that it "compresses" a very long stream of data bits into a short, nearly unique summary. This points out an important fact about interpreting the meaning of the bit difference between one checksum and another. It is not possible to determine where the error is by looking at where the checksums differ. A single-bit error in the data bits may cause a completely different checksum to occur, while a multiple-bit error may result only in a single checksum bit differing. All that can be said is that the checksums differ, so there is a difference in the bits that generated the checksums. The amount or location of the difference in data bits cannot be determined to any extent at all from the checksum difference.

Error Detection and Correction

Parity and checksums indicate that an error has occurred. The system protocol must then be used to call for a retransmission of the data bits that were received with error. In some applications, this is undesirable or impossible. Once again, a good example is a deep-space satellite. There may be no practical possibility of telling the satellite that the data were received with an error. In other cases the data may be so valuable that it needs some type of added insurance to make sure that any change in a bit is not only detectable, but correctable. Computer memories in critical systems are an example. If a bit in the computer memory changes due to noise or component failure, there is normally no way of finding out what the correct value was even when the memory is read again. The information is irretrievably lost.

For these situations, *error detection and correction (EDC)* was developed.

Multiple Error-Detection Levels

Some systems employ several levels of error detection. For example, the coding of letters that are typed at a keyboard may use seven bits of ASCII plus one parity bit. The next level, the format, may add a checksum field based on the bits that it receives, including the ASCII character bits and the appended parity bit. The formatting layer treats these simply as a group of eight bits, without realizing or caring that seven of the bits are data and one bit is parity. The parity bit becomes part of the bit stream that is used to generate the checksum, but this is not a problem, due to the transparency of data handling between layers of the system. At the receiving end, the checksum field is stripped off and compared for errors by the layer that corresponds to the transmitters' formatting layer. The remaining bits are passed to the decoding layer, which takes off the parity bit and sees if there is a parity error. The actual character data bits are then interpreted to see what character was actually sent.

EDC uses a group of bits that act as *checking bits* on a relatively small group of data bits. For example, one type of EDC for eight bits of data requires five checking bits in order to detect all one- and two-bit errors, most three-bit errors, and allow correction of all single-bit errors. The concept of CDC is to use the data bits, in various combinations, to produce the checking bits. The combinations that are used are worked out in the complex theory of EDCs. The checking bits are transmitted along with the data bits, and as before, the receiver performs the same operation and compares results. If they agree, the data bits are correct.

If these results disagree, though, the receiver can go one step further. By looking at the exact locations of the differences between the received checking bits compared to the recalculated checking bits, the receiver can actually determine which data bit was received in error. The solution: Simply invert the state of the bits in error. Once again, here is another advantage of digital systems. If you know that a bit is in error, then you know what its correct value must be, since there are only two possible choices. A binary 1 that is in error must really be a binary 0, and vice versa. Contrast this to an analog system, where you may receive a 1.27-V signal that you know that some noise or error: Knowing there is an error does not indicate what the correct value is. Coding for error correction is known as *forward error correction (FEC)* since the receiver does not have to go back to the transmitter to get a retransmission if it detects an error.

The combinations of data bits used to calculate the checking bits are relatively simple. In one form of EDC, called *Hamming code* after its inventor, the checking bits are formed by taking the parity (odd or even) of the data bits in combinations as shown in Figure 12.19. (Note that the five checking bits are numbered 0 through 4.) Despite their apparent complexity, these combination results are very easy to calculate using simple digital logic circuitry, similar to that used for parity generation earlier in this section.

Suppose that the eight data bits are 11000101. Checking bits CB0 through CB4 are then 00111 using the combination patterns. Next, let the received data bits have a single error in the LSB. The checking bits are recalculated at the receiver and are 11101. The single-bit error has caused three checking bits (CB0, CB1, and CB3) to change. The changed checking bits are called the *error syndrome* and are marked by 0s, while the unchanged bits are marked by 1s. The

		Data bits							
Check bits	7	6	5	4	3	2	1	0	(LSB)
CB0				X	X		X	X	
CB1		X	X		X	X		X	
CB2	X		X	X		X	X		
CB3	X	X				X	X	X	
CB4	X	X	X	X	X				

X means perform parity (odd or even) on these bits only

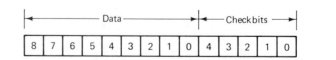

Figure 12.19 Development of Hamming checkbit patterns: eight data bits producing five checkbits.

syndrome bits are matched against a table that has been calculated in advance by the theory. This table, Figure 12.20, shows that this syndrome is the result of an error in data bit 0, the LSB. To correct the error, the receiver just inverts the LSB.

Note that the EDC can detect and correct an error that occurs in the five checking bits, not just the eight data bits. Errors do not distinguish between original data and checksums or checking bits, and can occur with equal likelihood in any bit.

Other EDC algorithms exist and each provides slightly different capabilities. For example, some EDC codes are better at detecting and correcting errors that occur together in *clusters*, such as when a severe noise pulse or atmospheric fade occurs. Others are optimized to detect and correct errors that are widely separated, as may occur from very sporadic random noise. Although these codes have some very different mathematical analysis behind them to determine the impact of different error patterns, the circuitry used is very similar. All that changes is the number of EDC checking bits required for the number of data bits, and the specific combinations of data bits used to calculate each of the checking bits. EDC codes have limits in the number of bit errors they can detect and correct per number of extra bits used; the different EDC algorithms provide a choice.

EDC codes provide many benefits but they have one major drawback. They require that many extra bits be transmitted (or stored in memory). For a five-bit checking field on eight bits of data; $5/(8 + 5) = 38\%$ of the bits contain no new information, only checking information. This is a high price to pay. Many applications cannot afford the circuitry, bandwidth, or reduction in overall throughput for this. For critical applications, however, EDC codes provide the answer to the problem of knowing there is an error and yet not being able to go back and get the message sent again.

Error location		Error syndrome bits				
		CB0	CB1	CB2	CB3	CB4
Data bit	0	0	0	1	0	1
	1	0	1	0	0	1
	2	1	0	0	0	1
	3	0	0	1	1	0
	4	0	1	0	1	0
	5	1	0	0	1	0
	6	1	0	1	0	0
	7	1	1	0	0	0
Check bit	0	0	1	1	1	1
	1	1	0	1	1	1
	2	1	1	0	1	1
	3	1	1	1	0	1
	4	1	1	1	1	0

→ Error syndrome in this example (Data bit 0 row)

Figure 12.20 Hamming error syndrome bits, formed by marking the changed checkbits by 0s, indicate where the error is located.

CD Players and EDC

The complex steps used in taking basic digitized analog signals and using them in a communications system may seem necessary and worthwhile only for space satellites, worldwide networks, and similar long-distance applications. In fact, however, these same techniques are used in the standard consumer compact disk (CD) music system for recording and playing music. These disks store the digitized audio as binary signals, represented by tiny pits (or lack of pits) on the surface of a plastic disk. The pits are read by a laser beam and photodetector in the CD player.

The analogy between the CD system and a worldwide communications system is this: Digitizing the music signal and storing the equivalent binary form on the disk is like transmitting a signal; the compact disk itself is the communications medium (like a cable, or air, or vacuum); and reading the disk surface and transforming the recovered bits back into the original audio is the receiver process. The recording industry has set standards for CD systems: the analog signal is sampled 44,000 times/s and digitized to 16 bits (96 dB of dynamic range). All CD systems adhere to this basic format, but there is much more to the CD format than simply performing these A/D conversions at the required rate and resolution.

The bits cannot be recorded directly onto the disk surface because they contain no framing or synchronization information, which is needed to retrieve them reliably. Even with bits added for framing and sync, errors caused by dirt or scratches on the disk, or by noise in the electronics, would cause annoying audio clicks or even loss of framing and sync. To prevent this, the 16-bit A/D conversion values from many samples are scrambled (following a specified set of rules), so that the bits from a single conversion are not in one physically adjacent section of the disk. To make the bit pattern useful and reliable, these scrambled patterns are transformed with error-correcting codes and sync bits. The final result is that an audio signal that was originally represented by 192 bits (12 samples × 16 bits/sample) is actually recorded onto the disk surface as a total of 588 bits: 27 sync-related bits and 561 data plus EDC bits (almost three times as many data bits are recorded as existed in the original digitized signal).

The CD player, of course, performs the reverse operation. Using the sync bits on the disk, it reads the pitting pattern and produces a 1,0 pattern. It then performs error correction on this bit pattern (using the EDC bits in the pattern), and unscrambles the bits to produce the original 16-bit samples. Finally, the 16-bit values are passed through a D/A converter to produce the analog sound that is the same as the original music.

This complex and costly scheme (in terms of bits, CD surface space required, and the algorithms needed) was developed by the engineering teams at Sony (in Japan) and Philips (Holland) with good reason. It was done to ensure that even severe scratches or dirt on the disk do not cause uncorrectable errors, which means loss of sync, irritating audio noise, loud clicks, or disk player malfunction (which might appear to be a problem in the mechanism, but is not). After all, a small scratch or dirt particle will cause hundreds of pits on the disk surface to be obscured and read back in error. By scrambling up the samples, the scheme makes sure that the bit errors caused by a scratch or dirt are, in effect, spread across many digitized samples; the use of EDC makes sure that virtually all these errors are not only detected but also corrected.

Bit Error Rate

The overall quality of a digital communications system is measured by several parameters. One of the most important is the *bit error rate (BER)*, which shows what fraction of the total bits are in error. A relatively low performance system will have a BER of 1 in a million (10^6), while better systems provide BERs that are a million times better (1 in 10^{12}). There are some communications systems that have very poor BERs, as low as 1 error in 10,000 bits, due to especially poor signal conditions.

The most important factor in determining BER is the system *signal-to-noise ratio (SNR)*. As SNR increases, BER drops dramatically. Most communications systems begin to provide reasonably good BERs when the SNR is above 20 to 30 dB. The performance of the system versus SNR is usually shown on a graph, such

12.7 Error Detection and Correction

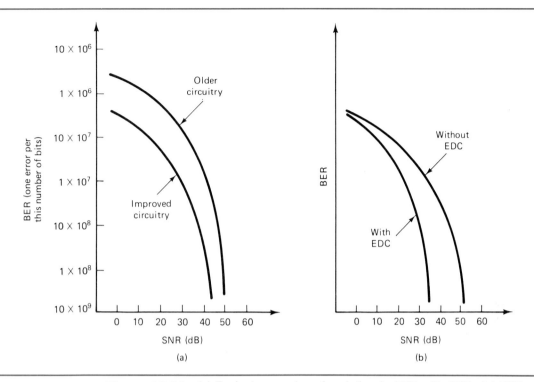

Figure 12.21 (a) Typical examples of variation in BER with SNR; (b) BER versus SNR for a system using both uncorrected and corrected (EDC) data bits shows improvement that error correction provides at any given SNR.

as Figure 12.21a. The SNR scale uses dB while the BER axis is labeled in powers of 10. As the system changes, such as may happen with improved circuitry (a new, low-noise front end, or more stable clocks and clock recovery, a better link, or better response to changes in signal level), the graph shows these results.

The BER versus SNR chart also shows the impact of EDC in systems that provide it. Figure 12.21b shows the BER for the ''raw'' uncorrected data bits, and then the BER that the message user sees after EDC. Note how under poor conditions with low SNR the beneficial effect of EDC is small, since the number of errors is greater than the EDC circuitry can handle; there is virtually no improvement. In contrast, as the SNR increases and the number of raw errors decreases, the EDC is able to become effective and dramatically reduce the BER. In the next chapter we discuss BER testing and test equipment in more detail.

Review Questions

1. Compare gross errors and single or several bit errors. Why are bit errors more subtle, yet sometimes more significant than the gross errors?
2. Why are error detection, and error detection and correction, possible for digital but not analog signals?
3. What is the basic idea behind all error-detection schemes, in terms of how the receiver used the received data bits to evaluate the errors in the data?

4. What is parity? Why is it useful?
5. How many data bits does a single parity bit usually cover?
6. What groups of errors will parity detect, and which will it miss?
7. What is the checksum concept? How does it differ from parity?
8. Compare the number of error detection bits versus the number of data bits for a checksum and for parity.
9. Does a checksum provide an absolute guarantee of error detection? About what percentage of errors will it detect?
10. Can a checksum indicate what the exact error is? What does a one-bit difference in the checksum indicate compared to several bits that differ?
11. How can a system use both parity and checksum error detection? Why is this not a problem?
12. Where is error detection and correction used? What is its efficiency? What is its reliability at detecting and correcting errors?
13. How is error detection and correction done? What types of errors will a five-bit Hamming code on eight data bits detect? Which will it correct?
14. What is the sequence of steps that detects and corrects an error? What is the syndrome? How is the table that links the syndrome to the bit in error devised?
15. When an error is located, what does the receiver do?
16. What is BER? What are key factors that determine BER?

SUMMARY

An effective and properly designed digital communications system makes use of a simple fact: Unlike analog signals, digital ones can be extensively processed and manipulated to provide improved performance. The information to be transmitted is coded into some character set that represents it; then the bits are formatted into larger groups with additional information that shows where the data begin and end, provides error detection and correction capability, and allows for a continuous, high-speed block of bits to be sent. The rules of the communications "give and take" are defined by a protocol that defines all states of the system and the many paths between states.

Asynchronous systems provide simple, low-speed performance with less complicated formats and usually simpler protocols, at rates up to about 100 kilobits/s. For higher speed, synchronous systems provide long streams of bits to be transmitted, judged for errors, and retransmitted efficiently. Error-detection schemes include parity, which uses one checking bit to indicate some errors in a small group of bits, and checksums, which can show if there is an error in a large block of bits. For absolute data integrity, error-correction codes send many extra bits along with the data and allow the receiver to detect errors and then correct them, without a retransmission from the source.

All this makes digital communications system more complex but very effective and reliable. Transparency between the various layers of the system means that a change in rules in one layer is irrelevant to the layers above and below it, which gives systems a great deal of flexibility. Each layer just follows the rules of coding, formatting, modulation, or protocol on the bits it receives from the previous layer. If all this seems necessary only for

Summary

large systems such as those that span the country, world, or space, that is only partially correct. The standard home audio compact disk (CD) system uses many of the same ideas in coding the digitized audio signal, formatting it on disk, and providing error detection and correction. The distances involved are only a fraction of an inch, but it is still a digital communications system (although there is no real protocol, since it is a simplex system!).

Summary Questions

1. Compare the complexity of analog versus digital communications systems. Why the differences? What are the benefits of the additional complexity?
2. What is the importance of transparency?
3. Compare and contrast modulation, code, format, and protocol. How do they interrelate? What is the role of each?
4. What is ASCII code? What are some alternatives? What are the drawbacks of numeric codes?
5. What is the meaning and importance of delimiters?
6. What information is typically contained in a preamble? In a postamble?
7. Compare the features and applications of fixed- versus variable-character-block lengths.
8. Compare digital gates and line drivers/receivers in their ability to drive a signal properly into a short line and a longer one at various data rates.
9. Why are bipolar signals often preferred over unipolar?
10. Compare conventional NRZ data bits to Manchester encoded bits. What are the differences and features of each?
11. Describe three exception conditions that a protocol must handle.
12. What do ACK and NAK messages indicate? What does a protocol typically do when it receives either? When it receives neither?
13. Compare the operation of stop and go versus continuous protocols. Compare the relationship of protocols with simplex, half-duplex, and full-duplex systems.
14. What are state diagrams? Why are they very useful?
15. Compare bit rate and character rate in asynchronous and synchronous systems.
16. What is the difference between an asynchronous system and a synchronous system?
17. Why do asynchronous systems need start and stop bits? What do these do?
18. How do ICs implement the rules of synchronous and asynchronous protocols and formats? Explain how a UART acts as the interface between the microprocessor bus and the serial communications link. What does the UART do for transmitting? For receiving? How does the microprocessor set the UART up and determine its condition?
19. How does a communications system recognize and respond to gross errors? To single- or several-bit errors?
20. What is the basic concept used in error detection, by either parity or checksum methods? What must the receiver do?
21. Compare the ability of parity and checksum schemes to detect errors. What does each require? What type of errors does each catch?
22. How is a checksum formed? What is the significance of the checksum bits?

23. What is the advantage and the disadvantage of error detection and correction?

24. How does a checksum or error detection/correction system handle the presence of a parity bit in the data?

25. What happens when the checksum bits differ in error detection and correction? What if there is an error in the checking bit, not the data bit?

26. Explain why error detection and correction is less effective at low SNRs and high BERs.

PRACTICE PROBLEMS

Section 12.2

1. What are the individual ASCII codes for S, Q, 7, 9, :, and /?

2. What is the sequence of bits, in order, when the word STOP is transmitted using ASCII?

3. How many bits does it take to transmit the number 64.76 in seven bit ASCII? How many does it require when each digit is represented by a four bit binary equivalent (0000 through 1010) and 1011 is the decimal point? What is the number of bits needed in the four-bit code compared to seven-bit ASCII?

4. A 10-bit binary field is used to represent numbers from 0000 to 1023. How many bits does it take to represent the number 1016? How many bits are needed in this code compared to the seven-bit ASCII representation?

Section 12.3

1. Sketch the format for the message NEXT TIME using four-bit binary fields for the preamble and postamble information, 1010 as the start delimiter, and 0101 as the end delimiter. The message is number 9, the block length is variable, and there is no receiver field.

2. A message uses ASCII codes for the preamble fields, with a start pattern using ASCII //, a terminating pattern of **, a message number from 00 to 99, and block length that ranges from 00 to 99. Sketch the frame for message number 23: SEE YOU LATER.

3. What is the frame when the example of problem 2 is expanded to include a special eight-bit field in the postamble for error-detection purposes?

Section 12.4

1. For a unipolar signal with 0 and +5 V representing binary 0 and 1, what is the average value after each bit when the bits consist of a stream of four 0s, then four 1s, then eight 0s?

2. Repeat problem 1 when the signals are bipolar, with −5 V for binary 0 and +5 V for binary 1.

3. What is the average signal value after each bit for the ±5-V bipolar voltages when the bit stream is 01110011?

4. Sketch the NRZ, bipolar version of the bit stream 10001001. Sketch the Manchester RZ version.

5. How many characters/s are transmitted using seven-bit ASCII at 2400 baud?

6. The seven-bit ASCII characters at 2400 baud are now formatted with a preamble that contains 32 bits and a postamble that contains 16 bits. What is the throughput in characters/s when a block has 16 characters?

7. What is the throughput for the situation in problem 6 when a block is lengthened to 128 characters? By what factor has it increased compared to problem 6?

8. What is the throughput for the communications in problem 6 when the data rate is increased to 19,200 baud?

Section 12.5

1. Show the timing diagram for the stop-and-go protocol when a total of six messages are sent and messages 2 and 5 have errors.

2. Repeat problem 1 for the continuous protocol.

3. Show the timing diagram for problem 1 when neither an ACK nor a NAK is received, and the timeout is set equal to one message length.

4. Draw a state diagram for a stop-and-go protocol where a message is retransmitted up to three times if an error is detected, and then a special "system recovery" mode is used to find the source of the problem. If the problem is cleared up, message transmissions resume; if it is not, a message to a system technician is generated.

5. Draw the state diagram for a continuous protocol that initializes the system with two test sequences before the first user message is sent and then begins actual user message transmission.

Section 12.7

1. Determine even parity for these bit patterns: 0110, 011001, and 11001011.

2. What is the parity (odd) for 1001, 100101, 0101001, and 11110000?

3. What is the parity for the bits of problem 1 when only the last bit changes? What is it when the last two bits change?

4. What is the parity for the bits of problem 2 when the line shorts out and all bits are seen as 0s?

5. Calculate the checksum for a digital stream of 16 alternating 1s and 0s using the shift register and XOR feedback shown in this section.

6. Repeat problem 5 but with a single-bit difference: the first bit is a 0, not a 1.

7. What are the Hamming checkbits for a 10101010 pattern? What are the checkbits when the pattern has a one-bit error in the MSB?

8. Using the results of problem 7, show how the syndrome points to the error.

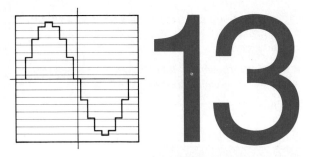

13

Digital Modulation and Testing

CHAPTER OBJECTIVES

When you have completed this chapter, you will understand:

- The unique requirements of modulation and demodulation for digital signals, and various methods used
- Digital modulation, especially quadrature modulation, for multiple-level signals, including constellation space, noise margins, and eye patterns
- Basics of communications testing for binary and multiple-level digital signals
- Generation and use of random bit sequences for both testing and data encryption (as protection against unauthorized eavesdroppers)

INTRODUCTION

Three types of modulation—AM, FM, and PM—were studied in early chapters of this book. Digital signals use these types of modulation following the same basic principles, yet with some new perspectives. In this section we look at how restricting the value of the modulating signal to just two values (binary) or a small group of values (multiple level) affects the results of modulation and the ability to recover the original data. Digital modulation is not restricted to using just one type of modulation at a time. Communications systems use combinations of AM, FM, and PM to achieve the necessary data rate, error performance, and ease of signal recovery at various signal-to-noise ratios.

We will also look at how digitally modulated systems are tested. To test the performance of a system carrying information in digital form, traditional instruments such as the oscilloscope are used in different ways so that the signal quality and the amount of signal noise can be measured and observed. The effect of adjustments or changes in circuitry can be watched as they take place. Specialized bit-error-rate instruments generate long series of bits, and then compare these generated bits with the bits as received. The result is a single number that precisely summarizes the performance of the digital system. Testing is much simpler for the technician or engineer when the transmitted data bits are looped back at the receiver and sent back directly to the source: A single person can perform all tests working from just one end of the system.

Special circuitry, based on the shift register used for checksums, generates the long, random bit patterns needed for testing. The same circuitry is used with small modifications to provide data security so that unauthorized eavesdroppers cannot make any sense of data bits they intercept, while authorized receivers can easily determine the correct bit values.

13.1 BASIC MODULATION AND DEMODULATION

The data bits of the entire message—including any start, stop, preamble, and postamble bits—interface to the communications channel at the *physical* level. The encoded and formatted data bits are now ready for actual transmission, but they must be in a form that the channel can accept. Amplitude, frequency, or phase modulation is used to transform the simple voltage levels that represent the data into signals that are compatible with the channel. In some applications, a combination of AM, FM, and PM provides the best results.

The problem of demodulation for digital signals is very different than it is for analog, continuous-value signals. The receiver and demodulation circuitry of a digital system are designed knowing that only specific signal values are allowed or expected. Instead of having to *reproduce* the transmitted signal exactly, the receiver must simply decide into which digital bin or subzone the received signal falls, corresponding to the original transmitted digital signal. What digital signaling does is to change the demodulation problem from one of accurately *estimating* the original signal value (1.25 V versus 1.26 V, for example) into a problem of *detecting* the received signal and *determining* the correct digital value.

This is not a minor point. The circuitry for modulating a signal to one of several allowed values is similar to analog modulation, although the trade-offs and alternatives that AM, FM, and PM offer must be understood in the perspective of the digital world as well as the analog one. However, although digital signal detection and determination at the receiver require more complex circuitry than analog signal estimation, it can be performed with much greater reliability and fewer errors than estimation, as we will see in this section.

Binary versus Multiple-Level Digital Modulation

In binary systems there are only two allowed signal values. From a demodulation perspective, this provides the greatest resistance to corruption by noise since the two signals can be separated by the maximum difference or distance in voltage, frequency, or phase for AM, FM, or PM, respectively. However, binary signals make poor use of the bandwidth of the channel. Many digital modulation systems use 4, 8, or 16 digital levels to increase the utilization of the available channel bandwidth. The technical trade-off is that the signals are closer to each other—

there is less *separation* in there voltage, frequency, or phase—and thus more susceptible to corruption by noise and subsequent misinterpretion. The system designer must analyze the expected noise and the acceptable error rates, along with the impact of any error detection or correction bits.

To take binary data and transform it into multiple-level digital signals for modulation, the binary bits are grouped together. For example, consider a binary stream of 16 bits:

$$0110001011010010$$

If simple two-level binary modulation is used, these 1s and 0s can directly modulate the carrier, and a total of 16 bit periods is required. When four digital levels are available, the bits are grouped into clusters of two bits (called *dibits*). Each two-bit group of the data stream will have one of four unique values (00, 01, 10, and 11):

$$01\ 10\ 00\ 10\ 11\ 01\ 00\ 10 \quad \text{(spaces for clarity only)}$$

Each dibit modulates the carrier to one of four values, and only eight signaling periods are needed. For example, 00, 01, 10, and 11 might correspond to amplitude modulation of a carrier with a 0-, 1-, 2-, and 3-V modulating signal (Figure 13.1a).

Example 13.1

What is the modulating voltage pattern for 01100100 in the binary case when 1 V is represented by logic 0 and 4 V is logic 1? What is it when the same bits are transmitted with four-level digital modulation, with 1-, 2-, 3-, and 4-V signals for the first through fourth levels?

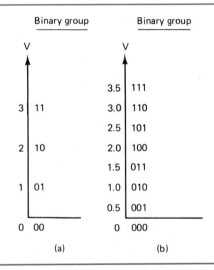

Figure 13.1 (a) Digital AM representing two-bit groups with one of four distinct voltage levels; (b) digital AM for three-bit groups requires eight distinct voltage values.

Solution

For the binary case, it is 1, 4, 4, 1, 1, 4, 1, and 1 V. For the four-level case, group the bits: 01 10 01 00. This corresponds to 2-, 3-, 2-, and 1-V modulating signals.

For systems where eight modulation levels are available, the sixteen bits are grouped into clusters of three bits (*tribits*) to indicate one of eight unique digital values (Figure 13.1b):

$$011 \quad 000 \quad 101 \quad 101 \quad 001 \quad 000$$

(the last two bits are 0s that are added to "even out" the final group). Note that only six signaling periods are needed.

Bit Rate and Baud Rate

The grouping of bits leads to the difference between bit rate and baud rate. For the two-level binary system, the number of bits represented by the signal is equal to the number of up-and-down signal transitions, or baud. Therefore, the bit rate and the baud rate are the same. For multiple-level digital signals, this is not the case. Consider the four-level digital signal where each digital level corresponds to a two-bit data group. Each complete signal transition now represents two bits of information, not one. The number of unique signal values is twice the number of transitions, so the effective information bit rate is twice the baud rate. A 1200-baud signal using four levels (two bits) is transferring information at 2400 bits/s. Similarly, if the system uses eight levels (three bits), the 1200-baud rate is equivalent to 3600 bits/s. Here we can clearly see how multiple-level digital is more efficient than simple two-level binary for transferring information bits at a given signaling transition rate—but at a cost of possibly greater error rates due to reduced distance between signal states.

Amplitude Modulation

The simplest form of modulation and demodulation is AM, where the signal voltage sets the carrier level to one of several distinct values corresponding to digital values. For binary digital signals, there are only two carrier levels. These two levels are usually set at half-amplitude (binary 0) and near full amplitude (binary 1). The distance (the spacing) between the two values would be greater—and thus more tolerant of added noise or distortion—if the binary 0 level is set at 0, which means that a binary 0 corresponds to no received signal level. This makes demodulation and troubleshooting difficult—no received signal indicates a binary 0, but it also might be due to a problem with the channel.

The receiver and demodulator for the binary AM signal is shown in Figure 13.2. After the modulating data signal is recovered from the carrier by standard AM demodulation techniques, it is passed on to a *comparator* circuit. The *threshold* of this comparator is set between the signal levels that represent binary 0 and binary 1. Any signal presented to the comparator with amplitude below the threshold is a 0, and any signal above the threshold is a 1.

The threshold is normally set exactly midway between the nominal 0 and 1 levels, but in some cases it may be offset from the midpoint. This is done when the noise is not constant versus modulated signal amplitude, as is sometimes the case.

406 Chap. 13 Digital Modulation and Testing

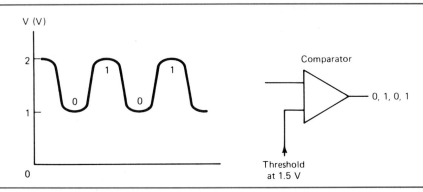

Figure 13.2 The basic demodulator and decision circuitry for binary AM uses a comparator to judge the input signal versus a threshold.

For example, if the transmission system and receiver has more noise at low signal levels and less at higher levels (due to internal circuitry "quieting" effects), the binary 0s will have more added noise and the 1s will have less. The threshold will be set farther from the 0 level and closer to the 1 level to accommodate this asymmetry.

The situation for the nonbinary digital case is slightly more complicated. Instead of a single comparator to determine 1s and 0s, the receiver uses multiple comparators to form a *window* that frames each of the digital levels, shown in Figure 13.3 for the two-bit, four-level case. Each digital level requires two comparators, one with a threshold at the high boundary and the other with the threshold at the low boundary. After demodulation, the signal is applied to all the window comparators. The window comparator output is *asserted* (goes active) only if the signal meets two conditions: It is both above the low limit and below the high limit. The other window comparator outputs are inactive. For any received signal value, only one window comparator output will be active, and this points to one of the four levels. The transformation from the one-of-four output lines of the bank of window comparators to the original binary pattern is done by ordinary digital logic circuitry. A single 74148 IC performs *8 line-to-3 bit* conversion needed for this.

Frequency Modulation

Amplitudes are often severely corrupted by noise, have varying levels due to fading, and suffer from misadjustment of gains in different amplifier stages in systems. Frequency modulation is much less sensitive to these problems. As an added benefit, AGC and limiters can be used to maximize the signal amplitude since the amplitude conveys no information. FM is used on signals that travel through links that have characteristics not under the control of the sender or receiver, such as telephone lines.

For binary FM, often called *frequency shift keying* (*FSK*), distinct tones represent the 1s and 0s. For example, a 1000-Hz tone can represent a 0, while a 3000-Hz tone is a 1. The receiver requires the basic FM demodulator and then presents the recovered modulating signal to two narrow-bandwidth *bandpass* filters (Figure 13.4), which pass only those signals within the frequency set by their design. The filter outputs then go to comparators with adjustable thresholds, which judge if either filter output is strong enough to qualify as a valid signal. Only

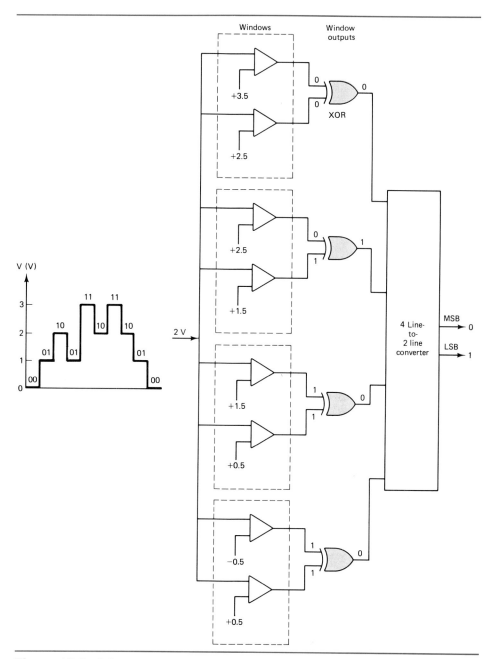

Figure 13.3 A four-level AM demodulator requires window comparators and output logic to reconstruct the two-bit pattern that produced the received voltage.

one filter output can be active at a time, representing either a 1 or a 0 received bit. If the overall signal level is too low due to line or channel problems, neither output will be valid and both comparator outputs will be inactive. Simple digital logic circuitry can convert the comparator outputs, when valid, to a single 1 or a single 0 corresponding to the data bit.

For multiple-level digital systems, a bank of narrow bandpass filters is re-

408　Chap. 13　Digital Modulation and Testing

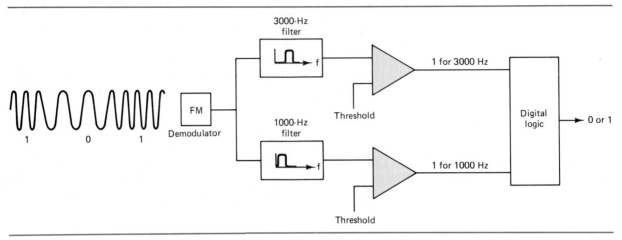

Figure 13.4　A binary FM demodulator has filters, threshold comparators, and output logic to judge if the received frequency strength is sufficient for a valid bit.

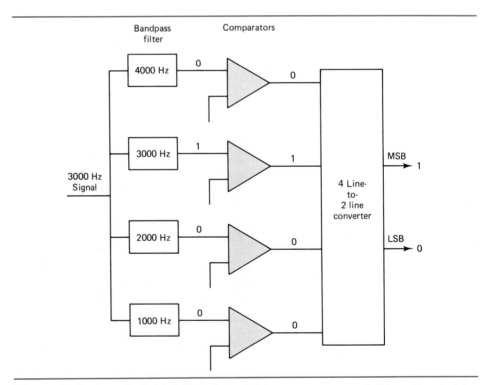

Figure 13.5　A multiple-level FSK demodulator uses a group of filters, threshold comparators, and logic circuitry to recreate the original dibit, here shown for a 3000-Hz received signal.

quired instead of the bank of window comparators of the AM system (Figure 13.5). Multiple frequencies, such as 1000, 2000, 3000, and 4000 Hz, provide four levels of digital representation for 00, 01, 10, and 11. Each filter provides an output if there is signal energy in its passband, and the filter outputs are checked by comparators with thresholds to make sure that noise does not cause a false output. The four comparator outputs are reduced to the correct two-bit pattern by digital logic, as in the AM case. The figure shows the signal values and logic states for a binary 1,0 data group (the third level).

Phase Modulation

Filters such as those used to recover FSK are relatively costly and need to be adjusted carefully. This is a system drawback that grows significantly as the number of filters increases for multiple-level digital signals. For this and other reasons, PM in the form of *phase shift keying* (*PSK*) is often used. PM and PSK are often used at higher signaling rates than regular FM or FSK allow on low- to medium-quality channels. A constant-frequency signal is phase shifted to represent the various digital values. For binary signals, only two phase values are needed (such as 0 and 90°). Multiple-level PSK uses multiple distinct phase shifts such as 0°, 30°, 60°, and 90°. Larger amounts of phase shifting separate the signals more and reduce the chance that noise and phase jitter will cause errors due to misinterpretation of the correct data bit.

The basic receiver must recover the modulating signal from the carrier using circuitry similar to an FM receiver. Circuitry to recover the PSK data uses a single phase detector, regardless of the number of digital levels (this is in contrast to FSK, which needs a filter for each FSK value). The phase detector compares the phase of the demodulated signal to a reference signal and produces an output whose amplitude varies in inverse proportion to the phase difference between the two signals. When the reference and the received signal are in phase (0° shift), the phase detector output is maximum and decreases as the phase difference (error) between the reference and the received signal increases. For example, the phase detector output may be 1 V for 0° error, and decrease to 0 V for 180° of error.

The phase detector output next goes to a bank of window comparators, as in the AM case. These comparators provide an output indicating the digital value corresponding to the phase shift. The window boundaries are normally set midway between the voltages corresponding to the phase shift values. As in the AM case, the window outputs are then converted by ordinary digital logic to show what the original, modulating data bit pattern was that caused the shift. Note that the top-to-bottom order is reversed for PSK, since a 0° shift produces the largest output, but this is easily taken care of by the digital logic. The figure shows the signal values and logic states for a four-level PSK system where 01 is the data pattern.

Of course, there is a technical difficulty with PSK compared to FSK. The reference phase signal must be provided with great accuracy and must track any changes (due to instability, channel noise, or jitter) in the 0° phase reference of the transmitted signal. Filtering and phase-locked loops are usually used to extract the unshifted signal reference from the received modulated signal.

Review Questions

1. In what way is the demodulation problem for digital signals very different than it is for analog signals?

2. Why is binary the simplest form of digital modulation, and why does it provide the best error resistance? Why does it use bandwidth poorly?

3. How are binary data bits transformed into multiple-level digital signals?

4. Why is the baud rate equal to the bit rate for binary signals? What is the relationship between the two rates for nonbinary digital signals?

5. Explain how dibits result in an effective bit data rate of twice the baud rate.

6. How are comparators used to decode multiple-level digital signals into one of N signal lines? How is the digital bin number converted back to the binary pattern?

13.2 QUADRATURE AMPLITUDE MODULATION

To achieve a good combination of high data rates, efficient use of available bandwidth, low error rates, and ease of demodulation, a combination of AM and PM is often used in communications systems. In *quadrature amplitude modulation (QAM)*, a single carrier is both amplitude and phase modulated, with a resulting signal $s(t) = \sin[2\pi ft + \phi(t)]$. Since AM is prone to corruption by noise, QAM is used on channels that have stable and predictable performance, such as the microwave links that relay wide bandwidth signals from tower to tower across country.

In actual implementation, QAM is generated from two carriers of identical frequency, 90° out of phase, by performing a 90° phase shift on a single carrier (Figure 13.6a). QAM is described as having two components: an *in-phase (I)*

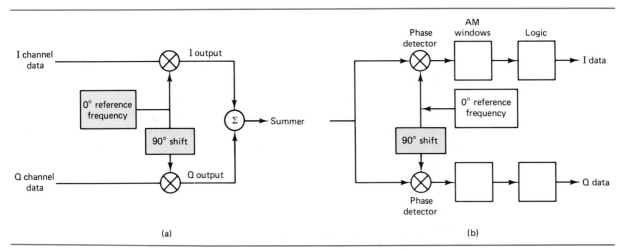

Figure 13.6 (a) The QAM modulator uses identical circuitry for *I* and *Q* channels, except that the *Q*-channel carrier is shifted by 90°. The *I* and *Q* outputs are summed to form the final QAM output. (b) The QAM demodulator is similar to the PSK demodulator, except that the *I* and *Q* signals are demodulated with 90° out-of-phase references.

carrier component and a *quadrature* (*Q*) carrier component at 90° phase shift with respect to the *I* component. The equation that describes the QAM signal $s(t)$ of frequency $2\pi f$ as a function of *I* and *Q* is

$$s(t) = i(t) \sin 2\pi f t + q(t) \cos 2\pi f t$$

where $i(t)$ is the amplitude of the in-phase component (the sine) and $q(t)$ is the amplitude of the quadrature component (the cosine). (Recall from trigonometry of signals that a sine and cosine of identical frequencies are 90° out of phase with each other.)

It may seem that we are really dealing with two separate signals here rather than varying the amplitude and phase of a single signal. Trigometry can be used to show that the (cos + sin) equation for $s(t)$ is equivalent to a single carrier with fixed frequency but variable amplitude and phase. Varying *I* and *Q* signal amplitudes with $i(t)$ and $q(t)$ is the same as performing AM and PM on a single signal, but the *I–Q* method is much easier to implement, demodulate, and analyze.

The *I–Q* signals of QAM can be modulated independently in amplitude. This modulation is often called *vector modulation* because the resulting time-varying signal voltage can be resolved into the two separate components of in-phase and quadrature signals, with $i(t) = \cos \phi(t)$ and $q(t) = \sin \phi(t)$. (This is just like resolving a force vector into two components at right angles to each other: A force at any angle can be expressed as the sum of an appropriate size horizontal force combined with a vertical force.) Demodulation of *I–Q* signals is shown in Figure 13.6b with circuitry similar to the modulation circuitry; it is discussed in more detail below.

In QAM, each carrier is modulated to one of several digital levels. With *n* distinct levels for each carrier, a total of $n \times n = n^2$ unique digital states can be represented by a QAM signal. For example, two amplitude levels and two phase levels provide $2 \times 2 = 4$ digital signal points. Each of these points represents a single two-bit pattern. For 4×4 modulation, there are 16 signal points identified by each grouping of four bits (a *quadbit*) since $2^4 = 16$.

QAM provides a good compromise in performance for multiple-level digital signals. Its vector modulation provides two independent channels—*I* and *Q*—of information in the bandwidth that a single channel would normally occupy and is relatively easy to implement and demodulate at the receiver. A QAM vector demodulator (Figure 13.6b) demodulates and decodes the *I* and Q signals separately. A single in-phase reference is used in a phase detector to extract the *I* component, which is then demodulated to determine $i(t)$. The same reference shifted by 90° is used to extract the *Q* component and determine $q(t)$. Signals $i(t)$ and $q(t)$ are then decoded into the original bits by window comparators and digital logic, operating on both the *I* and *Q* outputs. When the *I* and *Q* modulations are identical in $n \times n$ QAM, the decoders are also identical.

Systems in commercial use have 2×2, 4×4, and 8×8 QAM, equivalent to 4, 16, and 64 unique states and representing 2, 4, and 6 bits per signal point (since $2^2 = 4$, $2^4 = 16$, and $2^6 = 64$). The popular 8×8 QAM provides a bit rate that is six times the baud value. It is relatively efficient in using the available bandwidth, but has reduced distance between signal states and thus greater possibility for errors, depending on the noise levels.

Constellation Plots

The signal patterns can be graphed in a *constellation plot* (it resembles stars), shown in Figure 13.7a for 2×2 QAM and Figure 13.7b for 4×4 QAM. The horizontal axis shows the modulation values of the I carrier and the vertical axis shows the Q carrier values. At any instant, the received signal has a single amplitude and phase value, which can be demodulated and decoded into the corresponding multiple-bit pattern. The I axis represents the first n bits of the data group, and the Q axis represents the remaining n bits (thus defining $n \times n$ unique points in the constellation). For 2×2, the I axis can have two values (0 or 1) from one bit position, while the Q axis also has two positions (1 and 0) from the second bit of the dibit. In 4×4, there are four positions (00, 01, 10, and 11) derived from the two bit positions along each axis, for a total of 16 points defined by four bits.

The constellation diagram also shows the distance between each digital signal and its neighbors, and how much I and Q noise can be tolerated before errors in decision making occur. Noise and distortion on a received signal will move it closer to one or more neighbors. Note that signal points have different distances from other points; a signal is much more likely to be confused with a nearer neighbor than with a more distant one.

QAM does not have to be symmetrical, with n bits on each axis. It is technically possible to use formats such as 4×2 (representing 2 bits + 1 bit = 3 bits, or 8 states) or 8×4 (representing 3 bits plus 2 bits = 5 bits or 32 signal states). This may provide the right balance between bandwidth use and distance between signals. However, the symmetrical patterns are much easier to use and troubleshoot, since the I and Q channels are identical in modulation and demodulation (except for the 90° phase shift) and can even use duplicated circuitry. This is a major practical advantage.

The constellation diagram is very useful for judging the quality of the received signal and its demodulation (Figure 13.8). Noise in the I-channel component means that the constellation dots will not be points but will have left-to-right movement, whereas Q-channel noise will cause up-and-down movement. Noise problems in both channels of the signal make the points become fuzzy circles, and these will become larger as the noise increases. As the fuzzy circles of adjacent points get larger and overlap, the receiver decision-making circuitry will make errors. When the amplitude or phase levels are not evenly separated or there are amplifier nonlinearities (usually due to misadjustment at the transmitter or receiver), the constellation will be distorted. All these problems can quickly be seen by the eye, and the effect of any changes or adjustments can easily be seen as the adjustments are made.

In summary, the constellation diagram immediately shows the test technician or engineer several key things: where the digital signals are, how many states of digital signaling are being used, how close the states are, how much noise there is, how constant the channel performance is (since the dot circles expand, contract, and move as the channel characteristics change), and the effects of any tuning or adjustments in the system electronics. Some test instruments actually show the signal path (trajectory) from dot to dot. This *vector* plot is even more informative but harder to interpret.

Review Questions

1. What is QAM? Why is it a good choice for efficient use of bandwidth?
2. How is a QAM signal usually generated? Why is the implementation different than the classical AM + PM equation indicates, but still equivalent? What are the I and Q channels?
3. How does QAM provide multiple levels of digital modulation? Explain how n levels of I and Q modulation represent n^2 signal states or the equivalent number of bits.
4. What is a constellation diagram? What does it show in terms of QAM signal quality and problems?
5. How is QAM demodulated? When can the I- and Q-channel demodulation circuitry be identical?

13.2 Quadrature Amplitude Modulation

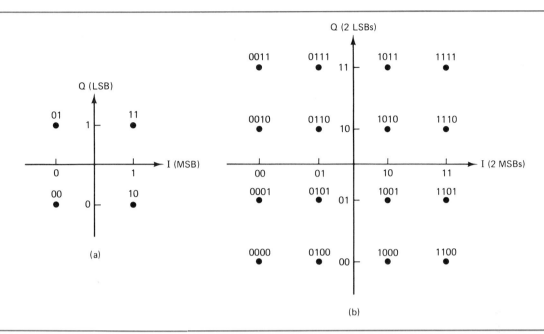

Figure 13.7 (a) The 2 × 2 QAM constellation shows how each of the four dibits occupies a unique point in the *IQ* graph. (b) The 4 × 4 QAM constellation shows 16 *I*, *Q* points, each representing a single four-bit value.

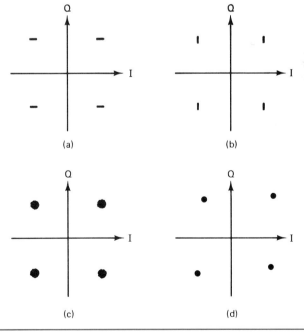

Figure 13.8 Noise and distortion effects are easily seen in constellations: (a) *I*-channel noise only; (b) *Q*-channel noise only; (c) both channels with noise; (d) system nonlinearities.

13.3 LOOPBACKS, ERROR RATES, AND EYE PATTERNS

The techniques used to test digital communications systems start with some of the same fundamentals that an analog system uses, measuring signal distortion, noise, frequency response, and linearity. But just as a digital communications system has layers of functionality—modulation, formatting, encoding, and protocol—a digital system can additionally be tested at different layers to provide a different and less ambiguous set of measures. The step above testing the fundamental characteristics of the channel using eye patterns (discussed below) is testing with known characters that have been encoded and formatted. Above this level of testing is the need to use random bit patterns for the characters, to see if the system can properly encode and format these characters.

Loopbacks

Many communications systems cover large distances and it is impractical for a technician working from the transmitting end to observe the results sitting at the receiver. Even if there is another technician at the receiving end, the two would have to coordinate their activities and talk to each other, often difficult or even impossible. A better solution is to have the receiver send the original data back to the transmitter. This is what *loopback* capability provides for half- and full-duplex systems (Figure 13.9).

Loopbacks echo the data back toward the data source. For full-duplex systems, the forward and reverse channels are set up so that the data makes a "U-turn" and is sent back to the transmitter immediately, except for unavoidable propagation delays. In half-duplex systems, a block of data is sent in the forward direction, then the channel is reversed and the data are sent back. In either case, the transmitter electronic test circuitry must compare the looped back data, as received, with the data that were originally sent. Any differences indicate the bit error rate of the two-way transmission.

The *bit error rate* (*BER*) for the forward direction cannot be precisely determined from the round-trip BER since errors can occur in either the forward or the looped back paths. The two-way BER does certainly set an upper limit on the system BER, since the BER in one direction cannot be greater than the two-way

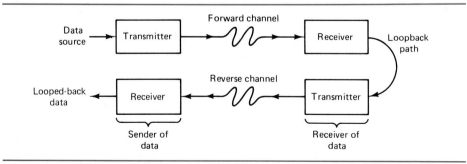

Figure 13.9 The loopback scheme returns received, unprocessed data to the sending point so that tests can be run by one person or be completely automated.

value. Thus the BER measured by loopback gives a *worst-case* value for the BER in the forward direction or in the reverse direction. Another solution is to assume that half the errors occurred in each direction. Although this is often not strictly the true situation, it is often a good estimate.

The loopback is switched on for the system in one of several ways. A person at the far end can physically connect the received signal output back to the transmitted signal input using a cable or special switch setting. Although effective under some circumstances, this approach is not effective for one person or unattended, automatic testing. Instead, most communications interface circuits have a *loopback mode,* where they are instructed by a special test command simply to echo back the data they receive. Many UARTs and synchronous protocol ICs provide this capability. A special instruction causes the IC simply to take the received data bits and send them back out. There is no need for the bits to pass through the un-formatting, decoding, or protocol stages of the receiver, only to have them go through these steps again as they are retransmitted back. The data transmitter simply wants to know if the bits as received agree with the bits as sent at the UART.

Any error detection and correction circuitry must be disabled or bypassed, or else some bits that are actually received with errors will be corrected before retransmission. This would indicate the overall system performance in the face of errors, but not the quality of the communications channel itself. A channel with absolutely no errors would have the same BER as one with a few errors that had been corrected, so the effect of any changes in channel performance would be obscured. This does not mean that the performance of the system with EDC is of no interest, simply that the BER test must be clearly specified to show the channel performance versus the performance of the system with EDC in place.

The bit patterns sent by the transmitter when in loopback and test modes are of interest. Most loopback testing includes sending special bit patterns to see if there are any special sensitivities in the system, its clocks, or its circuitry. Common patterns for the data include all 1s, all 0s, alternating 1s and 0s, and random bit streams. These patterns allow the test to show how well the circuitry handles different situations and maintains proper timing, as well as to measure the operation of the modulation and demodulation circuitry of the physical interface.

Error Rates

Error rates are a very clear indicator of system performance under various conditions. The error rate can be determined with or without the loopback scheme, although it is easier to implement with loopbacks. A bit error rate tester (or tester pair, for nonloopback situations) is designed for this purpose and makes the task easier, although it is not absolutely necessary.

For the system with loopback, the BER test unit transmits a known bit pattern (which can be one of several that it provides). The bits are received back at the source, while the BER tester receives them and compares them to what it sent. Since there is some delay between sending and receiving, the BER tester must align the received frame with the transmitted frame so that it compares bit values in the same relative locations in the bit stream (Figure 13.10). Otherwise, it would think that there were many errors when the real problem is that the BER tester is looking too early or too late at the received bits in comparison to the transmitted bits—comparing the first looped-back bit with the third transmitted bit, the second looped-back bit with the fourth transmitted bit, and so on.

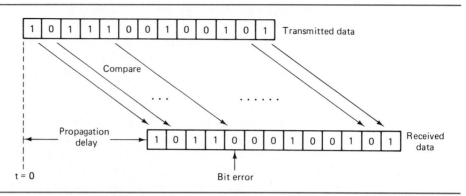

Figure 13.10 BER testing requires that the received data frame be put in time alignment with the transmitted data frame before the frames are compared for differences.

When there is no loopback and two BER test units are used—one at each end of the system—each BER unit must use the same pattern. The receiving BER unit must know which pattern is being sent so that it can compare bits as received with what they should be. This is obviously straightforward for all 1s, all 0s, and alternating 1,0 patterns, but it may seem impossible for a random 1,0 pattern. In practice, the solution is simple: Special circuitry is used which generates what may appear to be arbitrary, random patterns, but they are actually exactly reproducible patterns. The same circuitry produces the same pattern in each tester. The communications channel does not care that the pattern in not truly random (like flipping a coin) but instead meets all the normal tests of randomness, such as an equal number of overall 1s and 0s, a variety of patterns of 1s or 0s in a row, and no obvious repetitions within the bit stream. We discuss such random pattern generator circuitry in more detail in the next section.

Note that random bits cannot always be inserted into the communications system directly at the modulation point. Even a properly working system will have problems if the format fields of the preamble and postamble do not agree with the data fields they represent. If random bits are used to replace the entire bit stream (preamble, data fields, postamble), there will certainly be discrepancies and inconsistencies. Instead, the data bits can be random to check the ability of the system to handle any data patterns, but the preamble and postamble bits must be determined by the system based on its formatting rules.

Additional BER tests A typical BER test unit such as the one shown in Figure 13.11 does much more than send meaningless patterns of bits and determine the number of bits in error. A BER tester sends specific patterns that are compatible with various standard codes and formats. These include actual words and sentences using ASCII code (with start and stop bits), which is useful for someone watching the received characters on a terminal. More powerful testers also produce bits that follow formats such as the T-1 format (Chapter 12), and some common asynchronous and synchronous frames, with preambles and postambles.

There are two reasons why BER testers send bit patterns and frames that follow standard formats in addition to the more random patterns. First, for some systems to even accept and return the data, it must fit into the standard frame

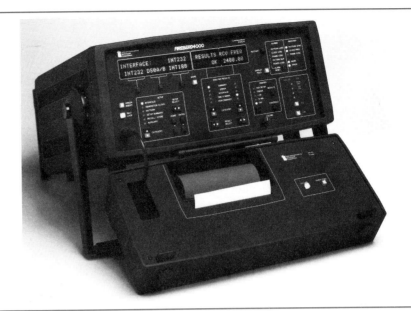

Figure 13.11 A typical BER test unit generates the data bit patterns and compares this pattern to the received bits with a variety of measures (courtesy of Telecommunications Techniques Corp.).

format of the system. Second, these BER testers can be set to analyze the types of errors and thus reveal more information about the kind of problems the system has. These include measurements of:

The BER as discussed so far, based on thousands of bits

The percentage of total seconds that have no errors at all

The percentage of frames that have errors, or the percentage that have no errors

The importance of these measures is explained with examples. Consider a system that sends data in frames of 100 bits, including preamble, message, and postamble bits, and each frame takes 0.5 s (a very slow rate of 200 bits/s, but this is for illustration only). When 1000 frames are sent, a total of 100,000 bits is transmitted, and this requires 500 s. Suppose that the system averages one bit error every 2 s, for whatever reason. The overall BER is 250 errors per 100,000 bits = 2.5 per 10^3 bits. The percentage of error free seconds is 50%, since there is one error every 2 s. Finally, since there is one error every 2 s, and four frames in that time period, the percentage of frames that have errors is $\frac{1}{4} = 25\%$.

Now, consider the case where the same number of errors (500) occur but they are concentrated with 100 errors/s spread over the first 2.5 s. The BER is still 2.5 per 10^3, but the percentage of error-free seconds has risen to (500 − 2.5/500 = 99.5%. The percentage of frames without error is also 99.5%, by similar reasoning.

There are many other possibilities for the spread of errors while the BER remains unchanged. For example, the first 250 bits could be in error, resulting in 1.25 s that are all error and (500 − 1.25/500) = 99.7.5% error-free seconds, and three frames that are completely in error, so the percentage without error is (1000 − 3/1000) = 99.7. Other examples will produce percent error-free seconds and percent error-free frames that differ significantly.

The interpretation of the three error measures is this: When both the raw bit error rate and the frame error rate are high, while the percentage of error-free seconds is low, the errors are scattered randomly through the total transmission. This is most likely due to continual severe noise in the channel or system, or circuit misadjustment or malfunctions. However, when the BER is high yet only a small percentage of the total transmission time and frames had errors, these errors are concentrated and probably caused by some severe noise but of short duration or by temporary synchronization problems. This information is very useful in beginning to troubleshoot the system.

BER and noise testing   The BER test unit is often used to measure the system performance as a function of SNR. In many situations it is difficult to vary or control the SNR as desired. For example, the SNR of a satellite link cannot be directly affected, since there is no way to exert control over the ambient electrical environment. The solution is to fool the system by deliberately corrupting the transmitted signal with precise amounts of noise added by the BER tester to the signal it generates. This noise adds to the unavoidable noise of the link itself. The test uses these injected noise values to judge the performance of the system at various SNR levels. The highest SNR that can be measured exists when the artificially induced noise is zero; SNRs below this value cannot be obtained.

To measure system performance below this SNR floor, a special test setup must be used. For the satellite case, for example, the electronics of the ground transmitter and the satellite are placed in a special chamber that excludes all external electrical noise. The transmitter power is reduced to simulate the effect of the signal loss over the distance of the real satellite link, and the tests are run with the injected noise and known SNRs. Other special corrective factors are used in the final calculations to compensate for differences between the real application and the lab. The result is that BER versus noise performance for a wide range of SNRs can be measured. These special chamber tests are complex and not done routinely. In contrast, BER and BER versus SNR are easily measured and done routinely or even automated, despite the unavoidable system noise, as long as this noise can be accepted as part of the test.

Eye Patterns

Although the communications signals represent digital information, the real world that they travel through is analog in nature. Even a "perfect" digital signal with absolutely square corners will look quite different in reality. The corners will be rounded by limited bandwidth, the signal will be distorted or tilted due to unequal propagation delays of the various spectral components, and the amplitude will be reduced or varying with time. Changes in the communications channel will result in changes in the digital signal shape and in the ability of the receiver to recover the data bits with little or no error.

Just as the constellation pattern shows the digital signals conveniently, the *eye pattern* is the indicator that summarizes all the imperfections and realities of the received data bits (Figure 13.12). To form an eye pattern, all the data bit

13.3 Loopbacks, Error Rates, and Eye Patterns

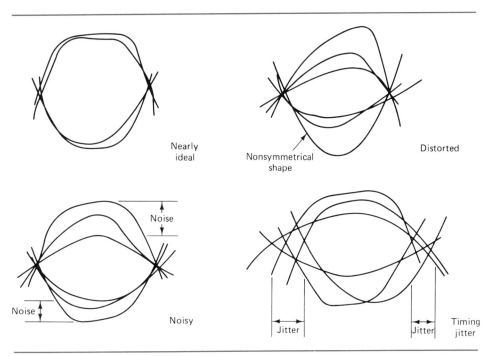

Figure 13.12 Eye patterns quickly summarize many characteristics of the communications channel and data bits, including noise, distortion, and timing jitter.

signals are "overlaid" on one another. If the bits were ideal, the eye pattern for binary signals would be a perfect rectangle. Instead, the overlap result looks more like an eye. The opening of the eye shows the distance between the voltages representing 1s and 0s. As this opening gets smaller, the margin or distance between the two signal states is less and thus more prone to errors in determining if the signal is a 1 or 0. Blurring, thick eye tops and bottoms mean that the signal levels have more noise. As conditions improve, the eye opening gets wider and the noise margin improves.

The receiver should ideally sample the data bits at the point where the eye is widest, which may not be the same as the midpoint of the eye. Some advanced communications receivers actually adjust their receiver clock timing to look at the bits at this optimum point rather than the midpoint. By advancing or delaying the judgement time at the receiver to this wider part of the eye, the BER at any SNR value is improved.

The eye pattern also shows how much clock jitter and undesired phase shifting the received bits suffer. Although the eye may be wide open, left-to-right blurring at the crossover points means that there is instability in the system timing. This causes errors because the system timing will not be stable: the signal decision points will change, or the crossover points will come close to the decision time rather than stay at the maximum opening part of the eye.

Like the constellation, the eye pattern changes immediately to show any effect of changes in the system or the channel as filters are varied, noise levels change, or adjustments in the circuitry are made. Fortunately, the eye pattern is easy to set up using a standard oscilloscope.

Setting Up an Eye or Constellation Pattern

Developing an eye pattern does not require special equipment, although many communications test instruments also provide eye pattern display capability. An ordinary oscilloscope is used with its time base is set to sweep across the CRT face in one- to two-bit periods. The data signal is connected to the vertical input of the scope. Most important, the trigger for the sweep is the clock signal used to receive the data bits. This must come from the part of the receiver that provides recovered clock (for synchronous systems) or the internal clock (for asynchronous systems) to the rest of the receiver.

As each bit is received, it triggers the scope sweep and is displayed on the scope screen. Each subsequent bit begins a new sweep across the screen, but the persistence of the scope display causes the viewers' eye to blend all of these together and integrate them into the overlaid image known as the eye pattern. When the data bits change in amplitude, noise, or timing, the eye pattern also changes. In effect, the eye pattern is a simple and accurate indicator of signal quality in noise and timing and a great aid for tuning and adjusting systems.

An oscilloscope is also the basic element for the constellation display (Figure 13.13a). The demodulated I signal is fed into the X (horizontal) input amplifier, while the Q signal is fed into the Y (vertical) input amplifier. The gain and position of each channel of the oscilloscope must be adjusted by the amplifier knobs, of course, so that the digital signal levels fill the scope face and are centered. As the I and Q signals take on their many values, the scope face will show the constellation dots as the X and Y voltage inputs.

Unfortunately, the scope will also show the transitions between digital states, and the screen will be a confusing array of dots interconnected by many lines where each line is the transition path from one digital state to another. A small additional circuit is needed to eliminate the interconnecting lines and provide just the dots. The scope Z-axis (usually an input on the back of the scope) controls the intensity of the display, with full intensity when the Z input is 1 V and no intensity (darkness) when the Z input is at 0 V. This occurs independent of what the X and Y inputs are.

This special circuit uses the data clock signal to develop a signal that is always at 0 V (darkness) but is 1 V (full intensity) at those times when the clock indicates that the data signal is valid (Figure 13.13b). This signal goes to the scope Z input. When digital values are input for display on the scope screen via the X and Y inputs, the display will be dark between the valid data times when the signals are at the constellation points. The transitions will thus draw no visible lines on the screen. When the data clock shows that the data signal is valid and meaningful (not in the transition time), the circuit lets the scope signal be visible. Thus only constellation points are displayed, and it is clear to the viewer if these points are sharp, fuzzy, or moving.

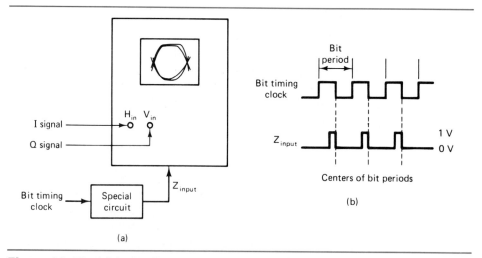

Figure 13.13 (a) An oscilloscope setup for constellation display requires an additional circuit to control display intensity; (b) timing of bits versus signal to control scope intensity for constellation.

13.4 Random Bit Generation and Data Encryption

For QAM systems, two eye patterns—one each for the *I* and *Q* channels—are usually displayed simultaneously, one above the other. This is very useful since QAM systems require precise and stable 90° separation between the two carriers, and any deviation from 90° appears as system noise, reduces SNR, and increases BER. The simultaneous display of *I* and *Q* eye patterns lets the technician judge this quadrature difference as well as make adjustments to the system to improve quadrature performance.

Review Questions

1. What is a loopback? What does it show?
2. Why must EDC be diabled for meaningful loopback results?
3. Why does the loopback test often use many specific data patterns?
4. Why must the BER tester align the received bit frame with the transmitted one? What happens if they are not aligned?
5. Explain why the BER tester uses some standard codes, or format and frame patterns, in addition to patterns of meaningless bits.
6. What are the three types of error measurements common in error testing? What does each indicate? What do all three, when interpreted together, tell about the system error source?
7. How can very high SNRs (low noise) situations be tested?
8. What is an eye pattern? What does it represent for a digital signal?
9. How is a scope used for displaying the digital constellation? Why is a special-intensity control circuit needed? How is this signal derived?

13.4 RANDOM BIT GENERATION AND DATA ENCRYPTION

In the preceding section we saw that there is a need in testing for apparently random bit patterns, yet this random pattern must be reproducible in different test instruments for use in comparing what was sent with what was received. This seems to be a pair of incompatible goals: randomness in a bit pattern but repeatability of the pattern when needed.

Random Bit Generation

Fortunately, there is an easy solution to the problem. The algebra theory which showed how to develop CRC checksum and EDC patterns (Chapter 12) also shows how simple circuitry can be adapted for generating bit patterns that appear completely random but are also recreatable as needed. The circuitry uses a variation of the shift register with XOR feedback paths. Recall that the CRC checksum is developed by setting the register to all 0s and then clocking in the data bits. When the last data bit is entered in the shift register, the remainder value in the register is the checksum.

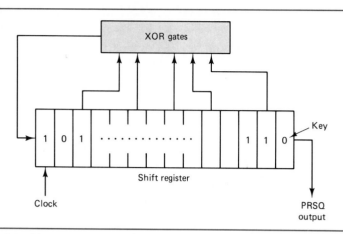

Figure 13.14 Circuitry for generating random-like bit pattern is based on simple modification of CRC checksum circuitry.

Figure 13.14 shows the circuitry difference for generating random-like patterns: Instead of using an initial state of all 0s, initialize the register with a binary pattern that we will call the *key*. Disconnect the input to the register where the data bits enter to have their checksum computed but leave the feedback path to the first flip-flop. Clock the shift register continuously and observe the rightmost bit as the shift register is clocked. What occurs is that for some combinations of register size, feedback taps, and initial binary patterns, the bit pattern emerging from the rightmost flip-flop resembles a random stream of bits.

This stream of bits meets all the traditional tests for random patterns. Over many hundreds or even thousands of bits, there will be an equal number of 1s and 0s. There will be some clumps of two, three, or even more 1s (or 0s), but the bit stream will not have any repetitious patterns. An observer will be unable to predict if the next bit is a 1 or a 0 from looking at the previous bits.

At the same time, this bit pattern is not truly random. Unlike coin flipping—which produces a random pattern of heads/tails (1s and 0s)—this random bit generator circuit produces the same pattern every time, as long as the circuitry is unchanged and the same key is used. We have solved the dilemma of producing an apparently random stream of bits, yet one that can be reproduced as needed. Since this pattern looks to an observer like a truly random pattern, yet can be recreated as needed, it is called a *pseudorandom* pattern or *pseudorandom sequence (PRSQ)*. Note that the pattern appears random and an observer cannot predict the next bit, but someone who has the "secret" of the initial state can calculate the entire pattern in advance and regenerate it as needed.

If you know the algorithm—the structure of the circuit and the initial state—you can reproduce the pattern as needed; if you do not know the algorithm, it is virtually impossible to recreate the bit pattern. For testing communications systems, there is no need to hide this algorithm and the key, but we will see elsewhere in this section where this becomes a powerful tool. The PRSQ does eventually repeat itself, depending on the circuit specifics. For testing purposes, repeat cycles of several hundred bits length are long enough, and the typical shift register has 32 or 64 bits.

13.4 Random Bit Generation and Data Encryption

> **Examples of Generated Pseudorandom Sequences**
>
> Use the shift register wired as shown in Figure 13.16, and set the initial pattern to 1101. Figure 13.15a shows the contents of the register at each clock cycle. The state of the last flip-flop in the register is the PRSQ bit pattern, in this case 1011 0010 0011 110 . . . . Using 0001 as the key pattern (Figure 13.15b), the same hardware circuitry generates a different PRSQ: 1000 1111 1011 00 . . . . Since this is only a four-bit register, it is not possible to get a PRSQ pattern longer than 15 bits without repeating—note that the first pattern repeated after 15 bits, while the second repeated after 14 bits. However, the shift register, feedback taps, and key principle applies to the use of longer shift registers to generate very long patterns before repetition occurs. Note that the PRSQ pattern repeats whenever the shift register contents become equal to the initial key value.

Data Encryption

One problem in communications systems is that unauthorized listeners may intercept a signal and find out critical information or data. Two basic defenses against eavesdropping are available. First, the communications link can be made physically secure so that an eavesdropper cannot intercept any signal. Of course, this is often difficult, since the link may be radio wave rather than cable or fiber optics,

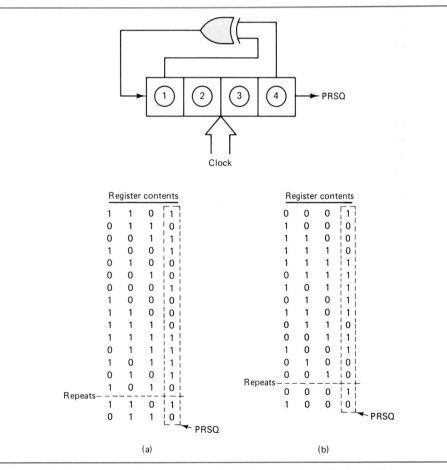

Figure 13.15 The PRSQ generated with (a) key 1101 and (b) key 0001.

and anyone can put an antenna in the air and pick off some of the signal. The second line of defense is to encode the signal in some way so that the eavesdropper can receive the signal perfectly but cannot make any sense of what it says. This is called *encryption* to distinguish it from conventional encoding, where the data uses some well-known code (such as ASCII) and the coding process is not an attempt to hide information.

It is difficult to encrypt signals that are in analog form, although schemes that play "tricks" with the modulation and way the modulating information signal is processed before it modulates the carrier are used. In fact, many cable TV services do this to prevent nonpaying viewers from using a homemade box to intercept signals via satellite dish antenna and so have free access to hundreds of channels.

A simple approach to encryption of digital symbols is to substitute one symbol for another. Children often use a simple code where the letter "a" is replaced by "d," "b" by "q," and so on, for example. In the real world, however, this simple substitution encryption is easy to break by studying the intercepted encrypted message. A practical substitution scheme with digital bits appears hopeless, since a fixed, unchanging substitution rule to change a 1 to a 0 (or vice versa) would be easily broken—simply invert the bits.

Use of PRSQs for data encryption When the information signal is in digital form, however, there is another scheme based on use of a PRSQ which allows simple encryption and nearly perfect security. The transmitter takes the information bits and adds these in sequence to the bits of a PRSQ, one bit at a time. Circuitry that implements this is shown in Figure 13.16, where the addition function is accomplished with a single XOR gate (an XOR logic operation is the same as binary addition). The user data bits now appear random as well, and there is no way for an eavesdropper to determine that an eight-bit pattern really represents 1.27 V, for example. In the simple substitution scheme, a specific symbol is always replaced by the same new symbol, but in this scheme the replacement symbol is continually changed, in effect, by the PRSQ. It is as if the child's scheme first substituted "d" for "a," then used "v" for "a," and then used "m" for "a"—this is much more difficult to break.

For the intended receiver, the seemingly senseless bits are easily restored to their original value. The receiver generates the identical PRSQ and then adds this,

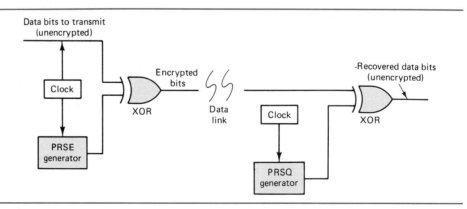

Figure 13.16 The use of PRSQ addition to original data bits for encryption, and addition again for decryption.

13.4 Random Bit Generation and Data Encryption

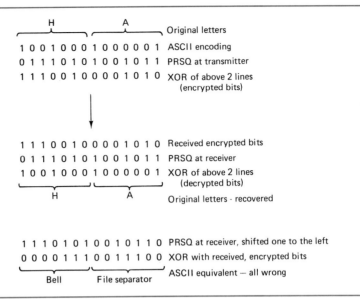

Figure 13.17 Result of encrypting a stream of bits (here, from ASCII encoding) at the source with a PRSQ, and then decrypting them at the receiver; also shown is the effect when the PRSQ is not properly synchronized.

bit by bit, to the received stream of encrypted bits. The result: the original, unencrypted bits emerge. Figure 13.17 shows the effect on a message with standard ASCII encoding of characters that is also encrypted with a PRSQ, then decrypted with the same PRSQ at the receiver. Anyone who attempts to recover the original bits without the PRSQ will be extremely frustrated.

From this we see that successful data security can be achieved by combining the data bits with a PRSQ. Both the sender and the receiver must use the same PRSQ generator circuit and key. As additional protection, the keys can be changed on a regular basis, in case an eavesdropper happens to have the correct circuitry and key (by theft, capture, or determined trial-and-error efforts).

The benefits of transparency are also apparent in this encryption scheme. The data bits are combined with the PRSQ before formatting bits, preamble, postamble, or EDC bits are added, and before final modulation is done. The encrypted bits that result are treated by layers of the system as any other data bits, and processed identically. The only difference is that the bits are meaningless without the proper decryption scheme, but the algorithms and circuitry used to develop the format, preamble, postamble, and EDC bits, or to modulate the bits onto the link make no judgment as to what the bits mean. They simply take bits as received and process them according to a prearranged recipe.

With the wide use of digital signals in many previously analog applications and the appearance of many new, all-digital applications, data security and encryption are being used more often to prevent eavesdroppers from making sense of the bits they intercept. These applications include digital telephones, computer-to-computer communication, facsimile machines, and military communications and pictures. Some compact disks for military applications are now produced with the bits (representing digitized voice, or ASCII coding of typed messages) passed through an encryption circuit *prior* to the deliberate scrambling and EDC bit algorithm used to generate the standard CD bit pattern (Chapter 12). An enemy who captures such a disk (and even the disk player) could not read it; only authorized personnel with the key sequence can see the contents. Encryption is also used to prevent unauthorized people from intercepting and changing the contents of a message, such as when a bank computer sends account-related information from a branch to the main office (tapping in to change the "$100 deposit" message to "$10,000 deposit").

Although this encryption easily fills the need for providing security, it has one technical difficulty. The PRSQ generator used at the receiver must be frame-synchronized with the one used by the receiver, or else the decryption process PRSQ will not line up properly bit by bit with the received bits and the PRSQ used to produce them, also shown in Figure 13.17. A one-bit mismatch in sync means that the supposedly decrypted bits are completely incorrect. Special synchronization techniques are required to accomplish this, and these are more complex than PRSQ generation or the encryption process itself.

Review Questions

1. Explain the apparent contradiction in testing needs for a random bit pattern that is also recreatable on demand.
2. How is the dilemma of question 1 solved? What is a PRSQ?
3. What is the need for data security? What are the two basic approaches? Why is one of these techniques for maintaining secrecy often impractical?
4. How are PRSQs used to solve the secrecy problem? How are the bits encrypted? How are they recovered by the intended receiver?
5. Why can't an eavesdropper make sense of the bits even if they are intercepted perfectly? What is a key? What is the effect of changing the key while leaving the circuitry unchanged?
6. Why is transparency between system layers necessary for ease of data encryption?

SUMMARY

Using data bits to modulate a carrier involves several technical choices. AM, FM, or PM can be used, either individually or in some combinations. AM is simplest but most noise sensitive. FM and PM offer better noise immunity and are less affected than AM by signal-level variations. Binary data are the easiest to modulate and recover at the receiver, as well as providing the largest margin against noise-induced errors; however, binary format makes poor use of the available bandwidth.

To increase the use of the bandwidth, multiple-level digital modulation with 4, 8, or even 16 distinct levels (two, three, or four bits) is used. Multiple-level schemes provide much better use of the channel bandwidth but require more complex circuitry at the transmitter and receiver to recover the data bits. In addition, the signal values are more closely spaced than they are in the binary case, so noise has a greater chance of causing an error. To use multiple-level digital signaling, the data bits are clustered into dibit, tribit, or quadbit groups, where the group value determines which level the signal will have. Comparators and digital logic at the demodulator ensure that only valid bit levels are used and reproduce the original binary data pattern from the multiple-level demodulator output. Quadrature amplitude modulation combines AM and PM by modulating two identical frequencies that are 90° out of phase. The resulting signal defines a signal constellation, with each unique point in the constellation representing a multiple-bit value. QAM is used where the channel has well-controlled characteristics and AM noise is low.

Testing digital modulation requires a bit error rate test unit, which produces a known data pattern for transmission. This pattern is compared to the received data pattern to measure the error rate. EDC and related functions must be disabled so that the errors are not corrected, which would obscure the true channel BER. A loopback allows the entire system to be tested from one end, by automatically returning the received data bits to the original source. In many systems the loopback function can be initiated entirely from the transmitter, which allows one-person testing or even unattended testing. An eye pattern overlays all data bits on top of one another and creates a picture that summarizes the general shape of the data signal. Changes in the shape and closing of the eye indicate how noise, signal distance, and channel distortion are affecting the data bit signals.

Testing often requires long streams of random bits, yet the bit pattern must also be reproduced in other test instruments. The solution to this dilemma is to modify the CRC checksum shift register so that it produces a pseudorandom sequence of bits. These appear random to an outsider but can be generated identically by anyone with the same circuitry and key value. When the pseudorandom bit pattern is added to the stream data bits, the resulting bit pattern is encrypted and senseless to anyone who receives them, unless they have the same circuitry and key for decryption.

Summary Questions

1. Compare the features of AM, FM, and PM for binary digital signals in terms of complexity, reliability, and bandwidth use.

2. Repeat question 1 for multiple-level digital signals.

3. Explain the concept of baud rate versus bit rate for binary and multiple-level digital signals.

4. What are the structures of receivers for binary signals using AM, FM, and PM?

5. Compare multiple-digital signal level receivers for AM, FM, and PM to binary signal receivers.

6. What is QAM in concept? In application? What do I and Q represent?

7. What are the advantages of QAM? Where is it most often used? Why?

8. What do QAM constellation points represent? Why is the constellation useful? How does QAM represent a multiple-level digital signal?

9. Why does QAM often use identical I- and Q-channel data recovery circuits?

10. Why is a loopback useful in digital testing? How is testing done without a loopback?

11. Compare loopback testing with full- versus half-duplex systems. What is the role of the BER test unit?

12. What are common meaningless patterns sent by BER testers? Why do testers also send bit in standard formats and codes?

13. Compare the interpretation of the source of errors when all three error measures are about the same percentage to situations where the percentages differ significantly.

14. What is an eye pattern? How is it displayed? What does it indicate, and what is its usefulness?

15. Compare what an eye pattern shows with what a QAM constellation indicates.
16. Explain why random bits are needed for testing and how they can be easily generated.
17. Why is data security needed? Why is simple substitution quickly ineffective for alphabet letters and useless for binary symbols? How can an eavesdropper be defeated even if he or she intercepts a message?
18. Explain how PRSQs are used for data encryption and decryption. Why is this scheme so hard to defeat?
19. What is the major technical problem with data encryption and decryption using PRSQs?
20. What determines the length of the PRSQ before it repeats? Why is the fact that the sequence is not infinitely long not a problem?

PRACTICE PROBLEMS

Section 13.1

1. What is the effective bit rate at 300 baud when bits are represented by a single voltage that can have one of 16 distinct values? What is the effective bit rate when eight values are used?
2. Voltages of 0 V and +3 V represent binary 0 and 1. What is the sequence of voltages when the bit pattern is 0110000011001001?
3. Repeat problem 2 when the bits are transmitted using four voltage values in each signaling period, with values 0, 1, 2, and 3 V for binary 00 through 11.
4. The 0 to +3 V span is now extended from 0 to 3.5 V and divided into eight parts. What is the distance between digital signal levels? What is the sequence of voltages for the bit pattern in problem 2?
5. Repeat problem 4 with the voltage span divided into 16 distinct values beginning at 0 V and separated by 0.25 V.

Section 13.2

1. Draw the constellation diagram for 2 × 4 QAM. How many bits does each point represent? How many points are there?
2. Use a constellation diagram for 2 × 2 QAM to indicate, graphically, the distance between closest states and the distance between the farthest states.

Section 13.3

1. A BER tester sends bits at 2000 baud for 10 s. There are a total of 50 errors. What is the BER?
2. The data bits of problem 1 are sent as 200-bit frames. All the errors are concentrated in the first two frames. What is the percentage of error-free seconds and the percentage of error-free frames?
3. Repeat problem 2 when the errors all occur immediately and the rest of the transmission is error-free.
4. Using 200-bit/frame at 2000 baud and 10 s of transmission numbers, what are the BER, percentage of error-free seconds, and percentage of error-free frames when

Practice Problems

there are 500 errors spread evenly through the 10-s period? When there are 1000 errors?

5. What are the three error measures when the 500 errors occur at the average rate of 100 errors/s for the first 5 s?

Section 13.4

1. A 4-bit register with feedback taps at locations 2 and 4 uses a key of 0011. What are the first 16 bits of the PRSQ? What are the next 16 bits?

2. What are the first 16 bits for the register of problem 1 when the key is 0101?

3. A PRSQ of 0110 0011 0101 01 is added to the data pattern for the ASCII representation of letters A and B (use seven-bit ASCII). What is the encrypted bit pattern? Show that the receiver can decrypt these bits and recover the original letters.

4. The PRSQ of problem 3 is not synchronized properly to the data bits but is positioned one bit to the left. Show the result of the decryption process.

14
TV/Video and Facsimile

CHAPTER OBJECTIVES

When you have completed this chapter, you will understand:

- How a two-dimensional image is converted to a signal voltage versus time through raster scanning in standard broadcast TV
- How and why synchronization is added to the video signal, and the types of modulation used for the video signal (and accompanying audio signal)
- What the TV receiver does to demodulate and recover the video information and recreate the original image
- How color information is added to the monochrome signal, and how it is recovered at the receiver so that a true color picture is seen
- How a facsimile machine uses digital techniques and data compression to transmit a nonmoving image in a few seconds over standard phone lines

INTRODUCTION

With this chapter we begin studying many specific applications of communications electronics and see how the fundamentals of the preceding chapters transform into systems used in real-world situations. Beginning with the common yet extremely complex system of video and TV, we will see how the many problems of converting a two-dimensional image into an electronic signal were solved between the 1920s—when TV was first conceived in modern form—and the 1940s and 1950s—when TV became a common reality. The resulting system can transmit live (real time) pictures in black and white or in color, yet is simple to use. TV and video have also served to advance the basic technology of electronic components and integrated circuits, since the enormous number of TV sets provides tremendous incentive for manufacturers to develop high-performance parts that replace many separate components at lower cost. These component advances have "spilled over" into the design of other communications systems and provided significant improvements in non-TV/video systems.

Many of the concepts of TV are also used in facsimile machines, which send a "still" image over standard telephone lines. Yet the TV signal is analog while the facsimile signal is digital, which provides a good opportunity to see the benefits of each form of signal transmission in its intended application. Digital techniques such as data compression and error correction are used in facsimile systems to send a single page image in less than a minute.

14.1 IMAGING BASICS

A video or television (TV) communications system must do much more than simply take a signal and modulate, transmit and receive, and then demodulate it. Somehow, it must convert the original information of the two-dimensional picture into a time-varying signal that completely captures the picture's information, and must also provide the necessary information so that the receiver can re-create the picture faithfully.

The key to this is the *raster scan* of the image to be transmitted (Figure 14.1) with a repetitive crosswise scan where each successive scan line is slightly offset from the previous line. The image is scanned by the TV camera beginning at its top left corner. The raster scan proceeds from this corner across the picture to the right-hand side, then goes back to the left side and begins a new line of scanning just below the first line. The scan is completed when the last line finishes at the bottom right corner and the scanning goes back to the starting point. As the image is scanned, a voltage signal proportional to the image brightness is generated by the camera, and it is this voltage that forms the heart of the transmitted signal. This brightness corresponds to the *gray-scale* intensity of the picture for monochrome (black and white) TV. The similarities and differences between monochrome and full-color are discussed in Section 14.3; for now, we will use monochrome to explore the basic concepts.

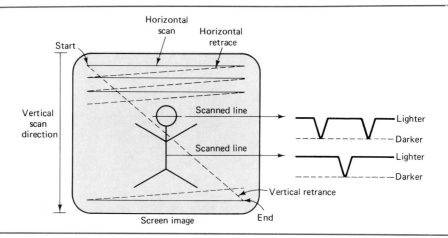

Figure 14.1 Raster scan and retrace shows how a two-dimensional image is converted to a signal that varies with time; detail shows signal for some portions of image scene.

The TV camera is a light-sensitive device which generates a signal voltage proportional to the light intensity it sees at any point on its surface. Traditionally, the camera was also known as a *vidicon,* after the vacuum tube which is the heart of the camera; more recently, vidicons are being replaced by *charge-coupled devices* (CCDs), solid-state equivalents that are smaller, use less power, and are more sensitive to low light levels. The signal from the TV camera (which we will study in greater detail in the next section) consists of a voltage representing the brightness along one line, followed by a short *retrace* time during which the scanning beam is retracing from the right to left and during which there is no image information, then the voltage corresponding to the second line, and so on, down to the end of the last line.

A complete raster scan of the screen is called a *frame,* and to create the illusion of a moving picture, at least 50 to 60 screen images per second must be transmitted. Below that rate the brain does not combine the rapid succession of images into a single smoothly moving picture. Instead, it sees *flicker* as a group of individual still pictures flashed quickly one after the other. We will see soon how this 50 to 60-image/second rate is achieved even though the TV signal bandwidth does not really allow for it.

The number of lines in the raster scan determines the vertical *resolution* of the final image. More lines produce higher resolution and better definition of details, but at the cost of sending more information. This, in turn, means that higher resolution requires scanning more lines per frame and wider bandwidths. Standard TV in the United States (and some other countries) uses 525 lines per frame.

At the receiver, the image corresponding to the TV camera raster scan voltage must be recreated. Basically, the method is this: a *cathode ray tube* (*CRT*) is the screen of the TV and produces an electron beam from its cathode gun, aimed at the front faceplate of the CRT (Figure 14.2). When the electrons strike a spot on the phosphor-coated faceplate, these phosphors glow; a stronger electron beam produces greater brightness of the small spot. The position of the electron beam on the faceplate can be steered in either the horizontal and vertical direction by a pair of magnetic or electric fields.

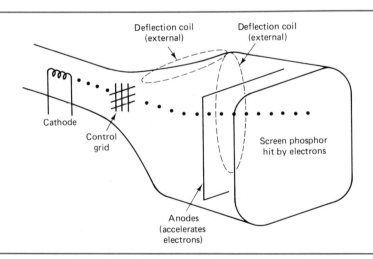

Figure 14.2 Basic design and operation of a CRT.

The TV or video receiver must develop a raster scan identical to the one used at the TV camera, to control the position of the electron beam of the CRT. The amplitude of the received signal determines the beam intensity and thus image brightness at any point on the screen. This leads to the next interesting point about transmitting the video image: There must be synchronizing signals indicating to the receiver when a new frame begins and when any new line within the frame begins. These are *vertical synchronization* (vertical sync) for the start of a new frame and *horizontal synchronization* (horizontal sync) for the new lines.

Several key frequencies and times periods occur in TV systems. The video images are presented 60 times/s, which is the same as the power line frequency. If a different frequency is used, such as 55 or 65 Hz, the unavoidable mixing of the power line and frame frequencies will produce a difference frequency that appears on the TV screen as a visible bar, moving at this difference frequency. (For this reason, countries with 50-Hz ac line power use a 50-Hz frame update instead of 60 Hz). The horizontal line sweeps are created by an oscillator at $(525/2) \times 60$ Hz = 15.75 kHz (the source of the $\div 2$ factor is discussed later in this section.)

The synchronization information is transmitted as part of the continuously varying voltage that represents the image itself. It is not sent along separate wires or on another frequency or channel. In the next section we will look at the complete transmitted signal and see how the sync information is encoded along with the image signal at the transmitter, yet can be identified clearly and separated out by the receiver.

The result of the raster scan and sync is that the TV camera and the TV receiver CRT are scanning the same image point at the same time: The camera is generating a signal level corresponding to the image brightness at the point, while the CRT is controlling an electron beam and the screen brightness at the same point. As the two move along in synchronization across every horizontal line and vertically from one line to the next line, a complete picture is re-created on the CRT face. Since this re-creation of the image is repeated at a high frame rate, the eye and brain perceives it as a moving picture, although it is really a rapid series of stills.

Interlacing The complete TV picture frame consist of 525 lines. Unfortunately, transmitting 60 frames of 525 lines each second would require too much bandwidth—TV stations are assigned 6-MHz channel slots. One alternative is to send fewer lines, but then the resolution is poor and the picture is coarse and grainy. The other choice, to send fewer frames/s, results in very unpleasant flicker since the brain can no longer combine the images into one moving picture.

The solution is *interlacing*, which "tricks" the mind into thinking that the video picture has more to it than it really does. Instead of transmitting a complete frame of 525 lines, each frame is divided into odd and even line numbers and now consists of two *fields*. The even-numbered lines are scanned and transmitted in one frame (the *even field*) and the odd-numbered lines are sent in the next (the *odd field*). Since each frame now has half as many lines as before, the bandwidth needed is cut in half (in theory; in practice it is slightly less than half). Since 60 images (fields) per second are sent, the flicker is not a problem and the eye integrates the alternating odd and even fields to form the complete picture as it combines all the fields to make up the moving image. (Note that 60 fields/s = 30 complete frames/s, the original rate at which the image was scanned.)

Integration of the odd and even is not difficult for the brain, since the odd and even fields are really very similar—there is not much difference in the contents of an odd field and an even field image if you think about what a typical picture looks like. Successive frames of a moving picture have a great deal of repetition, and only small parts of the overall image change from one frame to the next. This similarity between alternate fields—and between successive frames—plays a role in digital TV (Section 14.5).

If interlacing seems like "something for nothing," to some extent it is. Although the amount of information being transmitted has been reduced—which is the goal of interlacing—the viewer does not notice it and does not perceive that something is lost. In fact, less information is transmitted in each second, but this is obscured since there is so much redundancy in successive lines and between successive complete frames, and really not much new information anyway. If each line or frame were completely different from the preceding line or frame, the situation would be different.

Review Questions

1. What is a raster scan? How does it convert a two-dimensional image to a time-varying voltage? What is the retrace period?
2. Explain how horizontal lines make up a frame. How is the vertical resolution determined?
3. How does a CRT produce an image corresponding to the TV camera raster scan? How is beam intensity and position controlled?
4. Why are horizontal and vertical sync needed?
5. What is the standard number of lines/frame? How many frames/s are sent, and what is the relationship of this number to the line frequency? What does the brain perceive if the frame rate is too low?
6. What is interlacing? Why is it done? What is lost but not missed?

14.2 THE TV SIGNAL

The signal that is used to modulate the TV or video transmitter combines video brightness information, horizontal (line) sync and vertical (frame) sync information, and a sound (audio) signal. This *composite* signal format is defined for North America by Electronic Industries Association (EIA) standard RS-170, which precisely defines the voltage levels and timing of all aspects of the signal. Since all broadcasters use the RS-170 standard, any TV receiver can recover the image and display it, regardless of the transmitter or source. (The EIA is a voluntary standards-setting organization that defines many key communications standards for the industry.)

The signal amplitude versus time is shown in Figure 14.3. RS-170 defines amplitudes in IRE units, not volts, with 1 IRE unit equal to 0.0215 V. Each line of video information has four parts: the actual picture intensity, a *front porch*, a horizontal sync pulse for the line, and a *back porch*. Each of these four has specific IRE unit levels associated with it. (IRE stands for Institute of Radio

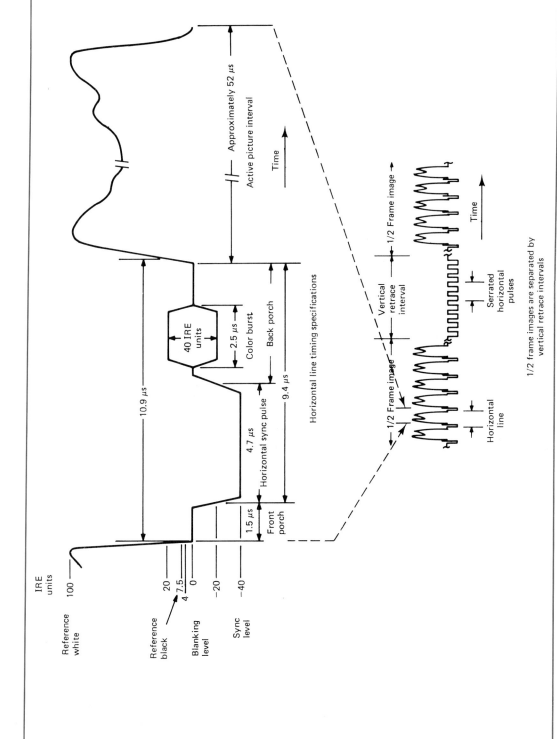

Figure 14.3 The RS-170 standard specifies the entire video signal that results from the raster scan, including timing and signal levels (courtesy of Analog Devices, Inc.).

Engineers, which no longer exists as an independent organization but merged with the American Institute of Electrical Engineers to form the Institute of Electrical and Electronic Engineers, the IEEE).

The front and back porches are at the *blanking* level of 0 IRE, which means that the screen is "blacker than black," and these porches serve as boundaries around the sync pulse. The sync pulse magnitude extends down to −40 IRE. This does not drive the screen any darker—that's not possible—but it does mean that the 4.7-μs-long sync pulse can easily be distinguished and separated from the video information at the receiver, since only the sync pulse is below 0 IRE. The raster scan retrace that occurs during the front porch, sync, and back porch time does not appear on the screen because the retrace signal levels effectively turn the beam of phosphor-exciting electrons off during this *blanking period*.

The actual video image intensity information for a single line takes 52 μs. The signal level for *reference white* (maximum normal brightness) is 100 IRE units, while *reference black* is 7.5 IRE units. The signal magnitude varies between these two values corresponding to the image brightness. Here we see how the originators of the RS-170 signal solved the problem of combining both image information and timing information into a single composite signal: levels between 7.5 and 100 IRE units are picture-related, while levels below 0 IRE units are for timing and sync. The receiver circuitry is designed to distinguish signals by their levels. A special *color burst signal* is transmitted with color broadcasts; it is discussed in Section 14.3.

There is still the problem of synchronizing the beginning of an entire field of lines. The two half-frames—the fields—that make up each complete image frame are separated by a group of pulses that trigger a vertical retrace and synchronization to the start of a new picture at the upper left-hand corner. The horizontal sync pulses are present throughout the retrace time (and the screen is blacker-than-black), so the retrace is invisible, but their duty cycle (the ratio of pulse-on to pulse-off times) is changed and these pulses are wider. This change in effective pulse width is how the receiver recognizes the end of a frame and the beginning of the vertical retrace period that starts a new frame.

Although the complete picture has 525 horizontal lines, visible picture information is contained in only 485 of these. Each visible field therefore has 242½ lines, interlaced to provide the 485-line total. The remaining lines are used for test and reference information that the TV station may need to transmit. Specially coded information is also located there, such as close-captioned subtitles for the hard-of-hearing. A special decoder at the TV receiver looks at these invisible lines, extracts the subtitle letters, and then displays them on the screen along with the picture.

Modulation and Bandwidth

TV stations are allocated 6 MHz of bandwidth by the Federal Communications Commission (FCC), with VHF channel 2 at 54 to 60 MHz, up to channel 6 at 82 to 88 MHz (there is a 4-MHz gap for other services from 72 to 76 MHz). Channel 7 begins at 174 MHz, and this part of the VHF TV band goes up through channel 13, which begins at 210 MHz. Channel 14 begins in the UHF band at 470 MHz; the last UHF channel is 83, which begins at 884 MHz. As you can see, TV is allocated a tremendous amount of the VHF and UHF spectrum, with the UHF stations ranging contiguously from 470 through 890 MHz. [Frequencies between 806 and

14.2 The TV Signal

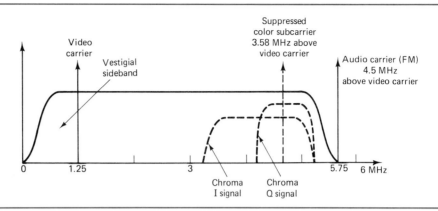

Figure 14.4 Use of the 6-MHz video signal bandwidth allocation for monochrome video, color information, and sound.

890 MHz (channels 70 through 83) were recently reallocated to mobile radio since they were not often used and there was tremendous need for additional mobile radio spectrum.]

Figure 14.4 shows how this bandwidth is used by the composite video signal. The actual video carrier signal, which is located 1.25 MHz ± 1000 Hz above the channel lower boundary, is amplitude modulated by a 4-MHz modulating signal using *vestigial sideband* techniques where one sideband and the carrier are transmitted but a large part of the other sideband is removed (Chapter 4). This modulation technique was chosen as the best compromise among three choices:

Complete SSB-SC, which needs the least bandwidth and makes better use of power but requires more complex and sophisticated transmitter and receiver circuitry, may have undesired phase shifts and is harder to demodulate

Double sideband with suppressed carrier, which occupies more bandwidth but does not waste too much carrier power, yet is difficult to demodulate

Conventional DSB-AM with full carrier power and full bandwidth, which wastes carrier power while permitting easiest demodulation

As a result of the vestigial sideband modulation, the upper sideband occupies the full 4-MHz bandwidth, but the lower sideband is only 1.25 MHz wide below the carrier.

TV Sound

A complete TV signal from a commercial station also must include the audio information. The audio signal is modulated using narrowband FM with ±25 kHz deviation, with the audio carrier 4.5 MHz ± 1000 Hz above the picture carrier. Since FM reception is not affected by amplitude modulation, this minimizes distortion that occurs if the video signal inadvertently amplitude modulates the sound signal as a result of circuit imperfection or nonlinearity.

Horizontal Resolution

The vertical resolution of the screen image is fixed by the number of lines displayed. However, determining the resolution across each line—the *horizontal resolution*—is not as direct. The displayed line does not have distinct dots, but is smoothly continuous from left to right. This does not mean that the resolution can be as fine as you desire: The bandwidth of the TV signal determines how much unique information exists, not how you choose to present the signal. The horizontal resolution of a TV line is between 350 and 500 unique *picture elements* (*pixels*) depending on how you define the smallest feature that can be represented and seen properly. One way to test this is to put checkerboard patterns with smaller and smaller squares in front of the TV camera and see where the squares can no longer be distinguished. In practice, special test patterns with lines at precise spacings are used to measure the effective resolution and signal quality.

Review Questions

1. What four information groups does the composite video signal contain?
2. What does the video signal look like versus time? What measurement scale is used?

Analog versus Digital TV

The video signal is an analog waveform that ranges from 0 IRE units (black) to 100 IRE units (white) in magnitude, and the complete TV signal—video and sound and sync—occupies 6 MHz of bandwidth. We can look at the signal bandwidth and data rates if the video information were sent as a digital signal so that the many benefits of digital systems would be available for TV (storing and recalling images as desired, perfect freeze-frame, and *image processing* to improve the picture sharpness, for example).

The vertical resolution is fixed by the number of lines, while the horizontal resolution is harder to define since each line is a continuous visual element and is not obviously divided into smaller units. However, the effective resolution within a line is determined by the available bandwidth, and each line can only show detail up to a certain fineness given that allocated signal bandwidth. Recall from Fourier analysis that sharper details and faster changes need more bandwidth.

For digital versions of the TV signal, divide each horizontal line into individual spots or picture elements called *pixels*. Typically, the line is divided into 384 pixels, although this number is not fixed and is sometimes changed in specific applications. This means that there are 525×384 pixels/frame. Digitizing a pixel to one of 64 levels of gray requires six bits ($2^6 = 64$), for a total of $525 \times 384 \times 6 = 1{,}209{,}600$ bits needed to represent the information contained in a single frame. In other words, by transmitting 1.2096 Mbits, the information of one frame can be sent (assuming that 64 levels of gray-scale intensity are sufficient for the application). However, to maintain the illusion of a moving image, 30 complete frames/s must be sent, or more than 36.2 Mbits/s! The approximate rule is that the bandwidth of a channel must be at least twice the bit rate, and is typically five times the bit rate, so a channel bandwidth of more than 72 MHz is needed. This is much greater than the 6 MHz allocated.

The bit rate and bandwidth increase even more if 256 levels of gray, corresponding to eight bits, are needed. The reason that TV can manage with only 6 MHz is that an analog signal voltage is used. As seen in previous chapters, digital signals use a lot of bandwidth, and two-level digital (binary) is the worst offender in this; 4-, 8-, and 16-level digital are progressively better. Digital signals do not make good use of the total available signal magnitude span, although of course they have many other benefits. The analog signal used for the video can be thought of as a signal with an infinite set of levels and is therefore very good at conveying lots of information in each specific signal value.

From these calculations it appears that transmitting digital video signals is hopeless given the available bandwidth and technical difficulty of handling so many bits/s. Fortunately, several solutions are available and will be discussed in Section 14.5.

3. What is the purpose of the blanking interval?

4. How are sync pulses distinguished from video information? How are vertical and horizontal sync distinguished from each other?

5. Why are the terms "line sync" and "frame sync" used in horizontal and vertical sync?

6. What bandwidth is allotted for the TV station signal? What frequencies are used for TV signal carriers?

7. How is the sound signal modulated, and where is it in the overall 6-MHz bandwidth?

8. What is the vertical resolution of a TV picture? What is the horizontal resolution, and why is it harder to define and measure?

14.3 COLOR TV

Television could provide only monochrome images at its beginnings in the 1920s and 1930s, although there were some early experiments with schemes to provide color. As TV became a household item in the 1940s and 1950s, engineers began to study how a low-cost, practical color system could be developed. The designers of color TV faced a tremendous challenge: the scheme for sending color pictures had to be fully compatible with monochrome in two ways. First, an existing monochrome TV receiver must be able to receive a color signal and still show a proper, equivalent gray-scale picture. Conversely, a color TV receiving a monochrome signal has to show a monochrome picture and not a variety of incorrect colors. The solution to this dilemma was devised based on understanding how the brain perceives large and small areas of color, available technology, and careful understanding of bandwidth and frequency spectral components. The RS-170 standard was modified for color in 1953 and is known as NTSC (National Television Standard Committee) color.

The method settled on encodes the color information and transmits this information as part of the complete transmitted signal (Figure 14.5). First, let's look at how the color information is extracted from the scene that the camera sees. Three *primary* colors—red, green, blue (R, G, and B)—mix in various proportions to create the entire spectrum of visible colors. You can see this yourself by taking three bright flashlights, placing red, green, and blue color filters in front of their lenses, and then mixing the colors on a white wall.

The color TV camera is like three cameras in one unit, as it takes the image from the lens and separates it into the three primary colors by passing the image through tinted filters. Each color component goes to a separate TV camera tube, identical to the one in a black-and-white camera. The three tubes scan the image in synchronism, and produce three signals—one each for R, G, and B—corresponding to the intensity of that color component in the original image at every point.

These three signals are next added together by resistor matrix circuits (using resistors to sum the signals in an op amp) to form several combinations. The combination of $0.30R + 0.59G + 0.11B$ is the *luminance signal Y* and corresponds to what we have called brightness in a monochrome picture. These particular

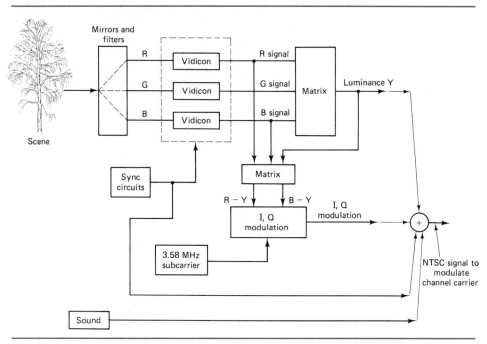

Figure 14.5 Block diagram of color transmitter takes three camera channels (for *RGB*) and combines them via matrix circuitry into special signals that are monochrome compatible.

coefficients of R, G, and B were chosen through an understanding of color perception to produce what appears to be pure white to the eye at the receiver, since the eye has different sensitivity to each of the three primary colors. The luminance Y signal has a bandwidth of 4.1 MHz, for sharp, detailed monochrome features.

Two other combinations of R, G, and B, called the *chrominance* or *chroma* signals since they convey the actual color information, are created by matrix circuitry. An *in-phase signal* I is formed by $0.60R - 0.28G - 0.32B$, and a *quadrature signal* Q is created by $0.21R - 0.52G + 0.31B$. These two signals are used to modulate a special *subcarrier* signal via a balanced modulator, which produces information sidebands but leaves no carrier that could interfere with existing signal. (*Note*: Since Y is a function of R, G, and B, the I and Q signals are also expressible in terms of $R - Y$ and $B - Y$, which includes the G factor as well.) The I and Q signals are of much narrower bandwidth than Y, with I limited to 1.5 MHz and Q to only 0.5 MHz.

This was done deliberately: There is a need to conserve bandwidth, but the question is how to do it. The eye is much more sensitive to the color detail represented by the I signal (yellowish red through greenish blue) than it is to the colors represented by the Q signal (reddish blue to yellowish green), and therefore more detail (and more bandwidth) must be presented for the I signal. This shows how a good understanding of color perception, not just simplistic application of technology, was part of the scheme for providing color TV that is compatible with monochrome.

The color subcarrier frequency is 3.579545 MHz ± 10 Hz, usually referred to as 3.58 MHz. This frequency was chosen because it is the 455th harmonic of one-half the exact horizontal scanning frequency of 15,734.26 Hz. Fourier and spectral analysis show that the monochrome information produces frequency components that are clustered at harmonics of the nominal 15.75-kHz line rate throughout the 4-MHz video bandwidth. The 3.58-MHz subcarrier puts the color information spectral components between the monochrome video information frequency components, so there is no interference or overlap between the color and monochrome information components. (*Note*: For a variety of circuitry reasons, the horizontal oscillator is designed to free-run at 59.94 Hz, not 60 Hz.)

The *I* and *Q* modulation of this subcarrier produces a *vector* or *phasor* whose instantaneous phase determines what basic color—the *hue* or *tint*—will be displayed. Its instantaneous amplitude determines how much of a given color—the *saturation*—will be displayed. For example, perceived colors from light green through deep green have the same pure green hue but with differing amounts of saturation (depth of color). Conversely, deep green and deep yellow have different hues but may have the same saturation.

At the receiver, the *I* and *Q* signals must be recovered as part of the demodulation process. We saw in phase modulation that the phase can be defined only relative to some reference signal, so for the *I* and *Q* demodulation a reference must be provided (the 3.58-MHz color subcarrier is supressed before transmission). Therefore, the color TV transmitter provides a short *color burst* of eight cycles of its 3.58-MHz subcarrier on the back porch of the horizontal sync pule of the transmitted signal, to define the reference at the receiver, which we study next.

Other Color Standards

The NTSC method of encoding color information on the monochrome signal was defined in 1953, and is used in the United States, Canada, Mexico, Japan and some other countries. There are alternative approaches that were developed later

---

RGB Color and NTSC

The process of combining *R*, *G,* and *B* color signals from the camera into the final NTSC format signal inevitably loses some of the color information and fine details. This is due both to circuitry imperfections and the bandwidth limitations of the NTSC format. In addition, the two-field interlacing per frame of the NTSC signal, necessary to conserve bandwidth, is very awkward when the image it represents must be processed or modified (zoom, special effects, and similar features). To minimize problems with signal quality and interlacing, many studios convey signals internally within studio equipment, and from one studio system to another, using the original *RGB* signals. Similarly, when a personal computer generates a color image on its screen, the color information is not encoded as a single NTSC signal but instead uses three separate wires, one each for *R, G,* and *B*. This *RGB* method is practical where the signal path is wire and distances are relatively short and bandwidth limitations or channel allocations are not a problem.

If the *RGB* signal must be broadcast, an *RGB-to-NTSC converter* or *scan converter* is used to combine the three independent color signals into the NTSC-compatible signal, including development of the interlacing from the noninterlaced *RGB* signal. This is usually done at the last possible point in the chain of equipment. In the reverse direction, *NTSC-to-RGB converters* (also referred to as *scan converters*) take a received NTSC signal and decode it into the constituent *RGB* signals, so further processing of the image can be done with the three primary colors. These scan converters also combine two successive fields of a frame to produce a single 525-line frame so that the resultant *RGB* signal represents a complete raster scan rather than two half-frames, to each signal manipulation and processing. When converting an NTSC signal to *RGB*, the image detail and color fidelity will not be as sharp as a signal that was never in the NTSC format since some picture information and detail are lost in the NTSC process.

and benefited from newer technology capabilities but still use the 6-MHz channel allocation. The PAL and SECAM standards are used in other parts of the world and have greater resolution and color performance, in theory. However, these systems are not at all compatible with NTSC, so a TV system designed for one signal cannot be used for the others. Special "universal" TV receivers are available that accept all formats, but these are costly.

Review Questions

1. How is the color information extracted from the scene the camera sees? Why are red, green, and blue used?
2. What is luminance? How is it formed? Why were specific coefficients chosen?
3. What is chroma? How are the chroma signals formed, and what do they represent? What are hue, tint, and saturation?
4. How are the I and Q signals developed? How do they represent color information?
5. What is the color subcarrier, and how was the subcarrier frequency selected? How is the subcarrier modulation?
6. What are the bandwidths of the Y, I, and Q signals? How does color perception play a role?
7. Why is a reference needed to demodulate the color information?

14.4 TV RECEIVERS

Do not be deceived by the low price and commonness of the standard consumer TV receiver: It is a marvel of complex circuitry that combines sophisticated signal processing, wideband stable high-frequency circuitry, and complex color-related features in a reliable, low-cost system. In this section we will see how a monochrome or color signal is received, demodulated, decoded, and used to recreate the original scene as seen by the TV camera.

Warning: Do not explore the inside of a TV set without proper documentation and supervision! Although many signal-processing circuits are low voltage and safe, there are some high-voltage sections (>20,000 V) in every TV set. TV capacitors can retain thousands of volts for minutes and hours after the set is turned off and can therefore deliver a lethal shock even when the set is off and unplugged. Finally, depending on how the ac cord is plugged into the wall outlet and the wiring of the outlet, the TV set chassis may be "hot" at line voltage. To avoid danger, special precautions must be used, including use of isolation transformers, discharging of capacitors, and proper procedures.

A receiver block diagram is shown in Figure 14.6. The circuitry within each block was originally designed with vacuum tubes, which gave way to transistors and then ICs. TV receiver design, especially for color, is the ultimate in refinement of many of the circuits we have studied so far. There are stringent requirements in terms of bandwidth; frequency response; filtering; low drift, high

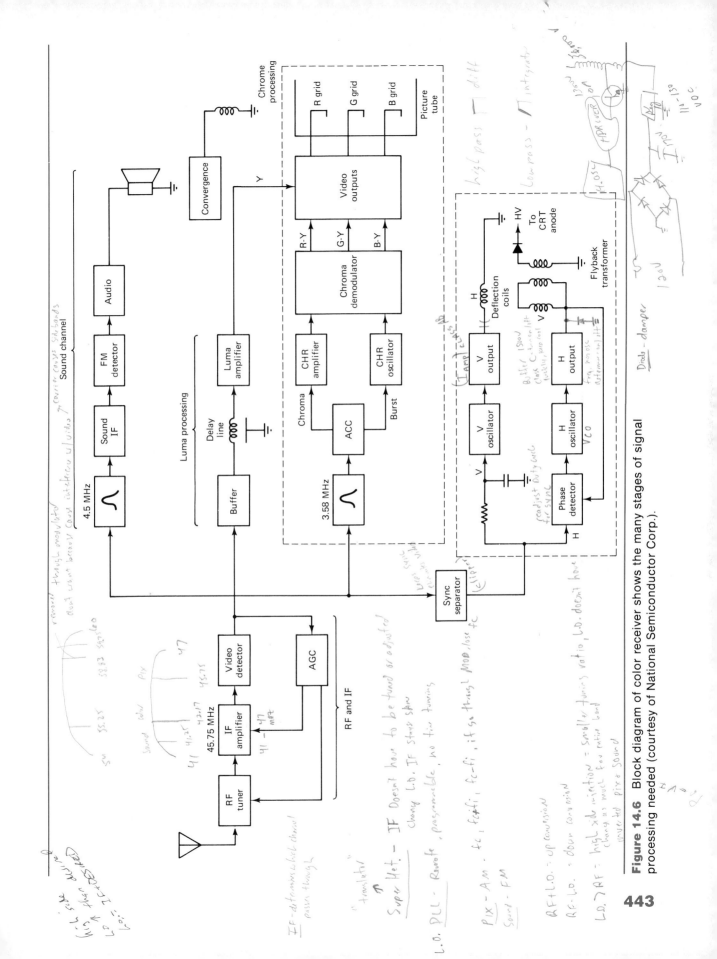

Figure 14.6 Block diagram of color receiver shows the many stages of signal processing needed (courtesy of National Semiconductor Corp.).

stability, and precision of signal and oscillator amplitude, frequency, and phase; plus matching of performance in various parts of the circuit. Today's TVs use extremely complex ICs that reduce the innards of the TV to a few packages, yet provide better performance, with few points that need readjustment.

RF and IF Sections

All received RF signals, from channel 2 at 54 to 60 MHz through channel 69 at 800 to 806 MHz, pass through a tuner, and are mixed to an intermediate frequency of 45.75 MHz, using standard superheterodyne techniques. A tuner cannot provide one-band, continuous tuning over such a wide range. Instead, the lower VHF, upper VHF, and UHF signals are handled as three different bands by switching in different circuitry. The 45.75-MHz IF signal, called the *video carrier,* goes to a *video detector,* which strips the video signal modulation from the carrier. It is this detected video information (which includes the audio signal as well) that must be decoded to re-create the picture.

The nominal output signal level from the video detector is around 3 V peak to peak. The signal levels at the antenna can range from as little as 10 μV rms to 0.5 V rms, depending on the transmitted power, the antenna type and size, and the distance between the transmitter and the receiver. For this reason a TV set has very effective AGC to compensate for variations in received signal level and adjust the gain of the tuner (RF stage) and IF stage to maintain a constant output level. The detected signal actually contains four groups of information:

1. Synchronizing signals for horizontal and vertical raster scanning directions.

2. Audio information, produced by frequency modulation of a 4.5-MHz subcarrier, which in turn modulates the video carrier at the channel frequency. The sound subcarrier is available at the video detector, but usually a separate detector is used to reduce crosstalk and interference between the audio and video signals.

3. Monochrome (luminance) information, which controls the instantaneous brightness of the electron beam as it scans across the screen. This signal is all the video information that is needed in a monochrome receiver. A negative-going video detector detects the luminance signal, with a more negative signal corresponding to a darker screen and a more positive signal corresponding to the brighter areas.

4. Color (chrominance) information, which contains the red, green, and blue information that has been combined and used to modulate the 3.58-MHz subcarrier.

The function of the rest of the TV receiver is to separate these four signal components and then send them to sections of the circuitry that use them to re-create the original picture on the screen. The amplifiers that process the video signals are wideband amplifiers, much wider in bandwidth and different in design than those for narrower-bandwidth audio signals. Wideband amplifiers must provide flat response over the entire bandwidth of interest, and for TV they must have carefully controlled phase characteristics. This is because the phase of the TV signal represents color information, and any undesired phase shifts cause

14.4 TV Receivers

color distortion. In addition, wideband amplifiers are usually ac-coupled between stages, but the TV signal contains information in its absolute dc level. TVs use a special circuit to provide *dc restoration*, which reinserts the dc level lost due to the ac interstage coupling.

Sync Separator

The *sync separator* circuitry that recovers the horizontal and vertical sync makes use of the fact that the sync signal magnitude is from 0 to −40 IRE, while the video information ranges from +7.5 to +100 IRE. The sync separator clips the complete video signal and passes only the video minus sync to the rest of the TV receiver (so that sync signals do not cause problems in other parts of the circuitry).

The total separated sync signal is then sorted into vertical and horizontal sync pulses. They are distinguished from each other by their period—the vertical synch pulses are relatively long, while the horizontal pulses are much shorter. The separated sync signal is applied to a low-pass filter (integrator) and a high-pass filter (differentiator). The wider, lower frequency (60 Hz nominal) vertical pulses produce a low-pass filter output at precisely the vertical frequency that the received signal indicates, and the high-pass filter output is the nominal 15.75-kHz horizontal sync pulse.

The vertical sync pulse goes to the vertical oscillator in the receiver, which free-runs at approximately 60 Hz, but must be made to run at exactly the same frequency as the camera scanned the original scene (which will differ by some small but still important amount). When the transmitter vertical rate and the receiver vertical oscillator rate differ, the picture will roll up or down (misadjusting the vertical hold control on the TV does this, as well). The recovered vertical sync triggers the vertical oscillator to begin another cycle of oscillation at the correct instant, so that each picture starts at the top of the screen. The output of the oscillator is a current that ramps up in time (a sawtooth waveform).

This ramp drives the *deflection coils* of the CRT to generate a magnetic field that goes from minimum to maximum value, and this magnetic beam deflects the electron beam from the CRT from top to bottom. [CRTs that deflect and steer the electron beam with electrostatic fields generated by voltage across flat plates (instead of magnetic fields and coils) require a ramping voltage, but otherwise the principle is the same.] At the end of the ramp, the oscillator output quickly returns back to the starting value, causing the fast retrace.

For the horizontal sync, a phase-locked scheme is generally used to maintain the correct frequency and phase of the receiver horizontal oscillator. The recovered horizontal sync pulse is the input to the PLL, and the PLL oscillator signal is the horizontal output. As with the vertical signal, a ramping current output is needed to drive the deflection coils (or ramping voltage for electrostatic deflection plates) that will steer the electron beam from left to right, and then cause a retrace during which the electron beam is blanked.

The high frequency of the horizontal output is also used to develop the 25-30 kV needed for CRT anode (which pulls the electron beam from its source at the cathode toward the CRT face at high velocity) by driving a *flyback transformer* primary side. Depending on design, this high-turns ratio transformer output is either directly rectified and used, or is put through a capacitive voltage doubler or tripler and then rectified.

At this point in the TV receiver, an electron beam in the CRT is being steered line by line from the upper left corner to the lower right corner, and retraces right to left after each line and bottom to top after each complete screen. The intensity of the electron beam is not yet controlled, however, so the screen is fully bright as its phosphor is hit by the full-power electron beam.

Luminance Processing

The Y signal is a regular amplitude-modulated signal and is recovered with a standard AM detector. It is used to control the amplitude of the electron beam reaching the CRT faceplate phosphors and so set the brightness of the screen at each point as the electron beam sweeps across each line. In a TV system that can only produce monochrome, this Y signal is applied directly to a *luminance amplifier,* which then controls the beam intensity when it is applied to a grid through which the beam passes between the electron emitting cathode and the screen. In a color TV set the recovered Y signal is first delayed by 0.8 μs before it goes to this amplifier. The delay ensures that black-and-white information does not reach the picture tube before the color information, which is unavoidably delayed by the extra circuitry it must pass through and by phase shifting in the narrow-bandwidth chroma section (recall that a frequency-domain phase shift is equivalent to a time-domain delay).

The luminance section also has controls for *contrast* and *brightness* of the screen image. A contrast control changes the maximum to minimum (dynamic) range of the Y signal, while the brightness control changes the dc level offset of the signal so that the entire intensity range is moved up or down.

Chroma Processing

A tuned circuit at 3.58 MHz extracts the reference color signal needed to demodulate the color information (I and Q). Accurate demodulation requires a continuous 3.58-MHz signal with both the correct amplitude and phase relative to the color burst (8 to 11 cycles) that was transmitted on the horizontal sync signal back porch. The amplitude and phase corrected 3.58-MHz signal is applied to a *chroma demodulator* whose outputs are color difference signals $R - Y$, $B - Y$, and $G - Y$.

Any errors, even slight ones, in the chroma subcarrier amplitude will cause the color information to be interpreted incorrectly. To prevent this, the chroma section has its own AGC, separate from the AGC of the tuner and IF. This *automatic chroma control (ACC)* stage compares the amplitude of the burst to a reference to keep the chroma output signal constant over an input range that can vary by as much as 20 dB.

The chroma signal is next divided (gated) into two time segments by a pulse derived from the horizontal section. During horizontal scans, the chroma subcarrier is sent to the chroma amplifier with a gain control to vary the saturation (color depth) of the picture. If there is no color burst—the picture as sent was monochrome, not color—the output of the chroma amplifier is "killed" so that is has no output, and the color TV screen is forced to monochrome operation. The reason for this is that a color TV set showing a monochrome picture would otherwise have colored specks caused by unavoidable noise, which would look like a small but legitimate color signal to the color demodulation and decoding circuits.

During the horizontal retrace period, the color burst is sent to a circuit that regenerates the color subcarrier to be like the one at the transmitter. This regenerator circuit is a crystal-controlled 3.58-MHz oscillator that is frequency and phase synchronized to the burst, and the circuit output is the reference output that recreates the transmitter 3.58-MHz subcarrier. The user's TV tint control, which shifts the final color of the picture, is a variable phase-shift network located here, and changes the color by shifting the phase of the regenerated reference output.

The chroma demodulator uses the outputs of the chroma amplifier and the color subcarrier regenerator, and consists of two synchronous detectors in quadrature, with one demodulating the I component of the chroma signal while the other demodulates the Q component. (Recall that I–Q demodulation requires knowledge of the carrier phase at the receiver.) These two recovered components are decoded to produce the original R, G, and B signals. The reverse of the equations that produced the original Y, I, and Q signals are implemented with resistor networks, as follows:

$$G = -I - Q + Y \qquad B = -I + Q + Y \qquad R = +I + Q + Y$$

where in each equation the I, Q, and Y factors have different numerical coefficients (not shown here) to combine these factors in the correct proportions.

The final step in color TV receiver electronics is to take the recovered R, G, and B information and use it to create the correct color on the screen. The CRT of a color TV has three electron guns, each raster-scanning identically across and down the CRT face but aimed slightly differently at the CRT faceplate. The inside of the faceplate is coated with *triads* of phosphors, in which one phosphor of the triad produces red light when hit, the second produces green, and the third produces blue. Each phosphor triad forms a single perceived spot on the screen. By careful design and aiming (*convergence*) of the electron beams, as well as use of a mechanical shield with holes called a *shadow mask,* the electron beam with red information can hit and excite only the red phosphor dot in each triad, the green beam excites only the green dot, and blue beam excites only the blue dot.

The tube faceplate has about 200,000 triads, equal to the number of displayed lines times the horizontal resolution. Developing manufacturing techniques for making the shadow mask and its holes, precisely placing three sets of color-producing phosphors on the inside of the CRT, and mechanically aligning the shadow mask and the phosphor triads were major impediments to producing practical color TV receivers, ones that had to be overcome reliably and at low cost. Most TVs have adjustments to make sure that the RGB beams *converge* only on their respective color dots and do not inadvertently excite one or both of the other dots in each triad.

The three *RGB* signals are amplified to about 100 V peak to peak in the video output stage and then applied to the R, G, and B grids to control the intensity of the three electron beams hitting their respective phosphors. In this way the three primary color components are re-created with the same intensities (relative to each other) that they had when they were separated at the TV camera.

Note that the entire chroma processing section is not needed in a monochrome receiver, which produces a gray-scale image from either a monochrome or a color signal. Color signal recovery adds immensely to the complexity of the receiver and CRT, as well as to the number of areas that have critical adjustments for phase, tint, convergence, and relative *RGB* intensities.

Audio

The 4.5-MHz sound subcarrier is amplified and limited as a standard FM signal. An FM detector recovers the original modulating signal and applies it to an audio amplifier. A typical home TV has audio output power between 1 and 5 W, although some of the new TV sets have much higher power amplifiers. Recently, TV with stereo audio has been introduced, in which the two channels of audio information are encoded in a manner similar to stereo FM radio, with an $L + R$ signal and an $L - R$ signal. A stereo TV receiver must decode these into the left and right channels.

Most of the functions of a color TV receiver are now implemented in ICs. Earlier ICs performed a single function, such as sync separator, horizontal or vertical oscillators, wideband chroma or luminance amplifier, or color information matrix. Advances in IC technology now produce ICs that incorporate many functional blocks into a single IC; thus some TVs now have only a half-dozen ICs plus some RF transistors for the front end and high-power transistors for driving the deflection coils.

Review Questions

1. How is the received video signal converted to a single IF frequency? What is the IF value used in standard TV?
2. What is the video carrier? What does the video detector do to the carrier?
3. Why does a TV need a very good AGC, encompassing both the RF and IF stages?
4. What four groups of information are in the detected video signal?
5. What are the characteristics of the wideband amplifiers in a TV receiver? What is dc restoration, and why is it needed?
6. How is the sync separated from the video? How are the horizontal and vertical sync signals separated from each other?
7. How do the vertical sync and oscillator cause scanning and retrace?
8. How is horizontal sync accomplished? How does the horizontal oscillator also develop the high voltage needed for the CRT?
9. How does a monochrome TV use a color TV signal?
10. Why is the Y signal delayed in a color TV?
11. How are brightness and contrast set?
12. Why is ACC needed? How does the TV re-create the exact phase and frequency of the transmitter 3.58-MHz subcarrier? How is tint manually adjusted?
13. What is the color killer? Why is it needed? How is it driven?
14. How do the recovered I and Q signals produce the RGB signals?
15. How do the RGB signals produce correct color and intensity on the CRT screen? What do the phosphor triads and shadow mask do?
16. How is TV sound recovered?

14.5 FACSIMILE

A *facsimile* machine (*fax*) (Figure 14.7) can send an image on a sheet of paper to another fax machine over regular phone lines. Unlike the analog TV signal, fax information is sent in digital format; practical bandwidth limitations of the system

Figure 14.7 A fax machine can scan and transmit a standard sheet of paper in only a few seconds, using digital technology and data compression (courtesy of Canon USA).

and communications channel (phone line) mean that a single image takes many seconds to transmit. (Recall that a digital version of a moving image requires tens of megabits/s.) However, there are similarities between a TV camera scanning a scene and the way that the paper in the fax machine is scanned. A fax machine also shows how digital technology can bring benefits, since information that has been digitized can be preprocessed to achieve specific goals.

The paper with the message written on it is inserted in the fax machine, where it is scanned, line by line and from top to bottom, using a special light and mirror mechanical assembly. A photocell senses the light reflected from the paper at any spot. Wherever there is a mark (writing, typing) on the paper, the photocell output voltage is very small and a binary 0 is generated; where there is no mark the photocell output is large and this generates a binary 1. The result is a series of 1s and 0s corresponding to blank areas and written areas of the single line. Each scan line is divided into 200 points per inch, so $8\frac{1}{2}$-in.-wide paper produces 1700 bits/line. Within the fax machine, the 1s and 0s are represented by digital IC voltages (typically TTL/CMOS values of 0 V and +5 V, nominally) but they are represented by two frequency tones when sent over the phone line, which is discussed further in Chapters 16 and 17.

At the receiving fax machine, a beam of light scans a specially treated drum surface in synchronization with the scan of the sending unit. The beam of light is turned on and off by the received 1s and 0s, and the drum turns slightly as each new line is presented. At the end of the scanned image information, the drum carries electrical charges on its surface that have the complete image, like a photographic negative, and then uses these charges to transfer toner ink to a piece of paper. This reproduces whatever was on the original paper. (Some machines use special heat-sensitive paper where a row of thermal printing dots is activated to darken the paper. This costs less but requires special paper.)

The resolution in lines per vertical inch is user selectable, with industry standard values of approximately 100, 200, and 400 lines/in. With these numbers, we see that a regular 11-in.-long piece of paper has 1100 scanned lines in the lowest resolution 100 line/in. mode. A complete sheet therefore produces 1100 lines × 1700 bits/line = 1.87 Mbits. Sending this number of bits at a 2400-baud rate (a standard telephone line value) will take about 779 s (nearly 13 minutes), which is a very expensive telephone call.

Here is where the potential of digital format is realized. The microprocessor in the fax machine can process the data bits before sending them and uses a previously agreed upon algorithm (Section 10.1) for *compressing* the data into fewer bits. The receiving fax is programmed to reverse this data compression scheme and reproduce the original bit pattern. Several different algorithms are available to compress and decompress the data.

In the simplest approach, the entire line is scanned and stored in the memory of the fax microprocessor, which carefully examines the string of 1s and 0s. Instead of sending each individual 1 and 0 over the communications channel, the fax machine sends information which indicates how many 1s in a row and how many 0s in a row are on the line. For example, consider a line that was scanned into 256 bits where the paper was almost entirely white, with just one pencil mark (Figure 14.8). In this example the scanned pattern is 20 1s, a single 0, and then 235 0s. Instead of sending these 256 bits, the format that is used says: Use an 8-bit field to indicate, in binary, the number of continuous 1s, followed by an 8-bit field to indicate how many 0s, and so on, until the complete line is defined.

Using this algorithm, the pattern is 00010100 (for the 20 1s), 00000001 (for the single 0), and 11101011 (for the 235 1s). Instead of sending 256 bits, only 3 × 8 = 24 bits have to be sent, which is less than 10% of the original number. A 13-minute

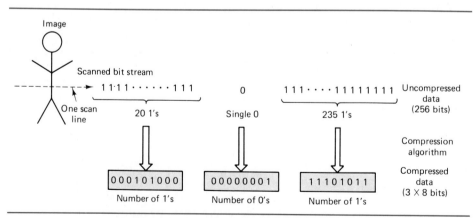

Figure 14.8 Data compression applied to one scanned line, showing the original image, the resulting binary pattern, and the compressed data.

message takes 1.3 minutes. Of course, this does not include bits that are used to synchronize the two machines, but synchronization must be done regardless of whether or not data compression is used. Some of the newest fax machines compress data so much that a full page takes only about 10 s to transmit in lowest-resolution mode, using a more sophisticated version of this algorithm, called *Group 3* or *G3*.

The data compression scheme is based on two observations. First, the image being scanned has broad areas where nothing changes (the white space of the paper) and sending the same information about the white space repeatedly is wasteful. Instead, the areas where nothing changes can be "summarized" by a data compression algorithm and then the overall description of this area is sent. Second, the data in digital format can be processed by the sending machine after they are scanned but before they are transmitted; similarly, the receiving fax machine can process the information after it is received but before it is used to put the image on the drum. Data compression for fax essentially sends a *summary description* of the line or of the page rather than every bit of the page, and this description is much shorter than the bit-by-bit method.

This is a major difference between information sent in analog compared to digital form: Processing of analog signals consists of amplifying, filtering, modulation, demodulation, and similar operations, but the basic information content is not changeable. An analog signal must be processed as it occurs (in *real time*), since it cannot be stored; digitized information is storeable and so can be processed and manipulated by various algorithms after it is generated. This processing can take as much time as needed—no data will be lost.

In this simple line-by-line algorithm, each line is compressed independently of the lines above and below it. More advanced data compression algorithms use the fact that adjacent lines are very similar to each other—there is large correlation and repetitiveness from one line to the next. Instead of sending the description of the pattern on each line, these algorithms send the description of one line, followed by only the descriptions of the differences between the first line and the

---

Data Compression, Coding, and Encryption

The idea of reducing communication time through data compression is not new. "Shorthand" codes were used from the first days of the telegraph to reduce the number of transmitted characters and save money. Standard shorthand books were published, and some people developed private codes to protect their message contents as well (a basic and somewhat effective form of encryption, Chapter 13). For example, ABC might mean "I will return tomorrow" while ABD means "I am delayed by a few days." With three-letter codes, a total of $26^3 = 17,576$ unique messages can be represented.

Data compression differs from simple coding to reduce message length in several ways. First, the intent of compression is not to conceal the message contents since the compression algorithm is published and known to all (although it can be done as part of the compression algorithm). But the most important difference is this: Compression as used in the fax machine applies to *all* scanned bit patterns. *Any* run of 1s and 0s that occurs from a line scan is compressed, and the compression activity is transparent with respect to the *meaning* of the image being scanned. The general-purpose compression algorithm used in fax systems really says: "For any 1,0 pattern from a line scan, here is a compressed version of it." As we have seen before, transparency to the message meaning is a critical feature in a communications system: It makes data compression a very powerful tool.

> **Video Data Compression and HDTV**
>
> The discussion of basic data compression and advanced algorithms is useful for more than just for fax machines. The ability to process video information in digital form and to perform operations such as picture enhancement, contrast improvement, removal of fuzzy edges on images, zooming, picture-in-picture, and other image manipulations means that there is a tremendous need for this digital video. However, as we have seen, the data rates required are extremely high. To reduce the amount of video data and the data rates, different types of data compression are used; some of these schemes reduce the data by factors of 50 to 100.
>
> All video data compression schemes use the basic fact that successive images on the screen have much in common, so that a lot of redundant information is sent. By using algorithms that remove this redundant information, more reasonable data rates and bandwidths can be used. Studies and proposals are now under way for various types of *high-definition TV (HDTV)*, which will provide much greater resolution and picture sharpness than NTSC TV, with about 1000 to 1500 horizontal lines. One problem is to do this in a way that uses only the available 6-MHz NTSC bandwidth, to remain compatible with existing TVs and spectrum allocations.
>
> Some HDTV proposals suggest setting up a new standard that is not NTSC compatible (which historically had to be monochrome TV compatible!) and thus make use of the many advances in digital signal, algorithms for signal processing, and manipulation of signals that have been digitized. Although no final decision on HDTV for the United States has been made, any final format will make use of the potential savings available through data compression to reduce bandwidth, data rates, and computer memory for image storage.

next line, and the one after that, and so on. This reduces the number of bits even further.

Unfortunately, there is a drawback to these advanced compression schemes. In the simple algorithm, a single bit in error affects only one line since each line is compressed without regard to other lines. Schemes that transmit only line-to-line changes are much more seriously affected by errors. Any bit error in the information corrupts not only the line that the bit is on, but also the subsequent lines. To compensate for this, advanced error correction techniques are used to detect errors and correct them (Chapter 12). The trade-off is increased complexity in the processing done by the fax machine, versus a significant saving in bits.

Before fax machines with built-in microprocessors and data compression schemes for shorter transmission time were readily available, many documents were sent by typing them on a keyboard (*telex* is a common name for this) and sending the ASCII code as a binary pattern over the same phone lines. A typed page typically has 1500 to 2000 characters, equal to 15,000 to 20,000 bits when the start bit, seven ASCII bits, stop bit, and parity bit are included. In contrast, the fax scanned page has nearly 2 Mbits. The disadvantages of keyboard entry are that the message must often be typed from the original page (unless the page was typed into a telex machine or computer to begin with), and only alphanumeric characters are acceptable. In contrast, a fax machine accepts an existing page—just insert it into the machine—and drawings and sketches can be used in addition to keyboard characters. When combined with data compression, fax is very convenient and quick.

Review Questions

1. What is a fax machine? How does it convert an image into electrical signals, and how does it perform the electrical to page image process?

Early Facsimile Machines

The first fax machines were mechanical implementations of a TV camera and CRT. The picture or paper to be transmitted was taped around a rotating drum, and a photocell scanned the drum surface as it turned. The mechanism holding the photocell also moved down alongside the drum as the drum turned, so that the entire drum surface was scanned. The photocell output, an analog voltage corresponding to the intensity of the image, was used to frequency modulate a tone sent over the phone line. An entire page/transmission scan (they occur simultaneously) took several minutes.

At the receiving end a similar mechanism was used. A sheet of light-sensitive paper was placed on a drum and the photocell was replaced by a focused light spot from a small bulb. The light intensity, controlled by the demodulated tone signal, exposed the special paper on the drum, which rotates in synchronization with the sending unit. After the entire image was scanned, the paper was removed and developed like a photo print.

Besides the obvious difficulty of using and developing light-sensitive paper, the mechanical scheme required careful adjustment for proper operation. Both drums had to rotate at the same speed or else the image was distorted and stretched; any noise in the phone lines caused speckles and spots on the final image at the receiver. The dynamic range of the photocell output and the intensity of the light bulb had to be carefully matched or the image was too light or too dark. Nevertheless, this mechanical fax was the major workhorse for transmitting images in a few minutes for many years, and was used extensively by business, police, and the armed forces. It is now obsolete, due to the sheet-fed fax with data compression, which is easier to use, more reliable, and transmits a page in a few seconds.

2. What are the horizontal and vertical resolutions of fax? How many bits does a single page require in lowest resolution? In highest resolution?

3. What is data compression? How is it applied to sending an image data by fax? What happens at the receiving fax unit?

4. Why does data compression reduce the number of bits sent?

5. Compare line-by-line data compression with compression based on comparing successive lines. What are the benefits and weaknesses of the advanced algorithm?

6. How does the processing of digital video differ radically from the processing of analog video?

SUMMARY

In this chapter we have studied an extremely advanced, yet common electronic communication system: TV/video. The first function of a TV transmitter is to raster-scan the image and produce a continuously varying signal that conveys the video brightness. Synchronization pulses for both horizontal lines and vertical frames must also be sent. These are differentiated from the video information by their levels. The TV receiver reconstructs the scanning of the TV camera and controls the brightness caused by an electron beam that excites the phosphors on the CRT faceplate.

Color TV is much more complicated. The color signal must be compatible with a monochrome signal, and a special color subcarrier is modulated with color information. The color TV receiver demodulates the color signal after recreating the color subcarrier precisely. The red, green, and blue signals control the intensity of three separate electron beams that excite phosphor triads on the CRT and mix R, G, and B light to reproduce the original color. A TV receiver combines sophisticated, precise circuitry with careful application of standard circuits; color TV uses all three types of modulation (AM for the video luminance and sync, PM for the color information, and FM for the sound).

A facsimile machine scans a printed image and sends it as a series of digital signals. Data compression algorithms are used to summarize and describe the bit sequence, thus saving a tremendous amount of transmission time. The receiving fax decompresses the data with the reverse algorithm and reproduces the original image. Unlike TV, which uses analog signals and so conveys a tremendous amount of information in a fixed bandwidth, the digital format requires more bandwidth. However, the digital information can be processed and manipulated, unlike analog signals.

Summary Questions

1. How does a TV camera capture the information of a scene and put it into electronic form?
2. What is the horizontal and vertical scanning of the raster scan? What are retrace and blanking? How many lines are in a frame? What are the line and frame frequencies?
3. What information is encoded in the video signal? How is sync differentiated from picture information?
4. What is the channel bandwidth for TV? What frequencies are available for TV station carriers?
5. What is the bandwidth of the modulating video signal? What type of modulation is used?
6. How is the audio part of the signal modulated? Where is it in the channel spectrum?
7. What is interlacing? Why is it used? What is lost in theory but not in practice?
8. Compare the bandwidth of regular TV with the bandwidth of a digital version? Why is the analog TV signal able to convey so much information?
9. How does a TV receiver recreate the monochrome image? What controls the position of the electron beam on the CRT? What controls its amplitude? How is the beam position aligned with the one at the TV camera?
10. How is the color information extracted by the TV camera? What are the primary colors? How is color information encoded onto the transmitted signal? What is the role of the color subcarrier?
11. How is the color signal made compatible with the monochrome one so that a monochrome TV receiver produces a correct picture?
12. Why is the bandwidth of the color information limited? How did an understanding of color perception and resolution affect these decisions?
13. Explain the operation of a TV receiver, in terms of IF, video detection, recovery of the sync information, dc restoration, color burst regeneration, recovery of the color and intensity information, and production of perceived color on the CRT screen.
14. How does a color TV receiver produce a correct monochrome picture for noncolor transmissions?
15. What is a fax machine? How is it similar to a TV system? How does it differ?
16. What is the idea behind data compression? How does it significantly reduce the amount of fax data to be sent? Why is this possible with digital signals but not with analog ones?

17. Compare data compression done one line at a time, versus data compression with one line compared to the preceding line. What are the advantages and drawbacks of each?

18. What do scan converters do? Why are they useful? What do they do with interlacing?

19. What information is lost or distorted when using NTSC signals compared to *RGB*? Explain why an *RBG* signal derived from an NTSC signal cannot recover the information loss but is still more useful than the NTSC version in some applications.

15
Frequency Synthesizers

CHAPTER OBJECTIVES

When you have completed this chapter, you will understand:

- Similarities and differences among direct, indirect, and direct digital frequency synthesis
- How a phase-locked loop generates a wide span of distinct frequencies based on a single master clock
- How the synthesis range and resolution can be increased using dual modulus synthesis
- The use of synthesizers in microprocessor-based systems

INTRODUCTION

The traditional tuning control of a transmitter or receiver is being replaced by the more precise and microprocessor-compatible scheme of indirect synthesis. Instead of varying a tuned circuit capacitance or inductance, all local oscillator or carrier frequencies needed are generated from a single reference oscillator, which can be as accurate and stable as required. The synthesizer blends analog circuitry and digital dividers in a phase-locked loop in several configurations to provide easily settable tuning. Synthesizers are an excellent example of the power of digital circuitry to improve analog communications systems operation, since the signal being tuned with the aid of the synthesizer can be a conventional analog signal (such as voice or music).

To overcome some limitations of this scheme, prescalers and dual-modulus counters are used, allowing use of high-speed but more complex digital components in only a few critical places and regular digital circuitry in the rest of the design. Besides providing precise tuning, synthesizers combined with a microprocessor can be preset to desired stations, are accurate across the whole band of interest, and provide features such as automatic scanning.

15.1
DIRECT AND INDIRECT SYNTHESIS

Traditional receivers provide continuous tuning across the entire frequency band. Like an analog voltage, the dial can be set to any frequency by adjusting a knob that varies a capacitor in a tuned circuit (mechanical tuning) or applies a dc bias to a varactor diode (electronic tuning). Even the traditional rotary TV tuner is really a special version of a continuous tuner, since the click-stops of the detent mechanism switch in discrete capacitors to tune the various carriers. While continuous tuning is the ultimate in flexibility (any frequency can be tuned), it is also costly, prone to drift and instability, imprecise and inaccurate in parts of the band, and not compatible with modern digital circuits and microprocessors. It usually needs calibration at the factory and some adjustment of trimming components to achieve acceptable accuracy.

An alternative to continuous tuning is *frequency synthesis,* which uses one or more precise frequency sources to provide the necessary local oscillator or similar necessary signals. In contrast to continuous tuning, synthesized tuning provides signals only at specific frequencies, separated by steps. The step size may be very small (25 or 100 Hz), which very nearly approximates continuous tuning, or it may be large to correspond with the 6-MHz channel separation of commercial TV signals.

In *direct synthesis,* crystals are used as the heart of the oscillator circuit. The advantages of crystals is that they are precise and stable, but using a crystal per channel is costly and bulky when there are many channels to tune. For example, using direct synthesis, a citizens' band (CB) *transceiver* (a receiver and transmitter integrated into one unit) for the 40 channels of the entire band requires 80 crystals (one for the transmit and one for the receive frequency of each channel). Schemes have been developed to reduce the overall number of crystals when a very wide range must be covered, as for a general-coverage 3- to 30-MHz "shortwave" receiver. By a combination of switching and intermediate signal mixing, for example, 100 crystals cover the shortwave band with a 1-kHz step size. Although precise, this is still not competitive with a continuously tuned system except where the stability and precision of crystals are the highest priorities of the receiver design.

An alternative to direct synthesis is *indirect synthesis,* where a complete range of frequencies is generated based on a single, precise reference frequency. The principle of indirect synthesis has been known for many years, but it required the low-cost, low-power frequency dividers available as digital ICs to make it practical. The basis of indirect synthesis is the phase-locked loop and some clever variations and enhancements to the basic PLL design. Virtually all frequency synthesizers in use today use the indirect approach.

The benefit of frequency synthesis compared to a continuously variable tuning scheme goes beyond its stable and precise tuning. Since a single precise crystal determines the tuning across the entire band, it can be calibrated to very high accuracy if required, and then the entire band is also calibrated. The frustrating problem of a nonlinear dial that is off-frequency by differing amounts at different points will no longer exist, since any error in the crystal shows up as the same amount of error everywhere. Some receivers for the most difficult applications such as deep-space satellite reception use atomic frequency standards in place of a crystal, or put the crystal in a temperature-controlled oven with carefully regulated power supplies for the oscillator circuit, and thus achieve outstanding stability and precision.

Pushbutton tuning of a synthesized system is very convenient: A keyboard can tune an FM station at 106.7 MHz (just key in 1, 0, 6, and 7) or a standard AM broadcast at 590 kHz (5, 9, 0). The tuning can also be set by a knob that looks like the control of a continuously tuned receiver but which really sets the synthesizer to different values. A digital display shows the exact frequency or channel number as the receiver is tuned, and there is no need for the user to try to figure out what the indicator line on a conventional analog dial really shows (in fact, most of these "line" indicators cannot be read very precisely anyway). This is a major advantage when trying to tune a weak signal in noise at a specific frequency. Remote control is practical for a digitally controlled tuning system as well.

The advantages of the synthesized approach go beyond the precision and convenience and stability. Some additional benefits when the system has a built-in microprocessor include:

The ability of the receiver to recall stations by a shorter code entered by the user, so that 1 is 106.7 MHz, 2 is 89.9 MHz, and so on.

Automatic search scanning up and down the entire band, where the receiver stops at any frequency that has a signal. This is useful in police and fire band receivers, where different transmitters are in use at many frequencies at the same time, or when using a radio in a new geographical region.

Automatic "turn on" of the receiver to the desired station frequency at a specific time, after the user programs the receiver with the necessary information.

In the next sections we look at how this indirect synthesis based on a PLL is accomplished and at some of the techniques used to extend performance. Most of the discussion centers around receivers, because a receiver usually has to be able to tune to many frequencies, while many transmitters operate at one fixed frequency. Recall that the key to tuning a receiver to the desired frequency is precise control of the local oscillator signal, which is mixed with the incoming signal to produce the desired intermediate frequency. The synthesizer output is used as the local oscillator signal, replacing the LC tuned oscillator or the oscillator with a varactor and applied dc voltage from a tuning knob.

However, the ideas of frequency synthesis are also used to provide the master oscillator carrier signal for transmitters that must have the ability to shift from one frequency to another, such as mobile phone, police, fire, or military units. Some synthesizers even incorporate additional circuitry specifically to be used for half-duplex transceivers where transmission and reception take place on frequencies that are slightly offset from each other.

Although a synthesized receiver can tune only to specific frequencies, typically spaced every 25, 50, or 100 Hz in the band, this is not a limitation. For many bands, there are prearranged frequency assignments at even spacing, so there is no need to tune continuously: Broadcast FM stations are every 200 kHz, TV stations are separated by exactly 6 MHz, and broadcast AM signals are spaced at 20-kHz intervals. Even for bands where there are no fixed assignments, users generally agree to even spacing to avoid interference and simplify initial setup and

15.1 Direct and Indirect Synthesis

Direct Digital Synthesis

Another technique for generating sine waves of desired frequency is *direct digital synthesis* (DDS) (Figure 15.1). A DDS system has three key elements: a binary counter, a read-only memory (ROM), and a digital-to-analog converter (DAC) (Chapter 11). The counter is driven by a clock, and so steps from 00...00, 00...01, 00...10, up to its maximum value of 11...11 and then "wraps around" to begin again at 00...00. The memory of the ROM is preprogrammed so that it contains the digitized values of the amplitude of a perfect sine wave: At each successive address of the ROM, there is the amplitude of the next point, in time, of the sine wave. The ROM output—the digitized value of the sine wave—is converted back to an analog signal by the DAC.

As the clock drives the counter, the counter output points to successive addresses of the ROM, which in turn deliver the digitized amplitude value to the DAC. The DAC produces a waveform that closely resembles a perfect sine wave but has small "stair steps" between the successive points (which are smoothed by low-pass filtering).

To produce sine waves of different frequencies, the frequency of the clock that drives the counter is increased or decreased. Then the counter steps through the ROM addresses more or less quickly, generating the amplitude points of the sine wave at this new rate. The resultant sine-wave frequency is controlled entirely by the clock rate.

To this point it seems that we have built a complex system which really offers no benefits beyond those of a good conventionally tunable oscillator or voltage-controllable oscillator. However, if the DDS system design removes the counter and replaces it with a *register*—a memory location—that is written to and updated continuously by a microprocessor, the frequency is determined by the values that the microprocessor writes to the register at any instant. Even more impressive, the values written to the register do not have to be in strict succession. Instead, the microprocessor can "jump around" to some extent and so directly produce frequency- and phase-modulated waveforms. This is a unique benefit of DDS: modulated waveforms under complete digital control (and therefore driven by a microprocessor program and software). The DDS system can generate FM with one deviation at one carrier, then quickly shift to another carrier, and finally instantly change the modulation index.

Despite the benefits of DDS, it is used only in specialized applications such as transmitters or test signal generators which must produce unusual types of modulation or rapidly change modulation type and value. This is due to the cost of the components needed and the present limitations on maximum frequency achievable (about 10 MHz).

calibration—it is awkward to work with channels at 101.5637 and 103.7013 MHz (approximately 0.2 MHz apart) compared to channel assignments spaced every 0.2 MHz beginning at 101.6 MHz. As long as the synthesizer step size is small enough for the band in use, the fact that frequencies that fall between the synthesizer outputs cannot be tuned at all is not a problem.

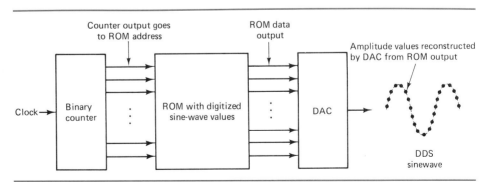

Figure 15.1 Direct digital synthesis uses a read-only memory (ROM) and DAC, with the ROM driven by the counter and presenting digitized amplitudes of successive points of the sine wave to the DAC.

Review Questions

1. Compare continuous tuning to synthesized tuning for precision, accuracy, linearity across the band, and the ability to tune any frequency.
2. What is direct synthesis? How does it achieve the precise tuning? What are the major drawbacks?
3. Why is indirect synthesis accurate across the entire band of interest?
4. How does indirect synthesis integrate with digital and microprocessor-based systems? What are some of the features that indirect synthesis provides?
5. What is the synthesized frequency used for in a receiver? In a transmitter?
6. What is direct digital synthesis? How does it generate a sine wave at one specific frequency? How is this frequency changed?

15.2 BASIC INDIRECT SYNTHESIS

We saw in Chapter 6 that the PLL can be used to track variations in the frequency of a received signal, and that the error signal which tries to force the VCO to follow the input represents the demodulated FM output. We have also seen how the PLL can be used as a narrow-bandwidth tracking filter to extract timing and synchronization information from a received stream of data bits. Now we will see how an enhancement to the versatile PLL is the basis for an indirect synthesizer.

The closed-loop design of the PLL and the feedback from the phase detector to the VCO act to force the VCO to track the phase—and therefore frequency—of the other input to the phase detector. If the external input of the PLL is a precise reference signal, the VCO will have the same frequency as the PLL locks onto the reference input. We can use the VCO output as the local oscillator signal (Figure 15.2).

Now insert a frequency divider with division ratio N between the VCO output and the phase detector input (Figure 15.3). (The term *modulo N* is often used for the divider, meaning that it counts to N and then resets and restarts from the beginning.) For the PLL to maintain phase lock with the reference input, it must operate at the reference frequency times division ratio N. Consider it from this perspective: The phase detector wants to see both of its inputs—the reference

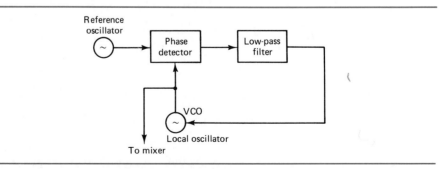

Figure 15.2 PLL VCO used as a local oscillator for a mixer.

15.2 Basic Indirect Synthesis

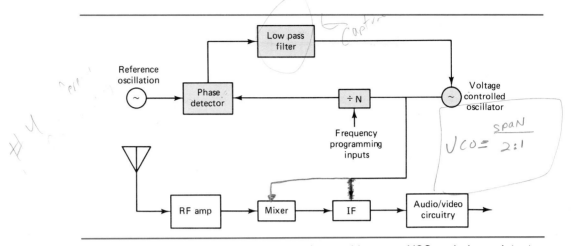

Figure 15.3 A synthesizer with divider inserted between VCO and phase detector can generate many LO frequencies from a single reference, resulting in a digitally tuned receiver.

and the VCO—at the same phase and frequency. When locked, $F_{ref} = F_{VCO}/N$, so $F_{VCO} = F_{ref} \times N$. The VCO does not "know" that its output is divided by N before it reaches the phase detector, and it must run at a faster rate so that its actual signal to the VCO matches the reference; the only way it can do this is to operate at $N \times F_{ref}$.

Frequency synthesis by this indirect method requires only a reference oscillator, a frequency divider that can be set to various values of N, a loop filter, and a phase comparator. All of these are easy to implement with modern ICs. For example, to provide VCO outputs that range from 1 to 2 MHz with 50-Hz steps, the synthesizer uses $F_{ref} = 50$ Hz, and dividers with modulo $N = 20{,}000$ to $40{,}000$. The VCO frequency will step through 1,000,000 Hz, to 1,000,050 Hz, to 1,000,100 Hz, and so on, as the division ratio N is set to 20,000, then 20,001, then 20,002, and so on.

We can now look at each element of the PLL-based synthesizer in more detail.

VCO

Any one of several standard VCO designs is used with a varactor diode as the circuit element that changes the VCO output frequency. The filtered phase detector output changes the varactor capacitance, which in turn changes the resonant frequency of the oscillator. Practical frequency spans of 2:1 (such as from 1 to 2 MHz, or 150 to 300 kHz) can be easily achieved; much wider spans are more difficult to realize reliably.

The VCO output is both the local oscillator signal and the input to the modulo N divider. For this reason its output is buffered by an amplifier that isolates it at the next stage (usually the mixer) so that its frequency and stability are unaffected by any changes in the mixer load or signal problems at the mixer.

Reference Oscillator

The reference oscillator must be stable and precise, since it determines the accuracy of the entire tuning. For higher-frequency references, a crystal oscillator can be used directly. Unfortunately, the reference frequency value is equal to the step size in this simple synthesizer, so reference oscillators of 50 or 100 Hz are needed for small steps. This very low frequency is not practical with most oscillator designs, so the actual reference oscillator operates at a much higher frequency and then is divided down by a fixed value before going to the phase detector. The phase detector sees the reference oscillator only after the division. Another advantage of this higher-frequency initial oscillator combined with the "divide down" scheme is that standard, low-cost crystals can be used for the reference, such as the 3.58-MHz color subcarrier crystal or the 32.768-kHz crystal of digital clocks. For larger steps such as the 200 kHz of the FM band, or the 6 MHz of TV, the reference frequency is much higher and the crystal oscillator output is used directly as the reference.

Phase Detector

The role of the phase detector is to indicate the phase difference between the reference output and the divided VCO output. The two inputs to the phase detector are "squared up" by a limiter so that their amplitudes are constant; this has no effect on operation because the phase detector is interested only in relative phase, not amplitude. Most phase detectors and PLLs actually operate so that they are in lock and synchronized when the two signals are in exact and constant quadrature (90°) with each other. Terms such as *no phase error* actually mean that the phase difference is a constant 90°, while *error* means that the phase difference between the signals is less than or greater than 90°. Of course, two signals that are always 90° out of phase with each other are at the same frequency, and two signals at different frequencies will have a changing phase difference with each other.

A phase detector for a synthesizer can be based on a simple digital exclusive-OR gate (Figure 15.4a). When the phase difference between the reference input and the divided VCO input is exactly 90°, the phase detector output consists of a square wave with a 50% duty cycle (Figure 15.4b). The average value of the output pulses is exactly one-half of the maximum amplitude of the square wave. When the difference becomes less than 90°, the output duty cycle decreases; therefore, the average of the output decreases. When the difference is 0°, the output becomes 0.

If the phase difference increases above 90°, the output duty cycle increases above 50% and the average is more than one-half of full scale. At the extreme where the difference is 180°, the output duty cycle is 100%, and of course, its average is then maximum.

Here we see how a very simple digital circuit functions as a phase detector for two square-wave inputs. The only restriction on this circuit is that the inputs—the VCO signal and the reference signal—have exactly 50% duty cycles, but this is the situation for any "squared-off and limited" sine-wave oscillation. The main point, then, is how the square wave with 0 to 100% duty cycle becomes a dc, steady-state voltage that is the error signal which controls the VCO. This is the function of the synthesizer filter.

Filter

A very critical element of synthesizer performance is the filter, which produces a signal whose average value corresponds to the duty cycle of the input square wave. Although it is essentially a low-pass filter, it determines *lock-in* time, indicating how quickly the VCO comes to its new frequency when the division ratio

15.2 Basic Indirect Synthesis

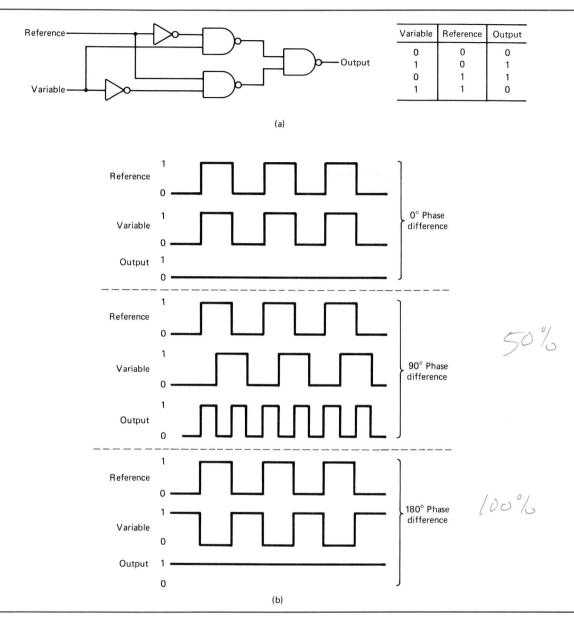

Figure 15.4 (a) Digital logic XOR as a phase detector; (b) key phase detector waveforms for 0, 90, 180° phase differences.

changes. It also determines the *trajectory* of the loop output frequency as it achieves lock-in: Does the VCO come smoothly to the new value, or does it overshoot and oscillate around the final value before it settles? Finally, the filter determines the stability of the VCO, and whether there is any phase jitter around the desired value even when the division ratio is constant.

When the division ratio changes, the phase detector suddenly shows a large phase error, which then pushes the VCO to the new frequency. The settling time of the filter response to this sudden change in phase detector output typically requires about one cycle time of the reference frequency for a *critically damped* filter design. This means that a 50-Hz reference takes about 0.02 s to settle; higher-frequency references which provide larger steps take correspondingly less time. If the filter is *underdamped,* the settling time is still longer and the filter output will actually overshoot beyond the correct value, then come back and possibly undershoot, before settling. The VCO driven by this signal will go too far, then come back too much, and finally, stabilize.

A filter time constant affects the overall stability of the synthesizer and its ability to respond to changes. A long-time-constant filter results in a VCO that is very stable at the various division ratios, since small jitter of the phase detector output caused by component variations will be ignored. However, there is a longer settling time for the filter output to reach a desired new value when the division ratio is switched. This is not a problem for synthesizers that are used primarily at one frequency, but is a problem for those that are rapidly scanning from one channel to the next or must switch back and forth between transmit and receive channels. The system designer must take this into account, together with the reference frequency value and the step size. Techniques for getting around this problem of small step size, low reference frequency, and long lockup time are discussed later in this chapter.

Divider

Integrated-circuit counters are used for the modulo N divider between the VCO and the phase detector. The modulo N ratio is preset into the divider either in binary format or *binary-coded-decimal* (BCD) format, depending on the design. BCD is convenient when manual switches are used to set the channel number, with one switch for the units, another for the 10s column, and so on, of the division ratio, while the binary format is more compatible where a microprocessor is the interface to the divider.

A *ripple counter* and comparator form one type of divider (Figure 15.5). The comparator is loaded with the desired factor, while the counter input is the stream of pulses from the VCO. When the counter value reaches the preset (modulo N) value in the comparator, the comparator generates a single pulse that both goes to the phase detector as the $\div N$ pulse and resets the counter to zero. This design is usable for VCO frequencies up to about 3 MHz, after which the internal propagation delays within the stages of the ripple counter are too great and the divider no longer functions.

For higher frequencies a *synchronous counter* design is used. This type of counter can operate up to 10 MHz for standard ICs, or as high as 100 MHz when faster IC technology is used. The synchronous counter does not have the problem of buildup of internal propagation delays that the ripple counter has, but its function in the synthesizer is otherwise the same. However, even synchronous programmable counters have speed limits: realize that a 100-MHz signal, for example, has a period of only 10 ns and that the ability of the digital ICs to operate with these fast signals is limited.

Some examples illustrate how this operates.

Example 15.1

A reference frequency of 100 Hz is used, with division ratios of $N = 100{,}000$ to $120{,}000$. (a) What is the step size? (b) What are the VCO output frequencies? (c) If

15.2 Basic Indirect Synthesis

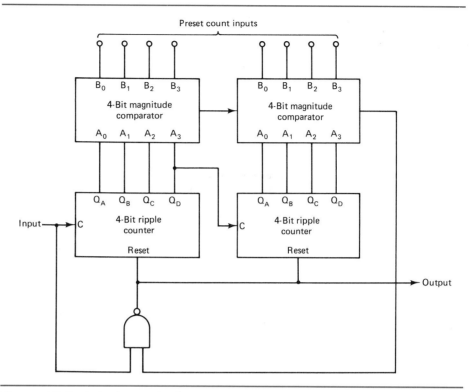

Figure 15.5 A ripple counter forms a programmable divider for digital pulses (courtesy of National Semiconductor Corporation).

the system IF is 5 MHz and the LO is above the received carrier, what frequencies can be tuned?

Solution

(a) The step size is equal to the reference frequency, 100 Hz.
(b) The VCO output at $N = 100{,}000$ is $(100 \times 100{,}000) = 10{,}000{,}000$ Hz = 10 MHz; at $N = 120{,}000$ it is $(100 \times 120{,}000) = 12$ MHz.
(c) The LO − the tuned carrier = IF; therefore, the carrier is 5 MHz when the LO is 10 MHz, and the carrier is 7 MHz for a 12-MHz LO.

Example 15.2

A LO must tune from 20 to 25 MHz in 1-kHz steps. What is the reference oscillator (VCO) frequency, and what are the division ratios needed?

Solution

The reference oscillator is 1 kHz (0.001 MHz). Since $F_{\text{VCO}} = N \times F_{\text{ref}}$, then $N = F_{\text{VCO}}/F_{\text{ref}}$. At 20 MHz, this means that $N = 20/0.001 = 20{,}000$; for 25 MHz, $N = 25/0.001 = 25{,}000$.

Review Questions

1. How does indirect synthesis use a modified PLL?
2. What element determines step size? What feature determines actual VCO frequency?
3. Why does the VCO operate at $N \times V_{ref}$?
4. What is the heart of the reference oscillator in most systems? Why does a small-step-size system usually use a reference oscillator that has been divided down from a higher frequency?
5. What is the average phase detector output, as a percent of full scale, when the phase difference is 0°, 90°, and 180°?
6. What is the low-pass filter? Why is it needed? How does the filter affect lock-in time, lock-in trajectory, and VCO stability? What is a typical lock-in time value?
7. What is the trade-off among lock-in time, step size, stability, and lock-in trajectory?
8. What does the modulo N divider do? How is it implemented?

15.3 EXTENDING SYNTHESIZERS

The major limitation to successful application of synthesizers at approximately 100 MHz and above is that modulo N programmable dividers are either not available or are very expensive and also consume more power than the circuit design can provide (hundreds of milliwatts or more, typically). The 100-MHz limit means that only TV channels 2 to 6 or the lower part of the standard 88- to 108-MHz FM band could be covered, for example. The solution to this limitation is to use a *prescaler* between the VCO and the programmable divider that feeds the phase detector (Figure 15.6). A prescaler is a fixed modulus divider which provides some convenient $\div M$ factor such as $\div 10$ or $\div 100$, or a binary factor like $\div 64$ or $\div 256$. This type of fixed modulus prescaler is implemented with a very fast digital logic family, such as *emitter-coupled logic* (*ECL*).

With a prescaler, the synthesizer output F_{VCO} is equal to $M \times (N \times F_{ref})$. The variable modulus $\div N$ counter which follows the prescaler can be fabricated using lower-speed, less costly, and less power-consuming digital logic technology, such as *low-power Schottky TTL,* since it sees the VCO output divided by the prescaler factor. The use of the prescaler greatly extends the achievable range of the frequency synthesizer concept.

Example 15.3

A 1-kHz reference is used in a synthesizer with N = 50,000 to 100,000, and a $\div 10$ prescaler is added. What is the range and step size without and with the prescaler?

Solution

Without the prescaler, the step size is 1 kHz, with F_{VCO} = 50,000 to 100,000 kHz.

15.3 Extending Synthesizers

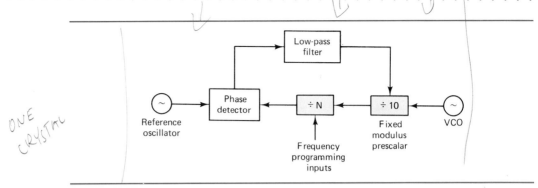

Figure 15.6 Use of a fixed-modulus prescaler in a PLL synthesizer to extend range.

With the prescaler, the step size is $M \times F_{\text{ref}} = 10$ kHz, and $F_{\text{VCO}} = 10 \times (1 \times 50{,}000) = 500{,}000$ kHz to $10 \times (1 \times 100{,}000) = 1$ MHz

The National Semiconductor Corp. prescaler DS8629 shown in Figure 15.7 is a good example of a prescaler IC. This device uses several internal stages to provide an overall fixed ÷100 modulus and can accept inputs signals from 0 Hz (dc) up to 120 MHz. The high-speed input stages are fabricated in ECL technology, while its lower-speed dividers and prescaled output use Schottky technology for lower power consumption and to provide TTL-compatible outputs to the phase detector. The input stage of the prescaler operates with signals as low as 100

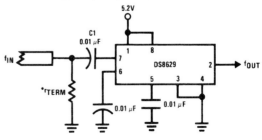

Figure 15.7 (a) Internal block diagram and (b) final circuit interconnection of National DS8629 ÷ 10 prescaler (courtesy of National Semiconductor Corp.).

mV and up to 1 V. Low input voltage ranges are important for two reasons: They consume less power, and they are more compatible with low-voltage VCOs [it is also more difficult to build VCOs that can provide the larger voltages (several volts) at these higher frequencies].

The major disadvantage of the prescaler method is that the step size has now increased by the prescalar modulus, if the reference is unchanged. This means that channel spacings are often too wide for the application. To maintain the same step size in a prescaler design, the reference frequency must be reduced by the prescaler factor. Using a simple PLL synthesizer design for 5-kHz spacing requires a 5-kHz reference, while a PLL with ÷10 prescaler design for 5-kHz spacing needs a 500-Hz reference. Unfortunately, low-frequency references are noiser than higher-frequency ones and require larger and more expensive components, and the associated PLL has a much longer lock-in time, which results in slower scanning or switching to a new frequency.

It appears that the alternatives are to use ÷N synthesizers up to about 100 MHz, then use a fixed ÷M modulus prescaler, which requires one of two unpleasant choices: larger step size and channel spacing, to continue using a higher reference frequency; or decreasing the reference frequency but accepting increased lock-in times, slower scanning, and more output noise. Some advanced systems circumvent this problem by shifting from one reference to another as the frequency of interest changes, or switching in the prescaler only at the higher frequencies. This provides an acceptable solution in many cases but also increases circuit complexity and cost. Fortunately, there is a very clever alternative, called the *dual-modulus prescaler*.

Dual-Modulus Prescalers

Replace the fixed modulus divider with a dual-modulus counting scheme (Figure 15.8) that consists of a main counter (÷M), an auxiliary counter (÷A), and a dual-modulus prescaler (N and $N + 1$) which divides by one of two numbers that differ

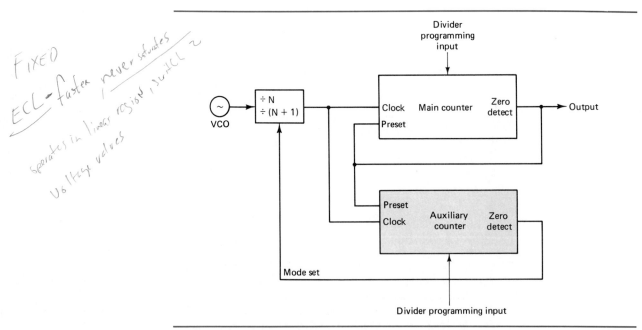

Figure 15.8 Block diagram of the dual-modulus scheme.

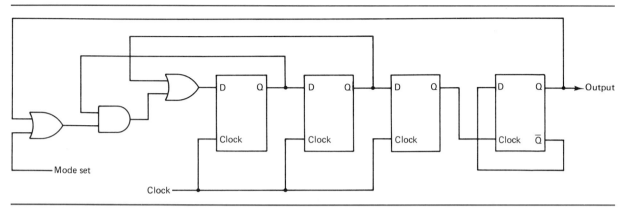

Figure 15.9 Logic circuit for a 10/11 prescaler.

by just 1 (such as 7 and 8, 15 and 16, or 63 and 64). The main and the auxiliary counters are preset independently of each other, but both are clocked from the same output pulses of the dual-modulus prescaler. A circuit for 10/11 prescaler is shown in Figure 15.9; its modulus 10 or 11 factor is set by a single control line.

In operation, the divide cycle begins with the dual-modulus prescaler counting by the larger of its two divisors ($N + 1$). When the auxiliary counter reaches zero, the mode of the prescaler is automatically changed to the smaller divisor N. The main counter continues to count down toward zero, while the auxiliary counter has stopped counting. When the main counter reaches zero, both it and the auxiliary counter return to their preset values and the cycle begins again.

We can see how many VCO cycles are needed to complete one cycle, assuming that the main counter is set to $\div M$ and the auxiliary counter is set to $\div A$ factors. Then $[(N + 1)A]$ cycles are needed in the first part of the cycle (until the auxiliary counter reaches zero) since the prescaler is set to $\div(N + 1)$. The prescaler mode switches to $\div N$ and the main counter counts down until it reaches zero, which requires $M - A$ counts or $[N(M - A)]$ cycles of the VCO output. The total number of VCO cycles required is

$$(N + 1)A + N(M - A) = NM + A$$

so

$$F_{VCO} = F_{ref}(NM + A) = (N \times M)F_{ref} + AF_{ref}$$

The effect of M on the overall division value is much greater than the effect of A, by the factor N. The step size is again equal to the reference frequency, while the prescaling is equal to the larger $N \times M$ factor. The dual modulus scheme allows the overall division factor to be set to smaller increments. Instead of simple prescaler factors of $\div M$, the prescale factor can be $\div M$, plus one or more steps of the reference. The prescale factor M sets the major steps, while the A factor determines the fine steps between the major steps.

Example 15.4

A 10/11 dual-modulus design is intended to provide LO signals from 200 to 210 MHz, with 1-kHz step size. What is the value of the reference needed? What is a possible combination of M and A values for the counters if $N = 10$?

Solution

The reference oscillator is 1 kHz (0.001 MHz). To generate 200 MHz, set A to 0. Since $N = 10$ and $F_{VCO} = (N \times M)F_{ref}$, then $M = 20,000$. To step from 200 MHz to 210 MHz, increment the count of A from 0 to 10 MHz/1 kHz = 10,000. With $A = 10,000$, $M = 20,000$, and $N = 10$, F_{VCO} (in MHz) = $(10 \times 20,000)0.001 + (10,000)0.001 = 210$ MHz

---

We see from this example how the prescaler factor $N \times M$ sets the initial range, and the A factor sets the number of steps, or fine-scale, after that value.

In actual circuitry, the prescale portion of the divider is designed for highest speed and is limited to a few simple integer values of M to simplify its design (such as 64, 128, 256, 10, or 100). The A factor then steps through all the intermediate values in the fully programmable divider. There is a restriction in using dual-modulus synthesizers: Since the auxiliary counter must reach zero before or simultaneously with the main counter, the value of M must be equal or greater than A. This limitation is generally not a problem and can be handled in the special cases when it occurs. (We will see one case of this in Section 15.4.)

For example, use a 1-kHz reference in the three cases of a simple synthesizer, the fixed-modulus ($\div 10$) prescaler, and a dual-modulus ($\div 10/11$) design. In the simple synthesizer, the steps are clearly 1 kHz, but the maximum VCO output is limited to about 100 MHz. A divider set to 100,000 produces this 100-MHz output. Suppose that we need to go to 200 MHz, so the $\div 10$ prescaler is used. Now, the presettable dividers only have to operate at 1/10 their previous rate, but the step size is 10×1 kHz = 10 kHz. When the programmable counter is set to 10,000, the VCO output is 100 MHz. As the divider is incremented by 1, the next VCO output is at 100.2 MHz, then 100.4, 100.6, and so on.

With the dual-modulus system, $F_{VCO} = (N \times M)F_{ref} + AF_{ref}$. For the 1-kHz reference and a $\div 10$ prescaler, this becomes F_{VCO} (in kHz) = $10M + A$. With M preset to 100,000, the F_{VCO} is 100 MHz and will increment by 1 kHz each time A is incremented by 1. This achieves both fixed high-speed divider operation and step sizes equal to F_{ref}. Figure 15.10 summarizes the main characteristics of the three synthesizer designs.

Review Questions

1. What is the performance limitation of the PLL-based synthesizer with a programmable modulo N divider?
2. What does a prescaler do? How does it affect step size?
3. How does a prescaler overcome the limitations of the modulo N divider?
4. What is the major disadvantage of the prescaler in a synthesizer? What are the practical problems of reducing the reference oscillator frequency?
5. What is a dual-modulus prescaler? How does it operate with two counters?
6. What benefit does the dual-modulus scheme offer in terms of achieved performance and step size? In terms of use of IC technologies?

	Basic indirect synthesizer	Synthesizer with modulo M prescaler	Synthesizer with dual-modulus prescaler
Step size	F_{ref}	MF_{ref}	F_{ref}
$F_{VCO} = F_{out}$	NF_{ref}	$(M \times N)F_{ref}$	$(MN + A)F_{ref} = MNF_{ref} + AF_{ref}$
Maximum frequency	100 MHz	500 MHz to 1 GHz	500 MHz to 1 GHz
Comments	Limited by $\div N$ speed	$\div M$ prescaler increases frequency and step size	Prescaler extends frequency; maintains step size

Figure 15.10 Summary of features of the simple, prescaler, and dual-modulus synthesizers.

15.4 SYNTHESIZERS AND MICROPROCESSOR SYSTEMS

The National DS8907 is a complete synthesizer for the standard AM and FM broadcast bands that uses a dual-modulus programmable divider to achieve the necessary range and channel spacing. Figure 15.11 shows it combined with a *microcontroller* (a specialized version of a microprocessor), keyboard for entering the desired frequency, and a display readout to indicate the frequency in an electronically tuned AM/FM radio. The figure does not show the RF, mixer, IF, or audio stages. The synthesizer has an internal Colpitts reference oscillator that

---

Synthesizers for Transmit/Receive Operation

In a receiver, the critical frequency for tuning is the local oscillator, which is mixed with the received signal carrier frequency and produces a fixed IF. Therefore, tuning involves setting the LO. For a transmitter, the operating frequency is determined by an oscillator at the carrier frequency, which is amplified and modulated as needed. From this we see that the frequency generated for a receive operation is different from the one for transmit, even if both are at the same frequency. Fast lock-in is needed when the user switches from transmit to receive mode, so that there is no irritating and interference-producing time lag when the synthesizer output is not at the correct frequency.

For example, consider a transceiver for 60 MHz, with a 10.7-MHz IF. The transmit carrier is 60 MHz, while the receiver LO is 60 + 10.7 = 70.7 MHz. A single frequency synthesizer is used to generate both. When the transceiver is set to transmit mode, the synthesizer (single or dual modulus) is preset to produce the 60-MHz signal, and in receive mode it is preset to yield a 70.7-MHz signal. The specific numbers to be loaded depend, of course, on the value of F_{ref} and the design of the counters.

Some frequency synthesizers have two internal registers that can store the appropriate numbers for transmit and receive operation, and then a single control line to the synthesizer indicates which mode is desired. This minimizes the interaction and data transfer between the transceiver microprocessor and synthesizer since actual numbers need be loaded only when a channel is changed, not when switching from transit to receive modes. In many transceivers the transmit and receive channels are offset from each other by a fixed amount; for example, transmit may be at 60.010 MHz and receive can be 10 kHz higher, at 60.020 MHz. The factors loaded into the synthesizer must take this into account, but this is just an additional minor calculation for the transceiver microprocessor.

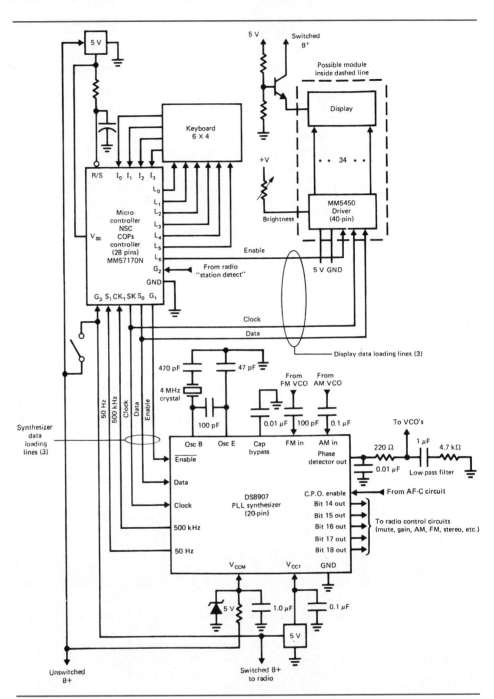

Figure 15.11 Complete electronically tuned AM/FM radio based on the National DS8907 synthesizer (courtesy of National Semiconductor Corporation).

needs only a 4-MHz crystal, which it internally divides down to provide either a 25-kHz reference for the FM band (88 to 108 MHz), or a 10-kHz reference for the AM band (550 to 1600 kHz). The phase detector output is filtered by components external to the IC, so the filter can be tailored to the application.

In operation, the user keys in the desired frequency and mode (AM or FM).

15.4 Synthesizers and Microprocessor Systems

The microcontroller calculates the correct counter values for the divider and prepares to load these into the DS8907. It also prepares a single *control bit* to set synthesizer operation for either AM or FM mode; this bit changes both the 4-MHz divide-down factor and also sets the dual-modulus divider to 7/8 operation for AM or 63/64 for FM. The entire AM band is tuned with 10-kHz resolution, while the FM band resolution is 25 kHz. This IC is capable of working up to 15 MHz in the AM mode—well beyond the standard broadcast band—while the FM mode output maximum is 120 MHz, which provides a local oscillator frequency high enough to mix with top of the band (108 MHz) and produce the standard 10.7-MHz IF. In addition, the microcontroller can be programmed by the user to tune desired stations simply by hitting a desired channel number, such as 1, 2, and so on, rather than entering the station frequency. For this mode, the microcontroller simply looks up a stored frequency value associated with the channel numbers with which it has been programmed.

To reduce cost and size, the entire synthesizer is packed in a 20-pin IC and requires 5 V at less than 160 mA. A *serial interface* instead of a *parallel interface* is used between the controlling microprocessor and the IC to reduce the number of IC pins and package size. This three-wire interface has one line for the actual data bits, one line as a clock for these data bits, and the third line as an *enable* signal indicating to the DS8907 that new, valid data are coming. The microprocessor sends the mode and counter data as part of an 18-bit field of data bits it transfers. The DS8907 ignores any clock or data bits until it sees the enable line go low (its active state), at which time it uses the clock signal to time and accept the 18 bits that are sent to it. When the 18-bit transfer is complete, the enable line is set high (inactive) by the controlling microprocessor, and the IC synthesizes the LO using the data it has received. Data are transferred as a binary number equivalent to the desired $\div N$ factor, expressed as a decimal number.

The advantage of the serial interface is that it uses only the three wires rather than the eight data lines that a parallel, byte-by-byte transfer would need, or 18 lines needed to transfer all the information as a single parallel word. This reduces the package size, although serial transfer takes longer than parallel transfer, which is not a problem in this application. The microcontroller also calculates the actual station frequency, and then sends bits representing the frequency digits to a digital display with the data and clock lines. Since different enable lines are used for the display and the synthesizer, each ignores data sent by the microcontroller to the other. The DS8907 also has some output bits that go to the rest of the radio circuitry, indicating the mode of operation (AM or FM) and other factors.

Setting the VCO Frequency

A total of 18 data bits are serially transferred to the DS8907. Five of these bits are for setting up the operating mode of the synthesizer, and the remaining 13 bits are for presetting the counters. The main counter is allocated seven bits and can be set from 0 to $2^7 = 127$. The remaining six bits are for setting the auxiliary counter to 0 to $2^6 = 63$.

For the AM mode, the reference frequency derived from the 4-MHz crystal oscillator is 10 kHz. With the modulo 7/8 prescaler, Figure 15.12a shows the values of M and A that provide tuning for the local oscillator across the entire standard broadcast band. Note that the IF is not the standard 455 kHz, but is 460 kHz, since AM channels are every 10 kHz beginning at 550 kHz. A 455-kHz IF with 10-kHz spacing produces channels every 10 kHz, but at frequencies ending in 5, not 0, kHz.

$$F_{VCO} = N \times M \times F_{ref} - AF_{ref} \quad N = 8, F_{ref} = 10 \text{ kHz}$$
$$= 8M(10) + A(10) \text{ kHz}$$

M	A	$F_{VCO} = $ LO(kHz)	$F_{carrier} = $ LO−460 (kHz)	Notes
12	0	960	500	
	1	970	510	
	:	:	:	
12	7	1030	570	A can also go to 12
13	0	1040	580	
	:	:	:	
13	7	1110	650	A can also go to 13
25	0	2000	1540	
	1	2010	1550	
	:	:	:	
25	7	2070	1610	

(a)

$$F_{VCO} = N \times M \times F_{ref} + AF_{ref} \quad N = 64, F_{ref} = 0.025 \text{ MHz}$$
$$= 64M(0.025) + A(0.025) \text{ MHz}$$

M	A	$F_{VCO} = $ LO(MHz)	$F_{carrier} = $ LO−10.7 (MHz)	Notes
61	0	97.600	86.900	
	1	97.625	86.925	
	2	97.650	86.950	
	:	:	:	
	48	98.800	88.100	First FM band channel
	:	:	:	
	60	99.100	88.400	
61	61	99.125	88.425	Coverage gap
62	0	99.200	88.500	
	:	:	:	
	60	100.700	90.000	
	61	100.725	90.025	
62	62	100.750	90.050	Coverage gap
63	0	100.800	90.100	
	:	:	:	
	62	102.350	91.650	
63	63	102.375	91.675	No gap
64	0	102.400	91.700	
	:	:	:	
	63	103.975	93.275	
64	64	104.000	93.275	Duplicate (overlap)
65	0	104.000	93.300	
	:	:	:	
	64	105.600	94.900	
65	65	105.625	94.925	
74	0	118.400	107.700	
	1	118.425	107.725	
	:	:	:	
74	8	118.600	107.900	Last FM band channel
75	63	119.975	109.275	A = 63 since it is a 6-bit counter

(b)

Figure 15.12 Settings of the DS8907 for: (a) AM operation: M, A, F_{VCO}, $F_{carrier}$; (b) FM operation.

15.4 Synthesizers and Microprocessor Systems

FM operation of the synthesizer, to tune station carriers spaced every 200 kHz (0.2 MHz) from 88.1 to 107.9 MHz, is more complex than AM operation. The prescaler operates in 63/64 modulus, with a reference frequency of 25 kHz. A standard 10.7-MHz IF is used, so F_{VCO} (which is the local oscillator signal) must range from 98.8 to 118.6 MHz. The M counter is loaded with 61, and A ranges from 0 to 63 (its maximum, since the A counter is six bits wide) (Figure 15.12b). The first F_{VCO} is then 97.600 MHz, corresponding to a carrier at 86.900 MHz.

Due to dual-modulus restriction that $A \leq M$, however, A cannot be >61 for $M = 61$. After $M = A = 61$, M is incremented to 62 and A begins again at 0, ranging to 62. Note that there is a small gap in coverage between $M = 61$, $A = 61$ and $M = 62$, $A = 0$. This gap means that some 25-kHz increments are not generated by the synthesizer, although all the 200-kHz station channels are synthesized. To cover applications that do need every 25-kHz increment (such as a signal generator or other test equipment) the DS8907 IC has special circuity designed-in which provides coverage for the gap values.

After $M = 62$ and $A = 62$, the counters are loaded with $M = 63$ and $A = 0$ to continue generating frequencies for the FM local oscillator. For the last standard FM channel (107.9 MHz) $M = 74$ and $A = 8$, although the DS8907 will go to $M = 74$ and $A = 74$ to provide F_{VCO} of 120.250 MHz (carrier of 109.550). Also note that some M and A pair combinations overlap and produce the same frequency output values (this can happen in AM mode, too). These overlaps are not a problem, since the system microprocessor performs the calculations and chooses, if possible, to change only A when switching to a new frequency, while leaving M unchanged as much as possible.

Example 15.5

What are the M and A values to generate a local oscillator signal of 1100 kHz for AM mode? What signal will be tuned with a 460-kHz IF?

Solution

To solve the equation for M, set $A = 0$ and see what value of M results: In this case 1100 kHz = $8 \times M \times 10$ kHz, so $M = 13.75$. Since M can only be an integer, truncate the result to 13, which is the major prescaler factor. To find A, use this truncated value of M in the equation and see what value A must have to bring F_{VCO} up to the required value: $1100 = 8 \times 10 \times 13 + 10A$, so $A = 6$. Therefore, to have F_{VCO} (the local oscillator) at 1100, use $M = 13$, $A = 6$. The signal tuned will be a $1100 - 460 = 640$ kHz (assuming a high LO).

Example 15.6

An FM station at 90.3 MHz must be tuned with a high LO and a 10.7-MHz IF. What is the needed LO value? What values of M and A must be set?

Solution

The LO must be at $90.3 + 10.7 = 101.000$ MHz. To get this value, set A to 0 and solve the F_{VCO} equation for M (with $F_{ref} = 25$ kHz). We get $101.000 = 64M(0.025)$, so $M = 63.125$. Once again, since M is an integer we truncate the decimal part and set $M = 63$ for the major prescaler factor. Next, we determine the fine steps needed, determined by the value of A by using $M = 63$ in the equation and solving for A: $101.000 = 64(63)(0.025) + (0.025)A$, which produces $A = 8$. Therefore, $M = 63$, $A = 8$ will produce $F_{VCO} = 101.000$ MHz and tune a carrier at 90.3 MHz.

Review Questions

1. What functions does an AM/FM synthesizer such as the DS8907 perform?
2. What are the reference frequencies for AM and FM operation? How are they derived? What are the dual-modulus values in AM and FM modes?
3. Why is a serial interface used between the microcontroller and the synthesizer ICs? How many signal lines are needed to transfer all the data bits? How is this done?
4. How many bits are in the M and A counters? What are the largest values in each?
5. How are M and A values calculated for AM operation? For FM operation?
6. Why does the FM tuning have gaps? Why is this usually not a problem? Why are overlaps in the FM not a problem either?

SUMMARY

Electronic, digital tuning using frequency synthesizers offers many advantages over traditional tuning with a continuously variable capacitor circuit. In indirect synthesis, a PLL VCO is forced to run at multiples of an accurate reference frequency by inserting dividers (counters) between the VCO and the phase detector. The VCO output serves as the local oscillator of a receiver or the master oscillator of a transmitter. Features of synthesized tuning include precision and accuracy over the entire band, compatibility with a microprocessor, programmability to desired stations, elimination of uncertainty on the correctness of the tuning, and versatility. The synthesizer loop filter plays a major role in the stability and settling, or lock-in, time of the VCO at any new value.

The simple synthesizer uses a divider (modulo N counter) and provides tuning up to about 100 MHz, with tuning step size is equal to the reference frequency. About 100 MHz, fully programmable dividers no longer function well, so a high-speed, fixed-modulus prescaler is inserted between the VCO output and the programmable divider. This reduces the speed needed in the programmable divider but increases the step size by the prescale factor. A dual-modulus prescaler overcomes this weakness by providing both a coarse and a fine scale setting through its two counters, so the step size is not increased but the operating frequency of most of the circuitry is reduced.

Summary Questions

1. What are the advantages of synthesized tuning over continuously variable tuning based on a capacitor? How is it microprocessor compatible?
2. Explain the operation of a direct digital synthesizer. What are the unique benefits it offers? What are the limitations?

3. Why does synthesized tuning have a fixed step size? Why is this not a problem? What is a modulus N counter?

4. What is the principle of indirect synthesis? What determines the accuracy of the tuning? What determines the actual frequency of the local oscillator?

5. What is the role of the phase detector in synthesized tuning? How does it work?

6. Why is the filter of the synthesizer critical? What performance characteristics does it determine? Why is a higher-frequency reference preferable, and what does it limit?

7. Why does a simple synthesizer become impractical beyond 100 MHz? How does a prescaler extend performance?

8. What are the features and drawbacks of the prescaler? How does the prescaler divider differ from the divider in the simple synthesizer?

9. What is a dual-modulus divider? How does it provide both prescaler functions and small step size?

10. Compare the simple synthesizer, prescaled synthesizer, and dual-modulus synthesizer in terms of step size, frequency range, lock-in time, and type of ICs used.

PRACTICE PROBLEMS

Section 15.2

1. A signal to be transmitted needs master oscillator frequencies of 50.0, 50.1, 50.2, 50.3, . . . , 51.9 MHz. What is the reference frequency? What are the divider factors?

2. A broadcast-band FM receiver (88.1 to 107.9 MHz, channel spacings of 200 kHz) has a 10.7-MHz IF, with the local oscillator above the carrier. What are the reference frequency and the modulo N factors needed?

3. A broadcast band AM receiver (550 to 1600 kHz, 20-kHz channel spacing) has a 450-kHz IF and a LO above the carrier. Determine the reference oscillator and N factors needed.

4. A two-way radio transmits and receives from 160 to 170 MHz on 25-kHz spaced channels. What is the reference frequency? What are the divider factors for transmitting? What are they for receiving if the IF is 20 MHz and the LO is below the carrier?

5. A system has a reference oscillator of 25 kHz and $\div N$ factors of 10,000 to 25,000. What distinct frequencies does the VCO generate? If this is the LO for a 25-MHz IF, and the LO is below the carrier, what frequencies will be tuned?

6. Sketch a reference signal and VCO input with 45° phase difference and the corresponding output of the exclusive-OR phase detector. What is the average value after filtering?

Section 15.3

1. Using a $\div 10$ prescaler to achieve a VCO output of 300 to 350 MHz with 100 kHz steps, what is the reference frequency? What are the values of the programmable divider?

2. A dual-modulus scheme is used for the VCO output of problem 1, with a 10/11 divider. What is the reference frequency? What counter values are needed?

3. A dual-modulus 64/65 synthesizer has a 10-kHz reference. For what counter values will this be a master oscillator from 128 to 136 MHz, and what step size will be achieved?

Section 15.4

1. Set the DS8907 for tuning an AM station at 1310 kHz. What is F_{VCO} if the IF is 460 kHz higher than the carrier? What are the values of divisors M and A?

2. The values $M = 15$, $A = 6$ are loaded into the synthesizer for AM operation. What is the generated F_{VCO}? What frequency would be tuned with an IF at 460 kHz if a high LO is used?

3. In FM operation, $M = 62$ and $A = 12$. What is F_{VCO}? What is the tuned carrier frequency with a 10.7-MHz IF above the carrier?

4. What values of M and A will tune an FM station at 101.7 MHz? What LO value is needed? Use a 10.7-MHz IF and a LO above the carrier.

16
The Telephone System

CHAPTER OBJECTIVES

When you have completed this chapter, you will understand:

- The basic operation of the telephone unit and its connection to the local telephone office
- How phones connected to the same central office are connected
- The interconnection of calls between local offices, and how connections are switched as needed
- The use of computer-controlled electronic switching at the local office and between offices for both analog and digital signals, and the features this adds beyond traditional electromechanical switching capability
- The problems caused by unavoidable signal echoes and various solutions to these problems

INTRODUCTION

The telephone system is a communications system that provides private two-way voice and data communication between virtually any locations, separated by a few yards to thousands of miles. A sophisticated network of switching circuitry allows a caller to dial and be connected to a desired party at the far end. Phone service at its simplest begins with a basic rotary dial phone, but the same connections can handle more sophisticated phones and pushbutton tone dialing.

Interconnections between phones are made at the local central office, either to other phones at the same office or to trunk lines that connect to other central offices. The central office switching used to be electromechanical, but solid-state switches are used in most new switches. Control of these switches is either by fixed-function digital logic circuitry or by computer-controlled electronic switching systems. The flexibility, reliability, and advanced features of electronic switching make it the best choice in many applications.

The trunks that interconnect switching locations are also advancing to computer control, providing new features that cannot be realized with traditional control. The computer-based systems allow the phone system to monitor its own performance and dynamically change routings and connections based on available signal lines and the amount of phone traffic. Both electronic switching at the central office and computer-controlled trunks base their operation on a computer program and a data base that contains information about all phone connections and lines and their status.

16.1 OVERVIEW OF THE SYSTEM

The telephone system is both the simplest and most complex communications system in general use. It is simple because the connection between the user's phone and the local phone switching office has just two wires, and the scheme that the office uses to interact with the phone is straightforward. At the same time, the phone system is complex because it can transmit voice and data over thousands of miles, through a mixed variety of links and communications channels, through analog and digital circuits, and through many adverse operating conditions, such as noise and errors. Most important, it does all of this without action by the user, who simply dials the number of the desired party. It differs considerably from systems with prewired links that forever connect specific users to other specific users: The phone system has extensive switching capability to connect any subscriber to any other subscriber. It is this switching that makes the overall phone system extremely flexible and complex.

The phone system manages itself, with switching systems and computers and providing alternative choices and redundancy, to produce a good, complete, reliable connection in most cases. Finally, the phone system is a marvel of electrical compatibility: The standard telephone plug that the user sees can accept basic voice phones, electronic phones, or computer and data interfaces, yet the system operates properly with each because careful technical decisions were made to ensure that newer developments do not make existing equipment obsolete.

In this section we look at the overall plan of the complete phone system, begin to understand the role played by each element, and see how the many elements integrate and tie together. In subsequent sections we will look in detail at the operation of each part of the phone system. As in many systems, the final performance that the telephone system can provide is closely related to how well each part performs, and it is essential to understand the capabilities and limitations of each part. The terminology used in the phone system is wide and varied, and much of it is migrating over to general communications systems.

The phone system begins with the *local loop*, which is the pair of wires that connects the *central office* to the user's phone. There is one central office for each *exchange*, which is all phones with the same first three digits in the seven-digit number. A single central office for exchange XYZ serves 10,000 phones with numbers from XYZ-0000 through XYZ-9999. In large cities, several exchanges are housed in the same building, but from a system perspective they are separate offices.

16.1 Overview of the System

The function of the central office is to *supervise* the local loop and the phone connected to this loop and to act as the interface between the phone and the rest of the system. This requires that the central office monitor the phone line to see if the phone is hung up and idle (*on-hook*) or in use (*off-hook*), to send a dial tone, ringing signals, and other tones to the phone, to listen for dialing information from the phone, and to meet other interface needs, such as providing power to the phone.

As the interface between the phone and the rest of the system, the central office makes the physical connection between one phone and another in the same exchange. If the call is to another exchange, the central office is responsible for connecting to a special link between exchanges called a *trunk*. A trunk can be a physical wire, a fiber optic cable, a radio link, or a satellite connection. Calls between adjacent towns usually involve only a single trunk between two exchanges, but a longer-distance call can be routed through several trunks and central offices. Looking at a diagram of central offices and trunks (Figure 16.1), we see that they form a *network* of interconnected *nodes*.

For heavily used long-distance paths, there are *supertrunks* that collect many trunk lines at regional switching centers and combine them together, often using multiplexing, and then go directly to the far end. For example, a call from New York to Los Angeles can go through many shorter trunks and their central offices (such as New York to Cleveland to Chicago to Denver to Los Angeles) or go on a supertrunk that connects New York and Los Angeles exclusively. A node that connects a supertrunk is different from a regular central office and is often called a *tandem switch* because it connects only to other central offices and does not have any local loops attached. It interfaces only those calls that already have been placed on a trunk for long-distance connection.

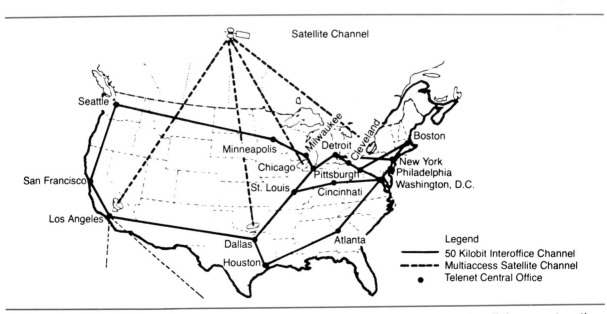

Figure 16.1 A network of nodes of the overall phone system links users together (Courtesy of Racal-Vadic, Inc.).

There is a carefully planned *hierarchy* of small, medium, and large switching centers. There are about a dozen of the largest switching centers in the United States. To provide maximum flexibility, a center at one level can connect to centers at other levels (Figure 16.2), and call routing does not have to go up to the top of one hierarchy chain and then start back down to the final destination on another. For example, there are a lot of calls between adjacent towns, and these are not routed up the hierarchy and then back down: A high single high-usage trunk directly links their central offices. (If this trunk is out of service, though, the calls are easily routed through another, more complex path.) The routing scheme adjusts dynamically for changes in circumstances, compensating for heavier traffic through some points, failed equipment, or different mixes of shorter- and longer-distance calls. The three-digit *area code* prefix of the regular seven-digit phone number indicates that the call is for a different major center.

The variety of network connections and options gives the phone system

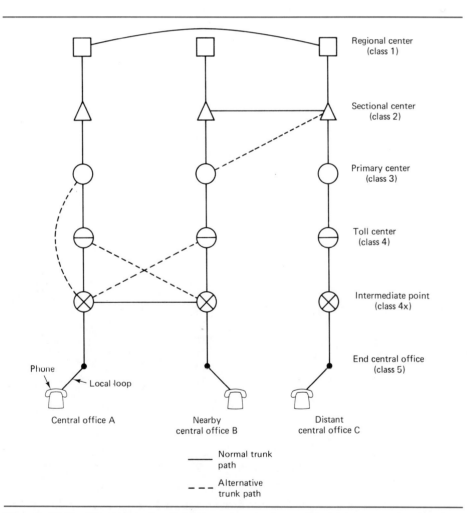

Figure 16.2 Hierarchy of switching centers and possible interconnection levels allows organized routing with flexibility.

tremendous flexibility to route calls through the shortest route or to provide alternative but perhaps longer routes when the preferred route is overloaded or has technical problems. The use of computers in recent years to control the central offices as well as to control groups of central offices has added to the ability of the phone system to find the best route and yet adapt and reconfigure the routing plan as needed to meet new situations and problems.

All of this is transparent (a concept we have seen before) to the user who is making the call. It is difficult to appreciate the tremendous "behind the scenes" complexity that makes efficient operation of the phone system possible. In the early days of telephones, callers were interconnected manually at switchboards by plugging wires into socket receptacles. Automated equipment soon replaced the manual operation, but these were still essentially mechanical switches that could only connect in the most straightforward manner and could not look at the many routing choices and pick the best one for a call at that specific time and circumstance. Computer-controlled central offices and network nodes now provide continuous updating and reevaluating of the network and routing, and can even change the route in midcall if the signal quality falls below standards. At the same time, the system accepts signals and routes calls from the simplest dial phone as well as high-speed computer interfaces.

Review Questions

1. What is a local loop? A central office? An exchange?
2. What is the function of the central office? What does it do for local loops on a single exchange?
3. How are central offices interconnected? When are trunks needed?
4. What is the hierarchy of switching centers? Why is there flexibility in the routing of calls via different levels of this hierarchy?
5. What are on- and off-hook conditions?

16.2 THE TELEPHONE INSTRUMENT AND THE LOCAL LOOP

The simplest form of telephone service that is provided by the central office to a basic phone with rotary dial or pushbuttons is known as *Plain Old Telephone Service* (*POTS*), representing the fundamental telephone instrument and local loop. A schematic of a basic phone is shown in Figure 16.3, along with Figure 16.4, showing loop voltage and current during different parts of the calling sequence. When the phone is on-hook, it appears as an open circuit on the loop and draws no current. (This is a major necessity: If phones that are not in use draw power, the central office would be supplying a considerable amount of costly power to the thousands of idle phones.) As soon as the phone goes off-hook, the local loop circuit is completed and about 20 mA of current flows, with power supplied by a -48-V battery at the central office. The presence of the current flow

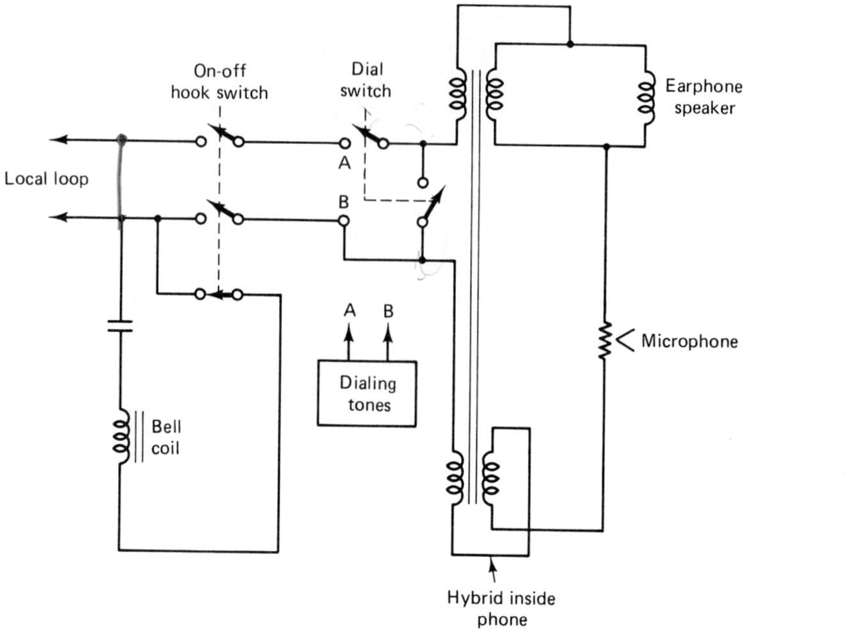

Ringer
80-120 V RMS
20 Hz
2 sec. ON
3 sec. OFF

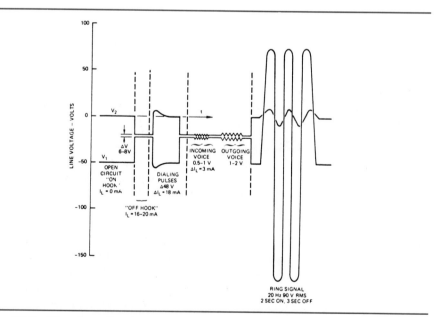

Figure 16.3 Schematic of basic telephone (both pulse and tone dialing) (courtesy of Monsanto Corp.)

Figure 16.4 Signals on local phone loop during phases of use (courtesy of Monsanto Corp.).

in the loop is detected by the central office and indicates that the phone has gone off-hook.

Although the phone is an open circuit with no dc current path when on-hook, there must be a way to ring the phone for incoming calls. An ac signal called the *ringing voltage* is sent from the central office to the phone, and this voltage activates the bell or electronic ringer in the phone. The bell circuit is across the local loop wires in series with a capacitor, which blocks any dc current and prevents the on-hook phone from drawing battery current, while allowing the ac ringing signal to pass.

Talking Signals

The voice on the local loop is an analog voltage. For outgoing voice, its signal level is about 1 to 2 V at approximately 3 mA, and the incoming level is about half the outgoing level. Since there are only two wires and the voice signal is full duplex (an incoming voice and outgoing voice can occur at the same time—and often do—instead of at separate transmit and receive times), a special transformer in the phone unit itself prevents most of the speaker's voice from feeding into the earphone speaker and blasting his or her ear. This *hybrid* transformer uses multiple windings in opposition so that the outgoing signal to the earphone cancels most of itself out (a little *sidetone* is needed so that you can hear yourself talking) but allows the full incoming signal to reach the ear. We will see more of the hybrid transformer later.

Loop Signal Levels

The signals on the loop must be large enough to be usable, but not so large that they overload the phone or amplifiers in the rest of the system. Phone company specifications detail the acceptable ranges. The outgoing power level of the "voice band" signal (averaged over a 3-s period to account for normal conversation ups and downs) from the phone to the central office is between -10 and 0 dBm into a 600-Ω load (900 Ω in some areas). Incoming signals have much greater variability since they pass through many stages, trunks, links, and switches, and are from -42 to -4 dBm. In addition, signal energy outside the voice band must be limited with the maximum out-of-band signal at 3995 Hz (and higher) at a level of -16 to -36 dBm or below, depending on frequency.

Note that both outgoing and incoming signals have a significant spread in acceptable levels, which means that phone system equipment must be designed to operate over a wide range. Impedances are also important: The central office terminates the local loop in a nominal 600-Ω (or 900 Ω) impedance, but due to unavoidable factors such as line length, temperature, humidity, and power levels, the equipment (the phone instrument or other unit on the loop) must be able to drive loads of 200 to 1200 Ω.

Dialing

There are two kinds of dialing: pulse and tone. *Pulse* dialing from a rotary dial phone is the original kind developed soon after the invention of the automatic switching equipment for the central office. When a digit is dialed, the loop circuit is opened/closed a number of times corresponding to the dialed digit, at a basic rate of 10 make/break cycles/s with approximately 50% duty cycle. The central office senses and counts these make/break sequences to recognize the dialed digit (Figure 16.5).

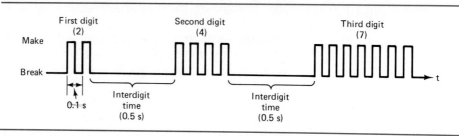

Figure 16.5 Signals on loop from pulse dialing, including interdigit time.

The individual digits of the complete number are separated by a slightly shorter period of time than a person requires to move his or her finger back to dial another number, approximately 0.5 s or more. This interdigit time is the key to recognizing that the following make/break pulses belong to a new digit, not the preceding one. The central office also has special timeout circuitry to make sure that the break part of the dialing pulse is not misinterpreted as the phone is being hung up at the end of a call, which also opens the local loop path.

Pulse dialing is very simple in concept but it has some major drawbacks: An electromechanical relay or its electronic equivalent is needed to generate the make/break cycles. Also, there is no way to send special controlling signals to the central office for special services, since any make/break while using the phone puts clicks on the line and could be misinterpreted as a dialed digit. Finally, pulse dialing takes a long time to complete.

This last point is the most significant. Consider dialing the number 742 with 10 make/break cycles/s and 0.5 s between digits. The total dialing time is then

$$0.7 + 0.5 + 0.4 + 0.5 + 0.2 = 2.3 \text{ s}$$

Of course, for a typical seven-digit number, the time will be much longer, and a complete long-distance number (1 + area code + phone number) can take many seconds. Yet time is a precious commodity on phone lines or any communications system, since central office equipment is allocated to sense the dialing sequence during the dialing period, and a long time is required to pass the dialed digit information on to the next central office for a long-distance (trunk) call.

Tone dialing (Touch-Tone), the alternative to pulse dialing, was developed in the 1950s by the Bell Telephone System to overcome the weaknesses of make/break digit signaling and is now in widespread use. In *tone dialing*, each digit is represented by a unique pair of tones, shown by the keyboard matrix in Figure 16.6. There are four low-group tones, one for each row, and three high-group tones, one for each column (actually, there is a fourth high-group column and tone but it is only used on special phones). When you "dial" the digit 5, for example, the corresponding low and high group tones are generated by the phone and sent to the central office: 770 and 1336 Hz. The # and * keys are for indicating special functions and are also sent as a tone pairs. With this *dual-tone multifrequency* (*DTMF*) dialing, all digits take the same amount of time and the tones are sent only as long as the pushbutton is held down by the person dialing, which is usually about 0.25 to 0.5 s.

16.2 The Telephone Instrument and the Local Loop

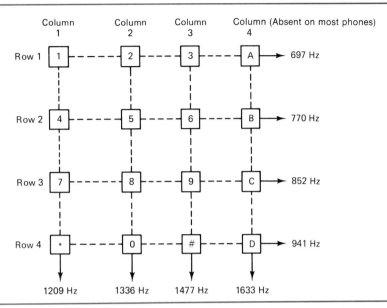

Figure 16.6 DTMF tones for each number, produced by keys arranged in a matrix.

Example 16.1

How long does it take to dial the number 784-3742 (a) using pulse dialing with 0.5-s interdigit time (b) with tone dialing where digits pushbuttons are held down for 0.25 s and the interdigit time is also 0.25 s?

Solution

(a) For the pulse dialing, the time is the sum of the digit times, plus six interdigit times:

$$0.7 + 0.8 + 0.4 + 0.3 + 0.7 + 0.4 + 0.2 + (6 \times 0.5) = 6.5 \text{ s}$$

(b) For tone dialing, the time is 7×0.25 (for the digits) + 6×0.25 (interdigit times) = 3.25 s.

From these typical values we see that DTMF dialing is much faster than pulse dialing in time required to send the dialed digit information from the phone to the central office. There are other advantages to tone dialing: Circuitry to send tones at the phone and sense them at the central office is easier and cheaper to build with modern integrated circuits compared to make/break pulse generation or detection. Most important, the tone pairs themselves can be used to signal the user's equipment at the receiving end and so initiate special action such as changing to a different mode of operation (or making a phone answering machine play back its accumulated recorded messages from a remote phone). This is because the tones are in the user's voice band of frequencies (*in-band* signals), and the phone system is designed to ignore any signals in the voice band after the initial

---

Choice of DTMF Tones

It may appear that the low and high tones assigned for DTMF are random and arbitrary, but the reality is the opposite—these tone frequencies were chosen only after careful study. Note that no tone is a harmonic, or even close to a harmonic, of another tone. Therefore, any harmonics due to distortion of the pure DTMF sine-wave pair (Fourier analysis, Chapter 2) will not produce a signal that could appear as another legitimate frequency. Second, the unavoidable mixing of the four tones (one tone pair plus the other tone pair), through circuit nonlinearities does not produce a sum or difference frequency that can be confused with a valid frequency. The designers of the system anticipated potential weaknesses, problems, and new frequencies resulting from signal distortion and mixing, then designed a scheme that is fairly resistant to being fooled into making errors.

Example 16.2

What are the second harmonics of the tones for the digit 3?

Solution

The fundamentals are 697 and 1477 Hz. The harmonics are 1394 and 2954 Hz. The 1394-Hz harmonic is midway between the column 2 and 3 values; the 2954-Hz harmonic is well beyond the 1633-Hz maximum of DTMF.

Example 16.3

What are the sum and difference frequencies generated through unavoidable mixing of the DTMF tones for the digit 3?

Solution

The sum is 697 + 1477 = 2174 Hz; the difference is 1477 − 697 = 780 Hz, which is far enough away from the 770-Hz "row 2" tone to be distinguished from it.

---

dialing sequence, since they may be generated by the user's voice. These ignored signals are passed directly through the system to the receiver, which may then take some prespecified action.

In contrast, the make/break signals cannot be used for anything except dialing. A break during conversation is normally interpreted as meaning that the phone has been hung-up, for example, so that the phone system is always monitoring the local loop to sense a line break during the talking cycle. In other words, in-band signals such as tones are transparent to the system after dialing, while make/break signals are not.

Note: Many phones with pushbutton keypads have a switch to set either tone or pulse dialing. When the selector is set to pulse mode, the person dialing pushes keys just like on a DTMF phone, but the phone electronically generates pulses like a standard rotary dial phone. This allows the convenience of buttons on a phone even where the central office does not support DTMF for that particular loop (or the user has decided not to pay the extra fee for DTMF capability). Although this gives the user the convenience and appearance of a tone phone, it is still a pulse phone and the dialing sequence takes the same amount of time to complete as it does on a dial phone.

Generation of DTMF Signals

The original DTMF generators used one of two schemes. Eight separate oscillators—one for each of the eight row and column tones—provided all required values, and pressing the number on the dial keypad connected the appropriate two

Input	Nominal (Hz)	Divider ratio N	Actual* (Hz)	Deviation* (%)
R_1	697	5120	699.13	+0.3
R_2	770	4672	766.17	−0.5
R_3	852	4224	847.43	−0.5
R_4	941	3776	947.97	+0.74
C_1	1209	2944	1215.88	+0.57
C_2	1336	2688	1331.68	−0.32
C_3	1477	2432	1471.85	−0.35
C_4	1633	2176	1645.01	+0.74

* f_{osc} = 3,579,545 Hz; f_{actual} = osc/N.

not harmonically related so as not to produce another tone the same as another

Figure 16.7 Division factors for MC14403 DTMF generator IC (courtesy of Motorola, Inc.).

oscillator outputs to a summing circuit. Alternatively, two oscillators were used, one for the rows and one for the columns. Pressing the key of the phone connected one of four different capacitors into each of the oscillator circuits, to produce the necessary output frequencies.

The drawback to these methods is that the precision and stability of the tone frequencies are dependent on the critical components in each oscillator. Each one has to be calibrated and adjusted for correct output frequency, and the oscillators and their capacitors will drift with supply voltage, time, and temperature. Two newer designs derive all DTMF frequencies from a single oscillator, which is simpler to adjust than many separate ones. The first uses a single very high frequency master oscillator (often based on a 3.579545-MHz TV chroma crystal) and then divides this down via fixed $\div N$ dividers to provide the correct output frequencies. For example, with $N = 4672$ the output frequency is 766.17 Hz, sufficiently close to the nominal 770-Hz value. Figure 16.7 shows the divisors used in the Motorola MC14403 DTMF generator IC and the resulting output frequencies. These divisor values are used because they provide sufficiently accurate results while minimizing critical factors such as IC complexity, cost, and power consumption.

The newest approach is shown in Figure 16.8 as implemented by the Motorola MC34013 telephone speech network and tone dialer IC. This single chip derives its 1.4-V operating voltage from the local loop and provides many of the functions needed in a telephone instrument, including interconnecting the local loop, telephone microphone, and ear speaker while providing DTMF tones and most of the hybrid function.

The DTMF tones are based on a 500-kHz master oscillator signal. When a key on the phone pad is pushed, logic circuitry within the IC senses the key row and column and sets the necessary $\div N$ factors in two separate 4-bit counters, one for the row tone factor and one for the column tone factor. These counters simply start at 0000 and count, driven by the 500-kHz clock, up to 1111 and then down back to 0000; then the count-up and -down cycle begins again and continues as long as the key is depressed. Each counter output goes to a digital-to-analog converter (DAC), which produces an analog voltage corresponding to the

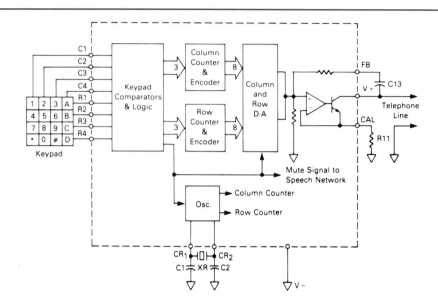

Figure 16.8 Synthesizing DTMF tones using counters and D/A converters in the MC34013. (a) DTMF block diagram; (b) frequency synthesizer errors (courtesy of Motorola, Inc.).

counter's value. The DAC output is then a "16-step" approximation of a sine wave, at a frequency determined entirely by the rate at which the counters increment. The approximations are low-pass filtered to smooth them, and then added together to form a DTMF signal.

The $\div N$ factors for the counters designed into the MC34013 produce tones with less than 0.15% error, which is very good. The only critical component is the 500-kHz crystal of the oscillator. Producing a tone this way is also called *frequency synthesis*, but it differs considerably in concept from the synthesis of local oscillator signals studied in Chapter 15 (it is actually a variation of direct digital synthesis).

Local Loop Characteristics

The local loop itself is thin wire, usually No. 22 to No. 28 AWG, and is anywhere from several hundred feet to several miles long, depending on the distance between the phone and the central office. This variation in distance means that signal levels can vary considerably from one loop to another. More important are the frequency characteristics of the line. A standard line is optimized for voice communication, with a passband that has about 10-dB variation from 300 to 2500 Hz. Signal attenuation outside the passband is much greater, and reaches about 30 dB at 100 Hz and 3500 Hz (Figure 16.9a). Similar characteristics apply to the trunk lines that interconnect central offices.

The time-delay characteristics of the phone line are also critical. Recall that a delay in the time domain is equivalent to a phase shift in the frequency domain. The time delay/phase shift introduced by the phone line, called *envelope delay*, causes distortion of the overall waveform as different spectrum components are delayed by differing amounts. For voice, this is not too serious, since the ear is relatively insensitive to phase shifts in the spectrum. However, this is a severe problem for digital signals sent over the phone lines, which we discuss in Section 16.7.

Telephone lines are available in several performance grades. A standard unconditioned voice-grade line is a *dial-up* or *switched line* since it is the line you

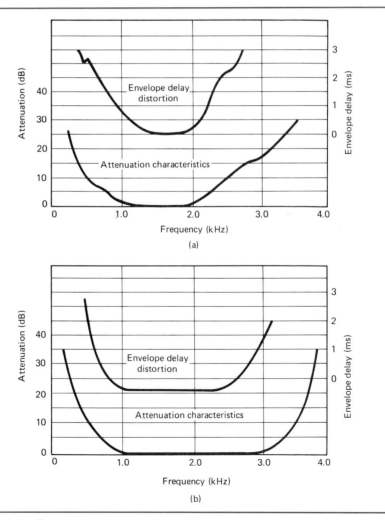

Figure 16.9 Frequency and envelope delay characteristics for: (a) ordinary unconditioned line (courtesy of Racal-Vadic, Inc.); (b) better-quality conditioned line.

randomly get when you pick up a phone and are switched to a trunk. In contrast, a user who needs wider bandwidth and less envelope delay can have a special line installed or reserved which is guaranteed to meet superior specifications; this is called a *dedicated, conditioned,* or *leased line* (Figure 16.9b). Five standard grades of *C-conditioning performance* are available, needed when digital signals at increasingly higher baud rates or analog signals of wider bandwidth are sent over the lines, and low distortion and error rates are required.

Review Questions

1. What does the phone look like electrically when it is on-hook? Off-hook?
2. What is the typical supply voltage for phone power?
3. How is the phone rung? How does the ringing signal get through the open loop?
4. What does the hybrid do in a phone instrument?
5. What are the nominal and actual phone impedances?
6. How is pulse dialing done? How are the digits recognized? How are the digits separated from each other?
7. What are the disadvantages of pulse dialing? What is the most serious problem?
8. What is tone dialing? Why is it called DTMF? How does it work? Why does a pushbutton phone sometimes look like a DTMF unit but really remain a pulse dialing unit?
9. How were the tones for DTMF selected?
10. What is the typical time to dial a seven-digit number with tone dialing?
11. Explain how DTMF tones can be synthesized by using a master oscillator, $\div N$ counters, and a DAC.
12. What are the loop characteristics for frequency response and envelope delay?
13. What are the differences between dial-up versus dedicated lines?

16.3 THE CENTRAL OFFICE AND LOOP SUPERVISION

The central office is the interface between the local loop, other local loops at the same central office, and trunks to other central offices. The interface to a local loop is called a *subscriber loop interface circuit (SLIC)* or *line card* and is responsible for providing signals to the loop, sensing loop activity, and sending control signals to the phone at the end of the loop. One SLIC is needed for each phone connected to the central office. The signals on the loop during on- and off-hook, dialing, talking, and ringing were shown in Figure 16.4. The two wires of the local loop are called the *ring* (−48 V) and *tip* (ground), from the early days when the connections were made at a switchboard with a two-contact plug similar to those used in today's audio equipment.

Operation Sequence

Since using a phone is so commonplace, we may not realize that a complex sequence of events occurs when a call is made. The *calling party* picks up the phone, the central office detects this change from on-hook to off-hook status, sends dial tone to the phone, and switches in circuitry to recognize the dialed digits (either pulse or DTMF). Only then can the dialing begin. As soon as the central office recognizes the first dialed digit, it disconnects dial tone from the loop.

After the complete number is dialed, the central office, trunks, and switching circuitry make the connection to the *called party* phone and a ringing voltage generator is connected to ring that phone (assuming that it is not busy). When the called party picks up the phone, the central office for that phone senses this and disconnects the ring generator; the regular talking circuits of the two phones are connected, so conversation begins.

The six functions provided by the SLIC for the loop are summarized as *BORSHT*, an acronym for "battery, overvoltage protection, ring trip, supervision, hybrid, and test." The specifics of these functions are described below.

Battery. The phone loop requires voltage when it is off-hook, both to drive the earphone/speaker and to provide a supply signal that the microphone can modulate with the talker's voice intensity and frequency. Most commonly, the battery is −48 V dc with respect to ground, provided by batteries at the central office. The batteries themselves are continuously charging from power lines so that the phone system operates even when there is main power failure; backup time is usually up to 24 or 48 hours.

Overvoltage. The phone lines are long and not shielded. They can easily pick up electrical interference of hundreds of volts from nearby equipment, power lines, and even lightning, as well as misconnection during installation. The overvoltage protection in the SLIC protects the SLIC and central office circuitry from these excessive voltages.

Ring trip. To ring the phone, the central office sends the 90-V_{AC} 20-Hz ac signal that is on for 2 s and off for 4 s. When the phone is picked up, the SLIC detects the off-hook via the ring trip, disconnects the ringing voltage source, and connects the normal talking circuitry in its place.

Ringback

It is a common misconception that the ringing voltage applied to the called phone is also the ringing signal heard by the caller. In fact, a separate and much lower-level local ringing signal is sent from the central office of the calling party back to the calling phone, so the caller can hear a sound which indicates that the far-end phone is ringing—but this signal does not come from the generator ringing the called phone. The called phone ringing and the calling phone *ringback* are each supplied separately by the central office. In a long-distance call that involves two new central offices, the ringing and ringback come separately from the central office responsible for each local loop, respectively, the ringback is supplied by the called loop office on most older offices.

If you have ever had the "strange" experience of someone picking up the phone before it apparently rings, this is why: The ringing signal generator and the ringback generator are not synchronized, and it is possible for the ringing signal to be sent to the called phone, and the phone picked up, slightly before the ringback signal is generated for the calling phone.

Supervision. The SLIC monitors the phone loop to detect when an on-hook phone is picked up, indicating that a user wishes to begin the calling cycle. As soon as the phone goes off-hook, the SLIC senses this and indicates to the central office that dial tone must be sent to the phone, so the user can begin dialing. Similarly, when the caller hangs up, the SLIC senses that the call is finished and indicates this to the rest of the central office.

Hybrid. The local loop consists of just two wires and carries full-duplex signals—the voice from the person at the phone to the far end, and the voice from the far end speaker to the ear of the person at the phone. For these two signals to be amplified and transmitted through the rest of the phone system, they must be split into incoming and outgoing signals, thus establishing a channel in each direction. A transformer-based circuit called a *hybrid* does this (Figure 16.10), sometimes called a *two-wire to four-wire converter* or a *four-wire terminating set*, since it splits the local loop into two wire pairs, one for each direction. The hybrid circuit uses a special set of transformer windings that operate so that the incoming (receive) signal can pass onto the two-wire loop, but is self-canceled by opposing windings so that it does not pass on to the outgoing (transmit) wires. Signals from the phone on the two-wire loop go only to the transmit pair, not to the receive wires, also as a result of the canceling action of the transformer windings.

Here is how it works: If the hybrid balancing network and the two-wire line have identical impedances, the signal power from the receive pair divides equally between the balancing network and the two-wire line; thus no power enters the transmit pair, where it would return to the signal originator. Similarly, if the

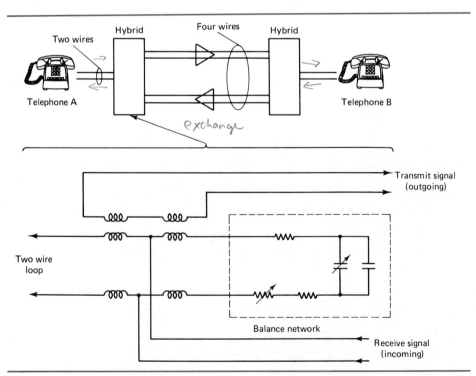

Figure 16.10 Location and schematic of hybrid transformer circuit (courtesy of Racal-Vadic, Inc.).

impedances of the receive and transmit pair are identical, the signal power that enters from the two-wire loop divides equally so that half goes to the transmit pair while the other half goes to the receive pair, where it is dissipated in the output impedance of the receiver pair amplifier. The hybrid introduces a 3-dB loss, unfortunately, each time signal power divides in half between a signal line and the balance network dissipation. The hybrid circuit is very similar to the one in the phone itself, which prevents the speaker's voice from blasting into the earphone while allowing the incoming signal to reach the earphone and then be heard.

The key to a successful hybrid is the balance network, with impedance that must match the local loop impedance. Any mismatch between these two results in some undesired signal feeding through when it is not wanted and line reflections and echoes (Section 16.6). A good hybrid circuit can produce 30 to 50 dB of rejection, or *transhybrid loss*, depending on the matching between the balance network impedance and the loop impedance. Of course, the wide span of loop impedances (due to loop-length variation, which affects resistance, inductance, and capacitance) means that a single standardized hybrid circuit will be matched for some loops, but not exactly right for others.

Each end of the calling-to-called phone connection requires a hybrid. The two-wire loop from the calling phone is converted to four wires (two pairs) for transmission and amplification, and then recombined into two wires at the far end to go onto the loop of the called phone.

Test. The test function allows the central office to put special test signals on the line, necessary to check the quality of the loop and phone performance. These signals must override regular use of the phone lines.

SLICs are now implemented with ICs that occupy less space, use less power, and provide more functionality than circuits made of discrete components. The Motorola MC3419 SLIC (Figure 16.11) provides all the BORSHT functions

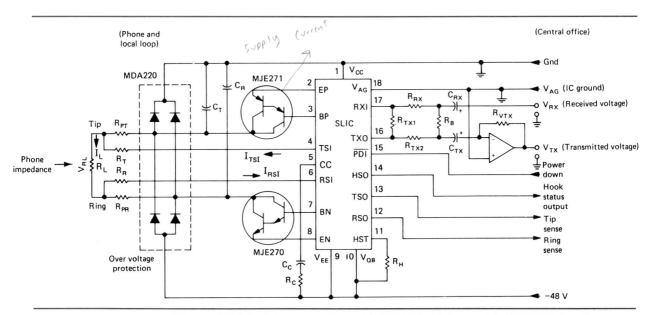

Figure 16.11 Interconnection of MC3419 SLIC to provide BORSHT functions (courtesy of Motorola, Inc.).

with a minimum number of external components. On the loop side, the MC3419 drives two transistors to provide up to 20 mA of loop current (its own internal transistors cannot supply this much current) and an external overvoltage protection diodes. The hybrid function is part of the IC itself, with several op amps to add or subtract the appropriate incoming and outgoing voltages as needed. This is much less expensive and more reliable than having a separate transformer component. Only the hybrid balance network—component R_B—is provided by the IC user, since it must be selected to match the loop impedance. R_B can be replaced by a reactive impedance if the line is not a pure resistance, for maximum hybrid performance.

Review Questions

1. What is the complete sequence of events from placing a call by going off-hook and completing a phone call when the far end goes off-hook?
2. What happens when the phone is busy?
3. Where does the ringing voltage originate? What is ringback? Where is it from?
4. What is a SLIC? How many SLICs are needed in a central office?
5. Describe each of the six BORSHT functions.
6. What is the two- to four-wire conversion function (and the reverse) of the hybrid? Why is it needed? Why does the hybrid introduce a 3-dB loss?

16.4 THE CENTRAL OFFICE AND SWITCHING

To connect the calling phone to the called phone, the central office must do two things: recognize the dialed digits of the phone number, and make a physical connection between the loops of the two phones (for calls within a single central office) or to the correct trunk (for calls to another central office). Pulse dialing make/breaks are recognized by circuitry that senses both the make/break cycle as well as the time between dialed digits, to recognize that the pulses from one digit are complete and a new digit is beginning.

For DTMF dialing, the central office must identify the two distinct tones that the telephone generates for each number depressed on the phone keypad. As soon as the phone goes off-hook, the central office does two things: It attaches the digit-recognizing circuitry to the loop and sends dial tone to the phone to indicate to the user that the special circuitry is connected and the central office is ready to accept dialing information. One way to identify DTMF is to use a bank of seven narrow-bandpass filters, with four filters tuned to the frequencies of the keypad rows (the low group) and three tuned for the column (high) group (Figure 16.12). As each DTMF tone pair is generated, only the two filters corresponding to the tone pair have significant outputs. Comparator circuits monitor the outputs of all seven filters, producing a logic 1 output if the filter output is above a threshold or a logic 0 if the filter output is below the threshold. Comparators with thresholds above

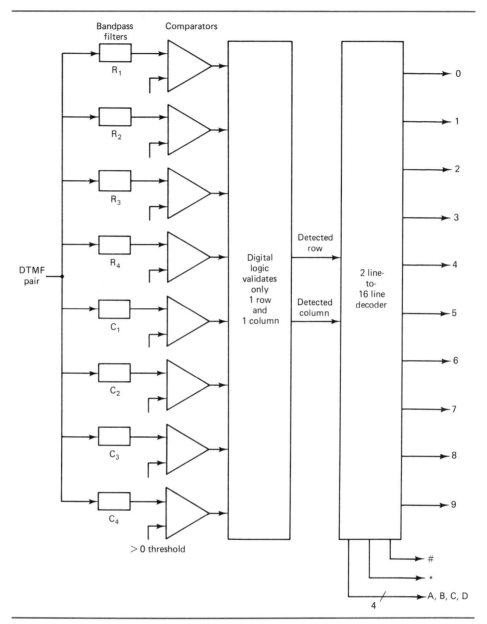

Figure 16.12 DTMF receiver using filters and digital logic circuitry to identify dialed digits.

0 V are used so that noise—which may have energy at any frequency—does not cause false responses.

The logic outputs of all seven comparators then go to additional digital logic circuitry, which decodes the two valid filter outputs (one from high and one from low) and produces an output on one of 10 lines, corresponding to dialed digits 0 through 9. The same logic circuitry is also designed to reject as invalid any condition other than one output from a low filter and one from a high filter.

Therefore, a single comparator 1 or three comparator 1s are ignored, as are two outputs from the high group or two from the low group of filters.

At this point the central office knows what digit was dialed. It then holds this digit in storage while the rest of the digits of the complete number are dialed. When all the digits have been received, the central office disconnects the digit recognition circuitry so that it can be used for another caller and local loop. Next, the central office must make the connection between loops. The complete dialed digit series goes to logic circuitry that associates these digits with a single local loop if the first three digits are the same as those of this central office. If the first three digits are different, the central office finds the trunk line that corresponds to the three digits that represent the other central office.

The calling cycle continues as the called party is connected to ringing, then answers by going off-hook. The central office then disconnects the ringing generator and, finally, makes the connection of the talking path.

Making the Connection

The physical path between the two phone loops must allow an analog signal voltage to pass. Earliest central offices made the connection by closing a mechanical switch contact or energizing a relay to close a contact. These were replaced by switching systems using contacts arranged in a *matrix* (also called a *crossbar* or *crosspoint switch*) that has incoming caller lines along the vertical side and outputs along the horizontal side. The control signal that energizes the contact is completely isolated from the talking signal path. The large matrix required is actually built from many smaller matrices, and some ingenious methods are used to reduce the total number needed in the central office. The contacts of the crossbar or rotary switch are exposed to air and vibration, which reduces their reliability. For much greater reliability, *reed relays* are now used, with contacts sealed in a glass enclosure and opened or closed by an electromagnetic coil wound around the enclosure.

Although switches and relays are effective, they have some drawbacks: They are fairly large, consume significant power, and will wear out like all electromechanical devices. The problem is especially serious when you consider the thousands used in a central office. In the early 1960s, central offices began to use *solid-state switches*, which are all-electronic equivalents of electromechanical switches and relays. Early designs used a single CMOS IC to replace a single switch crosspoint contact and external digital logic to point to the designed matrix connection point and turn it on or off. Present integrated-circuit technology fabricates many such switches of the matrix onto a single IC, along with the circuitry which decodes the address of the two lines that must be connected.

One such device provides a 12 × 8 matrix. This device has 12 ports (connection points) along one axis and 8 along the other, so it can steer any one of 8 lines to any one of 12 outputs. The fact that the matrix is not square is not a problem: The extra four lines can be for sending signals out to trunks for connection to other central offices, for example. The specifications of this device show how it replaces a mechanically completed switch connection: It has an on-resistance of just 35 Ω, off-resistance of several MΩ, can handle analog signals up to 6 V, and requires just 10 mA of current at 5 to 12 V dc for power. The IC can also be combined with other ICs to form larger matrix sizes.

In operation, the digit recognition circuitry of the central office produces a complete number of the desired called phone. The logic circuitry takes this num-

Pulse Dialing and Mechanical Switching

The original automatic central office was invented to ensure proper connection without the need for an actual operator. The *Strowger* automatic step-by-step switch (named after the undertaker who invented it in 1891 so that the local phone operator—his competitor's wife—could not connect calls for his services to the other mortician in town) served as the heart of virtually every central office, until tone dialing and electronic switching became common beginning in the 1960s. Now electronic switching has advanced to computer-controlled central office switches (Section 16.5).

Figure 16.13 shows in simplified form using the three-digit dialed number 184, how the automated crossbar system operates. A series of 10-position rotary switches is driven by the dial pulses, with the switch position setting a ratchet mechanism: As the first digit is dialed, the dial pulses (at a 10-pulse/s rate) cause the mechanism to step through from 1 to 10 steps (dialed numbers 1, 2, . . . , 8, 9, and 0). A special circuit recognizes the relatively long period between dialed digits (at least 0.5 s). When the last pulse from this first digit (1) occurs, this circuit directs the next digit to a second bank of rotary switches, which then step along with the pulses of this second digit (8). The time gap between digits 2 and 3 is next sensed, and the next group of pulses are directed to a third group of rotary switches. The third group is stepped by the dial pulses, and after the third digit (4), the calling phone loop is connected to the desired called party. Of course, the wiring must be set up so that dialed digits 0 to 9 correspond to rotary step positions, 1, 2, . . . , 9, 0, but this is just a matter of wire arrangement. The actual central office switches used provide the capability to step through and search for a free line to the next switch, so that a switch can serve for many users. This reduces the overall number of switches needed.

ber, converts it to the binary value that corresponds to the physical connection point of the phone, and sends two numbers to the matrix decoder and latches: one for the connection point of the loop that initiated the call, the other for the connection point of the called phone path. The decoding logic uses the binary values to activate a single internal switch, located at the correct crosspoint of the matrix, and the connection between phones (or intermediate links between phones) is made.

If the desired phone is at another central office, the sequence is slightly modified. As soon as the logic sees that the three-digit prefix of the phone number indicates a different central office, it uses these prefix digits to make the connection to a trunk to the other central office. It next sends the remaining digits (the last four digits of the seven-digit number) to the other central office, which can then connect to the correct loop and phone.

Trunk Characteristics

A trunk is almost always a four-wire path since a single-trunk wire pair carries voice in only one direction, so that a complete talking circuit requires two pairs (four wires). The transition from two-wire to four-wire, and the reverse, is made at the central office connected to each phone loop.

The signal characteristics of frequency response, envelope delay, and noise are much more tightly controlled on a trunk than on a local loop, for two reasons: First, since a trunk goes only between central offices and not to any local loops, it is easier to design and engineer it for the necessary performance. Each local loop, in contrast, has a different length, runs through a variety of electrical environments, may be moved as a phone is relocated or is exposed to weather, and so is usually different from other local loops.

Second, the trunk is a major link in the overall phone system. A local loop with only fair signal quality and characteristics affects only the user who is on that

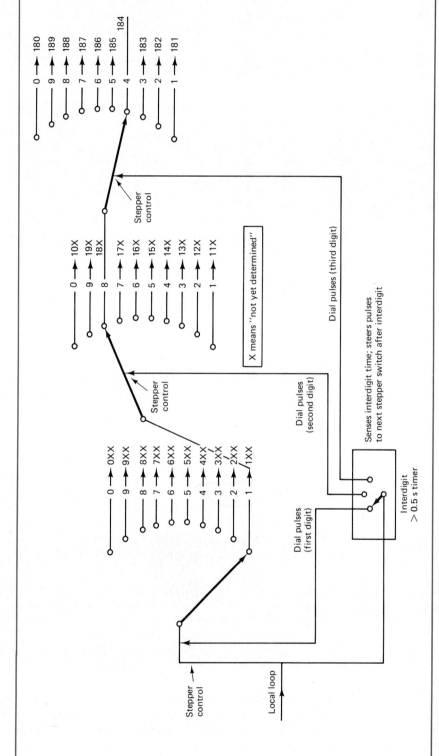

Figure 16.13 Switching connections for dialed number 184 in a simplified electromechanical switch.

loop. A trunk is used at different times by many users and has a broader overall impact. A call may involve many trunks connected in series, and the end-to-end performance is the sum of the performance of all these links in the chain. If each trunk's performance is only fair, the overall phone-to-phone connection will be poor. For example, if a trunk has a loss of just 6 dB and there are five trunks in the overall connection path, the total signal loss will be 30 dB.

For this reason, trunks have analog *repeaters* or *digital regenerative amplifiers* spaced at regular intervals to boost signal levels, reduce the effects of noise, and correct the frequency response of the trunk lines. For example, an analog trunk with excess rolloff between 2 and 3 kHz has an amplifier that equalizes all signals back to nominal value by providing more gain in the 2- to 3-kHz region, compared to the other frequencies. To reduce wiring or the number of individual trunk channels needed, most signals between central offices are actually multiplexed together into a single wider-bandwidth signal that is then transmitted over a single physical link or carrier (Chapter 22). At the receiving central office, the signal is demultiplexed into its original separate components and then switched to the called phone loop or another trunk.

Measuring Loop and Trunk Performance

A special test set such as that shown in Figure 16.14 is designed to characterize the loop and trunk completely by generating specified signals at various frequencies and levels. A loopback at the far end of the link returns the signal to the test set, which precisely measures amplitudes, time delays, and noise. By using this loopback, a single test technician can work from one end.

Of course, using the loopback means that the total test path is a round trip, so a single-direction path cannot be measured. However, there is usually a spe-

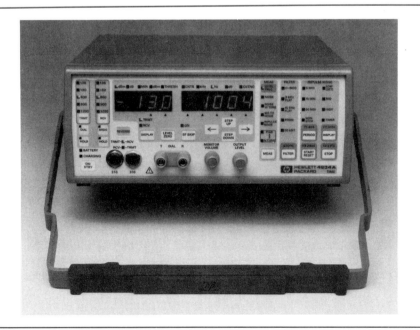

Figure 16.14 Typical line and trunk test set (courtesy of Hewlett-Packard Co.).

cial, dedicated test line with carefully controlled performance characteristics to be used for either the forward or reverse channel path. These known characteristics are then subtracted from the measured results, leaving the answer for the desired single-direction path only.

Review Questions

1. How are DTMF tones recognized at the central office with filters and digital logic? How many filters are needed?
2. What does the central office do with the recognized dialed digits? How does it establish a connection to the called party?
3. What is a matrix or crossbar switch? What are reed relays, and how do they provide an advantage over a conventional relay?
4. What are the advantages of solid-state switches over mechanical ones? How are they controlled?
5. What does the central office do when the first three digits of the phone number indicate a different exchange?
6. What are two reasons that trunk performance is more critical than loop performance? What is the role of the repeater?

16.5 ELECTRONIC SWITCHING SYSTEMS

The solid-state switches used to replace the crossbar mechanism and reed relays of the central office are controlled by *hardwired* digital logic, designed to initiate the same action each time in response to the same events. The next step in the evolution of the central office is the *electronic switching system (ESS)*, where the fixed-function logic is replaced by a computer acting as controller. ESS systems were first installed in the larger central offices and the tandems that connect only trunks and supertrunks. Smaller versions are now used in local central offices for towns and cities. An ESS is more capable yet smaller, less expensive, less power consuming, and more reliable than the digital-logic-controlled switching system. ESS was originally designed for reed-relay-based switching matrices, but it is even more compatible with the solid-state switches used in newer switching matrices. Note that even though the ESS switch control is a digital, computer-based system, it controls matrices that connect analog signals (as well as digital ones, depending on the crosspoint design).

But the most important benefit of the ESS is its flexibility. It can be programmed to provide many features that simply cannot be achieved in any other central office switch and can do advanced things such as checking its own performance and reconfiguring itself when it senses deterioration in signal quality or failure of some of its subsections. Let's see how the entire approach to managing the central office switching operation is very different with an ESS.

The heart of the ESS is the computer and program software that monitors all office activity, sees what service needs there are, and then initiates action to fulfill the need. The ESS controlling computer maintains in memory a *data base* of all phones it is connected to and responsible for, as well as all its trunks. With each phone number in the data base, the ESS stores various pieces of additional information, such as extra services which that phone may require (Figure 16.15). In operation, the ESS computer constantly scans the loop terminations and connects a digit receiver and dial tone (from its pool of available units) to any phone that goes off-hook and needs service. The dialed digits are received by the digit receiver and then passed to the ESS computer. Instead of 10 distinct lines indicating the dialed digit, the receiver uses a simple binary code of 0000 through 1001 for the digits.

So far, little seems new or different. But the ESS unit looks at these digits and determines via its computer program and stored information the status of all lines and equipment and the most effective path of connections to the called party. It then sends instructions to the matrices it controls to make that path complete, sees if the called phone is on-hook or busy, and sends the appropriate ringing or busy tone. Finally, when the called party phone goes off-hook in response to ringing, the ESS makes the actual connection between the two lines.

If there is any problem at the central office, such as equipment out of service, the ESS data base of available paths is changed so that a different routing is selected. The ESS-based central office requires fewer digit receivers than a traditional office, because its computer program is better at allocating them to the various phones that may need them.

The data base of the ESS also provides the basis for additional features. One such feature is *call forwarding* where someone wants all his or her calls temporarily forwarded to a different number. In a traditional central office this can be done only by physically moving wires. In an ESS office, this is done easily without touching any wires. The computer keeps a list of alternate numbers (provided by the phone users) that it associates with each number in the data base. When a phone number is dialed, the computer looks in its data base to see if there is an alternate number—the forwarding number—for the dialed number; if there is, the computer redirects the call.

Many other features are available. *Speed dialing* allows the user to dial an abbreviated code, such as *3, for frequently used numbers. The computer sees a

Phone number (logical)	Phone number (physical)	On hook	Off hook	Dialing	Talking	Incoming calls only	Call forward number (logical)	Blocked number (logical)	Camp-on number (logical)
0000	3714	Y	N	N	N	Y	0007	None	None
0001	4926	N	Y	Y	N	N	9823	None	None
0002	5701	N	Y	N	Y	N	None	None	None
0003	9006	N	Y	Y	N	N	None	3752	0392
⋮	⋮	⋮	⋮	⋮	⋮	⋮	⋮	⋮	⋮
9998	0134	Y	N	N	N	N	None	None	7412
9999	2287	N	Y	N	Y	Y	8475	None	0002

Figure 16.15 Data base table of ESS maintains current information on each phone line, status, user, and equipment, and is used as basis for software decisions.

call coming from this user and the abbreviated code and looks up in its memory the list of abbreviated codes and actual numbers that this user has previously stored. Another feature, *camp-on,* is used when a line is busy and the caller wants to be connected as soon as the called party hangs up and is available. The caller indicates this by keying a special code sequence on the phone, which the ESS sees and recognizes. Then the caller can hang up or use the phone for other calls. The ESS, meanwhile, keeps track of the number of the calling phone and the desired but busy called phone, and scans continuously to see if the called phone is busy or back on-hook. As soon as it goes on-hook, the ESS sends both it and the calling phone a ringing signal so that both parties are alerted that the desired connection can be made.

Other features include limiting calls to a certain area (the ESS connects only calls with the proper central office numbers); limiting a phone to receive calls only (the ESS will not respond to off-hook from this phone but will establish a talking path to it from another phone); and *call blocking,* where the user wants to block all calls except from a list of desired numbers. For call blocking, the central office keeps a list of acceptable calling numbers in the data base together with the called phone number. When someone dials the number, the ESS computer checks if the caller number is on the list. If it is, it completes the call; if it is not, it returns a busy signal to the caller. Call blocking works only within a single central office if the central offices are connected via conventional trunks and signaling (since the central office must know the calling phone number), but we will see in the next section how newer trunk interconnections and signaling allow call blocking to be used for calls from other exchanges as well. Of course, an ESS can easily restrict a phone to receive incoming calls only, or to make calls only within its own exchange or to selected exchanges, too, simply by noting these restrictions in its data base alongside the phone number.

In an ESS central office, the status of each phone loop (or trunk) is noted by *flag bits* in the data base alongside each phone number. There are bits for phone off-hook, phone dialing, phone ringing, conversation in progress, phone out of service, and many other factors. The decision as to what action to take next in the central office switching is made not by fixed digital logic or electromechanical action, but by the computer program of the ESS looking at the phone status flags and acting in accordance with the programmed rules and algorithms (Section 10.1). This gives the ESS tremendous power, because the actions taken can follow complex rules if programmed into it. In a non-ESS system, special features can be installed only by wiring special circuitry up for a specific phone, and the wiring must be changed whenever the features or phone numbers are changed as well. If someone has not paid his or her phone bill, the ESS can put the number out of service without touching a wire: The ESS operator simply uses a computer terminal to type in the delinquent phone number, and a flag bit next to the number is set indicating that this phone number is out of service due to too many unpaid bills. Then, when this phone goes off-hook to make a call or someone dials this number, the ESS program sees the bit and provides no service or connection.

As another example of how an ESS central office is easier to operate than a traditional one, consider someone at phone number 356-8978 who is moving from 100 Main Street across town to 25 Central Street and wants to keep the same phone number, possible since both locations have the same central office. In a conventional office, the phone company would have to actually move the physical connection of the local loop in the central office. The termination point for central office number 356-8978, for example, previously was the phone loop to 100 Main Street but now must be connected to the loop to 25 Central Street. When the 356-8978 number is dialed, the signal goes to the new address loop instead of the old.

In an ESS, there are *physical numbers* and *logical numbers*. The physical numbers correspond to the actual points where the local loops are terminated, numbered 0000 through 9999 in a single exchange central office. The logical numbers are the actual phone numbers associated with the physical numbers, and also range from 0000 to 9999. The data base list has a table of the physical numbers alongside the corresponding logical (user phone) numbers. For example, physical wire 4502 can be logical number 8978. When there is activity at a physical loop, such as wire 4502 going off-hook, the ESS sees this and then looks up in the data base to find that the actual caller is number 8978. When someone dials to call 8978, the ESS manages the entire call sequence and then rings the phone at physical location 4502 after looking in the data base and seeing that phone number 8978 is at physical point 4502.

Now, when the person moves to 25 Central Street, the ESS can solve the problem of keeping the same phone number simply by changing the entry in the data base of logical and physical numbers. The physical loop number for 25 Central Street replaces the physical loop number for 100 Main Street in the data base, alongside the existing logical number of this caller, 8978. The ESS manages everything as before when someone calls phone number 8978, but connects the ringing signal and makes the final connection to the new physical point, which has in effect been renamed by a change in the data base. No wires are actually moved, disconnected, or reconnected.

Common Channel Signaling

In the phone operation studied to this point, the trunk wires between central offices used for the final conversation—the talking wires—are also used as signaling wires to sense the phone activity, send ringing, and indicate what path or connection must be made. This is how the phone system operated for many years. When someone made a long-distance call that went through many central offices and trunks, the complete talking path was established as soon as possible after the number was dialed. This talking path serves both to connect the two phone users and also to connect the central offices as they supervise the establishment of the call itself.

Although this is a simple and effective approach, there is a drawback: Trunk lines and switching connections are tied up in establishing the call, even if the call

---

PBX and PABX

Many businesses have their own switchboards to route calls from one inside user to another as well as to handle calls to and from the outside. The switchboard acts as a scaled-down central office and is called *private branch exchange (PBX)*, capable of handling anywhere from dozens to hundreds or even thousands of phones. You have seen old movies with operators who ran switchboards using cables that plugged into a large interconnection board. Modern switchboards are highly automated, and place and route most calls automatically, unless an operator is needed for special situations. The newest of these *private automated branch exchanges (PABX)* are very similar to a regular telephone office ESS, providing many of the same functions for the phones in the office or company. They contain microprocessors which implement algorithms that operate based on information stored in the PABX data base of callers and phones.

is not completed because the called phone is busy or does not answer. For a single trunkline, this seems minor. But when you consider the thousands of trunks that interconnect major cities and central offices, it turns out that a large percentage of them and their switches are not actually carrying conversation but are supervising the setting up of calls. As a result, the phone company has to build extra trunks so that there are enough trunks for actual talking, which is wasteful.

At the same time, the use of ESS means that the central offices are much more sophisticated and capable of actions and decisions based on computer and software program control. Yet they are limited in what information they can pass to the next central office because there is a limit to the numbers and type of signals that can be sent via the talking path, and these signals can be confused with voice. The trunk voice connection is the path for both the data—the talking signal—and the supervisory signals that manage the path. This leads to conflicts and constraints in the kind of supervisory information that can be sent.

The solution developed in conjunction with ESS is *common channel signaling (CCS)*. In a CCS system, there are many talking path trunks between the central offices, and there is also a separate signal trunk called the *common channel* which is used only for signaling and supervisory information (Figure 16.16). In operation, the central offices pass information about what actions to take or the status of various phone lines via this common channel, which is a high-speed data path not used for any voice signal. Establishing the complete connection is managed entirely through the data on the common channel, which serves as a private communications link between ESS systems at the trunks. CCS between central offices is analogous to ESS within an office, since both separate the talking operation from the supervisory operation, and use computers, data bases, and sophisticated software programs to manage operation of the phone lines. Use of CCS is called *out-of-band* signaling, since the interoffice signaling is outside the talking band frequency and path.

For example, in establishing a call that goes to another central office, the calling phone central office sends the phone number via the common channel. The other central office finds the desired called phone in its data base and checks to see if it is busy or available. If it is not busy, the called central office rings the phone

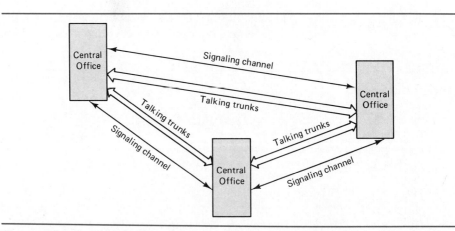

Figure 16.16 Talking and common channel trunk of common channel signaling shows separation of control path and talking circuit paths.

and reports back to the calling central office that it is ringing the called phone. The calling phone office sees this status message and sends ringback to the calling phone. If the phone is answered, the far end central office then signals, via the common channel, that the called phone has gone off-hook. Only at this point is the actual connection for talking made by connecting a free trunk (or trunks) using a path that has been calculated from the data-base listing of possible paths.

If the phone is busy, the far-end central office reports this to the calling exchange, which sends busy signal to the calling phone. No actual talking connection path is made while the phone is ringing or if it is busy. Therefore, no trunkline is needed until there is actually a need for it, when the calling party answers the phone.

Like ESS, common channel signaling provides new opportunities for efficient operation of the overall phone system. It does more than free up trunks from carrying signaling and supervisory information, leaving them for talking. It means that the same kind of sophisticated computer and program control can be used between central offices, as well as within the central office. CCS also provides more detailed messages about the central office activity and phone line status to be sent to other central offices that may need it. For example, if a central office is overloaded and cannot complete a connection to another central office in a call that involves many trunks, it can send a message indicating this. The calling office then looks for a path, perhaps a longer or more complex one, that uses other central offices.

Common channel signaling and ESS can be combined so that phone users can all block calls except from desired or selected phones. Once again, this is possible since the central office of the called phone can find out the number of the calling party that is trying to reach it, by sending a query via CCS to the central office of the originating call. The computer at that office knows the number of the

---

Call Tracing

In a CCS system, each central office is able to keep track of the phone numbers and activity of the calling and called parties, and use this information. Perhaps you have seen the old movies where the police are trying to trace a threatening call to see where it is coming from. They need to keep the caller on the line as long as necessary to follow the circuit path physically from loop to trunk to loop, which can take 20 minutes to an hour. If the caller hangs up too early, all the tracing effort is wasted. With a CCS system, tracing a call is a matter of typing a few commands into the terminal of the exchange. The command essentially requests information on the call activity associated with the phone number of the called phone. The central office has this information, of course, in its records of activity of the calling numbers and times of connections from any other phones in the same central office. But it also keeps records of the activity on the common channel, which has trunk numbers and times of connections from this called phone to other central offices. The central office can then send a query to the other central office, via CCS, asking for what called phone used a trunk at a specific time. The far-end central office, using ESS, has a record of all its phone activities with both that central office and with outside trunks. The ESS can piece this together to show what phone number at the other office used the trunk to call the target called phone.

All this can take place after the caller has hung up, since it is based on looking at software-maintained records of activities rather than the actual connection in place. This makes the difference between computer-controlled and regular systems clear: In a regular system, all calling information is lost after the call is disconnected (except for billing information); in an ESS and CCS system, the operation is managed by a computer that can also log connections in its memory and then onto a disk or tape for longer-term storage.

calling party and can pass the information over the common channel. CCS also means that call forwarding and camp-on can be extended to calls in other central offices, since additional information, requests, special conditions, and phone numbers can be transferred.

Review Questions

1. What is ESS? What is the fundamental difference between ESS and a traditional central office, even one with solid-state switches?
2. What is the ESS data base? What information does it contain? How is it used?
3. What are features such as call forwarding, speed dialing, call blocking, and restricted call access? How does an ESS implement them? How might it be done without computer control, and why is this impractical? Why does call blocking require that the central office know both the calling and the called phone numbers?
4. Compare the operation of changing phone locations while keeping the same number using non-ESS and ESS systems.
5. What are the benefits of CCS? Why does it integrate well with ESS?
6. What is the sequence of a call with CCS compared to non-CCS?
7. What can CCS do that conventional systems cannot? What are the benefits?
8. What is the fundamental difference in what happens to call-related information in a computer-controlled central office or switching system, compared to a hardwire system or electromechanical system?

16.6 ECHOES AND ECHO CANCELLATION

The two half-duplex signals on the incoming and outgoing paths of the phone connection are combined into a single loop signal by the hybrid transformer circuit. Ideally, the balance network impedance of the hybrid is exactly the same as that of the local loop, providing a perfect match and signal cancellation. In reality, there is some mismatch and therefore some reflection of an incoming signal back to its source. This is the cause of *echo*.

Echo occurs when a person talks into the phone and some of the signal energy that reaches the listeners' end of the link is reflected back to the talker. *Talker echo* is a problem for most people only when the round-trip echo time is greater than about 100 ms. Below 100 ms most talkers do not notice the echo, but above this time delay the echo annoys the talker and is very discomforting. It is very hard to talk coherently when you hear a delayed version of your voice. (In fact, this is a test to determine if someone is truly somewhat deaf or just pretending—a truly hard-of-hearing person will speak the same way with a delayed voice in their ear since they never hear themselves, while the pretender will have a very hard time talking with the delay in the ear and will trip up after a few words.)

Echo is not limited to just talker echo. Some of the returned signal energy

16.6 Echoes and Echo Cancellation

that reaches the talker end can be reflected back to the listener, so that the listener hears a reduced and delayed *listener echo* in addition to the original, louder voice. The effects of listener echo are generally different from talker echo: Because of its lower amplitude it is often just an annoyance to the listener. Both talker and listener echo can cause errors when the signal consists of data, not voice. The echo signal appears like noise that adds to the desired signal—and computers cannot normally distinguish between the original signal and its echo version while new signals are coming in simultaneously (unlike the brain, which can separate voices and echoes to a larger extent).

Echo Parameters

The key factor in echo is the signal travel time, which is determined by the travel distance. For talker echo, the distance is twice the distance between ends of the link, while for listener echo it is three times that amount. Echo time is calculated simply:

$$\text{echo time} = \left(\frac{\text{distance}}{\text{propagation velocity}}\right) \times [2 \text{ (for talker echo) or 3 (for listener echo)}]$$

The signal travel distance can be significant even when the callers are not far apart geographically. The actual signal between two callers only 100 miles apart may be routed over a complex series of trunks that stretches for thousands of miles, depending on the trunks available when the voice path is established. Echo potential is greater when calls are routed automatically via a satellite link. Communications satellites (Chapter 19) are in orbit approximately 22,300 miles above the earth, so the one-way signal distance is 44,600 miles; the talker echo distance is 89,200 miles (about $\frac{1}{2}$ s) and the listener echo distance is 134,000 miles (about $\frac{3}{4}$ s). Triangular geometry shows that the actual path distance is greater than the height to the satellite, because the ground points are not in the same location, but the basic 44,600-mile figure is a good first approximation.

Example 16.4

A signal travels via cables (propagation velocity $0.7c$) from New York to Los Angeles, a distance of 3000 miles. What are the talker echo and listener echo times?

Solution

The talker echo time is

$$\frac{3000 \text{ miles}}{186,000 \times 0.7 \text{ mi/s}} \times 2 = 0.046 \text{ s} = 46 \text{ ms}$$

and the listener echo time is

$$\frac{3000 \text{ miles}}{186,000 \times 0.7 \text{ mi/s}} \times 3 = 0.069 \text{ s} = 69 \text{ ms}$$

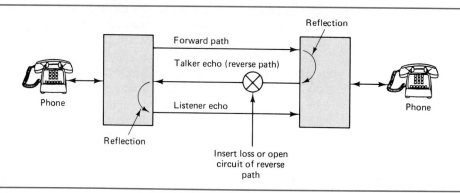

Figure 16.17 Echo suppression achieved by insertion of reverse path loss.

Solutions to Echo

Echo is a serious problem for reliable and useful long-distance communications. There are four solutions in general use: inserting loss in the reverse path, signal subtraction, digital signal processing, or a four-wire path.

The *reverse path loss* approach uses special circuitry to sense which ends' signal is larger and therefore which side of the connection is "talking," and then introduces significant loss in the return path between the listener and the talker or completely opens this path (Figure 16.17). About 100 ms after speech energy decreases, the circuitry "flips," turns on what was the reverse path and is now the forward path, while inserting loss in the former forward talking path.

Although this method is effective, it is very awkward for voice and inefficient for data signals. The reason is that the full-duplex path of the connection is really reduced to a half-duplex path, since only one person can be talking at a time. A conversation with the normal interruptions and "give and take" is impossible, especially if both parties occasionally try to speak at the same time. It is similar to a half-duplex two-way radio, where each party must say "over" to let the other party know that it is his or her turn to speak, except that circuitry decides automatically when one party has finished speaking.

For data communications, the situation is not socially awkward but severely reduces the capability of the system and the amount of data that can be transferred per second (throughput). A typical data communications sequence, even when the bulk of the data is going in one direction, involves many status and error-related messages from the receiver back to the sender (Chapter 12). These status messages let the sender know what has been received properly and what problems have been recognized. When done via full duplex, the status messages do not slow the system down, as the sender is interpreting them while sending new data.

When the system has a half-duplex link, the sender must stop sending, the link must switch directions by changing the forward and return path conditions, get status, switch back, and so on. A lot of time is lost since no overlap is possible between forward data transmission and returned status reception. In addition, the 100-ms time for sensing and switching direction is a significant loss in a data communications system—at 9600 baud, approximately 960 bits can be sent in that time.

The *signal subtraction* method uses circuitry that compensates for the echo by negating it (Figure 16.18). The original talker signal is fed into a *delay line* to generate a simulated echo which is then electronically subtracted from the signal plus real echo using an op amp configuration. When the electronically generated

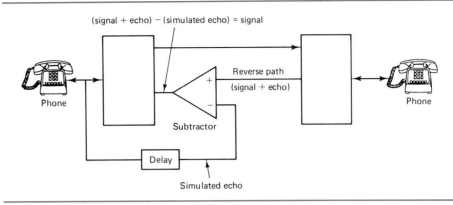

Figure 16.18 Echo suppression using signal subtraction method.

delay is equal to the signal delay due to travel distance, the echo effect is canceled out. The key to successful implementation of this circuit is that the delay time must be adjusted for the differing delays of each connection path. Also, the amplitude of the delayed signal must be equal to the amplitude of the echoed signal, so as to exactly cancel it when subtracted. These adjustments of time delay and amplitude are made by an *adaptive* circuit that electronically adjusts the delay to minimize the echo effect. The echo cancellation circuitry cannot "set and forget" the subtracter gain setting either, but must continuously adjust it as the echoed signal strength varies during a call.

The *digital signal processing (DSP)* echo cancellation scheme is an updated version of the subtracter method. Instead of using a delay line and analog subtraction, the signals (with echo) are converted to digital form and processed by sophisticated algorithms that can separate out the desired signal from its echo, which after all is similar to the desired signal but delayed and different in amplitude. The desired signal (echoless) is then converted back to analog format for further use. This DSP approach requires extensive arithmetic calculations on large sets of numbers, so it is becoming more common only as the capability and processing power of DSP microprocessors increases while their cost decreases.

Finally, the echo problem can be significantly reduced for voice, and completed eliminated for data, by the use of a complete four-wire circuit (Figure 16.19). Recall that the hybrid circuit which converts a full-duplex, two-wire loop into a pair of half-duplex two-wire paths (and the reverse) is the source of the problem. Instead, the two paths are extended all the way to the users at either end so that there is a completely separate and independent circuit path all the way through for each direction of signal travel. Now the only cause of echo for voice is any signal that comes out of the telephone earphone speaker and gets back into the microphone through air or the skull bones of the phone user. The amplitude of any such signal will be much less than is reflected back by an imperfect hybrid circuit. For data there is no echo at all, since the forward and reverse data paths are naturally electrically separate in the computer systems at each end. The only drawback to the four-wire end-to-end scheme is that special local loops must be used (ones that do not have the normal hybrid circuit at the central office). Such special loops cost more to set up, and they must be established and maintained throughout the complete signal path—any inadvertent hybrid along the way negates their benefit.

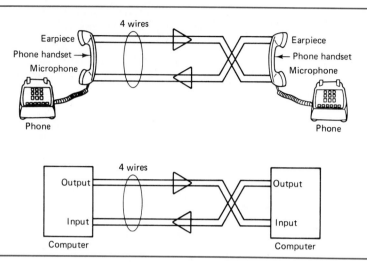

Figure 16.19 Connecting a signal with a four-wire path end-to-end eliminates echo by using two separate paths and no hybrids.

Review Questions

1. What is the basic cause of echo? Compare talker and listener echo causes and their effects on voice and data.

2. What are the minimum echo distances via satellite?

3. How does introducing loss in the reverse path reduce echo? Why is this scheme awkward for conversation and inefficient for data transmission?

4. How does signal subtraction cancel echo? What two factors are critical, and why must they be adjusted continuously?

5. How does the digital signal processing approach reduce echo?

6. What is the four-wire method of eliminating echo? Why is it so effective for both voice and data? What is the difficulty with it?

16.7 DIGITAL SIGNALS AND SWITCHING

The traditional telephone system is designed primarily to handle analog signals, but many of the newer trunks and central offices handle signals in digital format as well. These signals can be from inherently digital signal sources such as computers, or represent the digitized versions of analog signals. The voltages used for digital signals are not directly compatible with the telephone system and are converted to a compatible signal by an interface *modem*, discussed in Chapter 17.

A long-distance call routed through multiple trunks and central offices usually travels via a combination of analog and digital links, with the signal converted from analog to digital format, and vice versa, at special tandem switching centers.

There are also systems that are designed for digital signals only, without any path for analog signals (such as voice) which have not been digitized directly at the phone instrument and loop itself.

The telephone industry has a standard format for analog-to-digital conversion, with voice signals bandlimited to 3000 Hz and then digitized at 8000 samples/s to seven [8] bits (256 levels). (Recall from Chapter 10 that the sampling rate must be at least twice the bandwidth.) Twenty-four individual digitized signals—from 24 users—are then combined via multiplexing, have some signaling and framing bits added, and the entire resulting 1.544-Mbit/s stream of bits is used to modulate a carrier. (This T-1 format for combining digital signals for trunks is discussed in Chapter 22.)

There are two aspects to handling signals in digital format or having an all-digital telephone system. First, there is the overall set of protocols used to manage such a system and network. One such protocol (and related) format that is now being established is the integrated services digital network (ISDN), which is discussed in Chapter 18. The other aspect involves the routing and switching of digital signals, which is relevant to both the all-digital system such as ISDN and mixed (analog plus digital) systems. The trunklines and their repeater amplifiers, as well as the switching matrix, are different for digital signals as compared to analog signals.

Trunks and Amplifiers

The trunklines for digital signals must transmit the signals without excessive signal attenuation, of course, or noise in the lines and amplifiers will corrupt the signals enough so that a bit error will occur. There is also the problem of bit smearing or *intersymbol interference,* resulting primarily from the capacitance of the trunk wires. The high bit rate requires wide bandwidth, but capacitance, measured as envelope delay, causes the bits to spread out in time. The equivalent way of looking at this in the frequency domain is that the higher spectral components of the bits are increasingly phase-shifted by capacitance, so the final bit shape is distorted when all the spectral components come together at the far end. When bits begin to spread out in time, they reach a point where they actually overlap with subsequent bits and interfere with them. The receiver, of course, cannot distinguish one "stretched" bit from two overlapping bits, and may make an error in deciding whether a 1 or a 0 was sent. The eye pattern of Chapter 12 shows how to observe intersymbol interference as bits are overlapped on an oscilloscope screen, and see how they can be misinterpreted.

Amplifiers for digital signals are very different from the repeater amplifiers of analog signal trunks. The digital signal amplifier must judge and restore the smaller and noisier signal to its original digital value, and not simply amplify its level. For example, suppose that the signal is sent as -1 V for binary 0 and $+1$ V for binary 1, but is received at the amplifier as a much smaller signal plus noise corruption. The digital regeneration amplifier uses a comparator to take all signals ranging from 0 to -1 V and reestablish them at the -1-V level; all signals from 0 to $+1$ V are restored at $+1$ V. The original signal is thus perfectly recreated but without its noise or attenuation.

In contrast, the analog amplifier cannot make any judgment about what the correct signal value should be or how much noise has corrupted the signal. All the analog amplifier can do is boost the signal level, which means that the added noise is increased as well. Eventually, the noise becomes so excessive that it severely affects the quality of the received signal, making conversation difficult or causing errors. The analog signal amplifier is a linear amplifier, while the digital signal regenerative circuit is a limiting circuit.

Digital Switching

A switching matrix for digital signals such as those of ISDN—not digital information represented by analog signals—can use the same relay or solid-state circuitry that analog signals use, but this is not necessary nor the most efficient form. Instead, switches designed especially for digital signals provide better performance and more flexibility. One such device has 16 ports of four digital input/output bit connections, for a total of 64 bits. At each port, a group of four bits is directed to one or more of the other 15 groups, since each port can serve as an input or an output. This type of digital crossbar switch is used whenever data bits have to be transferred from one or more sources to one or more destinations.

The controlling information that directs the bits from an input to an output port is entered in advance under microprocessor control, and then the device is clocked. During the clock switching cycle, all of the four-bit port bits are passed onto a single 64-bit-wide bus within the IC. Then they are independently parceled out to the specified output ports. In effect, the switch matrix acts as the controller of an intersection of 16 4-bit highways, where the data bits entering from each highway can be directed to another exit, or to multiple exits at the same time.

At each clock cycle, the operation is repeated and four-bit chunks are transferred. The setup information is stored in internal registers within the device and does not have to be repeated prior to each transfer of bits. Only when there is a change in desired switching paths does the microprocessor that controls the IC have to issue new instructions. Special control lines allow all I/O connections to be forced to a high (binary 1) or low (binary 0) state, or to a "high-impedance" mode that makes it easier to connect multiple units together for larger matrices.

Review Questions

1. How does a phone system transmit a signal in both analog and digital formats?
2. What is the sampling rate used by the phone system when digitizing a single voice signal? How many bits of resolution are used? How many separate users are combined into a single digital trunk signal, and what is the final bit rate?
3. What type of regenerative amplifier is needed for digital signals? How does it differ in operation and output result from an analog signal repeater amplifier?
4. Explain the operation of a typical digital crossbar such as the S618840. How is the desired switching path chosen? When is data transferred? What is done when the switching path remains the same, and when it changes?

SUMMARY

The telephone system is remarkable because it is available virtually everywhere in modernized countries and is extremely simple to use, despite the incredible complexity and sophistication of its inner workings, with complex switching circuits allowing any user to be connected to any other. The world of telephones and signal switching and routing has a special language and unique communications needs. Beginning with the local phone loop and a basic telephone, the central office must supervise the phone operation, provide required tones, and sense phone dialing. The caller must be connected through to the desired called party, routed through just one central office or via one or more trunks and other central offices.

Pulse dialing has largely been replaced by tone dialing, which is faster and more compatible with electronic interfaces. The electromechanical switch for interconnections was supplanted by the reed relay, which in turn is being replaced by the solid-state switch matrix. Both relay and solid-state switching centers are now usually controlled by computer-based electronic switching systems, which decide what action should be taken by sensing and logging phone activity in a data base, and implementing procedures that are part of the computer software. Switching center interconnections have also advanced to computer control, with a common channel for signaling between offices that is separate from actual talking paths that are established between callers.

Echoes due to signal reflections at both ends of the connection can cause difficulty in conversation and data transmission. Various echo cancellation schemes range from minimizing the return path signal, to subtracting a simulation of the echo, to advanced digital signal processing of the signal and echo. The best method of eliminating echo provides two completely separate paths between the two ends of the system.

For digital signals, amplifiers regenerate the signal and actually eliminate any noise, something that cannot be done with analog signals and repeater amplifiers. The digital bits can be computer generated or a digitized voice signal. Many channels are usually multiplexed into a single wideband signal for greatest efficiency and sent over one link or channel.

Summary Questions

1. What is the local loop? What does it look like electrically when the phone is off- and on-hook?

2. What is the basic switching role of the central office? How does the phone numbering scheme indicate if a call goes to the same office or another one?

3. How are trunks used to interconnect local loops from different central offices?

4. How is the phone rung from the central office, even though it is on-hook? How does a calling phone know the called phone is ringing?

5. Why is a hybrid circuit needed? How does it work? What signal loss does it introduce? What is the function of the hybrid in the phone and in the central office? What is the balancing network of the hybrid, and what is the effect if it does not match the loop impedance?

6. Compare and contrast pulse and tone dialing in terms of what the signals look like, speed of dialing, and methods of signal generation. Why is pulse dialing simpler in concept? What are the benefits of tone dialing in terms of overall system capabilities? How and why do some phones have pushbuttons but actually emulate pulse dialing signals?

7. How is each dialed number signaled by a row and column tones in DTMF systems? How are eight tones needed for the 16 possible dialed numbers and symbols? Why were the apparently uneven frequencies chosen for DTMF tones?

8. What are the frequency and envelope delay characteristics of conditioned versus unconditioned lines? Why is a conditioned line better but more costly?

9. What are the six BORSHT functions? What is the role of each?

10. How does a central office make the physical signal connection? Compare crossbar, reed relay, and solid-state switch characteristics.

11. What are repeaters? What to they do for equalizing the signal frequencies?

12. What is an ESS? How does a data base and computer program control form the heart of it?

13. What things does an ESS do that a hardwired control system and central office cannot? How does an ESS allocate lines and central office equipment?

14. What is the logical number for a phone, and the physical number? What does having two numbers do in terms of capability?

15. How does an ESS change phone numbers and features without touching wires?

16. What is CCS? Why does having a separate channel for interoffice signaling offer flexibility and more efficient use of trunklines?

17. How are ESS and CCS similar? How do they combine for the best in flexibility and capability?

18. Why can ESS and CCS systems provide detailed histories of calls, switching, and performance?

19. Compare the mechanism of talker and listener echoes. What echo times cause problems? Why are talker and listener echo effects different? How can there be echoes when the two phones are connected to central offices only a few miles apart?

20. How can echo be reduced by a delay line and subtractive scheme? What must be carefully adjusted, and why?

21. Why does a complete four-wire path end to end eliminate data echo and minimize voice echoes?

22. Compare a regenerative amplifier for digital signals with a repeater for analog signals. What can one do that the other cannot?

23. Compare and contrast the features of in-band and out-of-band signaling.

PRACTICE PROBLEMS

Section 16.3

1. How long does pulse dialing take for the number 345-9876? What is the tone dialing time? (Use digit times of the examples.)

2. Compare the pulse dialing time for number 123-2211 versus 898-9898.

3. What are the tone dialing times for the numbers of problem 2?

4. The long-distance number 1-908-345-9876 (with area code) is dialed by pulse and tone dialing. How long does dialing take?

5. What two DTMF tones are generated when the digit 8 is pressed?

6. What are the harmonics and sum/difference frequencies for the digit 8?

7. How close to valid DTMF tone frequencies are the first harmonic and sum/difference frequencies of the digit 0?

8. What sum and difference tones occur when the digits 3 and 7 are pressed at the same time?

Section 16.6

1. What is the talker echo time when the path distance is 8000 miles via undersea cables with propagation of $0.6c$?

2. What is the listener echo time for the situation of problem 1?

3. A call is made from Los Angeles to London with a $0.7c$ propagation velocity cable between Los Angeles and New York and a satellite orbiting 23,000 miles up for the New York-to-London portion. What is the talker echo? The listener echo? (Ignore the additional path distance due to New York-to-London separation.)

4. The Los Angeles-to-London call is now made via a double-satellite link, with the first link from Los Angeles to New York and the second from New York to Los Angeles. What are the talker and listener echo times?

17

The RS-232 Interface Standard and Modems

CHAPTER OBJECTIVES

When you have completed this chapter, you will understand:

- The technical and operating requirements of the RS-232 standard
- Implementation of this standard and ICs that support it
- Troubleshooting RS-232 interfaces
- The role of the modem and several standard modems in general use
- Limitations of RS-232 and other standards that overcome these limitations

INTRODUCTION

Successful communications occur only when the equipment at both ends of a path use the same values for the signal voltages, frequencies, timing, and meaning of key lines. The RS-232 is a very common standard for communicating between a terminal, printer, test instrument, or similar device, and its associated communications interface device to the main communications link. The interface standard defines some things with great precision, but also leaves the user many choices. Handshaking signals allow each device to indicate to the other whether it is ready or not. The functions of the RS-232 interface are implemented in special interface ICs that minimize the time-consuming, critical attention to detail activity by the microprocessor within each system. Other special ICs act as the physical interface to the communications link, providing the necessary signal levels and protection against incorrect installation.

The modem is an interface device that takes data signals and modulates them into a form suitable for standard communications links. AM, FM, and PM—

individually or in combinations—provide varying levels of performance, errors, and speed. Additional modem performance improvement is achieved by using multiple-level modulation, error detection, and error detection and correction codes on the information bits. The interface between the information source, or the user, and the modem is also enhanced by using standards other than RS-232 for increased speed, noise immunity, and distance.

17.1 ROLE OF THE INTERFACE STANDARD

An interface standard allows communications equipment from different manufacturers, or various models from the same manufacturer, to be connected together and function as a team. Without standards, everything involved in the communications connection—the connectors, the signal voltages, the signal timing, and the sequence of signaling—would have to be designed especially for the particular system in use. No other equipment would be compatible, since there would be physical and electrical mismatches. It would be like having a different power line plug and voltage requirement for every appliance in a house.

There are many standards in use. Each serves a different application need: Some are low cost, some are high speed, some are simple but not flexible, others are complex but very versatile. Electronic Industries Association (EIA) standard *RS-232* (RS means "recommended standard") is one of the most common low-to-moderate performance standards in use. RS-232 has three levels of specification: Some things are defined with exacting precision, for other technical issues it provides several different options, and it does not define any specifics of some points. We will study this standard because it is so common and because many of the issues it addresses arise in other standards besides RS-232.

RS-232 is designed as the specification standard between two types of equipment (Figure 17.1). One is the *data terminal equipment* (*DTE*), which is the device that is generating the data to be sent, or using the data that are received. For a variety of technical reasons, as we will see, the DTE is not electrically compatible with the data link. A special interface device called the *data communications equipment* (*DCE*) takes signals to and from the DTE and makes them compatible with the physical link. The physical and electrical connection interface between

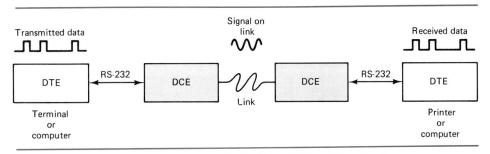

Figure 17.1 RS-232 used in DTE to DCE–link–DCE to DTE situation.

DTE and DCE is defined by RS-232, so that the DTE and DCE can plug directly into each other. At the other end of the communications link another DCE is connected, and it interfaces to the far-end DTE using the same RS-232 standard.

The DTE can be many things. Originally, it was a simple terminal (keyboard plus screen or printer), but now it can be a computer of any type or size. A DTE can also be a device that generates or accepts data, such as an electronic thermometer that produces digital temperature values to be logged in a data acquisition system, or a plotter that draws images on paper under the direction of a computer that calculates which lines should be drawn. The simple RS-232 standard is used in many ways beyond its original scope, which can cause technical problems that require careful understanding to overcome.

There does not even have to be a traditional "DTE to DCE–communications link–DCE to DTE" connection: Many systems use the DTE as one active end of the communications system and the DCE as the other, such as when a computer (DTE) is connected to its peripheral device, such as a printer functioning as a DCE even though it is not really data communications equipment.

The RS-232 standard defines a scheme for asynchronous communications, where there is a specified timing between data bits but no fixed timing between the characters that the bits form. In contrast, synchronous communications requires specific timing between bits and characters, and the clock for recovering data bits must be synchronized to or derived from the bit stream. Asynchronous systems are simpler but transmit data at much lower rates and are therefore less efficient than synchronous systems.

Communication defined by RS-232 is serial. There is a single wire, or link, for each direction of data flow, and the bits of the message are sent in sequence, one at a time. Compared to parallel systems where multiple lines carry many data bits simultaneously, serial communication requires less circuitry at either end of the data link, is easier to manage, and needs only a single physical link. Of course, serial systems have lower throughput than parallel systems, but in practice a parallel link is practical only over short distances of a few feet.

Although the RS-232 standard defines a serial system with just a single wire for each direction, the standard allows for other signal wires between the DTE and DCE as well. These additional signals are used to manage the data interface and to indicate communications status at any time to both the DTE device and the DCE device.

The RS-232 specification is intended to provide reliable communication up to a distance of 50 ft, at rates up to 20,000 baud. Although the RS-232 interconnection should not be longer than 50 ft, it has been successfully used for longer distance. Since the limiting factors of cable capacitance and noise are much less critical when the lower baud values are used, there are successful RS-232 links several hundred feet long, although this exceeds what the standard is designed to allow.

Finally, the phrase *serial ASCII format* is often used in conjunction with RS-232. The formal standard itself does not specify at all what bit patterns should be used to represent the actual information being transmitted. Most commonly, ASCII representation is used (Chapter 12). In special circumstances, however, non-ASCII formats are used, especially where only numbers are being sent. As we saw in the discussion of ASCII, it is relatively inefficient for numerical information since it requires a seven- or eight-bit field for each digit of the entire number. In contrast, a straight binary format can express values up to 255 in an eight-bit field. However, unless noted otherwise, we will assume that the RS-232 data are in ASCII format.

RS-232 is primarily a signal voltage and timing, DTE-to-DCE interface standard. Just as it does not discuss how the data should be represented by the bits, it does not define the overall message format and protocol. The definitions of these are left to the developers of the communications system. This means that RS-232 provides a great deal of flexibility, but also can have problems when signal levels and timing agree but message format, content, and protocol differ between the two DTE devices that it interconnects in a complete system.

Review Questions

1. What are the three levels of specification precision for RS-232?
2. What is a DTE? A DCE? Why is the DCE needed?
3. How is RS-232 used beyond its original scope?
4. What does "serial ASCII format" mean? Compare it to synchronous format and parallel format.
5. What is the maximum signal distance and speed under RS-232? When can these be extended?
6. Must RS-232 use ASCII representation? Explain. What are the alternatives? When is ASCII used?
7. What does the RS-232 specification say about message format and protocol? Where is RS-232 specific?

17.2 RS-232 OPERATION

There are three areas where RS-232 defines a communications standard: signal voltages, the use of signal lines, and signal and bit timing. These are the lowest levels of a complete communications interface protocol. By specifying these, RS-232 develops a basic building block for more complex protocols and message interchange. Without the basics such as those defined by RS-232, many sophisticated communications would fail because of simple incompatibilities at the lowest level (the screws fit in the holes, but the screws have slotted heads and you have a Phillips screwdriver).

Signal Levels

To represent a binary 1 or 0, two signal voltage spans are required (Figure 17.2). Any voltage between +3 and +25 V is a binary 0 (also known as a *space*, from historical use of mechanical teletypes), while a binary 1 (known as a *mark*) is any voltage from −3 to −25 V. The voltage span from −3 to +3 V is undefined (neither a 1 nor a 0) and should not exist in any system using the RS-232 standard. The signal transmitter sends any voltage between +5 and +25 V to indicate a 0; the receiver decides that it has received a 0 when it sees a voltage between +3 and +25 V. Note that the specification requires that the transmitted signal be either −5

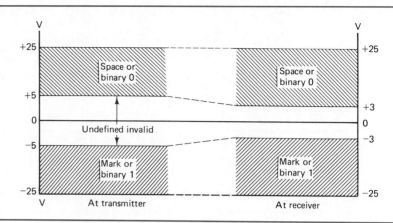

Figure 17.2 RS-232 specification voltage levels at the source and the receiver.

to −25 V or +5 to +25 V, but the received signal specification is relaxed to −3 and +3 V, respectively, to allow for line voltage drop.

In practice, the system has greatest resistance to noise or voltage loss in wires when the largest voltage possible is used at the transmitter. This noise margin is critical to reliable system operation. When a 25-V signal is sent but is corrupted by 10 V of noise or has 10 V of loss, it still arrives at 15 V, which is a valid binary 0. In contrast, when a +3-V signal is used, there is no margin: Any voltage decrease due to noise or line loss puts the signal in the undefined region.

For these reasons, RS-232 systems use a voltage as high as possible. The amount of noise margin needed depends on how much noise and line loss are realistically expected. For a very short interconnection, RS-232 levels of ±15 V or even ±5 V are sufficient and are easier to generate with most common digital logic circuitry and power supplies. If the conditions are more difficult, the larger voltages are used for additional "insurance."

Signal Lines

Twenty-five signal lines are defined by the RS-232 standard. This does not mean that every DTE-to-DCE interface requires 25 signal wires—many RS-232 interfaces use just a few. The lines are divided into four groups: *data, control, timing,* and special *secondary functions*. The signal lines, arranged by line number and by function, are listed in Figures 17.3 and 17.4. RS-232 specifies a signal ground, used for all signal wires, plus a chassis ground for protection and to make sure that the DTE and DCE chassis are at the same electrical potential. In practice, the signal ground and chassis ground are often the same wire or the chassis ground is omitted, which are really violations of the standard's requirements; fortunately, this usually works, but it can cause operational (or even safety) problems.

The *data lines* are the most important signals. Received data and transmitted data lines permit full-duplex communication between the DTE and DCE. To make sure that there is no confusion on the direction of data flow on the received data and transmitted data lines, the RS-232 standard specifies data flow directions from the perspective of the DTE: Data go from the DTE to the DCE on the transmit data line, and data go from the DCE to the DTE on the received data line. From the perspective of the DCE, then, data are transmitted on what is called the

17.2 RS-232 Operation

Line pin	Description
1	Protective ground
2	Transmitted data
3	Received data
4	Request to send
5	Clear to send
6	Data set ready
7	Signal ground (common return)
8	Received line signal detector (data carrier detect)
9	Reserved for data set testing
10	Reserved for data set testing
11	Unassigned
12	Secondary received line signal detector
13	Secondary clear to send
14	Secondary transmitted data
15	Transmission signal element timing (DCE source)
16	Secondary received data
17	Receiver signal element timing (DOE source)
18	Unassigned
19	Secondary request to send
20	Data terminal ready
21	Signal quality detector
22	Ring indicator
23	Data signal rate selector (DTE/DCE source)
24	Transmit signal element timing (DTE source)
25	Unassigned

Figure 17.3 Function of RS-232 lines, by pin number.

received data line and received on what is called the transmitted data line. This may seem confusing but it really avoids the more serious problem that will occur if both units try to transmit on the same line (signals clash) or receive on the same wire (there is no signal present, but there are two listeners on the same line and no signal source).

The *control lines* are next in importance (Figure 17.5). These lines are used for *handshaking*, so either the DTE or the DCE can signal to the other that there are data to be transmitted and can indicate to the other if either DTE or DCE is ready to accept new data. Four control lines are used most frequently:

1. *Request to Send* (*RTS*) from the DTE signals the DCE that the DTE has new data it would like to transfer.

2. *Clear to Send* (*CTS*) from the DCE indicates to the DTE that the DCE can accept the new data.

3. *Data Set Ready* (*DSR*) lets the DCE tell the DTE that it can accept new data.

	Line pin	Description	Gnd	Data From DCE	Data To DCE	Control From DCE	Control To DCE	Timing From DCE	Timing To DCE
Data	1	Protective ground	×						
	7	Signal ground/common return	×						
Control	2	Transmitted data			×				
	3	Received data		×					
	4	Request to send (RTS)					×		
	5	Clear to send (CTS)				×			
	6	Data set ready (DSR)				×			
	20	Data terminal ready (DTR)					×		
	22	Ring detector				×			
	8	Received line signal detector or data carrier detect (DCD)				×			
	21	Signal quality detector				×			
	23	Data signal rate selector (DTE)					×		
	23	Data signal rate selector (DCE)				×			
Timing	24	Transmitter signal element timing (DTE)							×
	15	Transmitter signal element timing (DCE)						×	
	17	Receiver signal element timing (DCE)						×	
(Secondary group)	14	Secondary transmitted data			×				
	16	Secondary received data		×					
	19	Secondary request to send					×		
	13	Secondary clear to send				×			
	12	Secondary received line signal detector				×			

Figure 17.4 RS-232 lines, organized by functional grouping together with direction of signal flow.

4. *Data Terminal Ready* (*DTR*) from the DTE is another indication that the DTE is ready.

Note that in the RS-232 standard the handshake lines are *asserted* or active (binary value = 1) when they are in the space zone of +3 to +25 V, which also corresponds to a binary 0 for data. This sometimes causes confusion when troubleshooting since an asserted control line is said to be in the binary 1 state, yet its

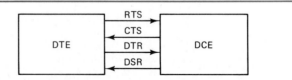

Figure 17.5 Four key control (handshaking) lines for DTE/DCE interfacing.

RS-232 Connectors

For many years, a 25-pin D-shaped connector with 13 pins on one row and 12 on a second row was the most common connector for RS-232. However, since many of the 25 wires specified by the standard are not used, this is wasteful and costly in terms of wire terminations and circuit board space. Newer systems often use a smaller connector such as a nine-pin D-shaped unit, or even a smaller type with just a few wires, at the DTE end of the cable. However, most DCE devices, such as printers, terminals, or modems (discussed later in this chapter), continue to use the 25-pin D-shaped connector, although this is changing. The figures show the pin assignments based on the 25-pin connector.

signal voltage is the same as a data bit that is binary 0. As long as everyone follows the standard, however, there is no operating problem.

Using these handshake lines, the complete data transmission sequence is a series of repeated requests to send new data, a handshake showing that the request can be honored, sending the data, a handshake indication that new data cannot be accepted because either the DCE or the DTE is full or the transmission link is unavailable, then watching for the "OK to send" signal finally appearing, and so on. The other control lines let the DCE indicate to the DTE that it has detected a ringing signal, or that the communications line signal quality is acceptable, among other functions.

Signal and Bit Timing

One more element is needed to complete the RS-232 picture: how the actual data bits are presented in the asynchronous environment. We studied this in Chapter 12; here, we will look at it with the specific values of RS-232 operation (see Figure 17.6).

Between data bits, when no signal is present, the RS-232 transmitted data line is at the marking state. When there are data to send, the line goes to the space state for one bit period called the *start bit*. This transition, indicating that a new series of data bits will follow, is sensed by circuitry in the RS-232 receiver as a warning to "wake up" and look for the incoming data bits.

The start bit is followed by five, six, seven, or eight data bits, depending on the design of the system (seven and eight bits are most common). The LSB of the data bits is sent first. After the last data bit, there is an optional parity bit for error

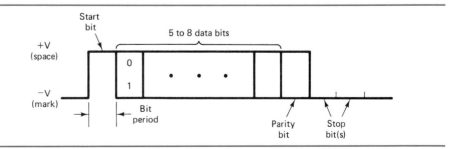

Figure 17.6 RS-232 signaling, with mark and space values for start, data, parity, and stop bits.

detection for systems that are using parity (Chapter 12). The entire sequence of data and parity bits is terminated by a *stop bit* (or bits), where the data line goes to the marking state for 1, 1½, or 2 bit periods. This indicates that the entire sequence of data bits is complete.

Let's look at how the letter B (1000010) looks in RS-232 format, using ASCII representation, along with seven bits, no parity, and one stop bit format (Figure 17.7a): 1 0100001 0 (spaces between bits are for printed clarity; also, the LSB of the character is sent first). For the digit 8, in ASCII with seven bits, no parity, and two stop bits, the pattern is 1 0001110 00 (Figure 17.7b). For the same example with even parity, it is 1 0001110 1 00 (Figure 17.7c).

From this we see that the RS-232 interface allows choices: number of data bits (five, six, seven, or eight) parity (none, even, or odd), and number of stop bits (1, 1½, or 2). Both the sender and receiver must be set to the same choices, or else all bits may be received perfectly but make no sense. For example, if the transmitter is set up for even parity and 1 stop bit while the receiver expects no parity and 2 stop bits, there will be confusion and apparent errors (''apparent'' because the bits are received as sent, but misinterpreted).

Bit Periods and Timing

The RS-232 standard allows many baud rates, up to a maximum of 20,000 baud (normally, 1 baud = 1 bit/s; see Chapter 13). Common rates are 110, 300, 600, 1200, 2400, 4800, 9600, and 19,200 bits/s. The 300-bit/s rate is used for mechanical teletypes or very low bandwidth, noisy channels, while most electronic systems

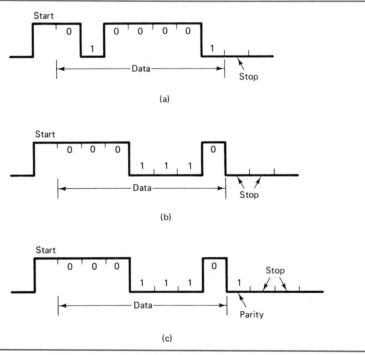

Figure 17.7 Three ASCII RS-232 examples: (a) seven-bit letter B, no parity, one stop bit; (b) 7-bit digit 8, no parity, two stop bits; (c) seven-bit digit 8, even parity, two stop bits.

17.2 RS-232 Operation

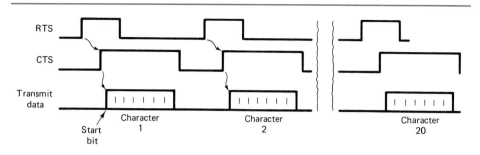

Figure 17.8 Timing diagram shows how handshake lines are used to control data flow and timing.

run at the higher rates of 1200 baud and above. A single bit period is the inverse of the baud value, for example, the bit period is 0.83 millisecond at 1200 baud. Although 19,200 baud is the fastest standard value within the 20,000-baud maximum value specified by RS-232, some systems communicate at up to twice that value (38.4 kilobaud). Once again, as with distances beyond 50 ft, this will work in a carefully controlled situation but may not work in the general interconnection and interface application.

RS-232 Operational Sequence

The interplay of data, handshaking, and bit transmission is seen in the *timing diagram* of full-duplex operation between a computer (DTE) and an interface to a phone line via a DCE, where a message consisting of 20 characters in ASCII representation is sent and a similar message received (Figure 17.8). The format is seven bits, no parity, and one stop bit.

The sequence begins when the computer asserts RTS, indicating to the DCE that it has characters to transmit. If the DCE can accept them, it responds by asserting its CTS. The DTE sees this, and begins sending the bits of the character at the specified baud value. After each character, the DCE uses its CTS handshake line to indicate back to the DTE that it is processing the character and

---

Break Detect

A unique signal indication called *break* is sometimes used with RS-232. In normal operation between characters, the RS-232 data line goes to the idle state (mark voltage), while during transmission of a character it makes transitions from mark to space as required by the character bit pattern. The line should, therefore, never have the space voltage for more than a few bit periods in succession.

This observation provides a method for interrupting the RS-232 receiver and getting its attention. When the transmitter holds the signal line in the space condition for longer than a complete character time period, it is called a *break;* receivers that are specially equipped to detect this have *break detect*. This break detect circuitry can be timing circuitry monitoring the line independent of the regular receiving circuitry, or can be built into the receiving ICs. The advantage of break detect is that it can be used to interrupt and get the attention of the receiver, regardless of what characters are being sent; in this way, it will almost always be seen by the receiver despite other problems (such as timing or parity) that may be present. Note that the break detect time is a function of the baud value in use: It must be greater than a complete character time period at that baud rate. In practice, the sender usually "breaks" the minimum acceptable value for two to three times the character period.

cannot accept the next one right away. The DTE continues to indicate that it has another character to send by asserting its RTS line. When the CTS line is once again asserted by the DCE to indicate that it can accept more bits, the DTE resumes sending them. After all 20 characters are sent, the DTE returns its RTS to the inactive state, showing the DCE that there are no more bits to transmit. Reverse direction activity is similar, except that the DTE uses the DTR line to show the DCE that it can accept new bits on its received data line.

Buffering

In the preceding discussion of sending 20 characters, the RTS/CTS handshake sequence was used between each character. This is inefficient, since valuable time is wasted asserting RTS and checking CTS. Even at high baud values, the actual throughput in characters/s will be much lower than the bit rate could achieve if bits were sent with no time gap between new characters. Continuous transfer and buffering are used to improve the efficiency and throughput.

To achieve continuous transfer, the DTE polls (checks) the status of the CTS handshake line while it is sending character bits, rather than asserting RTS and waiting for CTS after each character. If the CTS line is OK, the DTE sends the next complete character (stop, data, parity, and stop bits) immediately after the end of the present character. If, however, the CTS line indicates that the DCE cannot accept another character, the DTE completes the character it is sending, asserts its RTS line to show that it has more to send, and then polls the CTS to see when it is OK to send.

This more efficient operation is made possible by a *buffer,* which is an internal memory storage area in the DCE (and the DTE) (Figure 17.9). The buffer, also known as a *first in/first out (FIFO) memory,* allows the DCE to accept characters from the DTE even if it has not processed those already received and passed them onto the communications link. The DTE sends a stream of characters at high speed which fills the buffer in the DCE, and then the DTE goes on to doing other tasks. The DTE is not tied up in managing its communications interface, and there are no gaps between the stop bit of one character and the start of the next.

As long as the buffer has room, it indicates this by asserting the CTS line to the DTE. When the buffer is nearly full, it deactivates the CTS line so that the DTE sends no new characters. The CTS indication to stop is sent before the buffer is 100% full, so that any characters already in the process of being sent will be completed, and to provide a small cushion in case the DTE does not respond immediately to the change in CTS status. Then, as the DCE processes its received characters and empties the buffer by transmitting over the link, it signals once again to the DTE that it can accept new character bits.

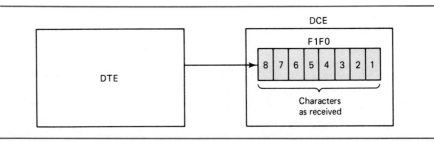

Figure 17.9 Use of FIFO buffer to increase efficiency of data link.

17.2 RS-232 Operation

For maximum throughput, the CTS trip point for reasserting CTS is set around the 5 to 20% full point, depending on system speed, expected number of characters, and other design factors. In this way, no data transmission time is lost in the unavoidable time delays caused by the handshake cycle and restarting of character transmission from the DTE to DCE. If the trip point for reasserting CTS is set where the buffer is completely empty, there is a small but significant period of time when no characters are sent from DTE to DCE, but some could have been since the buffer had space available for new data bits.

Software Handshaking

The handshake lines provide an effective means of indicating whether there are new characters to send, and if new characters can be accepted or not. There is a drawback to these extra lines, however: the DTE and DCE RS-232 interfaces need line drivers and receivers for the handshake lines, and the cable between the DTE and DCE needs additional wires. In some situations, this additional circuitry is not a concern, whereas in other applications it adds undesired complexity, cost, and weight to the system and cables. The alternative is a handshake scheme that uses special data characters on the data lines to indicate status. In *software handshaking,* the data contain the handshake information instead of separate hardware signals. Buffering is required for this scheme to work properly.

To implement software handshaking, two specific code patterns of the ASCII character set are reserved as handshake codes. One code, called *XON* (pronounced EX-ON), is sent from either the DTE or the DCE to the other device to indicate that it is OK to send new characters. A second code, *XOFF*, indicates that the buffer is nearly full and that no new characters should be sent.

In operation (Figure 17.10) the DTE (or DCE) does not transmit any data unless it has already received an XON code on its received data line. While it is sending data, special circuitry (and software) in the DTE (or DCE) continually monitors the received data line to see if an XOFF appears. If it does, the circuitry signals the software to immediately stop sending any new characters. When the buffer at the far end has room, that device sends an XON, which is received by the device that has data to transmit. The XON then enables the data transmission to resume.

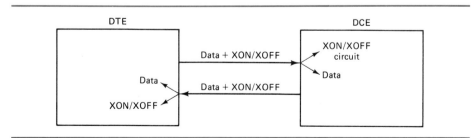

Figure 17.10 Use of XON and XOFF and circuitry to implement software handshaking.

Software Handshaking Requirements

Three things are needed to make software handshaking work. Each end of the interface must have a buffer so that any characters transmitted between the time the XOFF is sent back to the transmitter and it is recognized are captured, not lost. Next, special circuitry and supporting software must be designed into each end, to immediately recognize the presence of the XON/XOFF characters. Finally, there must be a full-duplex link between the DTE and DCE, so that the XON or XOFF character can be transmitted back to the data source while it is sending characters. If there is only a simplex (one-direction) data path, there is no way for the XON/XOFF character to go back to the data transmitter.

There is another subtlety to using software handshaking which precludes its use in all RS-232 applications in place of hardware handshaking. XON and XOFF are two specific characters in the ASCII character set, which is used to represent letters, numbers, punctuation, and special control functions (XON and XOFF fall into this last category). But if the data are being sent in non-ASCII representation, the data bit pattern itself may look like XON or XOFF. For example, suppose that the RS-232 interface is designed to send numbers only, with the eight bits of data representing 0 through 255. The decimal number 17 (0010001 in binary, listed as DC1 in ASCII tables) is the same as XON, while XOFF is decimal 19 (0010011 in binary, listed as DC3 on the ASCII table). Whenever either of these two numerical values is sent, the special circuitry and software will misinterpret them as XON or XOFF indications and believe that transmission can start or must stop! This causes all sorts of operating problems, since the start/stop will have nothing to do with the condition of the buffers or the transmission link. Instead, arbitrary and random data patterns (the numbers that the DTE or DCE are sending) that are unrelated to the communications status will have the unintended effect of controlling the interface "OK/not OK to transmit" status. For this reason, systems that use the data bits of the RS-232 interface for binary numbers cannot use XON/XOFF or any similar software handshaking.

Review Questions

1. What is the RS-232 mark state in binary? In voltage? What about "space"? What is the undefined region?
2. What are the four functional groupings of RS-232 signal lines? What are their roles?
3. From which device do received data flow? What about transmitted data? Why is the standard consistent, although it may seem contradictory?
4. What are the functions of the four handshake lines RTS, CTS, DSR, and DTR?
5. What is the state of an asserted control line? Compare this to a data bit.
6. What is a timing diagram? How is it used?
7. What is the interaction of handshaking when sending data from DTE to DCE?
8. Why is buffering useful, and how does it increase communications efficiency?
9. What is software handshaking? What are its benefits? What special circuitry is needed, and why?
10. What three things are needed for software handshaking?

17.3 RS-232 ICs

The simplistic way to generate the necessary start, data, parity, and stop bits for an RS-232 interface is to have the system microprocessor control one digital output bit of a multiple-bit port. The microprocessor must also generate the necessary handshake lines on other output bits of this port, sense incoming handshake signals on bit input lines, and receive information bits on an input line. However, such an approach is very inefficient. The microprocessor and software would have to devote a great deal of its time to servicing the RS-232 port, and perform the following operations to transmit:

Determine what the data bits are, as well as add the start bit ahead of them.

Calculate the parity bit (even or odd), if there is one.

Add a stop bit (1, $1\frac{1}{2}$, or 2) after the data bits.

Manage the handshake lines to generate a RTS signal, continually monitor CTS, and send out the bits only when the CTS line indicates that it is OK.

For receiving data, the task is equally complex:

Check the status of handshake lines for incoming data, and indicate to the remote device if it is OK to send (if the microprocessor can accept new bits).

Receive the bits, strip off the start and stop bits.

Strip off the parity bit (if any), calculate the parity of the received data bits, and check if the parity bit as received agrees with the parity bit as calculated.

All of these activities, for transmit and receive, would have to be done with precise timing at the specified baud value. Any deviation from the proper timing would cause an error: Transmitted bits would be sent out with the wrong timing, while received bits would be misinterpreted because they could not be checked at the right instants or the microprocessor would not be there to receive them. It is impractical to use the microprocessor and software directly to manage the RS-232 interface (or any other communications format) for any but the very slowest speeds of 100 to 300 baud.

Instead, a special IC is used which relieves the microprocessor from dealing directly with the RS-232 interface. The *universal asynchronous receiver/transmitter (UART)* IC is designed to provide the interface functions between the microprocessor and the RS-232 signal lines. One such UART is the Intel 8251 (Figure 17.11), connected to the microprocessor either as an input/output port or as a memory-mapped address of the CPU, with an interrupt line to let the UART signal important changes in status back to the microprocessor. Within the UART, several registers are available to let the CPU set up the desired mode of operation (parity, stop bits, baud value) in advance, and also receive characters and check

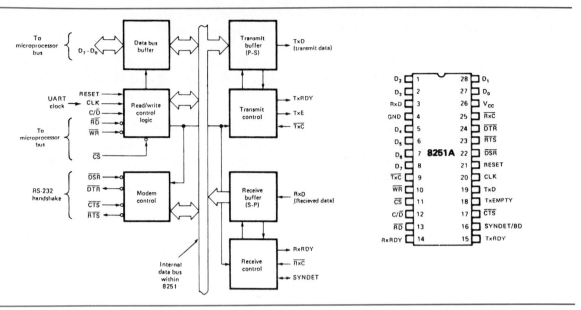

Figure 17.11 8251 UART IC block diagram and physical pinout (courtesy of Intel Corp.)

on status information details. The UART reduces the amount of microprocessor action required to send and receive RS-232 characters to an absolute minimum, while supporting baud values up to the maximum 19,200-baud value.

For transmitting data, the 8251 UART receives the character data bits in parallel from the system microprocessor. These bits may originate at a computer keyboard or be the result of some calculation within the system that produces a result to be transmitted; they can be ASCII representations of characters or binary numbers. The UART takes the data bits, adds a start bit, calculates and adds the parity bit (if any) and stop bit(s), and then checks the handshake lines to see if the receiver at the other end of the RS-232 interface can accept new data bits. If it can, the UART sends the bits out serially at a rate set in advance by the user. The 8251 electronically derives the bit timing for the rate from a master clock, and the microprocessor has only to select the desired rate. After the character is transmitted, the UART tells the microprocessor that it is ready to accept the next one for transmission.

For receiving, the UART manages the handshake lines of the RS-232 interface and indicates to the far end that it is prepared or not prepared to accept new serial characters. As these bits arrive, the 8251 captures them with its baud value clock, strips off the start and stop bits, calculates the parity of the received data, compares the data to the received parity bit for error, and signals to the system microprocessor that a new character has arrived and is stored within the UART. Handshake lines are then set to show that no new characters can be accepted. As soon as the system microprocessor has read the received character in the UART, the handshake lines are set by the UART to show that another character can be received.

A UART such as the 8251 does many of the small but essential and critical operations of RS-232 communications: adding, sensing, and stripping start and stop bits, calculating and comparing parity, managing handshake lines, and providing precise timing for transmitting or capturing bits. The system microprocessor involvement with these details is thus minimized. All that the microprocessor must do to transmit is write the data bit pattern to the UART, after checking that the UART can accept the next character; for received bits, the microprocessor has only to check that a new character is received and read it back.

The meaning of the data bits transmitted or received by the UART is totally transparent to the UART, which operates on any bit pattern. The UART can be used with any protocol or format since it makes no judgments on the data bits and does not study nor interpret the meaning of one character in relation to the previous ones or following ones. The UART acts as a servant in relieving the microprocessor of many details but makes no decisions beyond making sure that character bits are properly transmitted or received.

The interface between the 8251 UART and the microprocessor is a group of four registers, which are eight-bit memory locations within the UART (Figure 17.12). Two of the registers—a *command register* and a *write data register*—are for writing from the microprocessor to the UART, and two—a *status register* and a *read data register*—are for reading UART information back to the microprocessor. A single address line and the read or write cycle are used to distinguish these registers. Character bits to be transmitted are written to one data register, while received bits are retrieved by reading from the other data register. These registers are automatically cleared when the UART transmits the write data or the microprocessor retrieves the received data.

The command register lets the microprocessor set the 8251 to the specific operating modes: baud value, number of data bits (five, six, seven, or eight), number of stop bits, and parity (none, odd, even), as shown in Figure 17.13. Two bits set the desired baud value, derived by the UART from the master clock connected to this IC, to the clock frequency, or $\frac{1}{16}$ or $\frac{1}{64}$ of the clock frequency (called 1×, 16×, or 64× clocks on the IC data sheet). A 4800-baud interface rate is obtained with a 76.8-kHz master clock by selecting a 16× clock. There are two

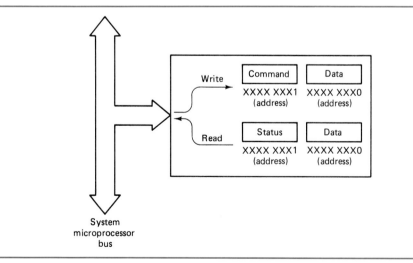

Figure 17.12 Microprocessor location of 8251 status and command registers and data registers.

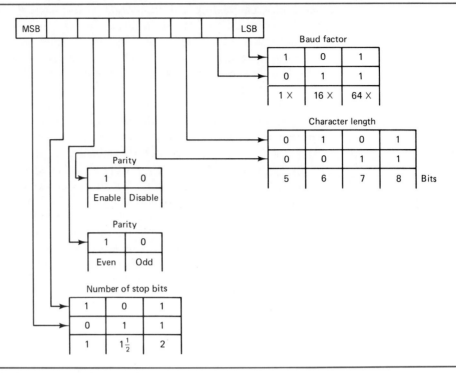

Figure 17.13 8251 command register bits.

bits for the parity mode, and two bits for setting the number of stop bits to 1, 1½, or 2. There are also two bits to set the number of data bits per transmitted character at five, six, seven, or eight.

Example 17.1

The UART is set to transmit at 16× clock, seven data bits, even parity, and two stop bits. What is the pattern that the microprocessor writes to the command register?

Solution

The pattern is 11111010.

The status register in the 8251 lets the microprocessor read and determine the exact status of the UART. The bits of the status register (Figure 17.14) tell the UART:

The status of the DSR handshake line

Whether a framing, overrun, or parity error has occurred (these errors are explained in detail below)

Whether the received data buffer is full or empty (if it is full, there is a newly arrived character and the microprocessor should read the received data register to retrieve it)

17.3 RS-232 ICs

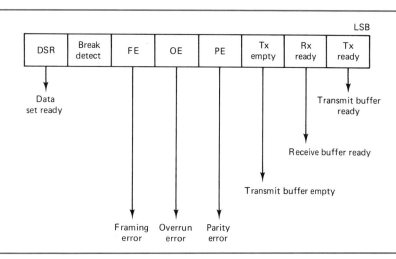

Figure 17.14 8251 status register bits.

Whether the transmit buffer is full or empty (if it is empty, the last character written was sent out and the UART can accept the next character from the microprocessor, to transmit as soon as possible)

(Note that a status bit is reserved to indicate "break detect," as discussed in Section 17.2.)

Receiving Cycle States

When the DSR line is ready and there are no errors and both buffers are empty, the status byte is 0000011X [X means "don't care" or "not relevant here"]. If a framing error and overrun error have occurred, the status byte is instead 0011011X.

When a new character is properly received by the UART, the status word is 1000010X. Note that the DSR indicates "not ready," since the receive buffer is full and therefore any external device must know that a new character cannot be accepted at this time. When the microprocessor reads the new character from the buffer, the status byte changes to 0000011X.

Transmitting Cycle States

Before transmitting, the microprocessor checks the status register to see if the transmit data buffer is empty. The status word should look like 000001XX. When the UART is loaded with the character, the status byte becomes 000000XX. After the character is actually transmitted over the RS-232 interface, the transmit buffer automatically empties, so the status byte changes to 000001XX.

RS-232 Line Drivers and Receivers

The UART is a sophisticated IC that provides management of the RS-232 interface bits and signals. However, it is not designed for actual physical connection to the data link, since any signal wire connected outside its original circuit board or chassis is subject to all types of abnormal conditions or miswiring. The complete RS-232 interface requires two types of ICs: a UART type, which provides management of the interface, and a more rugged physical interface IC. The physical

Error Conditions

Any communications interface must be designed to assume that errors will occur, detect them, and pass this information to the communication interface manager (microprocessor and software). A UART such as the 8251 detects three types of errors:

1. *Parity error*, meaning that the parity bit as received differs from what the UART calculates it should be, based on the received data bits.

2. *Overrun error*, which means that a new character has been received by the UART before the microprocessor has a chance to read the previous one from the read data register. This type of error should not occur if the handshake lines are operating properly, but the problem may occur anyway due to noise, mistiming, or glitches, and a communications system must be set up to deal with all foreseeable error situations without balking.

3. *Framing error*, indicating that the UART did not receive a valid stop bit at the end of the data bits and is thus unable to "frame" (align) the received bits in their proper position.

Note that there is a fundamental difference between a parity error versus overrun and framing errors. A parity error means that the UART apparently received all the bits properly but that a parity error was detected among them, like seeing a word and feeling that there is a typographical error (CET versus CAT). The overrun and framing errors indicate that there is a problem in actually capturing the received serial bits properly, which is a system error rather than an individual bit error (CAT versus C?T, where the bits of the middle character are unreadable).

---

interface IC is designed to withstand, without damage, voltages and currents that would damage an ordinary IC such as the UART. The UART IC is usually a +5-V component, while the RS-232 interface requires signals up to ±25 V and may be accidentally connected to higher voltages.

The physical interface IC must also deal with two very common situations: One end of the RS-232 link is powered up and delivering ±25 V signals, but the other end is not yet powered on, or a signal wire is accidentally shorted to ground—for example, the computer is on, but its RS-232 printer is not. The IC must be designed to remain undamaged in this situation, and also work properly as soon as it has its power applied. If any excessive external voltages or currents are received, the interface IC is designed to fail by blowing "open," thus acting like a fuse and protecting the more complex UART and microprocessor system from damage.

The most common physical interface pair of ICs in use are the 1488 *line driver* and 1489 *line receiver,* available from many manufacturers (Figure 17.15). A line driver such as the 1488 takes digital signals of 0 to +5 V (TTL levels) and translates them to signal levels up to the −25 and +25 V of RS-232. The driven line voltage is determined by the power supplies connected to the line driver IC: If a ±15-V supply is connected, the TTL signals are converted to −15- and +15-V signals. The 1489 line receiver is the complement of the line driver: It takes RS-232 level signals and translates them to 0- to +5-V signals to be used within the circuitry.

The use of a UART combined with a line driver/receiver pair is standard in RS-232 communications. The UART provides the interface management, while the line driver and receiver provide the higher voltages and appropriate interface to the physical link. In addition, by separating the UART functions from the physical interface, the communications system can switch to another type of physical link or to a more sophisticated UART device. For an optical fiber link,

17.3 RS-232 ICs

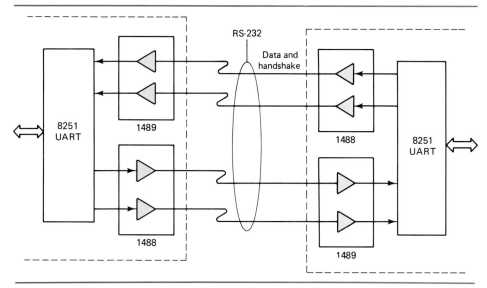

Figure 17.15 Use of line driver/receiver (1488/1489) with UART.

the 1488 and 1489 are replaced by a driver and receiver IC pair that is fiber optic compatible, but the same UART is retained. Similarly, if the RS-232 wire link is retained, the 8251 UART can be replaced by a UART that has additional internal memory registers (FIFO) so that many received characters can be stored before the system microprocessor must retrieve them. This greatly increases the efficiency of the interaction between the UART and the microprocessor, just as a buffer increases DTE-to-DCE efficiency. The line drivers and receivers, however, remain unchanged because they are appropriate for the interface despite the upgrade in UART capability.

Review Questions

1. What would a microprocessor and software have to do to transmit a character over RS-232 directly? What about receiving a character the same way? Why is this a problem?
2. What does a UART do?
3. How does a microprocessor interface to the UART? What are the roles of the command and status registers within the UART? What is set into them?
4. How does the microprocessor set the transmit and receive modes?
5. How is the baud value set? How does the UART derive it?
6. How does the status byte change during the receive cycle? During the transmit cycle?
7. What are parity, overrun, and framing errors? How does the parity error differ fundamentally from the other two error types?
8. Why are line drivers and receivers needed with the UART? What features do they offer? What circumstances must the line drivers and receivers deal with?
9. What does a 1488-type line driver do? What does a 1489-type receiver do?

17.4 RS-232 EXAMPLES AND TROUBLESHOOTING

In this section we look at some examples of RS-232-based interfacing, each of which highlights legitimate variations in using the standard. This is followed by discussion about troubleshooting an RS-232 interconnection that does not operate properly.

There are two important points to be aware of. First, RS-232 is not a "tight" standard. Although it does precisely define voltage levels and timing, it leaves many choices to the user besides the baud value, parity, and number of stop bits. These include which handshake lines should be implemented, how they are to be used, and exactly which type of connector is needed. Second, the RS-232 standard was originally designed for interfacing terminal equipment (DTE) to a communications device such as a modem (a type of DCE device discussed in the next section), which then transmits signals over the communications link to a similar DCE, connected to another piece of terminal equipment (another DTE).

Over time, however, the RS-232 standard has come to be used for direct interconnection of two electronic devices, neither of which is acting as a true DCE that is actually sending data over a communications link to another DCE. This was discussed at the beginning of this chapter and compared to the original RS-232 intention. It is in this area where most of the RS-232 difficulties arise: Without an intermediate communications link or actual DCE devices it is not immediately and unambiguously definable who is the DTE and who is the DCE when the standard is used as an effective, low-cost scheme to connect a computer to its printer, to connect a voltmeter to a computer that controls it and logs the voltmeter readings, or to connect two computers to each other directly. Let's look at some of these cases when there is no traditional "DTE to DCE–link–DCE to DTE" connection.

RS-232 Applications

One common application of the RS-232 standard is to interconnect an instrument such as a digital voltmeter with a nearby computer (Figure 17.16). In this application the voltmeter is monitoring some point in the circuit and reporting the voltage value in ASCII format at regular intervals, such as one reading every second. The role of the computer is to acquire all these readings, perform additional analysis on them, and then produce graphs and statistical summaries of the average voltage, largest and smallest values, rate of change, and similar factors. In this application the flow of data is entirely one way, from the voltmeter to the computer.

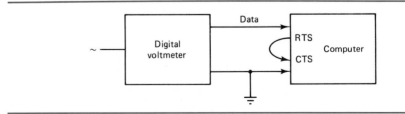

Figure 17.16 Simple RS-232 connection: digital voltmeter to computer.

The voltmeter's function is simple: Measure the voltage, display the value, convert the value to ASCII characters, and transmit the ASCII characters over the RS-232 interface. In many such installations, the voltmeter has only the transmit data line and ground, and no handshake lines at all. There is no point in having the computer tell the voltmeter not to send the next reading, since the voltmeter is not able to store those readings that it is not allowed to send. In other words, for the voltmeter there is no difference between continuing to send readings to a computer that cannot accept them, or being told to stop taking new readings altogether.

The interface is simple and effective, but it can produce a problem. Suppose that the RS-232 interface at the computer has some handshake lines and expects to see a handshake signal from the other device, indicating that the other device has new data. In this situation the voltmeter keeps sending, but the computer ignores the data because it has not seen the proper handshake line back to it asserted; in fact, the voltmeter will never assert the handshake line because it does not have one. The solution is to "fool" the computer into thinking that the voltmeter has asserted the handshake line, telling the computer that new data will be arriving. This *self-handshaking* is done at the computer, where its own RTS and CTS lines are wired together. Every time the computer thinks that it is signaling to the remote voltmeter device that it can accept new data, it is really telling itself that there are new data. If this looks like cheating, it is. While it will work in most cases, it also means that the handshake function at the computer is permanently defeated.

Another common example is a computer connected to a printer (Figure 17.17a). Here again, the flow in information is simple: The computer sends the printer the ASCII representations of the characters to be printed, and the printer

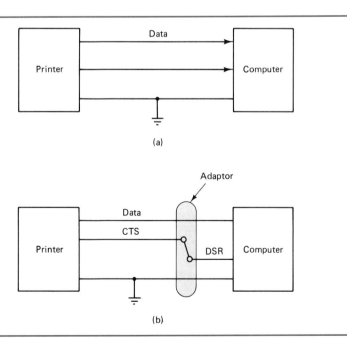

Figure 17.17 (a) Computer-to-printer RS-232 interconnection (single handshake line); (b) cross wiring of interconnect needed to make the interface work.

just prints them. However, handshaking is needed here, so that the printer can tell the computer if it is available for printing. A single handshake line from the printer to the computer is all that is needed. When this line is asserted, the computer can send the next character; when this line is inactive, the printer is saying to the computer that it is still printing the previous character, is out of paper, or is not yet turned on.

The handshake wire from the printer must arrive at the connection pin at which the computer is looking, to see if the printer is ready. If the printer is sending out a handshake line as CTS, but the computer is looking at DSR, the handshake information will be sent but never arrive at the correct RS-232 pin. The solution is to wire up a special adapter cable that brings the correct signal wires together (Figure 17.17b). If the handshaking line of the computer is simply defeated by the same scheme that was used in the voltmeter/computer example above, the computer will always believe that the printer is ready, and continually send characters to a not-ready printer, thinking they are being printed.

The special adapter can also be designed to handle the situation when one device (such as DTE) is looking for two handshake lines (CTS and DSR) but the other device is generating only a single line. The adapter branches out to two pins

---

Troubleshooting RS-232 Interfaces

There is a logical progression in actually getting two "RS-232-compatible" devices to communicate. Any troubleshooting should use this logical sequence or many frustrating hours will be spent in chasing misleading clues.

First, check the basics: Are both devices set to the same baud value, parity, and number of stop bits? These are set either by switches on each device (usually accessible from the outside so that no chassis disassembly is needed), or by user commands via the device keyboard (if it has one) in a special setup or initialization mode.

Next, the signal wires must correspond physically. The connectors must be compatible at each end. If one device has a 25-pin connector and the other has a nine-pin connector, a special cable must be used to bring the correct signal wires across. This cable carries the transmit data, receive data, ground, and any handshake lines. Even if the two devices use the same connector, such as the 25-pin D connector, they must have opposite senses (male and female) so that they can mate. A null modem may be needed to transpose signal wires, along with an adapter that changes gender of the connectors.

The most critical signal lines to identify are transmitted data, received data, and ground (normally pins 2, 3, and 7 on the 25-pin connector). The DTE end of the transmitted data line should have a voltage at the mark or space level, while the DCE end of the transmitted data line will normally measure near 0 V since it is an input and thus passive. Many installations use only one ground line rather than both chassis ground (pin 1) and signal ground (pin 7); unless both the DTE and DCE have at least one ground wire connected from end to end, errors and general malfunctions are likely.

Finally and most difficult, the operation of the handshake lines and sequence must be checked. This is done by looking at the manuals or documentation for each device, and answering these questions: Which is the DTE, and which is the DCE? What handshake lines does this device activate when it is saying that it can send or accept data to the other device? What handshake line does this device need to see asserted by the remote device in order to send or receive data? By studying the answer to these questions, the handshake lines between the two devices can be matched up by special jumpering of wires or by cross-wiring different signal leads to each other so that one device's request to send new data is seen by the other device, and then acknowledged by an "OK to send" signal that is returned. Of course, the source and the receiver must agree on the format of the data message. This is one step above the RS-232 interface itself, but disagreement on format can mean that characters themselves are received perfectly but are not understandable.

on the DTE connector, bringing the single handshake line to the two pins that must see a handshake for communications to proceed.

For both of these examples, the question arises: Who is the DCE, and who is the DTE? If this is not defined, one device will transmit on the wrong wires. This is best seen in a final example of two identical computers that are passing information to each other via RS-232, in full- or half-duplex modes. If both devices strictly observe the RS-232 standard, both will be transmitting data on pin 2, expecting to receive data on pin 3, and so on (assuming that both are set up as DTEs and using a 25-pin connector). In fact, no communication occurs, since the transmitted data bits of both collide, while the receive signal lines will have no activity! The same situation happens at the handshake lines, as one computer's outgoing handshake is wired to the other computer's outgoing handshake. The real source of the problem is that both are configured as DTEs (or DCEs), whereas the RS-232 standard is designed for interconnection of a DTE with a DCE.

The solution is a special interface cable called a *null modem*. This cable transposes signal wires pairs, so that one computer's pin 3 signal appears on the other's pin 2, for example. This problem and the null modem solution are separate from the incorrect handshaking problem of the previous examples, where the handshake lines were not used or were used differently.

RS-232 Test Equipment

It is difficult to check RS-232 operation with a voltmeter, because many lines must be monitored at once and these lines have signals that are not constant. A better solution is a *breakout box* (Figure 17.18), a simple device that brings all 25 wires of the common RS-232 connector out to accessible test points. These points can be monitored by a simple logic-level probe or an oscilloscope; a multichannel oscilloscope is most useful because it lets you see activity on several lines at the same time, and you can see the interaction between handshake activity and data flow. In addition, the breakout box has switches for every line (to let you disconnect any connection for test purposes) and any line can be interconnected to any other with jumper wires. This is essential when trying to make the null modem transposition of data lines, or trying to make the handshaking situation of the two devices correspond (or deliberately defeat it), as in the RS-232 examples above.

The weakness of the simple breakout box is that you can see signal-line activity but cannot actually interpret the data bits on the signal lines (start, stop, parity, XON, XOFF, character data bits) and their meaning. A more advanced instrument is the *line monitor*, which looks like a small terminal but sits passively on the RS-232 wires (Figure 17.19). The line monitor displays the actual characters on the lines, as well as special characters such as XON, XOFF, line feed, or carriage return. Using the line monitor, the technician or engineer can observe the characters on the line and see if they agree with what each RS-232 device is sending and what each device indicates that it has received. If there are errors due to incorrect settings of baud, parity, or stop bits, or noise and miswiring, the line monitor is very helpful. To show the data flow in each direction, most line monitors use black-on-white video for one direction and reverse video (white on black) for the other.

The line monitor is important because it lets users see the actual flow of data between devices, and see what messages are sent in response to other messages. One common troubleshooting dilemma is that everything seems OK, yet the device receiving the messages does not take the proper action upon receipt. In many

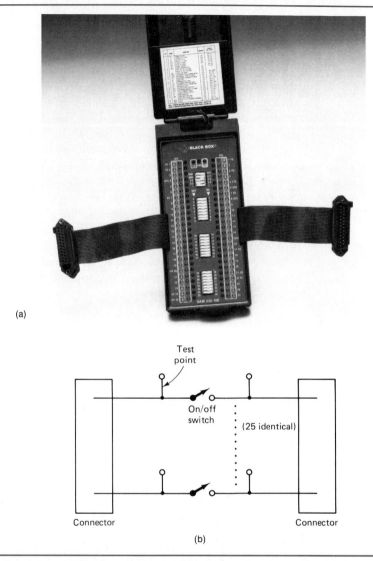

Figure 17.18 (a) Typical breakout box (courtesy of Black Box Corp.). (b) breakout box schematic.

cases the problem is not an error in the bits, but there is a message format difference. For example, the device sending characters may use a period (.) as the last character, while the device receiving the message may be designed to look for a semicolon (;) as the message terminator. Even when each character is received without error, the receiving device will not respond properly. The problem is not with the link, the bits, the parity, or the baud value: It is with the language—the two devices are using different languages. It is the same as a person reading a message in another language: Each letter is recognized perfectly, but the message cannot be interpreted or understood. The line monitor shows the actual characters, so the technician can check the system documentation to see what each device expects to receive and compare.

17.4 RS-232 Examples and Troubleshooting   543

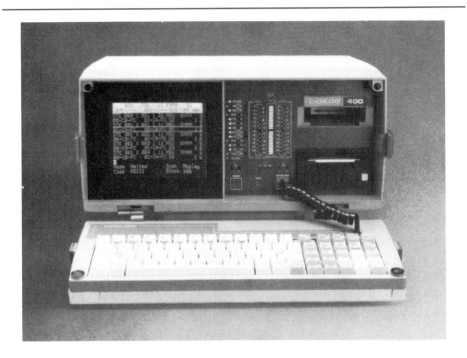

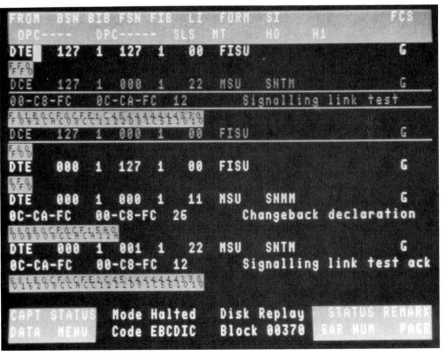

Figure 17.19 A line monitor is more complex but much more revealing than a simple breakout box (courtesy of Digilog, Inc.).

Review Questions

1. Why are there many legitimate RS-232 interface combinations?

2. Why does a simple voltmeter-to-computer RS-232 interface require no handshaking? How are the handshake lines of the computer interface fooled? Why is this done?

3. What happens when both devices are configured as DTEs (or DCEs) even if all handshake lines are implemented properly?

4. What is a null modem? What does it do?

5. What is the first step when troubleshooting an RS-232 interface?

6. What is a breakout box? What does it show? What can it not show? How does it allow RS-232 lines to be opened, closed, or wired to other lines?

7. How does a line monitor show format and protocol problems that occur even when every bit and character is received properly?

17.5 MODEM FUNCTIONS

The data signals used within electronic circuitry are generally not electrically compatible with the communications link. Standard TTL-level digital signal of 0 and +5 V cannot be directly connected to the phone wires, for example, because they have the wrong level, have excessive bandwidth, and can have a large dc component in their frequency spectrum (which is blocked by the frequency characteristics and circuitry of the phone or data link lines). The incompatibility between digital data signals and the communications channel is solved by a DCE device called a *modem (modulator—demodulator)* which takes the data signals and modulates them using AM, FM, or PM so that the signals can travel through the link. At the receiving end, the modem demodulates the signal to reproduce the original signal type and level.

Modems are available for many performance levels and types of applications and links. The simplest modems interface to their DTEs via RS-232 and can send data over standard phone lines at rates up to 2400 or 9600 baud with very few errors. More sophisticated modems can send data reliably at Mbit/s rates using synchronous communications over special, wideband lines or channels (wire, radio, fiber optic, satellite) while providing error detection and correction capability. In this section we concentrate on the lower-range modems because they are more common and they illustrate modem capabilities, and we will also look at some advanced modem features.

Modems are always used in pairs, with one at each end of the communications channel. The modems do not have to be from the same manufacturer, but they must have identical and complementary modulation and demodulation characteristics so that each can accept and understand signals from the other. For lower-performance ranges there are many companies which make modems that are identical in function, while the more advanced modems are usually unique to each manufacturer in modulation and error-handling techniques and so must be purchased in pairs. (There is less industry standardization for higher-performance modems.)

17.5 Modem Functions

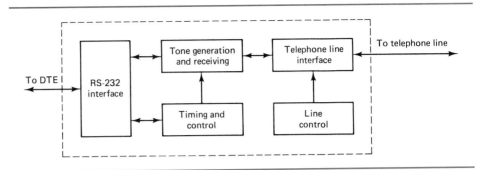

Figure 17.20 A modem (DCE) between a DTE and the phone line has five major functional blocks.

A block diagram of a modem is shown in Figure 17.20. The basic idea of a modem is to use tones to convey information, in contrast to the absolute voltage levels that convey information within a digital system. Depending on the modem design, the amplitude, frequency, or phase of the tone differentiates a binary 1 from a binary 0. As the modem receives data via RS-232, it generates the appropriate tone and then transmits it over the link; for received signals, it performs the reverse operation and produces 1s and 0s by decoding the received tones.

Modems can have functions beyond just tone generation. For telephone lines they must provide the correct signal levels and electrical appearance during dialing and calling phases. The modem (a DCE) must manage the handshake lines between it and its DTE and must also send status signals back to the DTE indicating what is happening on the phone line, using some of the other lines defined by the RS-232 interface.

For an example, we can use a modem that sends information with 1000 Hz for a 0 and 2000 Hz for a 1. Suppose that the DTE wants to send the letter F in eight-bit ASCII with no parity and one stop bit. It sets up its UART for the desired baud value and mode of operation, then passes 01000110 to the UART. In turn, the UART adds the start and stop bits and serially sends all 10 bits to the modem via RS-232 (remember that the LSB of the character is sent first): 1 01100010 0. The modem translates these to the correct frequencies as they arrive, and sends them onto the telephone lines as 2000, 1000, 2000, 2000, 1000, 1000, 1000, 2000, 1000, 1000 Hz (Figure 17.21). The receiving modem uses filters and comparators to decide if the frequency at any instant is 1000 or 2000 Hz, re-creates the binary data pattern, and then passes this entire pattern to the receiving DTE and its UART. This UART strips off the start and stop bits, tells the DTE circuitry that a new character has arrived, and then the DTE microprocessor reads the character. At higher data rates, the challenge for the modem is to decide, correctly and quickly, what frequency was received, despite varying signal levels, noise, and distortion.

Originate and Answer Modes

The simple example of two frequencies is effective for simplex or half-duplex communications, but it will not work for full duplex. If both modems transmit at 1000 and 2000 Hz, two things will happen: Their signals will interfere and conflict with each other, and the signals transmitted by each modem will overload the

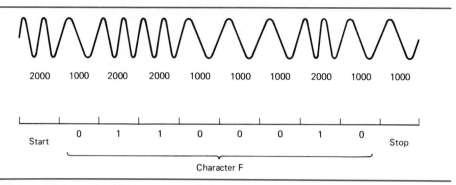

Figure 17.21 1000/2000-Hz tone sequence of a representative modem.

receiving circuitry within the same modem. The solution is to use two pairs of frequencies, with one modem transmitting at 1000 and 2000 Hz and the other at 3000 and 4000 Hz (Figure 17.22).

To make sure that one modem uses one pair while the other modem uses the second pair, modems operate in either *originate* mode or *answer* mode. The originate modem transmits the 1000/2000-Hz pair while the answer modem produces 3000/4000-Hz tones. The originate modem must detect 3000- and 4000-Hz tones, while the answer modem looks for the 1000- and 2000-Hz tones. Each modem is capable of operating in either originate or answer mode, as needed. The mode of operation is set when the initial connection is made: The modem that initiates the communications uses originate mode, while the one that responds uses answer mode.

There is another potential problem with full-duplex operation: The signal levels generated at the modem when transmitting are much greater than the received signal levels. This makes it extremely difficult to prevent transmitted tones from overloading the receiving circuitry front end within a modem, even though the frequencies are different. Along with channel bandwidth and noise, this is one of the factors limiting the maximum baud value for reliable communication. Lower baud values give the receiver circuitry more time to filter signals and settle despite high levels of nearby frequencies; at higher rates, this time is not available.

One solution to this problem is to operate in half-duplex mode (Chapter 12).

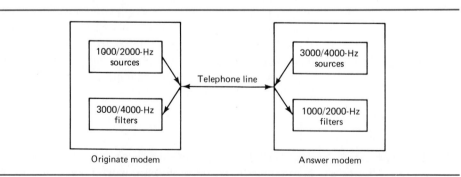

Figure 17.22 Originate/answer tone pairs avoid conflict and allow full-duplex communication.

Standalone and Integral Modems

The first modems were separate boxes acting as DCEs, connected by RS-232 to their DTE devices. These *stand-alone* modems are still used in many applications, because they have room for advanced circuitry to provide highest performance, can be changed and replaced if there is a problem, can be upgraded to newer designs if needed, and are independent of the DTE with which they connect.

However, stand-alone modems do have some drawbacks. Besides the obvious fact that they require more space and a cable to connect to their DTE, the RS-232 interface they use to the DTE does not provide flexibility. For example, suppose that you want the modem to produce the DTMF tones so that it can dial the other modem and automatically establish the call, without personal involvement. The RS-232 interface does not provide lines that specifically let DTE tell the modem what number to dial (the DTMF tone pairs to produce) or that it should begin dialing. To overcome this limitation, special character sequences that the modem uses only for instructions are defined by the modem manufacturers. The modem does not take these instruction sequences and transmit them; instead, it uses them to initiate dialing, for example.

The growth of personal computers and portable communications systems brings an alternative to the stand-alone modem. *Integral* modems are built into the computer system as part of the main processor circuit board or as an additional board that plugs into a socket within the computer. Unlike the stand-alone modem that can communicate with its DTE only via the RS-232 interface, the integral modem is a device on the CPU bus (Figure 17.23a). This type of modem has command and status registers like the UART and is easily set up for a desired mode of operation by the system microprocessor without having to work around the limitations of RS-232. The block diagram of a typical integral modem (Figure 17.23b) shows how it combines a modem IC (providing tone modulation and demodulation, filtering, and line supervision) with a UART and line interface circuitry to produce the complete modem function.

The drawback of the integral modem is that it has less potential performance than stand-alone units in terms of baud value and error rates, because its smaller size limits the complexity of the circuitry. A typical integral modem can achieve 2400 baud reliably, although advances in IC technology and digital signal-processing techniques are pushing this up to 9600 baud in newer designs. The choice between an integral modem and a stand-alone modem is dictated by the application: portability, the ability to change modems quickly and separate the modem from the DTE, and the level of communications sophistication needed.

Since there is data transmission in only one direction at a time, the circuitry does not have to contend with high-level, interfering tones. Half-duplex operation is awkward when there are many small blocks of data and lots of back-and-forth messages, such as when someone is typing on a terminal and expecting to see responses. It is acceptable and practical, however, when large blocks of data are being transmitted, since the need to reverse the channel direction is much less frequent and the time lost during channel turnaround is a smaller part of the overall time.

The four-wire connection (Chapter 16) is another effective solution to achieving high data rates in full-duplex mode. Since each direction has its own path from end to end (DTE to DTE), there is no interference. The receiving filter of a modem never sees the potentially overloading tones transmitted by that same modem.

Advanced Modems

The basic modem simply uses the binary 1,0 pattern to generate or distinguish signals that vary in frequency, as discussed above, as well as amplitude or phase. Beyond this, the basic modem does not add to the data pattern.

To improve performance—decreasing the error rate and increasing the baud value—the modem must do more than this. Advanced modems incorporate inter-

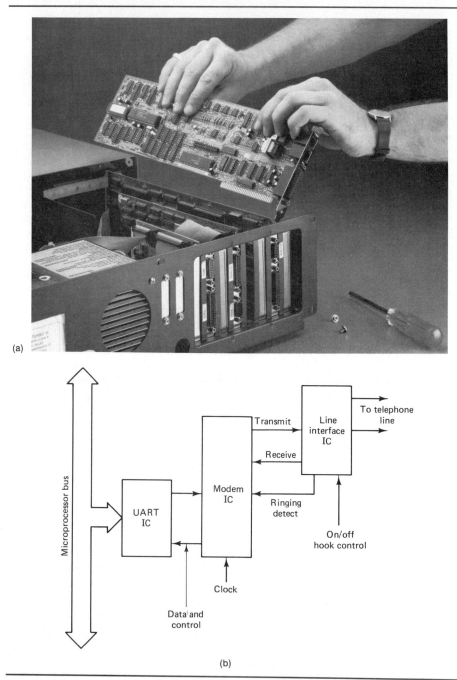

Figure 17.23 (a) Integral modem (courtesy of Black Box Corp.); (b) integral modem interfaces from microprocessor bus to phone line.

nal FIFO buffers to increase the achieved throughput at a given baud value. They also manipulate the data bits to develop error detection and correction codes, so that the receiving modem can reduce the bit error rate (BER) seen by its DTE. Finally, advanced modulation techniques are used to increase the bit rate while maintaining acceptable BER. Of course, synchronous communication is needed to go faster than about 20 to 100 kilobaud.

The FIFO improves performance because it reduces the amount of dead time between characters sent between the DTE and the modem. Without a FIFO, the DTE must check that the modem has sent the character over the link before it can send the next character to the modem. At the receiving end, when there is no FIFO, the modem must transfer each character as it is received to the DTE before it can receive the next from the link. Therefore, the baud value must be set low enough so that no characters are sent to the modem when they cannot be received (and therefore will be lost). By using the FIFO, a long stream of characters can be sent at higher speeds without time wasted for DTE-to-DCE handshaking after each character.

Smart modems also add error correction bits to the basic data bits. As we saw in Chapter 12, it is possible to take a group of bits and produce a checksum that summarizes this entire bit stream. A smart modem will generate a checksum on a block of bits and transmit this checksum after the block. The receiving modem calculates its own checksum based on the received bits and compares it to the received checksum. If they agree, there is no error; if they differ, the modem informs the DTE. The DTE then implements an error-handling protocol, such as sending a special character sequence back to the data source requesting retransmission of that block. Some modems can initiate the retransmission request on their own, without the DTE. In either case the addition of a checksum is transparent to the DTE. The data source DTE sends its bits, and the receiving DTE receives the same bits. Neither DTE "knows" that the transmitting modem has added checksum bits or that the receiving modem has removed them. The same transparency applies to the modems: Neither one attempts to know what the data bits represent. They only know that they receive a series of bits and must perform the checksum algorithm on these bits.

To operate at higher bit rates while maintaining acceptably low BER, there are even more advanced modems that perform error detection and correction automatically. The sending modem uses the EDC algorithms discussed in Chapter 12 to add a checking and correction field of bits to the data bits (arranged in groups of 8 or 16 bits most commonly). The receiving modem performs the EDC operation to see if the data bits received are error-free, and to correct them using the additional EDC bits if necessary. Because of this, the system can be run at higher rates while still providing low BERs. Like the checksum procedure, the EDC operation is transparent to the system and therefore not seen by either DTE. Only the results of EDC—received bits with fewer errors—are seen by the DTEs.

Many of the modems that use checksums for error detection, or EDC schemes, automatically set themselves to the highest baud value that provides acceptable performance. This approach works because the modem can tell how well the communications link is performing. For example, the modems can begin operation at 9600 baud and then both shift down to 4800 baud if they detect that too many errors are occurring. When conditions improve and the BER drops below some preset limit, the modems resume higher-speed operation, all without operator involvement. In this way, the modems can accommodate communications channels that get noisier or have varying amounts of distortion, envelope delay, and resulting variations in BER during the complete message time. This is a common occurrence in many channels, such as radio, microwave, and satellite links. The system baud value does not have to be set for the lowest baud value that produces acceptable performance under the worst conditions but adjusts dynamically to variations in channel characteristics.

Techniques such as internal modem FIFOs and adding error detection and correction improve modem performance by manipulating the basic digital bit information. Modem and communications system performance is also improved by changing the modulation scheme regardless of what the bits represent, and advanced modulation makes the best use of the channel capacity.

In the modem example earlier in this section we used a basic pair of frequencies to represent 1 and 0. However, we saw in Chapter 13 that multiple-level modulation (such as 4, 8, or 16 frequencies, or combined phase and amplitude modulation) makes better use of the available bandwidth and signal range than simple two-level modulation. To achieve values above 9600 baud over standard telephone lines, most modems use some type of multiple-level modulation. The bits from the DTE are grouped into 2-, 3-, or 4-bit clusters and then used to produce 4-, 8-, or 16-level modulation, respectively. Recall from Chapter 13 that when multiple-level modulation is used, the bit rate and the baud value are no longer the same.

However, these multiple-level methods reduce the distance, or spread, between signals that represent the bits, so that the noise margins are smaller and there is greater potential for error (due to noise and other channel characteristics) when the received signal is demodulated and decoded to produce the binary pattern. To compensate for this, modems that achieve higher bit rates through multiple-level modulation often also incorporate error detection (and retransmission protocols) or EDC to compensate for the large number of errors that may occur.

Modems for Non-Telephone Links

Modems designed for using the communications channel of the telephone system provide the greatest amount of flexibility since they can be used to connect any loops that the phone system interconnects, and worldwide communication is possible. A modem designed for interfacing to the telephone loop must be designed to meet phone system standards for signal levels, bandwidth, and absence of dc signal component, among others.

There are applications, however, when there is no need to use the phone system, such as when the two DTEs are located in the same building. The modems for these situations can be designed for a different set of rules than the telephone loop modem. Within the same building, a direct wire can be run between the modems. This wire provides wider bandwidth than the telephone loop and has no components to block dc or intermediate amplifier stages that the telephone system must provide to maintain signal levels. Such a *short-haul* or *direct-connect* modem is designed to provide high-speed, low-error performance on this type of dedicated path for a distance of several miles maximum. To the DTE, this modem has an RS-232 interface, but the circuitry within the modem that interfaces to the communications line is different from that for a telephone loop. This modem looks like a telephone-type modem to the DTE, but its circuitry interfaces to a different path.

Other modems are designed to interface to fiber optic lines instead of wire. The fiber optic link provides immunity to electrical interference, protection against someone tapping the line, and electrical isolation. Once again, the DTE does not see that the DCE—the modem—is connected to a different type of link.

17.5 Modem Functions

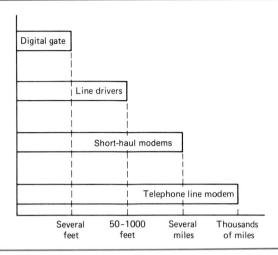

Figure 17.24 Comparison of relative data communications distance for a digital gate, line driver, short-haul modem, and telephone modem.

The signaling sequence and RS-232 interface remain the same. Some manufacturers offer several versions of the basic modem: one for telephone line interface, one for direct connect systems, and one for optical fiber. The user can choose the type that best suits the application, and change from one to another without making any changes in the DTE's RS-232 interface. Figure 17.24 summarizes the relative distances achieved with digital gates, line drivers, direct connect modems, and telephone line modems. Of course, higher rates are also easier to achieve at shorter distances as well.

Review Questions

1. Why is a modem needed to bridge the incompatibility between circuitry data signals and communications lines? What are three reasons that digital circuitry signals (such as TTL) are incompatible with telephone lines?
2. What does a modem do? Why are they used in pairs?
3. What are originate and answer modems? Why are they needed? How is it decided which modem is which mode?
4. Why is half-duplex an alternative to originate and answer modes? When is it a good alternative? When is it awkward?
5. How does a modem increase data rates while holding error rates low?
6. What does a FIFO do in a modem?
7. What is the role of error detection in a system? What is meant by saying that error detection and correction (or detection alone) is transparent to the DTEs and that data are transparent to the modem?
8. How can an advanced modem select the highest baud value automatically? Why is this useful?
9. How does advanced modulation improve modem performance? How does a modem implement multilevel modulation?

17.6 STANDARD MODEMS

Some basic modem designs have become so commonplace in the communications industry that many manufacturers use the same modulation and demodulation scheme. In this section we look at two such modems: the basic Bell 103 unit, providing operation to 300 baud, and the Bell 212 unit which operates at 300 and 1200 baud. Both were developed by the Bell Telephone System but became so popular that the terms ''103-compatible'' and ''212-compatible'' are now generic names for any modem, regardless of manufacturer, that provides the same operating specifications.

Bell 103 Modem

This modem provides full-duplex operation up to 300 baud. Frequency shift key modulation (FSK) is used, with one pair of tones for originate mode and another pair for answer mode. Figure 17.25 shows the frequency spectrum of this modem, with the outer envelope as the frequency response of a standard dial-up, unconditioned telephone loop. The frequency components of the modem signals must fall within this envelope.

In the Bell 103 modem in originating mode, a space uses 1070 Hz and a mark is 1270 Hz. For answer mode, these frequencies are 2025 and 2225 Hz. This spacing between originate and answer pairs allows properly designed filter circuitry within the modem to filter out the desired received tones even when the transmit tones at the modem are 50 dB greater.

The sequence of frequencies seen on the telephone line from a 103-type modem in originate mode for the letter E with one start and one stop bit, no parity, and seven-bit ASCII representation (1000101) is 1070, 1270, 1070, 1270, 1070, 1070, 1070, 1270, 1270 Hz, corresponding to a 010100011 (recall that LSB is sent first). For the same character sent by the modem in answer mode, the sequence is 2025, 2225, 2025, 2225, 2025, 2025, 2025, 2225, 2225 Hz.

Bell 212 Modem

The Bell 212 modem provides 300-baud operation identical to the 103-type modem, and 1200-baud operation using *differential* phase-shift key (DPSK) with four

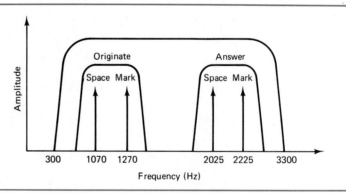

Figure 17.25 Bell 103 modem frequencies and bandpass of standard phone line.

17.6 Standard Modems

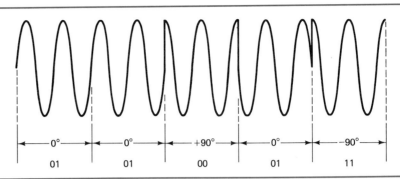

Figure 17.26 Differential phase shifting of the Bell 212 modem.

levels of phase shifting. Phase shifting is used because it provides better performance than frequency shifting with fewer errors at this higher baud value. In originating mode, a 1200-Hz carrier is phase shifted; in answer mode, a 2400-Hz signal is phase shifted. The four phase shifts are 0, 90, 180, and −90°. In operation, the data bits to be transmitted are grouped by the modem into clusters of two bits each (*dibits*), and then each dibit points to a specific phase-shift value relative to the previous bit waveform (this accounts for the phrase *differential*): 00 is 90°, 01 is 0°, 10 is 180°, and 11 is −90° shift from the existing waveform phase. For the letter E example above, the nine bits are grouped by twos, and an extra "mark" is added at the end since the line is idle. The grouping is 01 01 00 01 11 and the phase-shift pattern of the carrier is 0°, 0°, 90°, 0°, −90°, as shown in Figure 17.26. The demodulation circuitry determines the received phase and generates the corresponding two-bit code pattern. Note that twice as many bits of information are sent per second compared to the rate at which symbols are actually sent. This means that the modulation scheme sends more data in less time but at a cost of half as much margin against error-causing phase jitter and noise.

As modem speeds increase, modulation becomes more sophisticated. For highest speeds, assuming that the channel has the necessary bandwidth, quadrature modulation with four phase and four amplitude values is used for a total of 16 values/symbol (four bits, since $2^4 = 16$), or even eight phase and eight amplitude levels, yielding 64 values/symbol (six bits, $2^6 = 64$). Of course, noise protection schemes such as checksums or EDC are usually needed to maintain acceptably low BER with the reduced noise margins that these modulation systems also have unless specially shielded and conditioned low-noise lines are provided.

Review Questions

1. What are the originate tones of the Bell 103 modem? What are the answer tones? What is the maximum baud value for this modem?

2. What are the answer and originate carriers in the Bell 212 modem? What are the amounts of phase modulation in this device?

3. How do the bits determine the amount of phase shift? Why is this scheme used? What is differential phase-shift keying?

17.7 OTHER COMMUNICATIONS STANDARDS

The RS-232 interface standard is among the most common, but it does have some technical weaknesses. Most of these—except for low speed—arise not from the original application of connecting a DTE to its DCE device, but from the extended uses that have been added since the standard was developed. These include interfacing computers to printers, electronic instrumentation to other instruments, or using data systems in harsh electrical environments such as industrial plants.

The summarized limitations of RS-232 include:

50 ft is the maximum guaranteed distance from one end to the other.

The 20-kilobaud rate is too slow for many applications.

The standard is designed for "point-to-point" links, where there are just two users on the link.

The single ground wire of RS-232 that is used as the 0-V reference for all signals can pick up noise and suffers voltage drops and radiated signals, which in turn cause errors.

The -25 to -3 V and $+3$ to $+25$ V signal levels are not compatible with most newer electronic equipment which has ± 15-V or $+5$-V power supplies.

To overcome these weaknesses of RS-232 in some applications, other EIA standards were developed. We will look at three of these in more detail. No single standard overcomes all these limitations simultaneously, but each improves on RS-232 in one or more areas. The physical circuitry to implement these standards requires replacing only the line driver and receiver of RS-232, while the UART (or equivalent) can remain unchanged. Once again, this is the advantage of separating the formatting IC from the line interface IC.

RS-423 Standard

RS-423 allows for distances to 4000 ft, data rates to 100 kbits/s, and up to 10 line receivers driven by a single line driver (Figure 17.27) (so a single source can supply characters simultaneously to 10 video screens, for example). The standard achieves this distance and ability to support 10 receivers by using drive circuitry that produces more current along with line receiver input circuitry which draws less current from the transmission line.

The increase in speed is accomplished by lowering the voltages to -3.6 to -6 V for a binary 1 and $+3.6$ to $+6$ V for a binary 0. This reduces the maximum slew rate requirements on the line driver by a factor of 4 compared to the RS-232 standard (6 V versus 25 V) and assures that the line driver output can traverse the entire maximum voltage span in less time (12-V versus 50 V swings), yielding faster data rates. Like the RS-232 standard, RS-423 is a *single-ended* specification, with each signal represented by a single wire and one ground wire common to all signals.

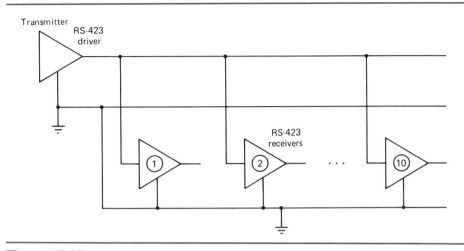

Figure 17.27 RS-423 with one driver, capable of driving 10 receivers.

RS-422 Standard

RS-422 improves on RS-423 by increasing the maximum data rate to 10 megabits/s while maintaining the ability to drive up to 4000 ft and 10 receivers. The signal levels are similar to RS-423: -2 to -6 V and $+2$ to $+6$ V. The major difference is that RS-422 is a differential standard (Chapter 7). Each signal is represented by a pair of wires, and the voltage difference across these wires is what is sensed at the receiver. This minimizes the effect of ground noise or the voltage drop along the signal leads. The signal arrives at the receiver with less noise since the ground wire noise is irrelevant, and with larger useful amplitude (the voltage difference between the wires, not the voltage between a signal wire and ground).

The penalty for using RS-422 is that two wires are needed for each signal, in contrast to the one wire per signal used in RS-232 and RS-423 (plus a single common ground for all). Line drivers for RS-422 take the single-ended 5-V signal within the circuitry and produce a differential output on two wires (Figure 17.28a). When the output is a binary 0, the A output is greater than B; the situation is

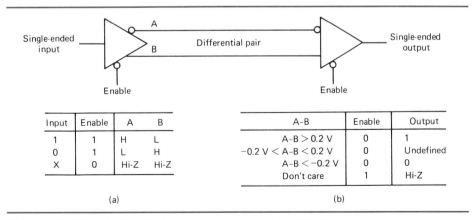

Figure 17.28 (a) RS-422 driver schematic and (b) RS-422 receiver schematic with both truth table showing voltage levels and logic states.

reversed when the output is a binary 1. (The role of the output enable pin is discussed below.) An RS-422 line receiver (Figure 17.28b) performs the reverse operation, taking the two differential wires and producing a 0-V or 5-V single-ended output based on the relative difference between inputs A and B. When the A-B difference is >0.2 V, the output is a binary 1 at 5 V, while a difference more negative than −0.2 V produces a binary 0 at 0 V.

RS-485 Standard

RS-485 is a standard that expands on RS-422 by allowing the same speed and distance, and up to 32 receivers, with signal levels of −1.5 to −6 V and +1.5 to +6 V. The major enhancement, however, is that RS-485 allows *multidrop* operation, where more than one transmitter can be connected on a line—though not active—at a time (Figure 17.29). Although only a single transmitter can be operating at any instant, the standard specifies that up to 32 transmitters can be connected on the differential wire pair at the same time. This is done by making sure that line drivers for RS-485 can be disabled, using a single control pin on the line driver IC. When disabled, the line driver goes to a high-impedance mode (sometimes called a *three-state driver*, for its 0, 1, and high-impedance states) where it draws virtually no current and produces neither a binary 0 nor 1. As far as the system is concerned, these disabled drivers are not present on the line.

The RS-485 standard is useful when many user points must be connected together, with a single user sending data at any time. To ensure that one and only one sender is active, the system circuitry must take special actions to instruct the users as to whose turn it is to become active. This forms the basis for the *backbone* of a *network* where messages can be sent by any user to any of other users. Receivers (users) who have no need to listen to the message can simply ignore the data since there is no electrical conflict in having many listeners. The only conflict occurs when there are two or more talkers at the same time, so

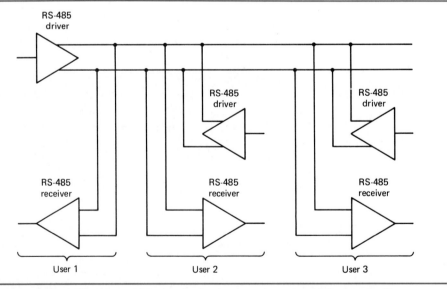

Figure 17.29 RS-485 multidrop standard can support multiple transmitters and receivers.

the use of the output enable and high-impedance mode on the line driver allows the system to specify which user will be the transmitter at any time period.

None of these standards allows achieving both the maximum distance and the maximum speed at the same time. A graph is published with each standard which shows the bit rate that is guaranteed at a given distance when using the appropriate line drivers, receivers, and signal levels.

Review Questions

1. What are some limitations of RS-232? In what situations are these a problem? How can RS-232 be changed to another standard?

2. What are the differences between RS-232 and RS-423 in distance, data rates, and number of receivers? How are each of these improvements accomplished?

3. Compare RS-422 and RS-423. What are the advantages and disadvantages of differential signals? How do the line drivers operate?

4. Compare RS-485 to RS-232, -423, and -422. What is multidrop? Where is it needed?

5. What does the output enable pin of a multidrop line driver do? Why is it needed? How is it controlled?

SUMMARY

RS-232 provides an effective way to transfer bits at rates up to 20 kilobaud, for distances up to 50 ft. It was originally designed as an asynchronous standard between a terminating device such as a computer, terminal, or printer, and the associated communications device, which then transmits over a data link to another peripheral device. Over time the RS-232 standard has come to be used for direct interconnection between one terminating device and another, which has caused some confusion and problems that require careful thought to resolve.

The functions of the RS-232 serial asynchronous interface are implemented in ICs which divide the overall responsibility. A universal asynchronous receiver/transmitter handles proper formatting of data bits from the microprocessor or data link and manages the RS-232 link handshake lines, parity, and error detection. Line drivers and receivers provide the muscle needed to drive the link while resisting fault, miswiring, or signal applied while power is off. Systems can be upgraded by replacing the UART with one that has a buffer, or the line driver and receiver can be replaced with a pair that is compatible with a different type of physical data link.

A modem is the device which converts data bits into modulated signals that are electrically compatible with the data link. Modems are available for a wide span of speeds, and for telephone lines, a direct wire connection, fiber optical links, and radio links. In each case, higher performance is achieved by advanced, synchronous, and multilevel modulation, as well as by adding error detection (and correction) to the basic data bit stream. All this is completely transparent to the source of the data bits or the user.

For applications where RS-232 is not capable of supporting the speed, distance, or number of users in the system, other line voltage standards are used. These require smaller voltage swings (thus increasing the potential data rates) and support more than one receiver on the line. A multidrop standard allows more than one transmitter, although only one can be active at a time. Special line drivers and receivers are used for these faster and longer distance standards.

Summary Questions

1. Explain where RS-232 is precise, where it allows a range of choices, and where it makes no restrictions. What are a DTE and a DCE?
2. What are the RS-232 voltage spans for mark and space? Which represents a binary 0 data bit? Which represents an asserted control line?
3. What are the four functional groups of RS-232 signal lines? What is represented by each?
4. Why does a DCE receive data on a line called transmitted data?
5. What are the four key handshake lines and their roles? What is break detect? How and why is it used?
6. What is the normal interaction of these handshake lines in communication?
7. Compare use of handshake lines with software handshaking in terms of number of signal wires, how each is done, benefits, and limitations.
8. What are the advantages of buffering in communications links? Where is it a necessity?
9. Why is using a UART much better than using a microprocessor directly to control an RS-232 communications port?
10. What does a UART do? How is it set up? What can be set? How does it indicate back to the system microprocessor what it is doing?
11. Explain parity, framing, and overrun errors. How are they handled?
12. What is the role of line driver and receiver ICs? Why not use the UART pins directly to interface to the communications link? What are the advantages of separating the UART functions from the line driving and receiving functions?
13. What happens when two identical RS-232 devices are connected to each other? How are the problems resolved?
14. What is the logical approach to RS-232 interface troubleshooting?
15. What role does the breakout box play? What does the line monitor do beyond the breakout box? Why is this useful?
16. What function does a modem perform? Why is this needed? Why are there many different types of modems?
17. Compare answer and originate modes in modems. Why are these modes needed? When are they not needed?
18. How do advanced modems improve performance with error detection and error detection plus correction codes?
19. Contrast the speed, modulation, and operation of the Bell 103 type of modem to the Bell 212 type of modem. How does each take data bits and present them to the communications line?
20. What are the performance limitations of RS-232? Why are these not a problem in low-speed, asynchronous systems?
21. Compare RS-232, −422, −423, and −485 in terms of signal levels, speed, distance, and number of users. Why are reduced voltage levels better for achieving higher speed? What is the potential drawback to these lower levels?

PRACTICE PROBLEMS

Section 17.2

1. What is the binary pattern for transmitting the digit 5 in seven-bit ASCII, with no parity and one stop bit? How long does this take at 1200 baud? At 19,200 baud?

2. What is the bit pattern for the symbol # in eight-bit ASCII with even parity and two stop bits? How long does this require to send at 9600 baud?

3. The number 15 is sent in eight-bit ASCII, even parity, one stop bit. What is the bit pattern? What is the pattern when 15 is sent as a single eight-bit binary number (LSB first) without use of ASCII? What percentage of bits (and time) is saved?

Section 17.3

1. What bit pattern to the command register sets the 8251 UART to 1× operation, eight bits per character, odd parity, and two stop bits?

2. For a master clock running at 614.4 kHz, what should be written to the command register for operation at 9600 baud, seven bits, even parity, and one stop bit?

3. What baud values are obtainable with a master clock at 38.4 kHz? What about when the clock to the UART is at 19,200 Hz?

4. What is the status byte when the transmit and receive buffers are ready, the data set ready line is asserted, and there are no errors detected by the UART?

5. How does the UART status byte change when a character is received but with framing and overrun errors?

Section 17.5

1. What is the tone pattern when a modem in originate mode is using 1000/2000-Hz originate frequencies and 3000/4000-Hz answer frequencies, and the letter B is sent in ASCII, eight-bit format, no parity, and one stop bit?

2. Repeat problem 1 when the modem is in answer mode.

3. Two modems are connected together. The originate modem is sending the message HI while the answer modem is responding with YES. Both use seven-bit ASCII, odd parity, one stop bit, and 1000/2000- and 3000/4000-Hz tone pairs. What frequency sequence is observed in each direction?

Section 17.6

1. What are the tones sent by the Bell 103, in originate mode, for the letter Q in ASCII with eight-bit format, no parity, and one stop bit?

2. Repeat for the case of odd parity.

3. What tones are sent in problem 1 when the modem is in answer mode but the format is two stop bits?

4. What is the phase-shift pattern for the Bell 212 modem for the letter Q as defined in problem 1?

5. Repeat problem 4 for the case of even parity.

18
Networks

CHAPTER OBJECTIVES

When you have completed this chapter, you will understand:

- Why a network is different from a fixed connection or path between two users
- The various choices for network topology, protocol, and access, with the benefits and drawbacks to each
- A variety of specific networks in use today
- Networks that cover small areas and wide physical areas, and how different networks may be interconnected to each other
- The characteristics of an inherently all-digit network called ISDN

INTRODUCTION

A network is a system that allows a group of users to communicate with each other. Behind this simple statement is a complex system that must have proper interconnection paths, appropriate physical connection of each user to the overall system, rules for handling messages and various situations, and a specific message format that all users agree to use. Standards define many, but not all, of the characteristics of the network operation and must be agreed and adhered to by all users if the network is to succeed.

Networks are used in industrial plants, offices, factories, scientific laboratories, and many other applications. The network can be local—within a section of a building—or spread over miles and even the world. Each network application has different requirements in terms of message speed, acceptable cost for interconnection, and ability to add new users. In this chapter we look at the basic structures of networks, the interaction among the many users, and the characteristics specific to networks in various applications. We also look at some specific networks in use today and the integrated circuits that support them. Finally, we look at a new type of all-digital telephone-based network, the integrated services digital network, which is being defined and installed now.

18.1
NETWORK APPLICATIONS

A network is a communications system that interconnects many users and is designed to let any user send messages to any and all other users on a common set of communication links. In contrast, a direct point-to-point system has only one user at each end, although there are often many intermediate links between the two users.

Network applications include offices, linking various personal computers to each other and to a larger computer; interconnecting larger computers located in different buildings or cities to let them share information, basic data, and results; and sending data collected at many locations to a single computer that integrates all the information. Here are some specific instances of networked communications:

1. An airline reservation system links together terminals that are spread across the country and the world. Whenever a reservation is needed, the list of available flights and seats maintained at a central computer is checked and then updated. This ensures that two people cannot be booked for the same seat, while reservation lists can be checked for multiple bookings by the same person.

2. A small office with 10 computers shares a single, high-speed printer. Any one of these computers can use the printer if it is available; if the printer is already in use, there is no undesired interference and the network automatically puts a would-be second user into a *queue* (an ordered waiting line). Memos and documents generated by these individual computers can be accessed electronically by users at other computers, if needed.

3. Tracking stations for space satellites are located worldwide, so that the satellites can be followed regardless of rotation of the earth with respect to the satellite. Each station receives general instructions (specifying which satellite to track, and when), acquires and performs initial data analysis on it, and then sends the data to the central point for final analysis.

4. A factory that manufacturers electric motors links together the many production machines. The central computer sends directions to the various machines on how many of each model to make. In turn, the machines report back what they have done. The central computer matches the production results with raw material inventory and incoming orders to see what has to be ordered from suppliers.

In each of these examples, the network users share a common need or goal and are working on various related aspects of a single project or system. They have a need to share information, either frequently or occasionally, with some or all other members of the network. For these reasons, a network is much more than a system that allows any user to send information to any other user. It must allow for constant interaction among the users, a carefully established standard for achieved performance, and a method to make sure that no foreseeable problem, including physical failure, power off, or noise-induced errors, can cause the entire network to fail.

The telephone system is often described as a network. It is a network in the sense that any user can pick up a phone (or computer) and contact any other user. But it is not a network in the sense that we are using in this chapter or in the standard industry definition. The many users who can access the phone system do not have a common project or goal that links them all together. The information that one pair of users pass between them is usually unrelated to the information between another pair of users: The conversations on different lines are generally independent of each other.

Interestingly, the telephone system is a network in another way, but unseen to the user. The computers and equipment that manage the central offices, trunks, and electronic switching systems make sure that information about activities and events at each point is available, as needed, to other parts of the system so that the overall telephone system operates reliably and efficiently. Of course, the telephone system can be a key part of a network that does link users who have common needs and goals.

An interface standard has four major elements: the mechanical portion, which defines the physical cable connection; the electrical part, which specifies the voltages and currents into and out of the nodes; the functional element, which details the interface signals that exist on the physical connection; and the operational portion, which discusses what messages are used.

Because a network links many users who must share information in a timely and reliable way, a network requires more than an interface standard such as RS-232. The interface specified by RS-232 (or similar standards) defines the physical interface: signal voltages, timing, and basic handshaking. This is an important and necessary first step to building an overall network, but much more is needed.

First, there is the physical interconnection schematic for all the users. This *topology* shows the path, or paths, between all the users. Many networks use a single path that is common to all users, and no other user can send a message when one user is already transmitting. Of course, one or more can be listening, if they need to. Some networks use many physical paths, so that more than one message can be passed at the same time, but this is quite expensive and complex.

Here again, this is a difference between the telephone system as a network and a conventional communications network. Many complete paths exist at any time in the telephone system, and the specific paths (connections) are changed as different calls are made. In contrast, a communications network usually uses one major path, and the users share this as needed. The physical connections do not change and are not switched, but the use of the communications link does vary with the network needs.

A network needs a *format*, or common language, for messages (Chapter 12). To make sure that each user can understand others as well as be understood by others, the way in which a complete message is built from individual characters is specified. This is one step above the specific bit pattern code used to represent the characters themselves (ASCII or similar). The format specifies how to begin a message, how to address it, how the information of the message must be presented, and how to indicate the end of the message. The network rules ensure that all users follow the same format and establishes a common language for all.

The format carried within each user does not have to follow the network standard, and usually does not. When a device must pass information among its own subsections, it often does not need the more elaborate format of the network. Also, the network standard may be inefficient for the device. For example, the computer-controlled machine tool for cutting metal needs many short commands to guide the cutting head and determine the exact status, at every instant, of the head. The summary information that is sent on the network, in contrast, describes how many pieces were cut successfully, and how many the machine tool could not complete properly, together with the reasons. Summary information can use longer messages and does not have to be as timely as the message used to control the actual operation.

A network *protocol* defines the rules of the conversation—the message passing—among the network users. Without a carefully defined protocol, users will conflict with each other as they try to send at the same time, and there will be confusion when unavoidable errors and equipment problems occur. A protocol sets up the rules to follow under any and all circumstances in the network, using a *state diagram* (Chapter 12), which shows all possible states of the network, what events can occur, what action to take for each circumstance, and the state that results from this action.

Local and Wide Area Networks

A network that connects users who are in the same general location is called a *local area network (LAN)*. A LAN can cover one corner of a building, a single building, or a cluster of buildings spread over a few miles. Beyond this distance, there are technical reasons that prevent a single LAN from doing the job reliably or at a low enough cost. A *wide area network (WAN)* is needed for areas that are spread out geographically. In some cases, the wide area network is built of smaller LANs that are closely linked, or made of mixed combinations of LANs and special longer-distance links.

Most LANs use a communications link—either wire, fiber optic link, or radio—that is installed specifically for that LAN and is not used for any other function. In contrast, a wide area network often uses a combination of private communication links combined with some other shared link, to avoid the expense of installing a long-distance private path. The other users of the shared link may be other networks, but they can be nonnetwork users who simply need an available path from one point to another. The way the physical path is connected among the many users has a major impact on network reliability, cost, efficiency, and throughput.

There is no single network topology, protocol, or format that is ideal for all situations. The best choice depends on the specific application, data speeds required, reliability, cost, and what other existing facilities are already in place. The different network specifics in topology, protocol, or format that we look at in the following sections are all used today in standard commercial systems and meet the needs of users with specific requirements. Other topologies, protocols, or formats might be poor choices, or more awkward ones, for the specific application.

Review Questions

1. Contrast a network with a point-to-point connection between two users.
2. What distinguishes a network, as used in this chapter, from the telephone network? In what way(s) is the telephone system a network?

3. Why does a network require more definition than just a specification such as RS-232?

4. What is topology? Protocol? Format? Code?

5. How can a network user have an internal format that differs from the network format? Why is this desirable in many cases?

6. What is a local area network? A wide area network? Compare their distances, links, and speed.

18.2 TOPOLOGIES

A network *topology* is a physical schematic that shows interconnection of the many users. There are four basic topologies: one to all, star, bus, and ring. Each has very different characteristics in amount of required cabling or paths, expansion potential, ease of management via its protocol, flexibility for sending messages, and reliability in case of problems. Some advanced networks use combinations of these topologies to provide the best overall performance. The point at which each user is connected to the network is called a *node*. Topology does not say anything about the type of physical link used between nodes: It can be coaxial cable, twisted pair, optical fiber, radio, or another type of media. It does not have to be the same between every node, but it usually is.

In the *one-to-all* topology (Figure 18.1) there is a path between every node and every other node. Although this is simple in concept, it is a very impractical scheme except when there are only a few nodes, up to a maximum of four in typical practice. The number of paths needed is defined by the equation $P = (n^2 - n)/2$, so a three-node system needs three paths, a four-node system has six, and a five-node system has 10. The number of paths required increases dramatically as the number of nodes increases. Each node, in addition, must be physically able to connect to paths to every other node, so a very large connector block and associated interface support circuitry is needed at each node. Adding a new node requires major rewiring and affects the software that manages communications for the system.

This topology does offer some advantages. First, the protocol is simple, since there is no chance of collisions between two transmissions on a link between nodes even when many users try to transmit at once. There is also reliability in the

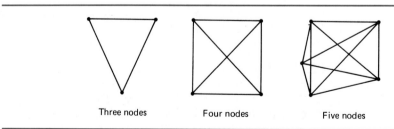

Figure 18.1 One-to-all network topology for three, four, and five nodes.

18.2 Topologies

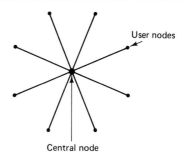

Figure 18.2 Star network topology has a central node.

system, since the loss of any single path affects only the two users who share it, and a problem at a single node (equipment or power failure) leaves the other nodes able to communicate completely. Still, the realities of this scheme make it impractical for most networks in use.

A practical improvement over the one-to-all topology is the *star* network (Figure 18.2). In this approach, all user nodes are connected to a central node point that interconnects all the individual user links and nodes. Data flows from one node, through the star center, to the desired receiving node. The central node is like a large switch that routes the data from the input line to the output line. More advanced central nodes can have multiple switching paths, so several paths exist at the same time. This topology is used in a telephone system central office (exchange) and it offers several advantages over the one-to-all; most obvious, it is practical when 10,000 nodes have to be connected.

Other features are: Adding a new user node requires running only a single link from the star center to that user, and does not disturb any other nodes; any problem with a single link affects only the node connected to the link; and it is easy to isolate users for test purposes as needed simply by disconnecting them at the star center. This can be done entirely from the star center, so the test technician does not have to travel from node to node to connect and disconnect users.

However, the star network has a major weakness. Although adding a new user is simple in concept, it does require running a completely new and separate link, which can be costly. This is especially aggravating when there is already another user node nearby yet the star requires that a completely new link be run alongside the existing one. There is no possibility of sharing of the physical link. Another weakness is that the central node is critical; if it fails, the entire network is out of service. Despite these drawbacks, the star topology is very useful when the distances are small (such as within an office) since it is relatively easy to expand to new users or move existing nodes to new office cubicles without disturbing other nodes.

One network topology that shares a single link or pathway among all users is the *bus* (Figure 18.3). In this topology, the link serves as a highway for all data signals, and users connect onto the bus at their node location. The bus is really an expansion of a multidrop standard such as the RS-485, studied in Chapter 17. Users can be added anywhere along the length of the bus by tapping into it, as long as the bus is nearby.

The reliability of the bus topology depends on how carefully the design is

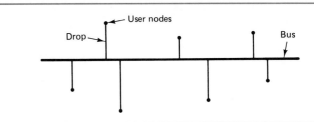

Figure 18.3 Bus network topology has a single path for all nodes to share.

done. If a single user fails, the bus normally can continue to function—unless the failure is the type that short-circuits or "locks up" the bus. To guard against this, most designs are very reliable, very rugged components such as transformers or optoisolators between the equipment at each node and the bus. In addition, *watchdog* or *timeout* circuitry within each device independently monitors the user device status. The device is designed normally to reset the watchdog circuitry frequently and periodically, say, every 100 ms. If this watchdog reset does not occur, it means that a failure or hangup of some sort has occurred in the node device. The watchdog circuitry is therefore not reset, and the watchdog automatically disconnects the device from the bus so that the bus itself is not disabled by this one failed device.

While the bus structure efficiently uses a single path (which saves significantly on cabling or channel costs), it does require more management than the star. In the star, the protocol is a minimum since the star center makes sure that the proper connections are made and that no collisions occur. In the bus there is no central management function. Therefore, each node must use some protocol to make sure that every user gets a turn, to ensure that messages are completed, and to handle the inevitable conflicts or errors that occur. This protocol uses some of the processing resource within each node, as well as occupying some of the bus time, which results in decreased efficiency and throughput.

The bus topology is very popular for communications systems networks; it is even used within a single chassis to interconnect various PC boards (i.e., processor, memory, interface, peripheral device). It provides a good combination of simple structure, ease of interconnection, and not too complex protocols in a network. The bus can be multiple wires or paths that connect in parallel, which is practical for short distances only (such as within a chassis or between subsystems).

The last major topology for networks is the *ring* (Figure 18.4). All user nodes are connected with the physical path acting as the links of a chain, and the last user node is connected back to the first node. A signal going on to the next node must be processed by the first node, which then passes it through to the next node. Adding a new user requires breaking the ring temporarily, inserting the new node, and then reestablishing the complete ring path, which is sometimes inconvenient or not acceptable. The ring topology, however, is useful for fiber optic network links, where the fiber acts as the point-to-point link between nodes and adding a new node does not require adding a new tap into the fiber optical cable. With present technology, splicing into a fiber optical cable is complex and difficult (although this is an advantage in terms of system security), whereas plugging in a new extension link is simple. In contrast, the bus topology requires that taps

18.2 Topologies

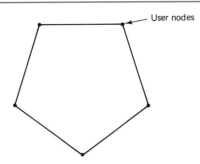

Figure 18.4 Ring topology forces messages to pass through each node.

(sometimes called *drops*) be used to connect to the main bus, which is simple with wire cable but awkward with fiber.

Rings also have simpler protocols than buses. In a bus configuration, each node must monitor the bus constantly to see if there are any messages for it or to see if the bus is clear before transmitting. (We will see these in detail in the next section.) The ring uses a simpler scheme: Each node accepts the message, processes it, extracts any pertinent data, modifies the message if needed, and then passes it on to the next node. This is simpler to implement and has more predictable performance.

There is an obvious drawback to the ring: The failure of any node effectively shuts the entire network down. Dual, redundant paths are used in critical applications, and watchdog circuitry at each node monitors the node performance. When the circuitry detects a problem with the node, it sets a switch which electrically bypasses that node and directs all input signals directly to the output and onto the next link in the network. (Note that in the bus, a failed node is disconnected, whereas in the ring it is bypassed.)

Review Questions

1. What is a node? How are nodes interconnected? Do all these interconnecting links have to be the same?
2. What is one-to-all topology? What are its key features, advantages, and disadvantages?
3. What is star topology? What is the role of the central node?
4. What are the pros and cons of the star?
5. What is a bus topology? What is the amount of interconnection path needed? How are users added or dropped?
6. What happens in the bus if a node fails? How does a watchdog operate to minimize the effect of a node failure on the bus network?
7. Contrast the management of the bus versus the star. Why is the bus protocol more complex?
8. Contrast the bus and the ring topologies and protocols. Why is the ring potentially simpler?
9. Why is a single-node failure in a ring a major problem? How is the problem solved?

18.3 PROTOCOLS AND ACCESS

The protocol defines the rules by which any node can access the network. In this section we examine four common protocols: command/response, interrupt-driven, token passing, and collision detect. Each of these can be used with the different topologies, although there are some topology and protocol combinations that work better than others.

Command/Response

A *command/response* protocol uses one node as the *master* and all other nodes are *slaves* (for this reason, it is also known as a *master/slave* protocol). All communication begins when the master sends an addressed message and the addressed slave responds to the master (Figure 18.5a). When a slave node has information that it wants to send to the master, it must wait for a command from the master saying: "It's your turn, what do you have?" The slave responds with its message—either new information or a message indicating nothing to report—at that time. The master can send messages to the many slave nodes in any sequence, or address the same slave node several times in a row. Most protocols require that a slave, upon receiving a message, provide a return receipt message to the master to confirm that the message was received properly.

The situation is more complicated when a slave node needs to pass a message to another slave node (Figure 18.5b). First, the originating slave node must be addressed by the master so that it can transmit its message to the master. The master then addresses the target slave node, sends the message it received from the originating slave, and awaits a confirmation of message received.

From this sequence we see that sending a message from a master to a slave requires two transmissions (one from the master to the slave, and one in the reverse direction confirming receipt) and occurs when the master dictates. Information from a slave to the master also requires two messages (one in each direction), but the slave cannot control the originating time of the message. Slaves can only respond to commands from the master, which accounts for the name of this protocol. Finally, a slave-to-slave message takes four complete messages (two in each direction).

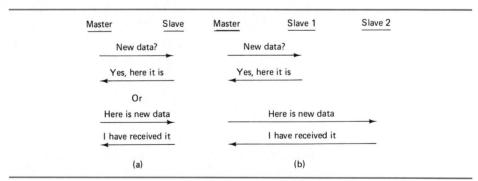

Figure 18.5 (a) Master-to-slave message protocol; (b) slave-to-slave message protocol.

information to a single, high-speed printer, a wait of a few seconds more is not a problem and CSMA/CD is a good choice. However, if the network links together factory production equipment to machine an auto engine block, maximum delays are critical and token passing is a better selection. It is slower on average but never has the unpredictable and sometimes very long message delay time of CSMA/CD.

Review Questions

1. What is command/response protocol? Where is it useful?
2. How is a message passed from slave to slave? How many messages are sent overall?
3. Why is command/response protocol simple and easy to follow? What are its drawbacks?
4. What is polling? What is its effect on the time for a slave to get a turn?
5. What is an interrupt-driven protocol? How does it improve on command/response?
6. What is token passing? How does it allow peer-to-peer communications? What fills the role of the master?
7. What is the advantage of continuous passing of the token in terms of longest possible wait for a turn?
8. What is the advantage of token passing? The disadvantage?
9. What is the basic idea of collision detect protocol? What causes collisions if each node checks the network before transmitting? How are collisions resolved? How is the back-off waiting time set?
10. How long is the wait when network traffic is low? When it is high? Why does performance change with network traffic levels? Why can't collision detect guarantee the maximum wait for user nodes?

18.4 NETWORK EXAMPLES

No single network meets the needs of all user applications. Different requirements in speed, cost, ability to handle smaller and larger volumes of traffic, expansion to accommodate new users, adequate performance despite errors, and predictability of overall performance affect which network is a good choice in the particular situation.

Just as in other communications activities, standards of interface are needed to ensure that all equipment connected to the network is compatible. This is especially true when the equipment is made by different manufacturers and may not be initially compatible. Network standards are either *open,* which means that a published standard available to all is used, or *proprietary,* where one manufacturer develops the standard and may offer it to others who agree to follow strict conditions on its use. Open standards are developed by industry groups comprised of manufacturers of equipment, users of the system, technical experts, and other interested parties. The final standard takes 5 to 10 years to develop under the guidance of a technical society such as the Institute of Electrical and Electronic Engineers (IEEE), a nonprofit organization devoted to advancing electronic technology. Among the many network standards they have published are:

IEEE 802.3 CSMA/CD for baseband and broadband systems

IEEE 802.4 token passing for baseband and broadband bus

IEEE 802.5 token passing for baseband rings

(*Note: Baseband* means that no carrier is modulated; *broadband* means that the data signals modulate a higher-frequency carrier.)

Note that a standard does not have to define every aspect of the network. Some standards let the user select the medium (the physical link) that provides the necessary bandwidth, data rate, and cost in the specific application, as well as the exact format of the messages between nodes. Other standards call out a specific medium and modulation scheme or have subsections for each of the different allowed variations; some standards do not specify these at all. Finally, some standards are very specific in format and protocol, whereas others provide just general guidelines.

In this section we look at four local area network standards, each designed primarily for a specific group of applications. We also look at some of the interface circuitry and ICs used for these networks.

AppleTalk

AppleTalk (also called LocalTalk) is a LAN designed to interconnect devices and computers in the Apple Macintosh family of personal computers and products. Most of the applications are in office environments, where a single high-speed laser printer is typically shared by many separate computers. Also, individual computer nodes may want to pass data files to each other (for example, a copy of a memo), so that there is a need to send messages from any node to any other node, in addition to all nodes sending to the printer node.

Low cost and simplicity are important for this network, since users of low-cost computers do not want the network node connection to cost as much as the computer itself. Fortunately, the LAN has relatively low to medium traffic, waiting times of several seconds are acceptable, and any errors due to noise can be handled by a simple scheme that detects an error and requests retransmission.

The topology of the AppleTalk network is a bus with up to 32 nodes maximum. Data is sent in serial format, with a single shielded twisted-wire pair as the medium. The wire is specified to have dc resistance of 17 Ω/300 m, a characteristic impedance of 78 Ω, and a capacitance of 68 pF/m. Some companies besides Apple offer fiber optic cables to replace the wire for longer distance and more noise resistance, and this change is transparent to the network nodes. A variation of the RS-422 standard is used for signal levels at 230,400 bits/s. The modulation is frequency shift keying (FSK), with frequency modulation between two values, and a maximum distance of 300 m (approximately 1000 ft) is guaranteed.

The most complicated portion of the AppleTalk standard is the message frame format, called *synchronous data-link control (SDLC)*. This format and its supporting protocol use a frame (Figure 18.8) that begins with an eight-bit synchronization field to indicate the frame starting point and to help the receiver

18.4 Network Examples

Sync	Address field	Control field	User bits	Error checking field	End field
01111110	8 bits	8 bits	Any number of bits	16 bits	01111110

Figure 18.8 SDLC frame shows format for key fields.

synchronize with the data bits. Following the sync field is an eight-bit address field for the receiving node address. An eight-bit control field lets the sending node indicate what kind of message this is: one of several special startup messages; a supervisory message which sets up the communications between the sending and receiving nodes and controls the flow of data on the network; or the actual information message itself.

The message is not restricted to a specific number of bits but can be anywhere from one to several thousand bits long. SDLC is a *bit-oriented* protocol, because SDLC considers the information it transmits as a single long string of bits. To the user, these bits can be a series of ASCII characters connected together, or eight-bit bytes of numerical values, or any other representation. The SDLC frame concludes with a 16-bit field for an error-checking code (cyclic redundancy codes, Chapter 12) and an end field that looks just like the sync field and terminates the frame.

Since low cost and ease of interconnection are important requirements in the AppleTalk network, the physical interconnection from the computer (or printer) to the network is made through a small-signal transformer at each node (Figure 18.9). This couples energy from the equipment to the cable, while providing isolation from miswiring (or connection to the power line!) and resultant damage, and makes it relatively easy to add or remove nodes on the bus. Transformers are very reliable and inexpensive; an AppleTalk node costs less than $50, and cabling is about 10 cents a foot.

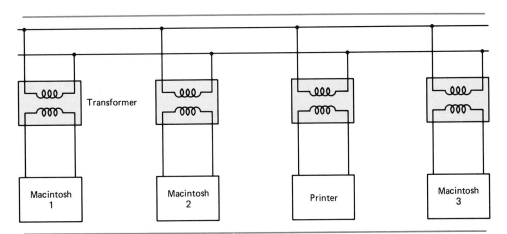

Figure 18.9 AppleTalk connection and transformer for each node and common bus.

Manufacturing Automation Protocol

Manufacturing Automation Protocol (MAP) is a token-passing bus standard (IEEE 802.4) intended to link together the various types of machinery used in a factory production area, such as presses, lathes, conveyors, and drills. It operates at either 5 or 10 Mbits/s, with FSK modulation. The digital signals control the carrier directly: When the data bit is a 0, one cycle of the carrier is sent in the bit time period; when the carrier is a 1, two complete cycles are sent (Figure 18.10). Therefore, the carrier is either 5 or 10 MHz for a 5-Mbits/s data rate, or correspondingly 10 or 20 MHz for the 10-Mbit/s rate. Token passing is used here because the maximum response time is critical in manufacturing and machinery operations, and the difficulty in guaranteeing the longest wait with a collision detect protocol is not acceptable for this application. Because factories are electrically noisy environments, the MAP specification requires that the receiver provide bit error rates of less than 1 in 10^9 with SNRs as low as 20 dB.

Ethernet

Ethernet (IEEE 802.3) is designed to link together a large number of users in office and engineering environments, especially those users who have large amounts of data to transfer among each other. For example, engineers using computers to do mechanical and electrical design need to send large volumes of information to drawing plotters, layout equipment, and machine tools for prototype fabrication.

Ethernet uses collision detect protocol in a bus topology and supports up to 1024 users on the bus. The bus medium is not specified but it is usually coaxial cable to support the maximum 10-Mbit/s data rate, although fiber optics can be used as well. Without signal repeaters, the maximum distance between any two nodes on the Ethernet bus is 500 m (1500 ft).

ICs are available to support the Ethernet protocol and provide the necessary electrical interface as well as bus management. Intel Corp. offers a set of ICs for the complete Ethernet interface, beginning with an 82586 LAN *coprocessor* (Figure 18.11). This coprocessor is a microprocessor specifically designed as an intelligent peripheral IC that manages the transmitting and receiving message frames, and so significantly reduces the communications burden on the system microprocessor. The 82586 sends and receives data, keeps track of message numbers, frames the outgoing data and strips framing from received data, and detects and identifies error conditions. It also sends test messages, if necessary, over the network and analyzes their results.

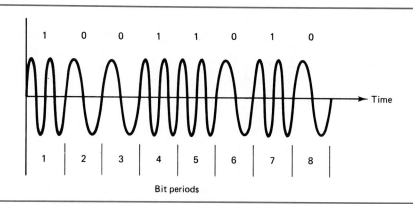

Figure 18.10 MAP modulation scheme.

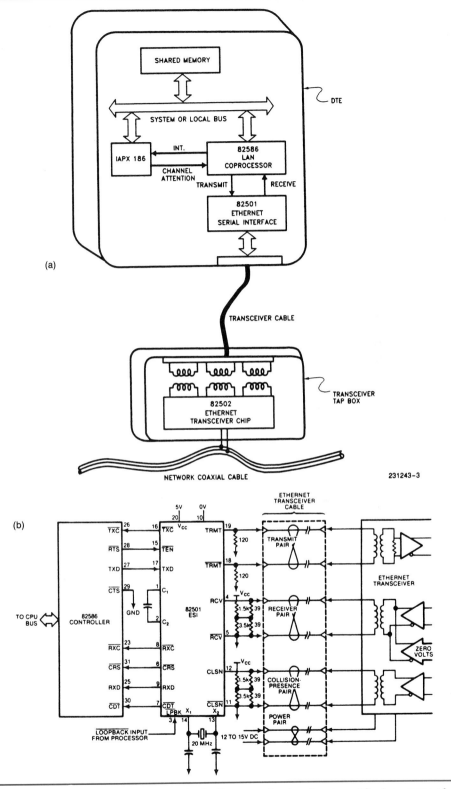

Figure 18.11 Ethernet ICs: (a) block diagram of role of various ICs (courtesy of Intel Corp.); (b) interconnection of ICs and bus (courtesy of EDN, *Build a VLSI workstation for the Ethernet Environment*, Cahners Publishing Co., Feb. 23, 1984).

The Intel 82501 Ethernet serial interface, which connects between the coprocessor and the Ethernet cable connection, is designed to work in conjunction with this LAN coprocessor. The 82501 provides a crystal oscillator for the Ethernet 10-MHz clock circuit, performs Manchester encoding and decoding which is required by the Ethernet specification (Chapter 12), and provides the electrical interface to the Ethernet line drivers and receivers. Also in the 82501 is a watchdog timer which independently monitors and checks if any transmission from the IC lasts more than the maximum allowed, meaning that for some reason the IC is stuck in transmit mode and locking up the bus. When the watchdog detects this condition, it issues a reset signal that clears and restarts the 82501.

Three signals interconnect to the Ethernet cable from this IC: transmit data, receive data, and collision detect (which indicates that a collision has occurred) and each uses one pair of wires. The signals are at the voltage level specified by the standard, and the line drivers/receivers within the IC are designed to be short-circuit-proof. The physical connection to the Ethernet cable is via transformer coupling for isolation and ruggedness.

IEEE-488

IEEE-488, also known as the *General Purpose Interface Bus (GPIB),* was defined in 1975 as a network standard for interconnection of engineering test instruments. Prior to this standard, test equipment instruments operated independently of each other. In such a typical test setup, a power supply is connected to the unit under test, while voltmeters, oscilloscopes, and frequency sources are connected as needed to key test points. The test itself is run by a technician or engineer, who adjusts signal levels and frequencies, measures the results, and writes them down. If a computer is used to analyze these results, this is done separately from the task of running the test and gathering the data itself.

As test instruments developed more capability using their own internal microprocessors to enable them to perform complex tests more easily and even process some of the test results, it became obvious that one thing was missing: a way to link all these instruments together with a central controller. The IEEE-488 bus interface standard is designed to fill this need. Any test equipment or instrument with this interface is designed to accept commands via the bus and to transmit results onto the bus as well.

When testing an audio amplifier with IEEE-488-compatible instruments (Figure 18.12), a personal computer acts as the controller for a test that requires producing a wide span of signal frequencies and amplitudes from a sine-wave generator, measuring the voltage at selected points in the amplifier, and measuring the distortion of the final output with a distortion meter. Although each instrument—sine-wave source, voltmeter, and distortion analyzer—can be operated manually from its front-panel controls, the entire test can be run automatically via the test controller and IEEE-488 interfaces on the various instruments. The controller tells the sine-wave generator exactly what frequencies and amplitudes to generate, tells the voltmeter to make a reading and report it back, and tells the distortion analyzer to make its readings and report those as well. When the test is complete, the controller does the necessary calculations, produces the final output graph, and stores the raw data and summary results on its internal disk.

The IEEE-488 standard is very specific in some areas and deliberately nonspecific in others. It allows up to 15 devices on the bus, each with its own device number (usually set by switches in the instrument itself); this allows two identical

18.4 Network Examples

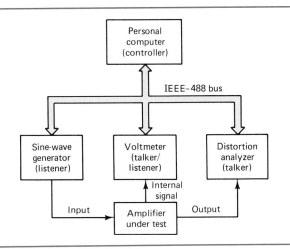

Figure 18.12 IEEE-488 bus set up for testing an amplifier with various instruments.

voltmeters to be used at the same time, just by assigning different addresses. Signals are transmitted at baseband (no carrier) with a maximum rate of 1 Mbit/s. The total bus length allowed is 20 m (60 ft), and the cable has 24 wires: eight for data (sent as eight bits in parallel, a single byte), eight for control and handshake, and eight for ground. The cable and instrument connectors are specified precisely so that connector mismatch never occurs.

Any IEEE-488 instrument is categorized as a *listener*, *talker*, or *controller*. A *controller* sends messages and receives them, and there can normally be only one controller for the system. A *listener* device can only receive messages, while a *talker* can only send them. Devices can be talker/listeners and so both send and receive. An example of listener-only instruments are power supplies, which get instructions on the desired output voltage or limits: a talker-only might be a voltmeter which reports voltages but accepts no instructions; an example of a talker/listener is a voltmeter that accepts instructions to change range and reports voltage readings. Most newer instruments are talker/listeners, because this mode provides maximum flexibility: Instruments can be remotely set so that there is no need for someone to use the front-panel controls.

Both command/response with polling and interrupt-driven protocols are allowed under the IEEE-488 specification. The entire specification comprises over 100 pages of detailed rules and state diagrams (Chapter 12) defining the allowed modes and how instruments should respond to instructions, together with the protocol for every conceivable situation. Any attempt to program a microprocessor to follow all these rules would be very time consuming, take significant amounts of memory, and probably have many errors (software bugs).

When the IEEE-488 specification was being defined, IC manufacturers began the design of components to support the standard. The TMS9914A from Texas Instruments is one such IC (Figure 18.13). Like the UART (Chapter 17), it communicates to its system microprocessor via command and status registers, and has internal data registers for data to transmit and for received data. It also implements, within its circuitry, all of the state diagrams and rules of the IEEE-488 specification protocol.

580   Chap. 18   Networks

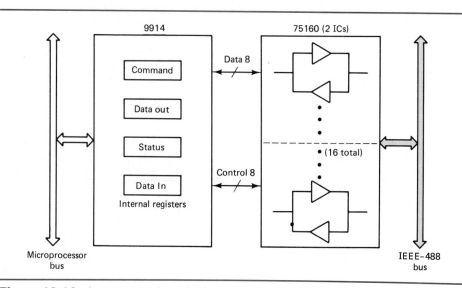

Figure 18.13 Interconnection of microprocessor bus, IEEE-488 TMS 9914A IC, line driver/receiver, and IEEE-488 bus.

Whenever an event occurs on the IEEE-488 bus, such as the controller receiving a talker request to send new data, the IC uses the state diagram rules to see what it should do next, depending on where it presently is. Figure 18.14 shows one such situation. The TMS9914A is either in the busy state or in the available state. When it receives a new data request, it takes the path dictated by the state diagram. If it is busy, it sends a message via the appropriate control line to show that it cannot accept new data; if it is available, it indicates that it can take the data. If the request does not have the proper timing, or has an error, the TMS9914A sends an error indication on the bus so that the requester knows there is a problem.

Special line drivers and receivers are needed for IEEE-488. These have the same function as RS-232 line drivers but meet different voltage and performance

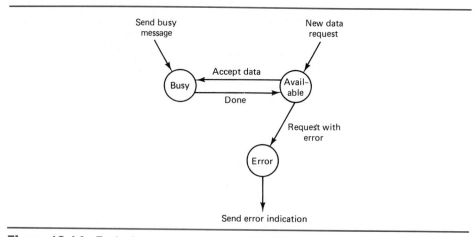

Figure 18.14 Typical state diagram for IEEE-488 operation.

IEEE-488 Format

While IEEE-488 is very specific about the physical interface and the protocol, it places virtually no restrictions on the message format itself. For example, a voltmeter from manufacturer A may send its data back to the controller using ASCII characters, in the format "10.93 V" whereas a voltmeter from another manufacturer uses a different format: "V = 10.93 #" (the # is a message terminator in this case). Both are allowed under the standard, but this is an area of great potential for confusion and errors. The person programming the IEEE-488 controller must know the specific format that each device uses and must read device manuals and instructions very carefully. There are some suggested format guidelines, but they do not have to be followed. IEEE-488 does not specify the format within the protocol (to allow maximum future flexibility since so many different types of electronic instrumentation can be connected to it), but this is sometimes a source of problems in operation. The problem often shows up only in special circumstances or error modes, when a new message appears that was not anticipated, in a format that is not the one in normal use. Because of this confusion, the IEEE is now developing format standards as well.

specifications. The Texas Instruments 75160 IC has eight bidirectional buffers (each buffer is a combined line driver and receiver) in a single package, and just two such ICs are needed to interconnect the TMS9914A to the IEEE-488 cable.

Test Equipment

Conventional test equipment such as voltmeters and oscilloscopes have very limited usefulness in troubleshooting networks. Most network problems do not involve the signal levels or shapes, but instead are due to different formats (where allowed) or subtle interaction in the give and take of the protocol. Specialized protocol analyzers are used to perform the following functions:

Monitor the bus while remaining passive and invisible, to show the messages on the bus, who sent them, and to whom. The information is displayed on the analyzer screen.

Show the responses to the messages and indicate messages that failed to get the required response.

Analyze the overall data traffic on the network to see the response time. Many network problems are not stated simply as "does not work" but involve slow response time, occasional errors or no response, or other symptoms that are not easily defined. Sometimes this is due to heavy overall traffic, or it could be due to noise that corrupts data and automatically causes repeated retransmission under the rules of the protocol.

Finally, the analyzer can become an active node on the network. It can generate single test messages and accept responses, or generate varying amounts of data to see how the network performs as the traffic level changes. It can even produce messages with deliberate bit errors, format errors, or protocol errors, to see if the network is handling these various error situations properly.

These network analyzers are made by many different manufacturers. Some analyzers are designed for one network standard only, such as MAP or Ethernet.

Others are more expensive but can be set to accommodate any one of many standards, and have different connection cables and kits for interface to the many types of network cabling and media.

Review Questions

1. What is the difference between an open versus a proprietary network standard? Do standards define all aspects of the network and interconnection? Explain.
2. What is the intended purpose of AppleTalk and what are its essential requirements? What are AppleTalk topology, number of users, and media? How is the user connected to the network at the node? What modulation is used, and what speed?
3. What are the format and protocol for AppleTalk? What does it mean to say that it is "bit-oriented"? What are the message fields?
4. How many nodes does Ethernet support? What are the topology and protocol?
5. Why is a standard such as IEEE-488 needed? How is it used in test situations?
6. How many devices does IEEE-488 support? How is each device uniquely identified?
7. How many wires are in the IEEE-488 bus? What is the maximum distance? What is a controller, a talker, a listener, and a talker/listener? Give examples of each.
8. What are the IEEE protocols and state diagrams? Why are protocol ICs required?
9. What does IEEE-488 say about specific message formats? Why can this be a problem? What is the method of preventing this problem?
10. Why is a protocol analyzer needed for network troubleshooting? What kind of problems does it have to handle? What does the analyzer do?

18.5 WIDE AREA NETWORKS AND PACKET SWITCHING

Unlike a local area network, which serves a small geographic area, a wide area network links nodes that span miles, states, countries, and continents. While a LAN uses a dedicated communications link that is installed specifically for the LAN, most wide area networks do not have this "luxury." Instead, they often use links that are available to others, and do not necessarily have the use of the path exclusively but share it with users who are not part of the network. There are applications that generate enough traffic to justify installing links—wire, fiber optics, radio, or satellite—dedicated to the application; examples include large corporations and the military services. Generally, however, wide area network users share their path and links with others.

18.5 Wide Area Networks and Packet Switching

Since the distances in a wide area network are large, propagation delay becomes an important factor in the overall performance. In addition, changes in the path that is set up for each message mean that there are large variations in this delay from message to message. There can even be differences in the propagation delay within a message when the communications path in use is changed "on the fly," which is done automatically in most cases when noise or errors increase past a certain limit.

For these reasons, wide area networks are not used for time-critical applications such as controlling the position of a motor from 1000 miles away. It would be dangerous and unreliable to do this, or very costly to ensure adequate performance. Instead, these networks are used primarily to pass information, such as how many parts to make, data that result from tests, airline reservation requests, and banking numbers. As long as the information reaches the destination in reasonable time, and without errors, the wide area network serves it purpose.

It is not economical to set up a single link for each user in a wide area network. For example, suppose that 1000 travel agent terminals are connected to a central reservation computer which keeps track of flights, available seats, and airline bookings. Setting up 1000 connections to this computer would be very costly. Instead, the data from each terminal are delivered to the wide area network by local links, similar to local loops of the telephone system. In the wide area network, the many data streams are combined together into a single wider-bandwidth signal, which is then sent over the network path. The combined signal is divided by the network management into groups called *packets*, and each packet is routed by the most economical and reliable path that is available at that instant. One moment it may be a coaxial cable, and the next moment it may be a fiber optic link that suddenly has some unused channels.

In a *packet-switched* network like this, the many packets are routed and handled independently of each other as they travel through the wide area network (Figure 18.15). Each packet has a destination address and a number, showing what position it had in the overall sequence of packets. When the many packets arrive at the destination, the network management system at that node reassembles the packets in correct order, since packets may arrive out of sequence due to the different routes each takes. By allowing the packets to take different routes,

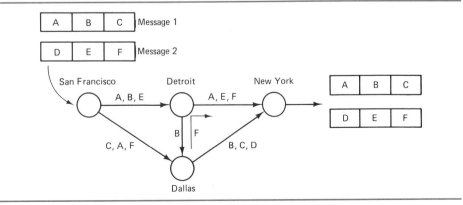

Figure 18.15 Packet switching and routing shows how long messages are split and recombined.

the wide area network can use standard data links despite constant changes in available channels, depending on the traffic levels.

Of course, keeping track of all the packets to make sure that they arrive where they should, and then reassembling them in sequence, is a major task. The wide area network protocols for packet switching are complex so that they can actually perform all these functions despite network link noise, errors, changes in the resources available for carrying messages, and sudden increases and decreases in the amount of data to be sent.

One consequence of a wide area network and packet switching and routing is that the link does not provide instantaneous communication. There can be delays of several seconds or even minutes for the complete message to reach the intended receiver. This occurs because a packet may be temporarily held up (*stored*) at a node along the way to the final destination—because no path is available due to technical problems or too much traffic—and then passed along (*forwarded*). For this reason, networks that switch packets are also called *store-and-forward* systems.

Some packet-switched systems are deliberately designed to hold data for longer periods, to take advantage of times when more links are available and line usage charges are lower. For example, a company that sends messages from its U.S. offices to Europe and Japan can use a wide area network that takes the various time zones into account. There is no point in rushing a message to the receiver if that user's office is closed. A store-and-forward system in this application send these messages at a lower priority, using cheaper communications links with "after hours" pricing. If, however, the message is sent during the business

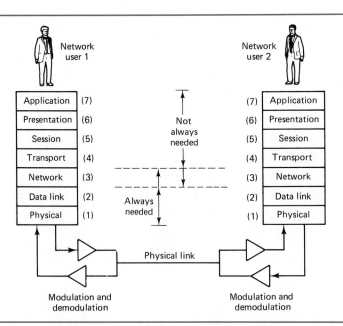

Figure 18.16 The complete OSI reference model for networks includes seven layers, ranging from the Physical layer up to the Application layer, and allows different types of users to share one network. Not all layers are needed in every application.

ISO OSI Model

To allow different types of users on the same network as well as different networks to communicate with each other, the International Organization for Standards (ISO using its French initials) has defined an overall communications framework called the model for *open systems interconnection (OSI)*, usually referred to simply as the *OSI reference model*. This model defines seven layers in a complete communication system (Figure 18.16).

The lowest layer, layer 1, is the actual node where the physical and electrical connection to the network is made. All the higher layers, from layer 2 up to the user (an individual or a computer actually using the data) at layer 7, are implemented by specialized electronic circuitry and software. Each layer handles accepts messages from the layer immediately below (or above), performs a precisely defined task, and passes the result to the layer immediately above (or below).

Although the exact function of each layer is difficult to describe, we can add an analogy to the formal definitions or the roles of each layer: a French sales manager dictating a message to a secretary who must send it as a letter to someone in the United States via a network:

Layer 7, the *Application* layer, provides a means to accept the message from the user, or to distribute the received message to the user. It is analogous to the sales manager dictating a message in French to the secretary.

Layer 6, the *Presentation* layer, makes sure that the information is in a form that the receiving system can use and translates, if needed. It may, for example, translate an ASCII code to an EBCDIC code. It is analogous to the secretary taking the dictation and translating into English.

Layer 5, the *Session* layer, manages and synchronizes conversations between different applications, including marking beginnings and ends of messages so that they are not cut off prematurely. This corresponds to an administrative assistant who adds the letter heading and closing address information, records the letter in a ledger, and finally puts it in an addressed envelope.

Layer 4, the *Transport* layer, combines multiple messages from different sources, if appropriate, and prepares them to go over the same link. It also guarantees the successful arrival data from one end of the link to the other. It is similar to a shipping manager who packs letters going to the same destination (city) into a larger mailbag, although envelopes within the bag may go to different people at the final destination.

Layer 3, the *Network* layer, routes communications and calls upon various routes needed to deliver the data. It acts as the network controller. A shipping manager performs this function by calling the destination office and assigning a route for the mailbag, attaching a routing slip, and putting the bag in a mailcart.

Layer 2, the *Data Link* layer, covers the synchronization and error control needed, regardless of the contents. It is equivalent to mailroom workers weighing the mailbag before shipping, or the receiving end mailroom workers weighing the bag when it arrives to make sure that no letters have been added or lost.

Layer 1, the *Physical* layer, includes circuitry and functions needed to interface to the physical connection. It does not say anything about the modulation or physical medium used. The mailroom loading dock workers who call the trucks and load or unload the mailbags have this type of function. The bag may subsequently travel by boat or airplane, but that is not the concern of these workers.

At the receiving end, the mail bag, envelope, and message enter at layer 1, and travel from layer 1 up to layer 7, where the received, checked, and decoded message is delivered in English to the intended recipient.

If all this seems extremely complicated, it is. The goal of the OSI model is to allow different types of users on one network, or networks with different formats and protocols to interconnect successfully, analogous to a French message from Paris arriving in the United States and being delivered in English. The layers assure that a message will arrive intact (no bit errors within the message or lost messages), and also be understandable by the recipient (English versus French). Fortunately, the operation of the layers is invisible to the actual users connected to layer 7 at each end. They simply put messages with desired destinations into the system (or receive messages) while the reference model handles the many details.

Many applications do not need to use all these layers. Within a single LAN, only layers 1 and 2 (and sometimes 3) are needed.

Gateways, Routers, and Bridges

User nodes can connect to both LANs and wide area networks. Consider a company with several facilities located around the country. There will be a lot of traffic within each facility, and there will also be traffic from location to location. It would be inefficient to have users within the same building use a wide area network to talk to each other just because they also need to communicate with company employees in other buildings as well.

A better solution is to use a separate LAN within each single location, and then have this LAN connect via a *gateway*, *bridge*, or *router* node to a wide area network whose nodes are the various facilities (Figure 18.17). In this way, messages among users within one building use the simpler and faster LAN, while only those messages between users in different locations have to use the slower, more complex wide area network.

A gateway interconnects two networks using different protocols and formats, and serves as a linkage at layer 4, 5, 6, or 7 (Figure 18.18). In contrast, a router is simpler and connects two networks at the network layer only (layer 3) while performing primarily protocol conversion. Finally, a bridge is the simplest, used to connect two networks with compatible protocols at layers 1 or 2. Although gateway, bridge, and router LAN interconnections paths represent interconnection at different levels, the three terms are often used interchangeably.

A gateway (or bridge, or router) node is relatively complex since it must support combinations of different protocols and/or formats at the same time and allow messages that originate under one protocol or format pair to make the transition smoothly to another protocol or format combination. The gateways nodes do not have to be the same everywhere, since the LAN that each supports may be different: The facility where the engineering design is done may be using Ethernet while the manufacturing facility many miles away can be MAP-based.

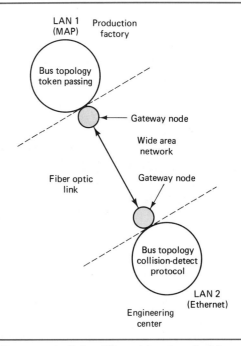

Figure 18.17 A gateway node links different local and wide area networks.

18.5 Wide Area Networks and Packet Switching

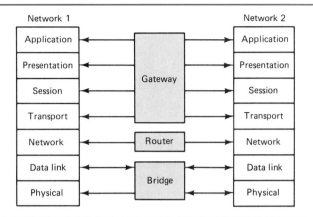

Figure 18.18 Gateways, bridges, and routers interconnect different networks at different layers, depending on the nature of the differences between the networks.

day, the system finds a channel that has space immediately available, but at a higher cost. In this way, the wide area network manages to provide the required level of message service, and tailors its performance to the varying needs of the users at different times of day and days of the week.

Several standard protocols are in use for wide area networks and packet switching. Most of these are published by the CCITT (International Consultative Committee for Telegraphy and Telephony), an international communications organization. The *X.25* protocol of CCITT provides specific rules for packet addressing, allows nodes that are adjacent to each other to send packets between themselves directly (bypassing the larger links of the network), and sets up procedures for moving packets from node to node when the sender and receiver are separated by several nodes. X.25 also specifies error detection rules, along with procedures for signaling that an error has been detected and for recovery from these errors.

Review Questions

1. What communications links do wide area networks usually use? What communications link issues are unique to wide area networks and do not occur for LANs?

2. What kind of information is typically sent over a wide area network?

3. What is packet switching? Why are its protocols complex?

4. What is a store-and-forward system? Why do delays occur, and why are they not a problem in many applications?

5. What is the OSI reference model? What are the roles of the seven layers? Why is such complexity needed?

6. What is a gateway? What is its role? Why is a gateway so complex?

18.6 INTEGRATED SERVICES DIGITAL NETWORKS

Today's conventional networks support digital signals with analog technology. To better meet the needs of digital systems, a completely new communications system called the *Integrated Services Digital Network (ISDN)* is being developed to integrate digitized voice, video, and data, using the same simple twisted-pair wire used as that used for telephone loops. It is integrated because all types of information—voice, data, packet-switched messages, and directly connected users—come to the user node on one line from one central source. ISDN standards for compatibility among pieces of equipment and the network are being developed by the CCITT.

Under the ISDN standard, there are five network access points, called R, S, T, U, and V (Figure 18.19). The R interface is the access point for existing, non-ISDN equipment (today's standard phone or an RS-232 interface, for example, and called TE2—Terminal Equipment 2—in the specification), which requires a terminal adapter to be compatible with the network. Different terminal adapters are needed for each device since an analog telephone and RS-232 interface differ greatly in physical, electrical, and control signal needs.

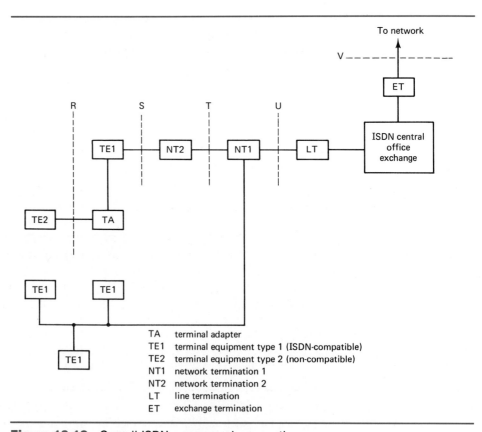

Figure 18.19 Overall ISDN access and connections.

The S interface connects ISDN-compatible equipment with the computer side of the network termination (an NT2, such as a local area network). The T interface is where ISDN switching equipment at the user location connects to the overall system side of the network termination NT1. The U interface is the connection between the customer premises and the central office equipment, while the V interface connects the central office to the rest of the network.

Both the S and T interfaces use two pairs of twisted wire, a total of four wires. The T interface uses one pair to transmit and one to receive, for full-duplex capability. The U interface requires just a single pair of wires because it uses echo cancellation techniques to permit full-duplex operation without separate transit and receive paths. The maximum distance specified for the U interface is 2500 to 6500 m (depending on wire type and number of connections) compared to 1000 m for the S and T interfaces.

ISDN Data Rates

Two classes of service, called the *basic rate* and the *primary rate*, are available with ISDN. The basic rate is known as 2B + D and consists of two 64-kbit/s B channels that can carry either voice or data, plus one 16-kbit/s D channel used for network signaling and control as well as user packet data. There is an additional 48-kbit/s channel used for other purposes, such as echoing the D channel back to the source, synchronizing the terminal units, and conveying other internal network information. The overall basic rate is (64 + 64 + 16 + 48) = 192 kbits/s.

In North America and Japan, the primary rate is called 23B + D and consists of 23 B channels at 64 kbit/s each and one D channel at 16 kbit/s. The primary rate is compatible with 1.544-Mbit/s transmission systems since it is the same as the Bell system T-1 trunk rate (Section 16.7) and is used to interconnect computer centers and to connect central offices to the network rather than to serve local users directly. [*Note:* Europe uses 30 B channels plus 1 D channel (30B + D) or 2 D channels (30B + 2D), which is compatible to transmission at 2.048 Mbits/s used by the CEPT system (Europe's approximate equivalent to T-1). Both North America/Japan and Europe speeds and channels are supported by ISDN standards.]

ISDN Format

The format used in ISDN depends on which part of the network is studied. Figure 18.20 shows the format for the S interface between TE and NT2, with data in both directions. Data transfers occur every 250 μs in 48-bit frames, with the data rate in each direction at 192 kbit/s. Note that the data bits of the two channels are offset in time from each other by two bits to avoid certain conflicts that might otherwise occur.

ISDN components, similar in function but more complex than Ethernet or IEEE-488 ICs, are now available for use in communications systems. Several companies offer complete circuit boards with ISDN interfacing, such as the one shown in (Figure 18.21), which plugs into an IBM personal computer and supports R, S, T, and U interface access.

Review Questions

1. What is ISDN? How does it differ from systems where analog signals are handled in digital form?

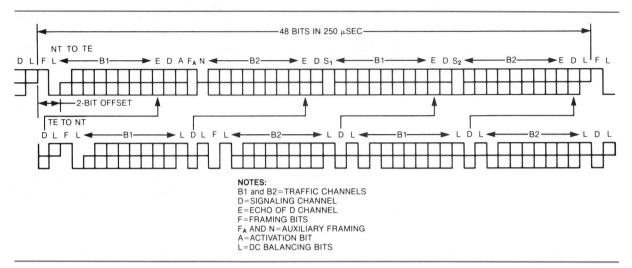

Figure 18.20 ISDN bit format at the *S* interface (courtesy of EDN magazine, Cahners Publishing Co.).

2. What are the ISDN access points? How are they used?

3. Why does the U interface require just one pair of wires for full duplex?

4. What is the ISDN basic rate in terms of channel types and data rate?

5. What is the ISDN primary rate in North America and Japan? How many channels are there, and of what types? What is the overall rate?

6. What is the frame length in ISDN format between TE and NT2?

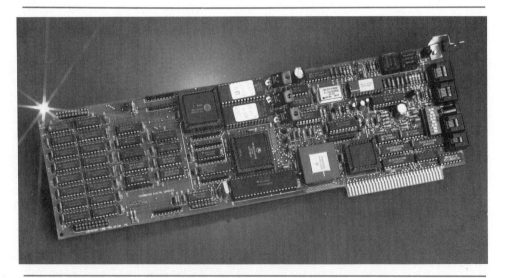

Figure 18.21 Plug-in board for IBM PC allows the PC to connect to ISDN and for study, investigation, or actual network use (courtesy of Motorola, Inc.).

SUMMARY

In this chapter we have studied networks, the most complex and wide-ranging type of communication systems. A network requires agreement by all users on virtually every aspect of its operation, and the rules they agree to follow are set, to a varying extent, by several industry standards. No single network standard is ideal for all applications, and the best network in a specific case is determined by required speed, distance, reliability, interconnection scheme, and general flexibility.

A network specification (or agreement among users) begins with the physical medium and the modulation used to put digital signals onto the medium. Next, all users must agree to the coding that takes the data bits and encodes them into a final 1,0 pattern. Coding forms the basic elements of the larger message format, or organization of individual message, including any heading or closing information on the message. Messages are sent following the protocol—the rules of conversation—which must allow each user a turn to transmit, accommodate requests for messages, and make sure that errors and problems are handled gracefully without stopping the network. Finally, the network interconnection topology shows the paths among users.

Many standards in use define the specifics of medium, modulation, coding, format, protocol, and topology from among the many board choices (including one to all, star, bus, and ring topologies and command/response, interrupt-driven, token-passing, and collision detect protocols). Some standards define only a few of these, and letting the users define the rest depending on their specific situations. Other standards define virtually all aspects of the network but allow users several choices in each category, such as the type of physical medium used.

Because of their complexity, testing of networks requires much more than voltmeters and oscilloscopes: It requires a sophisticated instrument such as a network protocol analyzer. This instrument can monitor the network, analyze the many messages and their flow and meaning, and even generate messages to test the network's ability to handle more traffic or various error conditions.

Summary Questions

1. What is a network? How does it differ from a simpler connection between two users? What makes a network more complex?

2. In what way is the telephone system a network? How is it not a network? When is it part of a network?

3. What is the meaning of coding, format, protocol, and topology in networks? Why are these needed? Do all network standards define all of these? Explain.

4. Compare local and wide area networks in terms of users, technical issues, and speed.

5. Describe one-to-all, star, bus, and ring topologies. Compare them in terms of handling small and larger number of nodes, ease of expansion, overall reliability if a node fails, links used, speed, and necessary protocols.

6. What is a watchdog timer? What does it do? Why does a network need one at each node? What does it do if a bus node fails? If a ring node fails?

7. Describe the four basic protocols and compare them with respect to simplicity, speed, advantages, and weaknesses.

8. When is the waiting time of a token-passing network shorter than that of a collision detect network? When is the situation reversed?

9. What are the typical primary applications for AppleTalk, Ethernet, MAP, and

IEEE-488 network standards? What does each define explicitly, and what does each leave to the user choice? How many nodes does each support?

10. Explain the function and need for network ICs such as those in Ethernet and IEEE-488.

11. Compare applications and links for local and wide area networks.

12. What is the ISO OSI reference model? What is the function of each layer? Why is this complexity sometimes needed? When is it not needed?

13. How does a gateway node allow interfacing between a local and a wide area network? Compare the functions of a gateway, a router, and a bridge with respect to their complexity and layer within the OSI model.

14. What is a packet-switched protocol? What must it do? When is it useful?

15. What is ISDN? Describe the five ISDN interfaces.

16. What are the channel types and rates for basic and primary rates in the United States?

19
Satellite Communication and Navigation

CHAPTER OBJECTIVES

When you have completed this chapter, you will understand:
- How communications satellites are used, their benefits and limitations
- The integration of many critical electrical and mechanical aspects of a complete satellite system design
- The role and operation of satellite ground stations
- The use of non-satellite-based electronic systems for navigation
- The application of satellites to navigation via the global positioning system

INTRODUCTION

We take for granted the reality of instant communication of video, audio, and data signals to or from virtually any place on earth. Such communication is made possible by systems based on satellites as the major link between users. Acting as space-based signal relay stations, satellites overcome the difficulties and limitations in broadcasting beyond the visible horizon.

A satellite system represents the careful and finely tuned integration of electronics, mechanical structure, rocketry, and antenna design, supported by a system of earth-based ground stations, computers, and radar for tracking satellite position precisely. Different orbits provide a selection of coverage of the earth surface, with orbital times ranging from 90 minutes to 24 hours. Using frequencies in the gigahertz range, satellites are effective and predictable in performance but have very large path losses from earth to satellite to earth. A complete system requires high-gain antennas, sensitive and low-noise front-end receiver stages, and careful planning.

In addition to handling messages, navigation is a vital role for communications. Simple triangulation, which has been used for many years, has limited position accuracy and range. A LORAN system uses precisely timed signals to provide greater accuracy in position, but has limitations in range and cannot cover the entire globe. The ultimate in navigation is a satellite-based positioning system which works with special ground stations and user receivers to provide outstanding positional accuracy anytime and anywhere on earth.

19.1 COMMUNICATIONS AND ORBITS

In the discussion of antennas in Chapter 9, we saw that the line-of-sight distance between two antennas was limited by the curvature of the earth and the horizon distance visible from the antenna height. Although higher antennas are the obvious way to increase the achievable distance, high antenna towers are expensive and impractical in many locations. There is an alternative: to use a space satellite as a relay station for signals to overcome the limits imposed by the earth's horizon. A *communications satellite* is designed to receive a signal from a transmitting station on the ground and retransmit it to a receiving station located elsewhere. (Although some satellites are used as relay points between other satellites or manned spacecraft, we will concentrate on their use as the intermediate point between two earth-based stations.) The *uplink* is the signal path from transmitter to satellite; the *downlink* is the corresponding path from satellite to earth-based receiver.

Satellites are placed in orbits around the earth at various angles and altitudes (height). Each angle and altitude provides some combination of desired and undesired characteristics. A satellite position is measured by its *elevation* angle with respect to the horizon and its *azimuth* angle measured clockwise from the direction of true north (Figure 19.1).

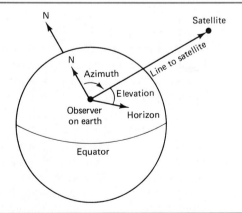

Figure 19.1 Elevation and azimuth angles of a satellite (courtesy of COMSAT).

19.1 Communications and Orbits 595

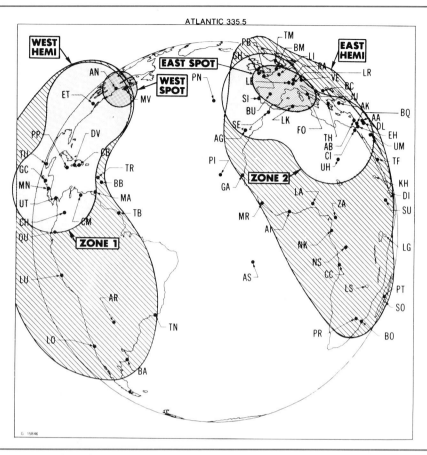

Figure 19.2 Footprint of INTELSAT V satellite (courtesy of COMSAT) (from *COMSAT Technical Review,* vol. 7, no. 1, p. 314, Spring 1977).

Footprint

The *footprint* of a satellite is the earth area that the satellite can receive from or transmit to. This is a function of both the satellite orbit and height and the type of antenna the satellite uses. As the satellite gets higher, it is in a line of sight to more of the earth surface. However, it is not always useful to have a satellite with a broad footprint for its uplink or downlink antennas. This wider beamwidth means that more interfering signals will be picked up at the satellite uplink receiver while the desired signal strength is lower; also, the transmitter downlink antenna will broadcast to areas besides the intended earth station of the downlink and so wastefully spread out its limited power.

By using a focused antenna pattern, the footprint of the satellite is more sharply restricted. Figure 19.2 shows the footprint of the general-purpose INTELSAT V satellite, orbited as a relay between North America and Europe. The figure shows the footprint of primary zones of coverage in each region, as well as secondary, weaker zones and the widest (but weakest) zones. The two letter abbreviations in the figure are the uplink and downlink ground station two-letter designations.

Altitude

The satellite orbit around the earth can be relatively low—50 to several hundred miles—or as high as 23,000 miles. The time for one complete orbit of the satellite is determined primarily by this height. A satellite at 100 miles takes 90 minutes for an orbit, while one at 10,000 miles takes approximately 12 hours for one orbit; in contrast, an orbit at 22,300 miles takes 24 hours. The 24-hour orbit, when combined with a specific elevation angle, is *geostationary* where the satellite appears to be stationary over one point on the earth's surface, since it and the earth rotate at the same rate. (Incidentally, the geostationary scheme was first proposed by science fiction writer Arthur C. Clarke in the 1940s.)

Lower orbits require less power at both the transmitter and the satellite, since the distances involved are much shorter. In addition, the power from the satellite that reaches the ground receiver is much greater than it is for a higher orbit satellite, so a smaller receiver antenna is needed and noise is not as serious a problem. However, the low orbits require that the transmitting and receiving antennas move physically (or be scanned electronically) to track the satellite. Although this tracking is possible, it means precise alignment of the antenna is needed to aim properly at all times. A complete control system with motors and angle-measuring indicators is needed to move the antenna smoothly and accurately as the satellite passes overhead.

An alternative is to use a receiving antenna with a broad footprint so that actual physical tracking is not needed. Unfortunately, these antennas provide low gain, so that signal levels and the signal-to-noise ratio at the receiver will be low. The transmitter antenna cannot have too broad a beamwidth since this will cause interference with other stations.

Orbits are generally not true circles but are elliptical. This complicates the tracking problem since the satellite will be closer to earth at some points than at others. As the satellite height increases, orbits that are more nearly circular are possible.

Although the gravitational pull between the earth and the orbiting satellite is the primary factor in determining the satellite movement, it is not the only one. For lower orbits (100 miles) the earth factor essentially overwhelms all others, yet even the earth is not a perfect sphere. The resulting orbital ellipse has imperfections that cause it to change over time. In addition, the atmospheric drag on the satellite causes its speed to slow down and the orbit to decay slowly.

At the higher altitudes, the sun and moon have small influence compared to the earth—about 1% of the earth's—but this is a significant impact on the orbital path. The sun and moon pull on the satellite and distort its path from a truly elliptical shape, and this distorting pull changes with the sun and moon positions with respect to the satellite. For the satellite to maintain its desired orbit consistently, it must therefore use small on-board rockets or thrusters periodically to correct its position against these corrupting influences.

Angles

An orbit is defined by the angle with respect to the earth equator, which is the 0° plane. Three types of basic orbital angles are possible: polar, inclined, and equatorial (Figure 19.3). In the polar orbit, the angle of the orbital plane is 90° and the satellite basically rotates over the North and South Poles as the earth turns below it. At some time, then, over the course of many orbits, the satellite will be in view of every point on earth. This polar orbit is used for low-altitude satellites (see Section 19.5) as well as spy satellites that take pictures or eavesdrop on ground

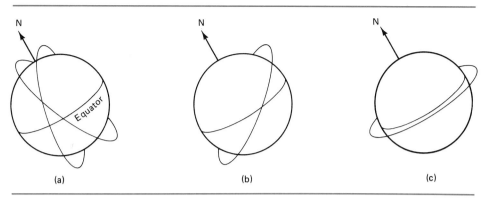

Figure 19.3 (a) Polar, (b) inclined, and (c) equatorial orbits.

radio signals. Most polar orbit schemes use many satellites, so that at any time there is at least one satellite in view (or that can view you!).

In an inclined orbit, the satellite orbit plane is more nearly parallel with the equator, in contrast to the polar orbit, where the satellite plane is at 90° to the equator. As the satellite inclination gets closer to the equator, it covers more of the central part of the globe on each pass. In this type of orbit, the satellite footprint is concentrated around the earth's middle, with much less coverage of the higher latitudes toward the two poles.

When the orbit inclination is 0°, the satellite plane and the earth equatorial plane are the same—the satellite plane cuts along the equator. This leads (at the 22,300-mile height) to the unique geostationary orbit, which is very useful in communications since the satellite takes 24 hours for a complete orbit and appears to be hovering over a fixed point on the earth's surface. From a single geostationary satellite, about 40% of the earth can be covered by a single satellite. Three geostationary satellites provide coverage for about 95% of the earth surface (with some unavoidable overlap); only the extreme polar regions are unreachable.

Satellites in geostationary orbit form an *equatorial belt* around the earth. Since these are very desirable locations for communications satellites, international telecommunications agreements have been worked out to avoid electronic interference and physical collision. Originally, the satellites were spaced at every 4° of the 360° complete circle, so only 360/4 = 90 "parking spaces" were available. Due to technology improvements with lower-noise receivers, better satellite electronics, and more precise tracking, these spacings have now been reduced to only 2°, and agreements are being established for 1° spacing. This, of course, allows for many more geostationary satellites.

Frequencies

Satellite communications usually use frequencies in the 1-GHz and higher ranges. There are several reasons for this. First, there is a great deal of available bandwidth at these higher frequencies, so large amounts of information can be sent. Second, the propagation characteristics at these frequencies are very consistent. Signals travel in perfectly straight lines undisturbed by geographical features such as mountains, water, and clouds (there is some attenuation due to water and clouds, but the line of travel is not affected). The loss in signal strength with distance is very predictable, so the system and circuitry designs can be worked

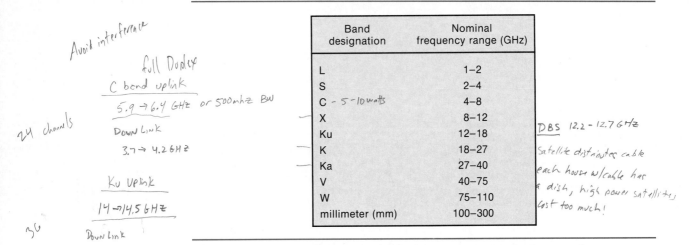

Figure 19.4 Frequency bands used in satellite communications, and letter designations.

out for consistent results. Noise levels from atmospheric and solar sources, as well as random movement of molecules, are lower than at the lower frequencies. Finally, the signal wavelengths are shorter, so effective satellite antennas of high gain and narrow beamwidth can be built, while reasonable in size and weight—a major practical consideration.

Not all satellites operate at such high frequencies, where standard electronic components and connectors are not usable, and where waveguides, stripline, and unique amplifiers and other electronic components are required. Some special-purpose satellites, such as for amateur radio signals, operate in the several-hundred-MHz range, but these are much less common for commercial applications because of limited distance, noise, difficulties in predicting performance, and less available bandwidth.

The available frequency bands are assigned letters (Figure 19.4). Each satellite uses a pair of bands, with one for the uplink and the second for the downlink. A satellite with an uplink frequency at 6 GHz and a downlink at 3 GHz is called a C/S band satellite. It is not desirable to use the same band for both up and downlinks since the signal broadcast from the satellite would overload the sensitive front-end circuitry of the satellite's own receiver. With two different bands, bandpass filtering is used in front of the receiver to filter out virtually all of the undesired downlink signal, which is many orders of magnitude stronger than the signal received on the uplink. There are two antennas on the satellite, as well. One is designed for the uplink frequency and footprint, and oriented toward the ground transmitter; the other is optimized for the downlink frequency and desired footprint, to point at the ground receiver. The satellite may also have other antennas for special, nonmessage channels such as its telemetry and control subsystem, discussed in the next section.

Review Questions

1. What is the basic function of the communications satellite? Why is it needed? Why are the uplink and downlink frequencies different?

2. What is footprint? Why is a larger footprint not always better? How is the footprint adjusted?

3. What is the range of orbital heights and times? What is the unique feature of a geostationary orbit? What is the geostationary height and inclination?

4. Contrast lower orbits versus the geostationary orbit for required transmitter power, receiver power, antenna size, and antenna control complexity.

5. How are orbital planes characterized? Compare polar, inclined, and equatorial orbits for position and features.

6. How are geostationary satellites spaced? Why is tighter spacing desirable and now practical?

7. Give four reasons why GHz frequencies are used. What are the satellite bands and their typical designations?

19.2 SATELLITE DESIGN

A satellite is an extremely complex and sophisticated electronic and mechanical device. Every aspect of its design must consider the many constraints and conflicting goals: allowable weight and size, available dc power, required receiver sensitivity, desired transmitter power, antenna size and beamwidth, the number of needed communications channels and bandwidth, plus the critical *telemetry* channel used by the ground station to keep watch over the satellite and issue essential commands for required physical position and internal configuration changes. The satellite also needs a guidance system to make sure that it maintains the proper angle to the earth, and rocket thrusters plus fuel so that it can make small orbit adjustments.

Of course, reliability is a major concern, since repair is impractical (it has been done for a low-altitude satellite using the Space Shuttle, but it is extremely costly), so redundant circuitry is included in case the primary circuits fail—but this adds weight and consumes power. A typical satellite weighs anywhere from several hundred to several thousand pounds and costs about 50 to 100 million dollars (*plus* the cost of actual launch). Construction of a satellite takes about two years, including testing, and the satellite is designed to function for at least 10 years in space.

A satellite such as INTELSAT V (Figure 19.5a) consists of many subsystem functions, all carefully integrated into a single system (Figure 19.5b). These include:

- The power subsystem
- The telemetry and control subsystem
- Main and auxiliary propulsion subsystems
- The communications channel subsystem
- Antennas

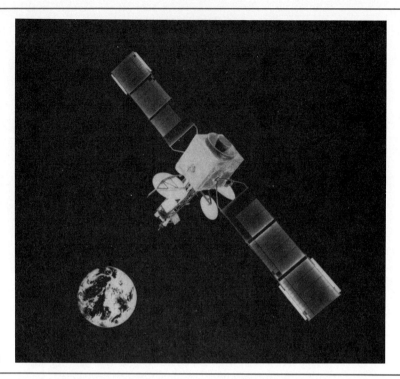

Figure 19.5 (a) INTELSAT V (courtesy of COMSAT).

Power Subsystem

The power subsystem is responsible for providing the primary dc power and the regulated, secondary supply voltages for the satellite circuits. Depending on design and complexity, a communications satellite requires several hundred to several thousand watts of electrical power.

Primary power originates from solar cells, which convert the sun's light into electricity. Since these cells have an efficiency of only about 10% in converting light to electricity, large panels of cells are required. These extend from the satellite like wings, or completely cover the outer surface of the satellite, and the cells must be oriented to face the sun directly for maximum output. Since the satellite is not in sunlight as it passes behind the earth, the solar cell power is not used directly. Instead, the illuminated cells charge storage batteries on board, which in turn provide the primary power to the satellite power converter. The storage batteries must be able to provide power for more than 12 hours of operation without recharging.

The internal circuitry of the satellite requires a variety of dc voltages from +5 and ±15 V for low-level logic and amplifiers, to several hundred volts for power amplifiers, with differing amounts of regulation. A dc-to-dc power converter produces these *secondary* voltages, with the required amounts of regulation, precision, and stability, by taking primary battery power and converting it to a high-frequency alternating current (frequencies between 20 and 100 kHz). This ac is then stepped up (or down) via transformers, rectified back to dc, and regulated as needed. The efficiency of the power conversion subsystem is critical,

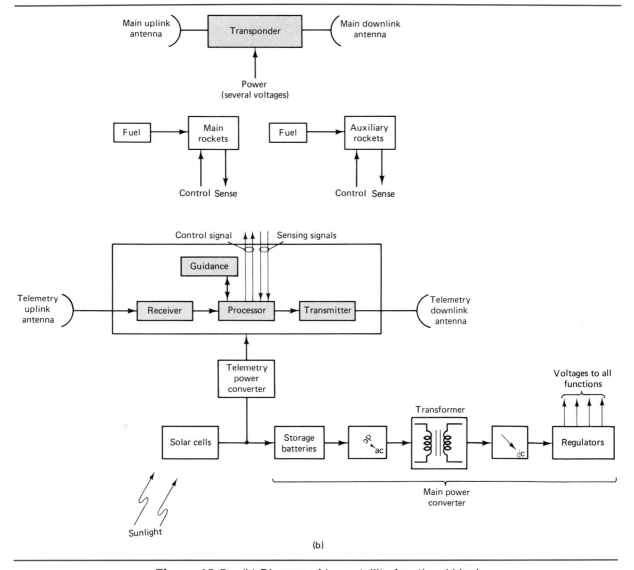

Figure 19.5 (b) Diagram of key satellite functional blocks.

since any power wasted in the converter means that more solar cells and larger storage batteries are needed. Power converter efficiencies of 80 to 90% are achieved by careful and expensive design.

Telemetry and Control Subsystems

Telemetry and control subsystems provide a way for the ground station to manage the satellite. The communications channels used for this function are entirely separate from the communications channels used to relay user messages from uplink to downlink (which is the satellite's primary purpose). Via *status telemetry*, the satellite transmits information back to the ground station about hundreds of vital internal voltages, currents, temperatures, and fuel levels, as well as positional and orientation information. This information is used to monitor the "health" and operation continuously, indicating both sudden failures and signs of potential problems (such as temperature too high in a high-power downlink amplifier).

There is also *command telemetry* for the ground station to command the satellite to activate its rocket thrusters to change position or orientation, to cause the electronics to switch to a different mode of operation (modulation, frequency, number of channels, and bandwidth), or to electronically reconfigure some of the internal interconnections. This last feature is especially useful when a problem is detected by the ground station. The uplink command telemetry instructs the satellite to rearrange the internal signal path to bypass failed units and switch in backup units that are available. In many cases, this capability has been used to save an otherwise-failing satellite.

Because the telemetry and control subsystem is so critical to ensuring that the satellite achieves proper orbit and maintains and corrects orbital position (called *stationkeeping*), as well as for monitoring the performance and even fixing the satellite, the telemetry plus control electronics are as separate as possible from the communications electronics. The only element they share in many satellite designs is the solar cell array for basic power. The telemetry and control subsystem has its own battery, power converter, uplink, downlink, and even antenna system. Unlike the very directional antennas of the communications links, the telemetry antenna has a broad beamwidth or is omnidirectional, so that an incorrectly oriented (or even tumbling) satellite can receive ground instructions and perhaps be saved. Similarly, any failure in the communications subcircuitry does not affect the telemetry and control circuitry, which is needed to reconfigure the rest of the satellite when problems occur.

Main and Auxiliary Propulsion Subsystems

The satellite is launched into its approximate desired orbit by a large rocket, which may be carrying several satellites at the same time. The satellite then uses its own main propulsion system to go to its final orbit, correcting for any unavoidable errors in the initial launch. Since only a limited amount of main rocket fuel is available on board the satellite, this orbit adjustment must be correct on the first or second try. Some surveillance (spy) satellites have additional main propulsion fuel so that they can actually shift from one orbit to another, via ground command, to take pictures of other parts of the earth or listen in to signals from different areas, but this naturally limits the overall useful life.

Once a satellite has achieved correct orbit, auxiliary rocket engines (thrusters) are used to turn the satellite to the correct orientation to the earth and sun, and to perform many small stationkeeping corrections for the orbital distortions and drift caused by the sun, moon, and the earth itself (this applies to all satellites, regardless of orbit type). The amount of auxiliary fuel is a limiting factor in the total useful life of the satellite, since the thrusters are used repeatedly. It is not possible simply to "set and forget" the satellite into the perfectly correct orbit and orientation—the irregular influences of moon, sun, and earth gravity are never absent and must be compensated for on a regular basis.

Commands to fire the thrusters come from the ground station, which performs the necessary calculations on how much thruster firing is needed and in what direction. Some advanced satellites can perform the calculations and thruster control as well, but the satellite itself has a poorer idea of where it actually is (and its orbital errors) than does the ground station, which is tracking it by radar.

Communications Channel Subsystem

The central purpose of the satellite is the communications channel subsystem—everything else exists to serve it. A *transponder* takes the received signal from the antenna at the uplink frequency, heterodynes (mixes) it to the downlink frequency, and amplifies it before transmitting. A complete basic transponder consists of an input filter, low-noise input amplifier, bandpass filter, mixer plus local oscillator, another bandpass filter, and the power amplifier (Figure 19.6). This type of transponder is open and transparent and handles any signal that fits into its bandwidth. The output signal format and modulation is identical to the input signal, and only the carrier frequency is changed.

The power amplifier is usually built from a special amplifier component known as a *traveling-wave tube (TWT)*, which can provide the necessary gain and output power (typically, 5 to 20 W) at the GHz frequencies of the downlink (Chapter 23). TWTs are wideband, high-gain devices with good efficiency at converting supplied dc power to RF, an essential requirement in a satellite. In addition, they are lightweight, reliable, and give consistent performance over a long life. In recent years, solid-state (transistorized) power amplifiers have been developed, and these are replacing TWTs in some satellites because of their potential for higher efficiency, greater reliability, and lower distortion.

A satellite is generally not limited to a single transponder channel. Multiple channels provide greater reliability in case one channel fails and allow the characteristics of each channel to be tailored to the type of signal, its modulation, and its bandwidth. Several standards are commonly used: 12 to 24 separate transponders each with bandwidth of 36 MHz, for frequencies below 10 GHz, plus wideband transponders with 80- and 200-MHz bandwidths above 10 GHz. Some advanced satellites provide an *RF switching matrix* so that any uplink frequency can be switched to any of the downlink frequencies instead of just being associated with a single downlink frequency. This provides additional flexibility in operation and in case of transponder failure, but does require complex GHz range switching circuits.

Transponders do not have to be transparent, simply translating the uplink signal to the downlink one. Advanced designs take the received signal, demodulate it to recover the modulating information, then remodulate this using another specific type of modulation. For example, FM of one modulation index can be transformed to another index (with corresponding change in signal-to-noise ratio), or FM can become PM. In addition, digital signals can be converted from binary FM modulation to quadrature phase shifting, for example. Remodulation is often

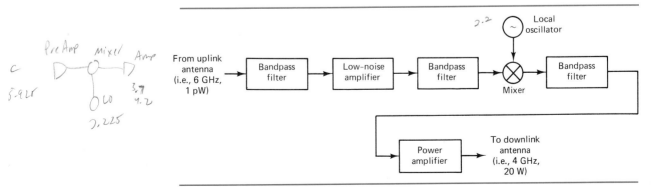

Figure 19.6 Block diagram of typical transponder.

Satellite Access

The goal of satellite transponder access is to ensure that the transponder channels are used as much as possible, since vacant channels represent wasted bandwidth and information transmission opportunity. Also, a single satellite usually must serve more than one uplink ground station or downlink ground station. Two common techniques to allow greater utilization of the satellite are *time-division multiple access (TDMA)* and *frequency-division multiple access (FDMA)* (studied further in Chapter 22).

In TDMA, a single transponder channel is shared among various uplinks by assigned time periods (time slices) to each one. Each ground station knows which time period it is assigned and transmits on the transponder frequency only during its period. In some communications systems this time assignment is fixed in advance (static); for example, four ground stations can share each 1-s period, with station 1 assigned the first 0.25 s, station 2 having the next 0.25 s, and so on. The disadvantage of this *static allocation* is that channel time is wasted if a station has nothing to transmit during its allocated time. Of course, all ground stations must have extremely precise clocks, and these must be synchronized so that all station clocks agree on when each second starts.

The alternative to static TDMA allocation is *dynamic allocation*, where the time slice assigned varies with the needs of each station, similar to the needs of a network with various nodes sharing a common medium (Chapter 18). This more complex arrangement allows the transponder to be used nearer to its maximum capacity, but the management of the ground stations is more complex. Each ground station constantly communicates with the others, to see who has information to send, and the assignments of available time periods must be carefully arranged. Instead of assigning a wideband transponder channel for carrying the ground station network management information, this information is combined by multiplexing with other unrelated information on a wideband channel or uses a lower bandwidth channel. For redundancy, the management information is carried by land and undersea cables as well.

FDMA is simpler in concept. One or more transponder frequencies are assigned to each ground station, and no two ground stations can have the same frequency at the same time. This is similar to assigning a broadcast band to various radio or TV stations so that there is no interference, except in this case the overall band (such as 88 to 108 MHz) is replaced by the multiple transponders within the satellite. The method of allocation can be static or dynamic; each allocation method has trade-offs in complexity versus maximum utilization of transponders in a manner similar to TDMA allocation.

---

used to provide more resistance to interference from downlinks of other satellites, or to provide security against unauthorized listeners. Finally, demodulated digital signals can be processed at the transponder to encode them for privacy and secrecy, or buffered and stored within the transponder for transmission at a later time when the desired downlink ground station is in view. Digital modulation also provides benefits for error detection and correction, and for improving error rates by different data modulation and encoding schemes.

Antennas

The satellite uplink and downlink antennas must combine the need for high gain (to increase effective received or transmitted power) with the beamwidth appropriate to the application. A high-gain antenna tends to have a narrow beam, which helps reduce received interference from other transmitters and provides a more focused footprint, but requires much more accuracy in aim.

Some early satellites used an array of small dipoles, electrically phased with each other to provide the necessary antenna field pattern. The gain achieved was relatively low, and most newer satellites use some type of parabolic dish antenna, or a horn antenna, for higher gain and narrow beamwidth. Typical communica-

Passive Satellites

A communication satellite is extremely complex and sophisticated. *Passive satellites*—signal reflectors without any amplification—were used before the technology and rocket power was available to provide all the needed subfunctions in a package that was light enough to lift into orbit. *Moon-bounce* experiments in the 1950s showed that a signal broadcast from earth could be reflected from the moon and back to another earth station.

The moon-bounce method required tremendous power at the transmitter, since most of the signal power was lost in the round-trip distance of nearly half a million miles, and the moon is a poor reflector of radio-frequency signals. Also, this scheme worked only when the moon was in sight, and could support only a few ground stations since the received signal footprint covered most of the earth (once again, due to the distance) even if the transmitted antenna beamwidth was very narrow. Finally, the uplink and downlink frequencies were the same (no frequency translation is possible from a passive reflector), so full-duplex communications was impractical unless each direction used a significantly different frequency from the other.

Another experiment in the 1960s used a large sphere as the satellite (Echo I), whose surface was covered with metallized Mylar to improve its reflection. Since this satellite was in a much lower orbit than the moon (several hundred miles) and was a better reflector, it provided better received SNR and a smaller footprint. However, this simple passive satellite still provided no gain (requiring powerful transmitters and large ground station antennas), had no directional antennas (so its uplink and downlink footprints were still relatively wide and uncontrollable), and suffered orbital decay since there were no rocket thrusters to correct any orbit disturbances or errors. Development of more powerful rockets combined with advances in high-performance, low-power, lightweight electronics have eliminated passive satellites and have completely replaced them with active satellites.

---

tions satellite antenna diameters are 2 to 6 m, with a gain of 10 to 40 dB. The antenna design must also minimize sidelobes, which will allow an interfering signal to get into the receiver front end or cause interference with another ground station.

Since the uplink and downlink frequencies and directions are different, there is a separate antenna for each. To allow more satellites with closer spacing, adjacent satellites use different polarizations (horizontal or vertical), since this further reduces interference. A vertically polarized signal will have little effect on an antenna that is designed for a horizontally polarized signal, and vice versa.

Review Questions

1. What are the many factors in satellite design? What is a typical satellite weight range, cost, and operational lifetime?
2. What does the power subsystem do? Where does primary power come from?
3. How much power is needed? What does the power converter do, and how are various voltages generated? Why are storage batteries needed?
4. What are the roles of telemetry and control subsystem in satellite management? What must it monitor and control? Why?
5. Why can the command telemetry force a reconfiguration of the satellite interconnections?
6. Why is the telemetry and control channel separate from the rest of the satellite electronics? Why does this channel use broader-beamwidth antennas rather than communications up- and downlinks?

7. What is the function of the main propulsion system? Why is there a need for an auxiliary propulsion system? Why is the available fuel a limiting factoring in the service life of the satellite?

8. What is a transponder? What are its main functional blocks? How does it perform its role? What is an open transponder?

9. How many transponders are used on a typical satellite? Why is there more than one? What are the typical number of transponder channels and their bandwidths?

10. What is access? What is its major goal?

11. Compare static and dynamic access of TDMA for features and drawbacks. Why is a separate management channel used?

12. What is FDMA? How does it increase access and use?

13. What are the roles of downlink and uplink antennas in gain, footprint, and focus? What are typical diameters and gain values?

14. Compare the moon and a metallized sphere for signal strength, availability, footprint, and long-term life, versus a modern satellite.

19.3 GROUND STATIONS

Although the communications satellite is the most sophisticated of communications systems in design and construction, the ground station (or earth station) is a critical part of a successful satellite communications system. A *ground station* (Figure 19.7) has five main subsystems: the interface to the earth-based signal source (the user who is sending and receiving a message), the transmit chain, the receive chain, the antenna, and the antenna control.

The baseband signal is modulated onto an intermediate-frequency carrier, typically 70 MHz, which is up-converted to the actual broadcast frequency (such as 6 GHz). It is then amplified as necessary before actually reaching the antenna. Transmitter power varies with the satellite distance and antenna gain. Tens of watts can provide reliable signals to low-level satellites, while power values of up

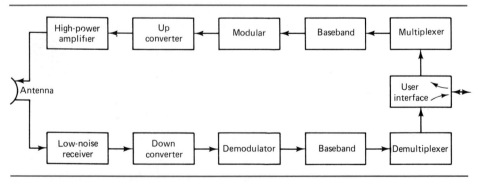

Figure 19.7 Ground station block diagram.

to 1 MW are used to reach deep-space satellites. Coaxial cable or waveguide is used to transfer the amplified signal to the antenna, depending on frequency and power level.

In the receive chain, a low-noise amplifier located at the antenna is critical since received power levels and SNR are low. The received frequency is separated from the transmit frequency to prevent the much larger transmitted signal (many orders of magnitude greater in power: watts versus picowatts) from interfering with the received signal or overloading the sensitive receiver front end. This separation is done with special filters (splitters and circulators) for microwave and millimeter signals, since traditional lumped-constant inductors and capacitors are nearly useless at GHz frequencies. The received signal is down-converted to an IF, again typically 70 MHz, and then demodulated.

The antenna is a critical part of the earth station. The uplink antenna provides the required gain to increase the effective transmitted signal strength (as seen by the satellite) and received signal strength (as seen by the low-noise receiver), while maintaining a narrow beamwidth to avoid interfering with adjacent satellites. The downlink receiving antenna must capture as much signal as possible from the desired satellite while minimizing signals from other sources. Antenna size and style affect the gain, the beamwidth, and the ability to avoid sidelobes which minimize interference with nearby transmitters. Depending on available power, diameters from 1 to 60 m are used, and beamwidths are about 1°.

Finally, the antenna control must be precise in positioning, since the narrow beamwidth means that even a small aiming error corresponds to a lost signal or missed target. For communication with a single geostationary satellite, the antenna orientation can be carefully measured and fixed. In practice, the larger antennas are steerable so that they can track different satellites. To achieve these difficult goals, the mechanism provides a position control mechanism that can point the antenna accurately at the desired angles, yet is strong enough to move the large antenna mass and hold it steady in wind and rain. This requires a unique combination of fine position control to extremely high angular accuracy and resolution, and brute strength for actual movement and holding. The circuitry, motors, and mechanical supports for the antenna require advanced mechanical and electrical control designs. The problem is even more difficult when the satellite is not geostationary, and the control mechanism must move the antenna smoothly to track the satellite course in the sky.

Link Budgets

A successful communication satellite system depends on two critical parameters that are closely related. The signal-to-noise ratio at the satellite uplink receiver must be large enough to ensure that the signal is useful and recoverable, and thus not buried in noise. The same applies to the signal as received by the downlink ground station antenna. Second, the actual signal power at these two points must be large enough for the respective receiver front ends to operate properly; if the power level is too low, the circuitry will not be able to perform the transponding function or the demodulation function.

To make sure that the SNR and power levels are sufficient, satellite system designers calculate the *link budget* that characterizes these two parameters at each point between the uplink transmitter output and the downlink receiver input. Each source of signal gain or loss, as well as the added noise, is put into the budget calculation. (We looked at this briefly in Chapter 9 as part of propagation and path loss.) If the results show that the power levels of SNR are too low, additional gain and noise reduction must be incorporated into the system design by increasing power and amplification, and lower noise components must be used as well. Let's look at a link budget for power levels.

A 6-GHz uplink ground station transmitter for a geostationary satellite (23,000 miles, 37,000 km) has a power output of 100,000 W, and the 4-GHZ downlink ground station receiver must see 1 pW at its input to operate. The transmitted signal has these gains and losses:

A transmitter antenna with gain of +50 dB.

The free-wave path loss for the uplink, discussed in Section 9.2, defined by

$$\text{path loss (in dB)} = 32.4 + 20 \log f \text{ (in MHz)} + 20 \log d \text{ (in km)}.$$

Note that the free-space path loss is determined by frequency and distance and increases with the log of these two factors. For this satellite, the loss is

---

The Voyager Spacecraft

The ultimate in long-range satellite communications was demonstrated in August 1989 when the *Voyager 2* spacecraft transmitted pictures to earth from the planet Neptune, a distance of 2.8 billion miles from earth (the travel time for the signal is 4 hours at this distance). *Voyager* was launched on August 20, 1977 to explore only Jupiter and Saturn, but continued to operate after passing these planets and sent pictures from Uranus in January 1986. To achieve this incredible performance, every aspect of the spacecraft, signal coding and modulation, and receiving stations has to use the most advanced techniques, many of which we have studied in this book.

The satellite. The satellite itself weighs 1800 pounds and has six computers, two transmit radios, and two receivers. Total power to operate the satellite is 400 W, but solar cells cannot be used at this distance because the sunlight intensity is so low: only $\frac{1}{1000}$ that of the earth. Instead, a small power generator that uses the heat produced by decay of radioactive plutonium develops the power needed (note that this is very different from a nuclear reactor power plant). The dish antenna on *Voyager* has a 12-ft diameter, and the actual transmitted power is just 22 W. Frequencies of 8415 and 2295 MHz are used.

Signal processing and encoding. *Voyager* has two monochrome cameras, one for wide-angle images and one for narrow-angle but more magnified views. Although each camera records pictures only in shades of gray, the cameras are equipped with color filters. By making and transmitting three images—based on red, green, and blue filtering—a complete and accurate color picture can be recreated.

Each image consists of a raster scan of 800 lines, and each line is divided into 800 pixels, for a total of 640,000 pixels/image. The intensity of each pixel is digitized with eight-bit resolution (256 distinct intensity levels) to produce a single byte per pixel, requiring a total of $8 \times 640,000 = 5.12$ million bits/image. The digitized image is enhanced on-board the satellite to improve its contrast and sharpness, since the very weak sunlight produces a relatively dark picture (like taking a picture of a room of people using only a small candle at the far end of the room for light). Data compression is then used by the satellite computers to reduce the number of bits by about 70%, using advanced algorithms: Each line is divided into five sections and the intensity of each pixel in a section is expressed in terms of its intensity relative to the first pixel in the section. Only the first pixel absolute value of intensity is transmitted in each section, with succeeding pixel intensities sent as difference information relative to this pixel's value. The earth-based receiving stations and computers use the reverse of this algorithm to reconstruct the original values.

Due to the distance and time lag between the satellite and earth, it is impractical to request retransmission if noise corrupts the received data. Many bits are

$$32.4 + 20 \log (6000) + 20 \log (37{,}000) = 32.4 + 75.6 + 91.4 = 200 \text{ dB}$$

(Recall that loss is subtracted, or expressed as negative dB and then added.)

Satellite receiver antenna gain of +20 dB.

Transponder gain of +90 dB.

Satellite transmitter antenna gain of +25 dB.

Downlink free-wave path loss governed by the same equation but using the 4-GHZ downlink frequency equal to $32.4 + 72 + 91.4 = 196$ dB.

Receiving ground station antenna gain of +25 dB.

By adding up all these figures, the total signal path gain is

$$+50 - 200 + 20 + 90 + 25 - 196 + 25 = -186 \text{ dB}$$

For a 100,000-W uplink transmitter output, the standard dB formula shows a 2.5×10^{-14} W signal at the ground station receiver front end.

---

added to the compressed data for error detection and correction, so that even received bit patterns with errors can be corrected. The data rate used for the satellite-to-earth path from Neptune is 21,600 bits/s. Although this means that a single picture takes several minutes to send, it allows the earth receiver to have a narrow bandwidth and so limit noise entering the system. This maintains an acceptable SNR, which is the most significant challenge for the data bits to be received and recovered. The original 22-watt signal from *Voyager* reaches the earth with a signal level of 10^{-14} W, a loss of 307 dB. (For the transmission from Jupiter, only 500 million miles from the earth, the data rate was 115.2 kbits/s, but the rate had to be reduced for the Neptune distance to maintain acceptable SNR.)

Earth tracking stations. The primary antennas used for tracking *Voyager*, transmitting commands to it, and receiving signals from it are part of the Deep Space Network operated by the Jet Propulsion Laboratory of the California Institute of Technology, and NASA. Ground stations are located in California, New Mexico, Spain, and Australia, to provide continuous coverage as the earth turns. Each ground station has one 70-m (230-ft) dish and two 34-m dish antennas. The three primary stations were aided by an array of 27 physically separate but electronically connected 25-m antennas in New Mexico. All ground stations were linked together and communicated both their tracking information and received data with each other continuously and immediately, so that any one could take over as dictated by earth rotation.

To communicate from the earth to *Voyager*, ground stations used 100-kW transmitters. The 70-m dish antennas were rebuilt for this mission to improve their beamwidth aiming accuracy and focus on the feedhorn, since every femtowatt (10^{-15} W) received is critical. The smoothness of the dish skin was improved from within 0.8 mm to within 0.1 mm of a perfect parabola, which increased the antenna gain by about 2 dB.

The result of the *Voyager 2* mission was reception of high-resolution, sharp, clear color pictures of Neptune and its captive orbiting satellites from as close as 3000 miles of their surfaces. Although the *Voyager 2* was launched in 1977 and designed only to function as far as Jupiter (500 million miles) and Saturn, it operated out to Uranus (1986, 1.8 billion miles) and Neptune (1989, 2.8 billion miles) before leaving our solar system. Engineers and scientists on the ground changed *Voyager's* course and data rates, as well as its algorithms, to achieve the extra, originally unplanned parts of the mission. Note how the success depended on analog-to-digital conversion of the video image pixel intensity, image enhancement algorithms, data compression, error correction and detection, low data rates, reception of 10^{-14}-W signals, and the largest dish antennas available.

This is far less than the 1 pW (1×10^{-12} W) needed by this receiver. Solutions to overcome this deficiency include increasing the transmitter power, antenna gains, transponder gain, and decreasing the signal needed at the receiver input by using lower-noise input circuitry. The path loss cannot be changed except by going to much lower frequencies (which have other types of losses in the atmosphere) or a low-altitude, nongeostationary satellite.

Similar but more complex calculations are performed for the SNR part of the link budget. The additional complexity arises because the cumulative effect of additive noise depends on where it is added to the signal—the SNR impact of 10-μW of noise on a 1-W signal is much less significant than on a 100-μW signal.

Review Questions

1. Why is the ground station a critical part of a satellite system?
2. How is the user baseband signal converted to the GHz range and higher power?
3. Why is precise antenna positioning important? Why are geostationary satellite ground station antenna steerable?
4. What is the link budget? What factors enter into it?
5. Explain how both signal power and SNR figure in the link budget. Why is the effect of noise more difficult to calculate?

19.4 LORAN NAVIGATION

Navigation is the art and science of knowing where you are. Behind this simple statement is a great challenge that has continuously faced explorers and sailors from before recorded history, and still is a major concern for modern ships and airplanes. Navigation on the open seas, without landmarks or recognizable markers is difficult: it is made much easier and more precise with the aid of communications systems; this is also true for aircraft which must determine their position while moving at hundreds of miles per hour. Any error in position results in wasted time while going in the wrong direction, can cause serious consequences if the craft runs out of fuel, or even crashes when the ship runs aground or the airplane crashes into a mountainside.

LORAN (Long-Range Navigation) was developed during World War II as a major electronic tool for navigation off the U.S. coasts, at ranges up to 800 to 1000 miles. The first version, LORAN-A, was discontinued by 1980 and replaced by LORAN C. Although originally intended for larger ships, LORAN is now used in many recreational boats and aircraft, due to the decreased cost and increased simplicity of the required electronic circuitry. Systems that are equivalent to LORAN are also available worldwide from many other countries, and LORAN is being expanded to cover the Great Lakes and the continental U.S. landmass. There is also some research investigating whether LORAN can be used to locate cars and trucks (with built-in maps) and trains.

19.4 Loran Navigation

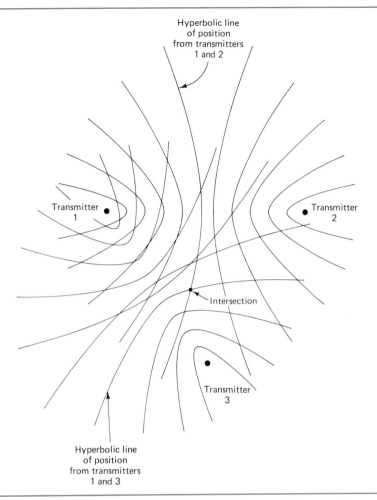

Figure 19.8 LORAN transmitters used to develop hyperbolas, used for position plotting.

The basic principle of LORAN is to have the ship or aircraft receive specially coded signals from pairs of powerful, shore-based transmitters at known locations (Figure 19.8). The time interval between the two coded signals in each pair is precisely measured by the receiver and is converted to distance using the propagation speed of radio waves (distance equals speed × time). From basic geometry it is known that the only set of points (the locus) that can have a specific time (and distance) difference from two sources is a *hyperbola*, and the specific hyperbola depends on the value of the time (distance) difference. Thus far, LORAN has placed the craft's location somewhere on a curved line of a hyperbola but has not identified the specific point.

The location is fixed more precisely by using another pair of transmitters (one can be the same as from the first pair). The two resultant hyperbolas intersect and define one or two unique points (depending on how they cross) which define the location(s) of the receiver. When there are two points, they are usually located far enough apart that the ship navigator can easily tell which one is not a possible location, but if there is uncertainty, a third transmitter pair is used to resolve the conflict. With a LORAN system under ideal operating conditions, absolute accuracy of better than $\frac{1}{4}$ mile is achieved and a change in position of about 100 ft is detected.

Transmitters and Signals

LORAN stations broadcast at carrier frequencies around 100 kHz, with the transmitter power in the multimegawatt range. A chain of stations is set up in each geographic area, and there are now 17 chains with a total of 50 stations. Within each chain, a master station transmits first, followed by the secondary stations in a specific order in an overall signal sequence called a *group repetition interval (GRI)* (Figure 19.9). The master station transmits eight pulses spaced 1 ms apart, followed by a ninth pulse 2 ms later. Each secondary station in the chain follows after a prescribed delay and transmits eight pulses spaced at the 1-ms interval. Phase modulation and a special phase scheme are used to make it easier for the receiver to distinguish master station pulses from secondary station pulses.

One key to the accuracy of the calculated result is that the transmitters must be precise in their coded signal transmission times. LORAN stations now use multiple *cesium clocks* (a type of atomic clock) for highest precision and for redundancy, in case one should fail. The clocks from the multiple stations in a chain are synchronized with each other once and then maintain time with such accuracy that they will differ by only 1 second after 1 million years.

LORAN systems errors are mainly due to propagation problems: The transmitted signals do not travel along the earth's curved surface in the shortest distance between source and receiver (in what is commonly called a "straight line" except that the earth is round). Atmospheric conditions, boundaries between hot and cold air masses, and land/air/water boundaries distort the travel path, so that the actual distance is greater than the timed value computation shows. In addition, the same signal may split and travel on more than one path, leading to two or more received signals and consequent ambiguities. These are reasons for using several fixes from more than two pairs of transmitters, if possible.

LORAN Receiver Circuitry

For navigation measurements, the beginning of the third cycle of the transmitted pulse is used as the reference point and a bandpass filter is used to extract the pulses from the carrier. Typically, this filter has a −3-dB bandwidth of 21 kHz

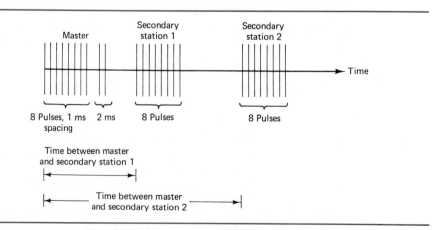

Figure 19.9 Loran GRI sequence.

around its 100-kHz center frequency, and its envelope distortion in the passband (equivalent to time delay), which would corrupt the accuracy of the received pulse timing, is kept to a minimum. One filter design is shown in Figure 19.10a, with the response shown in Figure 19.10b.

Review Questions

1. What does LORAN cover in the United States? What is typical accuracy?

2. How does LORAN develop the two lines of position that define a single point where they cross? Why are these lines hyperbolas rather than straight lines? How is the second hyperbolic curve determined? What if there are two points of intersection?

3. How does transmitter timing specify a unique hyperbola?

4. What is the LORAN carrier frequency and transmitter power? What do LORAN signals look like? What is the GRI? What is the sequence of signals for master and secondary stations in a chain?

5. Why is the synchronization accuracy of transmitters critical? How is this achieved?

6. What propagation problems cause LORAN errors?

7. How can an accurate clock and sextant be used for determining longitude? Assuming no error in sextant measurement, what clock accuracy is needed for results to be precise to 30 miles after 6 weeks (assuming that the clock was set perfectly at the start)?

8. Why is the clock and sextant method unsatisfactory? What can't it determine? When can't it work? What are its sources of error?

9. What are the sources of error in triangulation? Why does a small angular error correspond to a large positional error?

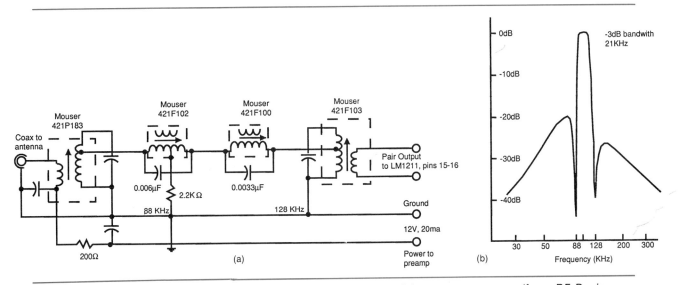

Figure 19.10 (a) Loran filter design; (b) frequency response (from *RF Design*, pp. 23, 25, August 1988).

Rudimentary Navigation Methods

The earliest form of navigation used the position of the stars to identify direction and distance. This requires complex star charts and tables, and clear night sky, of course, and cannot be automated easily. Nevertheless, it is believed that star navigation was used by ancient peoples who sailed across the Mediterranean Sea or across other large open bodies of water.

The sun is also used for navigation. A *sextant* identifies the angle of the sun above the horizon, and the difference between local time and some standard time gives the ship *longitude* (east–west position). Using the 0° longitude time, which goes through the historical Greenwich, England observatory as a reference, a ship has a clock set to *Greenwich Mean Time (GMT)*. At sea, the ship navigator notes the instant that the sun is directly above (12 noon local time) and reads the GMT time on the clock. The difference shows the longitude, with each hour corresponding to 15°, $\frac{1}{24}$ of the earth's 360° span. Position accuracy depends on the accuracy of the clock that is carried on-board and the precision with which high noon is detected using the sextant.

To indicate how valuable the clock and sun method is, in 1714 the British government offered a huge reward of 200,000 pounds—equivalent to several million dollars today—for a chronometer (a precise clock that could withstand the rigors of a sea voyage) that could determine the ship's longitude to 30 miles after a six-week voyage, equivalent to roughly 3 seconds per day time error. The prize was finally won in 1735. (A $20 digital watch today has the same accuracy.) However, an accurate clock does not solve the navigation problem entirely: The sun angle must still be determined (which is impossible in bad weather or at night), any error in measuring the sun's angle causes significant error in judging position, the method only provides longitude [not latitude (north–south) information], and it is difficult to automate.

Triangulation

An early scheme for radio-based navigation used the angle between the ship and two transmitters at known sites. In *triangulation* (Figure 19.11) a loop antenna (Chapter 9) is turned to find a null in received signal strength from one transmitter. This establishes an angle line between the ship and that transmitter. The antenna is then turned to locate the angle to a second transmitter, which determines another line of position. The intersection of the two lines is the present position.

This method requires manual operation (expensive automated nulling receivers have been built but are cumbersome) but does work regardless of time of day and weather. However, it is relatively inaccurate. An error of just $\frac{1}{2}$° in the measured null angle from a transmitter 500 miles away produces an error of approximately 4 miles; a 2° angle error has a 16-mile position error at the same distance (error is approximately equal to the distance between the transmitter and receiving antenna, times the sine of the angular error). Although this is not acceptable for many navigational needs, it complements the sun/time method or star method, both of which are weather-dependent and can provide only longitude.

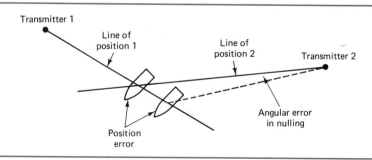

Figure 19.11 Triangulation fixes position based on angles to two known transmitters.

19.5 SATELLITE NAVIGATION

The *Global Positioning System (GPS)* and the similar *NAVSTAR* (*Navigation System using Timing and Ranging*) represent application of the extraordinary advanced technology of satellites to solve society's oldest problem of accurate, reliable, 24 hour/day navigation. This system uses a combination of ground stations, orbiting satellites, and special receivers to provide position information with accuracy to better than 20 m horizontally and 30 m vertically, and is designed for ships, aircraft, and even land vehicles. The entire system will have 18 satellites when complete (in the 1990s) to provide coverage virtually anywhere on the globe, and a full complement of 24 satellites to provide redundancy and a way to cross-check results.

GPS is based on accurate knowledge by the vessel of the position of each of four satellites, as calculated by the time with respect to the user. The user position is determined by extensive numeric calculations based on signals received from the four satellites. The differences in time of signal arrival at the user for the signals from these four satellites is used first to determine the distance between the user and the satellites. A subsequent set of calculations converts user position, with respect to four satellites, into position with respect to latitude and longitude (since the satellite orbits are known).

The basic scheme is shown in Figure 19.12. Each satellite continuously sends a precise time signal, indicating what time it has. These reach the user after various amounts of propagation delay, due to the varying distance between the user and each satellite. The user receiver compares its own clock time with the time signal as received from each satellite. Their time difference, when multiplied by the speed of light, indicates the range to the satellite.

Since there are three unknowns—position coordinates x, y, and z of the user—three equations from three satellites are needed for solution. However, there is a fourth unknown: the error in the user clock. Any error in this clock affects the accuracy of the time-difference measurement. To eliminate this clock bias as an error factor, the fourth satellite yields a fourth equation and provides for an overall solution for the four unknowns from four simultaneous equations: user position in x, y, and z coordinates, and the clock bias (which is effectively canceled out). (*Note:* The receiver clock is an inexpensive, crystal-controlled clock accurate to a few milliseconds. An atomic clock could be used by the user to eliminate the need for the fourth reading and equation, but this would make every user receiver much more costly.) The user x, y, and z position results are converted to latitude and longitude by additional calculation. (Note that if only three satellites are accessible or in range, the receiver can still produce a result using its own uncorrected clock, but with reduced overall accuracy.)

The equations shown with the figure are relatively simple and can be computed in a few milliseconds. In real systems there are many other correction factors that are applied to compensate for inaccuracies in the satellite orbits, Doppler effect on the received signal (Chapter 21), variations in propagation speed on the satellite signal as the signal passes through layers of the ionosphere, and similar, very subtle effects. With present technology, the GPS receiver requires between 1 and 3 minutes to acquire the time stamp, perform all calculations, and display the location information.

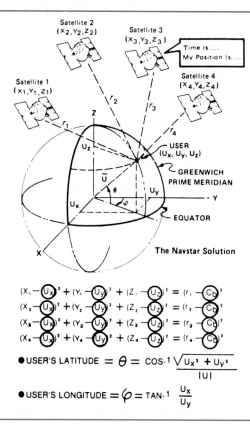

Figure 19.12 GPS satellite scheme and basic equations (from *Aerospace and Electronic Systems,* May 1987, copyright IEEE).

The complete GPS system has three segments:

1. The *control segment,* with master control stations (ground stations), which assures the overall system performance and accuracy.

2. The *space segment,* consisting of the satellites themselves, transmitting time codes and orbital position information to the users

3. The *user segment,* the actual user electronic circuitry, which must receive signals from the four satellites, compute the time differences, and determine position. There is, of course, no limit to the number of receivers that can use the system simultaneously, since the satellites' broadcasts are not directed to a specific user receiver but are broadcast to all.

We can look at each segment function in more detail:

Control Segment

The control segment uses fixed-location ground stations at precisely known latitude and longitude in Alaska, Guam (Pacific Ocean), Hawaii, Kwajalein Island (Pacific Ocean), Diego Garcia (Indian Ocean), Ascension Island (off West Africa), and California, among others. These stations monitor the signal from each satellite as it passes overhead and compare the actual station position to the GPS computed position. At the same time, each station synchronizes the satellite clock to the GPS time from the master station in California. Based on results of the calculations, the ground monitor station determines what errors the satellite has in its time and position information and sends the satellite a new navigation message with corrected information, which the satellite then incorporates into its message to users. This overall process continually checks the system accuracy and compensates for any errors, so long-term drift and errors do not accumulate.

Space Segment

GPS satellites are not geostationary but have an average height of 9476 miles and take approximately 12 hours to complete one orbit. The satellite orbits have an inclination of 55° to the equator, so each satellite covers most of the earth's surface as the earth turns under the satellite path during a 24-hour period. Within each satellite are two atomic (cesium) clocks and a third, spare clock.

Transmitters within each satellite operate in the L-band near 1.57542 and 1.2276 GHz. Both frequencies are transmitted simultaneously with nearly the same information, which permits corrections to be made for ionospheric propagation delays. Since the change in signal velocity depends on frequency, the user computer can correct for this source of error by making measurements at the two frequencies.

The signal transmitted by the satellite is not a simple message—a time stamp—such as "It is now 12:07:59" using ASCII characters. This message would be too easily affected by noise, and the time precision would be poor: taking a long time to say what time it is blurs the precision with which you mark the time point itself. Instead, a scheme based on a *pseudorandom sequences (PRSQ)* is used (Section 13.4). In a pseudorandom sequence, the bits all appear to be random 1s and 0s, as if generated by flipping a coin: 1 for a head, 0 for a tail. However, the pattern is generated by specially designed circuitry, and the same circuitry will always generate the same apparently random pattern. Thus the name *pseudorandom* means that the pattern appears truly random but is really not, and can be re-created as necessary with appropriate circuitry. Eventually, the pattern begins to repeat itself, after a number of bits, which depends on the complexity of the generating circuitry.

The satellite transmits a pseudorandom sequence, and the receiver has circuitry to generate the same pattern. In the discussion of the user segment (below), we will see how this is used to measure time differences precisely.

Each satellite in the GPS provides two unique pseudorandom patterns to allow the receiver to identify which GPS satellite's signal it is receiving. The *P pattern* is very long, sent at 10.23 Mbits/s, and can provide more precise timing (and therefore more accurate location information) but takes longer to manipulate at the receiver. The shorter but less precise *C/A pattern* code is sent at 1.023 Mbits/s; the accuracy it provides is sufficient for many applications.

User Segment

The user segment is the receiver for the GPS system. Its first function is to act as a front end for signal reception of the two GPS frequencies and to recover the P and C/A pseudorandom bit streams that have been modulated onto the carrier. The receiver also produces its identical pseudorandom patterns. There is a critical difference, however: the received bit streams and the locally generated bit

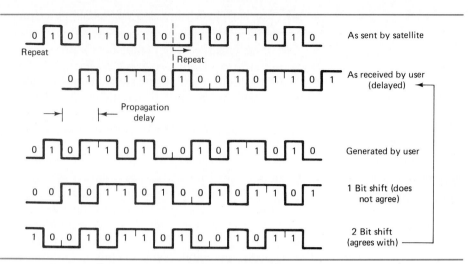

Figure 19.13 Alignment of psuedorandom code for precise time measurement.

streams are not aligned in time. In other words, the bits are the same if written out, but the received and local streams are out of phase with each other. What the receiver then does is delay (shift) its locally generated version until it aligns precisely with the received pattern. The amount of shifting it must produce for alignment tells the receiver what the time at the satellite was when it sent the signal.

Consider the situation of Figure 19.13: The satellite transmits the pseudorandom pattern . . . 01011010 . . . repeatedly beginning at time 0000. The receiver also generates the same . . . 01011010 . . . pattern beginning at what its clock says is time 0000. Due to the distance between the satellite and the receiver, the signal takes an amount of time equivalent to two bit periods to reach the receiver. Thus, when the receiver is locally producing . . . 01011010 . . ., at the same time it receives . . . 01101001 . . . (the same pattern but delayed by two bits). To match the two, it must adjust the shift of its own pattern by two bits. Once the match occurs, the receiver can say: "I had to shift my pattern by two bits to match the satellite signal, so the signal took two bit periods of time to reach me." In this way the precise time difference (and thus distance) between the transmitting satellite and the receiver is measured.

The pseudorandom pattern has several advantages over the more obvious scheme of sending a time marker and a message with the marker. The pseudorandom pattern can be picked up at any time and used for the pattern matching and time-shift determination. In contrast, a time marker can only be used at the times it is sent, and if the receiver begins reception in the middle of the associated message, it must wait for the next complete marker plus message. Second, the resolution of the pseudorandom scheme is to one bit period, while a time stamp can resolve only as precisely as the message it sends with the time marker. "It is now 12:00:01, it is now 12:00:02" has 1-s resolution. Finally, when the time stamp marker (and message) is corrupted by noise, it becomes almost useless. The pseudorandom sequence, however, has been shown by advanced analysis to be very resilient when noise corrupts some of its bits: It simply takes a little longer for the receiver to make the match, while the corrupted bits work their way through and out of the matching circuitry. As an additional feature, the chance of errors inadvertently causing a match at the wrong place in the two bit streams is extremely low.

Once the receiver of the user segment identifies the satellite and time difference for each of the four satellites, it begins the very involved calculations that produce the final position, with all corrections and compensations. This is all done by specialized microprocessor-based circuitry in the receiver.

Review Questions

1. How does GPS use the measurement of time at each satellite, as seen by the receiver, to solve the position problem?

2. Why are four satellite signals needed if there are only three unknowns? What information does a fourth satellite time signal provide? Why does this allow a simple crystal clock (not atomic) at the receiver?

3. Why are the calculations that convert time clock differences to position simple in concept but complex in practice? What other corrections must be applied?

4. What is the GPS ground segment? How many stations are there? What does each ground segment do?

5. What is the GPS satellite height and orbital period? What frequencies are used for GPS satellites? Why does each satellite use two separate frequencies?

6. Give two reasons why a pseudorandom bit pattern is used for time marking rather than a simple time message. How is this pattern generated?

7. How does the receiver determine the time delay using the received pseudorandom pattern and its own version? What are the advantages of the pseudorandom pattern in terms of noise tolerance, ability to pick up the pattern at any point, and resolution?

8. Why are two patterns (P and C/A) used? What are their rates?

SUMMARY

Satellite-based communications systems provide worldwide communications capability between two ground stations. Orbits can be polar, inclined, or parallel to the equator, proving different regions of earth coverage and footprints. The satellite itself combines directional antennas, a power subsystem, a sensitive receiver/frequency translator/amplifier transponder, and a control and telemetry system so that the ground station can manage (and correct) the satellite operation. There are also auxiliary rocket thrusters and control for stationkeeping, to compensate for the orbit distortions and decay caused by the earth, moon, and sun. Frequencies of operation are usually in the GHz range.

The complete signal path begins at a ground station, proceeds to an uplink, to the satellite itself, to a downlink, and ends at a receiving ground station. The sum of all signal gains and losses along the way constitutes the link budget. The final signal, as received by the ground station, must have enough power and a high-enough signal-to-noise ratio to be useful. Careful design is required to make this possible, since losses in the signal path are so high.

Radio signals are used for navigation as well as for communications. Triangulation fixes a user position at the intersection of direction lines to transmitters at known locations using a direction-finding nulling antenna. However, the accuracy is only fair and the range is only up to about 1000 miles from a transmitter. The LORAN system uses specially coded signals, broadcast by a chain by synchronized transmitters. The coding lets the user calculate a set of hyperbolic lines of position, and these determine the position where they cross. The newest approach to the age-old problem of navigation is a satellite-based system in which the user determines the time of signal travel for coded signals from four separate satellites to the user, and computes position based on these times. The overall calculation also corrects for the many small and subtle—yet significant—possible sources of error, to provide extremely accurate position data anytime, anywhere.

Summary Questions

1. What basic problem is solved by using a satellite? What are the alternatives? What are the roles of the uplink ground station, uplink, satellite, downlink, and downlink ground station in a satellite communications system?

2. Compare three types of basic orbital planes. What is the orbital plane in a geostationary orbit? What is the significant feature to the user of this orbit?

3. Explain why a satellite is an extremely sophisticated and complex design.

4. What are typical uplink and downlink frequencies? Why do they differ?

5. What is the function of the power subsystem? Where is power generated and stored? What are major concerns in this subsystem?

6. What is the role of telemetry and control? Why is this function made as independent as possible in the satellite? Compare its antenna to the regular uplink and downlink antennas. Why is it different?

7. What are the two main functions of the open transponder? What are typical output power levels?

8. What is the satellite footprint? Compare the features of a narrow footprint and a wider one. What determines the footprint?

9. What is satellite access? What is the goal of access schemes? Compare TDMA and FDMA in terms of operation, benefits, drawbacks, and static versus dynamic allocation.

10. What are the key functional blocks of a ground station?

11. Why is antenna positioning difficult, yet critical?

12. What is the link budget? What does it show, and what can it be used for? What factors enter into its calculation?

13. Explain how a precise pulse pattern and transmitter pair allows LORAN to determine a position hyperbola. Why is it a hyperbola and not a straight line? How is a position point determined? How does LORAN use time to calculate position?

14. Compare LORAN to triangulation in terms of accuracy, complexity of transmitted signal, receiver complexity, and sources of error.

15. What is the GPS concept? How many satellites are in the total system?

16. What raw data are converted into distance? How? How is distance information from the satellites converted into exact position?

17. What are the three GPS segments? What does each do?

18. Why can the GPS receiver use a crystal clock instead of an atomic clock, and why is this an advantage?

19. How is a pseudorandom bit pattern by the receiver used to calculate propagation time between satellite and receiver? What are three advantages of this scheme compared to a simple time message?

20. Why does GPS use two pseudorandom sequences? What are the GPS transmitted frequencies, and why are there two distinct frequencies?

21. Compare LORAN and GPS for achieved accuracy, transmitted signal type, overall system function, receiver operations to perform, and sources of possible error.

20
Cellular Telephone Systems

CHAPTER OBJECTIVES

When you have completed this chapter, you will understand:

- The way that cellular systems overcome the limitations of conventional mobile radios and radiotelephones
- The functions that a cellular system implements at both the base station and phone, along with some of the ICs needed
- The types of messages that a cellular phone and base station exchange between each other

INTRODUCTION

Tx 825-845
Rx 870-890

large subscriber capacity
efficient use of spectrum
nationwide capability
widespread availability
adaptable

Cellular systems allow many users to share the limited number of frequency channels available in a region. Although the amount of spectrum (and channels) available cannot accommodate all users if each user were assigned its own channel, a cellular system lets frequency channels be reused in the same area by dividing the entire region into many smaller cells.

Of course, this simple concept does not get implemented without a great deal of support. The central base station of each cell is not independent of other cells in the region: All cells are linked together and must carefully coordinate their activities, including allocation of channels and passing control of a mobile phone from one cell to another as the user moves and crosses cell boundaries. The phones used are much more sophisticated than combined transmitter/receivers. They use digitally controlled synthesis to set the operating frequency and must change frequency in milliseconds, based on commands from the base station. The phone unit also has its power level controlled by the base station, to keep signal strength to the minimum necessary for reliable communications while avoiding generating interference.

20.1 THE CELLULAR CONCEPT

It is straightforward to understand that hand-held transmitter/receiver combinations (*transceivers, walkie-talkies,* or *radiotelephones*) provide two-way communication between a pair of users. For example, standard 27-MHz citizens band (CB) units are available to let private individuals keep in touch over a range of up to about 20 miles (depending on conditions) with a transmitted power of less than 5 W. The number of pairs of users that any band supports is limited by the number of available *channels*— assigned carrier frequencies and bandwidth—and the allocated portion of the frequency spectrum, along with the mode of operation (half-duplex or full duplex). Forty channels, for example, support 40 half-duplex or 20 full-duplex conversations simultaneously. A channel is usually not permanently assigned to a specific physical transceiver: In a basic form of dynamic allocation, a pair of users takes an empty channel, and when they are done, another user pair can access the channel.

There are problems with this scheme of direct contact between pairs of users. First, the range is limited. To communicate over greater distances requires more power, larger antennas, and more sensitive receivers. Signal strength will be sufficient when the users are closer, but falls off as they move apart from each other. In fact, transmitter power may be wasted if the two users are very close, yet they will have too little power for other distances.

A better solution is to have a single, *base station* to serve as a central node for all users. Each user communicates with the base station, which acts as a higher-powered relay station (similar to a satellite but fixed on earth). The base station receives a signal and rebroadcasts it at higher power, often with a higher and more effective antenna than that of any hand-held (or vehicle mounted) transceiver antenna. A base station also makes the area of coverage uniform for all units. As long as the transceiver signal can reach the base station, the transceiver signal will be rebroadcast with enough power by the station to reach the users on the fringe of the coverage area. At the same time, the transceiver power does not have to reach as far—it only has to travel from an edge of coverage to the centrally located base station rather than to a user on the opposite edge of coverage.

A base station improves the achievable range but does nothing to increase the limit on the number of channels and users within a fixed frequency band. One method to improve the utilization of available channels is to use one channel as a *hailing* or *calling* channel. Using this one channel, the mobile phone user contacts the other desired party and then both switch to another channel for talking. This scheme means that only one channel is used for the inefficient waiting and setup process, and the remaining channels are used for actual conversation (somewhat similar to common channel signaling in telephone operation between central offices, Chapter 16). Although this increases the use of the available channels, there is still a fundamental limitation on the number of conversations that can be carried simultaneously. As a result, very few people who needed or wanted a mobile phone for their car or boat could get one. Even special services such as police and fire, which have separate frequency bands in the spectrum, had restrictions on the number of vehicle radiotelephones that could be handled.

This problem was studied for many years, and the most attractive solution was to divide the total area to cover into many smaller *cells*. The total area that is

Figure 20.1 Cellular phone (courtesy of Motorola, Inc.).

covered by the radiotelephones is actually comprised of a quilt of many smaller cells, each touching adjacent cells so that there are no areas without coverage. The only drawback to this solution is that it requires fairly sophisticated management of the cells and how they interact, and thus the cellular phones themselves (Figure 20.1) must be capable of some advanced operations and functions.

Cells expand the number of users and the coverage range, Figure 20.2, since the total area that must have cellular phone coverage is divided into smaller cells, about 2 to 10 miles in diameter depending on specific location. These hexagon-shaped cells provide complete signal coverage over the area of interest. At the center of each is a relatively low-powered transmitter base station, and different cells are assigned different groupings of channels in the overall band. Let's call these groups by the letters A, B, C, and so on, where each letter represents many channels but there are no channels in common between groups A, B, C, ...

Suppose that a cellular phone is in a cell with group A channels and communicating with the base station of that cell. In adjacent cell B, another cellular phone is in use. In another cell A, a third phone is in use. As long as the power used in the first A-group cell is low so that signals do not spill over to the second A-group cell, it is possible to have two different users on the same frequency in

20.1 The Cellular Concept

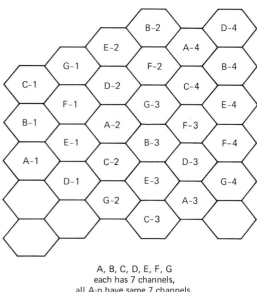

A, B, C, D, E, F, G
each has 7 channels,
all A-n have same 7 channels,
all B-n have same 7, etc.

Figure 20.2 Use of cells to allow frequency reuse: seven cells, repeated four times.

different cells, yet without any interference. However, this system appears very restrictive since users cannot talk outside their own cell, which is fairly small.

The next key step is to link the base stations of each cell together so that signals from one cell can be passed to another cell. These intercell links are not the channels used by the cellular phone to link to its base station. Instead, they are dedicated links of wire or optical fiber, and communication between cells in no way interferes with the cellular phone signals. Using these intercell links, any base stations can be connected, and thus two users in any of the cells can be connected to each other.

What we have done here is to subdivide the overall area so that frequencies can be reused many times in the larger area, without interference. To get around the small size of the cells, which is what allows this reuse, the cells themselves are linked together directly. A conversation between a user in an A-group cell and a user in a B-group cell actually is routed from the A-user to the A base station, to the B base station, and to the B user. A conversation between two different A-group users travels via the first base A-group station, over the hardwired link to the second A-group station, and then to the second user.

There is one problem: what to do when the car in the area covered by A-group channels crosses the cell boundary into a B-group cell. If the cellular phone is now useless, this is a poor system. Instead, a more advanced scheme is used for all the cells. As the car crosses from A to B, the cellular phone is switched to one of the available frequencies in the B group, as "control" of that phone is handed off to the B-cell base station. Meanwhile, the now-open A-group channel in the first cell is available to a new user in the A-cell area. This could be a new conversation just getting started, or someone who crosses into the A cell while continuing a conversation that began in another cell.

Thus we see the keys to a cellular system:

By dividing the total area into small cells, many users can actually be using the same channels *at the same time*, since adjacent cells have different frequency groups.

The base stations of each cell are linked together so that a conversation can pass from one cell to another. This links users in any two cells, regardless of their channel assignments.

A cellular phone changes from one channel to another as it crosses cell boundaries, even while the conversation is in progress. This is very different from simply using a different channel each time a conversation starts and staying with that channel until the conversation is over.

These principles are straightforward, but the real management of the overall cellular system is extremely complex. Among other functions, the base stations must be linked into a network, and each station must carry not only the user conversation but also information about what channels are available so that a cellular phone can be switched to an available channel as it crosses the boundary. A lot of detailed coordination is needed, since each base station must keep track of its open channels, ensure switchover to a new channel for units entering its boundaries, reassign channels that were in use but are now open, and also connect to other base stations to pass actual user signals.

The cellular phone system is really an advanced network with extremely dynamic allocation so that as many users have access as possible. These are concepts we looked at in connection with networks (Chapter 18) and satellites (Chapter 19). What makes the cellular system unique is that the user equipment—the phone themselves—are connected by radio links, and these phones are extremely sophisticated, yet fairly inexpensive (under $1000). In the next section we study the operation of a cellular phone unit in more detail.

[*Note*: A cellular phone and cellular phone system are very different from a portable phone ($50 to $200), used to give freedom in the home from the local loop phone wire. A portable phone extends the wired phone link using a radio channel. It is dedicated to that particular wired phone acting as its base station; it cannot function independent of the base station function provided by the wired phone, and no network protocol is needed to support it. The portable phone is really a limited-range wireless implementation of the last few feet of wire of the local loop (up to several hundred feet).]

Cell Size and Number of Possible Users

Suppose that there are 49 channels available in a part of the spectrum reserved for cellular phones. If the overall area is set up as a single cell (a single base station), only 49 users can be served at the same time. Now divide the overall area into many cells and assign channels to each so that no adjacent cells have the same channels. To ensure that no adjacent cells have the same channels assigned, seven groups of frequencies are used, A through G, each with seven channels. Suddenly, the total number of users who can be served is multiplied by the number of times that a group is reused.

If the overall area served with the original 49 channels is divided into 28 cells, each frequency group is available in four different cell locations. Therefore, the total number of channels apparently available is now $4 \times 49 = 196$, which is a significant increase. Of course, more cells provide a further increase in the number of conversations that can occur over the same number of channel assignments.

The cell size is a factor. If the cells are large and there are only seven channels in each cell, the total number of users per square mile is small. But as the cell size is decreased, the number of users that can be handled, per square mile, gets fairly large. Of course, there is a trade-off: Each cell requires a base station and links to the other cell base stations. At some point the network that links the base stations becomes very complicated and expensive to install and operate reliably. In heavily populated areas, though, it may be worthwhile since the demand for phones is so high.

Note that the required base station power and antenna size decrease as the cell size gets smaller, since the area to be covered is reduced. This means that base stations for smaller cells are less expensive, which partially offsets the cost of additional base stations and interstation links.

The size of all the cells does not have to be the same, and the original cell sizing and layout can be changed after the system is installed to accommodate new areas of population and need. In *cell splitting*, a single cell is subdivided into smaller cells. This provides the greatest number of users per square mile where it is needed, and less capacity in population outskirts. Of course, cell splitting means that new base stations must be built, but this is usually justified in terms of need and potential revenue.

Another advantage of cellular design is that the coverage area can be incremented and expanded easily. Suppose that business and user needs increase to the north of the original area that is split into cells. New cells are then built to the north, and the new cell map can follow the locations that have demand for services. In contrast, a single base station noncellular format expands its coverage only by increasing the transmitting power and antenna size. This fails to provide new channels and also aggravates the reliability of coverage problem, since mobile phones in these new fringe areas may not be powerful enough to reach the base station reliably, or there may be geographic obstructions over the longer distances.

Cell Shapes

The first analysis and experiments with cellular systems used circular cells not hexagons. Circles are easiest to visualize, but to provide complete coverage they must overlap at their edges; if there is no overlap allowed and the circle perimeters just touch, there will be coverage gaps. Hexagons provide complete coverage without overlap, yet are roughly similar to circles.

The actual cell shape is determined by the antenna pattern. Cellular systems use a combination of antennas to form the hexagonal pattern desired. In fact, other shapes, such as triangles or squares, can be used to provide 100% coverage without overlap, but it is much harder to shape the antenna pattern to these because of the sharp corners of triangle or square radiation patterns. All this discussion of ideal patterns, whether hexagon, circle, or other shape, must be viewed with the reality that cell shapes in real systems are not this simple. Local geographical features, including hills, small and tall buildings, and ponds and lakes, distort the pattern from its ideal shape. The cellular system company usually tries to correct and compensate by modifying the antenna elements to generate a nonhexagon pattern, which will end up as a hexagon after the propagation distortion. Alternatively, adjacent cells are shaped so that they compensate for each other, and a pattern weakness in one cell is covered by additional strength in its neighbor's radiation pattern.

Review Questions

1. How can using a base station improve the performance of a two-way radiotelephone system compared to having many users without such a station?
2. Why does the use of a separate hailing channel plus the talking channels improve the utilization of the system to a limited extent?
3. What is the concept of cellular systems? How does it provide for a greater number of simultaneous users despite the limited number of channels?
4. What is the function of the base station? Why is the base station in one cell connected to other cell base stations?
5. What happens when a cellular phone user crosses into another cell?
6. What are three key concepts in a cellular system?
7. Why does a cellular system have relatively complex management?
8. What is cell splitting? When is it done, and why? What happens to the base station when there is cell splitting?

20.2 CELLULAR SYSTEM IMPLEMENTATION

In the preceding section we saw that dividing the overall area into many smaller cells increases the number of users for a fixed number of channels (since frequencies can be reused in nonadjacent cells) but that the cell base stations and the cellular phones are fairly complex. All of the cell base stations are linked together by a *mobile telephone switching office* (*MTSO*), which acts as the central office and management node for the group. The MTSO is also the linking point to the regular phone system as well, since, of course, cellular phones need to connect to standard, wired phones. By going through an actual call, we can see how the system works in practice.

All base stations continuously send out identification signals (ID) of equal, fixed strength. When a mobile unit—the cellular phone itself—is picked up and goes off-hook, it senses these identification signals and identifies which is strongest. This tells the phone which cell it is in and should be associated with, since a cellular phone may pick up more than one base station signal. The phone then signals to that cell's base station with its ID code, and the base station passes this to the MTSO, which keeps track of this phone and its present cell in its data base. The phone is told what channel to use for talking, is given dial tone, and the call activity proceeds just like a regular call. All of the nontalking activity is done on a setup channel with digital codes.

During the call, the base station keeps track of all the active and unused channels in its cell and the six adjacent cells. As the mobile phone user travels, the relative signal strength received at the base station will vary. When the signal strength as received at the base station of its cell (call it cell A) decreases while increasing at the base station of adjacent cell B, this indicates that the phone is nearing the boundary between cells and will soon be crossing into the next cell. Note that cell B is monitoring the signal strength of the phone in cell A, although the phone conversation is linked not to it but to cell A.

As the phone crosses from cell A to cell B as indicated by the changes in signal strength, base station A hands the phone over to cell B and the phone is told to switch to a new frequency that is under control of cell B and available in that cell (this switchover involves no action by the phone user). The mobile phone switches to its new transmit/receive channel pair in about 100 ms, which is not noticed by the user in conversation. It is able to switch so easily because it uses frequency synthesis under microprocessor control (Chapter 15) to establish both its transmit and receive channel frequencies, and the microprocessor in turn is directed by commands from the base station. The data base showing which units are in which cells, and which channels are in use or available, is changed to show that the channel of cell A is free while the newly assigned channel in cell B is now taken.

The power output of a cellular phone is 3 W maximum. To reduce interference to other phones and to minimize overloading with base station receiver when the phone is close, this level can be varied by seven -1-dB steps, down to 0.7 W. The output power is controlled not by the user but by the base station, which issues a power up/down command depending on the signal strength it receives and other conditions. As with the channel changing, this is invisible (actually, "inaudible" is more appropriate) to the user.

Base Station

The base station is located at the center of the cell and has two antenna sets. A simple half-wave vertical element is used for transmitting to all the phones in the cell, with a power of about 5 W, depending on the cell size and terrain. This antenna provides uniform coverage in all directions (not considering distortion by geographical features and structures) and vertical polarization. Each mobile phone requires a vertically polarized antenna, too, usually a small "whip" mounted on the rear window of the car.

The receiving antenna at the base station is more complex. It is usually built of six separate half-wavelength vertical dipoles arranged symmetrically around a central pole, each dipole with a *corner reflector* behind it (Figure 20.3). The reflector provides directivity by shaping the dipole field pattern to a 60° beamwidth, corresponding to one section of the hexagon, while providing a gain of 15 to 18 dB.

There are two major reasons for this receiving structure. First, it shapes the base station field pattern to the desired hexagon. Second, it provides some direction-finding capability for the base station, which is needed to help the base station decide when and where to hand off the mobile unit to an adjacent cell. The six received signals, one from each segment, are compared against each other. By judging the relative strength of a mobile unit signal at these, the base station has a good indication of where the mobile unit is in the cell (Figure 20.4). For example, if the received signal is strong at segment 2 but equally weaker at segments 1 and 3, the mobile unit is probably in the center of the antenna 2 segment. In contrast, if the signal is equally strong in segments 1 and 2, the mobile phone is probably along the border between these two segments.

Note that this is an elementary form of direction finding, similar to triangulation (Chapter 19). In that scheme the direction to a transmitter was measured by a rotatable antenna to determine a line of position. By using measurements to two transmitters, two such lines are determined and their intersection identifies a unique position. In a cellular system, only one reading is used, to determine a rough line of position that corresponds to the sector of the cell. The physically

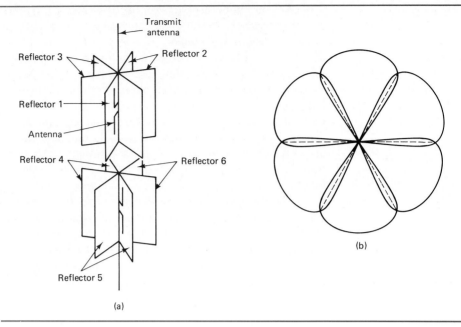

Figure 20.3 Base station receiver antenna (a) design and (b) pattern.

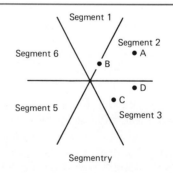

Vehicle location	Relative signal strength in segment					
	1	2	3	4	5	6
A	Low-medium	High	Low-medium	0	0	0
B	Medium	Medium	0	0	0	0
C	0	Low-medium	High	Low-medium	0	0
D	0	Medium-high	High	Medium	0	0

Figure 20.4 Localizing a phone to one cell segment via relative signal strength in receiver antennas, compared to each other.

turning antenna is replaced by a six-segment fixed antenna which provides a coarse angle measurement to 60°, compared to the typical 1° resolution of a direction-finding navigational antenna.

The base station also must serve as a frequency translator, similar to the satellite transponder. It takes the signal intended for a specific cellular phone from another phone in the cell or another cell via the MTSO, and transmits it on the channel assigned to that phone; it also receives on the specific channel assigned to the phone. It must do this for all the cellular units within its cell, and change transmit/receive channels as required by the situation in the cell.

Cellular Phone Details

The block diagram of a cellular phone is shown in Figure 20.5. Since the phone is full duplex, there is complete circuitry for receive and transmit channels. A special *duplex filter* connects the transmitter circuitry, receiver circuitry, and antenna and directs the flow of signal power so that the receiver front end is not overloaded by the output of the relatively powerful transmitter output; at the same time it directs the weak received signal from the antenna to the receiver front end [the overall duplex filter function is similar to the hybrid coil of the telephone (Chapter 16) and is discussed in more detail in Chapters 21 and 23].

The link between the voice transceiver circuitry and the microprocessor-based controlling circuitry occurs at three points: where the controller extracts base station messages from the received signal and also sends out its status and identification signals; where the frequency synthesizer sets the receive and transmit channels and changes them if directed to do so; and at the output power control, where the power level is adjusted by the microprocessor, once again acting on base station directives.

Cellular phones typically use frequencies from 825 to 890 MHz, which were previously assigned to UHF TV channels 73 through 83 but were reassigned by the Federal Communications Commission to mobile radios such as cellular phones. Phones transmit in the band from 825 to 845 MHz while receiving in the band from 870 to 890 MHz, with a 45-MHz separation between the transmit and receive channel pair. Channel 1 transmit/receive is 825.015/870.015 MHz.

The user voice signal (300 to 3000 Hz) is broadcast as narrowband FM, with ± 12 kHz deviation for 100% modulation. This corresponds to a bandwidth of 30 kHz, using Carson's rule (Chapter 6), which states that the bandwidth is approximately twice (the sum of the carrier deviation plus the signal bandwidth) = 2 (12 kHz + 3 kHz) = 30 kHz. Because of the high demand for cellular channels there is no guard band between channels, and channels are assigned every 30 kHz (unlike broadcast FM, for example, which uses a 50-kHz guard band between adjacent station assignments). The cellular receivers have a selectivity of 50 dB, to minimize interference from adjacent channels while providing 1-μV sensitivity (a moderate, not supersensitive figure, but superior sensitivity is not needed at these distances). A total of 665 transmit and 665 receive channels are assigned in the frequency spectrum that is allocated.

Cellular systems provide capacity for a large number of users, despite limited available signal spectrum. In many ways, they are a combination of common channel signaling and satellite systems. Instead of the space-based relay point (the satellite) and earth-fixed ground station, in a cellular system the base station is a fixed relay point and the ground stations are mobile. However, the goal is the same; to make use of a higher relay point for increased distance via the central

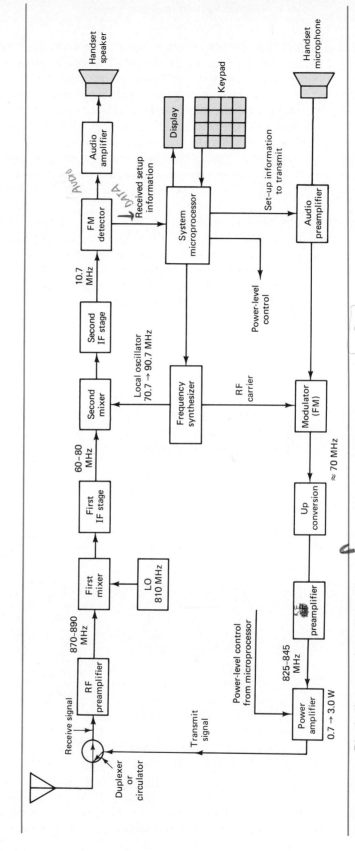

Figure 20.5 Block diagram of a cellular phone incorporates sophisticated analog circuitry, digital circuitry, and microprocessor control.

base station. The dynamic access of the cellular system is like the satellite system, as well. Each available channel functions as a transponder in the satellite, and the objective is to use all the transponders 100% of the time. To accomplish this, the access strategy must constantly allocate available channels, note newly freed channels, and manage overall use of the channels.

Like the common channel signaling scheme of the phone system, the cellular system also use a separate scheme for managing the network. The talking channels are used for conversation, but the setup and management are carried out by a separate set of links (in this case, the setup channels and coding of the cellular system). By doing this, very advanced messages and directives can be passed to the user phones without interfering with the actual conversation in progress. This is analogous to the use of a single common channel between central phone offices, where the common channel carries all types of information that helps the system make efficient use of the available resources.

Of course, all of this would not be possible if the cellular phone itself did not have a channel that allowed it to receive and interpret directives from the base station. This is possible due to the microprocessor and digital synthesizer within the phone itself, which acts as a sophisticated controller for the phone even while a conversation is ongoing, yet keeps all of this advanced activity inaudible to the person talking on the phone.

ICs for Cellular Phone

Cellular phone systems put a premium on small size and low power consumption as well as low cost. Integrated-circuit manufacturers have developed special ICs that incorporate many of the functions needed in a cellular phone. Philips Corporation (also known as Signetics in the United States) has a set of six ICs (Figure 20.6) which work together as the basis for an integrated set of cellular circuits and include almost all internal circuitry except the receiver and transmitter RF stages (the speaker, microphone, keypad, and display are external). All the ICs in this set are linked to each other and pass information among each other via what the manufacturer calls an I^{2}C (inter-integrated circuit) bus, which is a small-scale local area network (existing within a single circuit board) that uses bus topology, master/slave protocol, and requires just two wires as the interconnection among all the ICs (Chapter 18). This simple bus hardware saves board space and reduces the need for numerous line drivers/receivers, and although it is relatively slow compared to a more complex LAN, the speed of operation is sufficient for the application.

The six ICs consist of:

1. The PCB80/83C552 single-chip eight-bit Microcontroller, which is a microprocessor with additional circuity built in to optimize it as the ''nerve center'' and ''brain'' of a cellular phone. The CMOS microcontroller contains memory for the actual program that runs the phone, and this memory can be expanded with external memory ICs if needed. Also included are six digital input/output ports (eight bits each), an eight-bit analog/digital converter (Chapter 11), a timer circuitry to measure elapsed time periods, a UART (Chapter 17), and a watchdog timer.

2. The UMA1000 Data Processor for cellular radio handles all the signaling and supervision tasks associated with the phone operations, similar to the

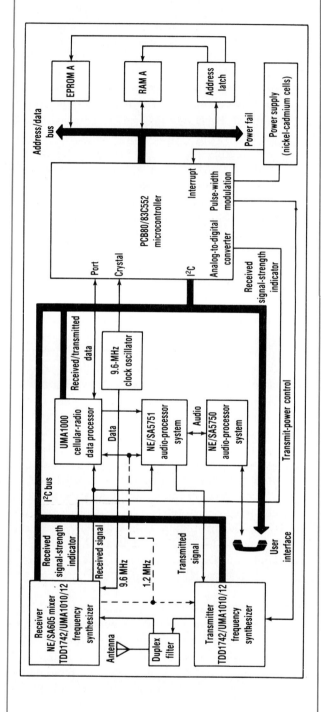

Figure 20.6 This set of six ICs use a mix of analog and digital ICs to provide a high degree of integration and nearly all signal processing functions needed in a cellular phone (courtesy of *Electronic Design*, Penton Publishing Co.).

BORSHT function of a central office (Chapter 16). This IC relieves the microcontroller of the many small but essential details of managing the phone as part of a cellular network (as a UART does for serial bit transmission). It consumes just 2 mA of current since it is built with CMOS technology.

3. The NE5750 and NE5751 Audio Processor System uses a pair of ICs for all the transmitted and received voice signal functions. The NE5750 provides a low-noise microphone preamplifier with adjustable gain, a voice-operated transmit/receive switch, a compressor to increase modulation index, a speaker amplifier, and an earphone amplifier. The NE5751 contains the I^{2}C bus interface for the pair, along with 300- to 3000-Hz bandpass filters for the transmitted and received audio signal, a deviation limiter, pre- and deemphasis circuitry (Chapter 6), a digitally controlled volume control with 30 dB of range, the DTMF generator, and power-up and power-down circuitry.

4. The NE605 Mixer/Oscillator and FM IF System combines the required second-stage mixer, local oscillator, amplifier, and demodulator functions on a single IC. The mixer can operate up to 500 MHz, and the internal oscillator can be used up to 150 MHz, or an external signal as high as 1 GHz supplied. The IF operates at 21 MHz and provides 102 dB of signal gain. A quadrature detector is built in for demodulation, along with a mute circuit that cuts off all audio output when the received signal is very weak or there is no received signal. A signal strength indicator is also included, with 90 dB of range.

5. The TDD1742 Frequency Synthesizer is a CMOS IC used for either transmitter or receiver frequency control, with an external prescaler. It is designed to switch frequency of operation in minimum time and without any false intermediate-frequency steps.

Although the IC set does not provide 100% of the circuitry for a cellular phone, it provides the most complex circuit functions and minimizes the number of additional components needed to complete the phone design. It shows how specially designed digital ICs and analog ICs can reduce the component count of a sophisticated design dramatically (although limiting the opportunity that a technician has to probe within a change the circuit operation, or troubleshoot below the major IC level). For perspective, the entire set sells for $60 to any phone manufacturer who orders 10,000 sets.

Review Questions

1. What is the role of the MTSO in a cellular system? How does a cellular phone know what cell it is in when it goes off-hook to make a call?

2. What is the sequence of events before actual dialing begins? How is call setup done?

3. How does the cellular phone cross from cell to cell? What does the base station direct in the cellular phone?

4. Why is the phone's output power level controlled?

> 5. What is the power level of a typical base station? What does its transmit antenna look like?
>
> 6. How does the base station find the approximate position of a cellular phone in its cell? Why does it need to know this?
>
> 7. Why does the base station do frequency translation for the signals under its control?
>
> 8. What is the frequency band where cellular phones operate? What modulation and deviation are used?

20.3 CELLULAR SYSTEM PROTOCOL AND TESTING

The cellular phone system uses analog signals to transmit voice but also has a separate digital channel to handle supervision and signaling information. [*Note:* A new standard is being developed for *digital* cellular phones, where the voice itself is digitized at the phone (or base station) and transmitted in digital format, but the final standards are not yet established.] In this section we examine the protocol and format used to manage activity between the phone and base station.

The 665 transmit and 665 receive channels are divided into two independent and coexisting cellular networks of 333 (or 332) channels, run by different companies by order of the FCC, which wants to promote competition in each region. Since each cell is surrounded by up to six adjacent cells, the 333 (or 332) channels are divided into 45 user voice channels/cell. The remaining 18 or 17 channels are reserved as control channels, needed for network access, call establishment, and call management.

A phone call over the cellular network actually requires simultaneous operation on two full-duplex channels, with one full-duplex channel for control and the other for the user voice. The base station transmits on the *forward control channel* and the *forward voice channel*; the mobile phone uses the corresponding *reverse control channel* and *reverse voice channel*. The cellular phone calling cycle begins when the phone is activated, equivalent to going off-hook for a wired phone. At this time, the phone unit reads its internal *numeric assignment module* memory location to determine which of the two cellular networks is its normal "home" network. Then the phone scans the control channels, locking onto the strongest received signal.

Next, the phone accesses the network. This is done by monitoring the busy/idle bit of the forward control channel and seizing the control channel when this bit indicates that the channel is free. The initial transmission over the reverse control channel includes registration information. This registration tells the network which cell the mobile phone is in, so that incoming calls can be directed appropriately.

This registration information is sent at 10 kbps and includes the telephone's *equipment serial number* (*ESN*), 32 bits; its *system identification number* (*SID*), 15 bits; and its *mobile identification number* (*MIN*), 34 bits. The registration process is repeated whenever a phone switches cells. The ESN identifies the manufacturer and specific phone unit. The MIN is a binary-coded version of the mobile unit's phone number. Finally, the SID indicate the home network for the mobile phone (the network to which that phone and its owner subscribe and where the bill for calls should be sent). The SID also identifies a phone that has gone beyond its own network to the foreign network into which it has traveled. (Just as a cellular base station is linked to the conventional phone system so that a cellular phone can contact a wired phone, cellular systems themselves are linked together via their base stations so that cellular system-to-cellular system communication is possible and so that phones which leave their home network can be served.)

At this point, the user can dial the desired phone number and press the SEND button. The dialing information travels over the reverse control channel as a data word in the *mobile station call initiation* message. When the dialed information is accepted by the base station, the phone is instructed by the base station to go to a specific voice channel for talking. Now, the caller can speak and listen through the phone. Once the call is established, the cellular network continues to exchange control signals with the phone via the control channel. It can tell the phone to increase or decrease power, or assign it a new channel before handing the phone off to a new cell.

The standard protocols used for the forward and reverse control channels are shown in Figure 20.7. Words within a message are repeated five times to help ensure successful reception. The forward channel also includes filler bits (similar to null codes) for maintaining a continuous data stream, thus easing synchronization by the phone. The forward channel *digital color code* (*DCC*) is added so that the control channel is not confused with a control channel from a nonadjacent cell that is reusing that frequency. A *discontinuous message* (*DM*) allows the base station to command the mobile phone to change its transmit power.

Reverse control channel messages handle registration, call initiation, and call reception. The mobile phone reads the base station's DCC and returns a coded version of the DCC, verifying that the phone is locked onto the correct signal. When the call is finished, the user presses the END button and the phone generates 1.8 seconds of signaling timeout over the reverse voice channel. The network monitors the voice channel so that if the user forgets to press the END button but still hangs up the phone, the system "times out" and ends the call anyway. An interesting note: In cellular systems, the user is billed for total time of accessing the network, including dialing, listening to far-end ringing, and listening to busy signals, since all of these activities tie up one of the limited number of channels. In contrast, with conventional wired phones, you are billed only for actual talking time.

Testing Cellular Phones

The block diagram of a cellular phone reveals two things: There is a lot of complex circuitry integrated together, yet many of the subsections perform standard communications functions such as modulation, mixing, and frequency synthesis. A test system for cellular phones can be built from standard communications test instruments that are traditionally used for noncellular systems, combined with a special interface specifically intended for cellular systems and a personal computer and software acting as the test controller.

The Hewlett-Packard HP8957S cellular radio test system (Figure 20.8) is designed to run a complete series of tests on cellular radios. This system consists of a *modulation analyzer,* for studying the signal produced by the radio; a *synthe-*

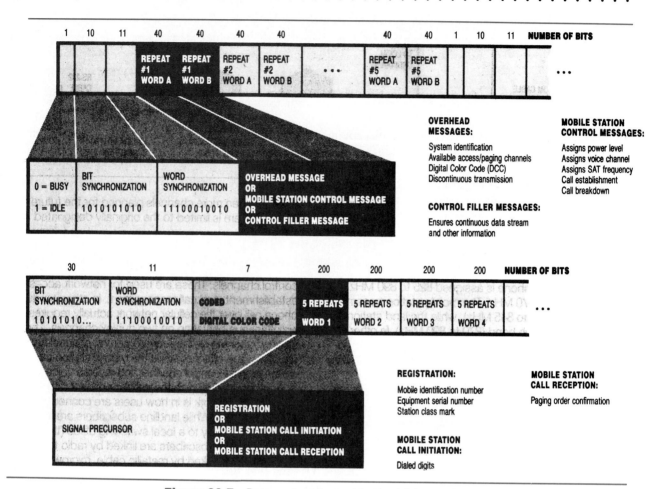

Figure 20.7 Data protocols used in the (a) forward and (b) reverse control channels show the information needed to set up and maintain a call (reprinted from October 1989, *Data Communications,* Copyright McGraw-Hill, Inc.).

sized signal generator, to produce signals which simulate those from the base station; and an *audio analyzer,* to perform measurements on the recovered audio signal. All of these standard instruments connect to a *cellular radio interface,* which simulates a cell base station and produces the signaling messages used in cellular systems. All instruments are linked together via the IEEE-488 bus (Chapter 18) under the control of a personal computer that runs a program designed specifically for this purpose.

The computer and program set the signal generator to necessary frequencies, levels, and modes of modulation, specify which readings the modulation analyzer and audio analyzer must make, and analyze the test results, all with a high degree of automation. There is also a power supply to provide nominal +12 V dc to the cellular phone, controlled by the program and computer to produce various power supply voltages and fluctuations similar to those that occur in use in a car's electrical system. Final results include specifications for interference from adjacent channels, frequency accuracy, carrier purity and undesired distortion, frequency response and sensitivity, and power measurements.

Figure 20.8 HP8957S test system designed for cellular phones (courtesy of Hewlett-Packard).

Review Questions

1. Explain how 665 transmit and 665 receive channels result in 45 user channels/cell for each cellular network. How many channels are available for control in each network?

2. Why are two full-duplex channels needed for each user? In what fundamental way does the duplex talking channel pair differ from the duplex control channel pair?

3. What types of registration information does the phone unit report when it starts the call sequence? What is the role of each?

4. What sequence of messages occurs when a phone call is initiated? Why is the message to the phone repeated five times?

5. What allows a cellular network to handle a phone which is not based in that network system?

6. What instrumentation is needed to test a cellular phone? What functions are measured? What results are produced? What are the roles of the personal computer and software in the testing operation?

SUMMARY

Cellular systems divide the overall area to be covered by mobile radiotelephones into smaller cells so that frequencies can be reused. The overall system combines networking of base stations, a mobile telephone switching office, and dynamic allocation of available frequencies with the sophisticated operation of a two-way radio that is digitally controlled by a microprocessor, which in turn takes direction from the base station. The key to the success of cellular operation is that the many base stations can share information about channel use and movement of the phones so that a phone which crosses from one cell to another is smoothly "handed off" to the new cell. A data base is maintained which identifies all units in use and their location, and is used as part of the information needed for the best operating decision. In many ways, a cellular phone system combines aspects of computer networks, common channel phone signaling, and satellite systems.

Each cell base station has a low power transmitter of about 5 W, which provides the necessary signal throughout the cell area. Besides providing more available channels in an area, using many base stations means that the mobile units can be lower-powered as well, since they do not have to transmit as far. Also, "dead spots" in coverage due to geography and terrain are more easily overcome. As the area to be covered by the cellular system changes, new cells can be attached to existing cells in the direction needed. Similarly, individual cells themselves can be divided into smaller cells to provide more channels in an area that has a lot of new demand.

Summary Questions

1. Compare the use of separate hailing channel/talking channels, a central base station for the entire region, and a cellular system in terms of efficient use of available channels, reliability of performance over a large region, number and power of base stations, and operating complexity.

2. How does a cellular system operate in concept? What allows the same frequencies to be reused, without interference, by other phones?

3. How is a cellular system expanded to new, adjacent areas? How is it modified to handle an excess of new users in existing areas?

4. What is the basic function of the base station? Of the MTSO? How does the cellular system ensure that calls continue without interruption as the phone user crosses cell boundaries?

5. What is the sequence of actions when a phone goes off-hook, before dialing begins?

6. How do the base stations know approximately where the phone user is upon off-hook occurring? Why do they need to know this? How do they know when a call is in progress and the user is near a boundary?

7. What is the power level of a cellular phone? Of a base station? How and why is the phone output power adjusted?

8. What frequencies are used for cellular phones? How is this set—by the user or by the system? When and how does it get changed? What are the transmit and received channel spacing?

9. What are the modulation and bandwidth of a cellular voice signal? When is the guard band used?

10. Explain how and why the base station transmit and receive antennas differ in design and function.

11. Why are hexagons used as the basic cell shape? What other shapes are possible? Why are real cells not perfect hexagons?

12. What are the similarities between a cellular system and a satellite system in concept? What about those between the cellular system and phone system common channel signaling?

13. Discuss the type of information that a phone must report to the base station when a call is initiated.

14. How does the cellular system maintain control between the base station and the phone before, during, and after a conversation?

15. What four channels are operating simultaneously during a cellular call? What is the role of each? What is the protocol of each? How do these channels differ?

21
Radar Systems

CHAPTER OBJECTIVES

When you have completed this chapter, you will understand:

- The principles of radar and the critical factors in radar system performance
- The signals used in radar systems (power, frequency, and shape)
- Circuitry and components for radar transmitter and receiver systems, including those unique to radar applications
- Advanced designs such as bistatic, multistatic, over-the-horizon, and synthetic aperture radar systems

INTRODUCTION

The development of radar has dramatically changed aviation, warfare, and scientific exploration because it allows us to "see" things electronically that are visually blocked or are too distant for conventional optical methods of naked eye, binoculars, or telescope. Based on the simple principle of measuring the time (and therefore distance) between a transmitted signal and its received reflection from the target, radar has evolved into an extremely complex and sophisticated system that can detect tiny objects hundreds or thousands of miles away.

Practical radar systems require high-powered transmitters operating at GHz frequencies to produce large enough reflections to be detected by the receiver, with sufficient precision in angle and distance. The shape and nature of the target also play a large role in the radar's ability to function over the required distances. Some radars are designed to detect the presence of all targets, while others are designed to show only targets that are moving. Computers now play a large role in improving radar performance, by varying key parameters during operation and optimizing the receiver so that it has the best opportunity to detect the target's reflection correctly. More advanced radars use transmitters and receivers that are not located at the same site, HF frequencies that refract from the ionosphere to sense targets thousands of miles away, and moving antennas to simulate a much larger physical antenna. All of these advanced radars need sophisticated signal processing to make their results meaningful, and faster computers with appropriate software are making this practical.

21.1 RADAR CONCEPTS AND DISPLAY

Radar, an acroynm for "radio detection and ranging," uses the reflection of electromagnetic waves to detect an object at a distance. From its first serious use in the 1940s (during World War II, where it allowed the British air force to detect German bombers before they became visible and so intercept them), modern radar is capable of detecting the presence, shape, and motion of ships, aircraft, land vehicles, and people at distances ranging up to hundreds and thousands of miles, depending on conditions.

Radar operates by transmitting a known signal and then detecting the echo of this signal as it reflects from any object in the signals' propagation path (Figure 21.1). The transmitter and receiver are at the same location in most radars and share the same antenna. The distance to the object is easily calculated from the time between the transmitted signal and the received echo and is displayed on a screen similar to that of an oscilloscope. Different schemes are used for the transmitted pulse and echo receiver so that *range* and *angle* to the target object are determined, or the motion and velocity of the target are measured.

Although the concept of radar is simple, the effective implementation is difficult. The transmitted beamwidth must be very narrow, so that the target angle is precise; a wide beamwidth will indicate the angle to a very coarse value. This requires antennas that are at least several wavelengths in diameter, to provide the sharper beamwidth. Since early radars operated in the range 50 to 100 MHz because the electronic circuitry was unavailable to provide power and precise signals at higher frequencies, the beamwidths were relatively wide and imprecise. Today's radar systems use the shorter wavelength frequencies, from the multi-hundred MHz to multi-GHz ranges (see the table of frequency bands and designations in Chapter 19), where a sharply focused antenna that is many wavelengths in size is much more practical.

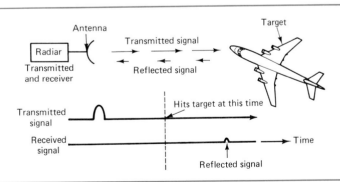

Figure 21.1 Basic radar concept involves detecting the echo of a transmitted pulse from the target while measuring the time between the transmitted pulse and the returned echo.

Chap. 21 Radar Systems

To make the situation more difficult, very little of the transmitted power reaches the target object because of the large distance and the unavoidable spreading of the propagating electromagnetic energy. In addition, the target itself reflects only a very small fraction of the energy that hits it, and this weak reflected signal is further attentuated as it travels back to the receiver over the same distance.

Radar Cross Section

The amount of power reflected by the target depends on many factors, including the size, shape, material (metal, plastic, wood, or water), and edges (sharp or rounded) of the target, as well as the frequency of the incident radar signal and the angle between the radar system and the target. Radar systems summarize all of these factors by the *radar cross section* (*RCS*), which has units of square meters and indicates the target size as seen by the radar. A flat piece of metal 7.5 cm (0.075 m) on each side (A = 0.075 m × 0.075 m = 0.0056 m²) has an RCS of 0.44 at 10 GHz, for example. A larger RCS means that the target reflects proportionally more of the transmitted beam that it intercepts.

A simple target such as a sphere will scatter the incoming energy in many directions, and only a very small fraction will be returned. In contrast, a large flat plate is a more effective reflector with much large RCS. Figure 21.2 gives the RCS of some basic shapes as a function of dimension and wavelength, as well as typical values for some common objects when viewed "head on" at 1 GHz.

Example 21.1

What is the RCS of the flat plate discussed above at 1 GHz?

Solution

From the formula, RCS = $4\pi A^2/\lambda^2$, where $A = (0.075 \times 0.075)$ and λ can be

Item	RCS (m²)
Sphere of radius a, $a \gg \lambda$	πa^2
Sphere of radius a, $a \ll \lambda$	$144\pi^5 a^6/\lambda^4$
Resonant dipole	$0.866\lambda^2$
Flat plate, broadside, area A	$4\pi A^2/\lambda^2$
Small airplane	1–2
Jet fighter	6
Large jet aircraft	40
Jumbo jet aircraft	100
Small pleasure boat	2
Cabin cruiser	10
Pickup truck	200
Car	100
Person	1
Bird	0.01

Figure 21.2 Typical RCS values show influence of shape, size, and wavelength.

calculated by $\lambda = c/f$, where $c = 3 \times 10^8$ m/s and $f = 1 \times 10^9$ Hz. Therefore, RCS = $4\pi(0.000032)/(0.3)^2 = 0.0044$ m², which is smaller by a factor of 100.

Example 21.2

What is the RCS of a sphere of radius 0.5 m when the frequency is 10 GHz ($\lambda = 0.03$ m)?

Solution
Since radius a is much greater than λ, we use the formula

$$\text{RCS} = \pi a^2 = 0.785 \text{ m}^2$$

The rudder of an airplane is a major factor in its RCS. A truck has a larger RCS than a large airplane, although the truck is physically smaller; this is due to the flat sides and sharp edges of the truck body. Poor radar targets (such as lifeboats) that do want to be visible often increase their RCS by carrying special lightweight signal reflectors designed for maximum RCS.

Radar Range

The range over which the radar system will reliably detect an object is determined by many factors. The reflected signal power must be large enough so that the receiver can detect it, and is therefore closely related to the SNR required for reliable detection. When the transmitting and receiving antenna are the same—which is usually the case—the signal power that reaches the receiver is

$$P_r = \frac{P_t G^2 \lambda^2 (\text{RCS})}{(4\pi)^3 R^4}$$

where P_r = received echo signal power (watts)
P_t = transmitted power (watts)
G = antenna gain (expressed as a ratio, not in dB)
λ = signal wavelength (meters)
R = range to the target (meters)

Example 21.3

What power is received when the transmitted signal is 1000 W at 1 GHz (0.3 m), the target is 10,000 m away (about 5.7 miles), the target RCS is 1 m², and the antenna gain is 40 dB?

Solution
First, convert the gain in dB to a ratio: 40 dB is a gain of 10,000.

$$P_r = \frac{(1000)(10,000)^2(0.3)^2(1)}{(4\pi)^3(10,000)^4}$$

$$= \frac{(1000)(1 \times 10^8)(0.09)(1)}{(1984)(1 \times 10^8)} = 0.045 \text{ W}$$

The *maximum range* that the system can achieve occurs when the received echo power is at least equal to the minimum signal that the receiver can detect despite unavoidable external and internal noise. There are several variables that increase this maximum range, and each has advantages and drawbacks. The transmitted power can be increased (more costly equipment and more power must be provided), the antenna can have higher gain (a larger, more complex design), the signal wavelength can be increased (but lower-frequency signals suffer more atmospheric noise and provide less range resolution, which is discussed in Section 21.2), lower-noise receivers can be used (costly or unavailable), or the RCS can be increased by changing target shape and size (of course, this is not often practical, and an enemy target will strive to decrease its RCS).

Note that the received power decreases with the fourth power of the range. This is because the transmitted beam spreads out and its power decreases with the square of the distance; similarly, the reflected power from the target spreads and also decreases with the square of the distance. The consequence is that to increase the range for fixed receiver sensitivity, signal wavelength, RCS, and antenna gain, the transmitted power must be increased by a significant factor. For example, doubling the useful range requires increasing the power by $2^4 = 16$; tripling the distance requires $3^4 = 81$ times the power. For these reasons, radar designs place a great deal of emphasis on higher-gain antennas and very low-noise receivers which can detect ever-smaller received signals.

Radar Displays

Early radar systems used *deflection modulation*, a simple variation of an oscilloscope screen to display the echo information (Figure 21.3). The horizontal trace, at a constant sweep rate, is started simultaneously with the transmitted signal, and the received signal is applied to the vertical input. The echo then produces a pulse or *blip* on the screen along the horizontal axis at a distance from the left-hand starting point that is proportional to the round-trip signal time. The horizontal axis is calibrated in range units proportional to the time units, since range is

$$R = \frac{c\,\Delta t}{2}$$

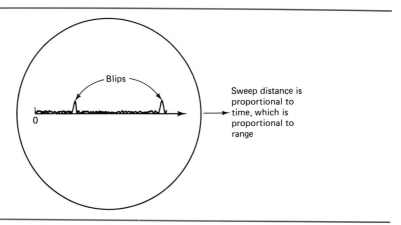

Figure 21.3 Deflection modulation screen indicates time as distance along axis.

where c is the speed of light and Δt is the total time between the transmitted signal and the received echo. This time is easily measured on the screen as the number of divisions between the sweep origin and the radar blip, divided by the known sweep rate (div ÷ div/s). The factor of $\frac{1}{2}$ is needed because the signal has completed a round trip to the target and back, covering twice the actual target range.

Example 21.4

What is the range when Δt is 15 μs?

Solution

Range = $(3 \times 10^8 \text{ m/s})(15 \times 10^{-6} \text{ s})/2 = 2250$ m, which is a little over a mile.

This type of display is adequate where the antenna does not change its direction. In many radar applications, however, the radar antenna physically or electronically scans the sky, rotating as a *search radar* around a complete 360° circle or sweeping back and forth through some smaller angle as an *aiming radar*. The single horizontal line display cannot directly show both the target angle and range, and although the angle could be displayed on a separate indicator, this requires that the operator look at two displays at the same time. Also, it is hard to determine movement of the target on successive sweeps.

In a *plan-position indicator* (*PPI*) the oscilloscope screen is modified to present a 360° image, with the radar system at the center of the screen (Figure 21.4). The fixed horizontal line of the simple display now rotates around the center, in synchronization with the antenna movement (a typical rate is 6 rev/min). In early radar systems this display rotation was accomplished by actually rotating the deflection coils (or plates) mechanically in sync with the antenna while power was fed to the coils or plates via slip rings. This relatively unreliable and expensive method has been replaced, and the scan is now done electronically by continuously varying the signals applied to the CRT x and y deflection circuitry in the correct mathematical relationship to produce the appearance of a rotating axis, or by a raster scan that creates the same visual effect.

While the axis rotates, the deflection circuitry pulls the CRT beam from the center outward at a known rate. The beam is cut off when there is no received echo, so the screen is dark; a received signal lets the electron beam hit the CRT face and causes a bright spot to appear. The screen display shows both the range and direction as a clear, easy-to-interpret image. As the target moves nearer or farther but at the same angle from the radar station, the change in range is shown by a line radiating out from the center at the correct angle. When the target moves crosswise with respect to the radar station, the echo points on each successive sweep of the screen generate a series of dots. The persistence of the CRT screen phosphor combines with the observer's eye persistence to form a single perceived line representing the change in target position.

Review Questions

1. What is the basic principle of radar?
2. What is RCS? What are some of the factors that determine it?

Figure 21.4 PPI screen shows a complete 360° sweep, with time (distance) measured from the screen center (courtesy of Sperry Marine Division of Tenneco).

3. What factors determine range? How can the maximum range be increased? Why is increasing transmitter power a difficult way to increase range?

4. How does simple deflection modulation operate? How does it show the target range?

5. What is PPI? How does it operate? How does it show both range and direction?

21.2 PULSE SHAPES

A *pulsed radar* system transmits a relatively narrow pulse by turning the high-frequency carrier on for a short time (Figure 21.5). The *pulse repetition frequency* (*PRF*) or *pulse repetition rate* (*PRR*) defines how frequently these pulses are sent. The *pulse width* indicates the precise duration of a pulse. The ratio of the pulse width to the total time between successive pulses (the inverse of the PRF) is the transmitter *duty cycle*. The radar receiver is designed to detect the reflected signal after each pulse is sent. The average power of the transmitter is equal to the peak power times the duty cycle.

As in most engineering situations, there is no single ideal value for the pulse

21.2 Pulse Shapes

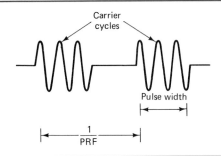

Figure 21.5 Radar pulse and carrier cycles.

width, PRF, or duty cycle. Instead, there is a trade-off in different aspects of achieveable performance. When the PRF is low, the radar will have to scan slowly to ensure that it does not miss targets. However, a high PRF brings with it more serious problems. In a basic radar system the *maximum unambiguous range* or *maximum usable range* is limited by the time between pulses. Consider what happens if a pulse is sent out every 1 μs, and the distance to target is such that the echo takes more than 1 μs to return (Figure 21.6). The receiver will not be able to tell if the echo is from the first pulse or from a successive pulse, and the target would appear closer than it really is. As the PRF increases, the maximum unambiguous range decreases.

Example 21.5

The PRF is 1000 Hz. What is the maximum unambiguous range?

Solution
Any pulse that returns after one time period between pulses will be confused with a return from the second pulse, so the maximum range occurs when $\Delta t = 1/\text{PRF}$. Using this in the range equation, we have

$$R = \frac{(3 \times 10^8)(0.001)}{2} = 0.0015 \times 10^8 = 150{,}000 \text{ m} \qquad \text{about 93 miles}$$

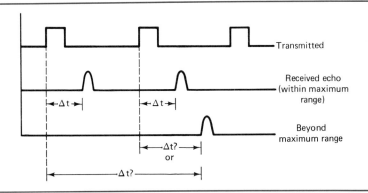

Figure 21.6 Maximum unambiguous range and false distance reading situation.

There is also a compromise in the pulse width. The receiver sees the echo if the returned signal energy (signal level) is above a threshold based on receiver sensitivity and noise. This energy is really the integral over time of the received signal power, and increasing the transmitter power will increase range by providing more returned power for a fixed pulse width. However, as we saw in the preceding section, this is a costly way to achieve more range, since just doubling range requires 16 times the transmitted power.

The alternative is to increase the width of the transmitted pulse (and therefore the duty cycle) so that more energy is transmitted although the peak power is unchanged. This improves the performance, but it requires a transmitter capable of providing more average power. More important, the wider pulse increases the radar *blind spot*, the time where no reflection can be detected because the transmitter is still on while a signal is already returning. Just as more power is needed to see targets farther away, too wide a pulse prevents the receiver from seeing targets that are close by since the receiver is blinded by the transmitter power. The radar ability to distinguish between two closely spaced targets along the same line is *range resolution* and is determined primarily by the narrowness of the pulse width.

Example 21.6

The pulse width is 0.01 ms. What is the blind-spot distance?

Solution

Any signal that returns within 0.01 ms will not be seen. Using the range equation yields

$$R = \frac{(3 \times 10^8)(0.00001)}{2} = 1500 \text{ m} \quad \text{about a mile}$$

Wider pulses do have other advantages besides offering more average power. The bandwidth of a wider pulse is less than that of a narrow pulse (see Chapter 2), so that the receiver can have a narrower bandwidth input filter. This reduces the amount of broadband external noise power received along with the desired signal, and such noise is one of the major limiting factors in the distance that can be achieved.

To summarize: In a pulsed radar system, a low PRF means that there will be long periods between the radar pulses, which limits the ability of the system to search quickly for unknown targets or track known targets. Increasing the PRF improves tracking and searching but reduces the maximum unambiguous range at which targets can be confirmed without confusion. Increased transmitted signal power increases the achievable range for a fixed pulse width but is a difficult way to achieve increased range. The alternative is to increase the pulse duty cycle, but this increases the short-range blink spot within which no reflection signal will be seen. Some advanced radars use computers that constantly vary the PRR and pulse width to overcome these constraints, based on sophisticated algorithms with which they have been programmed.

Doppler Radars

The pulsed radar shows echoes from any reflecting object in the signal path. In practice, this means that many of the echoes are of stationary objects (buildings, geographic features such as hills, and similar) that do not change on successive sweeps of the antenna. Generally, most of the interesting and important information occurs when the target (or targets) changes position, whether they be aircraft, airport landing strip, trucks, or other vehicles. These stationary objects are called *clutter* and can overwhelm the radar system. They do this in two ways: first, they provide many reflections that must be processed, and this can overload the signal-processing capability of the radar system; second, they are often larger than the moving targets, producing reflections that obscure the smaller reflections of moving targets.

To overcome this problem, radar systems based on the *Doppler shift* were developed. The Doppler shift in frequency occurs when there is relative movement between a signal source and the signal receiver as a fixed number of signal oscillation cycles is stretched over an increasing distance or compressed into a decreasing distance. (We already saw Doppler shift in satellite systems, Chapter 19.) Although speeds of even thousands of miles/hour cause only small shifts in the frequency of waves traveling at the speed of light, this shift is measurable and useful. The amount of change f_d, in hertz, in the original frequency is

$$f_d = \frac{2v \cos \theta}{\lambda}$$

where v is the velocity between the radar and the target (meters/s), θ is the angle between target line of movement and a line between the radar and the target, and λ is the signal wavelength (meters).

When the target is moving toward the radar, the velocity is positive and the signal shifts higher in frequency. For targets moving along the line between radar and target, θ is 0° and cos 0° = 1, so the Doppler shift is at its maximum; for targets moving crosswise to the line between radar and target, θ is 90° and cos 90° = 0, so there is no apparent shift in frequency. The relative velocity between the radar system and the target is $v \cos \theta$.

Example 21.7

What is the Doppler shift when tracking a car moving away from the radar at 100 mi/hr, at 1 GHz? (λ = 0.3 m, and 1 mi/hr is about 0.5 m/s.)

Solution

f_d = 2(100 mi/hr)(0.5 m/s per mi/hr)(cos 0°)/(0.3 m) = 333 Hz, which is a very small fraction of the frequency.

Example 21.8

Repeat Example 21.7 when the frequency is 10 GHz.

Solution

The shift f_d is now 10 times as great, or 3330 Hz. This is why higher frequencies are preferred in Doppler-based systems, although the carrier frequency is higher and the radar requires more complex circuitry.

In the *moving target indicator* (*MTI*) pulsed radar, the Doppler shift is used to minimize clutter effects and only see targets that are moving. This is done by using the phase shift of the received signal with respect to the transmitted signal, to sense where there has been a change (target movement) from the previous scan. (Recall that phase and frequency are intimately related, and a change in one produces a corresponding shift in the other.) The received echo passes through a discriminator circuit (see Chapter 6), which converts the phase information to amplitude variations. The signals from the first pulse is delayed by one PRR and then subtracted from the signal from a successive pulse, so that only the difference is seen. Figure 21.7 shows a typical example of this, where the successive signal echoes have already been delayed to line them up in time at the same starting point.

Note two important facts about MTI. First, it eliminates clutter signals which are present and may even be larger than the moving target echo, since stationary object echoes cancel themselves out. Moving targets that are smaller than the stationary ones can thus be seen. Second, it reduces the effect of noise and thus increases the useful range of the system for a given power, RCS, and related factors. This reduction occurs because noise is random, and on successive echoes the noise values do not simply add up as the desired echo signal does. Instead, noise power increases with $\sqrt{N}$ for N successive additions. Therefore, the signal power increases N times, whereas the noise power increases only $\sqrt{N}$ times, which is a much slower rate.

There is a problem in pulsed Doppler systems. When the PRF equals the Doppler frequency f_d or any integer multiple of it, the target velocity cannot be distinguished from the stationary clutter and will appear to have no Doppler shift at all. The relative velocity $v \cos \theta$ that results in a Doppler frequency equal to the PRF (or some multiple of it) is called the *blind speed,* since the radar system will not see the moving target at this speed and confuses it with clutter. Blind speed v_b is given by $v_b = n\lambda \text{PRF}/2$, where n is any integer (1, 2, etc.)(velocity in m/s; PRF in Hz).

Repetitive Addition of Signal plus Noise

The fact that most forms of noise corrupt the desired signal by adding to (or subtracting from) the signal is often used in communications systems that have unavoidably low SNRs, such as radar and deep-space satellites. The signal is sent N times, and all the received versions are added together. Since the noise is random, the total received noise increases $\sqrt{N}$ times, while the total received signal power increases linearly by a factor of N times, a much greater amount.

Suppose that the signal and noise powers are both equal to 0.01 W (SNR is 0 dB), and the signal is sent 16 times. The final signal power will be $16 \times 0.01 = 0.16$ W; the noise power is $\sqrt{16} \times 0.01 = 0.04$ W. The SNR is now 0.16/0.04 = 4, or 6 dB. As N is increased, the SNR improves even more. This technique is used to recover signals that are buried in noise, with SNRs below 0.

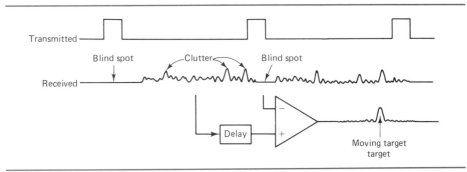

Figure 21.7 MTI radar operation to eliminate clutter and show only moving targets, which may also have smaller echoes than clutter.

Example 21.9

What are the first three blind speeds at 1 GHz when the PRF is 1000 Hz?

Solution

The first blind speed $v_b = 1(0.3 \text{ m})(1000 \text{ m/s})/2 = 150$ m/s (about 300 mi/hr); the next blind speed is 300 m/s (600 mi/hr), and the third is 450 m/s (900 mi/hr). The blind speeds form a "comb" of speed values at which the target will not be visible, evenly spaced out between the speed zones at which the target is visible.

Just as newer radar systems overcome near-range blindness (due to pulse width) and maximum usable range limitations (due to PRF) by varying these parameters, blind speeds that limit the usefulness of the pulsed Doppler system are avoided by varying the PRF over some range around a nominal value. This is part of the major trend in modern radar systems: to use the power of the microprocessor to improve performance by adjusting the principal performance specifications (PRF, duty cycle, pulse width) as needed, and then accounting for these changes in the final signal processing of the received echo. This is especially critical in military systems, where the enemy can exploit the weaknesses of the radar system and avoid detection by flying at the blind speed, just inside the blind spot distance, or just beyond the maximum range due to the PRF.

Another approach that uses the Doppler effect is *continuous-wave* (*CW*) radar, where there are no individual carrier pulses and instead the carrier is always transmitted (therefore, this radar has no blind speeds). In a CW system, the carrier is at frequency f, and the received signal is at frequency $f \pm f_d$ ($\pm$ depending on movement toward the radar or away from it). The receiver must filter out the relatively high-power transmitted carrier it sees (since it is shares the antenna) from the very low-power reflection it gets. A special component called *circulator* (similar in function to a hybrid in a telephone but of different construction) is used for this, reducing the undesired CW signal by 30 to 40 dB. The small amount of transmitted signal that does get past the circulator is mixed in the receiver as a reference to determine the difference in phase, and thus the velocity, between the transmitted signal and the received signal. (In a pulsed radar system, the receiver input is short circuited during the transmission so that its front-end stage is not overloaded; this also causes the blind spot.)

CW radars are used where only velocity information is of interest and actual range is not needed. A prime example of this is police radar for catching cars traveling above the speed limit; another example is measuring the motion of waves on the water. CW systems are characterized by low-power operation; since they are always on, the peak and average powers are the same. The presence of the transmitted signal while attempting to receive results in relatively low sensitivity and limited range, but this may not be a problem in the application.

Radar Pulses and Matched Filters

In pulsed radar systems, the transmitted signal has the very simple shape of a nearly rectangular envelope around a high-frequency carrier. Of course, this perfect envelope cannot be produced in practice, since bandwidth limitations, stray capacitance, and realistic (nonideal) components produce rounding of the corners to some extent. In the next section we look at some of the radar transmitter and receiver blocks in more detail. At this point, assume that the signal has been transmitted, reflected, received, and demodulated so that the circuitry examining the signal—which may be the reflection from a target—is making a decision: Is the echo signal large enough to constitute a valid target, or is it just noise? This is not a trivial point. The radar system should not miss real targets (a miss), but it should also not falsely indicate that there is a target when there is none (false hit). All the decision circuitry has to work with is the received signal power (and corresponding voltage), which is a greatly attenuated, distorted, and noisy version of the original transmitted pulse.

This problem was studied for many years, and for most envelope waveforms it turns out that one type of filter is the optimum choice to shape the received signal just before it goes to the yes/no decision circuitry. The *matched filter* takes the received signal and maximizes the ratio of the peak signal power to the average noise power. Of course, this maximum SNR produces the best chance of the circuitry making a correct decision rather than a miss or a false hit.

The matched filter is unique for each type of signal. It is a filter whose time response to an impulse (an ideal, infinitely short pulse, Chapter 2) is the time inverse of the transmitted waveform. Several matched filter examples are shown in Figure 21.8. The radar receiver decision problem now becomes a problem in designing a matched filter for the transmitted waveform and being able to build this filter. In many cases, the preferred transmitted signal envelope is modified so that a matched filter can actually be built for it. Unfortunately, it can be shown mathematically that the matched filter for many desirable or practical waveforms cannot be designed with conventional filters using resistors, capacitors, and inductors, even if ideal components existed. For many other waveforms, the filter can be designed in theory, but it cannot be realized with nonideal components, and yet no component is perfect: Every resistor and inductor has some capacitance, every inductor has some resistance, and so on.

Chirp Waveforms

The range resolution ability of a radar system to separate and distinguish two targets that are ΔR apart in range is related to the narrowness of the transmitted pulse. A narrower pulse allows greater resolution, and the radar can better distinguish the two targets. Unfortunately, we saw that for a given transmitter peak

Digital Filtering

The power of digital systems provides a solution to the non-realizable matched filter dilemma in many cases. The received signal is converted to digital form (Chapter 11), and then *digitally filtered* using extensive numerical computations on the digitized samples. The matched filter function is implemented (*emulated*) by the computation. The results of these calculations are signal values that can be interpreted as the voltages coming from the output of a conventional analog filter, if desired, although this is not often necessary. There are many benefits to using digital filtering: the real world of nonideal components is completely avoided since only calculations are used to provide the filtering function; and the filter parameters can be varied simply by changing the multiplication and addition factors in the calculations, rather than switching components. In fact, the radar system microprocessor can simultaneously change the transmitted signal shape and adjust the digital filter factors in the software to provide the correct values of the matched filter for this new waveform—all without actually changing a circuit component.

Although extremely important, these benefits are not the only unique features of digital filtering. Digital filters represent the possibility of building perfect matched filters numerically. However, there is another aspect to digital filtering: There are some matched filters which theory shows cannot be built with conventional components (even if they are ideal) but can be realized digitally. This removes a major constraint in the design of the transmitted pulse shape, since the matched filter can be "built" by the using appropriate computational sequence and numerical factors in a processor. This, in turn, allows radar systems to use the best pulse shape for the application and not be restricted to only filters that can be built out of discrete components.

power limitation these narrower pulses have less energy than broader ones, resulting in lower SNR and consequently less overall range, together with a greater chance of a missed echo.

To overcome this problem, many radar systems transmit a frequency *chirp* (Figure 21.9), where the carrier sweeps from a lower frequency to a higher one during the "pulse on" period rather than remaining at a fixed frequency (it is called a chirp because it sounds, in the audio range, like a rising whistle). This FM chirp is a special form of *pulse compression,* where a wider pulse is "coded" with

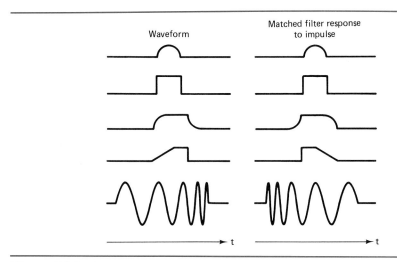

Figure 21.8 Matched filters for some radar signal envelope.

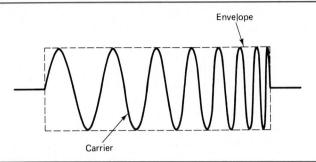

Figure 21.9 Chirp waveform achieves both higher energy and increased resolution.

a distinctive pattern that changes with time and which can then be recognized in the echo. Since the pulse is relatively wide, it has more total power than a narrow pulse. At the same time, the higher-frequency aspect at the end of the chirp gives it the ability to resolve closely spaced targets.

This seems to be an ideal way to solve the problem of constructing a powerful signal that still offers good range resolution. The matched filter for this signal, however, is extremely difficult to design "on paper" and requires many delay lines and numerous components; it is even more difficult to realize effectively with actual electronic circuit components, since all values must be precise. Despite this, many radar systems use the chirp and its matched filter built from conventional components because its benefits are so overwhelming in actual application.

However, the capabilities of microprocessors and digital signal processing circuitry make much more practical goals which were desired but hard to achieve. The received signal is digitized and then numerically processed by a sequence of calculations that implement the desired matched filter. For both conventional pulses and chirps, the system can dynamically vary the transmitted PRF, pulse width, and shape under software control to search for an unknown target and then track the target once located, all while dealing with conflicting targets, yet still provide the required matched filter for each pulse by software changes alone.

Review Questions

1. What is a pulsed radar system? What are PRF and pulse width?

2. How does a low PRF affect scanning speed and angular resolution? How does a high PFR determine the maximum range? How can a radar system think the target is closer than it really is?

3. How does the pulse width affect performance? What aspects are affected? Why is larger pulse width desirable for effective pulse echo detection? What are the drawbacks of this wider pulse? What is a blind spot?

4. What is clutter? What two problems may it cause?

5. What is the cause of Doppler shift? How large is it for typical radar frequencies and targets? What is the sign of the change it causes when the target and radar are closing on each other? What happens when the target is moving crosswise to the radar beam?

6. How is Doppler shift used to eliminate clutter in an MTI radar? How does the radar system physically implement the MTI scheme? How do Doppler and MTI increase noise slightly while increasing signal strength significantly?

7. What is blind speed? At what values does it occur?

8. How do modern radars avoid blind spots at fixed values?

9. What is CW Doppler? What radar information can it provide? What information does pulsed radar provide that CW Doppler cannot?

10. How does a CW system prevent overload of the receiver front end?

11. What is a false radar hit? A radar miss? On what basis is the presence or absence of a valid return signal decided?

12. What is a matched filter for a specific waveform? Why is it used, and where, in the radar system? Give two reasons why a perfect matched filter is not possible for many waveforms.

13. How is digital filtering used to provide the matched filter? What are the advantages of digital filtering?

14. What is a chirp waveform? How does it combine the benefits of wider and narrower pulses? What are some of the difficulties with the chirp waveform?

21.3 RADAR SYSTEM CIRCUITRY AND COMPONENTS

A radar system is like a conventional transmitter and receiver in many ways but with some significant differences, due to several unique features of radar: the relatively high power required at the transmitter, the basic on/off pulse shape of the modulating signal, the high frequencies used (GHz range), and the shared use of one antenna by a high-power transmitter and an extremely sensitive receiver.

Figure 21.10 shows the block diagram of a pulsed, non-Doppler radar system. There are four main functions: transmitter, receiver and display, duplexer, and antenna. The electronic and mechanical control for the antenna position is not shown. All of these elements operate as a coordinated and integrated system to provide the information on target range and position.

Transmitter

The radar transmitter develops high-powered pulses at carrier frequencies beginning in the several hundred MHz range, but more commonly in the multi-GHz range. This places extraordinary demands on the circuitry used, since conventional components—transistors or common vacuum tubes—cannot provide the power required at these frequencies. Special microwave components such as *magnetrons, traveling-wave tubes,* or *klystrons* (discussed in Chapter 23) are used, and these are keyed on and off by the pulse-forming circuit. An important aspect of the high-power, high-frequency oscillator is that it will be used in a pulsed on/off mode rather than a continuous modulation mode. Therefore, it must be capable of fast turn-on and turn-off while providing peak power that is far greater than the average power.

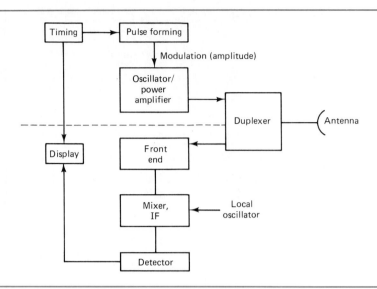

Figure 21.10 Pulsed, non-Doppler radar system block diagram.

The *pulse-forming* circuit, which produces a keying pulse of the desired shape and width, is driven by a timing circuit, which also determines the PRF of the system. In older systems the pulse width and PRF were either fixed or could be adjusted manually; newer systems use computers to adjust these dynamically to accommodate changing conditions, vary the blind spot and maximum unambiguous range, and minimize the ability of an enemy to fool the radar through *electronic countermeasures (ECM)*.

Receiver

A radar receiver is based on the standard superheterodyne configuration. The high carrier frequency requires a multiple-conversion receiver, with IFs at 500 MHz, 80 MHz, and 30 MHz in a typical receiver design. Klystrons or *Gunn oscillators* are used for the local oscillator, and the IF stages must have very wide bandwidths (called *video bandwidths* even though the signal is not a conventional TV signal) because of the narrow radar pulses. The detector is optimized for sensing the presence and level of a pulse waveform rather than determining the shape of a continuous signal (such as a voice signal) and may include a limiter to emphasize signal detection rather than trying to measure its exact magnitude. In radar, most of the information is contained in the presence or absence of the echo (and its precise time) rather than the shape of the echo. In recent years, the limiter has been supplemented by the matched filter, discussed in the preceding section, because it is optimum for the transmitted pulse shape and provides a more correct output voltage than the limiter under more conditions. Digital implementations of matched filters are accelerating this trend.

The receiver display is synchronized with the transmitter timing, so that it begins its sweep when the pulse is transmitted. Most radar systems use the amplified version of the detected output to control the screen intensity. The newest systems perform additional processing of this information with analog and digital circuitry and then enhance the image on the screen by intensifying some echoes

21.3 Radar System Circuitry and Components

and reducing others, depending on the application. For example, echoes from smaller, faster-moving targets can be amplified so that they are not obscured by larger, slower targets.

In addition, radar screens often add information labels to the screen blips, to help the observer sort out the picture. An air traffic control screen shows dozens of aircraft blips, which are impossible to make sense of without help. To make the display useful, each blip has an airplane number next to it, and the aircraft altitude. Newer systems add color to the blips, with blue for slower-moving aircraft ranging through orange for fastest. Now that computers are processing the information and controlling the display (rather having the echo go directly to the screen) the screen also shows the line of travel of faster aircraft (their vector) and flashes the blips of aircraft that are on collision paths if they maintain their present vectors and altitudes.

Duplexer

One of the difficulties that held up the initial application of radar in its early stages is that the powerful transmitter signal feeds right into the sensitive receiver input and overloads it. Even if this overload does not damage the front end (it usually does), the front end goes into saturation and takes many milliseconds to recover and return to normal operation. During this recovery period, the radar system is blind to any received echoes.

The solution requires shorting out the receiver front end while the transmitter pulse is one, thereby diverting any signal energy from reaching the front end. Of course, this cannot be done mechanically, since the many-millisecond switchover time of any mechanical switch would be far too long a period during which no desired echo signal would reach the receiver. Even if a fast enough mechanical switch could be built, it would not last long in service since the repeated switching of the large currents would burn out the contacts. The development of an electronic, self-actuating transmit/receive switch called a *duplexer* was a key step in practical radar implementation.

The duplexer consists of a sealed glass tube filled with ionizing gas and with two electrodes separated by a small distance, just like the spark plug in a car, and the size of the tube is designed to be resonant (like an *LC* circuit) at the carrier frequency. The spark gap connects the receiver input terminal and ground, and normally has nearly infinite resistance between the electrodes. As transmitted RF carrier energy passes through the duplexer, the duplexer is resonant and absorbs some of it, and ionizes the gas, which now acts as a very conductive, low-resistance path. This shorts out the receiver input, and the carrier signal cannot reach the receiver front end. As soon as the carrier is turned off, the gas deionizes and the duplexer returns to its high-resistance mode. The time for ionizing and deionizing is about a microsecond, far faster than that of any mechanical switch. The duplexer design has no moving parts and nothing to wear out, since the gas can ionize and deionize repeatedly without decay or wear.

For CW radars, a duplexer would not be useful since the carrier is always on. Instead, a *circulator* (Section 21.2) is used. This special type of waveguide sets up electromagnetic fields that cancel one another. The circulator ensures that most transmitted energy reaches only the antenna, while the echo signal energy goes to the receiver front end. The circulator is not perfect and reduces the transmitted signal as seen by the receiver by 30 to 40 dB, but this is enough since the CW transmitted signal is usually less than 10 W. In contrast, the peak power that a duplexer must shunt away from the receiver is usually thousands of watts and can be as high as megawatts.

Antenna

The antenna in a radar system can have fixed aim, be mechanically steered, or have electronic steering using a phased array (Chapter 9). The fixed antenna would be used in the nose of an aircraft, for example, to provide radar information on what is directly ahead. The more complex movable mechanical antenna allows the radar to search the sky to acquire a target and then track the target once it is acquired. Most movable antennas rotate or move in a constant path around a complete circle or back and forth over a limited arc, but the direction of a fully steerable antenna can be changed as needed to handle different target situations. The steerable antenna is more flexible in use but far more complex than a fixed antenna, and in many cases it is being replaced by the phased array design. Mechanical radar antennas are usually based on the parabolic dish or some dish variation.

In a *monopulse* radar, a single antenna with a complex feedhorn is used to provide a better idea of where the target is based on the return from a single transmitted pulse. Instead of a single feed at the focal point of the antenna, a group of four feeds is centered around the focal point. All four are driven simultaneously for the transmit portion of the radar operating cycle. When receiving, though, each feed is connected to a separate front-end and IF section, so that the signal strength of each is available independently. Since each feed is off the main axis of the antenna, each feed points to a slightly different angle from the normal axis. By comparing the signal received by each against one another, the receiver can locate the target with respect to the present antenna position and then direct the antenna to move in the correct direction toward the target position. [This scheme is similar to the coarse direction finding used in cellular phone base station (Chapter 20) but is more accurate.]

The drawbacks to this scheme are a more complex antenna feed structure, the need for multiple-receiver front ends and IF sections, and more duplexers to protect each front end. The benefits are that the antenna can find and then track the target more effectively, since it gets an immediate coarse indication of where the target is in relation to the present antenna orientation. Otherwise, the antenna would have to send out a pulse, get the echo, move to a new position, get the next echo, repeat this several times for slightly different directions, and then compare the results of this multiple-pulse operation. During this time, of course, a high-speed aircraft could move, significantly complicating the situation.

Solid-State Devices for Radar

The high frequencies and high power levels of radar transmitters can usually be realized only with special vacuum tubes or related devices such as the magnetron and klystron (Chapter 23). Unfortunately, these devices are not efficient at converting dc power into RF: Only 20 to 60% of the input power is transformed into useful RF output. Power semiconductors are solid-state transistors—fabricated as both bipolar devices and field-effect transistors (FETs)—that operate with greater efficiency, lower weight, and potentially higher reliability. Present technology limits these to operation between 20 MHz and a few hundred MHz, which are lower frequencies than commonly used in radar and at relative low power levels of only a few watts. This is sufficient for some radar applications, such as speed trap radar guns providing CW Doppler, but is not enough for many other applications.

Progress is being made in increasing both the operating frequency and power. For example, bipolar transistors now available can produce a 700-W pulsed output with 70% efficiency at 425 MHz. Although the peak pulse power that is needed is very high, the average power is much lower because of the low radar duty cycle, and this eases the design of these devices and systems. A 1000-W peak power transmitter with 1% duty cycle is far different than one with 100% duty cycle (CW mode). FET devices, especially, have an interesting characteristic: smaller, low-power modules can easily be combined in parallel to produce a single higher-powered signal, something that is very difficult to do with bipolar transistors. (There is an added advantage to this: If a single module fails, the output still works but provides reduced power. In a single-component design, a component failure takes the transmitter off the air.)

For radar receivers—where the need to provide power is not a factor—solid-state devices are now very common. Low-noise, front-end bipolar transistor amplifiers are available fabricated in silicon or gallium arsenide (GaAs) that can operate into the multi-GHz range while providing 5 to 25 dB of gain. Note that although radar receivers operate at the same frequencies as transmitters, they have an important difference: transmitters must produce relatively high power at these frequencies, while receiver front-end components are not used as power sources but must have low noise and provide sufficient gain to drive the subsequent mixer and IF stages.

Review Questions

1. How are radar transmitters similar to conventional transmitters? Where do they differ, and why?
2. What are the roles of the timing and pulse-forming circuitry?
3. What is the basic structure of the radar receiver? How is a received radar blip displayed directly? How is it enhanced in modern radar systems? What makes this possible?
4. What is the need for a duplexer? How is one built? Why is a mechanical switch not good for this function?
5. Why is a circulator and not a duplexer used for CW radar?
6. What does a movable antenna do? What about the capability of a steerable antenna?
7. What is monopulse radar? How does it work? When is it used? How does it add to system complexity?
8. What are the differences and similarities between solid-state device requirements for radar transmitters and receivers?

21.4 ADVANCED RADAR SYSTEMS

The basic radar system we have examined consists of a transmitter, an antenna, and a receiver at the same location and using frequencies from several hundred MHz up to the GHz region. There are several variations of this basic radar system in use, besides this basic design.

Bistatic and Multistatic Radar

Monostatic radar systems have their transmitter and receiver in the same location and usually share the same antenna. This provides one perspective on the target, but a limited one. In some situations more can be learned about the target if the transmitter and receiver, along with their antennas, are at different locations. In a *bistatic* radar system, the transmitter and receiver are separated by a *baseline* distance (Figure 21.11) with just one transmitter and receiver, for now. When the baseline distance is approximately the same or greater than the target distance, the overall performance of the bistatic system differs significantly from the monostatic one. It is equivalent to shining a flashlight on an object in a darkened room and then looking at the object from different positions. Each position in the room reveals another perspective on the object, which cannot be seen simply by standing at the flashlight position. In radar, the target has a different effective radar cross section, depending on the angle between the transmitter and receiver.

The bistatic radar system is often expanded to a multistatic configuration with multiple transmitters and one receiver (Figure 21.11). Each transmitter takes its turn in sending out the radar pulse, and the receiver observes all the echoes (knowing, of course, where each transmitter is and which echo is from which transmitter signal). This gives the single receiver many perspectives on the target. It also increases the likelihood that at least one path between a transmitter, the target, and the receiver will have the largest possible RCS—such as broadside to the airplane target—so that the largest echo will be returned.

Of course, the multistatic system could use more than one receiver instead of multiple transmitters. This is usually not the scheme used, however. The reasons are simple: Although transmitters are costly, they are fairly straightforward systems and designed basically to send out precise pulses at high power. In

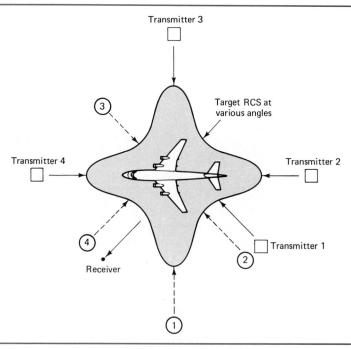

Figure 21.11   Multistatic radar configuration uses many separated transmitters and one receiver to generate multiple radar perspectives of the target.

contrast, modern radar receivers are extremely sophisticated: They use conventional signal processing techniques (low noise amplifiers, precise oscillators, mixers, IFs, and similar functional blocks), in combination with computers and digital signal processing, to take the received echoes and process them into useful information (to extract every subtlety and nuance from the signal). These sensitive receivers and advanced signal processing systems are actually more costly than the transmitters. In addition, the single receiver can combine all the echoes it receives from the many transmitters into a single, mosaic image of the target that is a fairly detailed and accurate composite.

Over-the-Horizon Radar

Radars using frequencies of hundreds of megahertz and above are limited to line-of-sight range because of the propagation characteristics at these frequencies (Chapter 9). Greater ranges can be achieved only with higher antennas, which is often impractical. An alternative is to use the principles of radar but with frequencies in the range 5 to 30 MHz (wavelengths of 50 to 200 ft), where the signals refract (bend) as they pass through the earth's ionosphere. The ionosphere acts essentially as a mirror that reflects radar signals from the transmitter over the horizon, to a target that returns some amount of the signal, and some of this signal is reflected back toward the radar site (Figure 21.12). Because this allows the radar to see targets, including low-flying aircraft, that are beyond the horizon at ranges of up to 2000 miles, it is called an *over-the-horizon* (OTH) radar.

An OTH system must contend with some very difficult technical problems. As there are already many users in the HF band that OTH uses, and lots of natural atmospheric electrical noise, the transmitter uses a continuous waveform (CW) rather than the harder-to-detect pulses of conventional radar. The entire OTH system is dependent on the propagation of HF waves and their refraction, which is a constantly changing situation and one that cannot be predicted in advance with high accuracy. In contrast, radars that use frequencies above several hundred MHz are fairly independent of the atmospheric conditions for propagation (rain and clouds have a small effect) but can only be used in line-of-sight situations.

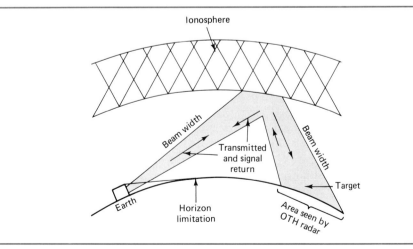

Figure 21.12 Over-the-horizon configuration and radar reflections.

To prevent the continuous transmitted signal from overloading and blinding the receiver, the OTH radar system receiver is separated by about 100 miles from the transmitter. This is not really a bistatic system, though, since the separation is much less than the target range; it is a modified monostatic system. In fact, if the transmitter and receiver are separated by much more than this distance, the transmit path and the receive path would be through different regions of the ionosphere and very little signal would actually return to the source (similar to using different mirrors for each signal, each aligned in a different direction). The 100-mile distance is a compromise between preventing blinding of the receiver by the transmitter using separation, and getting virtually no reflected signal since the transmit and receive paths are radically different.

The characteristics of the ionosphere and the way it bends the signals constantly change and must be carefully monitored by the radar system operators. Computers using advanced algorithms are needed to process the received signal, taking into account the bending effect to determine where the transmitted beam actually hit the target and was echoed back. An OTH radar covers a vast area and generates an enormous amount of data, so these computers must be able to process large amounts of data in a short time to be useful. A further complication is the dispersion (spread) of the return echo so that it is much smaller than a directly reflected one from a target "in view." High-powered transmitters, combined with very large antennas, low-noise circuits, and advanced computers, are the major characteristics of an OTH radar system.

Synthetic Aperture Radar

A narrow transmitted beamwidth is necessary to achieve a high degree of angular resolution in a radar system, to separate features and objects that are next to each other but at approximately the same range. For example, an airplane needs to see a map of the ground and features below, but while these rivers, roads, and land are at approximately the same range, it is their left-to-right angular separation that is of interest.

To achieve the desired fineness of angular resolution, the beamwidth must be narrow and therefore the antenna must be very large (have a diameter of many wavelengths). This is not possible in an airplane, of course. To overcome this limitation, *synthetic aperture radar* (*SAR*) was developed. In an SAR system, the appearance of a larger antenna is created by a clever and complex method that takes advantage of the movement of the airplane. A large number of radar pulses are sent as the airplane moves, and these are combined to simulate a signal from a single larger antenna. The effective size of this synthesized antenna is equal to the distance the airplane travels while taking these multiple radar images. SAR has also been used to map the surface of planets using aircraft or satellite-based radar systems. (The 27-antenna array in New Mexico used for deep-space satellites and planetary mapping (Section 19.3) is a land-based SAR, spread over 20 miles.)

The reason a SAR system is so complex is that the many separate "exposures" must be combined into one image. This is done by delaying each echo by the appropriate time and then performing complex signal processing on all of them simultaneously to produce the final radar image. The first echo is held the longest, the next one is held a slightly shorter period of time, and so on. The final result is that all the radar echoes, which were generated at different times as the aircraft (or spacecraft) traveled, are used to develop the signal's very high resolution image. SAR is used to generate maps with astonishing detail (Figure 21.13) using a

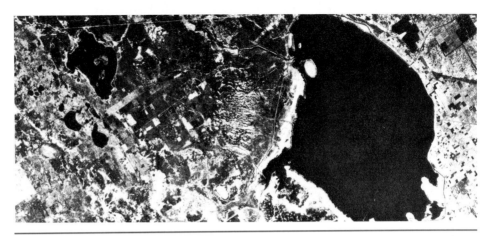

Figure 21.13 Mapping image generated by SAR [courtesy of NASA, (National Aeronautics and Space Administration)].

relatively small antenna with a physical diameter of only a few meters but a synthesized diameter of hundreds of meters. It has even been used to provide detailed maps of the surface of Venus, which is normally covered by clouds that make standard telescopes useless. (The delay lines are not needed with the ground-based array, since the antennas are not moving during their operation.)

Newer SAR systems replace the analog delay lines that hold the echo signal information with *digital signal processing* (*DSP*). The echoes are digitized by extremely high speed analog-to-digital converters, stored in a computer memory, and then numerically processed to produce the final result. This technique is costly but more accurate and flexible than the preceding method, which requires many precise delay lines to assure that the many SAR echoes all come together at the same time and is not easily adjusted for different aircraft speeds, PRF, and similar factors. [Combining many separate images of the same object is not limited to SAR. Medical instrumentation such at the CAT (computerized axial tomography) scanner and MRI (magnetic resonance imaging) take thousands of narrow-beam pictures around a patient which are then combined by a computer into a single, meaningful picture.]

Review Questions

1. What is a monostatic radar? What are bistatic and multistatic radars?
2. What can a bistatic radar do that a monostatic one cannot?
3. Why do multistatic radars use multiple transmitters, not multiple receivers?
4. What is OTH radar? What can it do? How is it implemented? What are some of the difficulties of OTH systems?
5. Why are the transmitter and receiver separated in an OTH radar? Why is an OTH radar not considered a bistatic design? Why not separate the transmitter and receiver by a greater distance in an OTH system?
6. Why is OTH difficult to implement reliably? What are the major characteristics of an OTH system?

> 7. What is SAR? How does it provide large antenna resolution in a physically small antenna?

SUMMARY

Radar provides us with the ability to see well beyond what are eyes, either naked or aided, can see. Many factors are intertwined in a radar system: the operating frequency, the radar pulse rate and width, the shape of the modulating envelope, the antenna gain, and the target distance and speed. The target's radar cross section is also a major factor and can seldom be changed by the radar system user. Radar displays show target distance, or target distance and angle, as a group of screen blips; more advanced displays annotate these blips and also show target directions and potential dangers.

The basic radar system transmits a pulse and measures the time of the reflected return from the target. This simple relationship establishes target distance. Doppler radars sense target motion while ignoring stationary targets, but are blind at some target speeds. Most radar systems require operation at frequencies in the hundreds of MHz to multi-GHz range, which in turn mandates special electronic amplifier components for the transmitter and receiver. These are very different from components used at HF and VHF frequencies, especially as transmit power levels increase and received signal strengths decrease. A special component called a duplexer (or a calculator) allows the high-powered transmitter to share the same antenna with the extremely sensitive receiver without damaging or overloading the receiver front end.

Computer-controlled radar systems provide higher performance because they allow operating parameters to be varied to fit the application. Traditional radar system design requires an inevitable compromise among pulse width, repetition rate, shape, and carrier frequency, where improving one aspect of performance degrades another. Newer systems can make continuous changes and then compensate for these by advanced signal processing. This is especially true in the use of digital filtering to provide the optimum analysis of the returned radar signal. Synthetic aperture radar uses a computer to combine many separate radar signals taken from a moving aircraft into a single high-resolution image, replicating a radar map taken with a physically enormous antenna. Similarly, radar systems with transmitters and receivers at different locations create a detailed composite image of the target based on the mosaic created by many separate echo returns. Radars operating in the HF band can "see" over the horizon, but managing the signal propagation strategy and analyzing the returned echoes is also complex.

Summary Questions

1. What is the basic principle of a radar system? How does it determine distance to a target?

2. What is RCS? Why is it so important? Why can't it be determined simply by looking at the apparent size of the target?

3. What affects radar range and resolution?

4. By what factor does maximum radar distance increase with power increase?

5. What is deflection modulation? Why is it supplanted by a PPI display in many applications? What do advanced PPIs show?

6. What are PRF and pulse width? What is the effect of increasing or decreasing either?

7. What is the blind spot? How is it minimized? What problems does this cause? Compare it to the blind speed.

8. Why is clutter undesired? What radar designs minimize it? What methods are used? What else is minimized along with clutter in some systems?

9. Explain Doppler shift and its role in pulsed radar and in CW radar. What is the magnitude of Doppler shift in typical applications?

10. What can a CW system indicate, and what can it not show? Why?

11. What is a matched filter? Where and why is it used? Why does it affect the shape of the transmitted pulses that can be considered for use in a radar system?

12. What is the chirp waveform? Why is it used? What is the difficulty with it?

13. What is a digital filter? What benefits does it bring to radar system design and performance? How does it do this?

14. Compare conventional VHF-band transmitters and receivers with radar transmitters and receivers. What are the principal blocks of a radar system?

15. What is a circulator? A duplexer? When is each used, and why? What is the construction of a duplexer, and why is it effective and reliable?

16. Compare the performance and capabilities of fixed, movable, fully steerable, and monopulse radar antennas. Why are phased array antennas becoming popular?

17. Contrast monostatic, bistatic, and multistatic radar systems in terms of performance, capabilities, number of transmitters and receivers, and locations. What are some of the technical problems with bistatic and multistatic systems?

18. What is OTH radar? How does it work? How does it differ from conventional radar?

19. What is a SAR? What problems does it overcome, and how? Why is it not easily done?

PRACTICE PROBLEMS

Section 21.1

1. What is the RCS of a flat plate 1 m on each side at 1 GHz? At 10 GHz?

2. Compare the RCS of a 0.1-m-radius sphere at 100 MHz and at 30 GHz.

3. What is the received power with a 1000-W transmitter, 80 dB of antenna gain, for the 1-m plate of problem 1 at 1 GHz with a range of 1,000,000 m (about 570 miles)?

4. The receiver in a 1-GHz, 1-MW radar requires at least 0.001 W to detect a valid signal properly. What is the RCS required if the target will be at a range of 1,000,000 m and the antenna gain is 40 dB? What would be the Δt between the transmitted and received signals?

5. A small airplane can fool a radar system into thinking it is much bigger by adding a radar reflector that increases the RCS. How large a flat plate should a jet fighter (RCS = 6) add to its side so that it looks like a large jet (RCS = 40) at 1 GHz?

6. What will be the ratio of received power to transmitted power (the signal path loss) from this apparently larger jet at 10,000 m if the antenna has a gain of 60 dB? What is the loss in dB? What is the round-trip signal time at this range?

Section 21.2

1. What is the Doppler shift for a 500-MHz signal when the target is moving directly toward the radar at 250 m/s (about 500 mi/hr)? What about when the target is flying directly away? What is the shift when the target is moving at 45° toward the radar?

2. A police radar to catch speeders operates at 2 GHz. What is the Doppler shift when the person is traveling at 20 m/s (about 40 mi/hr) toward the police radar unit? What is the shift when the person speeds up to 30 m/s?

3. What is the Doppler shift when the police unit is aimed at a curved part of the road, where there is a 30° angle of turn in the road, for the two speeds in problem 2?

4. What is the first blind speed for the 2-GHz radar if the PRF is 1 kHz?

5. What is the lowest value of PRF that can be used in the police radar system to ensure that the first blind speed is above 50 m/s (about 100 mi/hr)?

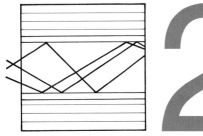

22

Multiplexing

CHAPTER OBJECTIVES

When you have completed this chapter, you will understand:

- The role of multiplexing in communication systems, and the relationship between multiplexing and modulation
- The key characteristics of space-, frequency-, and time-division multiplexing, together with the relative benefits and drawbacks of each
- How a communication system may use multiple stages of multiplexing to meet the needs of an application.

INTRODUCTION

Multiplexing is an important part of many communications systems. We have already seen it in stereo FM broadcasts as well as the standard TV signal. Two functions are accomplished by multiplexing: It allows an existing channel or link between a sender and receiver to be used for more than one message at a time (increasing capacity); it also allows related signals to be clustered together into a single entity which is then processed as one signal by the system.

Multiplexing makes sure that two signals do not occupy the same space, frequency, and time simultaneously. It does this by adding a new physical link (space division), sharing the overall bandwidth spectrum among several signals (frequency division), or allowing each user a turn at access to the link (time division). Each technique provides advantages and disadvantages in installation, cost, reliability, ease of troubleshooting, and the level of performance achieved. Multiplexing can be used for analog and digital signals, but time-division multiplexing is compatible with digital signals and makes good use of digital circuitry for these signals.

22.1 INTRODUCTION TO MULTIPLEXING

An electronic signal travels through the communications channel or link at a certain area in space, in a defined frequency band, and for a known period of time. When these three elements—space, frequency, and time—are all the same for two or more signals, interference and conflict result. *Multiplexing* is a technique for allowing more than one signal to coexist in the channel by developing a scheme to share the space, frequency, or time. With multiplexing, many signals can share an existing channel and make better use of the channel capacity.

There are several reasons to use multiplexing. The communications system can have several new, separate users who need to send messages between the same two points as a first user, and it is often impractical to run another physical wire between them or to establish a new transmitter and receiver pair. A good example of this is the trunk channel between telephone central offices, which carries dozens of conversations. Another reason to use multiplexing is that it allows several different signals to be clustered into a single group, so that they can be handled as a single entity from that point on in the overall system. As we saw in the discussion of TV (Chapter 14), the audio and video portions of the TV broadcast signal are multiplexed together and handled through the transmitter and receiver systems as one signal until they are split apart and separately demodulated.

There are three ways to increase the amount of information, or the number of signals that travel from a sending point to a receiving point. In the historical order of their development, they are:

1. *Space-division multiplexing* (*SDM*), where multiple physical paths are established by running new wires alongside the existing ones
2. *Frequency-division multiplexing* (*FDM*), in which each user signal modulates a different carrier frequency in the overall available bandwidth
3. *Time-division multiplexing* (*TDM*), where each signal is assigned a "time interval" or "time slice" and gets a turn, in sequence, at using the channel link and frequencies.

None of these three types of multiplexing is inherently better than the other two. The best choice for the application depends on many factors: available bandwidth, distance, number and type of signals, cost and complexity, and reliability. Many applications, in fact, multiplex a signal more than once in its course from sender to receiver, and the type of multiplexing used can be different at each stage of the total signal flow path from the signal source to the signal user.

Multiplexing allows unused capacity (usually available potential bandwidth) of the channel to be put to use by other signals. Of course there are limits. As more signals are multiplexed together, more and more of the available bandwidth is consumed. Therefore, there is a limit on how much multiplexing can be done; this becomes a practical limit on how many voice signals, for example, can be carried by a single coaxial cable, or how many TV signals can be supported by a wideband UHF-band frequency allocation.

Multiplexing and Modulation

Multiplexing and modulation—two vital elements in communications—are related but should not be confused with one another. Amplitude, frequency, and phase modulation (AM, FM, PM) are the three ways in which an information-bearing signal can be associated with a much higher carrier frequency, and modulation is the key to frequency-division multiplexing. However, already modulated signals can be multiplexed in space, frequency, or time; similarly, signals that are already multiplexed often modulate a carrier using AM, FM, or PM techniques. Modulated signals that are multiplexed can each use a different type of modulation, although they often have the same type. An unmodulated signal that has no carrier frequency is called a *baseband* signal, and that expression is also used to describe a signal before it undergoes FDM.

Digital and Analog Signal Multiplexing

The basic definitions of SDM, FDM, and TDM do not indicate whether the user signals that are being multiplexed are analog or digital. Multiplexing can be applied to either type of signal (recall that digital signals are really a special case of analog signals). In actual systems that implement multiplexing, though, the types of signals that are being multiplexed do have a large effect on the design of the circuitry. Multiplexing circuitry for analog signals must accommodate signals that can have any value within the overall range, and thus require linear circuitry that does not distort the signal. Digital multiplexing systems expect to see signals that have only a limited set of values (two values in the binary digital case) and therefore can use digital logic circuitry.

Recovering Multiplexed Signals

Multiplexing by itself is only half the complete story. To be useful, the multiplexed signal—which combines two or more signals—must be *demultiplexed* and separated to recover the original individual signals and then send them to the appropriate users' and circuitry for the next stage of processing (such as demodulation). The circuitry that performs the demultiplexing has some similarities to the multiplexing circuitry, but also some essential differences that we will discuss. In many ways, demultiplexing is more difficult than multiplexing since the received, multiplexed signal is corrupted by noise. Although most multiplexers deal only with relative clean and well-defined signals at their source, the demultiplexer must handle a composite signal that has distortion, undesired mixing between the signals, unknown variations in timing, and a wide range of signal levels.

Review Questions

1. What is multiplexing? Why is it needed?
2. What are the three types of multiplexing? How does each allow multiple signals to avoid conflict and interference?
3. Why is multiplexing needed to exploit available system bandwidth? Why is multiplexing used to form a cluster of signals that travel together?
4. What is the relationship between multiplexing and modulation? Does each type of mulitplexing require a specific type of modulation?
5. Compare multiplexing of analog versus digital signals in concept and in application.
6. What is demultiplexing? Why is it more difficult than multiplexing?

22.2
SPACE-DIVISION MULTIPLEXING

The first electrical communications system was the telegraph, invented and patented in 1840 by Samuel F. B. Morse. In the Morse system, message characters are encoded as a series of short dots and longer dashes, and sent by making/breaking an extremely simple circuit: a series combination of a battery, a sending switch (key), a complete loop of wire for the circuit (sometimes earth ground was used for the return part of the loop), and a receiving electromagnet which clicks as the circuit opens and closes. [Note that this is a binary code, but the length (time period) of the code element is critical, which makes it unsuitable for modern digital systems, which prefer bits of equal width.] When early telegraph systems needed to send more messages between two offices, the only available solution was also simple: Run another complete key, loop, and receiver circuit between offices (Figure 22.1). Now two telegraph operators could send messages at the same time to their counterparts at the receiving station.

This first instance of multiplexing shows both the good and bad points of space-division multiplexing. SDM systems are easy to build and add to, in concept, since each path is independent of all the others. They also provide some redundancy, since a failure in one path has no effect on the other paths, which continue to function normally. In addition, the many identical paths make it easy to troubleshoot if there are problems: Swap pieces from a good link with the failed link until the problem is located (change batteries, then telegraph keys, then the wire paths, then the receivers). The potential performance of the system is easily determined, even as more users (and their links) are added to the system, since each user has a dedicated line and never has to wait for the activity on another line to cease. Unlike some networks (Chapter 18), where adding new users causes overall throughput to decrease, adding new SDM links increases total throughput proportional to the number of links.

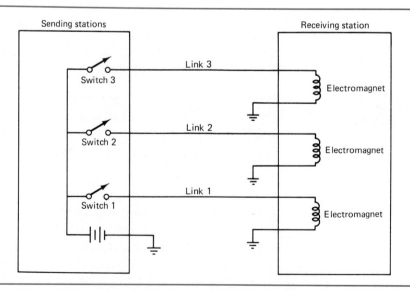

Figure 22.1 Telegraph circuit with two independent SDM loops.

In a long-distance SDM system, the initial cost depends almost entirely on the cost of the link, since only the basic transmitter and receiver are needed to establish a station-to-station connection. Of course, running the cable itself between stations is expensive, and this points to the first weakness of SDM. Adding another link later is equally costly, and especially painful if the bandwidth capability of the initial links is not being used fully. Even if many cables are installed at the same time, the bundle of cables quickly adds up in wire cost, size, and installation time. In the early days of telegraphy, telegraph poles (now called telephone poles) had crossarms carrying wires; as telephones became common, hundreds of wire blackened the sky.

There is another drawback to using an SDM system. The cost of station equipment for transmitting and receiving (in dollars, space, and power consumption) increases directly with the number of users. There is no possibility of sharing costs, since each link requires a complete transmitter, link, and receiver. It is like having a car for each person going from town A to town B and never being able to reduce costs by having the cost of one car shared among several users who are going between the same places. (However, a breakdown of a car stops only its driver; other cars and their drivers are unaffected.)

SDM Example and Circuitry

The most common example of an SDM system is the local telephone system (Chapter 16). Each phone is connected to the central office by a local loop that is not shared with any other user (common "party lines" were used in the early days of the telephone system to save wiring, and still exist in rural areas, but we will ignore these) and terminates at the central office at the *subscriber line interface circuit* (*SLIC*). These local loops are the SDM links. Note that in the SDM system, a failure in any loop does not affect other loops, and a problem with any SLIC affects only the loop connected to that SLIC.

The switching circuitry within the central office can connect any loop to any other, using a matrix of crossbar relays, reed relays, or solid-state switches, but the number of simultaneous conversations paths that can be complete at any time is limited by the number of switching elements in this matrix. Although 10,000 users can have 5000 conversations at the same time, it would be extremely expensive and unnecessary to provide this capability. Typically, the switching matrix is designed to provide connection paths for 5 to 10% of the loops, depending on the type of area that the central office serves (homes, small offices, large factories, for example) based on phone company historical data of phone usage for similar installations. At the central office, trunk lines connect to the adjacent central office, with each trunk carrying a single conversation. The number of available trunks is much smaller than the number of local loops, once again determined by the type of telephone traffic that the phone company expects in this region.

Demultiplexing SDM systems is straightforward. Since each signal has its own independent link and transmitter circuitry, demultiplexing simply requires that the switching matrix within the central office connect the appropriate loops together to complete a conversation path, or connect a loop and a trunk. The user signal itself—the voice—is a baseband signal and needs no processing or manipulation to make it usable at the far end, except perhaps for some filtering and amplification, which shape the frequency characteristics (correcting for distortion caused the line response) and boosts the signal amplitude to compensate for losses.

SDM is not limited to baseband signals. Specially designed switches can handle wideband coaxial or fiber optic links, carrying signals that have been modulated to much higher carrier frequencies. The principles of SDM remain the same, however.

Review Questions

1. What are the advantages and disadvantages of SDM in terms of ease of initial installation, initial space and equipment required, ease of expansion to new links, reliability and redundancy, testability, performance, and cost for one user versus cost for N users?
2. Give two examples of SDM.
3. How is SDM multiplexed? Demultiplexed?
4. Why is SDM often combined with a switching matrix?

22.3 FREQUENCY-DIVISION MULTIPLEXING

The congestion of telegraph wires and station equipment caused by space-division multiplexing led to the development of frequency-division multiplexing. FDM is possible when the bandwidth of the link—wires at first, then radio transmission links—is greater than the bandwidth of the message sent over the link. Another message can occupy the unused bandwidth if it is shifted away from baseband to a new carrier frequency. We studied this operation as modulation (AM, FM, and PM) where the information-bearing signal was impressed onto a carrier, varying the amplitude, frequency, or phase of the carrier.

In the earliest applications of FDM, the bandwidth of the wire link between telegraph stations was about five times the bandwidth of the telegraph signal itself. Therefore, five signals could be sent at once: one at baseband and the others evenly spaced throughout the remaining bandwidth (Figure 22.2). Modern coaxial

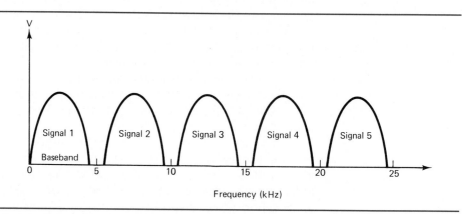

Figure 22.2 FDM spectrum combining five telegraph signals, with one remaining at baseband.

cables and fiber optic links have much more bandwidth, of course, so many more signals can be sent. (A No. 22 twisted-wire pair has a bandwidth of about 100 kHz, depending on wire twist tightness and spacing, wire gauge, insulation, and other factors.)

FDM is not limited to wires and cable links: It occurs when many broadcast radio and TV stations each use a unique assigned carrier within the overall band to avoid interference. None of the modulation signals remain at baseband; instead, each is translated to a much higher, and slightly different carrier frequency so they can be sent through the common channel and then sorted out at the receiver. This channel can be air, vacuum, or cable.

FDM and Grouped Signals

Our experience with FDM used for broadcast radio and TV stations shows that it allows independent signals to share the same time and space, at different frequencies, and that users pick out the signal of interest (the desired station) by tuning. The ability to provide this is a major feature of FDM, but there is another aspect that is important in communications systems applications. FDM allows several baseband signals to be grouped together around various *subcarriers*, and then be translated in frequency as a single cluster or entity to a final carrier value by a single multiplexing (modulation) operation. We have seen this in stereo FM (Chapter 6), where left and right audio signals are multiplexed into a single signal, and in broadcast TV, with the combination of an audio signal and a video signal into a single signal. This is done to make the rest of the communications system transmitter and receiver circuitry much simpler and more reliable and to ensure better performance. Instead of having to tune the audio and video signals of a TV channel separately, a single tuner (when switched) can locate the 6-MHz-wide segment in the spectrum used for this channel. This segment is then translated to a fixed IF frequency by the local oscillator and mixer, with the video signal and the audio signal at known frequencies relative to the assigned carrier. Since the baseband audio and video were originally multiplexed with precise subcarriers, they will always have that spacing, regardless of the final transmitted carrier frequency used. Note that the multiplexing operation with subcarriers did not require the same type of modulation—for standard TV, the video is vestigial sideband AM while the audio is narrowband FM.

Other FDM Applications

FDM is used extensively in areas besides commercial broadcast radio and TV. In *telemetry* many signals must be transmitted from a remote system under test (such as a rocket) to a base station. These signals represent the temperatures and pressures at key points in the rocket, along with directional and speed information. Each sensor signal is amplified as needed to provide a 0- to 1-V signal used for narrowband FM (NBFM), with a final bandwidth of 4000 Hz.

All these baseband NBFM signals are next multiplexed with various subcarriers spaced 4 kHz apart. The result is an entire cluster of telemetry signals, with bandwidth extending from 0 Hz to ($N \times 4$ kHz) for N telemetry channels. This cluster is next translated by standard AM, FM, or PM to the telemetry carrier channel assigned for this rocket. Other rockets (or aircraft) in the area are assigned different carriers, although they use the same subcarrier frequencies for their signals; similarly, any other cluster of subcarriers can be transmitted on a different channel simply by changing the final modulated carrier frequency. The

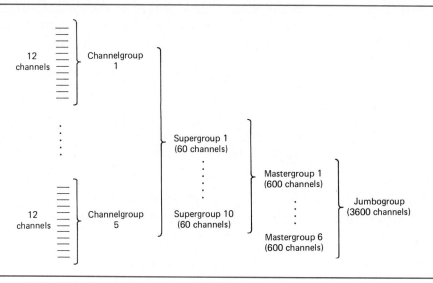

Figure 22.3 Telephone subcarriers: FDM from groups to jumbogroups.

receivers for any of the channels are identical, except for the front-end tuning section of each, which tunes the appropriate carrier signal. All telemetry channels have the same set of subcarriers, so the demultiplexing (demodulation) circuitry is the same.

The telephone system is a major user of FDM. Linking the central offices together with a transmitter and receiver pair and a link for each trunk required is very expensive and wasteful. Instead, the telephone system established a standard set of subcarriers (Figure 22.3). Twelve signals (each with bandwidth a little greater than 3 kHz) are frequency multiplexed with subcarriers spaced 4 kHz apart to form a *channel group*. Five such channel groups are next multiplexed with subcarriers to form a *supergroup* carrying 60 channels, and 10 supergroups are combined into a 600-channel *mastergroup*. Finally, six mastergroups form a *jumbogroup*, carrying 3600 conversations. This jumbogroup is then modulated onto the communications link, which can be a coaxial cable, radio, microwave, or satellite. All 3600 conversations travel together as a single 14.4-MHz-wide entity, regardless of the final carrier frequency used for the actual link between sending and receiving points. (*Note:* The actual jumbogroup bandwidth is slightly more than 14.44 MHz, to allow for 80-kHz guard bands between mastergroups.)

In this phone application, the subcarriers are evenly spaced since all conversations have the same bandwidth. However, subcarriers do not have to be spaced evenly and often are not in telemetry applications where different types of signals must be sent. If the telemetry system is sending back many low-bandwidth signals (from relatively slowly changing, narrow-bandwidth signals such as voice, temperature, or speed) and some wide-bandwidth signals (from video cameras), the subcarriers are spaced accordingly to make maximum use of the available total bandwidth. Spacing the subcarriers evenly for the largest bandwidth need is wasteful: Using the 6-MHz video bandwidth spacing to send 4-kHz bandwidth signals wastes $(6000 - 4) = 5996$ kHz, which is $5996/6000 = 99.9\%$ of the available spectrum.

22.3 Frequency-Division Multiplexing

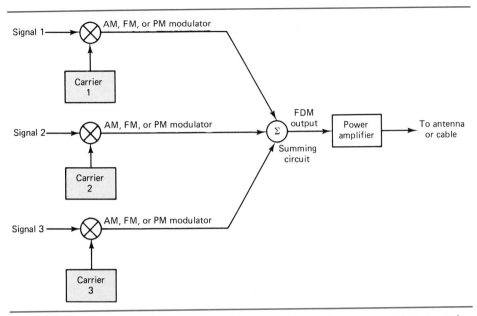

Figure 22.4 Basic FDM transmitter with subcarrier oscillators, modulators, and power amplifier.

FDM Circuitry and Performance

The circuitry for FDM operates at low signal power levels and amplifies only the complete multiplexed signal to the final power level to be transmitted. An FDM system is built from a bank of amplitude, frequency, and phase modulators with different carriers, all operating in parallel (Figure 22.4). The modulators do not have to be the same, as we saw in TV (video plus audio) example, but they certainly can be. The channel spacing is set by the carriers of each modulator, and the type of modulation is determined by the specific modulator used. After all the baseband signals have been modulated and summed together, the multiplexed group is amplified to the required power level.

At the receiver, demultiplexing begins by translating the multiplexed signal, if necessary, down to an intermediate frequency (IF) with standard superheterodyne techniques (Chapter 5). There is a local oscillator and mixer for each multiplexed channel, and the IF bandwidth is set equal to the bandwidth of an individual baseband signal after modulation (Figure 22.5). A receiver that demultiplexes 10 signals requires 10 separate local oscillators (LO) and mixers, with each LO operating at the frequency appropriate for its intended signal. This approach is effective but costly: Designing and adjusting a group of stable LOs that are precise and do not drift in time, and do not interfere with each other to produce *spurious* frequencies, is difficult. Tuning to another carrier (equivalent to switching from one TV station to another, for example) requires retuning all 10 LOs.

Another approach is to use a single LO and mixer, with an IF bandwidth equal to the overall bandwidth of the multiplexed signals. This is the method used in standard TV and stereo FM broadcasts, for example. The IF output is next tuned by a series of filters that separate the multiplexed signals, then pass each one on to appropriate detection circuitry to recover the baseband signal. Tuning the receiver for another multiplexed signal involves only varying a single LO; all subsequent stages have fixed values.

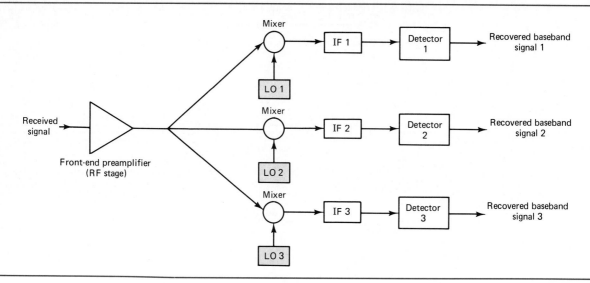

Figure 22.5 FDM demultiplexer with one LO per subcarrier.

FDM versus SDM

Clearly, an FDM system allows the communications system to increase the number of simultaneous signals between two points, up to the capacity of the link between these points. In contrast to SDM, which requires a new link for each signal, FDM continues to use a single link by squeezing the new user signals into the available bandwidth of the existing link. This results in major installation time and cost saving.

Like SDM, and FDM system requires circuitry for each channel that is being multiplexed. However, SDM requires a complete transmitter and receiver pair for each new channel, whereas FDM needs only the appropriate modulation and demodulation circuitry for each channel while sharing the final amplifier stages and receiver front ends. A single transmitter with the appropriate power output level, matched with a receiver with a sensitive front end, serve all the channels of a frequency-division multiplexed system. This means that the transmitter cost (in circuitry, dollars, power, and space) is shared, as is the cost of the front-end and tuning circuitry. Modulation is done at low power levels, and demodulation is performed on signals that have reasonable levels (100 mV to 1 V).

Of course, FDM does have some weaknesses compared to SDM. The main difference is in redundancy and reliability. If the single FDM transmitter or receiver fails, all users are affected; the same happens if the cable between them fails or there is interference and noise at the frequency in use. In an SDM system, the failure of any single transmitter, receiver, or link affect only that particular link. In practice, this is not as reliable as it appears: When one link fails, the others often also fail, because most link failures result from cable breaks and the various SDM cables all run together in a single bundle. (It is possible to guard against this by running the cables over different paths between the two endpoints, but this is more confusing and costly.)

Finally, SDM systems are easier to troubleshoot because each transmitter, link, and receiver path is identical to the others. This provides an easy way to troubleshoot simply by swapping and substituting adjacent operating units in the

22.4 Time-Division Multiplexing

failed link until the exact source of the failure is found. Like SDM, FDM uses essentially the same multiplexing and demultiplexing circuitry for analog or digital user signals.

Review Questions

1. Why is FDM possible over a link that is already in use?
2. What is the role of a subcarrier? Does the type of modulation of each subcarrier have to be identical? Explain.
3. Why and how is FDM used in telemetry?
4. How does the telephone system use FDM in interoffice communications? What subcarriers and clusters are standard in telephone applications?
5. What is the circuitry for producing an FDM signal? What are the relative power levels?
6. How can an FDM signal be demultiplexed? What are the two basic approaches and their pros and cons?
7. Compare SDM to FDM as to how each makes use of the available link, the required amount of circuitry for one and many users, reliability, redundancy, and ease of troubleshooting.

22.4 TIME-DIVISION MULTIPLEXING

The concept of TDM is simpler than FDM since it basically involves users taking turns in sequence, but practical implementation is not simple and presents new difficulties and issues. TDM has been used for many years where different users share turns at access to the communications link, but with each user taking a relatively long turn; there is no preset beginning or ending time for each user, and the order of turns is not fixed. For example, the trunk lines between telephone company central offices are time multiplexed: When one conversation over a trunk terminates, the trunk is free for another conversation to use it.

But modern TDM in communications systems really involves a plan and structure that provide access to the link for many user channels on a periodic and scheduled basis, such as where each of 10 users has access for 0.1 s during every second. In addition, unlike SDM and FDM, which use very similar circuitry for either digital or analog user signals, TDM designs are quite different for these two groups of signals. In fact, TDM is very suitable for and compatible with digital (and binary) signals, and for this reason, the overwhelming trend in recent years is to use TDM.

TDM for Analog Signals

The basics of a TDM system for analog signals are illustrated in Figure 22.6. In the heating, ventilation, and air conditioning (HVAC) system of an office building, the system that controls the overall building heating and cooling must monitor the temperature of each office and area. Based on these readings, the HVAC controller directs more hot or cool air to specific areas, while trying to minimize use of

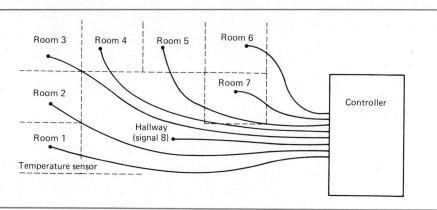

Figure 22.6 Routing of many HVAC signals to a single central controller, one wire per signal.

fuel (gas, oil, electricity). Each room and area has a sensor that produces a voltage proportional to the temperature, and each sensor has wires connecting it back to the controller location.

This system will work, but requires many long runs of wire throughout the entire building back to the controller. Yet the temperature information is a slowly changing signal and is not really checked by the controller more than once every few minutes. To minimize the many long wire runs, TDM is used. A multiplexer based on an analog solid-state switch (similar to the one in the telephone central office) between the HVAC controller and the temperature sensors lets the controller sequence through all sensors. The controller generates an address in binary format indicating which sensor signal should pass through the switch to the controller; all other sensor signals are blocked. The number of sensors that are addressed by N control lines is 2^N. Therefore, in a building with 64 sensors, the number of wire runs is reduced from 64 to just six address lines plus the one sensor signal. In operation, the controller sequences the multiplexer through all sensor channels at the appropriate rate.

This scheme is practical in a building but not in a conventional communications system where there is only a single message path—the channel—from the sender to the receiver of the message. There is normally no path for the receiving end—equivalent to the HVAC controller—to send out the address, indicating which user channel has the next turn, through the multiplexer. Of course, it is possible to establish this path, but this requires a full or half-duplex channel link, which is often not available. Even if such a link is available, it would be inefficient since much of the link message traffic is the user number rather than actual user messages. Also, as the rate at which the users get their turn increases (sometimes called the *scanning rate*), the system propagation delays, noise, the need for retransmission, and errors will combine to make the system ineffective.

There is another difference between this simple, yet useful HVAC system and a communications system. In the communications system, the many possible senders time-share the link, and their individual messages are re-sorted at the receiver and then distributed to the correct recipient. This is represented in Figure 22.7, where the sender and receiver each have identical multiple-position switches that control the signal flow path. At the sending end, the switch picks one of the N

22.4 Time-Division Multiplexing

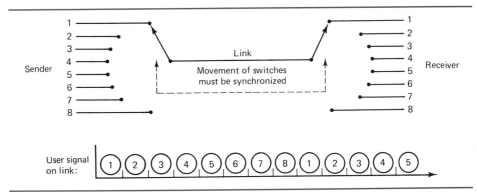

Figure 22.7 TDM system must scan, transmit, and distribute signals with proper channel synchronization.

users and puts that user signal on the link; at the receiving end the switch takes the signal from the link and distributes it to the corresponding point. In this way, N separate conversations share the link (apparently at the "same time" but not really), and each user is independent of the others. In contrast, the HVAC application has all user signals going to a single user, the controller.

Digital TDM In the design of the analog TDM scheme above, the biggest problems are identifying the start of a scanning sequence and maintaining proper stepping (clock) synchronization. We have seen these two points before in our discussion of digital

---

TDM Circuitry

A circuit that implements a communications TDM system is shown in Figure 22.8. A clock signal drives a two-bit counter, which produces counts from 00 through 11 and then back to 00. The counters are connected to decoding circuitry, which produces a single output at binary 1 but leaves the other three at 0, depending on the count value. This binary 1 output signal turns on an analog switch to allow the user signal to pass, while the three other switches are kept in the off state. At the receiver, the same action occurs, so that the signal from the first sender reaches the first receiver, and so on.

To make this scheme work, two forms of synchronization are required. The rate at which both switches step through the N positions must be the same, and the receiver must have some sort of marker that indicates when the switch is at the first position. At relatively low stepping speeds through the user channels, this is accomplished via accurate, preset clocks at each end, with the clocks synchronized to the starting points by a unique signal value on user channel 1 (this is similar to the line and frame sync signals on the TV signal, at <0 IRE levels). For example, the normal analog signal range might be -1 to $+1$ V, but channel 1 is fixed at $+2$ V. When the receiver switch senses the $+2$ V level, it automatically goes to position 1. The clocks are precise enough that they maintain the proper stepping relationship from positions 2 through N, and they are resynchronized when the transmitter goes from position N back to 1. (Alternatively, a clock signal can be sent with the user signal on another link as in the example, but then two lines in parallel are needed.)

At higher rates of thousands of steps/s, this type of sync does not work. Several factors work against it: As the number of users increases, the small difference between the two clocks is sufficient to cause them to fall out of sync with each other despite their re-syncing at position 1; link propagation delays and variations mean that the clock used by the sender may arrive differently at the receiver; also, noise can corrupt user signals so that they look like the special sync signal or corrupt the sync signal so that it looks like a user signal. Although at lower rates this is a workable design. Higher rates require a new approach.

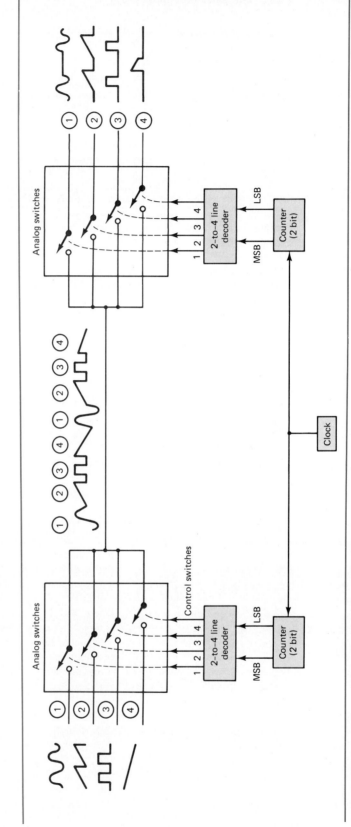

Figure 22.8 TDM circuitry, built using a two-bit counter, 2-to-4 line decode, and corresponding waveforms.

signals and transmissions. Clock sync is analogous to bit sync, while marking the start of the scanning is like frame sync. For analog systems, these are difficult to provide since the signals that are sent do not really provide any information about them. The analog signals from the many users of the TDM system are the user messages, not timing or sync information, and any timing or sync information sent along is difficult to separate from the user message signals. This is exactly the same situation that we saw in the phone system (Section 16.5), where using the talking path for signaling limited the effectiveness of system operation, while using a separate path (common channel signaling between central offices) offered much greater performance and flexibility.

If the TDM system is designed solely for digital signals, the design of the system changes radically because it is much easier to provide and maintain bit and frame sync. In the discussion of digital signals (Chapter 10) we saw how binary data bits themselves can be used to derive their underlying bit clock rate through the use of phase-locked loops (PLLs), regardless of the actual bit pattern of data. Similarly, a special bit pattern that never appears in the normal data is used for frame sync, so the receiving TDM switch can precisely and without ambiguity identify when data is coming from user 1.

With TDM for digital signals, even though the sync bits look like user data bits—the same voltage levels and width—they can carry unique information and must be separated out from routine message data bits by the receiver. This cannot be done as reliably or easily with analog voltages at higher stepping or scanning rates. In a digital system, the multiplexer clock rate for N users in just N times the clock of each user. Therefore, multiplexing 12 users, each at 19,200 bit/s requires a 230,400-bit/s clock.

Another difference is that the analog switching circuitry for TDM is now changed to standard digital logic gates (Figure 22.9). Since the only signals that appear at the receiver are binary 1s and 0s, there is no need for an analog switch that can handle any signal voltage in the overall range. When the decoder points to user 2, for example, the decoded controlling input to that AND gate is set equal to 1 and the gate output is the same as the user bit input. The decoded controlling inputs to gates 0, 1, and 3 are set to 0, so those gate outputs are 0 regardless of their data inputs. A complementary simple scheme demultiplexes the serial TDM stream of bits and distributes them to the correct users.

This multiplexing points out another subtlety and advantage of digital TDM. It does not matter at what precise point during a bit time period the TDM circuitry samples each bit, since the value of the bit is the same throughout the entire bit period. Therefore, the multiplexer circuitry that is scanning the bits in order to multiplex them can look at the bit from the first data source earlier in the bit period, look at the second source bit slightly later in the period, all the way to looking at the last data source toward the very end of the bit period. This is seen by looking carefully at the timing and inset part of Figure 22.9. The result of this flexibility is that the bit streams to be multiplexed do not have to be held in registers, since the multiplexing circuitry gets to all of them in a single bit period, although it samples each one at a different time *within* the period.

Bit versus One-Value-at-a-Time TDM

In the example above, the bits from each user were multiplexed with a single bit from each, in sequence. In many TDM systems, these bits represent voice or other analog signals that have been digitized, typically to seven or eight bits of resolution. The multiplexing and demultiplexing sequence are often slightly modi-

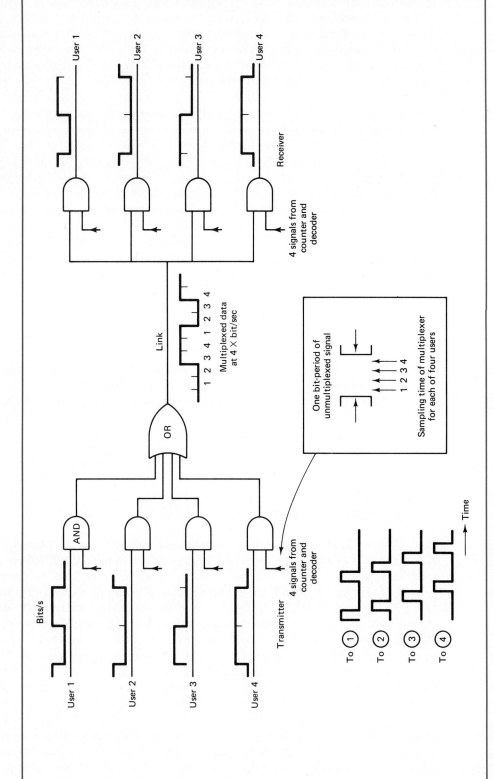

Figure 22.9 TDM circuitry for digital signals, built entirely from digital logic gates. Note how each signal is scanned in sequence and the resultant multiplexed bit rate is equal to the rate of one channel times the number of channels.

> **T-1 System**
>
> As communications systems were designed to make greater use of digital signals, the Bell Telephone System installed digital multiplexers called the *T-1 digital carrier communications system* beginning in the 1960s. This system, which is used extensively today, provides TDM connections for 24 voice signals from one central office to another, at a bit rate of 1.544 Mbits/s.
>
> The T-1 multiplexer circuitry samples each analog voice signal (bandwidth limited to 4 kHz) at a rate of 8000 times/s (to satisfy the Nyquist criteria for sampling), and digitizes each sample into a seven-bit value ($2^7 = 128$ levels). Each seven-bit value has an eighth bit added to it to serve as a marker for interoffice signaling information. The multiplexer design forms a long stream of bits, combining the eight bits of the first user, followed by the eight bits of the second user, all the way through the bits of the twenty-fourth user. A group of eight bits times 24 users is called a *frame* and contains 192 bits. Each such frame is preceded by a special *framing bit*, used to help establish frame sync at the receiver. Thus 193 bits must be sent 8000 times/s, for a total rate of 1,544,000 bits/s. The T-1 frame is the basic unit of this type of digital TDM.

fied from the one-bit-from-each-user method (selecting the first bit from user 1, then the first bit from user 2, all the way through the first bit of the last user and then to the second bit of the first user).

The TDM counting circuitry is changed so that the multiplexer sends the entire group of seven or eight bits from one user representing the complete digitized value bit pattern of one analog signal, then the group of bits from the second user, all the way to the bit grouping of the last user. After the last user it repeats the next seven- or eight-bit pattern from the first user, beginning a new sequence of scanning. The demultiplexer is designed for this sequence, of course, so there is no confusion. The difference in circuitry is relatively small: The same clock rate is used for the bits (and derived by a PLL at the receiver), but the counter that steps from user 1 to user 2 also acts as a divide by 7 (or 8) counter and steps to the next output position only after it has counted seven (or eight) clock pulses. Data buffers (registers) are used, as needed, to hold the data from the various users while they wait their turn to be multiplexed.

Number of Users and Bandwidth

Digital TDM seems an ideal way to put many individual users onto a single communications link. It may seem that there is no limit to the number of users that can share the link: Just increase the scanning rate at which the TDM switch steps from user to user. If the T-1 system presently handles 24 users at 8000 user samples/s, it appears straightforward to double the system clock and thereby increase the capacity to 48 users, at 3.088 Mbits/s. The bits representing the user signal will be narrower, but they will still have distinct 1 or 0 values. Unfortunately, this cannot be done.

Recall in the discussion of time and frequency domain of signals in Chapter 2 that we looked at the spectrum of digital signals. As the digital signal became narrower (a smaller period) in the time domain, the equivalent spectrum in the frequency domain broadened. Higher bit rates (narrower bits with shorter periods) simply require more spectrum. So we come up against the basic limitations of channel bandwidth and information rate: a link of specific, limited bandwidth cannot act as a conduit for bit rates with spectrum beyond its bandwidth capacity. As the bit rate increases—whether from one user at a higher bit rate, or more

TDM and Microprocessors

TDM is used not only in longer-distance communications systems. It is also very common in microprocessor-based circuitry, although with some technical differences. Every microprocessor has a set of address lines to indicate which memory location it is accessing, plus a set of data lines for actually transferring data between it and the memory, and a set of control lines that provide the handshaking and management of the microprocessor to memory communications. As microprocessors become more powerful, they are designed to address more memory locations and interface to wider memory words (more bits/address). Early microprocessors addressed only $2^{16} = 65,536$ (called 64K) locations of memory and so required 16 address lines, while data was handled in eight-bit bytes (eight data lines). More powerful microprocessors now address 2^{24} or more memory locations (>1677K) and the memory locations have 16 or 32 bits of information each. Therefore, the number of address and data lines has increased from just $16 + 8 = 24$ to at least $24 + 16 = 40$ (more if the microprocessor uses 32 bits for data). This means that 40 buffers and signal drivers are needed at the microprocessor and 40 corresponding receivers at the memory circuitry. This amount of circuitry requires considerable space, power, and expense—even running 40 tracks on the circuit board is space consuming.

The solution is to use TDM to minimize the problem. The digital lines between the microprocessor and the memory then represent, at different time, either address or memory data (Figure 22.10). A single control line from the microprocessor tells all the circuitry of the system whether the 24 lines are carrying address information or data, and the circuitry accepts the data or address bits accordingly. If it is address information, *latches* at the memory ICs accept and store this address information for when it is needed at the next part of the TDM cycle, when these same digital lines now present data for the memory. The memory ICs use the latched address information to identify where to store the data. (A similar operation is used to retrieve data at desired addresses in memory, to send back to the microprocessor.) To reduce the number of lines further, some microprocessors send all address and data in eight-bit bytes, so an address takes three bytes (24 bits) while the data needs two to four bytes (16 to 32 bits). The TDM signals on the eight paths represent, in sequence: address byte 1, address byte 2, address byte 3, data byte 1, data byte 2, data byte 3, data byte 4.

This provides tremendous savings in support circuitry cost, space, and power. Of course, there is a drawback to this use of TDM in the system. The overall *throughput* (the amount of data that can be transferred in any amount of time) is reduced. It now takes seven clock cycles (three for address, four for data) to send a complete data value to its memory location, whereas a nonmultiplexed system does it all in a single cycle (but with more circuitry) and is thus seven times faster.

This is a trade-off that the designer of the system must make based on the amount of data that must be handled and at what rates it must be processed. Multiplexing is very common in microprocessor-based systems, though, since the savings it provides are often essential and the slower performance that results is acceptable. In its most common form, the address and the data are multiplexed so that the system requires only one-half the circuitry, although the throughput is also cut in half. Synchronization for demultiplexing at the memory ICs is not a problem: The entire system shares a common clock (thus providing bit sync), while the multiplexer control line that the microprocessor generates to the rest of the system clearly indicates if the signal lines are address or data at any instant (equivalent to frame sync).

---

users each at their previous rate—the individual bits get shorter, while the required bandwidth to transmit them faithfully and with minimum distortion increases, and the channel limit is reached.

Although we normally think of digital signals and TDM in the time domain, their frequency-domain equivalent is the key. In FDM, in contrast, we are already thinking of signals in terms of their frequency-domain spectrum to see how many signals can be multiplexed into the available bandwidth. We more quickly and intuitively realize that the bandwidth can accommodate only a certain number of individual signals. When the sum of the bandwidths of the modulated signals equals the bandwidth available, this limit has been reached. The same relationship applies for TDM, only less obviously.

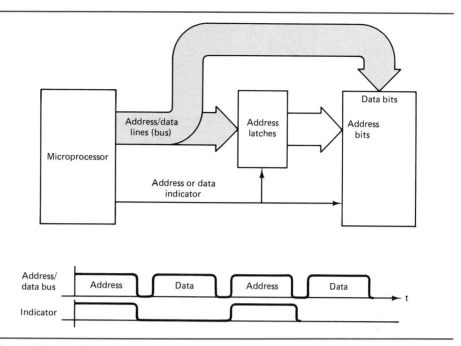

Figure 22.10 TDM as used in microprocessor circuitry.

TDM versus FDM and SDM

Unlike SDM and FDM, TDM handles digital signals differently than analog ones, but there are other differences among these three types of multiplexing. We saw that it is difficult to add new links in an SDM system, but that these links are independent of each other and add some amount of redundancy and reliability to the overall system. For FDM, a single-user system can often be improved (upgraded) to multiple-user support over an existing link by using FDM equipment at the transmitter and receiver, assuming that the link has excess available bandwidth. The modulators and demodulators of the FDM system are independent, so a failure of any one does not necessarily affect the others (unless it is off-frequency and causes interference). Therefore, the FDM system may continue to work reasonably well even with some hardware failures. Of course, a problem with the transmitter power amplifier, the link itself, or receiver front end affects all channels.

Compare this to the TDM situation. Like the FDM case, an existing link can be upgraded to multiple users by installing TDM multiplexing and corresponding demultiplexing circuitry, which allows many users to share the link. Distortion and signal quality on the existing link must be low enough that timing and sync can be established from the received signal. But there is more to using a system for TDM: In a TDM system, all channels are tightly linked through the multiplexing switch (analog or digital) that controls which signal is passed at any instant. Any problem with this critical element affects the entire system operation. Further, it is not possible to separate out one signal's circuitry and use it as a substitute for another, as is done for SDM, because there is very little circuitry belonging to that one user only.

In TDM the entire system is more closely integrated, which provides the greatest reduction in circuit complexity but the least amount of redundancy or extra subsections which can conveniently be interchanged. In summary, TDM systems either work very well or hardly function. The *bit error rate* (*BER*) (Chapter 10 through 13) indicates how well the TDM system is working, with typical values of less than 1 error in 10^8 bits (1 in 100 million) and better at signal-to-noise ratios of 20 to 30 dB and higher.

Review Questions

1. Why is TDM used increasingly in newer systems?
2. Why is the scheme where the multiplexing channel is selected by the receiving end of the link not practical for regular communications systems? Give two reasons.
3. How does TDM for digital signals overcome the limitations of TDM for analog signals in terms of sync problems and errors?
4. What circuitry is used for switching and selecting in digital TDM? Compare it to the circuitry for analog TDM. How is digital multiplexing and demultiplexing done?
5. Explain how digital TDM samples each of the various digital bit streams at different time within the bit period, but this still produces a valid multiplexed output.
6. Why can't a digital TDM system increase the number of users it sends onto the link endlessly simply by increasing the scanning rate? What is the basis for the limitation?
7. Why do microprocessor systems benefit from TDM? What is saved by using TDM in these systems? What is the drawback? How can the entire memory to microprocessor address and data bus be reduced to just eight lines plus control signals?

22.5 MULTIPLE-STAGE MULTIPLEXING

Many communications systems use several stages of multiplexing, in sequence, rather than just a single stage of SDM, FDM, or TDM. The successive stages of multiplexing do not have to be of the same type, although they may be. We have already seen two examples of multiple-stage multiplexing: in the phone system central office, where local loops (SDM) are interconnected to other central offices via T-1 trunk lines; and in telephone interoffice FDM, where the voice signals are combined into groups, groups into supergroups, and so on up to jumbo groups. In this section we look at some of the multiple stages of multiplexing.

TDM Followed by FDM

The 1.544-Mbit/s-bit-rate T-1 signal requires a link bandwidth of about 10 MHz for acceptably low error rates (narrower bandwidths have more errors since the digital signal suffers more distortion, and the receiver has a harder time making the correct 1,0 decision). Yet the bandwidth of a satellite communications link is much wider (Chapter 19), typically 36 MHz for the lower-frequency bands (below

10 GHz) and 80 to 200 MHz for the upper-frequency bands (above 10 GHz). Using one of these satellite links for a single T-1 signal is a waste of available bandwidth.

Instead of using the satellite for a single T-1 signal, three or four such signals are frequency multiplexed onto the 36-MHz-wide link (and more if the wider bandwidth link is used). The baseband T-1 format bit streams, each produced by 24 voice signals and independent of each other, are then amplitude modulated with subcarriers spaced 10 MHz apart to 60-, 70-, and 80-MHz carriers. The resulting cluster of three T-1 streams is translated up to the satellite carrier frequency, amplified, and transmitted. At the receiving ground station, the three signals are down-converted in a multiple-stage superheterodyne receiver back to the the 60-, 70-, and 80-MHz carriers, which are then demodulated (detected) to reconstruct the baseband T-1 signals.

Of course, the signals being combined by the FDM process do not have to be the same. A single satellite link can have any combination, such as a T-1 signal plus several standard 6-MHz TV signals (which are composed of the TV video and audio on their subcarriers) or analog voice groups and supergroups.

The communication system principle of *transparency* is essential to multiple-stage multiplexing. We have already seen this principle in many types of systems, such as when signals are encoded, formatted (even with added error correction bits), and sent with specific protocols in digital or network communications. Transparency means that each stage in the system performs specific, carefully defined operations on the signal it receives, and produces a predictable output for that input situation. But most important, the circuitry of each stage makes no judgment or decision based on what the data or signal actually represents. The bits can be ASCII characters of text, numbers in simple binary format, or digitized voice signals using seven bits per analog value, or even delta modulation (Chapter 11).

The corresponding stages in the receiver system perform the complementary operations and the reverse of the actions of the transmitter stages, following the opposite rules. Therefore, any circuit function in the multiple-stage multiplexing system treats input signals the same, whether they represent T-1 format bits of digitized voice or are just another continuous bit pattern with a different format. As long as both the multiplexer and demultiplexer each know what these rules are, the original signal that the sender generated is reconstructed and recovered at the receiving end.

TDM and TDM

Digital signals are often multiplexed, as in the T-1 format, where 24 user voice signals are digitized to seven bits, an eighth bit added, and then the digitized values are *interleaved* (which is a nice descriptive term for TDM) to produce a 1.544-Mbit/s bit stream. If this signal is sent over a wideband (bandwidth greater than about 10 MHz), bandwidth is wasted.

One solution is to use FDM to multiplex several such T-1 bit streams. However, an FDM solution requires subcarriers, oscillators, and precise filtering for proper operation. The alternative is to "stay digital" and combine, through TDM, several bit streams into a single higher-rate stream. This is easily done with digital signals and requires no analog circuitry at all. For N separate T-1 bit streams, a clock operating at N times the individual bit stream clock rates steps through all signals one bit at a time (with the appropriate multiplexing circuitry) and combines them into a digital supergroup (Figure 22.11). The final T-1 TDM signal is at

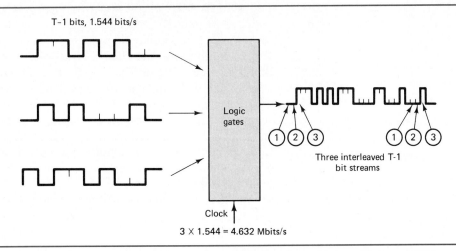

Figure 22.11 Digital supergroups: three T-1 signals multiplexed into a 4.632-Mbits/s signal.

$N \times 1.544$ Mbits/s and represents $N \times 24$ users. The fact that the T-1 signal represents 24 separate users rather than a single source of bits at 1.544 Mbits/s is transparent to this next stage of multiplexing and corresponding demultiplexing.

Digital signals are especially easy to time-division multiplex several times over since, as noted in the preceding section, the exact time for the multiplexer sampling the data from any source of data bits is not critical as long as the sampling is done sometime within the bit period. As additional stages of TDM are performed on various bit streams from different sources, the clock controlling the next stages' multiplexer has to be just accurate enough to make sure that all inputs are sampled within the bit period. It does not have to check at the exact center of the bit period, for example, since the bit value is the same throughout the period. This gives the multiplexing circuitry flexibility in operation and makes it easier for bits that are already the result of multiplexing to be combined with other multiplexed bits, to form yet another multiplexed bit stream at an even higher rate.

Review Questions

1. What is the purpose of using multiple stages of multiplexing?
2. Give two examples of multiple-stage multiplexing where the stages use the same type of multiplexing, and two where they do not.
3. Why is transparency an important feature of multiplexing?
4. Explain why the signals undergoing multiple stages of multiplexing do not have to be the same type or bandwidth, and can even mix digital and analog signals.
5. Why are digital bit streams that are already the result of TDM relatively easy to combine with other similar bit streams via another TDM operation for even higher data rates?

SUMMARY

To allow more users between two points, multiplexing is used. By adding new physical links, or making better use of the bandwidth of an existing link (sharing its frequency spectrum or dividing it into many time segments), more messages can be passed simultaneously. Space-, frequency-, and time-division multiplexing provide different amounts of additional capacity between two points and with differing amounts of new circuitry, reliability, and redundancy.

Space-division multiplexing is the oldest and simplest form: A new physical path is installed between the sending and receiving endpoints, along with new circuitry at each end. This is a costly way to increase the amount of message traffic, but it does make the circuitry of each user channel almost totally independent of the others, so that problems on one do not affect the others. Equipment can easily be swapped to help locate faults.

In frequency-division multiplexing, the many user signals are shifted to different parts of the link spectrum by various modulation schemes, so that the entire available bandwidth is used. Subcarriers are often used to establish precise frequency spacing among the various baseband signals. The type of modulation used for each signal does not have to be the same, nor does the bandwidth of each user signal—although they often are. Two examples of this multiplexing are the stereo FM signal and the TV signal of standard broadcast stations. As an added advantage, this type of multiplexing lets many signals be clustered together and then processed as a single signal, although composed of many separate signals in reality.

Time-division multiplexing requires sophisticated clock and synchronization circuitry to divide the overall time into many short periods, and each user signal gets its turn in sequence. This type of multiplexing can be used with analog or digital signals, but is especially useful with digital signals where the necessary circuitry, as well as synchronizing the clocks at both ends and the start of a sequence of users bits—a frame—is simplified. Time-division multiplexing is also used within microprocessor systems to reduce the number of signal lines between the microprocessor and its memory ICs.

Multiplexed systems are not limited to a single stage of multiplexing. Space-, frequency-, and time-division multiplexed signals can be multiplexed again, either by the same method or by one of the other two methods. For frequency and time multiplexing, the reverse process of separating the multiplexed signals into their original signals, called demultiplexing, is more complex than the multiplexing process because of signal distortion, noise, and propagation delays in the communications link.

Summary Questions

1. Give the two primary reasons for using multiplexing.
2. Compare SDM, FDM, and TDM in terms of basic operation and how each appears to an observer.
3. What is the basic circuitry for each of the three types of multiplexing? How is a single-user channel upgraded to multiple user for each? What new circuitry is required?
4. Compare and contrast multiplexing and modulation. Where do they overlap? Is one needed for the other? Explain.
5. What is the circuitry for demultiplexing SDM, FDM, and TDM? What are the key differences?
6. Compare the relative reliability, redundancy, and ease of troubleshooting of the three types of multiplexing.

7. How does the circuitry and dollar cost per user change with an increased numbers of users in SDM, FDM, and TDM? Where is the extra cost? Where is the extra savings? What changes in the physical link? The sending and receiving equipment?

8. What is a subcarrier? Where is it used? Why is it useful?

9. Give three examples of FDM. How many users can a spectrum allocation support? How is this number maximized in practical applications?

10. How is TDM used with a multiplexer controlled by the receiver? Why is this not useful in a general communication system?

11. Why are TDM multiplexer and demultiplexer sync difficult to achieve? What are frame and clock sync? How are they done at low speeds? Why don't these techniques work at higher speeds?

12. Compare TDM for analog and digital signals (and compare SDM and FDM for these as well). How is the sync (clock and frame) problem solved for digital signals? Why can't this technique be used for analog signals as well?

13. What is the factor that limits the number of digital TDM users rather than simply scanning faster to handle more users?

14. How is TDM used in microprocessor systems? Why? What are the benefits and drawbacks? Why is sync not an issue in this designs?

15. What is a multiple-stage multiplexing system? Give three examples. Why is transparency used as part of the design? What if the multiplexing was not transparent?

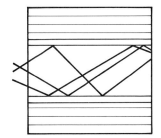

23

Microwave Equipment and Devices

CHAPTER OBJECTIVES

When you have completed this chapter, you will understand:

- The unique aspects of test, measurement, and design at microwave frequencies
- How and why vacuum-tube devices are used for signal generation and amplification at microwave frequencies
- Types of semiconductor devices for use at microwave frequency applications
- The operation and application surface acoustic wave devices at microwave (and lower) frequencies

INTRODUCTION

The microwave signal environment is very different from the lower-frequency world in terms of test equipment, available components, and the approach that is used for circuitry and systems. Some common and traditional types of test equipment are not useful at these short wavelengths, whereas others must be upgraded and new instruments and accessories developed.

Semiconductor devices are used extensively, except where they cannot provide the necessary power or have some other shortcoming that only vacuum-tube devices can overcome. Semiconductor diodes and transistors, fabricated in advanced structures from silicon and gallium arsenide, provide unique devices for microwave receivers and transmitters with good performance, efficiency, and low noise. For higher power levels, vacuum tubes are used in designs that combine electric and magnetic fields to control and manipulate electron beams. These tubes—klystrons, magnetrons, and traveling-wave tubes—overcome the limitations of conventional vacuum tubes at microwave frequencies and instead use these limitations as their basis for operation.

Surface acoustic wave devices are a solid-state approach to signal filtering. They convert electromagnetic energy to an acoustic wave, filter this signal, and then reconvert the acoustic wave to an electronic signal. They provide high-quality filtering in a single, low-cost, reliable device at frequencies where conventional filters are complex and difficult to manufacture.

23.1 TEST INSTRUMENTS AND METHODS

In previous chapters we saw some of the unique aspects of microwaves and microwave devices. Components such as waveguides (Chapter 8), microstrip and stripline transmission lines (Chapter 8), and klystrons and TWTs (Section 23.2) are very different from the conventional resistors, inductors, capacitors, wires, connectors, and transistors, and vacuum tubes of circuitry operating at lower frequencies.

The term *microwaves* has several meanings. In the earlier days of electronics, when it was difficult to build systems that operated reliably above about 100 MHz, microwaves indicated any frequency above 1 or 2 GHz (wavelengths less than 30 cm). Modern circuitry operates into the multi-GHz range, and *microwaves* is now used to indicate frequencies from 2 through about 18 GHz. Above 18 GHz, the term *millimeter* waves is used because the wavelengths are more easily specified in millimeters (at 30 GHz, one wavelength is 10 mm). Sometimes, however, the word *microwaves* indicates any signal frequency above 1 or 2 GHz, not just the more limited range up to 18 GHz. Occasionally, it is used to refer to frequencies below 1 GHz if the system is using waveguides.

Many things about the microwave and millimeter wave world are very different from the world at lower frequencies. Although the electromagnetic energy obeys the same laws of physics—Maxwell's equations—practical use of the frequencies and wavelengths requires a new type of component and a new approach to testing. Key aspects of microwave systems are:

1. Component values are smaller. capacitors are only a few picofarads, and even a $\frac{1}{2}$-pF change in value is significant. Placing a hand near the circuit or scope probe on the circuit affects everything.

2. Components are physically smaller, since their values are smaller, and this makes them harder to handle. The capacitance and inductance of the component lead wires is significant. For this reason, many microwave circuits use *chip components* that have no wire leads. Instead, the component has metallized end caps that are soldered directly into the circuit board.

3. *Stray* and *parasitic* capacitances and inductances, caused by the length of wire leads, circuit board tracks, and the physical assembly of the circuit, are insignificant compared to circuit component values at lower frequencies. But at microwave frequencies, they are significant and

sometimes larger than the component values themselves. The result: Board layout and grounding must follow special techniques, and measurements made of the circuit voltage, current, or power must also be done with great care. If not, the result may be misleading; more critically, the very act of measuring may cause temporary malfunction and even permanent damage.

4. Connectors used at microwave frequencies are very different from those at lower frequencies. These connectors must have low loss, provide precise matching and control of impedance [otherwise, standing waves and signal reflections (Chapter 8) will occur], and accept the unique types of coaxial cable used for microwaves.

5. The signal power levels that active microwave components such as transistors will handle is usually very low, since it is difficult to generate and control large amounts of microwave frequency power. The components used are often very sensitive to overloads, so *attenuators* are used in test to make sure that a 100-mW signal does not saturate or damage an amplifier that only works up to 10 mW. Of course, transmitter power levels are much higher.

6. Normally, routine electronic measurements such as voltage or current values at critical points cannot be made simply by connecting a multimeter. The meter's response at microwave frequencies, the type of connecting cable and connectors used, the effect of the meter impedance and capacitance on the circuit all affect the circuit operation and the measurement. In fact, most microwave measurements deal with power levels versus time or frequency, and voltage measurements are the exception (except for relatively static parameters such as device bias voltages).

7. Filtering and frequency measurement is more complex. Locating, measuring, and rejecting or passing a 1-MHz bandwidth signal at 100 MHz is much easier than doing the same at 10 GHz.

8. *Very important:* As noted in Chapter 8, microwave radiation can be **dangerous**! Special considerations and thought are needed when microwave energy leaks from a circuit or is deliberately transmitted by an antenna, so that it does not cause injury to the person standing nearby. **Caution** is required, and all **warning labels** must be followed. Those who might think to defeat or override safety interlocks on microwave equipment would be exposing themselves to a great deal of temporary and permanent body damage, including blindness, sterility, and genetic changes.

Based on these points, it may appear that working with microwaves is so difficult that it is not worth the effort. There are compelling reasons to use the microwave spectrum:

1. The lower frequencies are already in use. There is simply nowhere else to go if new communications bands are needed.

2. Many applications require transmitting large amounts of information, in both analog and digital format, in a short time. Because of this, a wide

bandwidth is needed. Only the microwave frequencies can support a 100-MHz bandwidth, for example, for a single user.

3. Even if a single user does not need the wide bandwidth, multiplexing (Chapter 22) of many users into a single wideband signal allows a single transmitter and receiver to be used. This allows sharing of cost among many users and a much lower cost per user. For this reason, the long-distance links used for high-capacity trunks between switching centers result in affordable long-distance phone rates for many.

4. The short wavelengths of microwaves (centimeters and shorter) means that high gain and very directional antennas (which require structures with dimensions several wavelengths long) occupy only a small space (Chapter 9).

5. The short wavelengths of microwaves are able to resolve small differences in distance in radar applications.

6. Microwave signals have propagation characteristics that are very predictable and repeatable, such as the free-space loss (Chapter 9). Unlike lower frequencies, which are severely affected by atmospheric noise and refraction—neither of which is controllable or accurately predictable—noise at microwave frequencies comes almost entirely from within the circuitry, and microwave propagation is line-of-sight (except when reflectors are deliberately used as in antennas). Microwave communications systems are reliable (this refers to the overall system, not the circuitry itself; lower-frequency circuitry is also very reliable, but its use in the overall application may not be).

Microwave Test Equipment

We have seen that measuring and testing components and circuits in the microwave region is not as straightforward as at the lower frequencies. There are no simple microwave tests, since each test setup requires careful understanding not only of what is to be measured, but also of the impact of the test equipment on the circuit or system under test. Unlike operation at audio or even the lower RF frequencies where measuring a voltage requires simply connecting a meter, or reducing a signal strength requires using a few resistors as a divider, microwave measurements require the right equipment used properly and connected correctly. There are no shortcuts or ''jury-rigged'' fixtures, because they usually give misleading or completely incorrect results.

Microwave test equipment is divided into four overall categories: basic connectors and signal path control, signal detectors, signal measurement and analysis, and signal sources. We will look at some units in each area. Note that microwave signals are carried both as currents/voltage in wires and cables, and as electromagnetic energy fields in waveguides. Most test equipment normally can interface to signals only when they are carried along in wires, but some equipment is able to handle waveguide signals as well. Impedances between source and load must be carefully matched so that reflections do not occur; 50 Ω is the most common impedance value in use.

Connectors and Adapters

Connectors and adapters are the basic mechanical interface between various cable assemblies, the system, and the test equipment. Many standard connectors are used in microwave systems (Figure 23.1). Adapters are small assemblies that

23.1 Test Instruments and Methods

(a)

Size	Series	Coupling	Impedance (Ω)	Frequency (GHz)	VSWR (max.)	Voltage (V)
Subminiature	SMA	Screw	50	12.4/18	1.3	500
	SMB	Snap-on	50	4	1.41	500
	SMC	Screw	50	10	1.6	500
Miniature	BNC	Bayonet	50	4	1.3	500
	TNC	Screw	50	11	1.3	500
	SHV	Bayonet	NC	NA	1.3	5000
	BN	Screw	50	0.2	1.3	200
	MC	Screw	50	0.5	1.3	200
Medium	C	Bayonet	50	11	1.35	1500
	N	Screw	50	11	1.3	1000
	SC	Screw	50	11	1.3	1000
	QM	Screw	50	4	1.3	5000
Large	QL	Screw	50	5	1.3	5000

(b)

Figure 23.1 (a) Appearance of typical microwave connectors (courtesy of AMP); (b) basic features of the most common connector series, (from *Machine Design,* p. 122, Feb. 9, 1989).

allow different connectors to be interconnected. For example, the test equipment may use SMA connectors, but the equipment under test may provide signals via an SMC connector. The adapter is either a single one-piece unit or a cable with corresponding connectors at each end to make the transition.

Attenuators

Attenuators reduce the signal amplitude. This is needed to match the signal amplitude to the acceptable input range of the test instrument, which is typically restricted to less than a few hundred mV or mW. Although this attenuation seems like a very easy task, it is not at microwave frequencies. The attenuator must be *flat*, providing the same attenuation factor (usually specified in dB) at all frequencies in order not to distort the signal. A *fixed* attenuator provides a single signal reduction, such as 1, 2, 5, or 10 dB, while *variable* attenuators allow the user to select any amount of attenuation in 1- or 2-dB steps, from 0 to 40 dB; the variable attenuators are easier to use but are much more complex and expensive.

It is difficult and costly to design a single attenuator that provides precise attenuation over a very wide range of microwave frequencies. Some manufacturers offer attenuators for specific bands, such as one for 10 to 20 GHz, another for 20 to 40 GHz, and so on. Others provide attenuators for a wider bandwidth but with a calibration chart showing the exact amount of attenuation at each frequency. The attenuation factor at each frequency must then be factored into any analysis of the data. Attenuators must also provide a good impedance match and low VSWR (<1.5 : 1); otherwise, they will cause reflections and standing waves that affect the unit under test and may invalidate the entire test procedure.

Attenuators are available for use with cable and connectors or with waveguides. A waveguide attenuator is an actual waveguide with some form of restrictive plate that impedes the flow of energy as it propagates through the waveguide, and it must be installed within the regular waveguides of the system. In contrast, attenuators for cables signals are connected much more easily. Incidentally, the signal "lost" through attenuation is dissipated as heat, so attenuators can become relatively warm depending on the input power and their attenuation factor.

Amplifiers

Amplifiers perform the opposite function of attenuators: They increase signal amplitude by known, fixed amounts so that the signal level is not too low for the test instrument to measure. Unlike attenuators, which are passive devices, amplifiers are active and more difficult to design to operate properly and faithfully over microwave ranges. Except for some klystron amplifiers (discussed in Section 23.2), which have waveguide input/output, they cannot be used with waveguides since signals enter and leave the amplifier via wires and connectors.

Amplifiers are specified by the frequency range they handle, the gain they provide (in dB), their noise figure, and their actual power output (dBm). Amplifiers for very low-level signals must be as low-noise as possible so that they do not significantly corrupt the desired signal with added noise. This additional noise is seen and analyzed by the test instrument as part of the original signal and will be amplified further by any subsequent gain stages in the system. Microwave test amplifiers have typical gains of +10 to +30 dB, power outputs of 10 to 100 dBm, and noise figures of several dB over the frequencies for which they are designed.

Switches and Splitters

Switches and splitters control the direction of flow of the microwave signal. Switches, of course, direct a signal to one of two or more paths. *Static switches* are set either by hand or by a digital control signal, and are intended to be switched infrequently. In contrast, *dynamic switches* are controlled only by a digital line, but switch from one position to the next in nanoseconds. They are used where the signal path must be rapidly changed or where two signals must be compared in rapid succession. The static switches are either mechanical or electronic; dynamic switches must use high-speed electronic components.

All switches have some *insertion loss*, which can range from a negligible 0.1 dB to a more significant 5 or 10 dB. If there is too much loss, an auxiliary amplifier may have to be used. As with attenuators, a switch must have flat amplitude versus frequency characteristics so that it does not attenuate some frequencies more than others. The overall performance of the switch is also measured by the amount of *crosstalk rejection* or *isolation* it provides between a selected (on) channel and the nonselected (off) channels; a typical value is 25 dB, which means that the unselected channel appears at the output 25 dB below its value on the input, while the selected channel suffers only insertion loss. Note that even a mechanical switch, which has virtually infinite resistance (and therefore infinite crosstalk rejection) at dc is far from perfect at microwave frequencies due to signal coupling within the switch via internal parasitic capacitances.

Splitters are needed when a signal must be directed to two or more points simultaneously, such as when the signal must be measured by an instrument and also go to another part of the overall system. The splitter is designed so that the signal source sees its intended impedance, even though driving two or more loads in parallel reduces the equivalent load impedance. Of course, in a *two-port* splitter the energy that appears at each port is only half (-3 dB) the original energy. Splitters are available both for waveguides and for insertion in cables. A good splitter is flat with frequency and has low insertion loss; some splitters are available that divide the initial signal not in half but by some other ratio, such as 10 : 1, so that a microwave signal can be picked off and monitored unobtrusively while most of it goes along the normal path.

Power Meters

Power meters are used at RF and microwave frequencies much more than at audio and medium frequencies. As their name states, they measure the amount of power in a signal. This is more useful than the actual signal voltage, which is not fixed in a circuit but is a function of the impedance the signal sees. Also, there is no single correct voltage: There is average voltage, rms voltage, and peak voltage, among others. These are difficult to measure at the higher frequencies. In contrast, power is defined uniquely regardless of the frequency. Also, the primary problem at microwave frequencies is the efficient amplification, manipulation, channeling, and sensing of signal power, since power is what drives circuitry.

A power meter consists of two parts: a *transducer* (sensor), which converts the signal power to its equivalent dc voltage, and a voltmeter, which measures and displays the dc value. The voltmeter is basically identical to an ordinary voltmeter, while the transducer is simple in concept but extremely sophisticated in realization (Figure 23.2). Different designs are used: The most common one takes the incoming microwave signal and uses it to heat either a *thermocouple* (two dissimilar wires connected together) or *bulk silicon*, which then produces a voltage directly proportional to the temperature. A well-designed transducer produces an output that is linear with the applied power, does not change with time or ambient temperature, and responds equally regardless of the microwave frequency. Many transducers come with a calibration chart to allow the user to correct for unavoidable imperfections.

Because of limitations in transducer physics, the power range of a meter with a single transducer is limited. A typical transducer can handle -30 to $+20$

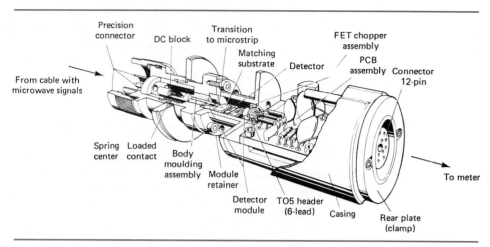

Figure 23.2 Construction of a power meter test head (courtesy of Marconi Instruments, Inc.).

dBm. For larger signals, attenuators are used. Alternatively, other transducers designed and optimized for different power ranges are substituted for the first one.

Depending on the microwave application, power meters must measure both continuous power (CW for *continuous wave*) and the power of pulsed signals. The power meter in Figure 23.3 indicates power for signals from 30 MHz to 40 GHz, at levels from −40 to +20 dBm. These signals can be either CW or pulses as short as 15 ns, which often occur in radar and related microwave situations. Using a built-in microprocessor and signal-processing algorithms, the power meter also indicates pulse rise time, fall time, and pulse width, all essential measurements in pulsed applications. An IEEE-488 interface (Chapter 18) is also built in so that the power meter can be part of an automated test station and be set up by a controlling computer and report measurements back to the computer; of course, it can be operated entirely from the front panel, if desired.

Figure 23.3 Microwave power meter (model 8500A) reads both CW and pulse peak power (courtesy of Wavetek Microwave, Inc.).

Spectrum Analyzers

Spectrum analyzers were discussed in Chapter 2 together with Fourier analysis and the concept of the equivalence of time and frequency domains. Signal frequencies and spectrums are often more revealing than signal voltage versus time in microwave applications, just as signal power is more important than voltage. Microwave spectrum analyzers are used to show the exact frequency of a waveform, its stability versus time, and to see if there are any undesired frequency components called *spurious* oscillations. Interfering signals are also seen. The spectrum analyzer shows signal power amplitude at various frequencies (which can be converted to voltage since the characteristic impedance is known), noise and distortion, and the effects of modulation on the carrier.

Spectrum analyzers for microwave ranges take the input signal and mix it with a local oscillator (LO) signal; the mixed output is then filtered by a narrow-bandpass IF to show the signal amplitude at a frequency equal to the difference between the LO and input frequencies (Figure 23.4). Sweeping the LO through the overall spectrum tunes the analyzer to cover the entire range of interest. More expensive and complex analyzers provide extremely fine frequency resolution: Some of the most advanced cover the range from a few GHz to over 100 GHz, with resolution of 10 Hz at the lower end of the range and 3 MHz at the upper. Another way to look at this is that 10 Hz at 1 GHz is just $\frac{1}{100}$ millionth of the nominal frequency; 3 MHz at 100 GHz is 3/100,000 of the frequency. Most spectrum analyzers are designed to handle more than just very small or very large signals but can accept a wide *dynamic range* of high to low, ranging from -120 to $+30$ dBm. This is necessary when looking for sidebands or spurious signal which may be 60 or 80 dB below the carrier itself—simply attenuating or amplifying the carrier can put the carrier in the acceptable range but still not allow the wide amplitude span to be seen along with it.

Counters and Frequency Meters

Counters and frequency meters are used to count the number of cycles of a signal and to determine the frequency by counting the number of cycles in a precisely known time period. They are needed to determine if equipment is operating at the correct frequency or if there has been signal drift, for example. Counters and

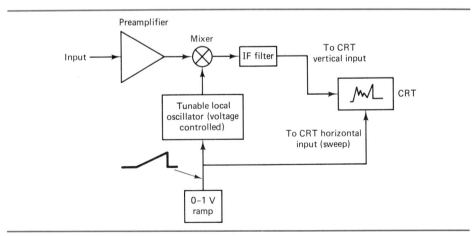

Figure 23.4 Block diagram of microwave spectrum analyzer shows use of tunable local oscillator and IF filter.

frequency meters are usually used in conjunction with the spectrum analyzer to make sure that there is only one signal for the meter to measure: It is meaningless to measure the frequency of two or three signals that are present together. Often, the spectrum analyzer shows that there are undesired signals that are large enough to confuse the frequency meter, or that the desired signal is too small for the meter to measure accurately (so that it requires amplification).

Oscilloscopes

Oscilloscopes are used in microwave applications, but not as much as at the lower frequencies. First, technology for microwave range oscilloscopes is very costly and not as accurate or wide-ranging as spectrum analyzers or power meters. Even more important, though, is that the "signal versus time" that a scope displays is not as useful as the signal's frequency spectrum or its power content.

Signal Generators, Sources, and Power Amplifiers

Signal generators, sources, and power amplifiers provide microwave signals to be injected or substituted into the system under test. A basic source produces *continuous-wave (CW)* carrier oscillation at a specified frequency and calibrated power level. Power levels are usually very low—from −100 dBm up to 0 dBm—since these are more critical in most applications. The flow of the signal through the system, and the various tranformations it undergoes as it passes through the stages of the system, are then measured. More advanced signal generators modulate the basic carrier with AM, FM, or PM, using a variety of analog or digital waveforms. The modulated signal is used in place of a signal at a receiver antenna and then tracked through the receiver itself, for example. *Sweep oscillators* generate a signal that begins at one frequency and smoothly and automatically sweeps through an entire preset range of frequencies. They are used for testing frequency response of amplifiers, attenuators, and filters, checking for linear performance at many frequencies, among other things.

Power amplifiers are broadband amplifiers that take the generator output and boost it to much higher levels, if needed, such as for testing transmitter antennas. A typical power amplifier output ranges from several hundred milliwatts to several watts (of course, larger ones are available, but these are much less common and must be used with extreme caution). The power amplifier output level is often not calibrated precisely but is measured with attenuators and a power meter. In most applications, however, only the lower power levels are needed at very precise levels for testing microwave systems.

A complete microwave test setup for checking the performance of an amplifier is connected as shown in Figure 23.5. The CW signal generator applies a precise input signal; the output is split so that the power meter can measure overall gain while the spectrum analyzer checks for distortion components produced by the amplifier. If the power is greater than the spectrum analyzer can accept, an attenuator is used between the amplifier output and the analyzer input. Modulated waveform performance of the amplifier is checked by switching in a generator that provides more complex waveforms than the CW source. The spectrum analyzer is especially useful in studying the output characteristics of the amplifier for non-sine-wave signals. The sweep oscillator replaces the simpler CW signal source for checking the linearity, performance, and performance of the amplifier across the entire band of interest.

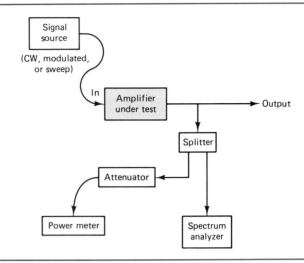

Figure 23.5 Setup for testing amplifiers.

Review Questions

1. Why are microwaves so different from lower frequencies, although they all follow the same laws of physics with respect to component values, size, stray circuit effects, connectors, power levels, and danger?

2. Why are microwaves used extensively in terms of information rate, available spectrum, and predictability?

3. Compare the use of microwaves in waveguides versus cables for test purposes.

4. Why are attenuators needed? Why are they not a trivial part of the test setup? What are the two ways that attenuator nonflatness over the range is accommodated? Compare fixed versus variable attenuators.

5. Why are amplifiers needed? Why are they more complex than attenuators? How are they specified? Why is noise a critical specification, unlike in attenuators?

6. Where are splitters used? Why not simply wire the signal to two points simultaneously?

7. Why are power meters more commonly used in microwave applications than at lower frequencies? Why not simply measure voltage? How is the power actually sensed, and then measured?

8. What is the resolution of a high-performance spectrum analyzer? What is dynamic range, and why is it important? Why are oscilloscopes less commonly used than spectrum analyzers in microwave work?

9. What does a CW signal source provide? What is added modulation used for? What is a power amplifier for a signal source? How is it calibrated, if necessary?

23.2 VACUUM-TUBE DEVICES

The first microwave transmitters, in the 1930s, used existing vacuum tubes (transistors did not exist yet) for their oscillators, amplifiers, and modulators, but could operate only at the lowest microwave frequencies. Increasingly smaller vacuum tubes were built, so that their dimensions would be compatible with the microwave wavelengths they were handling. However, when the wavelengths are only a few centimeters, the connecting wires within even a small tube look electrically like antennas and have considerable parasitic capacitance and inductance. The physical dimensions between the internal elements of even a small tube are such that electron *transit time* though the tube has the same approximate magnitude as one period of signal oscillation (period is the inverse of the frequency). This means that things no longer appear to occur "instantaneously" within the entire tube: during the time that the electron leaves the cathode on its journey towards the grid

---

Brief Description of Vacuum Tubes

The vacuum tube was the first active electronic device, capable of actually controlling and amplifying a small signal. It was invented in 1907 by Lee De Forest, who took the simple two-element vacuum tube diode and added one vital intermediate element—the grid—to control electron flow. The basic elements of a simple vacuum tube (Figure 23.6) are:

The *filament* (or heater), which heats up from applied ac or dc just like a light bulb filament and causes the surrounding *cathode* to emit electrons.

An *anode* or *plate,* which is charged to a high positive dc voltage and therefore attracts electrons emitted by the cathode. The flow of electrons—the plate current—is the tube's output into a load (an antenna, another stage of amplification, a filter, or a modulation or demodulation circuit, for example).

A *grid*, a screen between the cathode and anode plate through which the electrons must pass. By varying the voltage on the grid (usually from near zero to a negative voltage), the much larger flow of electrons between cathode and anode is controlled and modulated. This is the critical addition that revolutionized electronics, because the grid lets one small signal control a much larger one.

Although many circuit configurations are possible with vacuum tubes, in the most common one a small signal to be amplified is applied to the grid. This modulates the electron flow from the cathode to the plate, and the plate current is then a reproduction of the grid signal but with much greater current amplitude. The grid can also cut off the electron flow to the plate if it is driven negative so that it repels the negatively charged electrons emitted from the cathode. Like a standard light bulb, the entire assembly of elements is housed in an evacuated glass (or metal) envelope; this vacuum in the tube prevents the filament from oxidizing and burning out quickly. The vacuum also eliminates molecules in the tube which would interfere with the flow of electrons from cathode to plate.

This simple three-element *triode* vacuum tube (cathode, grid, anode; the filament does not convey signal but serves to heat the cathode and so is not counted) is an effective design that is still used; other refinements include additional grids for more control, as in the two-grid, four-element *tetrode* and the three-grid, five-element *pentode*. Small vacuum tubes were available for microwatt and millivolt signals but have been replaced by transistors; large tubes are still used extensively for kilowatt and higher signal power.

Note that when used as a diode, or rectifier, there are just two elements in a vacuum tube and no grid is needed. Electrons flow only in the direction from cathode to anode; an ac waveform applied at the cathode appears at the anode with only the negative half-cycles. Compare the complexity of this vacuum-tube rectifier diode—which requires a filament and filament power, plus high voltages for the plate—to the modern semiconductor diode, which just has two connections and is the essence of simplicity in use.

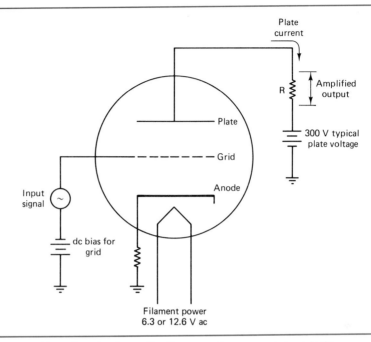

Figure 23.6 Functional schematic diagram of vacuum-tube triode.

and plate and actually reaches them, several cycles of the signal waveform occur. At the same time, higher power levels mean that a large amount of dc power must be supplied to the tube and dissipated heat must be conducted away, which gets more difficult as the size of the tube decreases.

The solution was to develop revolutionary new structures and designs that used electromagnetic principles in a different way from that of the conventional vacuum tube. Three important alternatives to the vacuum tube for microwave use are the *klystron, magnetron,* and the *traveling-wave tube.* Each turns the weakness of the conventional vacuum-tube structure to an advantage by exploiting the physics of the situation, manipulating and controlling the electrons with electric and magnetic fields. Figure 23.7 shows a "family tree" of electron tubes and the relationships between conventional tubes and various microwave tubes (note that we do not discuss some of these microwave tube variations but only the more significant and commonly used ones.)

Electron bunching is one physical characteristic that is used to generate power at microwave frequencies. When electrons move past a series of magnets (actually electromagnets), they are attracted in the direction of some magnet poles and repelled by the opposite poles, depending on whether they are north or south poles. As the electrons approach the attracting pole, they increase in velocity, and as they pass the pole they are pulled back to it and slowed (retarded). Similarly, for the repelling pole, the electrons are retarded as they approach the pole but pushed away after they pass it.

If the magnetic pole strengths are not large enough to completely repel, or hold the electrons, the electrons will arrive at their destination in bunches caused by the attract/repel action. The result is a series of high and low densities of electrons (bunches) in the stream that is flowing between the source and the

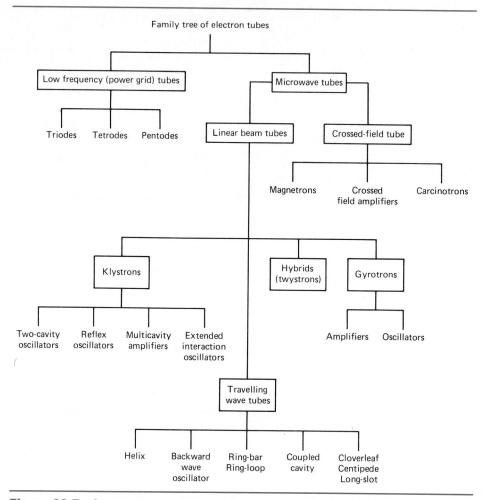

Figure 23.7 Conventional and microwave vacuum-tube family tree shows a wide variety of available devices (from *Microwave Systems News*, p. 20, June 1989).

destination. Varying the magnetic fields by changing the current in their electromagnet coils results in modulation of the electron beam current intensity. Electron bunching is also known as *intensity modulation* for this reason.

Klystrons

The klystron was developed in the 1930s and exists in both *multiple-cavity* and *single-cavity* (*reflex*) designs. The multiple-cavity klystron is an amplifier operating from about 250 to 1000 MHz, providing signal gain between 30 and 65 dB. It operates in a continuous or pulsed mode with 30 to 60% efficiency. In contrast, the smaller reflex klystron is an oscillator and is only 10% efficient, but it can operate up to about 100 GHz while producing power outputs from several hundred mW to several watts. It is used in test equipment, as the local oscillator in a receiver, and in low-power communications links.

The klystron requires no external components and uses a resonant cavity to determine the operating frequency and contain the microwave energy. In the

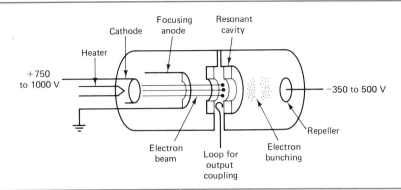

Figure 23.8 Reflex klystron block diagram.

simpler reflex klystron (Figure 23.8), electron emissions from the cathode are focused by the anode into a controlled beam of electrons, which pass through a *cavity gap*. Any initial disturbance in the smooth flow of electrons causes some electron bunching as they leave this gap. As these electron bunches approach the far end of the klystron, they are turned around by the high negative potential on the *repeller* located there and sent back through the cavity gap. If the electrons arrive at the correct time and phase relative to the oscillations within the cavity (when the oscillation within the cavity is passing through its zero point and just going positive), they will give up their own energy to the oscillations. The result is that the oscillations are enhanced while the electrons leave the gap as a dc current. (Unlike negative feedback, which serves to stabilize amplifiers and oscillators, this is positive *regenerative* feedback, which produces desired oscillations if properly controlled but results in wild and useless oscillation if not carefully done.)

The frequency of oscillation is determined by the size of the resonant cavity and can be adjusted by a screw that pushes on the wall of the cavity and changes the cavity's physical dimension. The success of the klystron operation depends on the repeller voltage, which controls the velocity at which the electrons are turned around and go back through the gap to sustain oscillation. This voltage must be set so that the electron traverses the gap, reaches the repeller, and goes back through the gap in three-fourths of a single cycle time of the RF frequency (plus any integral number of cycle times). By varying the repeller voltage slightly, it is possible to vary the oscillation frequency by a few megahertz for frequency modulation.

In multicavity klystrons (Figure 23.9) initial operation is similar to that of the reflex klystron. The RF signal in the first cavity intensity modulates (bunches) the electron beam as it traverses the gap. The electrons then travel through a space between cavities called a *drift* space, where the bunches tighten: Electrons that are accelerating continue to do so, while those that are slowed down also continue to slow down. The bunching action continues as the beam passes through a second cavity, and this activity is optimized by the appropriate values of applied voltage, the input frequency, and the length of the drift space. The bunched electrons excite the final cavity at its resonant frequency, and the output of this cavity is the amplified signal.

Klystrons range in size from about 6 in. long for the smaller reflex unit, to about 10 ft long for the higher-power multicavity designs. The entire assembly of a

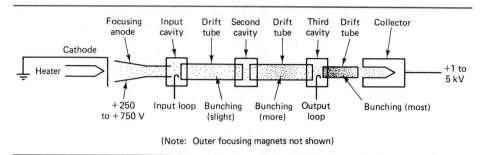

Figure 23.9 Multicavity klystron block diagram.

larger klystron weighs several hundred pounds and needs water cooling to keep the operating temperature within an acceptable range despite the large amounts of heat dissipated.

Magnetrons

The magnetron uses a powerful permanent magnet for its magnetic field (although a constant-current electromagnet can also be used) to bend the travel path of the electrons. It is a self-contained powerful oscillator that requires no external components and can provide up to 25 kW of continuous power or several megawatts of pulsed (low-duty cycle) power, operate at frequencies up to about 100 GHz, and yield relatively high efficiency of about 50 to 70% in converting applied dc power to microwave oscillation. [A special low-power (500 to 1000 W), low-frequency (2.5 GHz) version of the magnetron is used in home microwave ovens to heat food.]

The basis of the magnetron design (Figure 23.10) uses an electron-emitting cathode surrounded by the anode. Instead of a conventional grid, the magnetic field controls the electron flow and direction. Recall that the force on a charged particle—such as an electron—traveling through a magnetic field is at right angles to both the magnetic field and the electron path. In the magnetron, the emitted electron is attracted to the anode by the electric field that the anode voltage produces. If there is no magnetic field, the electrons travel directly to the surrounding anode.

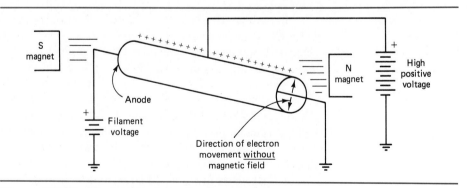

Figure 23.10 Electrical schematic of magnetron.

However, when there is a properly oriented magnetic field, it causes the electron path to curve as the electrons approach the anode; as the magnetic field strength increases, the curvature increases (its radius decreases). If the magnetic field is large enough, the electron path is curved so much that the electrons do not reach the anode but instead graze the anode while traveling in a circle back to the cathode. Further increases in the magnetic field make the electron travel in a smaller circle back to the cathode, never even touching the anode. The plate current due to electrons reaching it, therefore, is very large up to the critical value of magnetic field but then drops off to zero suddenly when the magnetic field is large enough to curve the electrons away before they ever reach the plate.

Now that we have electrons traveling in a circle, we must turn to the physical structure of the magnetron (Figure 23.11) before we see how this motion of the electrons is converted into microwave oscillation. The magnetron body consists of a hollowed-out block of brass or copper, with eight or more small cavities around the central chamber. It is within these cavities that the microwave energy is developed, and the cavity size determines the operating frequency. The alternate cavities are strapped together with metal bands to control internal phase and field relationships and effectively parallel the outputs of the cavities. A magnetron block for 25 W of continuous power is about 3 inches across and 1.5 inches thick; after the interval cavities are drilled and machined, the ends of the cylinder are covered and sealed with end plates. The magnetron block contains the electromagnetic energy, supports the leads for the cathode power and the microwave signal pickup probe loop that extracts the signal energy (see the discussion of waveguides in Chapter 8), and dissipates the excess heat that is generated. The magnetic field is along the central axis of the magnetron, like the axle of a wheel through a tire. The electric field within the magnetron extends out radially from the central axis, generated by the potential applied to the magnetron plate (the body) with respect to the central cathode.

If there is no applied magnetic field, the electrons generated at the cathode will travel directly to the anode in a straight line, under the influence of the electric field. The magnetic field, however, changes the situation completely. The combination of the electric field and the magnetic field causes the electrons to move in circles. As the electrons leave the cathode and move toward the anode, the magnetic field curves their path so they are traveling parallel to the anode. As they pass by the anode cavity opening (gap) when they are traveling in the same direction as the electric field in the gap, the electrons slow down. In doing so, they give up their energy of motion (kinetic energy), and this energy is not lost as heat but instead transfers to the electric field in the cavity.

The oscillations within the magnetron are the result of currents induced in the cavities by the moving electrons interacting with the electric field. The frequency of oscillation is determined by the size of the cavity and is therefore very stable, since these dimensions do not change, although they can be adjusted with small screws. The electron transit time is therefore used to advantage within the magnetron to allow the magnetic field to influence the travel path of the electrons, and divert them from the straight line between cathode and anode to curve up to the cavity gap.

Traveling-Wave Tubes

The traveling-wave tube (TWT) was perfected in the early 1950s and is used extensively today in many microwave systems, such as amplifiers in satellite

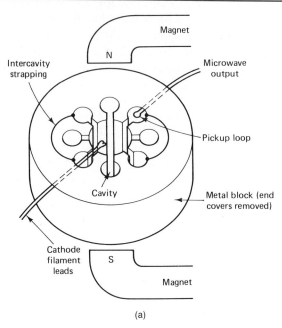

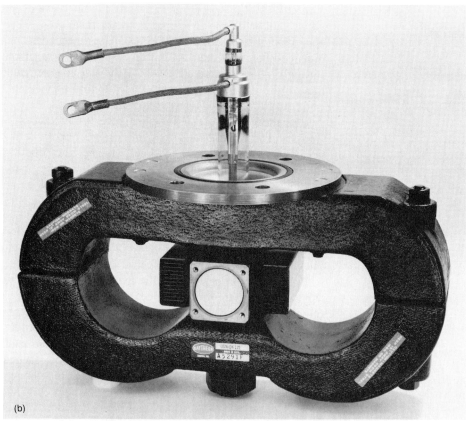

Figure 23.11 (a) Structural schematic of magnetron; (b) actual magnetron unit (courtesy of Raytheon Co., Microwave and Power Tube Division).

transponders (Chapter 19), because it offers a good combination of efficiency, reliability, and overall performance, along with wide bandwidth and low internal noise. Unlike the klystron or magnetron, which have only a brief interaction between the electron beam and the RF field in the resonant cavities, the TWT has nearly continuous interaction. In the TWT the electron beam and the RF field travel at approximately the same velocity, unlike in the klystron, where the field is stationary. The resonant cavities of the klystron and magnetron give both of them very high Q and narrow bandwidth, but the design of the TWT provides the wider bandwidth needed for applications such as transponders, radar transmitters, or general-purpose amplifiers. The electron bunching, which intensity-modulates the dc beam current in the TWT, occurs over the entire length of the device (at least 12 in. long) rather than just within a resonant cavity.

Low-power TWTs operate from 2 to 40 GHz, producing 25 to 50 dB of gain and output levels under 100 mW. These TWTs have been replaced almost entirely by solid-state devices (Section 23.3). Other TWTs operate to about 100 GHz while producing 2–5 kW of continuous power (and several times that in pulse power).

The problem that the TWT solves is this: The RF field from the signal to be amplified travels at the speed of light (c), but the electron beam travels much slower, at about 10% of c. Therefore, to have continuous interaction between these two elements requires that the RF field be slowed down. Of course, this cannot be done directly (since the speed of light is constant), so the TWT directs the RF field across and around the path of the electron beam in a helix (spiral) pattern. The wavefront of the RF field that is actually parallel to the beam therefore travels much slower than c, by a factor related to circumference of the helix and the pitch angle of the helix.

The TWT (Figure 23.12) uses an electron gun to shoot a stream of electrons down the center axis of the helical coil to an anode at the far end. The anode and helix are charged positively with respect to the cathode, which pulls electrons down the center path, and the anode, cathode, and helix are mounted in a vacuum tube. A microwave signal that is introduced into the helix at the cathode end travels at the speed of light, but its effective velocity along the axis is much less, as discussed above. This signal interacts with the electron beam which is going straight down the helix center axis, electron bunching occurs, and the electrons give up some of their energy to the wave on the helix. Bunching is maximized by adjusting the anode voltage so that electron beam travels slightly faster than the RF field propagates along the axial direction of the TWT.

The electron beam must remain focused at the anode and not be attracted to the helix (which is also at positive potential). External permanent magnets or electromagnets are used to develop a field within the TWT which keeps the beam focused.

In addition, since the RF field can propagate in either direction along the helix, oscillations may occur if some of the field reflects back from the helix end and constructively interferes (reinforces) with the forward wave. This is prevented by surrounding the glass TWT envelope with a lossy material (such as wire or graphite) that attenuates this *back wave* and so breaks the feedback path, although some gain is lost since the forward wave is also attenuated. (*Note:* This back wave characteristic is used, under carefully controlled circumstances, to turn the TWT into a *backward wave oscillator* instead of just an amplifier. The anode end of the helix is desired to have a high VSWR and large mismatch, while the back wave attenuator is removed.)

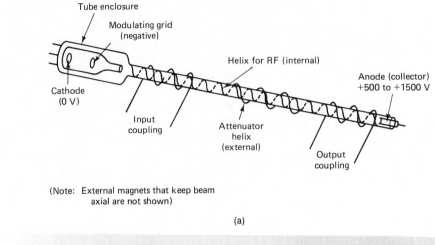

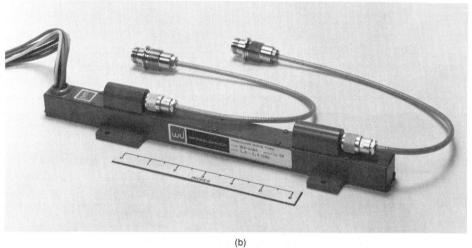

Figure 23.12 (a) Functional schematic of TWT; (b) typical unit (courtesy of Watkins-Johnson Co.).

Energy is coupled into and out of the TWT by several means. Most commonly, an input helix at the cathode end wrapped around the TWT main helix and combined with a similar output helix at the anode end is used, similar to coupling through adjacent, aligned transformer coils. Alternatively, special cavities are built into both ends of the TWT, with loops inserted to inject or extract the microwave energy. Other less common methods with other limitations are also used.

Figure 23.13 summarizes the power and frequency ranges for various microwave power sources in their more common models, along with a comparison to semiconductor (solid-state) devices, covered in the next section. Of course, for special application needs, many of these devices have been custom constructed in smaller and larger power versions and for higher and lower frequencies.

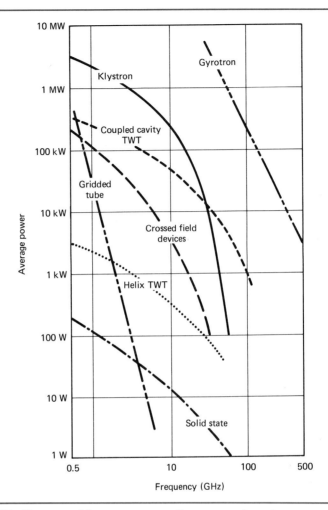

Figure 23.13 Power and frequency operating ranges for microwave vacuum tubes (along with solid-state devices) (from *Microwave Systems News,* p. 25, June 1989).

Review Questions

1. Why can't regular vacuum tubes be used for microwaves? Why does making the tube smaller help but still have some limits?

2. What is transit time, and what are its implications?

3. What is the basic operation of a vacuum tube and the roles of the filament, cathode, grid, and anode? How does a tube amplify? Why is the simplest amplifying tube called a triode? Compare a vacuum-tube rectifier to a semiconductor diode.

4. What is electron bunching? How does it occur? Why is it also called intensity modulation?

5. Compare the reflex and multiple-cavity klystron. How does the klystron use a resonant cavity? How does each type operate?

Peak versus Average Power

Communication systems often use measurements of *peak power* and *average power*. These are especially important in microwave systems, where the signal may be used in a pulsed mode rather than continuously transmitted. In the simplest case, where the carrier is turned either completely on or off, average power = peak power × duty cycle, where the duty cycle is the ratio of the "on" time to the "on + off" time. The difference between peak and average power can be substantial. A peak signal of 1000 W power that is turned on for only 10 μs every 2 s has a duty cycle of 0.000010/(2 + 0.000010) = 0.000005, so the average power is just 0.005 W. When the modulating waveform is not a simple on/off signal but has a complex shape, the actual average power is even less since the effective duty cycle is smaller still.

The implications of this for microwave devices such as the klystron, magnetron, and TWT are that the device must be designed for both the desired average and peak power of the application. For radar and similar low-duty-cycle applications, kilowatts or megawatts can be produced by a relatively smaller device, as long as the system can supply the necessary average dc power while dissipating the heat that unavoidable inefficiency produces. For CW applications, the device must be larger to handle the extremely large and continuous power supplied and heat generated. Some microwave devices (such as the magnetron) can be pulsed and provide high peak power values with acceptably low average ratings, while others such as the TWT are difficult to operate at a low duty cycle and really provide nearly identical average and peak power ratings. The type of microwave device chosen, therefore, depends on both the amount of power needed and the mode—continuous or pulsed—that is required.

---

6. How can klystron frequency be changed by a large amount? By a small amount? Why is the repeller voltage critical?

7. What is drift space? Why does the multiple-cavity klystron produce tighter bunching?

8. How does a magnetron use electric and magnetic fields to produce microwave power? What is the electron path in a magnetron under each field acting separately and under their combined effect?

9. What is the shape and construction of a magnetron? How does the physical arrangement make the magnetron work?

10. How does the magnetron plate current vary with applied magnetic field strength?

11. Why are TWTs used? What is the fundamental difference between the TWT operation and that of the klystron or magnetron?

12. What is the construction of a TWT and its principle of operation? How does the TWT allow continuous interaction of the slower electrons with the RF field traveling at c?

13. Why can a TWT have inadvertent oscillation? How is this controlled?

14. How is energy coupled into and out of the TWT?

15. Compare peak and average power needs for various applications. How does duty cycle relate these two parameters?

16. Why are peak and average power ratings important in selection of a microwave amplifier or oscillator?

23.3 SEMICONDUCTOR DEVICES

Early semiconductor (solid-state) devices operated only into the kilohertz (and soon the lower MHz) ranges and certainly could not provide any significant power, gain, or low-noise operation. Present devices—diodes and transistors—are now better than tubes in most performance aspects, and are used in place of vacuum tubes in virtually all microwave receiver applications, as well as lower-power transmitter systems. Vacuum tubes continue to be used for higher-power needs (beginning at about 10 W to 100 W, depending on frequency) where semiconductors cannot provide enough power. Vacuum tubes are also used when a combination of electrical and mechanical ruggedness is needed. Although in proper use semiconductor devices are as reliable and long-lived as tubes, the tubes retain one advantage: They can withstand amazing amounts of electrical abuse, overvoltage, electrical transients, radiation (from atomic reactions or in natural radiation of space), or excess heat buildup that would destroy a semiconductor device.

In this section we look at diodes and transistors used for microwave range oscillation and power. A wide variety of semiconductor process technologies are used, each with advantages and drawbacks. These include tunnel diodes, Gunn diodes, bipolar transistors, and field-effect transistors. Note that all these devices are pushing the limits of solid-state technology, and those who invent them are working at the fundamental levels of physics, including quantum theories of particle physics.

Advantages of Semiconductor Devices over Tubes

In general, the advantages of semiconductors are overwhelming tubes: no need for filament power, operation at lower voltages (12 and 24 V are common, compared to the hundreds or thousands of volts for tubes), smaller size (important in physical layout of microwave circuitry), and ease with which signals can be coupled into and out of the devices. The semiconductors can be used as *discrete* components that are soldered into a circuit board, and this circuit board can also incorporate microstrip and stripline tracks for carrying the microwave signal. This integration of the device, circuit board, and transmission line is low cost and reliable; this is another advantage of semiconductors over vacuum tubes, which often require separate waveguides to transfer energy into and out of them.

Discrete devices are also used as *chips*, which are then placed into a *hybrid* circuit. In a hybrid, all the components—semiconductors, capacitors, resistors, and so on—are mounted on a ceramic *substrate* and soldered to this substrate via their metallized end caps rather than by actual wire leads. Interconnections between devices are made by metal paths that are screened onto the substrate and fired at high temperature. The entire hybrid assembly is then put into a final package to be used as a single large-scale component. Hybrids are smaller, more reliable, and provide better performance than that of a traditional printed circuit board.

Advances in integrated circuitry have provided benefits to microwave systems as well. *Monolithic microwave integrated circuits* (*MMIC*) fabricate many microwave devices and their interconnections onto a single piece of semiconductor material [usually silicon, but other materials, such as gallium arsenide (GaAs) are also used]. This has the advantage of providing the lowest cost and greatest reliability, although presently MMIC performance trails that of hybrid devices and they cannot dissipate as much power, due to their smaller size. MMICs are available for amplifiers, op amps, mixers, local oscillators, front ends, modulators, and other stages of a receiver or transmitter. A great deal of research today is devoted to improving the discrete components in speed, power, noise, and efficiency for use in circuit boards and hybrids and in advancing the capabilities of MMICs.

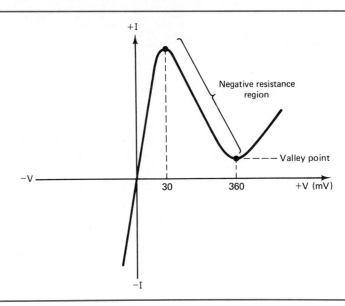

Figure 23.14 The negative resistance characteristic of the tunnel (Esaki) diode differs from the conventional resistor voltage–current graph.

Tunnel Diodes

Tunnel diodes, invented in the late 1950s by Esaki in Japan, can be used in oscillators or low-power (mW) amplifiers. The diodes have a *dynamic negative resistance* at certain operating conditions (Figure 23.14). Unlike a conventional resistor, in which the current flow increases as the applied voltage increases (although not always linearly), a negative resistance is one in which the current *decreases* as the voltage *increases* past a specific point. This characteristic allows the diode to be used in the feedback path of a circuit that can oscillate at frequencies up to tens of GHz.

The mechanism of the tunnel diode negative resistance is explained only by modern quantum physics: At certain energy values, some electrons on one side of the *p-n* junction within the diode actually migrate to the other side, despite not really having enough energy to do so. This is explained in quantum physics by showing that in special circumstances, some of the electrons "tunnel" through the energy barrier under certain unique conditions, an effect that cannot be explained by pre-quantum physics descriptions. Fortunately, the user of the tunnel diode does not have to understand any of this principle to use the device.

Gunn Diodes

Gunn diodes, invented in 1963, also have the negative resistance characteristic but develop it in an entirely different way. In contrast to the tunnel diode, which is a *junction device* dependent on activity at the *p-n* junction, a Gunn diode uses the basic properties of silicon (or other semiconductors such as GaAs) and so is a *bulk device* (Figure 23.15a). When a dc voltage is applied across a very thin slice of material so that the electric field has a high enough local intensity (>3300 V/cm) for GaAs), the material also exhibits negative resistance. A typical Gunn diode can produce several watts of power at 4 GHz with 20% efficiency, or 50 mW at 100 GHz, but with just 5% efficiency; Figure 23.15b shows a packaged Gunn diode assembly as it would be used in a circuit.

23.3 Semiconductor Devices   717

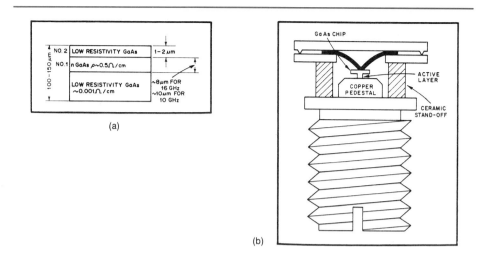

Figure 23.15 (a) Electronic structure of the Gunn diode (courtesy of American Radio Relay League, Inc. ARRL); (b) complete Gunn diode assembly (courtesy of American Radio Relay League, Inc., ARRL).

Other Microwave Diodes

Other diodes used in microwave applications include the three-layer *PIN diode* (for positive, intrinsic, negative doping of its three semiconductor layers), which switches from on (conducting) to off (nonconducting) in nanoseconds and maintains high off-impedance even to microwave signals. It is often used as a transmit/receive switch to direct transmitter signals to the system antenna while preventing them from reaching the receiver front end, and then directing received signal energy from this antenna to the receiver. Figure 23.16 shows a pair of PIN diodes in such an application. One diode is in series between the transmitter and antenna; the other is in shunt across the receiver input. A voltage at the bias terminal

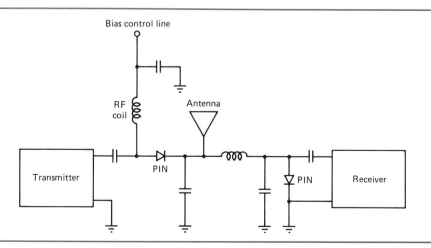

Figure 23.16 PIN diodes used as transmit/receive switch (courtesy of American Radio Relay League, Inc. ARRL).

(connected to the system transmit/receive control line) forward biases both diodes so that there is a low-impedance path from transmitter to antenna while shorting out the receiver input. When this voltage is absent, both diodes look like high impedances and the transmitter is effectively disconnected from the antenna while antenna signal appears across the receiver input terminals.

Other microwave diodes include the *IMPATT* ("impact avalanche and transit time") diode; and the *TRAPATT* ("trapped plasma avalanche triggered transit") diode, among others. They are fabricated in silicon, gallium arsenide, indium phosphide, and other materials.

The wide variety of microwave diodes provides many choices in design. They are available for different power levels, frequencies, and noise levels. Some diodes are low noise; others have very stable and repeatable noise characteristics and are ideal for noise sources in test equipment. However, diodes in general suffer from two problems. First, they are relatively inefficient at converting applied dc power to microwave signal power. Typically, only about 2 to 20% of the applied dc power is converted to microwave oscillation. Second, as *two-terminal* devices they are awkward to use in many circuit configurations. Unlike a *three-terminal* device (a transistor, for example), which has an input connection, an output connection, and a common connection for both input and output, diodes do not have separate connection points that distinguish input from output. In application, therefore, they are often used as feedback elements within a circuit to get some overall type of circuit operation.

Transistors

In contrast to the use of diodes in circuits, transistors form the basis of many well-understood circuit designs such as amplifiers. Transistors for microwave applications are usually fabricated from silicon or GaAs. Silicon is inexpensive since it is used for virtually all other semiconductors and ICs, and the equipment to process it into the final product (transistor or MMIC) is widely available. Silicon devices operate up to about the range 1 to 10 GHz; beyond that range, the more expensive and difficult to fabricate GaAs now allows construction of individual transistors and complete MMICs that operate at higher frequencies. GaAs operates at higher speeds because it has higher ion mobility than silicon, and higher peak electron velocity.

Although parasitic and internal capacitance and inductance affect transistor operation at microwave frequencies, and transit time is also a factor, the actual impact of these factors is not the same as in vacuum-tube devices. The dimensions of transistors are much smaller, and the manipulation and control of electron flow occurs within a solid substance where electrons travel slowly, rather than in a vacuum where an electron beam traveling at about 5 to 10% of the speed of light interacts with electromagnetic fields. Solid-state physics presents a very different and more complex environment than the electrons and fields of vacuum-tube devices. Appreciation of the parameters that limit the performance of microwave transistors require understanding of atomic structure, quantum physics, energy levels, materials science, and similar advanced subjects.

Two types of transistor structures are used at microwave frequencies, regardless of the basic bulk material—silicon or gallium arsenide—used. *Bipolar transistors* are current-driven *minority-carrier* devices that can provide significant power. However, as the power level increases, the bipolar device can actually work into a "vicious circle" of drawing yet more current, heating up even more, and then self-destructing. Bipolar circuits require additional components to limit the current and minimize problems from VSWR mismatch and reflection. The base and emitter of bipolar transistors are built as a series of interlocked fingers (*interdigitated*), to minimize parasitic capacitance and distribute the current flow more evenly.

Field-effect transistors (*FETs*) are now surpassing bipolar devices in many areas of microwave performance. FETs are voltage-driven, *majority-carrier* devices that tend to be self-controlled and limiting as the current they draw increases. In many situations, a FET is easier to bias and drive properly and is less sensitive to the type of load it is driving. FETs can also be operated in parallel (up to a limit) and so provide higher total power output than any single device can provide. FETs now produce gains and powers that equal or exceed those available from bipolar devices.

Noise is a major concern at microwave frequencies since it usually limits the useful sensitivity of the overall circuit. Presently, bipolar devices have lower noise at lower microwave frequencies, while FETs and bipolar device noise characteristics are about the same in the regions above 10 GHz. A small-signal bipolar transistor typically has a gain of 25 dB, an output of 100 mW, and a noise figure below 1 dB up to that frequency.

Advances in semiconductor devices for microwaves are occurring at a rapid rate. The physical packaging of the transistors must also be suitable for microwave frequencies (Figure 23.17). There are constant improvements in the power, frequency limits, efficiency, noise, and gain in both silicon and GaAs bipolar and FETs. Only a few years ago it was believed that the technology to replace vacuum tubes with semiconductors for VHF, UHF, and microwave transmitters delivering over a few watts output was a long way off in the future. This future has come much more quickly than most experts predicted.

Manufacturers also provide microwave amplifiers as complete modules, so that users do not have to design or build a circuit with transistors, resistors, and related components. Figure 23.18 shows an FET-based module that provides 22.5 dB of gain from 2 to 8 GHz and a 17-dBm output power level. The module has an input connector, output connector, and requires only that a power supply (± 12 V dc, 60 mA) be connected to operate. A module of this type (just 0.3×0.15 in.)

Figure 23.17 A variety of microwave transistor packages (courtesy of Phillips Components, Discrete Products Div.).

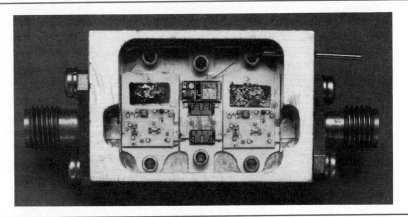

Figure 23.18 Microwave amplifier module using FETs is complete and easy to use, includes input and output connectors, and provides 22.5 dB of gain from 2 through 8 GHz (courtesy of Celeritek Inc.).

makes microwave amplification relatively easy since no circuit skills are needed by the user and all the intricacies of design and fabrication are already done and tested.

Review Questions

1. What are the advantages of semiconductor devices over tubes? Where do tubes still retain advantages?
2. What is a MMIC, and what are its advantages?
3. What are tunnel diodes? What is the negative resistance that a tunnel diode provides, and how is this used in oscillators?
4. What is a Gunn diode? What is meant by a bulk device rather than a junction device?
5. Explain the use of a PIN diode as a transmit/receive switch.
6. Why are diodes harder than transistors to design into circuits?
7. Compare bipolar transistors to FET devices. Why are FETs becoming more dominant in microwave applications?

23.4 SURFACE ACOUSTIC WAVES

Our entire study of communications systems has used electromagnetic energy—electric and magnetic fields and waves, voltage and current—to manipulate, transmit, and receive signals that carry information. There is one area where an alternative energy is used. *Surface acoustic wave (SAW) devices*, developed in laboratories in the 1960s and commercialized in the 1970s, use acoustic energy instead of electromagnetic energy. They are the modern equivalent of electromechanically resonant filters and provide excellent performance in precise bandpass filtering and oscillation applications. They can operate at frequencies of many hundreds of MHz and even low GHz values, providing stability and precision that conventional filter circuits (using resistors, capacitors, inductors) cannot easily achieve, yet with lower cost and greater reliability.

SAW devices are based on the combination of the *piezoelectric effect* and waveform interference. This same piezoelectric effect is what allows quartz crystals to be used as the frequency-determining factor in oscillators: When an ac voltage is applied across the crystal, the crystal mass physically vibrates. The frequency of this vibration is determined entirely by the physical properties of the crystal (its mass, dimensions, mounting, and internal atomic structure), and the vibrating crystal appears electronically as a very high-Q resonant circuit. In this way it acts as a precise bandpass filter, or it can be combined with a phase-shift network in the frequency-controlling circuit of an oscillator. Regular quartz crystals will function as precise frequency elements up to about 50 MHz; above that they get too small and fragile to be useful. Frequency multipliers (Chapter 6) can be used with standard crystals, but add to weight, cost, and power consumption.

In SAW devices (Figure 23.19) the same piezoelectric effect is exploited in a different way. An oscillating electrical signal is applied to a tiny piece of crystal that is formed as part of a larger, flat surface, and the piezoelectric effect causes the crystal to convert the electrical signal to vibration. These vibrations—which are therefore sound (acoustic) waves—travel across the surface until they reach a corresponding crystal at the other end, where their acoustic energy is converted back into electrical oscillations. [Note that although these truly are acoustic waves, their frequency is many orders of magnitude beyond what we normally consider as "sound" (up to 20 kHz).]

To make the SAW device useful as a filter, a precisely spaced row of metallic "fingers" is deposited on the flat surface (again an *interdigitated* structure). As the acoustic wave travels across the surface, it reflects back and forth as it impinges on the fingers. Depending on the wavelength of the acoustic signal and the interdigit spacing distance, some of the reflected energy will cancel out and attenuate the incident wave energy (*destructive interference*), while other frequencies will be aided by *constructive interference*, as standing waves form and reflections

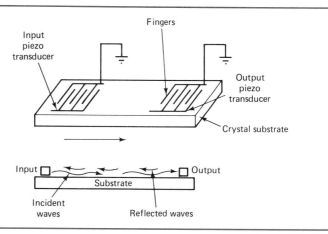

Figure 23.19 Functional and physical schematic of a SAW device.

add in or out of phase (Chapter 8). The exact wavelengths (and therefore frequencies) that are canceled or aided depend on the interdigit spacing, while the bandwidth of attenuated frequencies depends on factors such as the finger thickness and number of fingers.

The acoustic wave travels across the surface not at the speed of light (since it is not an electromagnetic signal) but at the much slower velocity of acoustic energy in the crystal, about 3000 m/s. The wavelength of the acoustic signal—the key factor in calculating the necessary interdigit spacing—is calculated using this velocity in the standard equation where wavelength = velocity/frequency. A 100-MHz electromagnetic signal travels in free space at 3×10^8 m/s and so has a wavelength of 3 m; the same frequency traveling as an acoustic wave has a wavelength of 0.00003 m (0.03 mm).

SAW devices can be fabricated for bandpass filters with center frequencies from 50 to 1000 MHz, with bandwidth (-3 dB) values around the center frequency of about 12 to 50 MHz. They attenuate the unwanted frequencies about 30 to 50 dB below the desired signal output level and are very useful for selecting a desired signal while rejecting others (Figure 23.20). They are very rugged and reliable, and there are no internal parts to be tuned or need adjustment since the entire SAW filter is fabricated from a single crystal (quartz, lithium niobate, or one of various other oxides) with carefully placed metallic lines. They are also small—a typical SAW device in its package is about 0.5×0.75 in. Contrast this to a filter built of discrete components, which requires a small circuit board or hybrid assembly.

The only major drawback to SAW devices is their high *insertion loss* of about 25 to 35 dB. This means that even the desired signal is severely attenuated and loses energy as it travels through the device (the undesired signal frequencies are attenuated even further below this, of course). For this reason, SAW devices

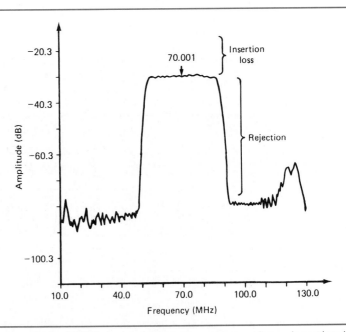

Figure 23.20 Typical SAW device frequency response shows insertion loss and out-of-band rejection (courtesy of Crystal Technology, Inc.).

cannot be used directly on low-level signals until they have been amplified up to several hundred mV, and another amplifier is needed after the SAW filter to restore the signal level. Despite this drawback, SAW filters are used extensively in communication systems (in single- and multiple-conversion superhet receiver IFs, and standard TVs to filter out the color subcarrier information) to extract some frequencies while rejecting others. They are also used as *delay lines* to delay an electronic signal precisely, which is often necessary in a circuit. A signal traversing across 1 cm of a SAW device takes about 2 μs; in contrast, an electronic signal traveling in a wire for 1 cm is delayed by only about 0.1 ns.

Review Questions

1. What is a SAW device? Explain its basic operation. How does a SAW device differ from an electronic filter?

2. Why does a SAW device provide such effective filtering? How is interference employed and exploited? What is the role of interdigitation?

3. What frequency ranges and bandwidths can SAW devices handle? What is the typical rejection of frequencies outside the passband?

4. What is the insertion loss of SAW devices? How is this accommodated?

5. Compare delay times in a SAW device to delay time in an equivalent length of wire. Why is there such a difference?

SUMMARY

The short wavelengths of microwave (and millimeter waves) renders many conventional components, circuits, and test techniques and instrumentation of little value. Measuring the magnitude of a microwave signal requires more than connecting a voltmeter: Constant impedance must be maintained, special connectors are used, power levels are critical, and attenuators and amplifiers are used to make sure that the signal under test and the range of the test instrument are compatible. The electromagnetic signal power is a vital parameter, and special power meters use sophisticated sensors to convert microwave energy into its equivalent dc value. Similarly, directing the flow of a signal in a cable and switching it from one location to another requires carefully designed power splitters and switches. The spectrum analyzer is a vital instrument for showing how much energy exists at various microwave frequencies, while the oscilloscope is used less than it is at lower frequencies.

Special vacuum tubes are required for generating higher-power microwave signals, since physical dimensions and stray capacitances makes most devices' conventional tubes useless at these frequencies. Electron bunching and transit time are two microwave region phenomena that these tubes make use of as they control the direction and shape of electric and magnetic fields, converting applied dc power to microwave signals. The frequency of oscillation is set not by discrete components such as capacitors and inductors, but by electromagnetic energy resonating inside precisely-sized cavities or traveling on carefully directed paths. Some applications require continuous power while others need short power pulses, and different tube structures are optimized for these different uses.

Semiconductor diodes and transistors are now used in place of tubes whenever possible, since they are much easier to interconnect and power, are smaller, and have better efficiency and lower noise. Special diodes such as the tunnel diode and Gunn diode have negative resistance characteristics and are used in microwave oscillators. Other devices, such as the PIN diode, switch very quickly and are used for transmit/receive switches. Transistors are more efficient than diodes, provide gain, and are easier to design into circuits. Microwave transistors are fabricated either in silicon or in gallium arsenide for even higher-frequency operation. Bipolar transistors can be designed using either material, but recent designs have resulted in silicon and gallium arsenide field-effect transistors with better performance and lower noise in many cases than is obtainable with bipolar ones. Physical packaging is crucial in microwave devices as well.

Solid-state materials are also used in surface acoustic wave devices, which use the piezoelectric effect inherent in a crystal to transform electromagnetic energy to acoustic energy and back. Specially placed fingers on the surface of SAW devices act as wave energy filters, yielding bandpass characteristics that cannot be obtained with RCL filters, and with greater precision and lower cost.

Summary Questions

1. Compare the difficulties of dealing with components and circuitry at microwave frequencies with those at much lower frequencies of operation.
2. What benefits does microwave operation provide?
3. Explain the role of the following test equipment: attenuators, amplifiers, switches, splitters, power meters, spectrum analyzers, counters, frequency meters, oscilloscopes, and signal sources. How is each used? What are the key specifications of merit of each?
4. Why is power measurement very common in microwave systems?
5. Why is flatness important? What are the two ways in which test equipment deals with nonflatness across the wide spectrum?
6. How does a basic triode vacuum tube operate? What are the key elements and their interaction? How does it provide amplification?
7. Why can't a conventional vacuum tube be used at microwave frequencies? What is done to extend the operating frequency as much as possible?
8. Explain transit time and its significance.
9. What is electron bunching? How is it caused? What is drift space?
10. Compare the structure and operating principles of the klystron, magnetron, and TWT. How do they differ? What is the role of a magnetic field, if any, in each?
11. What are the power levels, efficiency, and operating frequencies of the klystron, magnetron, and TWT? What determines the operating frequency of each?
12. Explain the difference between peak and average power and the significance of the difference.
13. Why are semiconductors preferred over vacuum tubes in most applications? Where are tubes still required?
14. Compare the use of semiconductors in printed circuit boards, hybrid devices, and MMICs.
15. Explain the construction and theory of the tunnel diode. How is the tunnel diode used?
16. Compare the tunnel diode to the Gunn diode, in basic structure.
17. What are PIN diodes? Where are they often used? Why?
18. Why are transistors preferred in many cases to diodes? What are the advantages and disadvantages of silicon versus gallium arsenide?

19. Compare bipolar transistors for microwave operation to FETs: What are the differences in operating theory? In performance? What are the benefits of each?

20. How are SAW devices used? What benefits do they provide? What are their drawbacks?

21. What is the operating principle of a SAW device? How does it exploit the piezoelectric effect? What are the interdigitated fingers used for?

22. To what frequencies can SAW devices operate?

23. How does a SAW device differ in operating principle from an electronic filter circuit?

24
Fiber Optics

CHAPTER OBJECTIVES

When you have completed this chapter, you will understand:

- The benefits and drawbacks of fiber optics systems compared to those of conventional wire cables, along with details of their construction
- Principles of optical fiber operation and the main types of fiber in use
- Types of components used as optical sources and detectors
- How a complete fiber optic system is characterized and how it performs
- Equipment and methods used for testing of fiber optic systems

INTRODUCTION

Fiber optics is the newest way of sending large amounts of data, although it is based on a principle of physics that has been known for several hundred years. Fiber optic links are used for applications ranging from short-distance local area networks to worldwide data communications. Hair-thin glass and plastic fibers act as conduits—pipes—for pulses of light generated by a light source. At the receiving end, an optical power detector senses the light pulses and converts them back to electronic data signals.

Practical fiber optics systems required the development of reliable sources and detectors, as well as low-loss optical fibers. These systems allow wideband communication within a single building or across the oceans, while providing benefits of no electrical interference, low error rates, security, and reduced bulk compared to standard wire cables or broadcast links. Optical fiber systems are most suitable for single point-to-single point applications, where there is a minimum of switching or routing, since optical energy is difficult to switch to other fiber paths.

Test instruments for fiber optics are very different from those for conventional electronic signals. Optical sources generate a light at a single wavelength (frequency) with imprecise power levels, so test procedures are changed to accommodate this situation and produce accurate results. Time-domain reflectrometry for electronic signals is modified for use with optical signals and can show nearly every subtlety of the optical fiber performance, working only from one end of the fiber.

24.1 FIBER OPTIC SYSTEM CHARACTERISTICS

In the late 1960s and early 1970s, scientists and engineers began exploring how to overcome some of the limitations of communications using metallic links (wires and cables). Using the principle of *total internal reflection* discovered over 200 years earlier (discussed in detail in Section 24.2), fiber-optic-based systems allow many benefits that cannot be achieved with any other type of medium. In such a system the electronic data signals are converted to light pulses and sent through a hair-thin glass or plastic fiber to a detector at the far end, where they are reconverted back to electronic signals. Although this is an extremely simple concept, many practical problems had to be solved to make fiber-optic-based systems reliable, practical, and low in cost. These problems have been solved well enough that fiber optic systems are being used in many applications instead of traditional wire and cable.

Both conventional electronic signals and light are electromagnetic waves and follow the same basic laws of physics. However, the extremely high frequencies and short wavelengths of light have dramatically different implications for signal transmission that do those of lower-frequency waves, including microwaves. Note that the frequencies in use are around 360 *terahertz* (3.6×10^{14} Hz), compared to the 1 GHz (1×10^9 Hz) at the beginning of the microwave region. At these frequencies it is easier to use the wavelength value as a measure. Fiber optic systems use wavelengths between 600 and 1500 nanometers (1 nm = 1×10^{-9} m), often referred to as 0.6 to 1.5 *micrometers* (μm) or *microns* (1 μm or micron = 1×10^{-6} m). Visible light spans 430 to 690 nm (violet to red), so fiber optic system wavelengths use part of the visible spectrum as well as the longer-wavelength, invisible infrared spectrum.

The benefits of using light and fiber optics are:

1. Tremendous bandwidth and consequent high data rates are easily achieved. An optical fiber system can easily support 100 Mbits/s; advanced systems are carrying beyond 1 Gbit/s.

2. The light pulses travel entirely within the fiber. They cause no interference, known as *electromagnetic interference* (*EMI*) and *radio-frequency interference* (*RFI*) in adjacent wire cables or optical fibers.

3. The optical fiber system is also immune to nearby signals and EMI/RFI, regardless of interference magnitude. An optical fiber placed next to an operating multimegawatt transmitter will perform as if the transmitter were turned off. There is also no interference with adjacent optical fibers.

4. There is complete electrical isolation between ends of the link. This eliminates *ground loops* (current that flows between two circuit grounds when they are not really at the same potential, although they should be so

ideally), which affect performance, as well as the danger of shock at one end if there is a misconnection or failure at the other end

5. Fiber optic systems are secure from unauthorized listeners. Since the light energy stays entirely within the fiber, the only way to intercept the signal is to tap physically into the line—there is no radiated energy field to intercept. Taps are difficult to accomplish physically, and a tap in the line causes a loss in signal power that is easily detected.

6. Since there is no electrical energy present, fiber optics can be used wherever there is danger of explosion from sparks.

7. The weight and bulk of a fiber optic cable is much less than the equivalent wire cable for the same effective bandwidth and number of users. A single coaxial cable weighs about 10 lb per 1000 ft; the optical fiber cable weighs about 4 lb. A single fiber optic system easily handles 1344 two-way conversations with one optical fiber for each direction in one standard telephone system implementation.

Communications links based on fiber optics are now used in many applications served previously by metallic cables, transmissions, and satellite links. They are used for long-distance trunks, radio area networks, and to carry signals within airplanes and ships (especially because of their weight savings and immunity to noise). Many of the new worldwide communications links being established use underground and undersea fiber optic cables to strongly supplement or even replace satellite links. This is because satellite frequencies and physical locations are being filled up; fiber optics offers better security against eavesdropping and deliberate interference; propagation delays and echo times in a fiber optic system are much less than in a satellite link (3000 miles from New York to California for a fiber optic link versus about 50,000 miles for a satellite path); and fiber optics are immune to natural radio interference (such as sunspots and atmospheric storms) that occasionally disrupts all atmospheric links.

Fiber optics systems do have some limitations. Many of these are changing as the technology advances, but some do not yet have solutions in sight:

1. The cost of the fiber is greater than that of basic copper wire in some configurations, although this is much less true today than it was 10 years ago. A basic optical fiber costs anywhere from $0.10 to $0.50 per foot, depending on the specific type of construction used.

2. It is difficult to splice optical fibers to make them longer or to repair breaks (both mechanical and chemical techniques are used to splice fibers). Copper wire is still easier to splice—connect the two wire ends in a mechanical crimp or solder them together. The type of temporary repair that is used for copper wire, where the wires are twisted and taped or quickly soldered, is impossible with optical fiber. The fiber optic splice must be done nearly perfectly using standardized procedures or it is virtually useless.

3. Connectors for fiber optics are more complex to attach to the cable and require precise physical alignment. In contrast, wire connectors are quickly crimped or soldered to the wire, and the actual mating connector

with its receptacle is tolerant of slight mechanical mismatch of misalignments.

4. Switching and routing of fiber optic signals is difficult. Taking a signal from a single fiber cable and distributing it to two cables requires a large, complex optical or electronic/optical assembly. Similarly, switching an optical fiber signal from one path to another requires some advanced and costly systems. In contrast, a single copper wire is easily connected to two other wires, and a simple, low-cost, reliable mechanical switch allows a signal on one wire to be sent to the desired path or used in a matrix.

5. The test equipment and techniques needed for fiber optics are different in many ways than they are for electronic signal paths.

Fiber optic systems are used almost exclusively for binary digital signals. There are some designs that use multiple-level digital signals and some that use analog signals, but these are rare. In all subsequent discussion of fiber optic systems and components, assume that data are being sent in binary form.

Warning: Never look into an optical fiber unless you can personally verify that the other end is also disconnected (not just connected to an "off" source). The light in the fiber, whether visible or invisible, may be intense enough or concentrated in a small spot to cause severe eye damage.

Review Questions

1. What is the basic principle of fiber optics? What are the frequencies used in fiber optics? What are the wavelengths in nanometers, microns, and micrometers?
2. What are the main benefits of fiber optics in terms of bandwidth, data rates, interference, isolation, security, safety, and bulk?
3. What are typical fiber optic applications? Why are fiber optic systems being used to supplement satellite links?
4. What types of signals are sent over an optical fiber? At what rates?
5. Why should a person never look into a fiber unless the other end is absolutely disconnected?

24.2 THE OPTICAL FIBER

The optical fiber is a thin strand of glass or plastic and consists of three parts (Figure 24.1). The *core* is a transmission area of the fiber, much like the hollow center of a pipe. A larger core gathers and transmits more light; typical core diameters range from 50 to 500 μm. Surrounding this core is the *cladding* which has a different *index of refraction* than the core. The cladding defines the optical boundary of the core and makes sure that total internal reflection occurs at the core outer skin; it is like the walls of the pipe, which contain the fluid within the pipe and prevent any leakage. A core and its cladding act as an *optical waveguide* or *light pipe*.

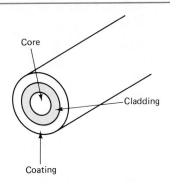

Figure 24.1 Optical fiber cross section: core, cladding, and coating.

Fibers are specified by the outer diameters of the core and cladding. A 50/150 fiber means that the core is 50 μm while the outside dimension of both the core plus cladding together is 150 μm. The core and cladding are surrounded by another layer, called the *coating*. This is a specially formulated plastic coating that provides a first level shock and abrasion resistance for the fiber. Coatings range in thickness from 250 to 1000 μm and are often stripped mechanically or chemically when two optical fibers are joined. The coating is not involved in the actual transmission of light, and its thickness is a function of the physical ruggedness needed. If the coating thickness is called out, it is the third number in the series of diameters, such as 50/150/400.

The coating is just the first layer of protection for the very thin optical fiber. Many types of outer layers formulated from various types of plastics, rubber, or metal are used as required by the installation. These coatings must protect the fiber against impact, normal abuse, and excessive bending which affect the fiber performance. A *tight* outer layer protects the fiber from crushing and impact but can protect the fiber only against some types of bending. To further protect against excessive bending, the optical fiber is often placed inside a *loose*, stiffer outer layer or jacket (similar in function to semiflexible metal conduit for power

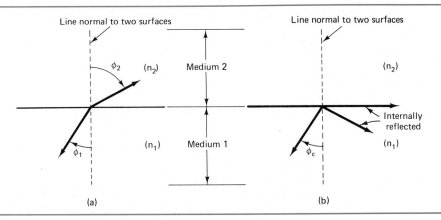

Figure 24.2 (a) Basic principle of the law of refraction; (b) total internal reflection prevents any light from leaving medium 1.

Total Internal Reflection

It may seem impossible that a perfectly clear fiber can constrain the lighwave rays to stay within the fiber as it travels many kilometers, yet not have these rays exit via the walls along the trip. The physics principle of *total internal reflection* ensures that the fiber acts as a pipe that does not let the light pass through the walls and so be lost. To understand how this works, let's look at what happens as a lightwave ray traveling through a medium with one *index of refraction* crosses into another medium with a different refractive index. These principles of refraction for lightwave rays were discovered independently by Willebrod Snell and René Descartes in the early seventeenth century, but they apply to all electromagnetic waves, including radio waves. Total internal reflection was demonstrated in 1870 when John Tyndall showed the Royal Society in London that light could be guided within a jet of water.

The index of refraction is inversely proportional to the speed of light through the medium; it is 1.00 for a vacuum by definition, 1.0003 for air, and varies between 1.2 and 2.00 for different glass and plastic formulations (when the index = 2, the speed of light in the medium is one-half of its speed in a vacuum). (The index of refraction also varies slightly with the wavelength of light, and it is this fact that causes a glass prism to split white light shining through it into its constituent spectrum of colors.) We have also seen the effects of refraction when an electromagnetic radio wave passes from one layer of the atmosphere to another and some bending occurs, since each layer has a very slightly different index of refraction from the nominal 1.0003 value of standard laboratory air.

Consider a ray of light with an *angle of incidence* ϕ_1 traveling in a medium with refractive index n_1 and existing as a refracted ray with angle ϕ_2 into a medium with refractive index n_2, Figure 24.2a (note that these angles are measured with respect to a line that is perpendicular, (normal) to the interface between the media.) The *law of refraction* states that

$$\frac{\sin \phi_1}{\sin \phi_2} = \frac{n_2}{n_1}$$

Refraction enables lenses to bend light to correct focus, magnify, and control images.

Example 24.1

What is ϕ_2 When $n_1 = 1.5$, $n_2 = 1.2$, and $\phi_1 = 45°$?

Solution

Using the formula, $\sin \phi_2 = \sin \phi_1 (n_1/n_2)$

$$\sin \phi_2 = 0.707 \left(\frac{1.5}{1.2}\right) = 0.884$$

so $\phi_2 = \sin^{-1} 0.884 = 62°$.

To see how total internal reflection reflection occurs, we assume a light-wave ray traveling from a medium with a higher index of refraction to one where the index of refraction is lower. As the angle of incidence ϕ_1 is increased, a situation is reached where the refracted ray points along the surface—its angle of refraction is 90° (Figure 24.2b). For angles of incidence larger than this *critical angle,* no refracted ray exists and the lightwave ray is prevented from leaving the first medium and instead returns. The critical angle ϕ_c is found by setting $\phi_2 = 90°$; then

$$\sin \phi_c = \frac{n_2}{n_1}$$

For a glass core with $n_1 = 1.5$ surrounded by cladding with $n_2 = 1.2$, $\sin \phi_c = 0.8$ and the critical angle is 53°. It is the principle of total internal reflection without light loss (in theory), which occurs by surrounding a higher index core with a lower index cladding, that makes fiber optics work: The lightwave ray bounces from one wall of the fiber to the other but never escapes. Total internal reflection cannot occur when the light travels from a medium with lower index of refraction to a medium with higher index.

You can see total internal reflection in a common real-world situation. When you are under water and looking up at the sky, you are in a medium with $n = 1.33$ (water) looking into one with $n = 1$ (air). At certain angles of looking up (the angle of incidence), there is total internal reflection and you see no sky at all, only your own darkness (unless you shine a light up at the sky and you will see some of it reflected back down; no light escapes and an observer above the water will not see you or your light at this angle). The critical angle at a water to air interface is

$$\sin \phi_c = \frac{1.00}{1.33} = 0.752$$

Thus $\phi_c = 48.6°$ or greater.

wiring). Of course, each layer of protection adds thickness and weight, and the final number and type of protective layers used are determined by the anticipated application and installation: A cable to be simply dropped into a trough between office partitions does not need as thick or rugged a jacket as one that will be pulled through ducts. Final fiber optic cable diameters, including all protective layers and jacketing, range from $\frac{1}{8}$ to $\frac{3}{4}$ in. and more.

Fiber Types

Fibers are identified by the type of paths, or *modes*, that the lightwave rays travel along within the fiber core (Figure 24.3). In multiple-pathway or *multimode* fibers, the lightway rays take many paths between the source and the far end of the fiber. As a result, the original sharp pulse at the source is spread out in time at the receiver, since some paths are longer than others. This spreading *dispersion* limits the maximum data rate and bandwidth that can be achieved, since adjacent pulses begin to overlap each other. Dispersion is much less critical at lower data rates, where the pulses are more widely spaced and overlap does not occur so soon.

There are two types of multimode fibers: step index and graded index. In a *step index* fiber (Figure 24.3a), there is a sharp step-like difference in the refractive index of the core versus the surrounding cladding. Step index fibers are relatively inexpensive to produce and useful at lower rates or shorter distances.

The *graded index* multimode fiber (Figure 24.3b) is more complex. Unlike a step index fiber, a graded index core contains many thin layers, each with a lower

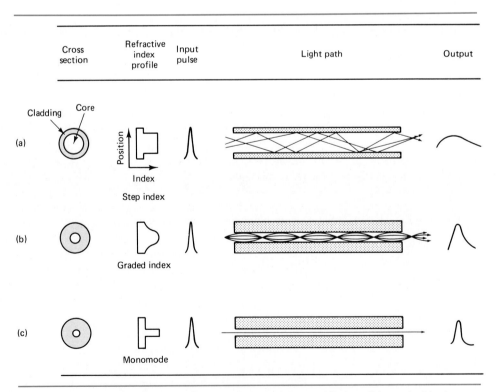

Figure 24.3 Light paths and dispersion in (a) stepped index, (b) graded index, and (c) monomode fibers.

index of refraction than the adjacent inner one. Since the speed of light increases as the index of refraction decreases, the effect of this grading is that lightwave rays that take the longer route to the edges of the core can travel faster than those that stick to the center path, so they arrive at the same time despite their longer route. The graded index fiber equalizes the overall travel time by compensating for all the possible paths or *modes* of travel (in many ways similar to the modes of energy travel in a waveguide, Chapter 8), since those that travel the longer paths also travel faster. The dispersion of graded index fibers is less than step index fibers and these fibers are useful at longer distances and higher rates, but the fiber is more costly to manufacture. Graded index fibers are available with core diameters of 50 through 100 μm.

The *single-mode* or *monomode* fiber (Figure 24.3c) allows only a single lightwave ray path or mode to be transmitted down the core, virtually eliminating any possibility of dispersion and overlap. The core diameter of a single mode fiber is extremely small (about 5 μm) and the light from the source must be aimed more precisely into the fiber to be effectively accepted and passed along the fiber. In addition, to compensate for any light that does not enter the fiber due to position misalignment, the light source must be more powerful (this will be discussed in more detail in Section 24.3). A properly installed single-mode fiber gives the highest rates and longest distances, but at the highest overall cost because the light source must be more powerful and precisely aligned; its light detector must also be aligned with greater precision to see the pulses as they exist the narrow fiber.

Optical Fiber Performance

The two most important parameters of fiber performance are *bandwidth* and *transmission loss* or *attenuation*. Overall bandwidth is usually expressed in the form of frequency times distance (MHz-km). A 300-MHz-km fiber can carry data with a 300-MHz bandwidth for 1 km, or data requiring 150 MHz for 2 km. Factors that limit the overall bandwidth are dispersion and transmission loss.

The attenuation due to transmission loss reduces the optical signal power as the light pulse travels through the fiber and is measured in dB/km. This attenuation is not caused by bandwidth limitations, but it does limit the bandwidth. There are five causes of this attenuation and loss of optical power within the fiber: optical fiber loss, microbending loss, connector loss, splicing loss, and coupling loss. Each cause is the result of a different mechanism.

Optical fiber loss Optical fiber loss varies with the specific wavelength of the lightwave and the exact composition and purity of plastic or glass used for the fiber and is shown for each fiber type by a graph (Figure 24.4). Note that certain wavelengths have much greater loss than the general loss for most other wavelengths. This is due to the physics of the situation, with losses due to internal fiber light absorption and scattering. *Absorption* is associated with the light energy causing energy shifts in the energy levels of atoms in the fiber as well as energy absorption by unavoidable impurities. For example, hydroxyl ions associated with water molecules in glass have strong absorption at 730, 950, 1380, and 2240 nm; conversely, many fibers have minimum absorption at 725, 820, 1300, and 1550 nm. *Scattering* is caused by imperfections such as tiny bubbles (visible only under a microscope) and very localized variations in the density and concentration of

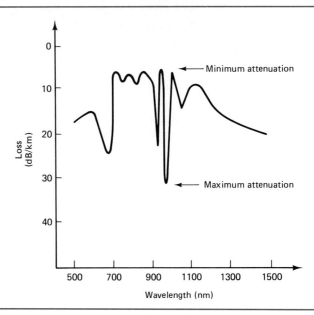

Figure 24.4 Power loss versus wavelength for a typical fiber; other fiber formulations have differing wavelengths of minimum and maximum loss (courtesy of Belden Wire and Cable Corp.).

glass or plastic in the fiber (imperfections in the internal uniformity). Extremely pure glass and plastic are needed for low attenuation.

Specially formulated fibers are used for best performance. Typical loss values for a 50/150-μm fiber operating at 1300 nm are less than 1 dB/km, and less than 3 dB (50% power loss) at 850 nm. (To dramatically illustrate the improvement in fiber technology in the last 20 years, early fibers had a loss of 200 to 700 dB/km!) The operating wavelength is chosen not just to minimize fiber loss; it must also be one that can be generated effectively by a source, and received effectively by a detector (these are examined in the next section).

Microbending loss Microbending loss is due to the minute fiber bends and deviations of the fiber. Even though the fiber may appear to be straight, the effect of the pull of gravity and resultant stress within the fiber causes tiny, localized bending. Some light escapes the total internal reflection at the bends and is absorbed by the cladding. In addition, the light-ray paths are changed from the theoretical ideal path so that dispersion increases. Microbending is reduced by the appropriate installation and jacketing. Step index fibers are more resistant to microbending losses than the graded index type.

Connector loss Connector loss is a function of the physical alignment of one fiber core to another fiber core. Three types of problems are possible: axial (angular) misalignment, transverse (also called lateral or radial) misalignment, and excessive separation between fibers (Figure 24.5). Of these, transverse causes the greatest loss. A displacement of just 10% of the core diameter produces a loss of about 0.5 dB. For a fiber with a 50-μm core, this 10% displacement corresponds to 5 μm, a very small distance.

24.2 The Optical Fiber

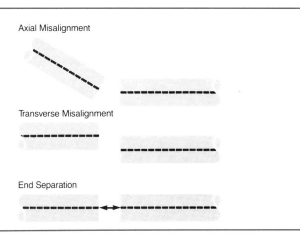

Figure 24.5 Three types of fiber misalignment, causing power loss at a splice.

The fiber optic connector is designed to minimize these problems, but connector loss is a major potential problem. Some connectors epoxy the fiber in place in an alignment sleeve, while others use rigid optical and mechanical alignment (sometimes aided by an optical fluid or gel that has a matching index of refraction, to avoid a "fiber to air to receptacle" set of interface surfaces and their losses). In both cases, the ends of the fiber must be carefully cut perpendicular to the axis and polished smooth. Typical connector losses are between 0.5 and 2 dB. (Contrast this to a metallic connection, which has close to 0 dB loss!)

Splice loss Splice loss occurs where two fibers are joined together to make a longer fiber. Splicing is done by melting the fiber ends, joining them, and then cooling the joined area (fusion and welding); by chemically bonding the ends (a form of gluing); or by precise mechanical fixturing and clamping (aided by an optical fluid or gel). Splice losses range from 0.1 to 1 dB. (Once again, compare with the loss when splicing two wires together.) For both splicing and connecting loss, alignment is critical and everything must be clean, smooth, and straight. A poorly made connector or splice has much greater loss than these typical values.

Coupling loss This loss between the fiber and the optical signal source or optical signal detector is a function of both the source or receiver device used and the type of fiber chosen. While the signal source emits the lightwave rays in a broadly spreading cone, to be properly *injected* into the fiber the optical fiber must capture these rays so that they can propagate properly within the fiber. The complementary situation holds for the rays leaving the fiber to be captured by the light detector: The rays inevitably spread out from the fiber end and must be captured by the detector. The size of the fiber core is a major factor in determining how much light is collected by the fiber from the source. We will study the nature of this coupling in detail in the next section.

Review Questions

1. What is the optical fiber core construction and function? What are typical diameters?

2. What is the role and diameter of cladding? What is the function of coating? What determines its thickness?

3. How many outer layers are there? How does the number and type vary with the application? What are the characteristics of a lightly wrapped fiber outer layer versus a tightly wrapped one?

4. What is a multimode fiber? What is dispersion? How does it limit the maximum data rate on the fiber?

5. How is step index fiber fabricated?

6. How does a graded index fiber reduce dispersion? Compare its construction and complexity with a step index fiber.

7. What are the features of monomode operation? Why is it the least dispersive? What is the core size?

8. What is the bandwidth of a fiber? How is it specified?

9. What is fiber attenuation? What are the causes of optical fiber loss? Why is it greater at some wavelengths than others?

10. What are absorption and scattering? What are typical attenuation values for optical fibers?

11. What is microbending loss? How is it reduced?

12. What are the three types of misalignment in connector loss? What are loss values for a connector? How is it minimized?

13. What is splice loss? What are three ways that splicing is done?

14. What is coupling loss at the signal source? At the signal receiver?

15. What is refraction? What is the law of refraction? What is the index of refraction for air, vacuum, and glass and plastics?

16. What is the relationship between index of refraction and speed of light in the medium? How are refractive angles specified? What is the critical angle? How is it calculated? How does refraction and the critical angle result in total internal reflection?

24.3 SOURCES AND DETECTORS

The emphasis in previous sections was on the principles of fiber optics and the optical fiber itself. Now we look at the light source and light detector, which are critical components of the overall fiber optic system. A complete fiber optic system requires an effective light source and light detector, matched both optically and physically to the fiber itself for the best optical power transfer.

Fiber Optic Sources

There are two devices suitable for generating light needed for the optical fiber: the *light-emitting diode* (*LED*) or *laser diode*. The light does not need to be in the visible range, and both infrared (frequencies below visible light, with longer wavelengths) and portions of the visible spectrum are useful in fiber optics. Like conventional diodes and transistors, both devices are solid-state and very reliable in

Optical Measurement Terminology and Parameters

When light is used for conveying information, the parameters that are measured and of most interest differ from those we use with radio waves, although they both are electromagnetic energy. *Radiometry* is the science of measurement of electromagnetic radiation, and *photometric* terms are those used for lightwaves in radiometry. The amount of light radiated by a source in all directions is the total power, also known as the *radiant flux* or *luminous flux*. It is measured in watts or *lumens* (a term in photometry that is equivalent to watts) and is usually simply called the power. *Radiant intensity* is the power in a given direction and is measured by the power/solid angle [a sphere encompasses a solid angle of 4π steradians (solid angle units, the three-dimensional equivalent of the conventional circle of 360° or 2π radians)]. The *flux density* shows how much radiant flux passes though a "window" of a specific area; its unit is watts/m² for the power/unit area.

These parameters matter because the light source must not only produce enough light for a usable signal, but the light must be in the correct direction to enter the optical fiber. One watt of light radiated evenly in all directions from a perfect point source (equivalent to an isotropic antenna, Chapter 9) puts much less light into the fiber than the same watt tightly directed and focused into the fiber end.

Numerical aperture (*NA*) is a mathematical measure of the fiber's ability to capture lightwaves from various angles (Figure 24.6). The *acceptance cone* half-angle ϕ that NA defines is calculated from the indices of refraction and is

$$\text{NA} = \sin \phi = \sqrt{n_1^2 - n_2^2} \text{ for a step index fiber}$$

(For graded index fibers, the NA is calculated using a much more complex equation that incorporates the rate at which the index of refraction changes with distance from the center axis.)

Example 24.2

A step index fiber core ($n_1 = 1.4$) is surrounded by cladding ($n_2 = 1.1$). What is the cone half-angle? What is the NA?

Solution

$$\text{NA} = \sin \phi = \sqrt{1.4^2 - 1.2^2} = \sqrt{0.75} = 0.87$$

Therefore the cone half-angle $\phi = \sin^{-1}(0.87) = 60.5°$.

A larger NA means that the optical fiber collects light from wider angles and is a better light gatherer for a given core and cladding diameter. A larger difference between the core and cladding indices results in a larger NA: A power increase of about 2 (3 dB) is achieved by increasing the NA from 0.20 to 0.30.

Numerical aperture is also used to indicate the sharpness of the cone of light emitting from the source. A smaller NA is usually desirable for the light source, since this indicates that the source output power is focused more tightly and can be aimed at the fiber core with less of this power missing the fiber's cone of acceptance. However, it must also be aimed precisely or all the light will miss the fiber end.

The other key factor in how much optical power is coupled into the fiber is the fiber diameter. Obviously, a larger core exposes more surface area at its end and gathers more light. The light-gathering ability increases with the square of the diameter: A 100-μm fiber has four times the cross-sectional area of a 50-μm fiber and intercepts four times as much light. Designers of fiber optic systems must trade off various combinations of NA and fiber diameter to ensure that enough light is gathered from the source. Thicker fibers are more costly, but stronger; higher NAs gather more light but are more costly to manufacture. The NA and thickness effects are often combined into an *optical collection factor* that can be considered a measure of the fiber efficiency for gathering optical radiation. Figure 24.7 shows collection factors (as a simple multiple and in dB) relative to a standard 100-μm core fiber for a variety of diameters and NA values.

Source and detector efficiency. A light source converts electrical input power to optical radiation, and a detector performs the reverse function of converting light input power to electrical signal power. How well input power is converted to output power is the *efficiency* of the source or detector. Higher efficiencies are desirable in a source—which is like a transmitter—since less power must be supplied by the system to the source and less heat is generated by the source (any electrical power that is not converted to light is transformed into wasted and damaging heat). At the detector—which is analogous to a receiver antenna—high efficiency is desired, so the relatively weak optical signal is converted into a large enough electrical signal to be used in the system. If the signal is too small, excessive errors will result from the poor SNR and extra-low-noise electronic amplifiers (equivalent to receiver front ends) must be used. An inefficient detector needs some combination of a more powerful source, lower-loss optical fiber, and shorter distance (with resultant lower attenuation) to achieve acceptable overall performance.

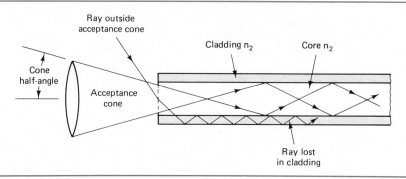

Figure 24.6 Geometrical meaning of numerical aperture and acceptance cone (courtesy of Belden Wire and Cable Corp.).

proper use. To be useful in a fiber optics data communications application, the light source must be capable of being pulsed at high rates by data bits (a continuous or CW source carries no information).

The light-generating process in both LEDs and laser diodes results from the recombination of electrons and holes inside a *p-n* junction, called *electroluminescence*. The recombined electron and hole pair has less energy than each constituent had separately before the recombination. When the hole and electron recombine, they give up this surplus energy difference, which leaves the site of recombination as a *photon* (the basic unit of light). The wavelength associated with this photon is determined by $\lambda = hc/E$, where h is Planck's constant (a fundamental constant in physics, equal to 6.63×10^{-34} joule·s) and E is the *bandgap* energy of the *p-n* material combination.

Note that the wavelength depends solely on the bandgap energy. For pure gallium arsenide (GaAs), λ is 900 nm. By adding minute amounts of aluminum (Al), the wavelength can be lowered to 780 nm. For even-lower wavelengths in the

Fiber core diameter (μm)	NA	Collection factor Relative to 100-μm standard	dB
300	0.27	14.1	+11.5
200	0.27	6.2	+8.0
200	0.18	1.6	+2.2
100	0.50	3.2	+5.0
100[a]	0.28	1.0	+0.0
85	0.26	0.62	−2.1
62	0.29	0.4	−3.8
50	0.20	0.13	−8.9

[a] Standard core fiber.

Figure 24.7 Optical collection factor for various fiber diameters, relative to a 100-μm fiber (courtesy of Belden Wire and Cable Corp.).

24.3 Sources and Detectors

visible-light region, other materials, such as gallium arsenide phosphide (GaAsP), indium gallium arsenide phosphide (InGaAsP), or gallium phosphide (GaP), are used.

A LED *p-n* junction emits light equally in all directions, with no preferred direction or focus except for where it is affected by the overall mechanical construction such as light reflection at the base and absorption within its semiconductor material. The numerical aperture—the angle over which the LED radiates—varies from 0.9 for a wide-angle LED to 0.2 for a narrow-angle LED with a focusing lens built into its assembly. This relatively wide NA means that much of the emitted light will miss the end face of a narrow fiber, and LEDs are therefore used with thicker fibers.

LEDs are low cost and produce output that is nearly linear with the applied drive current; current greater than a very low threshold level must be applied to them before they begin to operate. They require only simple digital drive circuitry in pulsed mode and have efficiencies between 30 and 70%. They are used for shorter-distance fiber optic systems, up to several tens of kilometers.

However, LEDs have some drawbacks. They can achieve pulse rates only up to about 100 MHz (which is certainly sufficient for many applications). The spectrum of their output light is not very sharp, but instead spreads about the center frequency; a typical spectrum is shown in Figure 24.8, where there is a 40-nm spread (at half-power compared to peak) about an 850-nm center. As this bandwidth of frequencies travels through the optical fiber, each sees a slightly different index of refraction (since index of refraction varies with wavelength in any given material). The resulting signal at the far end is no longer a sharp pulse but is dispersed. This, in turn, limits the maximum useful pulse rate, since the dispersed pulses overlap as their skirts get closer. The relatively large numerical aperature of LEDs requires larger fibers for effective coupling of the emitted light into the fiber; otherwise, much of the light never enters the fiber properly and is wasted.

Laser diodes were first built in the early 1960s. They are very different in size, required power input, and potential power output than glass-tube-based lasers, which were invented first (and which you may have seen in labs or mov-

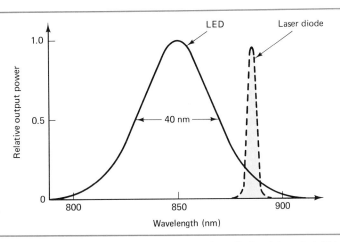

Figure 24.8 Relative output power versus wavelength (and bandwidth) of LED and laser diode.

ies). The light-generating process of a laser diode is similar to that of an LED, and the materials used are often the same. The difference is that the laser diode uses a much smaller junction area, and the concentration of injected carriers (holes and electrons) is much higher. The first laser diodes were unreliable, operated only at extremely low temperatures (77 K, −320°F), and required a large current but had low efficiency. Modern laser diodes are reliable enough to be used in commercial systems (although not yet as long lived as LEDs), operate at room temperature, and have reasonable operating currents and much higher efficiency.

The active region of a laser diode is enclosed by two aluminum-enriched layers of lower refractive index to act as optical reflectors. Because of the confinement caused by these reflectors, light can exit only from the front or back faces of the laser diode. These faces are semitransparent and form a resonator cavity for the light (not too different from resonant cavities for microwaves!). At a certain current density within the active region of the junction, the *optical gain* of photons generated exceeds the losses from the faces and the operating mode of the junction changes from the random type of the LED output to an organized, coherent, stimulated emission of a laser.

The *threshold* at which this change occurs is between 50 and 150 mA, depending on the laser diode material (in contrast to the LED, which has a very low current threshold of operation). The amount of emitted light decreases with increasing temperature, so laser diodes must be kept cool through the use of heat sinks and other mechanical schemes. Laser diodes can produce large amounts of optical output power (20 to 100 mW), with a very narrow output spectrum and hence little dispersion and overlap. They produce a very tightly directed beam with small NA which can be directed into the fiber with little loss. This makes them suitable for use with the thin monomode fibers. Laser diodes can operate at rates exceeding 1 GHz but require more complex circuitry to drive them than that of LEDs.

Modulating the Source

The LED or laser diode must be on/off modulated to produce a useful signal, with the on-state representing a binary 1 and off as binary 0. At lower rates—up to about 100 Mbits/s—this is done by fast but conventional circuitry which turns either type of diode on and off (Figure 24.9). This is especially practical with LEDs since they produce the same type of output spectrum whether they are on continuously or pulsed.

For laser diodes, the situation is somewhat different. Modulating the laser diode drive current causes its spectrum and nominal wavelength to change, which then causes system performance (attenuation, received power) to vary with the data rate. Therefore, the modulating circuitry is more complex, to avoid or compensate for these changes. At the highest rates—exceeding 1 Gbit/s—direct electronic modulation is replaced by optical modulation, where the laser diode light output passes through an optical "gate" that can be opened and closed in response to a controlling electronic signal. These gates are fabricated next to the diode output using exotic techniques of advanced physics such as the *Pockels cell, birefringence* (where an applied electric field varies the polarization of the light), or the *electro-optic effect* (where the index of refraction changes with voltage). Even the piezoelectric effect (SAW devices, Chapter 23) is used to allow an electronic signal to modulate the light energy wave.

All of these techniques are used, but they are costly, complex, and unsuit-

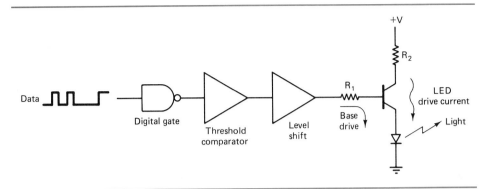

Figure 24.9 LED modulating circuit is relatively simple to interface to data.

able for widespread application at this time. Research is advancing the rates at which directly controlled and modulated laser diodes can be used, as well as improving the speed of LEDs.

Optical Detectors

The role of the optical detector is to efficiently convert the small amount of light energy received (photons) into an electrical signal. At the same time, the detector must be a low-noise device (just like a radio receiver front-end preamp), so it does not contribute excessively to low SNR and high errors. Three types of devices are used as detectors: photoconductors, PIN diodes, and avalanche photodiodes.

Photoconductors are fabricated from compound semiconductor material such as indium gallium arsenide, but not as junction devices. Basically, the resistance of the photoconductor is high when it is dark and low when it is illuminated. This resistance change occurs when photons hit the device and generate electron/hole pairs that facilitate the flow of externally applied current. Photoconductors are inexpensive but suffer from several drawbacks. Some current flows even when the device is dark, and this *dark current* translates into noise that reduces the overall performance of the system. Second, the range of "on to off" resistance is determined by the device *gain*, the internal relationship between number of incident photons to number of hole and electron pairs generated. The gain–bandwidth achieved is constant, so a high-gain device (easiest to use, with greatest difference between on and off performance) has low bandwidth; for higher bandwidths the gain is proportionally less.

PIN diodes are the most popular type of detector [they were used in microwave systems, (Chapter 23)]. They are relatively inexpensive, have low noise, and interface easily with conventional electronic circuitry. PIN diodes are fabricated from silicon for maximum sensitivity at 850 nm, and germanium or indium gallium arsenide phosphide (InGaAsP) for operation with wavelengths of about 1000 nm. The PIN diode is reverse biased in operation, so there is very little dark current. Bandwidths of greater than 10 GHz have been achieved with PIN diodes, and unlike the photoconductor, this bandwidth is essentially independent of the gain (ratio of incident photons to resulting electrons).

The output of the PIN diode—the electron current—cannot be used to directly drive digital logic circuitry: it is too weak, lacking enough current drive capability. A variation of the standard op amp, called a *transimpedance* amplifier,

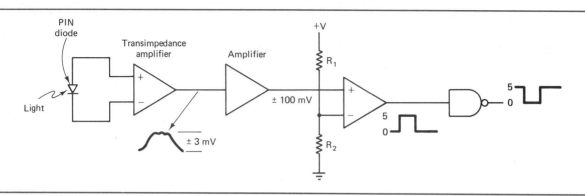

Figure 24.10 PIN diode receiver circuit includes a transimpedance amplifier, secondary-stage amplifier, threshold comparator, and digital circuit at receiver.

is used as a preamp and current-to-voltage converter (Figure 24.10). The output of this transimpedance amplifier is a small voltage (approximately 3 mV) when the diode is illuminated; this signal is then further amplified and then "squared off" by a pair of comparators that set thresholds of acceptable signal level. Signals below the threshold are considered to be the result of a dark diode (a binary 0), while those above the threshold are 1s. The comparator output then drives the digital circuitry that passes recovered data bits to the rest of the system, which determines what information these bits convey (data, headers, sync bits, error correction bits). Depending on speed, the comparator output and subsequent digital circuitry are usually TTL (transistor-transistor logic) or ECL (emitter-coupled logic) structures; TTL is useful up to about 100 Mbits/s, but ECL is required beyond that speed. Silicon is the preferred material for the circuitry at the lower speeds, and GaAs-based ECL is used for highest rates.

Avalanche photodiodes (*APDs*) are similar to PIN diodes but have one advantage: Each incident photon produces more than one electron/hole pair as a result of an internal "cascade" or chain reaction (hence the name "avalanche"). This gain means that the APD is more sensitive than the PIN diode to low light levels. Unfortunately, this avalanche effect also produces relatively high noise, which limits the usefulness of the device since the "off" signal level and the SNR are both higher compared to a PIN diode. In some cases, though, the overall result is a net improvement in the performance that can be achieved with the PIN diode.

Optical Wavelengths

Optical fibers have minimum attenuation at some specific wavelengths and more severe attenuation at others. The light source output and the light detector sensitivity also are not uniform across all the wavelengths, but have peaks and valleys. Finally, the numerical apertures (cones of light transmission and acceptance) vary for different implementations of each source, fiber, and detectors. The overall optical system must match the source, fiber, and detector characteristics so that there is maximum light transmission and detection from end to end. Otherwise, too much light will be lost in the fiber and the detector will not fully sense and convert the received light, so that the overall output will be low in amplitude and noisy.

The designer of a high-performance fiber optic system (greater than about

100 Mbits/s) weighs various trade-offs in choosing the type of source, fiber, and detector. For example, it may be better to use a source that has a little less output at a wavelength where the optical fiber has attenuation of 10 dB/km, compared to a wavelength where the source is slightly more powerful but the fiber attenuation is 20 dB/km. The overall output of the fiber going into the detector will be greater in the first case. We will explore this topic of overall link power losses further in the next section.

Review Questions

1. What is radiometry? What is flux? How is it measured? Why is direction and not just the amount of power critical in a fiber optic system?
2. What is numerical aperture? What does it indicate for a source? For a fiber? Explain NA in terms of an acceptance cone.
3. How does the light gathered by a fiber vary with fiber diameter? What is the optical collection factor? What is source efficiency? Detector efficiency? Why is high efficiency important for each?
4. How is an LED used as a source? What are typical LED wavelengths? What causes an LED to emit light? What materials are used to fabricate an LED?
5. What is the direction of LED light emission? Compare the NA of a naked LED with the NA of a LED plus lens. Why do LEDs need thicker optical fibers?
6. What is a laser diode? How is its construction similar to and different from a LED?
7. What is the laser diode threshold of operation, and the threshold for an LED? What is the operating current for a laser diode, and the effect of temperature on operation? What drive circuitry is used for lower-frequency operation and modulation?
8. What are the difficulties with laser diode modulation at higher rates?
9. What characteristic of the photodetector changes with incident light? What is dark current and the effect of too much dark current?
10. How are PIN diodes used as detectors? Why do they have low dark current? What are the bandwidth and gain of a PIN diode?
11. How is the output current of a PIN diode transformed into a useful binary signal?
12. What is an avalanche photodiode? Compare it in operation to a PIN diode.
13. What is the overall goal of a fiber optic system in matching the wavelength of the source, fiber attenuation versus wavelength, and detector sensitivity versus wavelength? What compromises must be made, and why?

24.4 COMPLETE SYSTEMS

Now that we have examined the individual components of a fiber optic system—source, cable, and detector—we can put them together into a complete system. Like any system made up from a series of individual links, there is a *link budget* or *link analysis* which shows how much signal loss or gain occurs in each stage (see Chapters 9 and 19, where this was discussed for satellite links). The goal of the link budget is to determine the signal strength at each point in the overall system and calculate if the power at the receiver (the detector) is sufficient for acceptable performance and error rates. If it is not, each stage is examined and some are upgraded (usually at higher cost) or guaranteed performance specifications—distance, speed, error rates—are reduced.

For fiber optic systems, the link budget also includes unavoidable variations in performance that occur with temperature as a result of component aging and from manufacturing tolerance differences between two nearly identical devices. In this respect, fiber optic systems need more careful study than all-electronic systems: There is greater device-to-device variation together with larger performance changes with time and temperature.

Figure 24.11 shows a basic point-to-point fiber optic system, which consists of an electrical data input signal, a source driver, an optical source, a 1-km optical fiber with maximum attenuation of 4 dB/km, the optical detector, and the receiver electronics. Shown with this system is the typical link budget analysis. It begins with optical output power of the source (−12 dBm in this case) and ends with the power that is seen by the detector. This analysis looks at each stage in the system and shows both the best- and worst-case power loss (or gain) for each link as a result of various factors, such as coupling losses, path loss, normal parts tolerance (best and worst for a specific model), temperature, and time. The analysis also allows for an additional 5-dB signal loss that will occur if any repairs or splices are made over the life of the system.

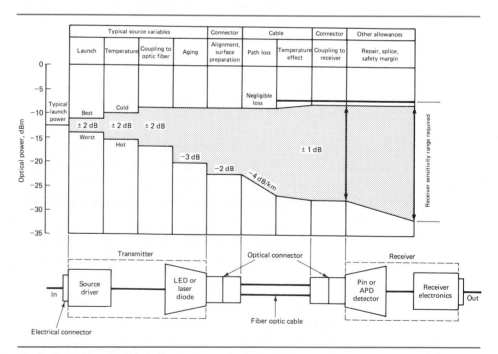

Figure 24.11 Link function schematic and budget analysis shows all factors and resultant maximum and minimum loss and signal power from source to receiver (courtesy of Belden Wire and Cable Corp.).

In this example the received optical power for a fully "on" signal can be anywhere between +7 dB (best case) and −23 dB (worst case) compared to the nominal source value; therefore, the detector must handle a dynamic range of optical signals from (−12 dBm + 7 dB = −5 dB) to (−12 dBm −23 dB = −35 dBm) representing a binary 1. Of course, when the source is dark for a binary 0, the received signal is also virtually zero except for system noise.

Example 24.3

The optical fiber is replaced by one that has loss of only 1 dB/km. What is the affect on link budget and received optical signal power?

Solution
Every stage is unchanged, but the loss for the 1-km fiber now can be between 0 dB (best case) and 1 dB (worst case); the received power will therefore be from +7 dB (best case) to −20 dB (worst case) compared to the −12-dBm source.

Standard Fiber Optic Components

When fiber optics first became available for data communications systems, each installation was custom built. There were no standard sources or detectors or their corresponding drive and receiver circuitry. This is in sharp contrast to all-electronic systems, where the component parts for nearly any application (except the most advanced) are available from many sources. This includes the mundane but still critical parts of any system, such as the connectors, splices, protective jacketing, and supports for mechanical integrity of the pieces, such as physical *strain reliefs*. In addition, careful link budget calculations had to be done by the installer of the system, using either the figures supplied by vendors for each part or measured by the installer for those unique parts that were custom built.

Although this is still true for fiber optic systems that perform at the higher rates—above several hundred MHz—or longer distances of many kilometers, it is not the case for applications that use fiber systems at lower rates and shorter distances. Many manufacturers now offer low-cost, easy-to-use fiber optic components. These include:

> A variety of LED and laser diode sources mounted in housings that easily accept a terminated fiber optic cable without special splicing
>
> Detectors that are also in housings that connect to a terminated cable directly
>
> Fiber optic cables in standard lengths (10 ft, 100 ft, 1 km) with the ends properly cut and finished, and with an attached *ferrule* that slips into the mating connector, just like an electrical connector pair

The optical fiber can be connected and disconnected like an ordinary connector, and the user does not have to worry about optical alignment or attenuation. The mechanical design of the connector forces the fiber and the actual source or detector into very accurate, close alignment with minimum loss. Similar connectors are also now used for splicing, where optical fibers that are fitted with ferrules can be joined in a mechanical splice using a double-sided connector that mechanically aligns and holds the fiber ends. Although the loss through such a connector is about twice the loss in a fused or bonded fiber splice, it is much simpler and quicker, requiring no tools or training. The difference is like using an extending cable that plugs into the existing cable to produce a longer test lead to a voltmeter, versus buying wire, connectors, and jacketing and soldering them together. The first method takes almost no time and skill, although it may result in slightly inferior signal quality than that obtained with the much more time consuming and skillful approach.

To make the overall fiber optic system design easier and more reliable, integrated sources and detectors are now available which incorporate the source drive electronics for modulation and the detector receiver electronics for recovering pulses as a single device or in a single enclosure. Figure 24.12 shows a monolithic IC that incorporates a lens plus photodiode, amplifier, and output transistor in a TO-18 style transistor package. This receiver has a sensitivity of 3 μW and achieves data rates up to 5 Mbits/s, using 50/125- or 200/300-μm multimode fibers, or 8/125-μm monomode fibers.

For higher performance, complete optical transmitter and receiver circuit units to mount on the main circuit board or connected via a short cable are available. All the designer has to do is supply power and apply the data bit pulses from the circuit to the transmitter, or connect the recovered pulses to the circuitry of the receiver. The electronic interface is usually TTL for data rates up to about 100 Mbits/s and ECL for higher rates.

Figure 24.13a shows one transmitter and receiver combination that sells for less than $20 per pair. The transmitter board has digital interface, amplifier, and LED driver which supplies up to 60 mA to the LED. In turn, the 850 nm LED provides -18 dBm typical output power into the user-supplied fiber. The receiver board (schematic shown in Figure 24.13b) has an 850-nm PIN diode followed by a transimpedance amplifier, additional amplifier, and a trigger circuit to form sharp digital pulses. A single +5-V supply is required. The overall sensitivity of the

Figure 24.12 LED and amplifier in single TO-18 package (courtesy of Microswitch Div. of Honeywell).

24.4 Complete Systems

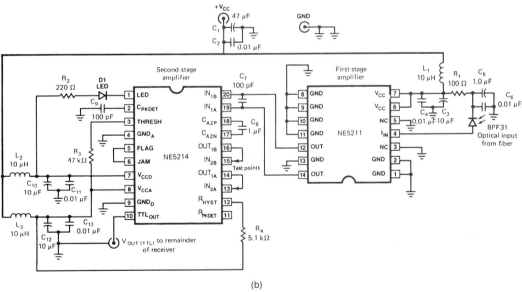

Figure 24.13 (a) Transmitter and receiver circuit board assemblies, complete with fiber optic drivers, receivers, and necessary electronics (courtesy of North American Philips Corporation); (b) schematic of receiver circuit board, showing two IC amplifiers which are the key elements (courtesy of the North American Philips Corporation).

receiver is −32 dBm (thus allowing for 14 dB of loss in the fiber and connectors) with operation up to 100 Mbits/s.

These integrated fiber optic components—which incorporate the drive circuitry, source, and connector housing, or detector, receive electronics, and connector housing in a single package—make using fiber optics in a system as straightforward as connecting systems with conventional metallic cable. A major benefit of this level of integration is that the performance is specified by the manufacturer for the entire unit, so the link budget analysis is simplified and has only to include the loss at each connector and within the fiber itself. They also simplify servicing in the field, since a defective unit can be removed and replaced easily.

For applications that are already installed using metallic wires, fiber optic replacements that can be dropped in are also available. One popular unit available from several companies replaces the very common RS-232 metallic interface link with fiber optics. This allows the user easily to get the fiber optic benefits of noise immunity, data security, and elimination of ground loops, although the data rate is of course limited to whatever baud value the RS-232 source is supporting. In these replacement units, the existing RS-232 connector on the computer or terminal connects to a small fiber optic interface box instead of a mating connector and metallic cable. This interface box converts the RS-232 voltages to signals that drive the source, which has a fiber connected to it. At the receiver end, the optical signal is detected and converted back into RS-232 voltage levels to directly connect to the far-end unit. The entire installation, of course, is *transparent* to the equipment at either end; neither the data source or user can tell that a traditional wire link using RS-232 voltages has been replaced by a fiber optic link.

Long-Distance Fiber Optic Links

Because of relatively low optical source power, low detector sensitivity, and high losses in fiber optic cables, the first installations of fiber optic systems were over short distances, less than a few dozen kilometers. The vast improvements in fiber optic components are demonstrated by the completion in 1988 of the first transatlantic optical fiber link, running from Tuckerton, New Jersey to Widemouth, England and Penmarch, France. (The fiber optic cable splits to England and France at the European continental shelf.) This TAT-8 fiber optic cable adds to the trans-Atlantic capacity provided by the existing copper-wire-based TAT-7 and similar undersea cables. It is 6740 km (3500 nautical miles) long and carries the equivalent of 40,000 two-way conversations that have been digitized and multiplexed into a pair of full-duplex digital bit streams.

The TAT-8 cable consists of three optical fibers (one for each direction plus a spare) encased in several layers of specially formulated polyethylenes for protection; the cable portions in shallower waters are further protected against sharks bites, dragged anchors, and other undersea hazards by two layers of armored steel tape wrapped around the inner layers. It uses a single-mode fiber with 1300-nm wavelength signals operating at 280 Mbits/s. The bit error rate is less than 4.4 errors per 100 million (10^8) bits.

This cable uses electronic *repeaters* every 80 to 100 km to restore (regenerate) the digital bits and boost their signal levels (*regenerators* would be the more correct term, but the two terms are now used interchangeably in industry). Each repeater is a miniature "optical to electronic to optical" converter. The repeater detects the attenuated optical signal (before it is so small or noisy as to be useless), converts it to an electronic signal, uses comparators and digital circuitry to restore the original signal shape and levels. It then uses this restored signal to drive another fiber optic source for the next segment of the journey. DC power for the 108 repeaters (7000 V, 1.6 A to minimize loss due to *IR* drops along the cable length) is provided by shore stations via conductive wrappers surrounding the cable's glass fiber/steel strength core.

24.4 Complete Systems

Fiber Optics and Networks

Fiber optics are used for the connecting links in local area networks (Chapter 18). Unlike the undersea cable, which is a custom-designed, one-of-a-kind installation, such networks require a high degree of standardization among the many users so that they can successfully and easily interface with each other, and connect/disconnect quickly as needed. The electronics industry is developing standards for fiber-optic-based networks that take into account the special characteristics of fiber optics and how they can be used most efficiently. The *fiber distributed data interface* (*FDDI*) is one standard that is getting a great deal of support from both network users and the manufacturers who provide the physical components for systems.

The FDDI standard does more than define the data rate, source and receiver characteristics, or the fiber and its connector. It provides a set of rules for the lower levels of the network so that a user can take any pattern of data bits (already formatted, with headers, error correction, and other fields), put them into a FDDI interface circuit, and have them sent over the fiber to the other end of the link. There, the formatted data bits are recreated and passed to the far-end user. There are many similarities and differences between the FDDI standard and the other IEEE 802 series of standards we looked at previously in Chapter 18. FDDI sends the data bits at 100 Mbits/s, but the particular fiber used (monomode, stepped index, or graded index) and the fiber size are not specified; they are chosen by the user to give the required performance for the distances involved.

Each FDDI interface node (Figure 24.14) contains a crystal-based clock oscillator, which is resynchronized precisely (*reclocking*) by using a phase-locked loop or SAW filter to extract basic timing information from the incoming data stream. As each node receives data bits from the link, it stores them temporarily in a *first-in first-out buffer* (*FIFO*); the data in the buffer are then "clocked out" to the circuitry using the resynchronized crystal clock. In effect, the FIFO acts as an "elastic" buffer to take up the small amounts of timing jitter in the received bits and then resend the bits with corrected timing.

In addition, the FDDI specification does not allow a string of binary 1s or 0s on the fiber, which would appear as a constant signal and thus not have the transitions needed for reclocking. To provide sync information, FDDI uses Man-

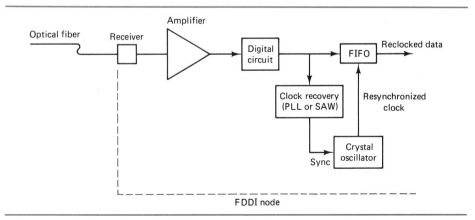

Figure 24.14 The block diagram of an FDDI interface node block shows how it restores signal timing using its own precision clock and a FIFO data buffer.

chester encoding (Chapter 12) to convert a conventional 1,0 binary stream into an alternating sequence, even if there is a run of two 1s or 0s in a row. The drawback to Manchester encoding is that the transmitter bit clock and Manchester bits must be at 200 MHz to support a bit rate that provides 100 Mbits/s of user information (since each original bit is now represented by a 1,0 or 0,1 pair); this is an acceptable technical trade-off for the benefits of Manchester encoding in effective and simplified reclocking.

The FDDI specification defines a ring topology with a token-passing protocol. To ensure operation even with hardware failures at a node or in the link, the FDDI standard requires dual rings (Figure 24.15a), where one ring is the primary path and the other is called the secondary path. The secondary path is not simply idle, to be used only if there is a problem with the primary path. Instead, if desired it can be used to supplement the primary path and so increase the total data throughput, but still be prepared to take the entire communications load if the primary path fails. The FDDI specification allows up to 500 nodes and up to 2 km between nodes. The optical output power into a fiber link must be between -14 and -20 dBm; optical detectors should operate with received power levels of -14 to -31 dBm. The overall bit error rate of a FDDI system is 2.5×10^{-10}, equivalent to a single error for every 4 billion bits (the inverse of the BER), using signal powers in the specified range and optical fiber with a 500-MHz-km bandwidth.

The FDDI dual rings are used for redundancy and reliability. When a failure occurs, the ring must be switched to bypass the failed link or node. A *dual-bypass* switch designed expressly for this purpose is shown in the figure at one of the nodes, along with a photo of the switch (Figure 24.15b). The normal straight-through path can be switched to bypass the node (and associated links) as shown in the figure. This switching is done under command from some sort of watchdog circuitry that activates a pivoting spherical mirror within the unit, and switching from normal to bypass paths takes just 10 μs. The dual-bypass switch is housed in a 5 in. $\times$ 5 in. $\times$ 1 in. box complete with fiber optic connectors, and signal power loss due to the presence of the switch is less than 1 dB.

Review Questions

1. What is a link budget and link analysis? Why is it needed?

2. Why do fiber optic systems require wider tolerances than electronic systems in the link analysis? What is done if the analysis shows the received signal power is too low?

3. What standard components are now available for fiber optic systems? What are the benefits of using such components? Compare practicality and performance of a slip-on fiber plus connector to a specially fabricated one.

4. What are integrated sensors and detectors? How can a system use standard fiber optic components and assemblies for maximum convenience and performance? What performance is achieved with an integrated receiver and with a circuit board transmitter/receiver pair?

5. Compare the TAT-8 link with a basic optical fiber. How is it protected? Why are repeaters used? How do they work?

6. What is FDDI? What network topology and protocol does it use? What does FDDI define, and what does it not specify?

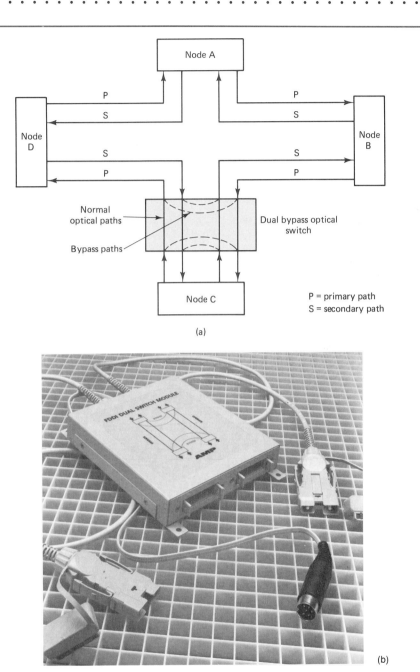

Figure 24.15 (a) An FDDI network schematic for four nodes, showing the dual optical paths (primary and secondary) and a bypass switch at node C; (b) a dual-bypass optical switch, using mirror pivoting, allows quick switching between the normal fiber paths and a path that bypasses the node (courtesy of AMP, Inc.).

Fiber Optic Switching and Multiplexing

As we saw in previous chapters on telephone systems, multiplexing, and other applications, the ability to switch signals from one path to another is the key to many communications systems, as is the ability to take a signal and distribute it simultaneously to several users. This capability is straightforward for electronic signals and is done with mechanical contacts or solid-state switches, plus amplifiers if needed to make up any signal loss.

For optical signals, the situation is much more complex, and at present there is no single solution to the problem of switching and routing. For this reason, most fiber optic links are used in a *point-to-point mode* as a *dedicated* high-performance link between two users, an application that does not require switching or routing. Examples include interoffice telephone trunks (the routing is done within the central office, but the trunk paths are fixed in place), connections between computers in adjacent buildings, and the links between a large computer and many smaller computers that it serves.

Nevertheless, we can look at some of the approaches that are presently being used or explored for switching and routing, to get an understanding of both the problems and potential. These methods include optical parts combined with electronic circuitry, mechanical devices, electro-optical devices, and twisted fibers.

1. *Combined electronics and optics* (sometimes called a *hybrid* design) use a detector to convert the fiber optic signal to conventional electronic form and do all manipulation of the signal when it is in electronic form using conventional high-speed circuitry. The signal is then reconverted to optical form and sent on its new fiber optic path. The repeater of the undersea electronic cable is one example of this hybrid approach, although it is used primarily to restore signal levels rather than to switch or route the signal.

Although this method is the most straightforward, it is also very costly since each switching node needs a large number of components and circuitry for converting between optical and electronic forms. Also, since electronic circuitry that can switch at the higher speeds of fiber optic data links (1 GHz) is complex, the overall link is often, instead, restricted to slower speeds (100 Mbits/s) so that simpler electronics can be used. Nevertheless, since signals in electronic form can be switched and amplified in many ways, the hybrid approach offers a large amount of flexibility in what it can achieve.

2. *Optical-mechanical devices* are available which control the path of optical signals. Using carefully built and aligned mirrors and prisms, with their motion controlled by small and very accurate motors, an optical signal can be steered to one of several paths. Half-silvered mirrors in some designs reflect some light but allow the rest to pass straight through (a *beam splitter*), so that a single-lightwave ray is divided and sent along two paths, although the signal power on each path is of course split as well.

This optical-mechanical design is effective for switching, and the final performance and cost compared to the hybrid approach depend very much on the specifics of the application. Certainly, there is no limitation on data bandwidth, since the mirrors and prisms handle the lightwave ray regardless of the data rate. The FDDI dual-bypass switch is a good example of such a switch optimixed for a particular application.

There are some drawbacks to this design approach, however. First, optical power is lost with each reflection since no reflector is perfect (typically about 1 dB is lost at each surface); when an optical signal is split

---

7. What are the FDDI clock rate and the effective bit rate? Why do they differ?

8. What does each FDDI node do? What is the function of the FIFO? How does the FDDI standard ensure reliability through dual rings?

9. Compare the difficulties of switching and routing electronic signals compared to fiber optic signals.

10. Why is a point-to-point link most compatible with fiber optics? Give two examples.

11. What is a hybrid fiber optic switch technique? What are its advantages and drawbacks?

for routing to two paths simultaneously, the power loss is at least 3 dB (this differs from the FDDI application, where the signal is routed to only one path at a time). (Unlike electronic signals, which are easily amplified to their original magnitude, optical wave amplifiers are only laboratory devices at this time.) The speed at which a signal can be switched from one path to the other is fairly low (microseconds and slower); in contrast, an electronic system can switch in nanoseconds. Finally, like all mechanical devices with moving parts, the reliability and ruggedness of the design is not as good as a properly designed all-electronic system that has no moving parts to break or be jarred by an impact. Nevertheless, the optical-mechanical design is good for applications where a new path is established for a relatively longer time period, broken, and then another path later established (as in a switching center or FDDI node).

3. *Electro-optical* or *photonic systems* use the concepts of electromechanical systems but implement them in a very different way. Physicists have found that there are certain crystals with optical properties that change when an electric field is applied. For example, the index of refraction of lithium niobate $LiNbO_3$ (which is transparent to light) changes sharply when the electric field exceeds a threshold value. Electronically controlled optical switches have been built from this crystal, fabricated as part of a larger device that also incorporates the optical energy couplers that interface to optical fibers. Lightwaves exit the source fiber and go into the crystal and exits through one of two paths to other fibers. The particular path that the light takes is determined by the specific index of refraction of the crystal, bending the light at one of two angles, depending on the index in effect at the time; the index of refraction in turn is controlled by the applied electric signal.

Other exotic methods are being investigated, such as ones that control the polarization of the light and then direct the light based on its polarization. Although these are not yet commercially available, they have the potential for highest-performance, very reliable systems. They have no moving parts, can be fabricated in solid-state devices (like integrated circuits) for mass production, and provide widest bandwidth since they handle the signal entirely in optical mode without lower bandwidth electronics.

Other research along these same principles is trying to develop better *optical modulators* which allow an electronic signal to impose information on lightwaves at rates exceeding 10 Gbits/s, for example. The next step is *integrated electro-optics*, which combines the source (or detector) with the modulator as a single monolithic manufactured structure: The semiconductor that forms the laser diode also has the modulator for the diode output fabricated alongside it.

4. *Twisted fibers* are being used to divide a single-fiber optical signal into several paths simultaneously, by upsetting some of the preconditions of internal reflection. The basic principle of total internal reflection shows that in a properly designed and installed fiber, virtually all the light will travel through the fiber without escaping. Besides attenuation within the fiber, there is some loss due to microbending and imperfections in the cladding which allow light to escape.

In a twisted-fiber scheme, a source fiber and several receiver fibers are twisted together in a carefully controlled shape and stress pattern. At this twisting node, light leaks from the source fiber at the extreme bend points (basically, there are many fine cracks in the cladding) and is captured by the receiver fibers through their bending cracks. For this to work, there must be precise alignment and control of the microbending crack patterns and fiber positions with respect to each other. This experimental technique has the drawback that only about 1% of the light energy from the source fiber is coupled into a receiver fiber, which is too low for practical use.

24.5
FIBER OPTIC TESTING

Testing of fiber optic systems differs in many ways from testing of nonoptical electronic systems. Although the overall goals of testing are the same—to ensure reliable performance and low error rates over a variety of operating conditions—the equipment and methods used to measure the performance of the fiber optic components of the system are unique and require new approaches. In this section we look at some of the basic instrumentation unique to fiber optic system measurements, along with some commonly performed tests.

Of course, conventional data testing instruments such as bit error rate (BER) testers (Chapter 12) are used to evaluate the final overall performance of any digital communications system. Since the BER test signal patterns are usually applied to the communications system as known substitutes for the regular electronic data signals, the use of an optical fiber medium in place of the metallic or radio link does not affect the test procedure. The link medium is transparent to the communications protocol in many of these situations, except that the fiber optic system can run at higher bit rates.

Signal Sources

Signal, waveform, or pulse generators are commonly used in electronic systems, with tunable frequency of operation. A signal source is also needed in fiber optics testing, but it differs in two significant ways from an electronic source. First, it is not tunable but instead operates only at one or a few distinct wavelengths because present technology does not offer reliable and affordable light sources that can be tuned in output wavelength. Wavelengths of 820, 850, 1300, and 1550 nm are most common in fiber optics because these are the wavelengths of least attenuation in standard fibers. Both LEDs and laser diodes are used as sources.

Second, the absolute power output calibration level of many optical sources is difficult to maintain with temperature changes and component aging. This power output may vary by as much as ±3 dB over all operating conditions of a typical source instrument. Instruments that are more stable in output power require advanced designs and are much more costly; they often use a detector installed right next to the source LED or laser diode to monitor the actual output power and develop a correcting signal to increase or decrease the source drive current (note that detector characteristics are more stable than source characteristics).

To overcome the difficulties of obtaining precise optical signal levels, many of the tests are arranged to look for differences, thereby canceling out the absolute source level value (plus any loss in the connectors). For this reason, fiber optic source instruments often use a beam splitter to take the single optical signal from the source diode and split it equally between two output connectors. Any change in absolute output level appears at both output connectors identically, to aid in implementing these difference tests.

Both continuous-wave (CW) and pulse sources are available. CW sources are needed to measure attenuation, losses, detector performance, and effects of connectors and splices. The pulse source simulates actual data signals in the fiber optic system and performance with dynamic signals and operation must be determined.

Power Meters

Power meters are used to measure the optical signal power. They usually provide a reading in both absolute units such as mW, and dB readings (relative to 0 dBm, as well as relative to another signal), which are generally more useful than absolute readings. Some power meters have two input connectors and two corresponding matched detectors so that the received optical power levels from two fibers can be measured and compared simultaneously. This is useful for many test setups.

Power meters use two types of detectors to sense the optical signal. *Thermal sensors* are similar to microwave power sensors and convert the optical energy

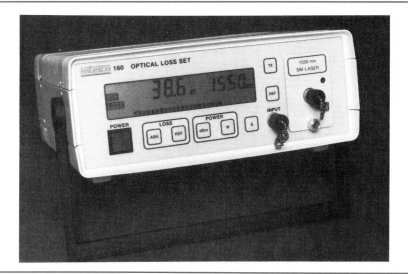

Figure 24.16 A fiber optic test set incorporates source and detector (courtesy of Inteleco Corp.).

into heat and then to a dc voltage. They are sensitive to a very broad spectrum of light wavelengths. In contrast, *quantum* sensors are basically specially calibrated versions of the fiber optical signal detectors studied previously, with maximum sensitivity only around one specific wavelength. This is not a shortcoming in many tests, since the source itself is also only providing an optical signal at one wavelength, is not tunable, and there are no other wavelengths to be concerned with (in contrast to radio signals, where there can be interference and undesired signals). Care must be used not to overload the power meter with too much optical signal power, which can temporarily blind the sensor or even cause a permanent shift in its sensitivity, and consequent loss of calibration.

For convenience, a source and a power meter are often combined into a single instrument called a fiber optic *test set* (Figure 24.16). The test set is very useful for portable work since it is smaller and easier to handle than two separate boxes, and some of the internal electronics (power supply, digital readout circuitry) are shared.

Attenuators

Attenuators are available for both fixed and variable amounts of attenuation and are needed to reduce signal levels to make them compatible with the range of test equipment optical inputs. Fixed attenuators use a piece of gray-coated (*neutral density*) glass filter with a precise amount of light transmission; for example, glass that transmits 50% of the light provides 3 dB attenuation. The filter is mounted in a special housing with an optical fiber connector on each side, so it can be connected in line with the signal fiber. The advantage of this type of attenuator is that it is broadband and can be used for any optical wavelength, requires no dc power, and requires no recalibration.

The same concept is expanded to use with continuously variable attenuators. A glass wheel is gray-coated with varying density to provide transmission from 0 to nearly 100%, and the resulting attenuation is marked in dB along the

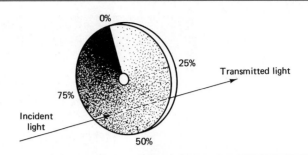

Figure 24.17 A 0 to 100% variable density glass attenuator works for any wavelength of light.

edge (Figure 24.17). Turning the wheel, which is mounted in a housing like that of the fixed attenuators, allows the user to select any attenuation, ranging from 0 dB to typically 40 dB. These mechanical attenuators are effective, broadband, inexpensive, and reliable, but they are not compatible with computerized testing equipment since they are not electronical settable (unless a motor drive is used).

Electronic attenuators use a variety of methods. A pair of *liquid crystals* control light polarization to act as variable darkeners that can be controlled by a voltage to have transmission between 10 and 90% (these liquid crystals are similar to those used in the digital displays of watches). These attenuators have the advantage of being electronically controllable, but they do not provide as wide a range of attenuation, are more costly, and do not provide the same attenuation at all wavelengths (which the variable density glass does). For this reason, therefore, manufacturers provide electronic attenuators that are calibrated at the standard fiber optical wavelengths.

In addition, the characteristics of the liquid crystal vary with temperature and aging, so some scheme must be used to accommodate this. One approach is to build a source, beamsplitter, and two detectors into the attenuator to monitor the actual transmission value continuously and thus maintain calibration. After the split, one beam proceeds directly to a detector while the other passes through the liquid crystal. The difference indicates the actual amount of attenuation provided, and the control voltage can be adjusted automatically to maintain the correct value based on this reading.

Measuring Cable Attenuation

One of the most common measurements is to determine the loss in an optical fiber to see if it is within specifications or if there is a crack in the fiber or a poor splice. A source and a power meter are the basic instruments needed for this test. If this were an electronic circuit, the easy method would be to send a known signal voltage into the cable (or system) and measure the output voltage. But with optical signals, the source value is often not accurately calibrated, so a *differential* or *two-point* measurement is used.

The test setup (Figure 24.18) first connects the source to the power meter through a known length (usually 1 km on a spool) of cable identical to the type of cable under test, and a reading P_1 is taken (in dBm) to act as a reference. This cable is then replaced by the cable under test and another reading P_2 is taken. The cable attenuation is then

24.5 Fiber Optic Testing

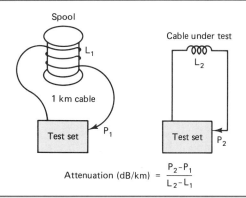

Figure 24.18 A cable loss test setup requires two steps to cancel out absolute power level of the source.

$$\text{attenuation (in dB/km)} = \frac{P_2 - P_1}{L}$$

where L is the difference in cable lengths, in km. Note that this technique does not rely on the absolute power output or long-term stability of the source, nor on the absolute accuracy of the power meter, and is not affected by loss at the test set connectors.

Example 24.4

The measured power through the 1-km test cable is 2 dB, while it is 8 dB through the 3-km cable under test. What is the attenuation factor for this cable?

Solution

Attenuation = $(8 - 2)/(3 - 1) = 3$ dB/km, which indicates an internal break in the cable under test.

To save time, the dual source output and dual power meter inputs are used. The split optical signal travels through two paths simultaneously, via the short known cable and the longer unknown cable. The power meter automatically shows the difference in dB. This technique assumes that the connectors for each cable have the same loss, which can be verified by swapping the cables and seeing if the received power difference changes.

Other techniques and instruments are available for measuring cable loss. These advanced methods rely on controlling the exact mode of light propagation through the optical fiber and are much more sophisticated and prone to errors. However, they do not require using a separate path and a known cable for comparison, which may be impractical in some cases. We will look at one of these later.

Measuring Connector Loss

Another common measurement is the amount of loss in a fiber optic connector, which is more difficult than a metallic connection to make properly; losses can be several dB or higher for a poor connection. Measuring the loss through the connector uses a differential technique similar to measuring the cable attenuation. A single optical signal is split; one path goes directly to the power meter while the other goes through the connector, using the same type of cable with total length equal to the direct cable (Figure 24.19). The power meter measures the difference, which is the loss in the connector alone. This same test can be done without a split output source and dual input meter, but it requires more connecting and disconnecting of cables.

Optical Time-Domain Reflectiometry

Optical time-domain reflectometry (OTDR) operates similar to time-domain reflectometry (TDR) used for wire cables (Chapter 7). OTDR sends a pulse into the end of the optical fiber and detects the returned light energy versus time, which corresponds directly to distance of light travel. Like TDR, it can show subtle characteristics of the cable and requires connection at only one end. OTDR actually shows more than obvious discontinuities in the optical path, such as splices, breaks, and connectors. It uses the physical phenomena called *Rayleigh backscattering*, which occurs within all fibers, to show the signal attenuation along all the fiber length. As a lightwave travels through the fiber, a very small amount of incident light in the cable is reflected and refracted back to the source by the atomic structure and impurities within the optical fiber.

Backscatter is different from the regular reflection of signal power that occurs at tangible discontinuities. Unlike OTDR, TDR for conventional metallic cables cannot show loss (attenuation) since the attenuated power dissipates as heat in the cable, whereas in OTDR a large part of the attenuation returns to the source as backscatter. Both TDR and OTDR, however, do show energy that reflects from discontinuities in impedance (TDR) or the optical path (OTDR).

An OTDR instrument block diagram is shown in Figure 24.20. The pulsed

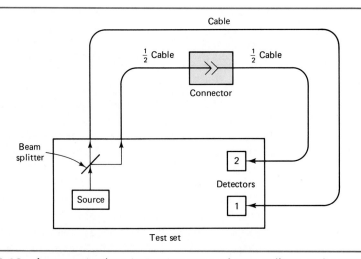

Figure 24.19 A connector loss test setup uses a beam splitter at the source and dual detectors to eliminate the effect of loss in the fiber itself.

24.5 Fiber Optic Testing

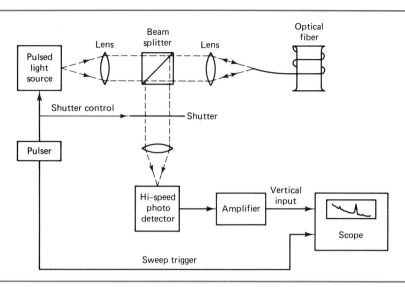

Figure 24.20 Block diagram of OTDR instrument for sensing and measuring reflected signal power versus distance along the entire cable length.

light source (an LED or laser diode) is split and one part injected into the optical fiber. The other part of the split optical signal goes to circuitry that shuts off the sensitive detector from being overloaded by the relatively large initial light pulse. The beam splitter serves another purpose: It directs the backscattered and reflected light that returns from the fiber end to the detector, where it is amplified and measured. The display screen horizontal sweep begins coincident with the pulse, and the amplified detector output controls the vertical position on the screen.

A typical ODTR display is shown in Figure 24.21. Note that the signal backscatter amplitude (vertical scale) declines steadily with distance (horizontal scale), and this slope can be calibrated to indicate loss/km. Any discontinuities show up as major increases in returned amplitude. OTDR is very useful for locating small breaks or problems in buried cables; even unburied fiber optic cables are hard to diagnose because many of the problems cannot be seen by inspection with the naked eye.

Three critical OTDR specifications are distance resolution, dead zone, and dynamic range. *Distance resolution* depends on the length of the source pulse and the speed of the associated electronics. A shorter pulse allows to define the distance to the reflection with greater precision. The *dead zone* is the time (and therefore distance) during which the detector is blinded after it receives any large returned signal, such as the one that comes from a sharp break in the cable. This reflection overloads the detector, which then takes some time to recover; during this time it may miss reflections from small faults (to avoid dead time after the relatively large injected pulse, the detector is protected during that pulse period). *Dynamic range* indicates how sensitive the OTDR system is, since the reflected and backscattered optical signal is orders of magnitude smaller than the initial pulse (these returns also undergo backscatter along with reflections at discontinuities as they return to the source, which reduces their amplitude even more).

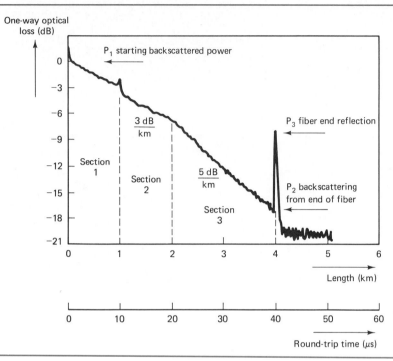

Figure 24.21 OTDR instrument screen showing what OTDR reveals about a 4-km fiber (courtesy of Hewlett-Packard Co.).

No single best combination of OTDR characteristics exists, since improvements in one area hurt performance in another. Increasing the pulse amplitude to increase dynamic range also increases the dead zone; decreasing the pulse width to improve resolution decreases the dead zone but produces a much smaller backscatter and reduces range.

To meet the varying needs of different applications, OTDR units are designed for either short-distance or long-haul operation. Users who test long lines (tens of kilometers) need systems with long range but are satisfied with resolution to a few meters. Those who are testing fiber optic systems in local area networks or even within ships and aircraft prefer high resolution (to a few cm) and smaller dead zone but need less dynamic range.

Review Questions

1. Where is a CW fiber optic signal source used? What is the use of a pulsed source?
2. In what two ways does a fiber optic source differ from a tunable RF source? Why is this not a problem?
3. How are tests designed to ignore absolute power levels or variations?
4. What does a power meter measure? Why have two inputs in parallel?
5. Compare the broadband power meter to the single-wavelength meter. How does the sensor in each differ? Why are one-wavelength units useful in fiber optics testing?

6. Why is power meter input overload a concern?

7. How is glass used to build a fixed or variable attenuator? What are the advantages and drawbacks of these gray-scale units?

8. How is cable loss determined differentially? Compare the measurement technique to electronic circuit loss measurement.

9. How does a dual output source and a dual input power meter save test time?

10. How is connector loss measured? Why is this a common measurement in fiber optics connectors but not for metallic connectors?

11. What is the principle of OTDR? What can it show? What is Rayleigh backscatter?

12. What does "distance resolution" of OTDR mean? How is it affected by pulse width? What causes dead zone? Why is dead zone a problem in some tests?

13. What is dynamic range? When is large dynamic range needed?

14. Why is there no available "best" combination of instrument settings for maximum distance resolution, minimum dead zone, and maximum dynamic range?

SUMMARY

Fiber optical systems provide many performance benefits that cannot be achieved with conventional electronic systems. Using a strand of glass or plastic, with suitable cladding, the physics principle of total internal reflection is employed to let the fiber act as a waveguide for the light energy. Fiber optic systems are reliable, lightweight, immune to causing or being affected by interference, and cannot be "tapped" easily. Although well suited to short and very long distance communications, they are most useful for links that do not need to switch or route the signals from one path to another.

The three key elements of a fiber optic system are the light source, the fiber, and the light detector. Light-emitting diodes and laser diodes are both used as sources, producing specific wavelengths of light. Laser diodes are more costly and difficult to modulate but achieve higher speeds with much smaller numerical aperture and are best used for launching optical power into a long-distance monomode fiber.

Fibers are made of glass and plastic with precise control of the thickness and index of refraction of both the core and its cladding. The outer coating and protective jacketing of the fiber protects it against abuse and bending, which cause attenuation of the optical signal; modern fibers have losses below 1 dB/km at their optimum wavelengths of transmission (and much greater at other wavelengths). Causes of fiber attenuation include impurities, energy absorption by atoms within the fiber, minute cracks in the cladding, and splices. Techniques and problems with optical splices and connections are very different than they are for conventional metallic cables, and these can be a source of excessive loss if not done with near perfection.

The optical detector can be a photodiode, PIN diode, or avalanche photodiode. These provide a choice of sensitivity, noise, and bandwidth; their tiny current outputs must be amplified to be useful, and then compared to acceptable signal thresholds. Individual components and integrated assemblies for sources, fibers, connectors, and detectors are now available as standard items; these make the design, installation, and repair of systems much easier.

Fiber optic tests center around an optical source and optical detector to measure cable and connector loss. Although the source output is usually not precise or stable with time and temperature, differential and two channel techniques are used to get necessary measurements. Optical time-domain reflectometry is a very powerful, sophisticated method for analyzing both gross cable discontinuities and subtle losses; unlike source and detector techniques, it requires connection to only one end of the cable.

Summary Questions

1. Compare fiber optics as a signal medium with a metallic medium in terms of bandwidth, distances, isolation, security, and interference.

2. Repeat question 1, comparing fiber optics to waveguides and to radio transmissions in air or vacuum.

3. What is refraction? What causes it? How is the angle of refraction measured? What is total internal reflection? How is it used in fiber optics?

4. Why are fiber optic links attractive for both shorter-distance links, such as local area networks, and long-distance, worldwide links?

5. What are the three layers of an optical fiber? What role does each play? Why must the refractive index of the second layer be less than that of the innermost layer?

6. Compare step index, graded index, and monomode fibers in terms of construction, thickness, performance features, and dispersion. What are typical diameters?

7. What bandwidths are achievable with fiber optics? What is the effect of dispersion on bandwidth?

8. What is the attenuation of a modern optical fiber? Explain how attenuation varies with wavelength.

9. What are some of the causes of attenuation? Which are inherent in the fiber, and which are due to installation or application?

10. How do the indices of refraction of the core and the cladding determine NA? What is the significance of NA (and angle of acceptance) for sources? For the fiber entry point? For the receiver?

11. How is light power measured? Why is the direction and spread of the light important in application?

12. Explain the basic operation of an LED. What determines the operating wavelength? What is the bandwidth of the LED spectrum? What is the threshold of operation? How is an LED modulated with the data signals?

13. Answer question 15 for laser diodes. What are the differences in operation and performance between LEDs and laser diodes? Why are laser diodes used for longer distances and monomode fibers?

14. Compare and contrast the bandwidth, noise, sensitivity, gain, and operation of photodetectors, PIN diodes, and avalanche photodiodes.

15. Explain a fiber optic system link budget. What does it show? Why is it needed? What are the sources of fiber optic performance variation? Why is this usually greater than the variation in conventional electronic components?

16. Explain the role of standard fiber optic system components: sources, fibers, connectors, and detectors. How do these ease analysis, installation, and repair of fiber optic links?

17. Explain the operation, topology, maximum node separation, and protocol of FDDI. What are the source and detector power levels? What does each node do? How is FDDI used? How are the data encoded, and why? What are the performance

"cost" and benefit of this encoding? How is reliability assured? What is the expected BER for FDDI?

18. Explain why fiber optics are used primarily for point-to-point links. Give examples.
19. How can fiber optic signals be switched and routed? What are some techniques? What are some of the difficulties?
20. Compare a fiber optic signal source to a conventional electronic source in tunability and stability. Why is this not a problem in many cases? Compare the use of a CW versus a pulsed source.
21. Explain the operation and benefits of a dual output/dual input test set. Is there another way to achieve the same results?
22. How is optical power measured? Why are single-wavelength power meters useful? Compare a single-wavelength sensor to a broad-bandwidth power meter sensor.
23. How is cable loss measured with a test set? What about the measurement of connector loss?
24. What is OTDR? What are the main functional blocks of an OTDR instrument? Compare OTDR to TDR for metallic cables: what does each show? What does OTDR show that TDR cannot?
25. What are the three critical performance parameters for OTDR? Why are they in conflict with each other to some extent? Which applications require which capabilities to a greater or lesser amount?

PRACTICE PROBLEMS

Section 24.2

1. What is the angle of refraction when the angle of incidence is 35° traveling from glass ($n = 1.4$) to air?
2. A lightwave ray travels from plastic ($n = 1.4$) to glass ($n = 1.6$). What are the angle of refraction when the angles of incidence are 30°, 45°, and 60°?
3. Calculate the critical angle for the figures of problem 1. Will total internal reflection occur?
4. Repeat problem 3 for $n_1 = 1.8$ and $n_2 = 1.3$.
5. A lightwave is incident at 50° going from a medium with $n = 1.7$ to one with $n = 1.4$. Will total internal reflection occur, and at what angle or greater?

Section 24.3

1. Calculate the NA and cone half-angle for a step fiber with core $n = 1.5$ and cladding $n = 1.3$.
2. What happens to NA when the cladding in problem 1 is changed to have $n = 1.2$?

Answers to Selected Problems

Chapter 1
Practice Problems

Section 1.2

1. wavelength = $3 \times 10^8/55{,}000$ = 5455 m
3. $5455/2$ = 2727 m
5. wavelength = $1.5 \times 10^8/1 \times 10^6 = 1.5 \times 10^2$ m = 150 m
7. velocity = $0.82 \times (3 \times 10^8) = 2.46 \times 10^8$ m/s.

Section 1.3

1. 25.1 Hz = 0.0251 kHz; 15.75 MHz = 15,750 kHz

Section 1.4

1. Ratio is (3000-300) MHz/(30-3) kHz = 100,000 (convert MHz to kHz or vice versa!)
3. HF = 27 MHz = 27,000 kHz; (audio) 3 kHz/27,000 = 0.0001 = 0.01%; (video) 6 MHz/27 MHz = .22 = 22%
5. capacity = $2000 \log_2 (1 + 150/12)$ = 7510 bits/s
7. capacity = $75{,}000 \log_2 (1 + 2000)$ = 822,488 bits/s

Chapter 2
Practice Problems

Section 2.3

1.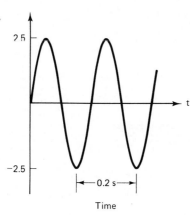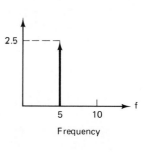

3.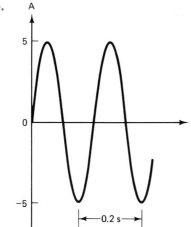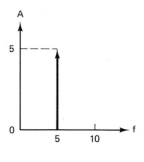

5. $V = \dfrac{4}{\pi n}$, for $n = 7$, $V = \dfrac{4}{\pi 7} = 0.18$

Section 2.6

1.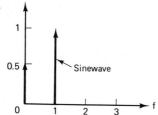

Chapter 3
Practice Problems

Section 3.2

1. 21.9 dB, −7.4 dB, 2.36 dB

3. $dB(V) = 20 \log \left(\frac{2 V_0}{V_0}\right) = 20 \log 2 = 6.02$

5. $P = 0.1$ W

7. $dB = 20 \log \frac{2.5}{1} = 7.96$ dB

Section 3.3

1. 0 dBW = 1W, 0 dBm = 0.001 W

   value in dBm = $10 \log \frac{1}{0.001} = 30$ dBm

3. 0, −1.37, 29.8 dBW

5. 23 dB, 53 dB

7. 85 dBm = 3.16×10^8 mW, 3.16×10^5 W

Section 3.4

1. gains are 5, 13, 13, 12, 11, 10, 10, 9, 9, 9, dB

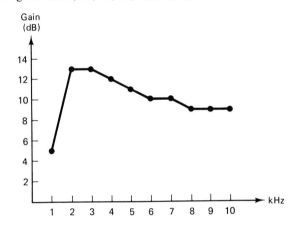

3. Output = input + gain
   30 + 5 = 35, 30 + 13 = 43, 30 + 13 = 43, 32 + 12 = 44, 32 + 11 = 43, 32 + 10 = 42, 29 + 10 = 39, 29 + 9 = 38, 29 + 9 = 38, 29 + 9 = 38 dB at each kHz point.

5. Output power is (input − loss)

0 kHz	10 dBm −20 dB = −10 dBm
1	−3 = 7 dBm
2	−3 = 7 dBm
3	−20 = −10 dBm
4	−20 = −10 dBm
5	−20 = −10 dBm
6	−30 = −20 dBm
7	−30 = −20 dBm
8	−30 = −20 dBm
9	−30 = −20 dBm
10	−30 = −20 dBm

7. Total gain is 50 + 40 + 20 + 25 + 10 = 145 dB
   After 2 stages, gain is 90 dB so level is −100 dBm + 90 dB = −10 dBm
   After all stages, −100 dBm + 145 dB = +45 dBm + 45 dBm $\Rightarrow .001 \times 10^{45/10} =$ 31.6 W

Section 3.8

1. $P = kT \Delta f$
   $= 1.38 \times 10^{-23} (273 + 25)(3500) = 1.44 \times 10^{-17}$ W

3. $V_N = \sqrt{4kT \Delta f R}$
   $= \sqrt{4 \cdot 1.38 \times 10^{-23} (273 + 100)(3000)(500)}$
   $= 1.76 \times 10^{-7}$ V

5. $R = \dfrac{(1 \times 10^{-9})^2}{4(1.38 \times 10^{-23})(273 + 25)(25 \times 10^3)} = 0.002 \, \Omega$

Section 3.9

1.
Noise (mV)	Square	# of occurances	Total
20	400	6	2400
−17	289	2	578
−35	1225	5	6125
25	625	3	1875
10	100	4	400
			11378

   $11378/20 = 568.9$; $\sqrt{568.9} = 23.85$ mV rms

3. 30 dB SNR

5. NF = 10 log (10450)/(11.75) = 1.76 dB

7. T = 290 (20 − 1) = 5510°

9. NT = 290 (NR−1) so NR = $\dfrac{NT}{290} + 1 = \dfrac{80}{290} + 1 = 1.27$

Chapter 4
Practice Problems

Section 4.2

1. 1 MHz ± 0.1 MHz = .9 MHz, 1.1 MHz

3. 10 ± 1 MHz = 9, 11 MHz

5. 100 kHz ± (1 to 10 kHz) ⇒ 90 to 99 kHz, 101 to 110 kHz

7. 995 to 1000, 1000 to 1005; 1095 to 1100, 1100 to 1105; 1195 to 1200, 1200 to 1205; 1295 to 1300, 1300 to 1305; 1395 to 1400, 1400 to 1405 kHz

9. 0.94 to 1.0, 1.0 to 1.06; 1.04 to 1.1, 1.1 to 1.16; 1.14 to 1.2, 1.2 to 1.26; 1.24 to 1.3, 1.3 to 1.36; 1.34 to 1.4, 1.4 to 1.46 MHz
   overlaps (interference): 1.04 to 1.06, 1.14 to 1.16, 1.74 to 1.26, 1.34 to 1.36 MHz

11. 150 ± 25 kHz = 125 to 175 kHz,
    200 ± 35 kHz = 165 to 235 kHz overlap from 165 to 175 kHz

Section 4.3

1. $m = \dfrac{100 - 80}{80} = 0.25$; $m = \dfrac{125 - 80}{80} = 0.56$

3. $m = \dfrac{1 - 0.1}{1 + .1} = 0.818$

Answers to Selected Problems

5. $\dfrac{P_T}{P_c} = \dfrac{1 + m^2}{2} = 1 + \dfrac{0.3^2}{2} = 1.045$
   for $m = 0.7$, $P_T/P_c = 1.245$

7. $P_T = 1000 + 300 = 1300$ W

   $m = \sqrt{2}\ \sqrt{\dfrac{1300}{1000} - 1} = 0.775$

Section 4.5

1. $P_T = P_C\left(1 + \dfrac{m^2}{2}\right) =$ a) 1125 W

   b) $P_C = 1000$, P sideband = 125 W

   c) 1000 W reduced by 20 dB is 10 W, therefore
      $1000 - 10$ W = 990 W suppressed

   d) $125 + 990 = 1115$ W

   e) dB = $10 \log \dfrac{1115}{125} = 9.5$ dB

3. $P_C = \dfrac{P_T}{1 + \dfrac{m^2}{2}} = 468$ W

   power = 468 reduced by 25 dB = 1.48 W

5. dB = $20 \log \dfrac{1}{.15} = 16$ dB → not enough — No

Section 4.6

1. suppression = $20 \log \cot\left(\dfrac{\phi}{2}\right)$; for $\phi = 0.25° = 53.2$ dB

3. $\phi = 2 \tan^{-1} \dfrac{1}{\left(10^{\frac{dB}{20}}\right)}$; for >40 dB, $\phi < 1.15°$

Chapter 5
Practice Problems

Section 5.2

1. $Q = \dfrac{10 \text{ MHz}}{30 \text{ kHz}} = 333$

3. BW = $\dfrac{5 \text{ MHz}}{250} = 20$ kHz; $\dfrac{6.5 \text{ MHz}}{250} = 26$ kHz; $\dfrac{8 \text{ MHz}}{250} = 32$ kHz

Section 5.3

1. Original SNR is $20 \log \dfrac{20}{2} = 20$ dB

   Output Signal is $(20 \times 1000) = 20{,}000$ μV
   Output Noise is $(2 \times 1000) + 50 = 2050$ μV

   Output SNR is $20 \log \dfrac{20{,}000}{2050} = 19.8$ dB

3. stage 1: Signal Out = (20 × 10) = 200 μV
Noise Out = (2 × 10) + 10 = 30 μV
SNR = 16.48 dB

stage 2: Signal Output = 200 × 100 = 20,000 μV
Noise Output = (30 × 100) + 50 = 3050 μV
SNR = 16.33 dB

Section 5.4

1. LO at (7.7 ± 0.8 MHz) = 6.9, 8.5 MHz

3. Signals at 4.995, 5, 5.005 MHz mixed with 3.5 MHz:
1.495, 8.495; 1.5, 8.5; 1.505, 8.505 MHz
IF is 1.5 MHz

5. $\dfrac{f_n}{f_o} = \dfrac{\sqrt{C_o}}{\sqrt{C_n}}$, so $f_n = \dfrac{f_o \sqrt{C_o}}{\sqrt{C_n}} = \dfrac{5\sqrt{100}}{\sqrt{90}} = 5.27$ MHz; $f_n = \dfrac{5\sqrt{100}}{\sqrt{110}} = 4.767$ MHz

Section 5.5

1. Low tracking LO is 9 − .455 = 8.545 MHz
image frequency is 8.545 − .455 = 8.090 MHz

3. a) LO #1 is 50 + 10.7 = 60.7 MHz
LO #2 is 10.7 + .455 = 11.155 MHz

b) output of mixer #1 is 50 ± 60.7 = 10.7, 110.7 MHz

c) into 2nd mixer is 10.7 MHz only.
output of 2nd mixer is 10.7 ± 11.155 = 0.455, 21.855 MHz

5. a) LO #1 = 110 − 10.7 MHz = 99.3 MHz
LO #2 = 10.7 − 0.455 = 10.245 MHz

b) 110 ± 99.3 = 10.7, 209.3 MHz

c) Input to mixer #2 is 10.7 MHz

d) Output of mixer #2 is 10.7 MHz, 10.7 ± 10.245 = 0.455, 20.945 MHz

Chapter 6
Practice Problems

Section 6.2

1. $m = \dfrac{60}{10} = 6$

3. $f_m = \dfrac{150 \text{ kHz}}{8} = 18.75$ kHz

5. BW = 2(125 + 10) = 270 kHz

7. m = 50 kHz/0.3 = 166.7; $m = \dfrac{50 \text{ kHz}}{5} = 10$

Section 6.3

1. Max Deviation = 65 × 3.5 = 227.50 MHz

3. Use: ×2 × 2 × 3 × 3 × 3 = ×108
Nominal frequency = 101.7/108 = 0.941666 MHz

Answers to Selected Problems

5. 1.2 MHz with 75 Hz deviation
   (× 36)
   43.2 MHz with 2700 Hz deviation
   (mix with 43 MHz)
   (86.2 MHz) + 0.2 MHz with 2700 Hz deviation
   (filter out)   (× 81)
   16.2 MHz with 218,700 Hz

Section 6.5

1.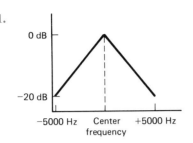

   $\dfrac{-20 \text{ dB}}{5 \text{ kHz}} = -4 \text{ dB/kHz}$

   at 2500 Hz deviation, response is -10 dB
   if 1V = 0 dB, then -10 dB = 0.316 V

   at 5000 Hz deviation, response is -20 dB
   when 1V = 0dB, then -20 dB = 0.1V

3. $\dfrac{20 \text{ mV}}{2 \,\mu\text{V}} = \dfrac{20{,}000}{2} = 10{,}000 = 80$ dB

Section 6.6

1. $\dfrac{10 \text{ kHz}}{1500 \text{ Hz/mV}} = 6.67$ mV

3. VCO range is $\dfrac{1.5 \text{ MHz}}{10 \text{ MHz}} = 0.15 = 15\%$

   can not capture (only 10%) but can lock (20%)

Section 6.8

1. 3 dB: $f = \dfrac{1}{2\pi RC} = \dfrac{1}{2\pi(50\mu S)} = 3183$ Hz

   $R_1 C_1 = 75;\ R_2 C_2 = 50$ but $C_1 = C_2$

   so $\dfrac{R_1}{R_2} = \dfrac{75}{50}$ so $R_2 = \dfrac{50}{75} R_1 = 0.667\ R_1$

Chapter 7
Practice Problems

Section 7.2

1. Differential = $10 - 5 = 5$V

3. CMRR = $20 \log \dfrac{10}{0.15} = 36.5$ dB

5. a) Diff. Voltage = $5 - (-5) = 10$ V
b) $V = 15 \cdot 10^{-80/20} = 0.0015$ V
c) $5 + 0.0015 = 5.0015$
$-5 + 0.0015 = -4.9985$

Section 7.3

1. $SR = \dfrac{6 - (-3)}{25 \ \mu s} = \dfrac{0.36 \ V}{\mu s}$

3. $SR = A \ 2\pi f = 5 \cdot 2\pi \cdot 1 \times 10^6 = 3.1416 \times 10^6$ V/s
$= 3.14$ V/μs

5. $I = C \ (SR) = (175 \times 10^{-12}) \dfrac{0.36 \ V}{10^{-6} \ sec} = 63 \times 10^{-6} A = 0.063^{-3}4 = 63$ mA

7. Differential $= (-2 - 1.5) = -3.5$ V = logic 0
Check: $CMV = \dfrac{-2 + 1.5}{2} = -0.5$ within CMV Limit

Section 7.5

1. $t = \dfrac{5m \cdot 2}{(3 \times 10^8)(0.85)} = 3.9 \times 10^{-8} = 39 \times 10^{-9} = 39$ ns

3. $D = \dfrac{v \cdot t}{2} = 3 \times 10^8 \dfrac{(.95)(1 \times 10^{-9})}{2} = 0.1425$ m = about 5.5″

Chapter 8
Practice Problems

Section 8.1

1. $Z_0 = \sqrt{L/C} = \dfrac{\sqrt{100 \times 10^{-9}}}{\sqrt{150 \times 10^{-12}}} = 25.8 \ \Omega$

3. $Z_0 = \sqrt{L/C} \Rightarrow C = L/Z_0^2 = \dfrac{55 \times 10^{-9}}{(120)^2} = 3.8 \times 10^{-12} F = 3.8$ pF

5. $Z_0 = \dfrac{149}{\sqrt{\epsilon}} = \dfrac{149}{\sqrt{2.3}}$

7. $Z_0 = 276 \log 2s/d \Rightarrow \dfrac{Z_0}{276} = \log \dfrac{2s}{d}$
$10^{Z_0/276} = \dfrac{2s}{d}$
$10^{\frac{200}{276}} = 5.3 = \dfrac{2s}{d} \Rightarrow s = 0.106''$

Section 8.2

1. $Z_0 = \dfrac{87}{\sqrt{E + 1.41}} \ln \left(\dfrac{5.98h}{0.8b + c} \right)$
$= \dfrac{87}{\sqrt{1.8 + 1.41}} \ln \left(\dfrac{5.98 \cdot 0.095}{(0.8)(0.15) + .008} \right) = 7.24 \ \Omega$

Answers to Selected Problems

3. $Z_0 = \dfrac{60}{\sqrt{E}} \ln \dfrac{4t}{.67\pi b \left(0.8 + 6/h\right)}$

   $= \dfrac{60}{\sqrt{2}} \ln \dfrac{4(0.05 \times 2)}{.67\pi\, 0.15 \left(.8 + \dfrac{.005}{.05}\right)} = 14.5\ \Omega$

5. $Z_0 = \dfrac{120}{\sqrt{2.3}} \ln \dfrac{\pi(0.1)}{0.15 + 0.008} = 54.4\ \Omega$

Section 8.3

1. $\lambda = \dfrac{2x}{m}$ mode 1 $= \dfrac{2 \times 0.85}{1} = 1.7$ cm

   mode 2 $= \dfrac{2 \times 0.85}{3} = 0.567$ cm

3. $\lambda = 1.1x = (1.1)(.55) = 0.605$ cm (rect)
   $\lambda = 2.8r = (2.8)(.55) = 1.54$ cm (circular)

5. $\lambda = \dfrac{2x}{m}$ $0.075 = \dfrac{(2)(0.75)}{m} \Rightarrow m = \dfrac{(2)(0.75)}{(0.075)} = 20$

7. coupling $\propto \sin \phi \propto \sin 35° = 0.573 = 57.3\%$

Section 8.4

1. elec. length $= \dfrac{5}{(3 \times 10^8)\, .75/(125 \times 10^6)} = 2.78$ wavelengths

3. $0.25 = \dfrac{\text{physical length}}{(3 \times 10^8)(.8)\, \dfrac{1}{98 \times 10^6}} \Rightarrow$ physical length $= 0.612$ m

5. $Z_0 = 100$ r, $Z_L = 300$ r

   $T = \dfrac{300 - 100}{300 + 100} = 0.5$

   SWR $= \dfrac{1 - .5}{1 + .5} = 0.333$; invert so SWR $= 3.33$

   power reflected $= 0.5 = 50\%$
   power absorbed $= 0.5 = 50\%$

7. $Z_0 = 50$ $Z_L = 150 + j100$

   $T = \dfrac{(150 + j100) - 50}{(150 + j100) + 50} = \dfrac{100 + j100}{200 + j100} = 0.6 + j0.2$

   $|T| = 0.632\ \underline{/18°}$

9. VSWR $= \dfrac{V_{max}}{V_{min}} \Rightarrow V_{max} =$ VSWR $V_{min} = (3.2)(300) = 960$ V ok

Section 8.5

1. $a_1 = 1.7$ $b_1 = 0.7$

   $S_{11} = \dfrac{b_1}{a_1} = \dfrac{0.7}{1.7} = 0.412$

   $S_{21} = \dfrac{b_2}{a_1} = \dfrac{a_1 - b_1}{a_1} = \dfrac{1.7 - .7}{1.7} = 0.588$

3. $a_1 = 1.7$
   What is b_1?

   $S_{11} = \dfrac{b_1}{a_1} \Rightarrow b_1 = S_{11}a_1 = (.3)(1.7) = 0.51$

Chapter 9
Practice Problems

Section 9.2

1. $d = \sqrt{2(1000)} = 44.7$ miles
   $d = \sqrt{2(1000)} + \sqrt{2(30)} = 44.7 + 7.8 = 52.5$ miles,
   additional distance is $52.5 - 44.7 = 7.8$ miles

3. a) path gain $= 32.4 + 20 \log f + 20 \log D$
   $= 32.4 + 20 \log 50 + 20 \log 10,000 = 146.4$ dB
   b) at 500 MHz, loss is 166.4 dB

5. $G_T + G_R -$ path loss $=$ total loss
   $20 + 10 - 120$ W $= 90$ dB

7. $P_T = P_R \, 10^{dB/10} - 10^{-6} (10^9) = 10^3$ W $= 1000$ W

Section 9.3

1. $F/B = F - B = 20 - 15 = 5$ dB
   $F/B = 14 - 9 = 5$ dB (same as above)

3. $200 - 175 = 25$ MHZ
   $175 - 150 = 25$ MHZ

   $$\frac{25}{175} = .143 = 14.3\%$$

   $$\text{Ratio} = \frac{200}{150} = 1.33 : 1$$

Section 9.6

1. at 220 MHZ: $\lambda = \dfrac{3 \times 10^8}{220 \times 10^6} = 1.36$, so $\dfrac{\lambda}{2} = .682$ m (driver)

   $\dfrac{\lambda}{2} + 5\% = .716$ m (reflector)

   $\dfrac{\lambda}{2} - 5\% = .648$ m (director)

   Boom $= (.18)\lambda (2) = .49$ m

3. $54 - 88$ MHZ
   at 88 MHZ, $\lambda = 3.4$ m, $\lambda/2 = 1.7$ m
   at 54 MHZ, $\lambda = 5.5$ m, $\lambda/2 = 2.777$ m $= l_1$

   $l_1 = 2.777$ m
   $l_2 = rl_1 = .9(2.777) = 2.5$ m
   $l_3 = rl_2 = 2.25$ m
   $l_4 = rl_3 = 2.025$ m
   $l_4 = rl_4 = 1.82$ m
   $l_6 = rl_5 = 1.64$ m
   $d_1 = .08\lambda = (.08)(5.5) = 0.44$ m
   $d_2 = .9(.44) = .396$ m
   $d_3 = .9 \, d_2 = .356$ m
   $d_4 = .9 \, d_3 = .32$
   $d_5 = .9 \, d_4 = .288$

   Boom $=$ sum of d_1 to $d_5 = 1.8$ m

Answers to Selected Problems

Section 9.7

1. $G = 15 \left(\dfrac{\pi D}{\lambda}\right)^2 \dfrac{NS}{\lambda}$

   $= 15 \left(\dfrac{\pi(\lambda/4)}{\lambda}\right)^2 \dfrac{20(\lambda/4)}{\lambda} = 15 \dfrac{\pi^2}{16} \dfrac{20}{4} = 46.2$

   gam of $46.2 = 16.6$ dB

   Beamwidth $= \dfrac{52}{\dfrac{\pi D}{\lambda} \sqrt{\dfrac{NS}{\lambda}}} = \dfrac{52}{\dfrac{\pi(\lambda/4)}{\lambda} \sqrt{\dfrac{20(\lambda/4)}{\lambda}}} = \dfrac{52}{\dfrac{\pi}{4} \sqrt{20/4}} = 29.6°$

   at 250 MHZ, $\lambda = \dfrac{3 \times 10^8}{250 \times 10^6} = 1.2$ m

   $D = \dfrac{1.2}{4} = .3$ m; $S = \dfrac{1.2}{4} = .3$ m, $L = NS = 20\left(\dfrac{1.2}{4}\right) = 6.0$ inches

3. $G \propto S \Rightarrow S/2$ implies $G \div 2$
   In dB: -3 dB

   Beamwidth $\propto \dfrac{1}{\sqrt{S}} \Rightarrow S/2$ implies $\dfrac{1}{\sqrt{S/2}} = \sqrt{2} = 1.4$ times beamwidth

   $L = NS = \dfrac{1}{2}$ previous

   for N doubled, S in half:
   $G \propto NS$ so its unchanged

   Beamwidth $\propto \dfrac{1}{\sqrt{NS}}$ so new beamwidth is unchanged

5. $G = \dfrac{GD^2}{\lambda_2}$, for $D = 10\lambda$ so $G = \dfrac{6(10\lambda)^2}{\lambda^2} = 6 \cdot 100 = 600$

   in dB: $10 \log 600 = 27.8$ dB

   Beamwidth $= \dfrac{70\lambda}{D} = \dfrac{70\lambda}{10\lambda} = 7°$

   at 250 MHZ, $D = 10 \cdot \dfrac{3 \times 10^8}{200 \times 10^6} = 15$ m

7. $G \propto \dfrac{1}{\lambda^2}$, at $2 \times$ frequency, $\lambda = \dfrac{1}{2}$

   $G = \dfrac{1}{(1/2)^2} = 4$ times

   Beamwidth $\propto \lambda \Rightarrow$ at $7 \times$ frequency, $\lambda = \dfrac{1}{2}$

   $B \propto \dfrac{1}{2}$

9. Beamwidth $= \dfrac{70\lambda}{D}$, so $D = \dfrac{70\lambda}{\text{Beamwidth}} = \dfrac{70\lambda}{1} = 70\lambda$

   $= 70 \left(\dfrac{3 \times 10^8}{500 \times 10^6}\right) = 42$ m

   $G = \dfrac{6D^2}{\lambda^2} = \dfrac{6(70\lambda)^2}{\lambda^2} = 29400$

   $= 44.7$ dB

Chapter 10
Practice Problems

Section 10.2

1. 4 V ± 2% = 3.92 to 4.08 V, or ±0.08 V

3. 2^{14} = 16,384; 2^{18} = 262,144.

5. Dynamic range = 20 log (3/.001) = 69.5 dB. For 6-V span, dynamic range is 20 log (6/.001) = 75.5 dB (doubling span increases dynamic range by 6 dB). Bits needed = 75.5/6 = 12.6, so 13 bits are needed.

7. First three: 0000, 0001, 0010; last three: 1101, 1110, 1111.

Section 10.3

1. 2 × 12 kHz = 24 kHz

3. Nyquist rate = 3300 × 2 = 6600 Hz or samples/sec. Bits/sec = 6600 × 10 bits/sample = 66,000 bits/s. Bandwidth is about 5 times that, or 330 kHz

5. 50 kilosamples/s are transmitted; each has 8 elements. 50,000 × 8 = 400,000 elements/sec are transmitted

7. Use 0 V for binary 0, 7 V for binary 1; noise immunity is (7-0)/2 = 3.5 V. For 4-symbols, use 0, 2.33, 4.67, and 7 V, noise margin is (2.33-0)/2 = 1.165 V.

Chapter 11
Practice Problems

Section 11.2

1. 01010101 and partial overlap with 10000000 is 11010101 (binary) which equals 213 (decimal). Each count is 5.12 V/2^8 = 0.02 V, therefore the voltage is 213 × 0.02 = 4.26 V. It should have been 10000000 = 128 (decimal) = 128 × 0.02 = 2.56 V.

3. The 0.03% error factor corresponds to 0.0003; T × 0.0003 = 5 µs, therefore, T = 5 µs/0.0003 = 16667 µs. At 100 kHz, each bit occupies 10 µs; therefore, 16667 µs/10 µs/bit = 1666.7 bits.

5. Using the figures in the problem, we find: Vout = 10 ln [1 + (10/100) Vin]/[ln (1 + 100)]. Here are Vin and corresponding Vout pairs: 0-0, 1-5.12, 2-6.6, 3-7.4, 4-8.0, 5-8.5, 6-8.9, 7-9.2, 8-9.5, 9-9.77, 10.0-10.0.

Section 11.3

1. Shift = 7 (Hz/mi/hr) × 25,000 mi/hr = 175,000 Hz or 175 kHz. 175 kHz/2450,000 kHz = 0.00007 = 0.007%

3. 1/(1 + 8) = 0.11 = 11%; 4/(64 + 4) = 0.058 = 5.8%

Chapter 12
Practice Problems

Section 12.2

1. S is 1010011; Q is 1010001; 7 is 0110111; 9 is 0111001; : is 0111010; / is 0101111 (MSB on left)

3. 64.76 in 7 bit ASCII takes 5 characters × 7 = 35 bits
In 4-bit Binary: 5 × 4 = 20 bits
20/35 = 57%

Section 12.3

1.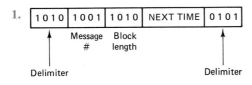

3. As above, but with 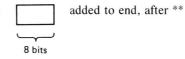 added to end, after **

Section 12.4

1.
Bit #	1	2	3	4	5	6	7	8	9	10	11	12	13	14	15	16
Voltage	0	0	0	0	5	5	5	5	0	0	0	0	0	0	0	0
Average	0	0	0	0	1			2.5					1.25			

$$\frac{0+0+0+0+5}{5}$$

$$\frac{4 \times 0 + 4 \times 5}{8}$$

$$\frac{0+0+0+0+5+5+5+5+0+0 \cdots 0}{16}$$

3.
Bits	0	1	1	1	0	0	1	1
Voltage	−5	+5	+5	+5	−5	−5	+5	+5
Average	−5	0	1.67	2.5	1.0	0	0.71	1.25

$$\frac{-5(1) + 5(2)}{3}$$

$$\frac{-5(3) + 5(5)}{8}$$

5. $\dfrac{2450 \text{ bits/sec}}{7 \text{ bits/ch}} = 343$ char/sec

7. $32 + (128 \times 7) + 16 = 944$ bits/128 char

$\dfrac{2400 \text{ bits/sec}}{944 \text{ bits/128 char}} = 325$ ch/sec

Compared to prob. 6: $\dfrac{325}{240} = 1.35$ times as many

Section 12.5

1.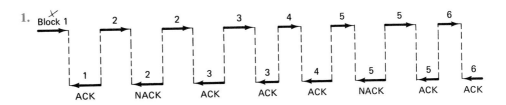

3.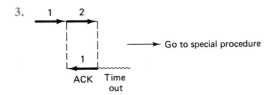

Answers to Selected Problems

5.

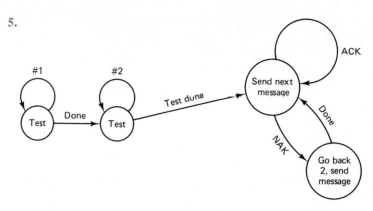

Section 12.7

1. even parity: 0110–0; 011001–1; 11001011–1

3. even: 0111–1; 011000–0; 11001010–0
   even: 0101–0; 011010–1; 11001000–1

5.

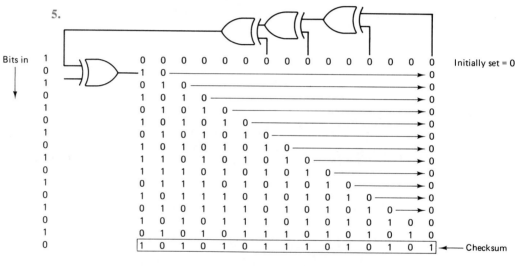

7.

	1	0	1	0	1	0	1	0	← Input bits	Odd homing bits
CB0			X	X		X	X			1
CB1	X	X		X	X		X			1
CB2	X		X	X			X	X		0
CB3	X	X				X	X	X		1
CB4	X	X	X	X	X					0

	0	0	1	0	1	0	1	0	← One bit error	
CB0			X	X		X	X			1
CB1	X	X		X	X		X			1
CB2	X		X	X			X	X		1
CB3	X	X				X	X	X		0
CB4	X	X	X	X	X					1

Hamming bits ① → 1 1 | 0 | 1 | 0 |
Hamming bits ② → 1 1 | 1 | 0 | 1 |
Syndrome → 1 1   0   0   0 → error in MSB (bit 7) see Fig. 12.19

Chapter 13
Practice Problems

Section 13.1

1. 16 levels = 4 bits (2^4 = 16)   4 × 300 = 1200 bits/sec
   8 levels = 3 bits (2^3 = 8)   3 × 300 = 900 bits/sec

3. 00 10 00 00 11 00 10 01
    1   2   0   0   3   0   2   1 V

5. 0110 0000 1100 1001
    1.5   0   3 $2\frac{1}{4}$ V

   since values are 0, $\frac{1}{4}$, $\frac{1}{2}$, $\frac{3}{4}$, 1, $1\frac{1}{4}$, $1\frac{1}{2}$, $1\frac{3}{4}$, 2, $2\frac{1}{4}$, $2\frac{1}{2}$, $2\frac{3}{4}$, 3, $3\frac{1}{4}$, $3\frac{1}{2}$, $3\frac{3}{4}$

Section 13.2

1.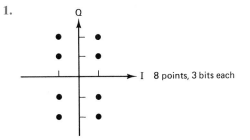
   8 points, 3 bits each

Section 13.3

1. Total bits = 2000 bits/sec × 10 sec = 20,000 bits

   $$\text{BER} = \frac{50 \text{ errors}}{20{,}000 \text{ bits}} = 5 \text{ in } 2000 = 1 \text{ in } 400 \text{ bits}$$

3. All 50 errors occur in first frame. Therefore:

   % EFS = $\frac{9}{10}$ = 90%

   % EFF = $\frac{(100-1)}{100}$ = 99%

5. BER = 500/20,000 = 5 in 200 = 1 in 40 bits

   % EFS: First 5 sec. have errors, the last 5 do not
   % EFS = $\frac{5}{10}$ = 50%

   % EFS: at 100 errors/sec, and 10 frames/sec, each frame has 10 errors. Therefore, all frames in first 5 sec have errors, while second group of 5 sec have none.

   % EFS = $\frac{5}{10}$ = 50%, total frames = 100, error frames = 50 (first 5 sec)

Answers to Selected Problems

Section 13.4

1.

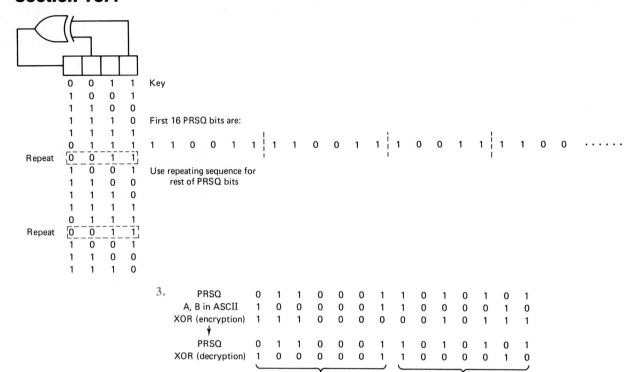

3.
```
            PRSQ              0 1 1 0 0 0 1  1 0 1 0 1 0 1
    A, B in ASCII             1 0 0 0 0 0 1  1 0 0 0 0 1 0
    XOR (encryption)          1 1 1 0 0 0 0  0 1 0 1 0 1 1
            ↓
            PRSQ              0 1 1 0 0 0 1  1 0 1 0 1 0 1
    XOR (decryption)          1 0 0 0 0 0 1  1 0 0 0 0 1 0
                              _____/ _____/
                                    A                B
```

Chapter 15
Practice Problems

Section 15.2

1. $N = \dfrac{F_{vco}}{F_{ref}} = \dfrac{50}{0.1} \rightarrow \dfrac{51.9}{0.1} = 500 \rightarrow 519$

3. $LO = (550 \rightarrow 1600 \text{ kHz}) + 450 \text{ kHz} = 1000 \rightarrow 2050 \text{ kHz}$

   ref = step size = 20 kHz

   $N = \dfrac{1000}{20} \rightarrow \dfrac{2050}{20} = 50 \rightarrow 102.5$ (use 102 or 103)

5. $F_{vco} = NF_{ref} = 25 \text{ kHz} \times (10{,}000 \rightarrow 25{,}000)$
   $= 250 \text{ MHz} \rightarrow 625 \text{ MHz}$

   Received frequency = $(2.50 \rightarrow 625 \text{ MHz}) + LO$

   $LO = 25$

   $LO = 275 \rightarrow 650 \text{ MHz}$

Section 15.3

1. a) step size = $Mf_{ref} \Rightarrow F_{ref} = \dfrac{\text{step size}}{m} = \dfrac{100 \text{ kHz}}{10} = 10 \text{ kHz} = 0.01 \text{ MHz}$

   b) $F_{vco} = M \times (N \times F_{ref}) \Rightarrow N = \dfrac{F_{vco}}{F_{ref} \cdot M} = \dfrac{300}{(0.01)(100)} \rightarrow \dfrac{350}{(0.01)(100)}$
   $= 300 \rightarrow 350$

Answers to Selected Problems

3. (64/65) desire Fvco from 128 → 136 MHz

   step size = ref = 10 kHz = 0.01 MHz
   N = 64

   Fvco = M · 64 · Fref

   $128 = M \cdot 64 \cdot 0.01 \Rightarrow M = \dfrac{128}{64 \cdot (0.01)} = 200$

   Fvco = (M × N × Fref) + A Fref = (200 · 64 · 0.01) + A Fref
   for Fvco = 136, A = 800

Section 15.4

1. Fvco = 1310 + 460 = 1770
   Fvco = (8 · M · 10) + A(10)

   determine M, set for A = 0 (provides Fvco just below desired value, without exceeding it)

   $M = \dfrac{1770}{80} = 22.125$—use M = 22

   Fvco = 8·22·10 + A·10
   desire Fvco = 1770 so:
   1770 = 8 · 22 · 10 + A(10) ⇒ A = 1

3. Fvco = 64 M (0.025) + A(0.025) MHz = 99.5
   M = 62   A = 12
   Fvco = 99.5 − 10.7 = 88.8 MHz

Chapter 16
Practice Problems

Section 16.2

1. Pulse: .3 + .4 + .5 + .9 + .8 + .7 + .6 + 6(.5) = 7.2 sec
   Tone: (7 × .25) + (6 × .25) = 3.25 sec

3. Both 3.25 sec

5. 8: 852, 1336 Hz

7. 9:Harmonics: 1704, 2672 Hz   Sum: 2188 Hz, Diff: 484 Hz

Section 16.6

1. $\dfrac{8000}{186{,}000 \times .6} \times 2 = .143$ sec

3. Talker echo $= \left[\left(\dfrac{3000}{186{,}000 \times .7}\right) + \left(\dfrac{23{,}000 \times 2}{186{,}000}\right)\right] \times 2 = .541$ sec

   Listener echo $= \left[\phantom{\dfrac{3000}{186{,}000 \times .7}}\right] \times 3 = .811$ sec

Chapter 17
Practice Problems

Section 17.2

1. 1 1010110 0   9 bits at 1200 baud = 0.0075 s = 7.5 ms
   at 19,200 baud = 0.00047 s = 0.47 ms

3. As 2 ASCII characters: 1 10001100 10 1 10101100 00
                                 1   5

   As a binary value   1 11110000 00
                         ‾‾15‾‾

   Time saved is ½ total

Section 17.3

1. 11011101

3. with clock at 38.4 kHz, baud values are 38.4 kbps, 2400, 600 bps. At 19,200 Hz, baud values are 19,200, 1200, and 300 bps

5. XX11XX0X
    ↑↑   ↑
   errors not
         ready

Section 17.5

1. B = 01000010
   Pattern is: 2000, 1000, 2000, 1000, 1000, 1000, 1000, 2000, 1000, 1000 Hz

3. H 1001000 1 ← odd parity
   I 1001001 0   "
   Y 1011001 1   "
   E 1000101 0   "
   S 1010011 0   "

   ORIGINATE:
   H: 2000, 1000, 1000, 1000, 2000, 1000, 1000, 2000, 2000, 1000 Hz
   I: 2000, 2000, 1000, 1000, 2000, 1000, 1000, 2000, 1000, 1000 Hz
   Y: 4000, 4000, 3000, 3000, 4000, 4000, 3000, 4000, 4000, 3000 Hz
   E: 4000, 4000, 3000, 4000, 3000, 3000, 3000, 4000, 3000, 3000 Hz
   S: 4000, 4000, 4000, 3000, 3000, 4000, 3000, 4000, 3000, 3000 Hz

Section 17.6

1. 1070, 1270, 1070, 1070, 1070, 1270, 1070, 1270, 1070, 1270 Hz

3. 2025, 2225, 2025, 2025, 2025, 2225, 2025, 2225, 2025, 2225, 2225

5. 0 1   0 0   0 1   0 1   0 1   1   1
   ↑   ↑   ↑   ↑
   Start   Parity Extra Stop
   0°   90°   0°   0°   0°   −90°

Chapter 21
Practice Problems

Section 21.1

1. 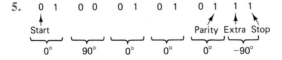 = 139.6 m² (1 GHz)

   $$= \frac{4\pi(1)^2}{\left(\frac{3 \times 10^8}{10 \times 10^9}\right)^2} = 13963 \text{ m}^2 \text{ (10 GHz)}$$

3. $P_r = \dfrac{P_T G^2 \lambda^2 \, RCS}{(4\pi)^3 \, R^4}$

at 80 dB, $G = 10^8$ $\lambda = \dfrac{3 \times 10^8}{1 \times 10^9} = 0.3$ m

$R = 10^6$ m $RCS = 140$

$P_T = 10^3$ W

$\Rightarrow P_R = 6.35 \times 10^{-8}$ W $= 63.5$ nanowatts

5. $RCS = 40 - 6 = 34$ m^2 $RCS = \dfrac{4\pi A^2}{\lambda^2} \Rightarrow A^2 = \dfrac{RCS \, \lambda^2}{4\pi}$

$\Rightarrow A^2 = .2435$ so $A = .49$

so side $= \sqrt{A} = .7$ m

Section 21.2

1. $fd = \dfrac{2 \, V \cos \phi}{\lambda} = \dfrac{2(250) \cos 0°}{\dfrac{3 \times 10^8}{500 \times 10^6}} = 833$ Hz

(away) $\phi = 180°$ so $\cos \phi = -1 \Rightarrow -833$ Hz

45°: 833 Hz $\cdot \cos 45° = 589$ Hz

3. 30°: $267 \times \cos 30 = 231$ Hz

$400 \times \cos 30 = 346$ Hz

5. $PRF = \dfrac{2 \, V_b}{n\lambda} = \dfrac{2 \cdot 50}{(1) \dfrac{3 \times 10^8}{2 \times 10^9}} = 667$ Hz

Chapter 24
Practice Problems

Section 24.2

1. $\dfrac{\sin \phi_1}{\sin \phi_2} = \dfrac{n_2}{n_1}$; $\phi_1 = 35°$, $n_1 = 1.4$, $n_2 = 1.0$

$\sin \phi_2 = (\sin \phi_1) \dfrac{n_1}{n_2} = 0.803 \Rightarrow \phi_1 = 53.4°$

3. $\sin \phi_c = \dfrac{n_2}{n_1} = \dfrac{1.0}{1.4} = 0.71 \Rightarrow \phi_c = 45.2°$

no, $\phi_c > \phi_1$ (45.2° > 35°)

5. $n_2 = 1.4$, $n_1 = 1.7$ $\sin \phi_c = \dfrac{1.4}{1.7} = 0.82 \Rightarrow \phi_c = 55.4°$

no, ϕ, must be $\geq \phi_c$

Section 24.3

1. $NA = \sqrt{n_1^2 - n_2^2} \cdot$; $n_1 = 1.5$, $n_2 = 1.3$

$= \sqrt{1.5^2 - 1.3^2} = \sqrt{0.56} = 0.748 = NA = \sin \phi$

$0.748 = \sin \phi \Rightarrow \phi = 48.4°$

Appendix A

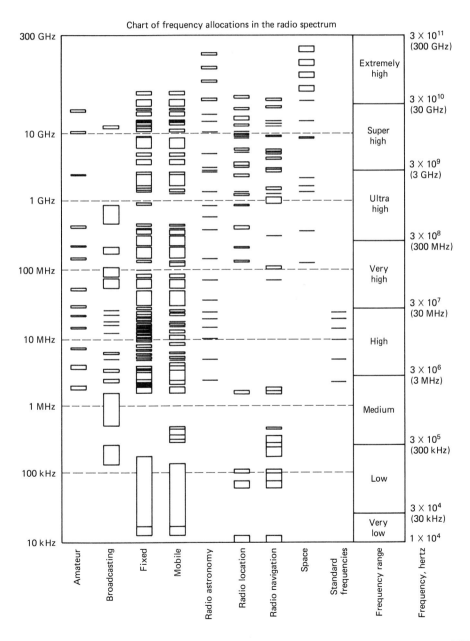

Appendix B

dB Table

dB	Voltage Or current ratio	Power ratio	dB	Voltage Or current ratio	Power ratio
0.0	1.000	1.000	26.	19.95	398.1
0.1	1.012	1.023	27.	22.39	501.2
0.2	1.023	1.047	28.	25.12	631.0
0.3	1.035	1.072	29.	28.18	794.3
0.4	1.047	1.096	30.	31.62	1000.
0.5	1.059	1.122	31.	35.48	1259.
0.6	1.072	1.148	32.	39.81	1585.
0.8	1.096	1.202	33.	44.67	1995.
1.0	1.122	1.259	34.	50.12	2512.
1.5	1.189	1.413	35.	56.23	3162.
2.0	1.259	1.585	36.	63.10	3981.
2.5	1.334	1.778	37.	70.79	5012.
3.0	1.413	1.995	38.	79.43	6310.
4.	1.585	2.512	39.	89.13	7943.
5.	1.778	3.162	40.	100.0	10000.
6.	1.995	3.981	41.	112.2	12590.
7.	2.239	5.012	42.	125.9	15850.
8.	2.512	6.310	43.	141.3	19950.
9.	2.818	7.943	44.	158.5	25120.
10.	3.162	10.000	45.	177.8	31620.
11.	3.548	12.59	46.	199.5	39810.
12.	3.981	15.85	47.	223.9	50120.
13.	4.467	19.95	48.	251.2	63100.
14.	5.012	25.12	49.	281.8	79430.
15.	5.623	31.62	50.	316.2	100000.
16.	6.310	39.81	51.	354.8	125900.
17.	7.079	50.12	52.	398.1	158500.
18.	7.943	63.10	53.	446.7	199500.
19.	8.913	79.43	54.	501.2	251200.
20.	10.000	100.00	55.	562.3	316200.
21.	11.22	125.9	56.	631.0	398100.
22.	12.59	158.5	57.	707.9	501200.
23.	14.13	199.5	58.	794.3	631000.
24.	15.85	251.2	59.	891.3	794300.
25.	17.78	316.2	60.	1000.0	1000000.

Appendix C

ASCII Character Set

ASCII Character	Meaning	(MSB) B7	B6	ASCII Bit Pattern B5	B4	B3	B2	(LSB) B1
NUL	Null	0	0	0	0	0	0	0
SOH	Start of heading	0	0	0	0	0	0	1
STX	Start of text	0	0	0	0	0	1	0
ETX	End of text	0	0	0	0	0	1	1
EOT	End of transmission	0	0	0	0	1	0	0
ENQ	Enquiry	0	0	0	0	1	0	1
ACK	Acknowledge	0	0	0	0	1	1	0
BEL	Bell	0	0	0	0	1	1	1
BS	Backspace	0	0	0	1	0	0	0
HT	Horizontal tabulation	0	0	0	1	0	0	1
NL	New line	0	0	0	1	0	1	0
VT	Vertical tabulation	0	0	0	1	0	1	1
FF	Form feed	0	0	0	1	1	0	0
RT	Return	0	0	0	1	1	0	1
SO	Shift out	0	0	0	1	1	1	0
SI	Shift in	0	0	0	1	1	1	1
DLE	Data line escape	0	0	1	0	0	0	0
DC1	Device control 1	0	0	1	0	0	0	1
DC2	Device control 2	0	0	1	0	0	1	0
DC3	Device control 3	0	0	1	0	0	1	1
DC4	Device control 4	0	0	1	0	1	0	0
NAK	Negative acknowledgment	0	0	1	0	1	0	1
SYN	Synchronous idle	0	0	1	0	1	1	0
ETB	End of transmission block	0	0	1	0	1	1	1
CAN	Cancel	0	0	1	1	0	0	0
EM	End of medium	0	0	1	1	0	0	1
SUB	Substitute	0	0	1	1	0	1	0
ESC	Escape	0	0	1	1	0	1	1

ASCII Character Set (cont.)

ASCII Character	Meaning	(MSB) B7	B6	ASCII Bit Pattern B5	B4	B3	B2	(LSB) B1
FS	File separator	0	0	1	1	1	0	0
GS	Group separator	0	0	1	1	1	0	1
RS	Record separator	0	0	1	1	1	1	0
US	Unit separator	0	0	1	1	1	1	1
SP	Space	0	1	0	0	0	0	0
!	Exclamation point	0	1	0	0	0	0	1
"	Quotation mark	0	1	0	0	0	1	0
#	Number sign	0	1	0	0	0	1	1
$	Dollar sign	0	1	0	0	1	0	0
%	Percentage sign	0	1	0	0	1	0	1
&	Ampersand	0	1	0	0	1	1	0
'	Apostrophe	0	1	0	0	1	1	1
(	Opening parenthesis	0	1	0	1	0	0	0
)	Closing parenthesis	0	1	0	1	0	0	1
*	Asterisk	0	1	0	1	0	1	0
+	Plus	0	1	0	1	0	1	1
,	Comma	0	1	0	1	1	0	0
-	Hyphen (minus)	0	1	0	1	1	0	1
.	Period (decimal)	0	1	0	1	1	1	0
/	Slant	0	1	0	1	1	1	1
0	Zero	0	1	1	0	0	0	0
1	One	0	1	1	0	0	0	1
2	Two	0	1	1	0	0	1	0
3	Three	0	1	1	0	0	1	1
4	Four	0	1	1	0	1	0	0
5	Five	0	1	1	0	1	0	1
6	Six	0	1	1	0	1	1	0
7	Seven	0	1	1	0	1	1	1
8	Eight	0	1	1	1	0	0	0
9	Nine	0	1	1	1	0	0	1
:	Colon	0	1	1	1	0	1	0
;	Semicolon	0	1	1	1	0	1	1
<	Less than	0	1	1	1	1	0	0
=	Equals	0	1	1	1	1	0	1
>	Greater than	0	1	1	1	1	1	0
?	Question mark	0	1	1	1	1	1	1
@	Commercial "at"	1	0	0	0	0	0	0
A	Uppercase A	1	0	0	0	0	0	1
B	Uppercase B	1	0	0	0	0	1	0
C	Uppercase C	1	0	0	0	0	1	1
D	Uppercase D	1	0	0	0	1	0	0
E	Uppercase E	1	0	0	0	1	0	1
F	Uppercase F	1	0	0	0	1	1	0
G	Uppercase G	1	0	0	0	1	1	1
H	Uppercase H	1	0	0	1	0	0	0
I	Uppercase I	1	0	0	1	0	0	1
J	Uppercase J	1	0	0	1	0	1	0
K	Uppercase K	1	0	0	1	0	1	1
L	Uppercase L	1	0	0	1	1	0	0
M	Uppercase M	1	0	0	1	1	0	1
N	Uppercase N	1	0	0	1	1	1	0
O	Uppercase O	1	0	0	1	1	1	1
P	Uppercase P	1	0	1	0	0	0	0

ASCII Character Set (cont.)

ASCII Character	Meaning	(MSB) B7	B6	ASCII Bit Pattern B5	B4	B3	B2	(LSB) B1
Q	Uppercase Q	1	0	1	0	0	0	1
R	Uppercase R	1	0	1	0	0	1	0
S	Uppercase S	1	0	1	0	0	1	1
T	Uppercase T	1	0	1	0	1	0	0
U	Uppercase U	1	0	1	0	1	0	1
V	Uppercase V	1	0	1	0	1	1	0
W	Uppercase W	1	0	1	0	1	1	1
X	Uppercase X	1	0	1	1	0	0	0
Y	Uppercase Y	1	0	1	1	0	0	1
Z	Uppercase Z	1	0	1	1	0	1	0
[	Opening bracket	1	0	1	1	0	1	1
\	Reverse slant	1	0	1	1	1	0	0
]	Closing bracket	1	0	1	1	1	0	1
^	Circumflex	1	0	1	1	1	1	0
_	Underscore	1	0	1	1	1	1	1
`	Grave accent	1	1	0	0	0	0	0
a	Lowercase a	1	1	0	0	0	0	1
b	Lowercase b	1	1	0	0	0	1	0
c	Lowercase c	1	1	0	0	0	1	1
d	Lowercase d	1	1	0	0	1	0	0
e	Lowercase e	1	1	0	0	1	0	1
f	Lowercase f	1	1	0	0	1	1	0
g	Lowercase g	1	1	0	0	1	1	1
h	Lowercase h	1	1	0	1	0	0	0
i	Lowercase i	1	1	0	1	0	0	1
j	Lowercase j	1	1	0	1	0	1	0
k	Lowercase k	1	1	0	1	0	1	1
l	Lowercase l	1	1	0	1	1	0	0
m	Lowercase m	1	1	0	1	1	0	1
n	Lowercase n	1	1	0	1	1	1	0
o	Lowercase o	1	1	0	1	1	1	1
p	Lowercase p	1	1	1	0	0	0	0
q	Lowercase q	1	1	1	0	0	0	1
r	Lowercase r	1	1	1	0	0	1	0
s	Lowercase s	1	1	1	0	0	1	1
t	Lowercase t	1	1	1	0	1	0	0
u	Lowercase u	1	1	1	0	1	0	1
v	Lowercase v	1	1	1	0	1	1	0
w	Lowercase w	1	1	1	0	1	1	1
x	Lowercase x	1	1	1	1	0	0	0
y	Lowercase y	1	1	1	1	0	0	1
z	Lowercase z	1	1	1	1	0	1	0
{	Opening brace	1	1	1	1	0	1	1
|	Vertical line	1	1	1	1	1	0	0
}	Closing brace	1	1	1	1	1	0	1
~	Tilde	1	1	1	1	1	1	0
DEL	Delete	1	1	1	1	1	1	1

Index

A

−3 dB, 61
A-law, 342
Absorption, 733
ACC (*see* Automatic chroma control)
Acceptance cone, 737
Accuracy, 316
ACK (*see* Acknowledge)
Acknowledge (ACK), 377
Acquire range, 185
Active element, 290
Adaptive circuitry, 511
ADC (*see* Converter, analog-to-digital admittance), 224
AFC (*see* Automatic frequency control)
AGC (*see* Automatic gain control)
Algorithm, 24, 314, 315
Aliasing, 324
Alignment, 131
AM receiver, 119
AM (*see also* amplitude modulation)
AM transmitter, 91
American Wire Gauge (AWG), 204
Amplitude modulation (AM), 78
Analog, 6
Analog drivers, 211
Analog input filter, 335
Analog signals, 309
Anechoic chamber, 275
Angle modulation, 190
Angle of incidence, 731
Angle of radiation, 270
Anode, 704

Answer mode, 546
Antenna, 604
 arrays, 290
 bandwidth, 278
 couplers, 288
 dish, 1, 300
 dummy, 279
 electronically steerable, 295
 folded dipole, 285
 grounded vertical, 286
 half-wave dipole, 276, 280
 helical, 298
 log-periodic, 293
 long-wire, 284
 loop, 285
 Marconi, 286
 multiband, 282
 parabolic, 300
 phased-array, 295
 scale, 275
 slot, 297
 stacked, 300
 Yagi-Udi array, 291
Antialiasing, 324
Antinodes, 239
Apple Talk, 574
Application layer, 585
Area code, 482
Armstrong. E. H., 124, 160
ASCII, 362
Asynchronous communication, 380
Attack, 108, 129, 142
Attenuation, 344
Attenuators, 695

Audio frequency (AF), 124
Automatic chroma control (ACC), 446
Automatic frequency control (AFC), 144, 177
Automatic gain control (AGC), 127, 141
Avalanche photodiodes, 742
Average power, 713
AWG (see American Wire Gauge)
Azimuth, 594

B

Back porch, 434
Backbone, 556
Backwave, 711
Balanced modulator, 96
Balanced ring modulator, 96
Balanced signals, 206
Bandgap energy, 738
Bandpass filter, 406
Bands, frequency, 12
Bandwidth, 4, 14, 733
Base boost, 191
Base station, 623
Baseband, 671
Baseband AM, 107
Basic rate, 589
Battery, 493
Baud, 374, 526
Baudot, 365
BCD (see Binary coded decimal)
Beamwidth, 277
Beat frequency oscillator (BFO), 143
BER (see Bit error rate)
Bessel functions, 162
BFO (see Beat Frequency Oscillator)
Binary, 34
Binary coded decimal (BCD), 463
Binary system, 310
Bipolar, 370
Bit:
  error rate (BER), 396, 414
  stuffing, 347
  sychronization, 347
  vs. baud rate, 405
Bits, 16, 310
Blanking level, 436
Blanking period, 436
Blind speed, 652
Blind spot, 650
Block length, 367
BORSHT, 493
Bounce, 268
Break, 527
Break detect, 527
Breakout box, 541
Bridge, 586
Brightness, 446
Buffer, 110, 209, 384, 528
Buffer circuit, 311
Bulk silicon, 699
Bus driver, 209

C

C conditioning, 492
C/A-pattern, 617
Call blocking, 504
Call forwarding, 503
Call tracing, 507
Calling channel, 623
Camp on, 504
Capture effect, 177
Capture range, 185
Carrier, 4, 32, 79
Carrier sense multiple access/collision detect
    (CSMA/CD), 571
Carson's rule, 165
Cascade amplifiers, 125
Cassegrain feed, 303
Cathode, 704
Cathode ray tube (CRT), 432
Cavity gap, 707
CCD (see Charge coupled device)
CCS (see Common channel signaling)
CD (see Compact disk)
Cell:
  shapes, 627
  size, 626
  splitting, 627
Cellular concept, 623
Cellular phone details, 636
Central office 480
Cesium clocks, 612
Channel capacity, 15
Channel group, 676
Character-by-character transmission, 366
Characteristic impedance, 204, 224
Charge coupled device (CCD), 432
Checking bits, 314, 394
Checksum, 388, 389, 391, 421
Chip components, 694
Chirp waveforms, 654
Chroma processing, 446
Chrominance (chroma signal), 440
Circulator, 653, 659
Class A operation, 93
Class B operation, 93
Class C operation, 93
Clear to send (CTS), 523
Clock, 336
Closed loop, 350
Cluster, 345
Clutter, 651
CMRR (see Common mode rejection ratio)
CMV (see Common mode voltage)
Coaxial cable, 215
Codec, 341
Coding, 360
Coherent detector, 185
Collision detect, 571
Color burst, 436, 441
Color TV, 439

Index

Command register, 384, 533
Common channel, 506
Common channel signaling (CCS), 506
Common mode rejection, 207
Common mode rejection ratio (CMRR), 207
Common mode voltage (CMV), 207
Communications receiver, 149
Compact disk (CD) players, 396
Compand, 341
Comparator, 348
Comparator circuit, 405
Complex conjugate, 224, 237
Composite signal, 434
Compression, 341, 450
Conditioned line, 492
Connector loss, 758
Constellation pattern, 420
Constellation plot, 412
Constructive interference, 295
Continuous tuning, 457
Continuous wave (CW), 108, 700
Continuous wave (CW) demodulation, 143
Continuously variable slope delta modulation (CVSD), 351
Contrast, 446
Control codes, 363
Control segment, 616
Controller, 579
Convergence, 447
Converter:
    analog-to-digital (ADC), 333
    digital-to-analog (DAC), 132, 333
    scan, 441
    serial-to-parallel, 337
    tracking, 350
Copper wire, 1
Coprocessor, 576
Corner reflector, 629
Counters, 701
Coupling, 235
CRC (*see* Cyclic redundancy code)
Critical angle, 731
Critical frequency, 270
Crossbar, 498
Crosspoint switch, 498
Crosstalk, 69
Crosstalk rejection, 699
CRT (*see* Cathode ray tube)
Crystal filter, 103
Crystal radio, 142
Crystals, 145
CTS (*see* Clear to send)
CSMA/CD (*see* Carrier sense, multiple access/collision detect)
Cutoff frequency, 233
Cutoff wavelength, 233
CVSD (*see* continuously variable slope delta modulation)
CW (*see* Continuous wave)
Cyclic redundancy code (CRC), 391

D

D-layer, 269
DAC (*see* Digital-to-analog converter)
Dark current, 741
Data terminal ready (DTR), 524
Data communication equipment (DCE), 519
Data compression, 451
Data link layer, 585
Data set ready (DSR), 523
Data terminal equipment (DTE), 519
Database, 503
dB (*also see* Decibels):
dBc, 52
dB charting, 57
dB estimating, 57
dBm, 51
dB and power, 47
dB reference values, 51
dB and voltage, 47
dbW, 51
dc component, 371
dc offset, 39
dc restoration, 445
DCC (*see* Digital color code)
CDE (*see* Data communication equipment)
DDC (*see* Direct digital synthesis)
Dead zone, 759
Decay, 108, 129, 142
Decibels (dB), 46
Decoder, 341
Dedicated channel, 375
Dedicated line, 492
Deemphasis, 193
Deflection coils, 445
Deflection modulation, 646
Delay lines, 723
Delimiters, 364
Delta modulation, 348
Demodulation, 69, 79, 90, 120, 140
Destructive interference, 295
Detection, 108, 140
Detector stage, 124
Deterministic, 571
Deviation, 161, 165
Dial-up line, 491
Dibits, 404, 553
Dielectric, 224
    constant, 225
    heating, 226
Difference frequency, 130
Difference signal, 81, 172
Differential phase, 553
Differential signals, 206
Diffraction, 266
Digital, 6, 34
    cellular phones, 636
    color code (DCC), 637
    drivers, 211
    filtering, 655

Digital (cont.)
  signal processing (DSP), 511, 665
  signals, 310
  switching, 514
  TDM, 681
Diode detector, 140
Direct FM, 168
Directivity, 277
Director, 290
Discontinuous message (DM), 637
Discrete values, 310
Dispersion, 732
Distance resolution, 759
DM (see Discontinuous message)
Dominant mode, 232
Doppler shift, 346, 651
Double conversion, 139
Double-sideband, suppressed carrier AM (DSB-AM), 95
Downlink, 594
Drift space, 707
Driven element, 290
Driver, 110
Drops, 567
DSP (see Digital signal processing)
DSR (see Data set ready)
DTE (see Data terminal equipment)
DTMF (see Dual tone multifrequency)
DTR (see Data terminal ready)
Dual bypass switch, 750
Dual tone multifrequency (DTMF), 486, 487
Dual-modulus prescaler, 468
Dummy load, 256
Duplex, 18
Duplex filter, 631
Duplexer, 659
Duty cycle, 648
Dynamic negative resistance, 716
Dynamic range, 318, 701, 759
Dynamic switches, 698

E

E-layer, 269
EBCDIC code, 365
Echo, 508
Echo cancellation, 509
ECM (see Electronic countermeasures)
EDC (see Error detection and correction)
Efficiency, 324
Electrical length, 238
Electromagnetic energy, 7
Electromagnetic interference (EMI), 727
Electromagnetic spectrum, 1, 11
Electron bunching, 705
Electronic countermeasures (ECM), 658
Electronic switching systems (ESS), 502
Electronic tuning, 132
Elements, 320

Elevation, 594
EMI (see Electromagnetic interference)
Encoder, 341
Encryption, 423, 424
Energy spectrum, 43
Envelope, 34, 81
Envelope delay, 491
Equatorial belt, 597
Equatorial orbit, 596
Equipment serial number (ESN), 636
Equivalent noise resistance, 69
Equivalent noise temperature, 68
Error detection, 388
Error detection and correction (EDC), 393
Error detection and correction (EDC), 421
ESN (see Equipment serial number)
ESS (see Electronic switching systems)
Estimation, 108
Ethernet, 576
Exchange, 480
Exciter, 110, 171
Expansion, 341
External noise, 66
Eye pattern, 418, 420

F

F-layer, 269
Facsimile (fax), 448
Fading, 120
Far-field measurement, 274
Fast Fourier transform (FFT), 24
FDDI (see Fiber Distributed Data Interface)
FDM (see Frequency-division multiplexing)
Feed, 300
Feedhorn, 303
Feedsystem, 303
Feedthrough, 98
FET (see Transistor, field effect)
FFT (see Fast Fourier transform)
Fiber Distributed Data Interface (FDDI), 749
Fiber optic attenuation, 756
Fiber optic attenuators, 755
Fiber optic cables, 748
Fiber optic cladding, 729
Fiber optic coating, 730
Fiber optic components, 745
Fiber optic core, 729
Fiber optic multiplexing, 752
Fiber optic power meters, 754
Fiber optics, 1, 727
Fiber optic signal sources, 754
Fibe optic testing, 753
Field strength meter, 274
Fields, video, 433
FIFO (see First in/first out buffer)
Filament, 704
Filter, 23, 54, 462
Filtering, 43
First in/first out (FIFO) buffer, 528, 749

Fixed block length, 367
Flag, 569
Flag bits, 504
Flashover, 244
Flat line, 238
Flat response, 53
Flicker, 432
Flip-flop, 349
Flow chart, 379
Flux density, 737
Flyback transformer, 445
FM demodulators, 178
FM receivers, 175
FM (*see also* Frequency modulation)
FM testing, 198
FM transmitters, 167
FM, indirect, 168
FM, narrowband (NBFM), 167
FM, narrowband (NBFM), 675
Focal point, 300
Footprint, 595
Format, 328, 360, 366, 562
Forward control channel, 636
Forward error correction (FEC), 394
Forward power gain, 249
Forward transmission coefficient, 249
Forward voice channel, 636
Foster-Seeley detector, 178
Fourier analysis, 22, 24
Fourier transform, 24
Frame, 368, 432, 685
Frame check, 386
Frame synchronization, 348
Framing bit, 685
Framing error, 536
Framing pattern, 348
Free-space path loss, 271
Frequency, 7
   domain, 24
    meters, 701
   modulation (FM), 33, 160
   multiplication, 169
    multiplier, 113
   response, 53
   shift keying (FSK), 186, 406
   translation, 82, 103
Frequency-division multiple access (FDMA), 604
Frequency-division multiplexing (FDM), 674
Frequency-sensitive voltmeter, 53
Front end stage, 66, 125
Front porch, 434
Front-end overload, 67
Front-to-back ratio, 277
FSK (*see* Frequency shift keying)
Full duplex, 18, 377

G

Gain, 276
Ganged capacitors, 131

Gateway, 586
General Purpose Interface Bus (GPIB), 578
Generator, 237
Geostationary, 596
Gigahertz (GHz), 11
Glitches, 336
Global Positioning System (GPS), 615
GMT (*see* Greenwich Mean Time)
GPIB (*see* General Purpose Interface Bus)
GPS (*see* Global Positioning System)
Graded index fiber, 732
Gray scale, 431
Greenwich Mean Time (GMT), 614
GRI (*see* Group repetition interval)
Grid, 704
Ground plane, 228
Ground station, 606
Ground wave, 265
Group 3 compression, 451
Group repetition interval (GRI), 612
Guardband, 166
Gunn diode, 716
Gunn oscillators, 658

H

Hailing channel, 623
Half duplex, 18, 377
Half-power points, 49
Hamming code, 394
Handshaking, 523
Harmonics, 30
HDLC (*see* High level data link control)
HDTV (*see* High definition TV)
Header, 367
Header block, 328
Hertz, 7
Hexadecimal, 328
Hierarchy, 482
High beat, 131
High definition TV (HDTV), 452
High level data link control (HDLC), 386
High tracking, 131, 137
High-level modulation, 109, 172
High-pass filter, 67
Hop, 268
Horizontal resolution, 438
Horizontal synchronization, 433
Hue, 441
Hybrid, 485, 494
Hyperbola, 611

I

I (*see* In-phase)
IEEE, 574
IEEE-488 format, 581
IEEE-488 standard, 578
IF (*see* Intermediate frequency)
IF transformers, 138

Image enhancement, 314
Image frequency, 137
Images, 136
IMPATT diode, 718
Impulse noise, 66
Impulse, spectrum, 37
In-band signaling, 487
In-phase (I), 410, 440
Inclined orbit, 596
Index of refraction, 266, 729, 731
Input reflection coefficient, 248
Insertion loss, 699, 722
Integrated Services Digital Network (ISDN), 588
Integrator, 349
INTELSATV, 595
Intensity modulation, 706
Interference, 12, 720
Interlacing, 433
Interleaved multiplexing, 689
Intermediate frequency (IF), 124, 129
Intermodulation distortion, 114
Internal noise, 66
Interrupt, 382
Interrupt-driven, 569
Intersymbol interference, 513
Ionosphere, 268
IR drop, 204
IRE units, 434
Isolation, 699
Isotropic, 276

J

Jitter, 113, 336
Johnson noise, 68
Jumbogroup, 676

K

Keying, 108
Kilohertz (kHz), 11
Klystron, 706
Klystrons, 657

L

LAN (see Local area network)
Laser diode, 736
Laser diodes, 739
Layers, 361
Leased line, 492
Least significant bit (LSB), 336
LED (see Light-emitting diode)
Light-emitting diode (LED), 736, 738
Lightpipe, 6
Limiter, 143, 176
Line card, 492
Line driver, 209, 370, 536
Line monitor, 541
Line receiver, 209, 370, 536

Line-by-line transmission, 367
Linear scale, 57
Link analysis, 744
Link budget, 607
Liquid crystals, 756
Listener, 579
Listener echo, 509
LO (see Local oscillator)
Loading coil, 289
Lobes, 276
Local loop, 480, 490
Local oscillator (LO), 124, 129
LocalTalk, 574
Lock condition, 185
Log-linear scale, 59
Log-log scale, 59
Logarithmic scale, 47, 57
Logical numbers, 505
Logic analyzer, 325
Logic probe, 325
Look-up table, 343
Loopbacks, 414
LORAN, 610
Loss, 53
Low beat, 131
Low level modulation, 172
Low tracking, 131
Low-pass filter, 67, 183, 336
LSB (see Least significant bit)
Luminance amplifier, 446
Luminance signal Y, 439
Lumens, 737

M

Magnetron, 657, 708
Manchester encoding, 372
Manufacturing Automation Protocol (MAP), 576
MAP (see Manufacturing Automation Protocol)
Mark, 521
Master, 568
Master oscillator, 113
Mastergroup, 676
Matched filter, 654
Matching, 245
Matching transformer, 245
Matrix, 172, 498
Maximum range, 646
Maximum unambiguous range, 649
Maximum usable frequency, 270
Maximum usable range, 649
Medium, 8
Megahertz (MHz), 11
Microbending, 734
Microcontrollers, 315
Microstrip line, 228
Microwave amplifiers, 698
Microwave attenuators, 698

Index

Microwave connectors, 696
Microwave test equipment, 696
Microwaves, 6, 694
Millimeter waves, 694
MIN (*see* Mobile Identification Number)
Mismatch, 224
Mixer, 124, 129
Mixing, 91
Mobile Identification Number (MIN), 636
Mobile Telephone Switching Office (MTSO), 628
Mode, 232
Modem, 544
  Bell 103, 552
  Bell 202, 552
  direct connect, 550
  integral, 547
  null, 541
  short haul, 550
  standalone, 547
Modulation, 4, 32, 79, 740
Modulation index, 81, 84, 162, 165
Modulation index and power, 86
Modulo-n counter, 460
Monochrome, 439
Monolithic microwave integrated circuit (MMIC), 715
Moonbounce, 605
Morse code, 144, 311
Most significant bit (MSB), 336
Moving target indicator (MTI), 652
MSB (*see* Most significant bit)
MTI (*see* Moving target indicator)
MTSO (*see* Mobile Telephone Switching Office)
Multidrop, 556
Multilayer, 228
Multilevel digital, 320
Multimode, 732
Multimode fiber, 732
Multiple-stage multiplexing, 688
Multiplexing, 6, 173, 670
Multiplier, 93, 170

N

NA (*see* Numerical aperture)
NAK (*see* Negative acknowledge)
NAM (*see* Numeric Assignment Module)
NAVSTAR, 615
NBFM (*see* FM, narrow band)
Near-field measurement, 274
Negative acknowledge (NAK), 379
Network, 377, 481, 560
  baseband, 574
  broadband, 574
  local area (LAN), 563
  n-port, 247
  open, 573
  proprietary, 573
  two-port, 247
  wide area (WAN), 563, 583
Network analyzers, 328
Network applications, 561
Network layer, 585
Nodes, 239, 379, 481
Noise, 16, 63, 126, 311
Noise and FM/PM, 193
Noise blanker, 150
Noise figure, 71
Noise figure and noise temperature, 73
Noise immunity, 313
Noise limiting, 149
Noise margin, 313
Noise measurement, 70
Noise sources, 69
Noise voltage, 69
Non-printing codes, 363
Nonresonant line, 238
Nonreturn to zero (NRZ), 372
Normalized values, 251
Notch filter, 67
NRZ (*see* Nonreturn to zero)
NTSC, 441
Null, 276, 363
Numeric Assignment Module (NAM), 636
Numerical aperture (NA), 737
Nyquist sampling theorem, 323

O

Off-hook, 481
Omnidirectional, 276
On-hook, 481
Open wire transmission line, 226
Optical collection factor, 737
Optical detectors, 741
Optical time domain reflectometry (OTDR), 758
Originate mode, 546
Oscilloscopes, 702
OSI reference model, 585
OTDR (*see* Optical time domain reflectometry)
OTH (*see* Over the horizon)
Out-of-band signaling, 506
Output reflection coefficient, 249
Overmodulation, 90
Overrun error, 536
Overshoot, 212, 326
Overvoltage, 493

P

P-pattern, 617
PABX (*see* Private automatic branch exchange)
Packets, 583
Packet switching, 583

Parallel-to-serial converter, 337
Parasitic elements, 290
Parity, 388, 389, 526
Parity error, 536
Path loss, 271
PBX (*see* Private branch exchange)
PCM (*see* Pulse code modulation)
Peak detector, 141
Peak power, 713
Pentode, 704
Phase, 29
Phase and Fourier analysis, 29
Phased array, 289
Phase detector, 183, 462
Phase discriminator, 178
Phase jitter, 69
Phase modulation (PM), 160, 190
Phase shift keying (PSK), 409
Phase-lock loop (PLL), 182, 346, 445
Phase-lock loop (PLL) as filter, 185
Phase-shift SSB generation, 105
Phasor, 441
Phosphor triad, 447
Photoconductors, 741
Photon, 738
Physical layer, 369, 585
Physical level, 403
Physical numbers, 505
Pickup, 113
Picture elements (pixels), 438
Piezoelectric effect, 720
Pilot carrier, 144
PIN diodes, 741
Plain old telephone service (POTS), 483
Plan-position indicator (PPI), 647
Plate, 704
PLL (*see* Phase-lock loop)
PM (*see* Phase modulation)
Point-to-point, 377
Polar orbit, 596
Polarization, 279
   circular, 279
   horizontal, 279
   vertical, 279
Polling, 569
Postamble, 367
POTS (*see* Plain old telephone service)
Power meter, 254
Power spectrum, 42
PPI (*see* Plan position indicator)
Preamble field, 367
Preamplifier, 55, 127
Preemphasis, 193
Prescaler, 466
Preselector, 127
Presentation layer, 585
PRF (*see* Pulse repetition frequency)
Primary colors, 439
Primary power, 600
Primary rate, 589

Printable codes, 363
Printed circuit board, 228
Printed wiring board, 228
Private automatic branch exchange (PABX), 505
Private branch exchange (PBX), 505
Probabilistic, 572
Product detector, 143
Propagation, 266
Propagation delays, 328
Propagation velocity, 8
Propulsion subsystems, 601
Protocol, 6, 328, 360, 376, 563
   bit-oriented, 575
   command/response, 568
   continuous, 377
   dedicated, 385
   master/slave, 568
   peer-to-peer, 569
   software, 385
   stop and go, 377
PRR (*see* Pulse repetition rate)
PRSQ (*see* Pseudorandom sequence)
Pseudorandom sequence (PRSQ), 422, 617
PSK (*see* Phase shift keying)
Pulse code modulation (PCM), 335
Pulse compression, 655
Pulse dialing, 485
Pulse repetition frequency (PRF), 648
Pulse repetition rate (PRR), 648

Q

Q (*see* Quadrature), 122
Quadbit, 411
Quadrature (Q), 180, 410, 440
Quadrature amplitude modulation (QAM), 410
Quadrature detector, 180
Quantization error, 317
Quantization noise, 317
Queue, 561
Quieting, 176

R

Radar, 643
   aiming, 647
   bistatic, 662
   continuous wave (CW), 653
   monopulse, 660
   multistatic, 662
   over the horizon (OTH), 663
   pulsed, 648
   search, 647
   synthetic aperture (SAR), 664
Radar cross section (RCS), 644
Radar display, 646
Radar range, 645
Radar solid-state devices, 660
Radiant flux, 737

Index

Radiation, 264
Radiation angle, 266, 277
Radiation loss, 226
Radio frequency (RF), 124
Radio frequency interference (RFI), 66, 727
Radiometry, 737
Radome, 301
Random bit generation, 421
Range resolution, 650
Raster scan, 431
Rayleigh backscattering, 758
RCS (*see* Radar cross section)
Read data register, 533
Read only memory (ROM), 343
Receiver:
   AM, 119
   bus, 209
   communications, 149
   FM, 175
   line, 209
   superheterodyne, 124
   TRF, 122
   TV, 442
Receiver testing, 154
Reciprocity, 265
Reed relays, 498
Reference black, 436
Reference oscillator, 461
Reference white, 436
Reflecting plane, 287
Reflection, 9, 266
Reflection coefficient, 243
Reflector, 290
Refraction, 9, 266
Regeneration, 212, 311
Regenerative amplifier, 313, 501
Remainder, 391
Repeaters, 501
Repeller, 707
Request to send (RTS), 523
Resistive heating loss, 226
Resolution, 26, 316
   video, 432
Retrace, 432
Return to zero (RZ), 372
Reverse control channel, 636
Reverse isolation coefficient, 249
Reverse path loss, 510
Reverse power leakage, 249
Reverse voice channel, 636
RF saturation, 67
RF (*see* Radio frequency)
RF switching matrix, 603
RFI (*see* Radio frequency interference)
Ring, 492
Ring trip, 493
Ringback, 493
Ringing, 212, 326
Ringing voltage, 485
Ripple, 62

Ripple counter, 463
ROM (*see* Read only memory)
Root mean square (rms), 70
Router, 586
RS-232, 345
RS-232, 519
RS-232 connector, 525
RS-232 control lines, 522
RS-232 data lines, 522
RS-232 secondary functions, 522
RS-232 timing, 522
RS-422, 555
RS-423, 554
RS-485, 556
RTS (*see* Request to send)
RZ (*see* Return to zero)

S

S-meter, 152
S-parameters, 247, 253
Sampled, 6
Sampling rate, 323
SAR (*see* Radar, synthetic aperture)
Satellite:
   communication, 594
   passive, 605
Satellite access, 604
Satellite frequencies, 597
Satellite wave, 265, 270
Saturation, 40, 54, 441
SAW (*see* Surface acoustic wave)
Scaling factor, 94
Scattering, 733
SDLC (*see* Synchronous data link control)
SDM (*see* Space-division multiplexing)
Selectivity, 122, 135
Self-handshaking, 539
Semilog scale, 59
Sensitivity, 127
Separation, 404
Serial ASCII, 520
Serial bits, 337
Session layer, 585
Shadow mask, 447
Shannon, 15
Shape factor, 135
Shielding, 205
Shift register, 337, 338, 391
SID (*see* System Identification Number)
Sidebands, 81
Sideband suppression, 106
Sidetone, 485
Signal detection, 403
Signal distance, 313
Signal estimation, 403
Signal strength indicator, 152
Signal subtraction, 510
Signal-to-noise ratio (SNR), 71, 396
Simplex, 17, 377

Single sideband suppressed carrier (SSB), 98
Single-ended signals, 205
Skin effect, 226
Skip conditions, 270
Skip zone, 268
Skirts, 135
Sky wave, 265, 268
Slew, 209
Slew rate, 210, 326
SLIC (*see* Subscriber loop interface circuit)
Slope detection, 178
Slope overload, 251
Slot coupling, 236
Smith chart, 246, 250, 253
SNR (*see* Signal-to-noise ratio)
Soft limiting, 176
Software handshaking, 529
Space, 521
Space segment, 617
Space wave, 265, 266
Space-division multiplexing (SDM), 672
Spectrum analysis, 22
Spectrum analyzer, 24, 701
Spectrum, digital signal, 34
Speed dialing, 503
Spillover, 302
Splice loss, 735
Splitters, 699
Spurious oscillation, 701
Spurious signals, 67
Squelch, 151
SSB demodulation, 143
SSB (*see* Single sideband suppressed carrier)
Stages, 55
Standing waves, 240
Start bit, 381, 525
Start of message, 367
State (transition) diagram, 379, 563, 580
Static switches, 698
Station keeping, 601
Status register, 384, 533
Status telemetry, 601
Step index fiber, 732
Step size, 316
Stereo FM, 172
Stereo FM decoding, 174
Stereo FM demodulation, 188
Stop bit, 381, 526
Store and forward systems, 584
Stratosphere, 269
Striplines, 228
Strowger switch, 499
Stub, 246
Subcarrier, 172, 440, 675
Subscriber line interface circuit (SLIC), 492, 673
Sum signal, 81, 172
Sunspot cycle, 67
Sunspots, 266

Supergroup, 676
Superheterodyne, 121
Superheterodyne receiver, 124
Superposition, 38
Supertrunk, 481
Supervision, 494
Surface acoustic wave (SAW) and devices, 103, 720
Surface waves, 266
Sweep oscillators, 702
Switched line, 491
SWR (*see* Voltage standing wave ratio)
Symbols, 320
Sync separator, 445
Synchronization, 64, 344
Synchronous communication, 380
Synchronous counter, 463
Synchronous data link control (SDLC), 386, 574
Synchronous systems, 328
Syndrome, 394
Synthesis:
   direct, 457
   direct digital (DDS), 459
   frequency, 457
   indirect, 457, 460
Synthesizers, 471
System Identification Number (SID), 636

T

T-1 format, 369
T-1 system, 685
T/R (*see* Transmit/receive)
Talker, 579
Talker echo, 508
Tandem switch, 481
TDM and microprocessors, 686
TDM (*see* Time-division multiplexing)
TDMA (*see* Time-division multiple access)
TDR (*see* Time domain reflectometry)
Telemetry, 186, 599, 675
Telephone signal levels, 485
Telex, 452
Terahertz (THz), 727
Terminating character, 367
Terminating set, 494
Tetrode, 704
Thermal noise, 68
Thermocouple, 699
Three-state driver, 556
Threshold, 405, 740
Throughput, 375
Time constant, 62
Time domain, 23
Time domain reflectometry (TDR), 216
Time-division multiple access (TDMA), 604
Time-division multiplexing (TDM), 679
Timeout circuitry, 566

Index

Timing diagram, 527
Tint, 441
Tip, 492
Token, 570
Token passing, 570
Tone dialing, 486
Topology, 562, 564
   bus, 565
   one-to-all, 564
   ring, 566
   star, 565
Total internal reflection, 727, 731
Touch-Tone, 486
Tracking ADC, 350
Transceiver, 195
Transducer, 264, 699
Transhybrid loss, 495
Transimpedance amplifiers, 741
Transistor, bipolar, 718
Transistor, field effect (FET), 719
Transmission loss, 733
Transmit/receive (T/R) switches, 283
Transmitter, 209
Transparency, 310, 360, 689
Transponder, 603
Transport layer, 585
Transverse electric mode, 232
Transverse magnetic mode, 232
TRAPATT diode, 718
Traps, 282
Traveling-wave tube (TWT), 603, 657, 709
TRF (*see* Tuned radio frequency), 122
Triangulation, 614
Triode, 704
Triple conversion, 139
Troposphere, 269
Trunk, 481, 499
Tuned radio frequency (TRF), 121
Tuned radio frequency (TRF) receivers, 122
Tunnel diodes, 716
TV receiver, 442
Twisted-pair cable, 214
TWT (*see* Traveling-wave tube)

U

UART error conditions, 536
UART (*see* Universal asynchronous receiver/transmitter)
Unbalanced signals, 205
Undersampling, 324

Unipolar, 370
Universal asynchronous receiver/transmitter (UART), 382
Universal asynchronous receiver/transmitter (UART), 531
Uplink, 594
User segment, 617

V

Vacuum tube, 94, 147, 704
Varactor, 132
Variable block length, 367
Variable gain amplifier (VGA), 141, 343
Variable reactance, 168
Varicap diode, 132
VCO (*see* Voltage controlled oscillator)
Vector, 441
Vector modulation, 411
Velocity, 7
Vertical synchronization, 433
Vestigial sideband (VSB), 103, 437
VGA (*see* Variable gain amplifier)
Video, 431
Video carrier, 444
Video detector, 444
Vidicon, 432
Voltage controlled oscillator (VCO), 183, 461
Voltage follower, 211
Voltage standing wave ratio (VSWR), 242
Voltage variable capacitance diode, 132
Voyager spacecraft, 608
VSB (*see* Vestigial sideband)
VSWR (*see* Voltage standing wave ratio)

W

Walkie-talkie, 195
WAN (*see* Network, wide area)
Watchdog circuitry, 566
Waveguide, 231
Waveguide losses, 237
Waveguide, circular, 235
Wavelength, 7
White noise, 68
Window, 406
Write data register, 533

X

XOFF, 529
XON, 529
μ-law, 342